HANDBOOK OF

NUCLEAR, BIOLOGICAL, and CHEMICAL AGENT EXPOSURES

HANDBOOK OF
NUCLEAR,
BIOLOGICAL,
and
CHEMICAL AGENT
EXPOSURES

Jerrold B. Leikin, M.D.
Robin B. McFee, D.O.

CRC Press
Taylor & Francis Group
Boca Raton London New York

CRC Press is an imprint of the
Taylor & Francis Group, an **informa** business

CRC Press
Taylor & Francis Group
6000 Broken Sound Parkway NW, Suite 300
Boca Raton, FL 33487-2742

First issued in paperback 2019

ISBN-13: 978-1-4200-4477-5 (hbk)
ISBN-13: 978-0-367-38892-8 (pbk)

Library of Congress Cataloging-in-Publication Data

Handbook of nuclear, biological, and chemical agent exposures / editors, Jerrold B. Leikin and Robin B. McFee.
    p. ; cm.
"A CRC title."
Includes bibliographical references and index.
ISBN-13: 978-1-4200-4477-5 (hardcover : alk. paper)
ISBN-10: 1-4200-4477-X (hardcover : alk. paper)
    1. Environmental toxicology--Handbooks, manuals, etc. 2. Poisoning--Handbooks, manuals, etc. I. Leikin, Jerrold B. II. McFee, Robin B.
    [DNLM: 1. Environmental Exposure--adverse effects--Handbooks. 2. Antidotes--therapeutic use--Handbooks. 3. Poisoning--therapy--Handbooks. WA 39 H23559 2007]

RA1226.H362 2007
615.9'02--dc22   2006039730

Visit the Taylor & Francis Web site at
http://www.taylorandfrancis.com

and the CRC Press Web site at
http://www.crcpress.com

TABLE OF CONTENTS

Section I - Antidotes

TABLE OF CONTENTS

Section IV - Laboratory Analysis

TABLE OF CONTENTS

Appendix

Patient Information Sheets

PREFACE

In virtually any field in medicine, each disease process is approached by the clinician one patient at a time. Diagnosis and therapeutic approach are thus individualized as is monitoring of the patient's status. Protocols serve as an inexact guide to aid in this process.

However, the approach to nuclear, biological, or chemical agent exposures presents different and unique challenges to the clinician. In these scenarios, the clinicians are usually encountering multiple exposures (sometimes even mass exposures) from a single, often poorly defined, event. Early presenting symptoms are not distinct and can often be quite variable. Laboratory analysis may be required from environmental (often nonbiologic) specimens. Scene evaluation and prehospital decontamination may turn out to be the most important intervention. Hospital resource utilization must be a consideration. Even the pathologist performing autopsies requires adequate preparation.

It is with these considerations that this book was created. While the book is divided into agent monographs, the clinician is advised that oftentimes multiple agents or even classifications may be responsible for a certain exposure. Toxidrome (or symptom complex) recognition provides the most important clinical clues as to the etiologic agent when multiple exposures are analyzed as a group. We have made every attempt to update this data regularly and welcome the reader's comments.

Jerrold B. Leikin, MD

Robin B. McFee, DO

EDITORS

Jerrold Blair Leikin, MD, FACP, FACEP, FACMT, FAACT, FACOEM

Dr. Leikin is director of medical toxicology at Evanston Northwestern Healthcare-OMEGA, Glenbrook Hospital, located in Glenview, Illinois. He is associate director of the Toxikon Consortium based at Cook County Hospital in Chicago. He is also professor of medicine at Rush Medical College and professor of emergency medicine at Feinberg School of Medicine at Northwestern University in Chicago, Illinois.

Dr. Leikin received his medical doctorate degree from the Chicago Medical School in 1980 and completed a combined residency in internal medicine and emergency medicine at Evanston Hospital and Northwestern Memorial Hospital in 1984. He completed a fellowship in medical toxicology at Cook County Hospital in Chicago in 1987. He is also board certified in the above specialties.

Dr. Leikin was the associate director of the emergency department from 1988–2001 at Rush-Presbyterian–St Luke's Medical Center in Chicago. Dr. Leikin was medical director of the Rush Poison Control Center for eleven years. He was also medical director of the United States Drug Testing Laboratory for five years. He is currently a consultant with the Illinois Poison Center, Wisconsin Poison Center, and PROSAR Product and Drug Safety Call Center located in St. Paul, Minnesota.

Dr. Leikin has presented over 100 abstracts at national meetings and has published about 200 articles in peer-reviewed medical journals. He is coeditor of the *American Medical Association Handbook of First Aid and Emergency Care*, and the *American Medical Association Complete Medical Encyclopedia*, both published by Random House. Dr. Leikin is also editor-in-chief of the journal *Disease-A-Month*. He has written several chapters on the subjects of toxicology, emergency medicine, critical care medicine, internal medicine, and observational medicine textbooks. Dr. Leikin is an active member of the American Academy of Clinical Toxicology, American College of Medical Toxicology, the American College of Emergency Physicians, the American College of Occupational and Environmental Medicine, and the American Medical Association. He is married and has two children.

Robin B. McFee, DO, MPH, FACPM

Dr. Robin McFee is a nationally recognized expert in medical toxicology, adolescent health, substance abuse, bioterrorism, and weapons of mass destruction. In addition to extensive expertise in emerging infectious diseases, she has experience in hazmat, search and rescue, as well as search and recovery. Dr. McFee has given approximately 300 invited lectures since September 11, 2001, regionally, nationally, and internationally. She conducts research in preparedness including the development of new interventions against WMD and has presented research at numerous scientific conferences. In 1997, she co-founded Seabury Adolescent Health; in 2001, she co-founded the Center for Bioterrorism and Weapons of Mass Destruction Preparedness (CB-PREP) in Ft. Lauderdale, and in 2004, Threat Science™.

Dr. McFee received her medical degree from the New York College of Osteopathic Medicine and a master's degree in public health (MPH) from The Mailman School of Public Health, Columbia University. She is board certified and completed a preventive medicine and public health residency at the State University of New York at Stony Brook. She is a fellow of the American College of Preventive Medicine. In addition, Dr. McFee completed post-residency training in medical toxicology at Winthrop University Hospital and has received extensive advanced training in WMD, public preparedness, and emerging global threats. She is a well-respected medical educator, having received numerous teaching awards including Innovations to Curriculum Award from the American Medical Association (AMA), an Excellence in

Teaching Award, the Aesculapius Medical Educator Award from SUNY/Stony Brook, and a Pfizer/AAMC Humanism in Medicine award. She is on the faculty of the medical schools at the State University of New York (SUNY) Stony Brook and NYCOM at the New York Institute of Technology. She has taught Basic Disaster Life Support (BDLS) and is an Advanced Hazmat Life Support (AHLS) instructor. During her career as clinician, teacher, subject matter expert, and thought leader, Dr. McFee has gained experience across the continuum of health care – from "in the trenches" emergency response to clinical medicine, program leadership, academia, public health, and health policy.

Dr. McFee consults with numerous federal agencies in addition to state and local governments. As an adolescent expert as well as toxicologist, she works with prevention programs, most recently as a speaker on Medscape's live conference on binge drinking. As a toxicologist, she consults with industry to develop newer interventions against WMD agents and consults for the Long Island Regional Poison Control Center in New York. Dr. McFee is on the National Nerve Agent Advisory Group. She has worked with various private industries, public agencies including the HRSA Bioterrorism Emergency Preparedness in Aging project, schools, and emergency medical systems. In addition, Dr. McFee has designed drills for such agencies as the Office of Emergency Management (OEM) and has been a designer, facilitator, and observer of more than 40 drills and exercises since September 11. Dr. McFee also has been a part of Region 7 preparedness efforts of the Florida Regional Security Task Force and is on several national/international task forces and advisory boards such as the Radiation Antidote Working Group and ASIS International Council for Global Terrorism, Political Instability, and International Crime.

In addition to approximately 50 adolescent health, bioterrorism, toxicology, and cardiac related publications in a wide range of peer reviewed medical literature, Dr. McFee was an associate editor of the *Johns Hopkins Journal Advanced Studies in Medicine* and serves as an advisor and editorial board member of other journals. She has authored several poisoning and toxicology related chapters in leading textbooks, and is co-developing two textbooks on WMD with anticipated publications dates in mid-2007.

CONTRIBUTORS

Judy Aberg, MD

Director, HIV Bellevue Hospital Center
Associate Professor of Medicine
New York University School of Medicine
New York, New York

Steven E. Aks, DO, FACMT, FACEP

Director, The Toxicon Consortium
Department of Emergency Medicine
John Stroger Hospital
Chicago, Illinois

Yona Amitai, MD, MPH

Director, Department of Mother, Child, and Adolescent Health
Ministry of Health
Jerusalem, Israel

Britney B. Anderson, MD

Physician
Department of Emergency Medicine
Northwestern Memorial Hospital
Chicago, Illinois

Carl R. Baum, MD, FAAP, FACMT

Director, Center for Children's Environmental Toxicology
Yale-New Haven Children's Hospital
Yale Medical School
New Haven, Connecticut

Rhonda M. Brand, PhD

Adjunct Associate Professor of Biological Systems Engineering and Environmental Health
Evanston Northwestern Healthcare
Northwestern University
Evanston, Illinois

Anthony M. Burda, RPh, DABAT

Chief Specialist
Illinois Poison Center
Chicago, Illinois

CONTRIBUTORS

Rachel Burke, MD

Physician
Department of Emergency Medicine
Rush-Presbyterian–St. Lukes Hospital
Chicago, Illinois

Keith Edsel, MBBS, MRCS, LRCP, MSc

Consultant in Radiation Medicine
Ryde Isle of Wight
England, United Kingdom

Christina Hantsch, MD

Physician
Loyola University Medical Center
Maywood, Illinois

Edward Krenzelok, PharmD, FAACT, DABAT

Director, Pittsburgh Poison Center
Children's Hospital of Pittsburgh
Pittsburgh, Pennsylvania

J. Marc Liu, MD

Physician
Department of Emergency Medicine
Medical College of Wisconsin
Milwaukee, Wisconsin

Mark B. Mycyk, MD

Physician
Department of Emergency Medicine
Northwestern Memorial Hospital
Chicago, Illinois

D. Robert Rodgers, MD

Physician
Department of Emergency Medicine
Northwestern Memorial Hospital
Chicago, Illinois

Bijan Saeedi, MS

Head Chemist
SET Environmental, Inc.
Wheeling, Illinois

Todd Sigg, PharmD, CSPI

Certified Specialist
Illinois Poison Center
Chicago, Illinois

Chad Tameling, BS

Emergency Response Coordinator
SET Environmental, Inc.
Wheeling, Illinois

Richard G. Thomas, PharmD, ABAT

Senior Pharmacist
Phoenix Children's Hospital
Phoenix, Arizona

Michael Wahl, MD, FACMT

Medical Director
Illinois Poison Center
Chicago, Illinois

Frank G. Walter, MD, FACEP, FAACT, FACMT

Chief, Division of Medical Toxicology
University of Arizona College of Medicine
Tucson, Arizona

REFERENCES

The following is a list of references utilized in the production of the individual monographs contained in this book.

AMA: Basic Disaster Life Support Provider Manual, Version 2.5, *AMA*, 2004.

Anderson PO, Knoben JE, and Troutman WG, *Handbook of Clinical Drug Data*, 9th ed, Stamford, CT: Appleton and Lange, 1999.

Auerbach PS, *Wilderness Medicine: Management of Wilderness and Environmental Emergencies*, 4th ed, St Louis, MO: Mosby, 2001.

Auerbach PS, *"A Medical Guide to Hazardous Marine Life"*, 2nd ed, St Louis, MO: Mosby, 1991.

Bartlett JC and Greenberg MI, *PDR Guide to Terrorism Response*, Thomson PDR: Montvale, NJ, 2005.

Baselt RC, *Drug Effects on Psychomotor Performance*, Foster City, CA: Biomedical Publications, 2001.

Baselt RC, *Disposition of Toxic Drugs and Chemicals in Man*, 7th ed, Foster City, CA: Biomedical Publications, 2004.

Benjamin DR, *Mushrooms: Poisons and Panaceas*, New York, NY: W.H. Freeman, 1995.

Billups NF, *American Drug Index*, 38th ed, St Louis, MO: Facts and Comparisons, 1994.

Bleecker ML and Hansen JA, *Occupational Neurology and Clinical Neurotoxicology*, Baltimore, MD: Williams and Wilkins, 1994.

Blumenthal M, Busse WR, Goldberg A, et al, eds, *The Complete German Commission E Monographs. Therapeutic Guide to Herbal Medicines*, Austin, TX: American Botanical Council; Boston, MA: Integrative Medicine Communications, 1998.

Bove AA, *Diving Medicine*, 3rd ed, Philadelphia, PA: W.B. Saunders, 1997.

Brent J, Wallace K, Burkhart KK, et al, *Critical Care Toxicology*, Elsevier Mosby: Philadelphia, PA, 2005.

Briggs GG, Freeman RK, and Yaffe SJ, *Drugs in Pregnancy and Lactation*, 6th ed, Baltimore, MD: Williams and Wilkins, 2002.

Bryson PD, *Comprehensive Review in Toxicology for Emergency Clinicians*, 3rd ed, Washington, DC: Taylor and Francis, 1996.

Ciottone GR, *Disaster Medicine*, Boston, MA: Elsevier Mosby, 2006.

Cooney DO, *Activated Charcoal in Medical Applications*, New York, NY: Marcel Dekker, Inc, 1995.

Dart RC, ed, *Medical Toxicology*, 4th ed, Philadelphia, PA: Lippincott, Williams, and Wilkins, 2004.

Derelank MJ and Hollinger MA, *CRC Handbook of Toxicology*, Boca Raton, FL: CRC Press, Inc, 1995.

Dorr RT and von Hoff DD, *Cancer Chemotherapy Handbook*, 2nd ed, Norwalk, CT: Appleton and Lange, 1994.

Drugdex System: Greenwood Village, CO: Intranet Database, Thomson Micromedex, Version 5.1, 2007.

Drummer OH and Odell M, *The Forensic Pharmacology of Drugs of Abuse*, London: Arnold, 2001.

Edmonds S and Stather R, *Reactions Weekly*, Langhorne, PA: Adis International Limited, 1995.

Ellis MD, *Dangerous Plants, Snakes, Arthropods, and Marine Life Toxicity and Treatment*, Hamilton, IL: Drug Intelligence Publications, Inc, 1978.

Emergency Response Guidebook, Washington DC: U.S. Department of Transportation (DOT), Label-master, 2000.

Fernandez H, *Heroin*, Center City, MN: Hazelden, 1998.

Ford MD, Delaney KA, Ling LJ, et al, *Clinical Toxicology*, Philadelphia, PA: W.B. Saunders, 2001.

Gardner DE, *Toxicology of the Lung*, 4th ed, Boca Raton, FL: CRC Press, Taylor & Francis Group, 2001.

Gilman AG, Rall TW, Nies AS, et al, eds, *Goodman and Gilman's The Pharmacological Basis of Therapeutics*, 8th ed, New York, NY: Pergamon Press, 1990.

Goldfrank LR, Flomenbaum NE, Hoffman RS, et al, *Goldfrank's Toxicologic Emergencies*, 8th ed, New York, NY: McGraw-Hill, 2006.

Goldstein SM and Wintroub BU, *Adverse Cutaneous Reactions to Medication*, CoMedica Inc., 1994.

Greenberg MI, Hamilton RJ, Phillips SD, and McCluskey GJ, *Occupational, Industrial, and Environmental Toxicology*, 2nd ed, St Louis, MO: CV Mosby Co, 2003.

Haddad LM, Shannon MW, and Winchester JF, *Clinical Management of Poisoning and Drug Overdose*, 3rd ed, Philadelphia, PA: WB Saunders Company, 1997.

Halstead BW and Halsted LG, *Poisonous and Venomous Marine Animals of the World*, revised edition, Princeton, NY: Darwin Press, Inc, 1978.

Hansten PD and Horn JR, *The Top 100 Drug Interactions: A Guide to Patient Management*, H & H Publications, 2000.

REFERENCES

Hartman DE, *Neuropsychological Toxicology*, 2nd ed, New York, NY: Plenum Press, 1995.

Hayes WJ and Laws ER, *Handbook of Pesticide Toxicology*, San Diego, CA: Academic Press, Inc, 1991.

Henderson DA, Inglesby TV, and O'Toole T, eds, *Bioterrorism: Guidelines for Medical and Public Health Management*, Chicago, IL: American Medical Association Press, 2002.

Hodgson E and Levi PE, *A Textbook of Modern Toxicology*, New York, NY: Elsevier, 1987.

Huang KC, *The Pharmacology of Chinese Herbs*, Boca Raton, FL: CRC Press, 1993.

Janicak PG, Davis JM, Preskorn MD, et al, *Principles and Practice of Psychopharmacotherapy*, 4th ed, Baltimore, MD: Lippincott Williams & Wilkins, 2006.

Katzung BG, *Basic and Clinical Pharmacology*, 8th ed, New York, NY: Lange Medical Books, 2001.

Klaassen CD, *Casarett and Doull's Toxicology: The Basic Science of Poisons*, 6th ed, New York, NY: McGraw-Hill, 2001.

Koren G, *Maternal-Fetal Toxicology: A Clinician's Guide*, 3rd ed, New York, NY: Marcel Dekker, Inc, 2001.

Lampe KF and McCann MA, *AMA Handbook of Poisonous and Injurious Plants*, Chicago, IL: American Medical Association, 1985.

Lewis RJ, *Sax's Dangerous Properties of Industrial Materials*, 8th ed, New York, NY: Van Nostrand Reinhold, 1992.

Litt JZ, *Physician's Guide to Drug Eruptions*, New York, NY: The Parthenon Publishing Group, 1998.

Liu RH and Goldberger BA, *Handbook of Workplace Drug Testing*, Washington, DC: AACC Press, 1995.

Mandell GL, Bennett JE, and Dolin R, *Principles and Practice of Infectious Diseases*, 4th ed, New York, NY: Churchill Livingstone, 1995.

Marzulli F and Maibach HI, *Dermatotoxicology*, 5th ed, Washington, DC: Taylor and Francis, 1996.

McCunney RJ, ed, *A Practical Approach to Occupational and Environmental Medicine*, 3rd ed, Philadephia, PA: Lippincott Williams & Wilkins, 2003.

McEvoy GK, *AHFS 2001 Drug Information American Hospital Formulary Service*, Bethesda, MD: American Society of Health-System Pharmacists, 2001.

Newall CA, Anderson LA, and Phillipson JD, *Herbal Medicines, A Guide for Health Care Professionals*, London, England: Pharmaceutical Press, 1997.

Noji EK, Kelen GP, and Goessel TK, *Manual of Toxicologic Emergencies*, Chicago, IL: Yearbook Medical Publisher, Inc, 1989.

Olin BR, *Drug Facts and Comparisons*, St Louis, MO: Facts and Comparisons Inc, JB Lippincott Co, 2001.

Olin BR, *Lawrence Review of Natural Products*, St Louis, MO: Facts and Comparisons Inc, 1989 with updates to 2001.

Penney DG, ed, *Carbon Monoxide Toxicity*, Boca Raton, FL; CRC Press, 2002.

Poisindex System: Intranet Database, Thomson Micromedex, Inc: Greenwood Village, CO; Poisindex Information System, Version 5.1, 2007.

Reigart JR and Roberts JR, *Recognition and Management of Pesticide Poisonings*, 5th ed, 1999. Available at: http://www.epa.gov/pesticides/safety/healthcare.

Reynolds JE, *Martindale The Extra Pharmacopoeia*, 32nd ed, London, England: Council of the Royal Pharmaceutical Society of Great Britain, 1999.

Russell RE, *Snake Venom Poisoning*, Philadelphia, PA: JB Lippincott Co, 1980.

Ryan RP and Terry CE, *Toxicology Desk Reference*, 4th ed, Taylor & Francis, 1997.

Schlesser JL, *1991 Drugs Available Abroad*, Detroit, MI: Medec Books/Gale Research Inc, 1991.

Segen JC, *Dictionary of Alternative Medicine*, Stamford, CT: Appleton and Lange, 1998.

Semla TP, Beizer JL, and Higbee MD, *Geriatric Dosage Handbook*, 8th ed, Hudson, OH: Lexi-Comp Inc, 2003.

Shulman ST, Phair JP, and Sommers HM, *The Biologic and Clinical Basis of Infectious Diseases*, 4th ed, Philadelphia, PA: WB Saunders Company, 1992.

Shults TF and St Clair S, *The Medical Review Officer Handbook*, 8th ed, Research Triangle Park, NC: Quadrangle Research, LLC, 2005.

Sidell FR, *Management of Chemical Warfare Agent Casualties: A Handbook for Emergency Medical Services*, Bel Air, MD: HB Publishing, 1995.

Sidell FR, Patrick III WC, Dashiell TR, et al, *Jane's Chem-Bio Handbook*, 2nd ed, Jane's Information Group: Alexandria, VA, 2002.

Smith RP, *A Primer of Environmental Toxicology*, Philadelphia, PA: Lea and Febiger, 1992.

Spandorfer M, Curtiss D, and Snyder J, *Making Art Safely*, New York, NY: Van Nostrand Reinhold, 1993.

Spencer PS, Schaumburg HH, and Ludloph AC, eds, *Experimental and Clinical Neurotoxicology*, New York, NY: 2000.

Spoerke DG and Rumack BH, *Handbook of Mushroom Poisoning, Diagnosis & Treatment*, Boca Raton, FL: CRC Press, 1994.

Spoerke DG and Smolinske SC, *Toxicity of Houseplants*, Boca Raton, FL: CRC Press, 1990.

Strange GR, Ahrens WR, Lelyveld S, et al, *Pediatric Emergency Medicine*, 2nd ed, American College of Emergency Physicians, McGraw-Hill, 2002.

Sullivan JB and Krieger GR, *Clinical Environmental Health and Toxic Exposures*, 2nd ed, Philadelphia, PA: Lippincott Williams and Wilkins, 2001.

Sweetman S, ed, *Martindale: The Complete Drug Reference*, Intranet Database, Verison 5.1, Lodon Pharmaceutical Press Electronic Version: Greenwood Village, CO: Thomson Micromedex, 2006.

Swotinsky R and Smith D, *The Medical Review Officer's Manual*, 3rd ed, Beverly Farms, MA: OEM Press, 2006.

Taketomo CK, Hodding JH, and Kraus DM, *Pediatric Dosage Handbook*, 9th ed, Hudson, OH: Lexi-Comp Inc, 2002-2003.

Tarcher AB, *Principles and Practice of Environmental Medicine*, New York, NY: Plenum Medical Book Company, 1992.

TOMES® System Intranet, Version 5.1, Thomason Micromedex: Greenwood Village, CO, 2006.

Trissel LA, *Handbook on Injectable Drugs*, 11th ed, Bethesda, MD: American Society of Hospital Pharmacists, 2001.

Turner NJ and Szczawinski AF, *Common Poisonous Plants and Mushrooms of North America*, Portland, OR: Timber Press, 1991.

Tyler VE, *Herbs of Choice: The Therapeutic Use of Phytomedicinals*, New York, NY: Pharmaceutical Products Press, 1994.

Tyler VE, *The Honest Herbal: A Sensible Guide to the Use of Herbs and Related Remedies*, 3rd ed, New York, NY: Pharmaceutical Products Press, 1993.

Upfal MJ, Krieger GR, Phillips SD, et al, *Clinics in Occupational and Environmental Medicine*, Vol 2, No 2, WB Saunders Co: Philadelphia, PA, 2003.

Walter FG, Klein R, Thomas RG, *Advance Hazmat Life Support: for Chemical Burns and Toxic Products of Combustion*, Arizona Board of Regents, 2006.

SECTION I
ANTIDOTES

Pharmacy Preparedness for Incidents Involving Nuclear, Biological, or Chemical Weapons

Anthony M. Burda, RPh, DABAT and Todd Sigg, PharmD, CSPI

Illinois Poison Center, Chicago, Illinois

Revised and updated, originally published in the *Journal of Pharmacy Practice*, April 2000, 13(2):141-55.

High-Yield Facts

1. Until delivery of large amounts of medical supplies from the Strategic National Stockpile (SNS) and other state emergency agencies, hospitals should be prepared to treat mass casualties for 4 to 24 hours using in-house inventory and/or supplies rapidly procurable from other institutions and pharmaceutical suppliers.
2. Nerve agent antidotes, such as injectable atropine sulfate, in various sizes, pralidoxime chloride in 1-gram vials, and injectable diazepam, are available through pharmaceutical manufacturers and wholesalers. Hospitals and government emergency agencies may now obtain military-style auto-injectors of these antidotes. Some hospitals have chosen to participate in the Chempack program, which prepositions large amounts of nerve agent antidotes on site.
3. The cyanide antidote kit containing amyl nitrite pearls and injectable sodium nitrite and sodium thiosulfate is manufactured by one U.S. company. Hydroxcobalamin, approved by the Food and Drug Administration (FDA) in December 2006, offers an improved safety and efficacy profile.
4. Pharmaceuticals used for exposures to radioactive isotopes, eg, Ca-DTPA/Zn-DTPA, Prussian Blue, and potassium iodide, are not typically stocked in hospital pharmacy inventories. They may be obtained through the SNS, state emergency agencies, or REAC/TS, and their manufacturer.
5. Many local, state, and federal domestic preparedness agencies are stockpiling antibiotics, eg, ciprofloxacin and doxycycline, for the treatment and prophylaxis of large numbers of patients, healthcare workers, and emergency personnel. These antibiotics will protect against many strains of anthrax, pneumonic plague, and other biological threats.
6. The Centers for Disease Control and Prevention (CDC) controls the supply and distribution of smallpox vaccines and vaccinia immune globulin.
7. The CDC controls the supply and distribution of botulinum antitoxin. Currently, there are no vaccines or antitoxins for other biotoxins such as ricin, abrin, staphylococcus enterotoxin B, and trichothecene mycotoxins.

Introduction

Events such as the sarin gas attack in a Tokyo subway station in March 1995, the September 11, 2001, attacks, several anthrax hoaxes, and the actual anthrax and ricin letters in the United States, have heightened concern among public health and law enforcement agencies that a real nuclear, biological, or chemical (NBC) attack may occur in this nation. The potential for a terrorist attack using a chemical or biological agent has many individuals involved in public health and safety equipping themselves with information, contingency plans, and procedures to cope with such threats. Information concerning preparedness for a terrorist attack involving NBC weapons is available from various governmental agencies and other organizations. Instruction in this area is available via journal articles, Web sites, and on-site and internet training programs, seminars, and conferences. Most references will provide practical discussion on issues such as local and statewide planning, on-site and hospital decontamination procedures, recognition and detection of NBC agents, diagnosis and pathophysiology of disease states, protocols for first responders, and a variety of other public health issues.

The objective of this chapter is to provide the practitioner a concise summary and description of the types of pharmaceutical products that a healthcare facility pharmacy may be asked to provide as part of an overall response to an incident involving weapons of mass destruction (WMD). These products include antidotes, antibiotics, antitoxins, and other agents used in the symptomatic and supportive care of the poisoned patient. Pharmacy managers are urged to check inventory for these products, and know where the nearest supplier (e.g., wholesaler, pharmaceutical manufacturer, etc) is located for each agent. It is important to know how supplies can be obtained quickly in emergencies; many healthcare facilities are unprepared for poisoning emergencies. Small nonurban hospitals are more likely to have fewer antidotes in stock than larger urban, tertiary care facilities. In addition to monitoring inventory, pharmacy managers and pharmacy and therapeutics (P&T) committee members should be aware of their local or state governmental agencies that may support a depot of some pharmaceuticals. Often this information is classified and not readily available to individuals not serving on WMD readiness task forces or committees.

The Nunn-Lugar-Domenici Domestic Preparedness Act of 1996 established a $250,000,000 program to train 120 cities in the U.S. in the preparedness of emergency and medical response to chemical and biological agents. Headed by the Department of Homeland Security (DHS), federal agencies involved with domestic preparedness programs include the Department of Energy (DOE), Federal Bureau of Investigation (FBI), Federal Emergency Management Agency (FEMA), Environmental Protection Agency (EPA), and CDC. Local government bodies created metropolitan medical response systems (MMRS), whose mission is to create a multiple level, technically diverse, professional response to any deliberate or accidental act involving an NBC agent within that jurisdiction. Antidote caches funded by federal grants to MMRS will be intended for use by Emergency Medical System (EMS) first responders for self-administration, to treat a number of casualties at the site of the NBC incident, and to provide pharmaceuticals to treat large numbers of hospitalized patients and asymptomatic outpatients. The first 4 to 12 hours of emergency response will need to be managed by local resources prior to the arrival of the DHS and other federal assistance, such as delivery and distribution of the SNS formerly known as the National Pharmaceutical Stockpile (NPS). Two primary missions of the Homeland Security Act of 2002 are to reduce the vulnerability of the U.S. to terrorism and to minimize the damage, and assist in the recovery, from terrorist attacks that do occur within the U.S.

Strategic National Stockpile

"The CDC, in consultation with other partners in chem/bio preparedness, has developed a stockpile to respond to biological or chemical terrorism emergencies. To determine and review the composition of the SNS, CDC considers many factors, such as current biological and/or chemical threats, the availability of medical materiel, and the ease of dissemination of pharmaceuticals. One of the most significant factors in determining SNS composition, however, is the medical vulnerability of the U.S. civilian population. SNS depots are stored at strategic locations throughout the United States to assure the most rapid response possible. CDC ensures that the medical material in these SNS storage facilities is rotated and kept within potency shelf life limits."

More information about the SNS program is available at www.bt.cdc.gov/stockpile, or by calling the SNS operations center at 404-639-7120.

The Veteran's Administration (VA) is a part of the National Disaster Medical System and can be asked to aid local civilian responders. The VA has a role in maintaining the SNS. VA procures pharmaceuticals for the CDC and manages contracts for the storage, rotation, security and transportation of these items. It is also responsible for the deployment of up to five "push packages" of pharmaceuticals in an emergency. VA has stockpiled 4 packages of pharmaceuticals and medical surgical supplies for the National Disaster Medical System and has a fifth package that is placed on site at high-risk national events such as a presidential inauguration.

Chempack Project

In 2003, the DHS/CDC Strategic National Stockpile Program began pilot testing a project that "forward" positions nerve agent antidote and symptomatic treatments at the local hospital and EMS level. The Chempack Project is a sustainable resource that would enhance local response to nerve agent attacks. As a cost-avoiding measure, the Chempack pharmaceuticals are kept fresh with respect to expiration dating via participation in the Food and Drug Administration Shelf-Life Extension Program (FDA SLE). It is projected that FDA SLE participation will deliver 97 million dollars of cost savings to taxpayers over 10 years; the alternative being outright replacement of product when it reaches the expiration date. Hospitals and EMS participation in this project was voluntary. The pilot test was successful, demonstrating project implementation and maintenance procedures to be feasible. The project expanded throughout 2004, with SNS selecting more localities nationwide to position Chempacks. Where Chempacks have been positioned, hospital pharmacists play a lead role, working with SNS and the local health department to receive, secure and maintain these assets.

Every hospital should have a WMD plan as part of their disaster readiness. The U.S. Army Edgewood Chemical Biological Center and the Army Corps Department of Justice (DoJ) Office of Domestic Preparedness offers training and technical assistance to jurisdictions to help them respond to, and mitigate the consequences of, domestic terrorism.

Chemical Agents

As defined by the U.S. Army, a chemical warfare agent is a "chemical substance intended for use in military operations to kill or seriously injure or incapacitate humans or animals through its toxicologic effects." Terrorists would find chemical agents attractive to use for several reasons, such as they are extremely toxic and readily available or easily synthesized.

In many respects, emergency response systems and healthcare facilities need to respond to a WMD chemical attack in the same fashion as a hazardous materials incident. The same principles regarding triage, decontamination and allocation of resources that shape disaster plans related to hazardous materials go into action during a WMD chemical attack.

Chemical agents are generally classified into several groups: blood agents, nerve agents, choking agents and blistering agents. Each agent has been designated its own North Atlantic Treaty Organization (NATO) designation symbol (a military symbol), which is not its chemical formula. This may be confusing at times. For instance, the NATO designation symbol for cyanide is AC, while its chemical formula is CN. Whereas, the NATO designation symbol for Mace® is CN.

An important principle to recognize concerning chemical agents is that the onset of symptoms is very rapid, typically within minutes. Therefore, prompt initiation of rescue, decontamination, medical attention, and antidotal therapy is critical in minimizing casualties. The Chemical and Biological Hotline (1-800-424-8802) based in Aberdeen, MD, serves as an emergency resource to all healthcare providers for technical assistance. Much of the following information has been adapted from the U.S. Army handbook, USAMRIID's Medical Management of Chemical Casualties Handbook. This reference is located online at www.usamriid.army.mil.

Cyanide

In the form of an NBC agent, cyanide would most likely be encountered as hydrogen cyanide (AC), cyanogen chloride (CK), and cyanogen bromide or cyanogen iodide gases. An explosion of an industrial storage tank containing acetonitrile or acrylonitrile would pose a high risk of delayed cyanide toxicity. Cyanide toxicity is characterized by a rapid onset of dizziness, confusion, dyspnea, tachycardia, and hypertension followed by coma, convulsions, bradycardia, hypotension, arrhythmias, and metabolic acidosis. Death may occur within minutes following significant exposures. Cyanide is classified as a blood agent by some WMD references. A cyanide antidote kit containing amyl nitrite pearls, injectable sodium nitrite, and sodium thiosulfate is available. Nitrites in this kit convert RBC hemoglobin into methemoglobin. Methemoglobin combines with cyanide to form cyanomethemoglobin, which theoretically decreases the amount of free cyanide available. In synergism with 100 percent oxygen, thiosulfate combines with cyanide via the rhodenese enzyme to form less toxic thiocyanate, which is eliminated in the urine. Two other pharmaceuticals, which serve as adjunctive therapy for cyanide poisoning, are injectable sodium bicarbonate to correct metabolic acidosis and benzodiazepines (e.g., diazepam or lorazepam) as anticonvulsants. Their availability is discussed later in the chapter.

Hydroxocobalamin, Vitamin B12A, (Cyanokit®) was approved by the FDA on December 15, 2006. It is manufactured by Dey, L.P. Dey, L.P. is located at 2751 Napa Valley Corporate Drive, Napa, CA 94558. The phone number to the company is (707) 224-3200. Its web site address is www.dey.com. The price of a Cyanokit® is currently unavailble and has a shelf life of 30 months. This chemical binds with cyanide *in vivo* to form nontoxic cyanocobalamin (Vitamin B12).

Another antidote, cobalt edentate (Celocyanor®), is used in Great Britain and France. A more rapid methemoglobin inducer, 4-dimethylaminophenol (4-DMAP), is used in some European countries. Stroma-free methemoglobin has been studied in animals and is not available for human use.

Eli Lilly & Company no longer manufactures the cyanide antidote package. It is now available from Akorn, Inc. Each package contains 12 amyl nitrite pearls, two-10 mL vials of 3 percent sodium nitrite for injection, and two 50-mL vials of 25 percent sodium thiosulfate for injection.

Akorn, Inc. is located at 2500 Millbrook Drive, Buffalo Grove, IL 60089. The phone number to the company is (800) 535-7155. Its Web site address is www.akorn.com. The price of the kit is $274.56 and has a shelf life of 24 months.

Three (3) percent sodium nitrite for injection (300 mg/10 mL vials) is available from Hope Pharmaceuticals, located at 8260 East Gelding, Suite 104, Scottsdale, AZ 85260. The phone number to the company is (800) 755-9595. Its Web site address is www.hopepharm.com. This product costs $42.48 and has a shelf life of 24 months.

Packages of 12 amyl nitrite pearls (0.18 mL or 0.3 mL) is available from James Alexander Corporation, located at 845 Route 94, Blairstown, NJ 07825. The phone number to the company is (908) 362-9266. Its Web site address is www.james-alexander.com. The products costs $3.96 and has a shelf life of 4 years (refrigerated).

This product also is available from Pharma-Tek, Inc, P.O. Box 1920, Huntington, NY 11743-0568. The phone number to the company is (800) 645-6655. Its Web site address is www.pharmatek.com. Please note that for maximum product stability, amyl nitrite inhalant should be packaged in unit-dose containers wrapped loosely in gauze or other suitable material and stored at 2-15°C.

Sodium nitrite powder USP can be used to extemporaneously prepare 3 percent sodium nitrite solution for injection, although no referenced source could be found for compounding instructions. Suppliers will add additional charges for special packaging and shipping. Sodium nitrite powder USP is available from Spectrum Chemicals and Laboratory Products, located at 14422 South San Pedro Street, Gardena, CA 90248. The phone number to the company is (800) 813-1514. Its Web site address is www.spectrumchemical.com. The product costs $21.00/50 grams or $72.00/2500 grams, and has a shelf life of 3 to 5 years.

The product also is available from Integra Chemical Company, located at 710 Thomas Avenue SW, Renton, WA 98055. The phone number to the company is (800) 322-6646. Its Web site address is www.integrachemical.com. The product costs $21.00/500 grams or $72.00/2500 grams.

The product also is available from Ruger Chemical Company, Inc, P.O. Box 806, Hillside, NJ 07205. The phone number to the company is (800) 274-7843. Its Web site address is www.rugerchemical.com. The cost of the product is $3.50-$7.00/pound in bulk.

Fifty (50) mL vials of 25 percent sodium thiosulfate solution for injection are available from American Regent Laboratories, Inc., a subsidiary of Luitpold Pharmaceuticals, Inc. The company is located at One Luitpold Drive, Shirley, NY 11967. The phone number to the company is (800) 645-1706. Its Web site address is www.luitpold.com. The cost of the product is $22.50/vial and has a shelf life of 36 months.

Nerve Agents

Nerve agents are organophosphates, which are potent inhibitors of acetylcholinesterase enzymes. Some examples of this group are sarin (GB), soman (GD), tabun (GA), GF, and VX. Signs and symptoms of nerve agent or organophosphate poisonings include muscarinic, nicotinic, and CNS findings. Muscarinic symptoms include "SLUDGE BAM": salivation, lacrimation, urination, defecation, gastric secretions, emesis, bronochspasm, bronchorrhea, bradycardia, abdominal cramping, and miosis. Nicotinic symptoms may include tachycardia, mydriasis, hypertension, muscle weakness, fasciculations, and respiratory paralysis. CNS symptoms range from blurred vision, restlessness, anxiety, and headaches to seizures and coma. Serious poisonings are managed with 3 antidotal agents. Atropine sulfate blocks muscarinic receptor sites, thus reversing all SLUDGE BAM complications. Glycopyrrolate (Robinul®), a quaternary ammonium anticholinergic medication, has been proposed as an adjunctive agent in the management of organophospate insecticide poisoning; however, it is not routinely administered. Pralidoxime chloride (Protopam®), also known as 2-PAM, regenerates cholinesterase enzyme activity. Administration of 2-PAM reverses nicotinic complications and works synergistically with atropine sulfate to correct muscarinic and CNS symptomatology. Obidoxime dichloride (Toxogonin®, LUH6) is an alternative agent to pralidoxime chloride that is available in some European countries but is not FDA-approved for use in the U.S. HI-6 (asoxime chloride) is an oxime currently under investigation. It is important to note that both atropine sulfate and pralidoxime chloride must be stocked for adequate antidote preparedness; atropine sulfate alone will correct muscarinic signs and symptoms, but 2-PAM is necessary to correct nicotinic signs and symptoms. Diazepam, lorazepam and other benzodiazepines as anticonvulsants are adjunctive measures to treat nerve agent-induced seizures. Barbiturates (eg, Phenobarbital) may be considered for seizures refractory to benzodiazepines. Topical ocular homatropine or atropine can relieve miosis, pain, dim vision, and nausea. Pyridostigmine bromide (Mestinon®) dosed at 30 mg orally every 8 hours serves as a "prophylactic antidote" or "antidote enhancement." Based on animal studies, pretreatment provides some protection against nerve agents, especially soman, which demonstrates rapid aging of the cholinesterase enzyme. On February 5, 2003, the FDA approved the use of pyridostigmine bromide to increase survival after exposure to soman nerve agent poisoning. Pyridostigmine bromide binds reversibly to 30 percent of the cholinesterase enzymes, thus, temporally protecting the enzyme from the nerve agent. Tablets are provided to military personnel in combat zones; pyridostigmine stockpiling in civilian hospitals is unnecessary and not recommended.

Pralidoxime chloride (2-PAM, Protopam®) was previously available from Wyeth, but as of August 2003, this product was sold to Baxter, Inc. is located at 95 Spring Street, New Providence, NJ 07974. The phone number to the company is (908) 286-7000. Its Web site address is www.baxter.com. The cost of the product is $108.38/vial and has a shelf life of 5 years.

Atropine sulfate for injection is available as a generic product from a variety of manufacturers. Some common forms are 0.4-mg/mL, 20-mL multidose vials (8 mg/vial) with a shelf-life of 24 to 36 months; syringe sizes (0.1-mg/mL, 5-mL or 10-mL syringes); and 1-mg/mL, 1-mL single-dose vials.

Stability of expired injectable atropine sulfate solutions was addressed in one study. The authors tested several samples ranging from in date to 12 years beyond expiration. They noted high concentrations of atropine sulfate in clear colorless solutions along with an absence of breakdown products in the test samples. These findings suggest their potential utility in times of emergency. Note: Readers are advised to refer to the Drug Topics Red Book or nearest pharmacy wholesaler for a complete list of drugs.

Atropine sulfate USP powder is available from several suppliers. This pharmaceutical grade powder may be used to prepare atropine sulfate for injection extemporaneously in large amounts. Two references were identified that give instructions for preparing intravenous (IV) solutions extemporaneously. The manufacturers of this pharmaceutical grade chemical have provided no shelf-life guidelines. One study of extemporaneously prepared atropine sulfate 1 mg/1 mL, in 100-mL multidose bags demonstrated stability of over 95 percent at controlled temperatures of 22.2°C (72°F) and 37.8°C (100°F) for periods of up to 72 hours. When ordering product, additional charges are added for special packaging and shipping. The product is available from Ruger Chemical Company, located at 83 Cordier Street, Irvington, NJ 07111. The phone number to the company is (800) 274-2636; fax (973) 926-4921. Its Web site address is www.rugerchemical.com. The cost of the product is $11.20/5 grams or $44.80/125 grams.

The product also is available from Spectrum Chemicals and Laboratory Products, located at 14422 South San Pedro Street, Gardena, CA 90248. The phone number to the company is (800) 772-8786. Its Web site address is www.spectrumchemical.com. The cost of the product is $20.45/5 grams or $80.60/25 grams.

The Letco Companies also has the product available. The company is located at 1316 Commerce Drive NW, Decatur, AL 35601. The phone number to the company is (800) 239-5288; fax (256) 353-7237. Its Web site address is www.letcoinc.com. The cost of the product is $16.00/25 grams or $415/1 kilogram.

Diazepam for injection is available as a generic product from a variety of manufacturers. Some common forms are 5-mg/mL, 2-mL prefilled syringes (10 mg/syringe) with a shelf life of 24 months; 5-mg/mL, 2-mL ampoules (10 mg/amp); and 5-mg/mL, 10-mL multidose vial (50 mg/vial).

Lorazepam for injection is available as a generic product from a variety of manufacturers. Some common forms are 2-mg/mL, 1-mL single-dose vial (2 mg/vial) with a shelf life of 18 to 24 months; 2-mg/mL, 10-mL multidose vial (20 mg/vial); 4-mg/mL, 1-mL single-dose vial (4 mg/vial); and 4-mg/mL, 10-mL multidose vial (40 mg/vial).

Military style autoinjectors containing atropine sulfate, pralidoxime chloride, and diazepam are available from Meridian Medical Technologies, Inc, the only FDA-approved manufacturer in the U.S. These autoinjectors are manufactured in large quantities of 25,000 and 40,000. Smaller orders may be filled if the product is in company inventory; however, the shelf life may be shorter. In 2003, the FDA approved the marketing of infant and pediatric strength AtroPen® autoinjectors; atropine sulfate 1 mg/0.7 mL, 0.5 mg/0.7 mL and 0.25 mg/0.3 mL. The doses approved for use in children and adolescents with mild symptoms of nerve agent poisoning include: 0.25 mg for infants weighing less than 15 lbs (generally less than 6 months), 0.5 mg for children weighing between 15 and 40 lbs (generally 6 months to 4 years),

1 mg for children weighing between 40 and 90 lbs (generally 4 to 10 years of age), and 2 mg doses for adults and children weighing over 90 lbs (generally over 10 years of age). For infants and children with symptoms of severe nerve agent poisoning, doses up to 3 times these doses may be given.

Each Mark I Kit (Nerve Agent Antidote Kit) contains one AtroPen® (atropine sulfate; 2 mg/0.7 mL) and one ComboPen® (pralidoxime chloride; 600 mg/2 mL).

The diazepam autoinjector contains 10 mg diazepam in 2 mL. Both the Mark I Kit and the diazepam autoinjector are available from Meridian Medical Technology, Inc., located at 10240 Old Columbia Road, Columbia, MD 21046. The phone number to the company is (800) 638-8093; fax (443) 259-7801. It Web site address is www.meridianmeds.com. Per Meridian Medical Technologies (MMT): "Please note that MMT requires the following information to accompany a purchase order (PO), excluding the training kits which do not contain drugs: a physician's prescription for all items; a copy of the DEA registration certificate is required if the PO includes the diazepam auto-injector; the PO must include the following wording: 'We certify that the items purchased under PO # .... will be used only by ....{company name}. The material will not be sold to a third party, distributed or used for any other purpose.'"

Nerve Agent Autoinjectors by Meridian Medical Technologies						
Autoinjectors	Item Number	Shelf-Life	Units/Box	Price/Box	Unit Price
Mark I Kit (NAAK)	NSN 6505-01-174-9919	5 y	30	$916.500	$30.55
Atropen® 2 mg	NDC 11704-106-01	3 y	12	$201.84	$16.82
Atropen® 1 mg	NDC 11704-105-01	3 y	12	$201.84	$16.82
Atropen® 0.5 mg	NDC 11704-104-01	3 y	12	$201.84	$16.82
Atropen® 0.25 mg	NDC 11704-107-01	3 y	12	$241.20	$20.10
ComboPen® (2Pam C1) 600 mg	NSN 6505-01-125-3248	5 y	100	$2175.00	$21.75
Diazepam (CANA) 10 mg, C-IV	NSN 6505-01-274-0951	4 y	15	$375.00	$25.00

As of February 2004, Bound Tree Medical, Inc., became the exclusive distributor of all Meridian autoinjector products to hospital pharmacies. Bound Tree Medical, Inc is located at 6106 Bausch Road, Galloway, OH 43119, and can be reached at (800) 533-0523 or by fax at (800) 257-5713. Their web address is www.boundtree.com.

Diomed in Istanbul, Turkey, manufactures DIO-ATRO®, an autoinjector containing atropine sulfate 2 mg and pralidoxime chloride 600 mg and has a NATO stock #6515-27-013-3995.

Pulmonary or Choking Agents

Chlorine (Cl), phosgene (CG), diphosgene (DP), and chloropicrin act primarily as pulmonary irritants causing cough, shortness of breath, and dyspnea. However, it may be several hours before serious complications become evident (e.g., pulmonary edema). No specific antidote is available for treatment of these exposures. Symptomatic and supportive care may include administration of oxygen, ventilatory support, and bronchodilators such as albuterol sulfate. Nebulized 3.75 percent sodium bicarbonate has provided dramatic symptomatic improvement in chlorine exposures, as noted in several anecdotal case reports. This may be prepared by mixing 2 mL of 7.5 percent sodium bicarbonate for injection with 2 mL of sterile 0.9 percent sodium chloride. Antibiotics should be reserved for those patients with an infectious process documented by sputum gram staining and culture. Parenteral steroids may be indicated in those patients demonstrating latent or overt reactive airway disease. Ipratropium bromide (Atrovent®) may be used adjunctively following chloropicrin exposure.

Albuterol sulfate is available as a generic product in a variety of forms such as a solution for nebulization, 0.09 mg/inhalation (17 grams) with a shelf life of 18 to 36 months; a solution for inhalation, 0.083 percent 3 mL vials; and a solution for inhalation, 0.5 percent 20 mL multidose vial.

Sodium bicarbonate for injection is available in concentrations of 4.2 percent, 7.5 percent, and 8.4 percent in syringes and vials in sizes of 10 mL or 50 mL, and has a shelf life of 18 months.

Methylprednisolone acetate for injection is available as a generic product in several concentrations, including 20-mg/mL, 10-mL vials (200 mg/vial) with a shelf life of 24 to 36 months; 40-mg/mL, 5-mL vials (200 mg/vial); and 80-mg/mL, 5-mL vials (400 mg/vial).

Blister Agents

A number of potent alkylating agents may be used as chemical warfare agents. Examples include nitrogen mustard (HS), distilled mustard (HD), mustard gas, phosgene oxime (CX), and Lewisite (L). Toxicity produced by these agents includes blisters, vesiculations, eye injury, airway damage, vomiting and diarrhea, and bone marrow stem cell suppression. Blisters may form several hours after contact with the skin. Erythema may be treated with calamine or other soothing lotion or cream. Denuded skin areas should be treated with topical antibiotics such as silver sulfadiazine or mafenide acetate. Systemic analgesics should be used liberally. For eye exposures, 2.5 percent sodium thiosulfate irrigations, homatropine ophthalmic ointment (or other mydriatics), topical antibiotics and topical steroids may be indicated depending on the severity of injury. Atropine or other anticholinergic agents or antiemetics should control the early nausea and vomiting. Antibiotics are necessary to treat infections, which are usually the cause of death. Colony stimulating factors [e.g., filgrastim (Neupogen®), sargramostin (Leukine®)] may be considered for patients demonstrating serious cytopenia. Lewisite is the only blistering agent for which an antidote may be useful since it is an arsenic derivative. The antidote dimercaprol (British anti-Lewisite; BAL) is a chelating agent for arsenicals and other heavy metals. BAL administration may reduce systemic toxicity of Lewisite. Although not commercially available, 5 percent BAL skin and eye ointments may reduce the severity of lesions when applied soon after decontamination. Since BAL is formulated in peanut oil, it must be given intramuscular (IM). BAL is available as 300 mg/3 mL vials in quantities of 10 from Akorn, Inc, located at 2500 Millbrook Drive, Buffalo Grove, IL 60089. The phone number to the company is (800) 535-7155. Its Web site address is www.akorn.com. The cost of the product is $716.45 and has a shelf life of five years.

Dimercaptosuccinic acid (DMSA, succimer, Chemet®) has been used experimentally in animals for the treatment of Lewisite exposures. It may be preferred in the treatment of multiple exposures because it is administered orally and exhibits fewer adverse reactions. Chemet® is supplied in 100-mg capsules and is available in quantities of 100 from Ovation Pharmaceuticals, Inc, located at Four Parkway North, Deerfield, IL 60015. The phone number to the company is (847) 282-1000, fax (847) 282-1001. Their web site address is www.ovationpharma.com. The average wholesale price of the product is $663.21 and has a shelf life of 24 months.

DMPS, which is the sodium salt of 2,3-Dimercapto-1-propanesulfonic acid, is a chelating agent available in Europe demonstrating some efficacy in treating Lewisite-exposed experimental animals.

The colony stimulating factors are available as: (1) Leukine® in 250-mcg/mL, 1 mL, multidose vial ($154.91) and 500-mcg/mL, 1 mL, multidose vial ($309.82). This product is available from Berlex Laboratories Inc, located at 1191 Second Avenue, Seattle, WA 98101-2120. The phone number to the company is (888) 237-5394. Its Web site address is www.berlex.com. The shelf life is 18 months for the liquid product, and 36 months for the powder product; and (2) Neupogen® in 300 mcg/0.5 mL ($236.28), 480 mcg/0.8 mL ($376.44), and 480 mcg/ 1.6 mL ($343.20). These products are available from Amgen, Inc, located at One Amgen

Center Drive, Thousand Oaks, CA 91320. The phone number to the company is (805) 447-1000. Its Web site address is www.amgen.com. The product has a shelf life of 24 months.

Based on experimental studies, it has been proposed that administration of sodium thiosulfate 12.5 grams may act as a "mustard scavenger" following exposure to mustard agents. Potential benefits of sodium thiosulfate therapy following mustard agent exposure are not known. See cyanide section for suppliers of sodium thiosulfate.

Incapacitating Agents

BZ, also known as QNB (3-quinuclidinyl benzilate), and Agent 15 (the Iraqi equivalent of BZ) are anticholinergic agents, which incapacitate victims by causing delirium and hallucinations. BZ and related anticholinergic compounds can be synthesized in clandestine laboratories. Patients demonstrate anticholinergic signs and symptoms (e.g., mydriasis, tachycardia, dry flushed skin, urinary retention, etc). Symptoms of agitation and hallucinations may be managed with benzodiazepines. Serious symptoms may be reversed by IV physostigmine salicylate (Antilirium®), a reversible carbamate cholinesterase inhibitor. Neostigmine methyl-sulfate (Prostigmin®) and pyridostigmine bromide (Mestinon®) are quaternary amines that do not cross the blood-brain barrier, and therefore, will not reverse CNS symptomatology caused by these agents. Antilirium® is available in 2-mL vials (1 mg/mL) in quantities of 10 vials from Akorn, Inc, located at 2500 Millbrook Drive, Buffalo Grove, IL 60089. The phone number to the company is (800) 535-7155. Its Web site address is www.akorn.com. The cost of the product is $48.60 and has a shelf life of 24 months.

On October 26, 2002, the Russian government used aerosolized fentanyl as an incapacitating agent during a terrorist attack in a Moscow theater. This incident resulted in the death of over 100 hostages and terrorists and critically injured many more. Fentanyl is a very potent opiate narcotic with significant exposures causing sedation, miosis, bradycardia, hypotension, and respiratory depression leading to coma and respiratory arrest. Fentanyl toxicity may be completely reversed by adequate doses of the opiate antagonists naloxone (Narcan®) or nalmefene (Revex®). Injectable naloxone is available generically in 0.4-mg/1 mL ampoules, 0.4 mg/1 mL syringes, 1-, 2-, and 10-mg vials, and 1-mg/1 mL in 2-mL ampoules. Nalmefene is available in ampoules of 1 mg/2 mL and 0.1 mg/1 mL from Baxter, Inc.

Riot Control Agents

These agents are commonly known as CN (alpha-chloroacetophenone, Mace®), CS (ortho-chlorobenzylidene malononitrile), or CR (dibenoxazepine) tear gas. Diphenylaminochloroarsine (adamsite) and diphenylchloroarsine are riot control/vomiting agents that are organic arsenicals; however, they do not cause systemic arsenic poisoning. Some "pepper spray" products contain hot pepper extracts such as oleoresin capsicum. Usually, the exposure effects of these agents are self-limiting. They include burning, itching, and watering of the eyes, burning and tingling of the skin, and respiratory discomfort. In most cases, no specific therapy is indicated other than basic measures such as movement of the patient to fresh air, eye irrigation with water and skin washing. More pronounced symptoms such as broncho-spasm and pulmonary edema are possible with significant exposures (ie, in enclosed spaces). Symptomatic patients may require supplemental oxygen, ventilatory support, and inhaled beta agonist with or without systemic steroids.

Nuclear Agents

Although the detonation of a nuclear weapon is a concern with respect to global and international conflicts, it is considered to be very difficult for a terrorist group to obtain, build, conceal, or deliver such a weapon. It is believed that a "dirty bomb" is more likely to be deployed. "Dirty bombs" are conventional explosives combined with radioactive material and are designed to frighten the populace and cause a major hazardous material cleanup rather than many injuries.

Radiation emitted by radioactive materials can be characterized into three types: alpha particles, beta particles, and gamma rays. The health hazards of radiation are divided into acute or chronic exposure risks. Chronic exposures (i.e., a low dose over a long period of time) increase the risk of cancers and cataracts; while acute exposures produce nausea, vomiting, blood dyscrasias, and death. Since alpha and beta particles do not travel great distances, materials such as protective clothing easily block them. Inhalation or ingestion exposures pose the greatest potential for harm. Gamma rays are the most harmful because they travel great distances and require dense materials, such as lead, to block penetration of tissues.

After exposure to a radiologic agent, patients may require treatment with either a chelator or a radionuclide blocker. Chelating agents for radionuclides are available through Radiation Emergency Assistance Center/Training Site (REAC/TS). Insoluble Prussian Blue (ferric hexacyanoferrate) is indicated for cesium and thallium chelation therapy. Prussian Blue is a pharmaceutical grade product obtained from Germany under the brand name Radiogardase®. It is available in 500-mg capsules, manufactured by Heyl GmbH, located in Berlin, Germany. On October 2, 2003, the FDA approved a New Drug Application for Radiogardase® to treat patients exposed to harmful levels of radioactive cesium or radioactive and nonradioactive thallium. Prussian Blue is now a component of the Strategic National Stockpile. Radiogardase® is available from Heyltex Corporation in Katy, Texas, located at 1800 South Mason Road, Suite 260, Katy, TX 77450. The phone number to the company is (281) 395-7040. Its web site address is www.heyltex.com. It has a price of $ 80/30 capsule bottle with a minimum order of 25 bottles, and has a shelf life of three years. Bottles may be purchased individually from the company with a physician's prescription.

Zinc-diethylenetriamine pentaacetic acid (Zn-DTPA), also known as pentetate zinc trisodium, and calcium-DTPA (Ca-DTPA), also known as pentetate calcium trisodium, are chelators for radioactive transuranic elements (see Table 2). On May 20, 2005, Ca-DTPA and Zn-DTPA were approved by the FDA for treatment of individuals with known or suspected internal contamination with plutonium, americium or curium to increase rates of elimination. They are currently available to hospitals and state and local health agencies through Akorn, Inc. www.akorn.com. Both of these drugs are supplied as 200mg/mL 5 mL, 1 gram ampuls.

REAC/TS trains, consults and assists in the response to all types of radiation accidents or incidents. The center utilizes physicians, nurses, health physicists, radiologists and emergency coordinators to provide 24-hour assistance at the local, national or international level.

Other pharmaceutical chelators may be used for a variety of radionucleotide exposures and would be found in any well-stocked hospital pharmacy. These include d-penicillamine (Cuprimine®), calcium disodium EDTA (Versenate®), dimercaprol (BAL®), succimer (Chemet®) and deferoxamine (Desferal®).

Chelating Agents for Radionucleotides	
Chelating Agent	**Radionucleotide**
Calcium and zinc – DTPA	Americium (^{243}Am), californium, cesium, curium, lanthanum, lutetium, plutonium, promethium, scandium, uranium, yttrium, zinc, rare earth metals
Calcium disodium EDTA	Copper, lead, and uranium
D-penicillamine	Copper, lead, and mercury (^{203}Hg)
Deferoxamine mesylate	Iron
Dimercaprol	Copper, mercury, and polonium
Succimer / DMSA	Lead and mercury

Colony stimulating factors, filgrastim (Neupogen®) and sargramostin (Leukine®), may be considered in patients experiencing significant bone marrow suppression. See blister agents section for suppliers of colony stimulating factors.

Radionuclide blocking agents saturate tissues with a nonradioactive element, which reduces the uptake of radioactive iodine. The most commonly used agents are potassium iodide (KI) tablets (130 mg and 65 mg), saturated solution of potassium iodide (SSKI) (300 mg/0.3 mL), and Lugol's solution (10 percent potassium iodide, 5 percent iodine), which reduce uptake of Iodine-131 into the thyroid tissue. Most states with departments of nuclear safety will stock enough quantity of potassium iodide tablets to protect workers and emergency personnel involved in a nuclear reactor incident. Some states offer supplies of KI to the general public residing in areas close to nuclear reactors. Sodium iodide (NaI) can theoretically be used instead of potassium iodide; however, no such pharmaceutical product is available in the United States. Sodium iodide USP powder is available from several sources listed in the 2005 Drug Topics Redbook. ThyroShield® is a recently approved KI solution for ease and use in pediatric administration; it is available in 30-mL bottles, in a concentration of 65 mg/mL. Several other examples of substances employed as radionuclide blockers include: sodium alginate for strontium, chlorthalidone for rubidium, and Lugol's for technetium.

The following table is provided by the U.S. Food and Drug Administration, Center for Drug Evaluation and Research (www.fda.gov/cder/guidance/4825fnl.htm). It gives the recommended doses of KI for various age and risk groups.

Threshold Thyroid Radioactive Exposures and Recommended Doses of KI for Different Risk Groups				
	Predicted Thyroid Exposures (cGy)	KI Dose (mg)	# of 130-mg Tablets	# of 65-mg Tablets
Adults over 40 y	≥500	130	1	2
Adults over 18-40 y	≥10	130	1	2
Pregnant or lactating women	≥5	130	1	2
Adolescents over 12-18 y*	≥5	65	1/2	1
Children over 3-12 y	≥5	65	1/2	1
Over 1 month to 3 y	≥5	32	1/4	1/2
Birth through 1 month	≥5	16	1/8	1/4

*Adolescents approaching adult size (≥70 kg) should receive the full adult dose (130 mg).

Note: KI from tablets (either whole or fractions) or a fresh saturated KI solution may be diluted in milk, formula, or water, and the appropriate volume administered to babies. A home preparation procedure for emergency administration of KI tablets to infants and small children using water, milk, juice, syrup, or soda pop can be found at www.fda.gov/cder/drugprepare/kiprep.htm. The KI prepared in these liquids will keep for up to 7 days in the refrigerator. The FDA recommends that the potassium iodide drink mixtures be prepared weekly; unused portions should be discarded.

Lugol's solution and SSKI are available from several suppliers; they have shelf lives of 2 to 5 years; see *2005 Drug Topics Red Book*.

FDA-approved manufacturers of KI tablets are: (1) Anbex, Inc, located at 15 West 75th Street, New York, NY 10023, which manufactures Iosat®. The phone number to the company is (212) 580-2810. Its Web site address is www.anbex.com. The product is available as 130-mg tablets with 14 tablets to a package. The cost of the product is $10.00/package and has a shelf life of 4 to 5 years; (2) Medpointe Inc. (www.nitro-pak.com or www.medpointeinc.com) manufactures Thyro-Block®, which is distributed by Nitro-Pak, Inc; their phone number is (800) 866-4876. The phone number to the manufacturer is (732) 564-2200. Thyro-Block® is available as 130-mg tablets with 14 tablets to a bottle; a case is 100 bottles. The cost of a single bottle is $9.95; a case of 100 bottles costs $799. Medpointe, Inc, markets potassium iodide tablets directly to nuclear power plants and utility companies. (3) Reciep AB (www.thyrosafe.com) manufactures and distributes Thyrosafe®. The phone number to the company is (866) 849-7672. The product is available as 65-mg tablets with 20 tablets per package. The cost of the product is $9.95 per package. (4) Fleming & Company Pharmaceuticals, 1733 Gilsinn Lane, Fenton, MO 63026, manufactures and distributes ThyroShield® liquid. The phone number to the company is (636) 343-8200, fax (636) 343-8203. The cost is $13.25 per each-ounce bottle and has a shelf life of 5 years.

For immediate assistance regarding a nuclear or radiological agent incident, contact the REAC/TS, located at P.O. Box 117, MS-39, Oak Ridge, TN 37831-0117. Their phone number is (865) 576-3131 (business hours); (865) 576-1005 (24-hour emergency line). Also contact the Department of Nuclear Safety - for the state in which the incident occurred - and the health physicists affiliated with hospital nuclear medicine departments, who can serve as expert consultants for radiation incidents.

Biological Agents

As NBC terrorist weapons, biological agents may be encountered as bacteria (eg, anthrax), viruses (eg, smallpox), or toxins (eg, botulinum or ricin). This fact will explain the striking differences in the manner that victims present to healthcare facilities. A catastrophe caused by detonation of a chemical weapon (eg, nerve agent or cyanide) would be characterized by immediate death or severe disablement of individuals at the site of the attack. First responders to such an incident would be paramedics, police, and other emergency personnel.

In contrast, with respect to biological agents, there is a delayed onset of signs and symptoms since incubation may take days. Emergency department clinicians and primary care practitioners would be the first to recognize and manage exposed patients. Knowing what medical and pharmaceutical interventions may be requested in a mass casualty event is crucial. It is beyond the scope of this chapter to delineate all the diagnostic clues, pathophysiology, laboratory monitoring, infectious disease control, guidelines for patient isolation, epidemiologic procedures, and public health ramifications of a bioterrorism event. Clinicians, however, should take note of this extensive list of pharmaceuticals (both oral and parenteral) that may be required in extraordinarily large amounts during an outbreak involving a biological weapon. Additionally, the duration of illness may be weeks to months, thus creating an additional stress on the healthcare infrastructure. Once the cause of illness in a large number of patients has been identified as a biological agent, prompt availability and distribution of appropriate medication can greatly mitigate the destructive impact of the act of terrorism. It should be noted that special products, uncommonly used vaccines and antitoxins would be provided via federal government storage and distribution programs. Pharmacist and public health personnel in the drug delivery system must be aware that it may require 24-48 hours or more for material in the strategic national stockpile to be transported, broken down, and delivered to local distribution centers. See discussion of the SNS in the introduction section. Adjunctive medications such as analgesics and antipyretics should be readily available to manage symptoms such as headache, fever, and myalgias. New therapies are under development. Detailed information on diagnosis, patient management, vaccines, etc, may be obtained through COMMANDER U.S. Army Medical Research Institute of Infectious Diseases (USAMRIID) at (301) 619-2833 during business hours, or at (888) USA-RIID, 24 hours a day. In the event of an emergency, contact the National Response Center at (800) 424-8802. Also, contact the CDC emergency operations center at (770) 488-7100.

The following information is a very brief synopsis of 14 possible threats as biological agents. Much of this information has been adapted from an article in JAMA: "Clinical Recognition and

Management of Patients Exposed to Biological Warfare Agents," and *USAMRIID's Medical Management of Biological Casualties Handbook*, 6th edition, April 2005.

Bacteria

Anthrax The etiologic agent causing anthrax is *Bacillus anthracis* (*B. anthracis*), a gram-positive spore-forming bacillus. As a biological weapon, the spores of these bacteria would be aerosolized, with inhalation being the primary route of exposure. The clinical course is characterized by a necrotizing hemorrhagic mediastinitis. Initially, symptoms may resemble a flu-like illness with fever, fatigue, malaise, vague chest pain, and nonproductive cough. Initial symptoms are followed by abrupt progression to dyspnea, stridor, diaphoresis, and cyanosis. Systemic complications of sepsis, shock, and meningitis may occur in up to half of the cases.

Unfortunately, once symptoms occur, treatment is usually ineffective. Intravenous ciprofloxacin should be initiated at the earliest sign of anthrax. Other fluoroquinolones may be substituted; however, no animal studies exist for quinolones other than ciprofloxacin. All natural strains of anthrax have been found to be sensitive to erythromycin, chloramphenicol, and gentamicin.

Historically, penicillin G has been the drug of choice for anthrax. An alternative regimen is doxycycline and one or two other antibiotics with *in vitro* activity against *B. anthracis*, such as rifampin, vancomycin, penicillin, ampicillin, chloramphenicol, imipenem, clindamycin, and clarithromycin. Ampicillin and penicillin should not be used alone because of possible beta-lactamase production. Chemoprophylaxis with oral ciprofloxacin, doxycycline, or amoxicillin, if the strain of *B. anthracis* is proven susceptible in exposed individuals, should be initiated and continued for at least 60 days or until 3 doses of anthrax vaccine are administered. A licensed attenuated vaccine, anthrax vaccine adsorbed (BioThrax®) in 5-mL multidose vials, is available for prophylaxis. This vaccine stock is owned by the Department of Defense. In order to obtain it, contact USAMRIID at (301) 619-2833. It is manufactured by BioPort Corporation, located at 3500 North Martin Luther King Jr. Boulevard, Lansing, MI 48906. The phone number to the company is (517) 327-1500. Its Web site address is www.bioport.com. The cost of 5 mL (10 doses) of the product is $1331.19 and has a shelf life of 18 months. The product must be stored between 2°C to 8°C. Per BioPort Corporation, as of February 2006, the only way one could obtain BioThrax® was by writing a letter describing the reason for the request, the number of people to be vaccinated, and contact information. Once this letter is received, it would go through an approval process at the Department of Defense. Depending on the level of risk, they would either accept or reject the request.

BioPort Corporation currently has vaccine available for civilian personnel and is very close to signing a domestic distributor for the management of the sale of BioThrax®. Once this becomes official, a physician will be able to contact the distributor to find out if he/she is eligible to receive the vaccine. The supply of the vaccine is still low in comparison to the very high demand, but BioPort will be selling the vaccine to those civilian markets most at risk and most vital to our country.

Brucellosis Human infection may be caused by four species of *Brucella*, which is a nonspore-forming gram-negative aerobic coccobacillus. Clinical manifestations include fever, chills, and malaise, which may lead to cough and pleuritic chest pain. Other complications may include osteomyelitis, genitourinary infection, hepatitis, endocarditis, and CNS infections. To prevent the possibility of relapse, combination therapy is advised. Various antibiotic regimens have been proposed from the following antimicrobials: doxycycline, gentamicin, streptomycin, rifamin, ofloxacin, sulfamethoxazole/trimethoprim (SMZ/TMP). There are no approved vaccines or chemoprophylaxis treatments.

15

Cholera This infection is caused by *Vibrio cholerae*, a gram-negative nonspore-forming bacillus. Clinical manifestations include vomiting, abdominal distention, and pain, with little or no fever, followed rapidly by diarrhea. Fluid losses may be excessive with death caused by dehydration and shock. Antibiotic treatment may include tetracycline, ampicillin, and SMZ/TMP. Intravenous fluid/electrolyte solutions are necessary to treat dehydration. At the present time, the manufacture and sale of the only licensed cholera vaccine in the U.S. (Wyeth-Ayerst) has been discontinued. Two recently developed vaccines for cholera are licensed and available in other countries [Dukoral, ® Biotec AB (www.activebiotech.com) and Mutacol, ® Berna (www.berna.org)]; however, neither of these two vaccines are available in the U.S.

Glanders Glanders and Melioidosis are caused by *Burkholderia mallei* and *Burkholderia pseudomallei,* respectively. Both are gram-negative bacilli with a safety pin appearance on microscopic examination. Both pathogens affect animals (e.g., horses, mules, donkeys) and human beings. Symptoms of inhalation exposure include high fever, rigors, sweating, myalgias, headache, pleuritic chest pain, cervical adenopathy, hepatosplenomegaly, and generalized papular/pustular eruptions. Pulmonary disease may progress to potentially fatal bacteremia and septicemia. Oral antibiotic regimens include amoxicillin/clavulanate, tetracycline or sulfamethoxazole/trimethoprim given for 60-150 days. For serious systemic disease, administer parental Ceftazidime and SMZ/TMP for two weeks followed by oral therapy for six months. There are currently no available vaccines for human use. Chemoprophylaxis may be considered using SMZ/TMP.

Pneumonic Plague The gram-negative, nonspore-forming bacillus, *Yersinia pestis*, is responsible for both pneumonic and bubonic plague. Patients exposed to pneumonic plague as a biological weapon will present with high fever, chills, malaise, cough with bloody sputum, headache, myalgia, and sepsis. Late in the course of illness, dyspnea, cyanosis, and respiratory failure may be noted. Effective antibiotic therapy includes streptomycin or gentamicin. Alternative drugs are ciprofloxacin, chloramphenicol, and doxycycline. In the U.S., a licensed, killed, whole bacilli vaccine was discontinued by its manufacturer in 1999, and is no longer available. Exposed individuals may be treated with doxycycline, ciprofloxacin, or chloramphenicol for 7 days for chemoprophylaxis.

Q-Fever Q-fever is a rickettsial disease caused by *Coxiella burnetii*. The most common symptoms of Q-fever are fever, chills, headache, fatigue, diaphoresis, malaise, anorexia, and myalgias. In some cases, cough with chest pain may be noted. Rare complications include hepatomegaly, splenomegaly, and jaundice. Effective therapies for Q-fever include tetracycline or doxycycline for 15 to 21 days; alternatives are ofloxacin or pefloxacin. Hydroxychloroquine added to antibiotic therapy has increased the effectiveness of therapy, especially in patients with chronic Q-Fever endocarditis. There is currently no commercially available Q-fever vaccine in the U.S. One vaccine, Q-VAX®, has been successfully tested and is commercially available for use in humans in Australia. Tetracycline or doxycycline may be given as chemoprophylaxis to exposed patients.

Tularemia Tularemia is caused by *Francisella tularensis*, a gram-negative aerobic coccobacillus. It is known as "rabbit fever" or "deer fly fever." Inhalation of tularemia organisms produces a typhoidal tularemia. Patients present with fever, weight loss, substernal discomfort, and nonproductive cough. The drug of choice is streptomycin. Other treatment options include gentamicin, fluoroquinolones (ciprofloxacin, norfloxacin), tetracycline, and chloramphenicol; however, high relapse rates are associated with these treatment options. Although not commercially available, a live attenuated vaccine is available under Investigational New Drug (IND) status for prophylaxis (available through USARMIID). Doxycycline or tetracycline may be used for chemoprophylaxis.

Viruses

Smallpox The etiologic agent that causes smallpox is the *variola major* virus. Smallpox was declared eradicated by the World Health Organization in 1980. Much concern exists regarding the stockpiling of this infectious agent as a weapon of bioterrorism due to its high morbidity and mortality. Patients infected with variola present with fever, malaise, rigors, vomiting, headache, and backache. Dermal manifestations include appearance of a rash followed by lesions, which appear as macules, then papules, then eventually form pustular vesicles. By the second week, scabs form, which leave depigmented scars upon healing. Patients are contagious until all scabs are healed. All patients exposed to variola virus must be immediately vaccinated. Those U.S. citizens who were vaccinated against smallpox in the 1950s and the 1960s are no longer protected against the virus. The only smallpox vaccines available in the U.S. are Dryvax® (Wyeth Laboratories, Inc) and WetVax® (Aventis Pharmaceuticals, Inc), available by calling the CDC at (404) 639-2888. Both smallpox vaccine products are components of the SNS. The CDC has contracted with Acambis to develop a new smallpox vaccine (ACAM 2000), which may have a more acceptable safety profile. In early 2003, the CDC made smallpox vaccine available through state health departments to civilian healthcare workers on a voluntary basis. Dryvax® is available as one vial of dried smallpox vaccine and one container of diluent (0.25 mL) with 100 sterile bifurcated needles. The manufacture of Dryvax® was discontinued in 1981. The U.S. Army and the CDC maintain a supply of vaccinia immune globulin (VIG), which is used for the treatment of complications due to the vaccinia vaccination. Limited data suggests that VIG may be of value in postexposure prophylaxis of smallpox when given within the first week following exposure, and concurrently with vaccination. Contact the CDC at (800) CDC-INFO or USARMRIID at 301-619-2833 to obtain VIG. There is currently no chemotherapeutic agent proven effective against smallpox. Cidofovir (Vistide®) is not a licensed treatment for smallpox; however, it is being studied under an FDA investigational protocol. Cidofovir is a nucleoside analog DNA polymerase inhibitor, which has been demonstrated *in vitro* studies to inhibit poxvirus replication and cell lysis. Cidofovir is currently licensed for the treatment of CMV retinitis and has demonstrated antiviral activity against poxviruses *in vitro*, and against cowpox and vaccinia viruses in mice. However, its use for the treatment of vaccinia adverse reactions is restricted under an IND protocol. Under the IND, cidofovir would only be used when VIG was not efficacious. Renal toxicity is a known adverse reaction of cidofovir. Although the CDC makes no recommendations for the use of antivirals at this time, it is recommended that healthcare providers continue to consult the CDC at 800-CDC-INFO to obtain updated information regarding treatment options for serious vaccine complications. Many states have surveyed hospital pharmacies to identify local sources of this product if a smallpox outbreak is suspected.

Cidofovir (Vistide®) is available from Gilead Sciences, located at 333 Lakeside Drive, Foster City, CA 94404. The phone number to the company is (800) 445-3235. Its Web site address is www.gilead.com. The product is available as 75-mg/ml, 5-ml vial for $888.00. It has a shelf life of 3 years.

Venezuelan Equine Encephalitis (VEE) Members of the alpha virus genus of the Togaviridae family produce this encephalopathic syndrome. The usual mode of transmission is mosquitoes; however, aerosolization makes those pathogens a very effective WMD. Alpha virus will produce neurologic syndromes noted by fever, headaches, confusion, drowsiness, seizures, dysphasia, ataxia, myoclonus, cranial nerve palsies, photophobia, myalgias, and vomiting. No specific chemotherapeutic agents are indicated. Treatment is symptomatic and supportive care. Antipyretics and anticonvulsants may be used in severe cases. A live attenuated vaccine for VEE TC-83 is available for prophylaxis, while inactivated vaccines are under IND status. A monoclonal antibody has been developed and is in the animal test phase for protection against infection and disease when given before or up to 24 hours after an airborne challenge with virulent virus.

Viral Hemorrhagic Fevers (VHF)   The most widely known examples of this group are the Ebola and Marburg viruses; these belong to the family Filoviridae, which are enveloped, nonsegmented, negative-stranded RNA viruses. Common features of VHF are myalgias, fever, and prostration. Mild symptoms include conjunctival injection, mild hypotension, flushing, and petechial hemorrhaging that may progress to shock. More severe symptoms include mucous membrane hemorrhage with maculopapular rashes and disseminated intravascular coagulation (DIC). Other VHF agents include Crimean Congo hemorrhagic fever (Bunyaviridae family), Hanta virus (Bunyaviridae family), and Lassa fever (Arenaviridae family). No specific antiviral agents are effective against Ebola or Marburg viruses. Other related strains (eg, Crimean Congo, Lassa) may respond to ribaviran. Postexposure prophylaxis may be considered with oral ribavirin. Many different pharmaceutical agents may need to be employed in the supportive management of hypotension, shock, and DIC. No vaccines or medicinals exist to protect against these viral illnesses.

Ribavirin (Virazole®) is available from ICN Pharmaceuticals, Inc, located at 3300 Hyland Avenue, Costa Mesa, CA 92626. The phone number to the company is (800) 548-5100. Its Web site address is www.valeant.com. The product is available as 6-g vials of lyophilized powder for $1573.80. It has a shelf life of 5 years.

Ribavirin (Rebetol®) is available from Schering Plough Corporation, located at 2000 Galloping Hill Road, Kenilworth, NJ 07033. The phone number to the company is (800) 222-7579. Its Web site address is www.schering-plough.com. It markets 200-mg capsules, available in bottles of 42, 56, 70, and 84, which cost $10.60/tablet. Its shelf life is "proprietary information."

Toxins

Botulinum   Botulinum toxin is a protein exotoxin produced by *Clostridium botulinum*, an anaerobic gram-positive bacillus. There are 7 types of botulism neurotoxins known as types A-G. Botulism poisonings are more commonly associated with improperly processed or canned foods. As a WMD, botulinum toxin may be inhaled from an aerosol or ingested in the form of sabotaged food. These toxins are the most toxic of all the NBC weapons. Clinical manifestations include blurred vision, mydriasis, diplopia, ptosis, photophobia, dysarthria, dysphonia, and dysphasia. Skeletal muscle paralysis follows, which presents as a symmetrical and descending progressive weakness resulting in respiratory failure. Patients are typically awake, alert, and afebrile. A trivalent equine antitoxin (types A, B, and E) for food-borne botulism is available from the CDC at 8 urban quarantine sites: Atlanta, Chicago, Honolulu, Los Angeles, Miami, New York City, San Francisco, and Seattle. To obtain these antitoxins, contact the CDC 24 hours a day at (404) 639-2888 or during business hours at (404) 639-2206. Connaught Laboratories, Ltd, one of only three suppliers in the world, manufactures this antitoxin for the CDC. A less immunogenic despeciated equine heptavalent antitoxin against all 7 types of botulinum is available from the U.S. Army. The U.S. Department of Health and Human Services ordered 200,000 doses of this product from cangene corporation for delivery into the Strategic National Stockpile (SNS) beginning of 2007. It should be noted that all horse serum-based antitoxins pose the risk of anaphylaxis and serum sickness; therefore, skin testing is advised. A pentavalent (types A-E) toxoid also is available under IND status.

Ricin/Abrin   Ricin is a biological toxin that derived from the plant *Ricinus communis*, commonly known as the castor bean. After inhalation exposure, victims may experience fever, weakness, cough, necrosis of upper and lower airway, and pulmonary edema followed by hypotension and cardiovascular collapse. Ingestion of castor beans or ricin may cause esophagitis, abdominal pain, nausea, vomiting, profuse bloody diarrhea, shock, and delayed cytotoxic effects to the liver, kidney, CNS, and adrenal glands. Abrin is a similar biotoxin, which is found in the seeds of the *Abris precatorius*, commonly known as the jequirity bean or rosary pea. There are no available antitoxins for either ricin or abrin. Treatment is supportive care only. Also, there are no commercially available vaccines or other chemoprophylactic agents; however, candidate vaccines for ricin are under development that are immunogenic and confer protection against lethal aerosol exposures in animals. The most promising of these is RiVax®, which has successfully completed a FDA Phase 1 clinical trial. Phase II human trials began in February 2005. This vaccine is manufactured by DOR BioPharma, Inc, located at 1691 Michigan Ave, Suite 435, Miami, FL 33139. The phone number to the company is (305) 534-3383. Its Web site address is www.dorbiopharma.com.

Staphylococcus Enterotoxin B (SEB)   SEB is an exotoxin produced by *Staphylococcus aureus*, a gram-positive cocci. SEB is most commonly recognized as a cause of food poisoning as it is produced by bacterial growth in improperly handled foods. Inhalation of SEB in a biological warfare scenario may rapidly incapacitate its victims. Signs of exposure may include fever, chills, headache, myalgias, and nonproductive cough with severe problems including dyspnea, retrosternal chest pain, vomiting, and diarrhea. No specific antitoxin is available. Supportive therapies are directed toward adequate oxygenation and hydration. Antipyretics and antitussives may provide symptomatic relief. The value of steroids is unknown. No vaccines for SEB are currently available; however, several vaccine candidates are in development.

Trichothecene (T-2) Mycotoxins   Fungi of the genera *Fusarium, Myrotecium, Trichoderma,* and *Stachybotrys* produce these compounds. Clinical manifestations of exposure include skin irritation, pruritus, redness, vesicles, necrosis, sloughing of the epidermis, nose and throat pain, nasal discharge, fever, cough, dyspnea, chest pain, and hemoptysis. Serious cases are associated with prostration, weakness, shock, and death. There is no antidote or antitoxin. Treatment is supportive care. No vaccine or chemoprophylactic agent exists for T-2 mycotoxins.

The following are some of the aforementioned pharmaceutical products available for the treatment of patients exposed to biological warfare agents. If a variety of sizes and formulations are available for a particular pharmaceutical product, only the largest dose forms are listed. (Readers are advised to refer to the *2005 Drug Topics Red Book* or nearest pharmacy wholesaler for a complete list of companies that manufacture or distribute the following products.)

- Ciprofloxacin is available in several different oral and parenteral dosage forms.
- Doxycycline hyclate is available from a number of generic suppliers in 100-mg tablets or capsules.
- Erythromycin is available in a large variety of salt forms and strengths: 500-mg tablets or capsules.
- Penicillin V and G are available in a variety of formulations, both oral and parenteral: 500-mg tablets, 10-million-unit vials.
- Gentamicin sulfate is a powerful aminoglycoside antibiotic that is only used for systemic infections. It is available in a variety of formulations: 40-mg/mL, 20-mL multidose vials (800 mg/vial), 40-mg/mL, 2-mL single-use vials (80 mg/vial).
- Streptomycin is a rarely used antibiotic; however, it has efficacy against several of the biological warfare agents. It is supplied in 1-gram vials for injection. Streptomycin is available from X-Gen Pharmaceuticals, Inc, PO Box 445, Big Flats, NY 14814. The phone number to the company is (607) 732-4411; fax (607) 732-2900. The cost of the product is $9.10/vial, available in units of 10 vials. It has a shelf life of 24 months.
- Chloramphenicol is another rarely used antibiotic. Smaller quantities can be obtained in vials and larger quantities may be obtained in powdered form. This may be required to treat a large number of casualties: 1000-mg vial, 25 grams USP powder.
- SMZ/TMP is another antibiotic that has efficacy against several of the biological warfare agents. The most commonly used formulation contains 800 mg of sulfamethoxazole and 160 mg of trimethoprim (Bactrim DS®): SMZ/TMP (800/160 tablets)

- Rifampin is available in 300-mg capsules or in large quantities in powdered form, 300-mg capsules, 500 grams USP powder.
- Tetracycline is available in 500-mg capsules or also in powdered form, 500-mg capsules, 100 g USP powder.

Conclusion

With the increasing probability of an incident involving a WMD agent, many local, state, and federal agencies have initiated plans for appropriate and effective emergency medical response. Experts in the area of EMS, emergency medicine, infectious disease and public health are becoming trained in the medical management of exposure to NBC agents.

Any large mass casualty scenario will demand the expertise and professional services of a hospital pharmacy. Therefore, clinicians should equip themselves with knowledge of antidotes, antibiotics, antitoxins, and other supportive agents used to treat casualties, and how they may be obtained quickly in the event of an act of terrorism. Currently, there are no guidelines mandating minimum hospital inventory of the pharmaceutical products that may be needed. Pharmacy managers, poison center personnel and pharmacy and therapeutics committee members are urged to participate in, or at least be familiar with, plans coordinated through local domestic preparedness programs.

Selected References

1. CDER KI Taskforce. "Guidance Potassium Iodide as a Thyroid Blocking Agent in Radiation Emergencies. U.S. Department of Health and Human Services Food and Drug Administration Center for Drug Evaluation and Research (CDER)," 2001 (http://www.fda.gov/cder/guidance/index.htm) and (http://fda.gov/cder/drugprepare/kiprep.htm)
2. Franz DR, Jahrling PB, Friedlander AM, et al, "Clinical Recognition and Management of Patients Exposed to Biological Warfare Agents," *JAMA*, 1997, 278(5):399-411.
3. Hendee W, Palmer R, Hall AH, et al, *Poisindex®*, Radiation (monograph on CD-ROM), Englewood, CO: Micromedex® Healthcare Series, Vol 127, 2006.
4. Inglesby TB, Henderson DA, Bartlett JG, et al, "Anthrax as a Biological Weapon: Medical and Public Health Management. Working Group on Civilian Biodefense," *JAMA*, 1999, 281(18):1735-45.
5. Kales SN, and Christiani DC. "Acute Chemical Emergencies," *N Engl J Med*, 2004, 350(8):800-8.
6. Ricks RC, Lowry PC, and Townsend RD, "Radiogardase-Cs Insoluble Prussian Blue (Ferric Hexacyanoferrate, Fe_4 [Fe $(CN)_6]_3$), Ca-DTPA (Trisodium Calcium Diethylene-triaminepentaacetate), Zn-DTPA (Trisodium Zinc Diethylenetriaminepentaacetate)," Oak Ridge: TN: Radiation Emergency Assistance Center/Training Site, November 2002 (www.usamriid.army.mil/).
7. The Centers for Disease Control (http://cdc.gov) Guide B - Vaccination Guidelines for State & Local Health Agencies, November 2002 (http://www.bt.cdc.gov/agent/smallpox/response-plan/index.asp).
8. The Centers for Disease Control (http://cdc.gov) Strategic National Stockpile, December 2002 (http://www.bt.cdc.gov/stockpile).
9. *USAMRIID's Medical Management of Biological Casualties Handbook*, 6th ed, Fort Detrick, MD: Operational Medicine Department, *USAMRIID*, 2005. (http://www.usamriid.army.mil/).
10. *USAMRIID's Medical Management of Chemical Casualties Handbook*, 4th ed, Fort Detrick, MD: Chemical Casualty Care Division, *USAMRIID*, 2001.

Poison Antidote Preparedness in Hospitals

Anthony M. Burda, RPh, DABAT

Illinois Poison Center, Chicago, Illinois

Michael Wahl, MD

Advocate Illinois Masonic Medical Center and Illinois Poison Center, Chicago, Illinois

Christina Hantsch, MD

Loyola University Medical Center and Illinois Poison Center, Chicago, Illinois

The quantities of medications listed in the IPC's (Illinois Poison Center's) antidote list are suggested guidelines; the amounts may be adjusted based on factors such as anticipated usage in the hospital's local area, the nearest alternate sources of antidotes, the distance to tertiary care institutions, etc. Keep in mind that some antidotes (eg, the cyanide antidote kit) must be immediately available on-site when a patient arrives at a hospital. Inadequate antidote preparedness may lead to increased morbidity or mortality.

All healthcare professionals should become familiar with regional poison control center poison prevention and treatment services. Staffed by pharmacists, nurses, physicians, and poison specialists, poison centers are available 24 hours a day, 7 days a week to all residents and health professionals of each state for consultation on the treatment of poisonings, medication interactions, occupational exposures, hazardous material incidents, envenomations, and other poison-related concerns.

The following statement can be attributed to Darryl S. Rich, PharmD, JCAHO Associate Director, Home Care Accreditation Services.

"The Joint Commission, in moving towards a functional and nondepartmental approach to its standard manuals, no longer has such a specific standard related to antidotes. However, there are two standards that would apply to the need for a pharmacy to maintain a supply of common antidotes in stock. Standard TX 3.1. The organization identified an appropriate selection of medications available for prescribing and ordering. This standard specifically required hospitals to develop criteria for the selection of products maintained in stock by the pharmacy. Those criteria must address patient need, given the diseases and conditions treated by the hospital and its emergency room.

The second relevant standard is TX 3.5.5. Emergency medications are consistently available, controlled, and secure in the pharmacy and patient care areas. Although this standard usually refers to the control and security of medications in emergency medication carts on the patient units, it can be used if appropriately selected antidotes (which the Joint Commission considers emergency medications) are not readily available.

Thus, from TX 3.1, it is incumbent upon the medical staff and the pharmacy, through its Pharmacy and Therapeutics Committee or other medical staff committee responsible for formulary selection, to select which antidotes the pharmacy should stock. Under standard TX 3.5.5, the pharmacy must then make sure the selected antidotes are readily available."

Uses and Suggested Minimum Stock Quantities for Various Poison Antidotes Used for Treatment of Poisonings Illinois, Poison Center 24-Hour Hotline: 1-800-942-5969

Illinois Poison Center Antidote List

Antidote	Poison/Drug/Toxin	Suggested Minimum Stock Quantity	Rationale/Comments
N-Acetylcysteine (Mucomyst®)	Acetaminophen Carbon tetrachloride Other hepatotoxins	600 mL in 10-mL or 30-mL vials of 20% solution	Acetaminophen is the most common drug involved in intentional and unintentional poisonings. This amount (600 mL/120 g) provides enough antidote to treat an adult for an entire 3-day course of therapy, or enough to treat 3 adults for 24 hours. Several vials may be stocked in the ED to provide a loading dose and the remaining vials in the pharmacy for the q4h maintenance doses.
Amyl nitrite, sodium nitrite, and sodium thiosulfate (cyanide antidote kit)	Acetonitrile Acrylonitrile Bromates (thiosulfate only) Chlorates (thiosulfate only) Cyanide (eg, HCN, KCN, and NaCN) Cyanogenic glycoside natural sources (eg, apricot pits and peach pits) Hydrogen sulfide (nitrites only) Laetrile Nitroprusside (thiosulfate only) Smoke inhalation (combustion of synthetic materials)	One to two kits	Stock one kit in the ED. Consider also stocking one kit in the pharmacy. **Note:** This has a short shelf-life of 18 months.

Table   (Continued)

Antidote	Poison/Drug/Toxin	Suggested Minimum Stock Quantity	Rationale/Comments
Antivenin, Crotalidae polyvalent	Pit viper envenomation (eg. rattlesnakes, cotton-mouths, and copperheads)	Ten vials	Stock in pharmacy. Advised in geographic areas with endemic populations of copperhead, water moccasin, or eastern massasauga rattlesnakes. In low-risk areas, know nearest alternate source of antivenin. **Note:** 20-40 vials or more may be needed for moderate to severe envenomations. The antivenin must be administered in a critical care setting since it is an equine-derived product.
Antivenin, Crotalidae polyvalent Immune Fab - Ovine (CroFab®)	Pit viper envenomation (eg. rattlesnakes, cotton-mouths, and copperheads)	Four to six vials	Stock in pharmacy. Recently FDA-approved product that is a possible alternate to equine product. May have lower risk of hypersensitivity reaction than equine product. Average dose in premarketing trials was 12 vials, but more may be needed. **Note:** Store in refrigerator. See equine antivenin also.
Antivenin, *Latrodectus mac-tans* (black widow spider)	Black widow spider envenomation	Zero to one vial	*Latrodectus* envenomations are very rare in Illinois. This product is only used for severe envenomations. Antivenin must be given in a critical care setting since it is an equine-derived product. Know the nearest source of antidote. **Note:** Product must be refrigerated at all times.

23

Table   (Continued)

Antidote	Poison/Drug/Toxin	Suggested Minimum Stock Quantity	Rationale/Comments
Atropine sulfate	α_2-agonists (eg, clonidine, guanabenz, and guanfacine) Antimyasthenic agents (eg, pyridostigmine) Bradyarrhythmia-producing agents (eg, β-blockers, calcium channel blockers, and digitalis glycosides) Cholinergic agonists (eg, bethanechol) Organophosphate and carbamate insecticides Muscarine-containing mushrooms (eg, Clitocybe and Inocybe) Nerve agents (eg, sarin, soman, tabun, and VX) Tacrine	Total 100-150 mg Available in various formulations: 0.4 mg/mL (1 mL, 0.4-mg ampuls) 0.4 mg/mL (20 mL, 8-mg vials) 0.1 mg/mL (10 mL, 1-mg ampuls)	The product should be immediately available in the ED. Some may also be stored in the pharmacy or other hospital sites, but should be easily mobilized if a severely poisoned patient needs treatment. Note: Product is necessary for adequate preparedness for a weapon of mass destruction (WMD) incident.
Calcium disodium EDTA (Versenate®)	Lead Zinc salts (eg, zinc chloride)	One 5-mL amp (200 mg/mL)	Stock in pharmacy. One vial provides one day of therapy for a child. More may be needed in lead endemic areas.
Calcium chloride and Calcium gluconate	Beta-blockers Black widow spider (Latrodectus mactans) envenomation Calcium channel blockers Fluoride salts (eg, NaF) Hydrofluoric acid (HF) Hyperkalemia (not digoxin-induced) Hypermagnesemia	10% calcium chloride: fifteen 10-mL vials 10% calcium gluconate: five 10-mL vials	Stock in ED. More may be stocked in pharmacy. Many ampuls of calcium chloride may be necessary in life-threatening calcium channel blocker or hydrofluoric acid poisoning.
Deferoxamine mesylate (Desferal®)	Iron	Twelve 500-mg vials	Stock in pharmacy. Note: Per package insert, the maximum daily dose is 6 g (12 vials). However, this dose may be exceeded in serious poisonings.

Table   (*Continued*)

Antidote	Poison/Drug/Toxin	Suggested Minimum Stock Quantity	Rationale/Comments
Digoxin immune fab (Digibind®)	Cardiac glycoside-containing plants (eg, foxglove and oleander) Digitoxin Digoxin	Ten vials	Stock in ED or pharmacy. This amount (10 vials) may be given to a digoxin-poisoned patient in whom the digoxin level is unknown. This amount would effectively neutralize a steady-state digoxin level of 14.2 ng/mL in a 70-kg patient. More may be necessary in severe intoxications. Know nearest source of additional supply.
Dimercaprol (BAL in Oil®)	Arsenic Copper Gold Lead Mercury	Two 3-mL ampuls (100 mg/mL)	Stock in pharmacy. This amount provides two doses of 3-5 mg/kg/dose given q4h to treat one seriously poisoned adult or provides enough to treat a 15-kg child for 24 hours.
Ethanol	Ethylene glycol Methanol	4 L of 10% ethanol in D_5W and 1 pint of 95% ethanol	Stock in pharmacy. This amount (4 L) provides enough to treat an adult (70 kg) with a loading dose followed by a maintenance infusion for 12 hours during dialysis. **Note:** Ethanol is unnecessary if fomepizole is stocked.
Flumazenil (Romazicon®)	Benzodiazepines Zolpidem	Total 1 mg: two 5-mL vials (0.1 mg/mL)	Suggested minimum is for ED stocking. Due to risk of seizures, use with extreme caution, if at all, in poisoned patients. More may be stocked in the pharmacy for use in reversal of conscious sedation.
Folic acid Folinic acid (leucovorin)	Methanol Methotrexate, trimetrexate Pyrimethamine Trimethoprim	Folic acid: three 50-mg vials Folinic acid: one 50-mg vial	Stock in pharmacy. For methanol-poisoned patients with an acidosis, give 50 mg folinic acid initially, then 50 mg of folic acid q4h for 6 doses.

Table   (Continued)

Antidote	Poison/Drug/Toxin	Suggested Minimum Stock Quantity	Rationale/Comments
Fomepizole (Antizol®)	Ethylene glycol Methanol	One 1.5-g vial **Note:** Available in a kit of four 1.5-g vials	Stock in pharmacy. Know where nearest alternate supply is located. One vial will provide at least one initial adult dose. Hospitals with critical care and hemodialysis capabilities may consider stocking 1 kit of 4 vials (enough to treat 1 patient for up to several days). **Note:** Product has a two-year shelf life, however, the manufacturer will replace expired product at no cost.
Glucagon	Beta-blockers Calcium channel blockers Hypoglycemia Hypoglycemic agents	Fifty 1-mg vials	Stock 20 mg in ED and remainder in pharmacy. This amount provides approximately 5-10 hours of high-dose therapy in life-threatening beta-blocker or calcium channel blocker poisoning.
Hyperbaric oxygen (HBO)	Carbon monoxide Carbon tetrachloride Cyanide Hydrogen sulfide Methemoglobinemia Brown recluse spider (*Loxosceles reclusus*) envenomation	Post the location and phone number of nearest HBO chamber.	Consult the regional poison control center to determine if HBO treatment is indicated.
Methylene blue	Methemoglobin-inducing agents including aniline, butyl nitrite, nitrates, nitrites, dapsone, dinitrophenol, local anesthetics (eg, benzocaine), metoclopramide, monomethylhydrazine-containing mushrooms (eg, *Gyromitra*), naphthalene, nitrobenzene, phenazopyridine	Three 10-mL ampuls (10 mg/mL)	Stock in pharmacy. This amount provides 3 doses of 1-2 mg/kg (0.1-0.2 mL/kg).

Table   (Continued)

Antidote	Poison/Drug/Toxin	Suggested Minimum Stock Quantity	Rationale/Comments
Nalmefene (Revex®) Naloxone (Narcan®)	α_2 agonists (eg, clonidine, guanabenz, and guan-facine) Angiotensin converting enzyme (ACE) inhibitors Coma of unknown cause Imidazoline decongestants (eg, oxymetazoline and tetrahydrozoline) Opioids (eg, codeine, dextromethorphan, diphenoxylate, fentanyl, heroin, meperidine, morphine, and propoxyphene) Tramadol Valproic acid	Nalmefene: None required Naloxone: Total 40 mg, any combination of 0.4-mg, 1-mg, and 2-mg ampuls	Stock 20 mg naloxone in the ED and 20 mg elsewhere in the institution. Note: Nalmefene has a longer duration of action but it offers no therapeutic advantage over a naloxone infusion.
D-Penicillamine (Cuprimine®)	Arsenic Copper Mercury	None required as an antidote. Available in bottles of 100 capsules (125-mg or 250-mg capsule)	D-Penicillamine is no longer considered the drug of choice for heavy metal poisonings. It may be stocked in the pharmacy for other indications such as Wilson's disease or rheumatoid arthritis.
Physostigmine salicylate (Antilirium®)	Anticholinergic alkaloid-containing plants (eg, deadly nightshade and jimson weed) Antihistamines Atropine and other anticholinergic agents Intrathecal baclofen	Two 2-mL ampuls (1 mg/mL)	Stock in ED or pharmacy. Usual adult dose is 1-2 mg slow I.V. push. Note: Duration of effect is 30-60 minutes.
Phytonadione (vitamin K₁) (AquaMEPHYTON®, Mephyton®)	Indanedione derivatives Long-acting anticoagulant rodenticides (eg, brodifacoum and bromadiolone) Warfarin	Two 0.5-mL ampuls (2 mg/mL) and two 5-mL ampuls (10 mg/mL)	Stock in pharmacy. Note: Menadione (vitamin K₃, Synkavite®) is ineffective and cannot be stocked as a substitute.
Pralidoxime chloride (2-PAM) (Protopam®)	Antimyasthenic agents (eg, pyridostigmine) Nerve agents (eg, sarin, soman, tabun and VX) Organophosphate insecticides Tacrine	Six 1-g vials	Stock in ED or pharmacy. Note: Serious intoxications may require 500 mg/h (12 g/day). Product is necessary for adequate preparedness for a weapon of mass destruction (WMD) incident.

27

Table (Continued)

Antidote	Poison/Drug/Toxin	Suggested Minimum Stock Quantity	Rationale/Comments
Protamine sulfate	Enoxaparin Heparin	Variable, consider recommendation of hospital P&T Committee Available as 5-mL ampuls (10 mg/mL) and 25-mL vials (250 mg/25 mL)	Stock in pharmacy.
Pyridoxine hydrochloride (Vitamin B₆)	Acrylamide Ethylene glycol Hydrazine Isoniazid (INH) Monomethylhydrazine-containing mushrooms (eg, Gyromitra)	Four 30-mL vials (100 mg/mL 3 g/vial) or ten 10-mL vials (100 mg/mL, 1 g/vial)	Stock in ED or pharmacy. Usual dose is 1 g pyridoxine HCl for each g of INH ingested. If amount ingested is unknown, give 5 g of pyridoxine. Repeat dose if seizures are uncontrolled. Know nearest source of additional supply.
Sodium bicarbonate	Chlorine gas Hyperkalemia Serum alkalinization: Agents producing a quinidine-like effect as noted by widened QRS complex on EKG (eg, amantadine, carbamazepine, chloroquine, cocaine, diphenhydramine, flecainide, propoxyphene, tricyclic antidepressants, quinidine, and related agents) Urine alkalinization: Weakly acidic agents (eg, chlorophenoxy herbicides, chlorpropamide, phenobarbital, and salicylates	Twenty 50-mEq vials	Stock 10 vials in the ED and 10 vials elsewhere in the hospital.
Succimer (Chemet®)	Arsenic Lead Mercury	One bottle of 100 capsules (100 mg/capsule)	Stock in pharmacy. FDA approved only for pediatric lead poisoning, however it has shown efficacy for other heavy metal poisonings.

© Illinois Poison Center

Chempack

Rachel Burke, MD

Department of Emergency Medicine, Rush Presbyterian-St Lukes Hospital, Chicago, Illinois

Steven Aks, DO, FACMT, FACEP

Department of Emergency Medicine, John Stroger Hospital, Chicago, Illinois

Introduction

The Strategic National Stockpile (SNS), previously known as the National Pharmaceutical Stockpile, was established in 1999 by the Department of Health and Human Services (DHHS) and the Centers for Disease Control and Prevention (CDC) in an effort to supply states with the medical resources needed to adequately respond to large-scale emergencies including natural disasters and acts of bioterrorism. The SNS is controlled by the Department of Health and Human Services (DHHS) and is intended to provide states and communities with necessary supplemental medical supplies within 12 hours of a Federal decision to deploy to national sites of large-scale emergencies and disasters. Materials contained in the SNS include medical supplies such as IVs and airway equipment and medications such as antibiotics and antidotes.

In the event of certain chemical emergencies or terrorist acts, specifically a nerve agent release, a 12-hour response time to deploy appropriate antidote is insufficient to adequately and effectively treat chemically exposed casualties. Nerve agent victims require timely treatment within minutes to hours of exposure for several reasons. First of all, nerve agents are extremely toxic chemicals that can result in severe illness or death within minutes of exposure unless immediate treatment is administered. Furthermore, antidote effectiveness may drastically decline if administration is delayed due to nerve agent aging, a process in which the bond between the nerve agent and its target, acetylcholinesterase, becomes irreversible ultimately rendering the antidote ineffective. The delayed response time required to deploy the SNS is a pitfall in the management of nerve agent mass casualty emergencies.

This potential pitfall in emergency preparedness was addressed by the CDC in 2002 with the Chempack Pilot Project, which was developed to determine the usefulness of more readily available stockpiles specifically for use in emergencies involving nerve agents. Chempacks are stockpiles of nerve agent antidotes that are strategically placed throughout participating regions to expedite deployment of antidote within 6 hours in response to a nerve agent event. After the Chempack was successfully piloted in rural South Dakota, a combined urban and rural area of Washington, and urban New York City, the DHHS implemented the plan throughout the United States and the surrounding territories, including Puerto Rico and the U.S. Virgin Islands.

Historical Threat Perspective

Chemical agents have been used in terrorism tactics and as warfare throughout history, including the use of cyanide during ancient Greece, irritant gases and blister agents in WWI, and nerve agents in Iraq. The nerve agents were first developed for use as pesticides by the Germans in the 1930s followed by the British in the 1950s. Due to their rapid onset of action, high volatility allowing easy dispersal, and potent toxicity, nerve agents are ideal terrorism and warfare agents. Despite their mass production during World War II, they were not utilized due to fear of retaliation, and although readily available, nerve agents have not been used as warfare agents to date with the exception of the Iran-Iraq War.

Nerve agents have been utilized in several terrorist activities over the past several decades. On a small scale, the Matsumoto incident in Japan in 1994 involved a sarin release carried out by the Aum Shinrikyo sect intended as revenge against presiding judges anticipated to rule unfavorably in a land dispute. Liquid sarin was vaporized by dripping it onto a heater and dispersing the vapor by fan toward the residential area housing the intended targets. This terrorism incident resulted in greater than 250 victims of which over fifty required hospitalization with seven deaths.

On a larger scale, the Tokyo Subway incident in 1995 also involved a sarin release carried out by the Aum Shinrikyo sect in an attempt to deter police from conducting a future raid on their chemical warfare development plant. In this incident, liquid sarin was carried onto commuter trains in plastic bags, which were then pierced with umbrella tips releasing liquid sarin into a confined space. The event was carried out during rush hour at the convergence point of several heavily populated trains located closely to the police department headquarters and several government buildings. This terrorism incident resulted in greater than 5000 people seeking medical care with 1000 mildly to moderately affected victims of which over 500 required hospitalizations and twelve died. Lack of appropriate decontamination and personal protection implementation resulted in secondary contamination of approximately 250 medical personnel.

Personal Protective Equipment and Decontamination Considerations

The Chempack is limited to pharmaceutical assets and does not include personal protective equipment (PPE) or decontamination supplies. To ensure adequate preparation for a mass casualty nerve agent event, each facility will need to inventory available PPE supplies and assess available personnel trained in decontamination.

As demonstrated in the Tokyo incident, first responders and emergency personnel are at high risk for becoming casualties. Due to threat of secondary contamination, healthcare providers must consider themselves potential victims and take appropriate personal protection precautions before attempting decontamination and treatment of nerve agent exposed victims. No decontamination or medical treatment should be rendered prior to donning appropriate PPE.

All chemically exposed patients must be appropriately decontaminated prior to entry into the treatment facility. Water irrigation is the most simple and effective initial method of decontamination. While the addition of soap, or dilute bleach solution may enhance the efficacy of decontamination, attempts to obtain bleach or soap should not delay the initiation of decontamination.

Resource Requirements

Nerve agent intoxication requires specific antidote administration and medical intervention to reverse toxicity, decrease severity of illness, and improve the likelihood of survival with a favorable outcome. Medications necessary for treatment of nerve agent toxicity include pralidoxime or 2-PAM, atropine, and diazepam. Pralidoxime supplies at most hospitals are limited and insufficient to treat large numbers of nerve agent exposures likely to be encountered in a mass casualty incident. Furthermore, certain nerve agents permanently bind acetylcholinesterase rapidly and require timely antidote intervention to reverse the binding process and prevent irreversible toxicity. Sarin, for instance, permanently binds or ages in approximately 5 to 6 hours.

In addition to a specific antidote requirement, nerve agent intoxication frequently requires large doses of atropine to adequately counteract the toxic effects. Although atropine is readily available in the hospital setting, a single severely exposed nerve agent casualty may require large amounts of atropine and numerous casualties could easily deplete hospital atropine stores. Urgency of administration and limited hospital supplies of nerve agent antidote and supporting medications necessitates readily accessible supplemental supplies in preparation for a mass casualty nerve agent event.

Chempack stockpiles of nerve agent antidotes are strategically placed throughout the United States for deployment in a mass casualty nerve agent event. Participation in the Chempack project is voluntary and the specific locations of the Chempack are determined by the individual participating states and regions based on hazard vulnerability analysis. In the circumstance of a special event or gathering, Chempack stockpiles can be mobilized to different locations temporarily to cover the event. Requirements for Chempack deployment include a working diagnosis of nerve agent exposure, a large scale emergency threatening the medical security of the community, and imminent depletion of hospital supplies of medically necessary life-dependent materials.

The Chempack is housed in a wire mesh and Plexiglas storage crate on wheels that weighs over 500 pounds. The contents must be kept within a certain temperature range to allow retention for use beyond their usual expiration date as sample tested and approved through the Shelf Life Extension Program (SLEP) developed by the Food and Drug Administration (FDA). The Chempack must be stored in a secure environment under lock and key and is wired with a Sensaphone device, which is directly connected to the CDC resulting in immediate notification if the temperature goes out of range or the contents are breached.

Contents of the Chempack are sufficient supplemental resources to adequately treat approximately 1000 mildly to moderately affected nerve agent exposures with the Hospital container and 454 mildly to moderately affected nerve agent exposures with the Emergency Medical Service (EMS) container. Two forms of the Chempack exist, the hospital Chempack and the EMS Chempack. The EMS Chempack is intended to be field-based and largely contains autoinjector forms of medications ideal for immediate use by emergency first responders. The hospital Chempack contains fewer autoinjectors and larger amounts of multi-dose vials for titrating medication dosages for prolonged inpatient care. The containers include pralidoxime chloride (2-PAM), which reactivates acetylcholine esterase (true antidote); atropine sulfate, which blocks effects of excess acetylcholine at its site of action; diazepam, which reduces the severity of acetylcholine-induced convulsions; auto-injectors for nonadjustable rapid IM dosing, ideal for emergent field use; and multidose vials for adjustable precision IV dosing, ideal for urgent hospital use and long-term care.

EMS Chempack Container for 454 Casualties			
	Qty	Unit Pack	Cases
Mark 1 autoinjector	240	5	1200
Atropine sulfate 0.4 mg/mL 20 mL	100	1	100
Pralidoxime 1 g inj 20 mL	276	1	276
Atropen 0.5 mg	144	1	144
Atropen 1.0 mg	144	1	144
Diazepam 5 mg/mL auto injector	150	2	300
Diazepam 5 mg/mL vial, 10 mL	25	2	50
Sterile water for injection (SWFI) 20 mL vials	100	2	200
Sensaphone® 2050	1	1	1
Satco B DEA Container	1	1	1

Hospital Chempack Container for 1000 Casualties			
	Qty	Unit Pack	Cases
Mark 1 autoinjector	390	240	2
Atropine sulfate 0.4 mg/mL 20 mL	850	100	9
Pralidoxime 1 g inj 20 mL	2730	276	10
Atropen 0.5 mg	144	144	1
Atropen 1.0 mg	144	144	1
Diazepam 5 mg/mL autoinjector	80	150	1
Diazepam 5 mg/mL vial, 10 mL	640	25	26
Sterile water for injection (SWFI) 20 mL vials	2760	100	23
Sensaphone® 2050	1	1	1
Satco B DEA Container	1	1	1

The CHEMPACK Project Guideines, CHEMPACK Project Office, Strategic National Stockpile Program, Centers for Disease Control and Prevention.

Toxicokinetics/Pathophysiology

Nerve agents refer to the category of toxins that specifically inhibit the acetylcholinesterase enzyme and include GA (Tabun), GB (Sarin), GD (Soman), and VX. By inhibiting acetylcholinesterase, these agents alter the homeostatic balance of the amount of acetylcholine in the nerve terminal resulting in a state of cholinergic excess. The cholinergic effects will be manifested by the degree of action at muscarinic and nicotinic receptor sites in the nervous system.

Clinical Manifestations/Lab Studies

Nerve agent victims will exhibit a classic toxidrome of cholinergic signs and symptoms. The list below breaks down the symptoms into muscarinic and nicotinic effects.

Muscarinic

Salivation
Lacrimation
Urination
Gastrointestinal distress (nausea, vomiting)
Bradycardia, bronchorrhea, bronchospasm
Abdominal cramps
Miosis

Nicotinic

Muscular fasciculations
Weakness
Tremors
Hypertension
Tachycardia

The most important muscarinic effects are the excessive respiratory secretions and bronchospasm. Antidotal administration of atropine should be administered to overcome excessive bronchial secretions. This is the most important endpoint of treating with atropine.

32

The nicotinic receptors are located at the neuromuscular junction, which accounts for fasciculations and weakness. There are also nicotinic receptors at the terminus of the preganglionic sympathetic nerve to the adrenal gland. This nicotinic stimulation will lead to tachycardia and hypertension, or precisely the opposite hemodynamic effect one may see after muscarinic stimulation. The degree of nicotinic versus muscarinic effects depends on the agent of exposure.

The most timely and reliable way to diagnose nerve agent exposures is toxidrome recognition on the part of the treating clinician. Appreciation of the signs and symptoms listed above will be critical to the recognition and treatment of nerve agent casualties in a WMD setting. In addition to clinical recognition for the diagnosis of nerve agent toxicity, there are several detectors that may be employed by EMS or crime scene investigators to aid in the diagnosis, but communication of these results to healthcare providers may be delayed.

There are two laboratory blood tests that can be obtained to confirm nerve agent exposure, the erythrocyte cholinesterase activity level and the serum or pseudocholinesterase activity level. Reliance on laboratory diagnosis of nerve agent toxicity is not practical because these tests will require several days for completion, and are not readily available in most clinical settings. (While the pseudocholinesterase activity level is more readily available, the erythrocyte cholinesterase level provides a more direct inference of nervous system cholinesterase activity than does the pseudocholinesterase activity. A depression of a cholinesterase activity below 10% suggests severe poisoning, 10-20% depression suggests moderate toxicity, and 20-50% suggests mild toxicity. Additionally, without baseline cholinesterase activity, the interpretation of these tests can be problematic.

Early Interventions

Following evacuation and decontamination, nerve agent antidote should be administered as soon as possible. In severe toxicity, antidote administration may be required prior to or during the decontamination and evacuation process. Antidote administration may include self aid, buddy aid, or third-party administration of autoinjectors, namely Mark I kits and CANA kits. Mark I kits contain two autoinjectors, a 2-mg atropine autoinjector and a 600-mg pralidoxime autoinjector, and CANA kits contain a single 10-mg diazepam autoinjector. The atropine should be administered first, followed by the pralidoxime. Nerve agent victims should be treated based on extent of exposure and severity of symptoms. Mildly symptomatic exposures should receive a single Mark I, moderately symptomatic should be administered two doses, and severe exposures should receive three Mark I doses plus a single CANA regardless of the presence or absence of seizure activity.

Consultation and Referral

Presentation of numerous patients exhibiting symptoms consistent with the cholinergic syndrome should create suspicion of a possible mass casualty chemical event involving organophosphate or nerve agent poisoning. Suspicion of a mass casualty incident should trigger activation of the internal hospital disaster plan and immediate inventory of hospital supplies of pralidoxime and atropine to determine resource availability and anticipate potential supply depletion.

Imminent depletion of nerve agent antidotes and supporting pharmaceuticals should activate deployment and utilization of Chempack resources. Local protocols developed through collaboration of the department of public health (DPH), the emergency management agency (EMA), and the emergency medical services (EMS) should be followed to mobilize and access medically necessary stockpiles. Supplemental sources for nerve agent management consultation include the poison control center and CDC Emergency Preparedness & Response website.

Suggested Readings

The working draft of the WHO publication "Health Aspects of Biological and Chemical Weapons," 2nd edition, 2001.

http://www.cbaci.org/pubs/fact_sheets/fact_sheet_2.pdf. – Matsumoto incident

Okumura T, Suzuki K, Fukuda Atsuhito, et al, "Tokyo Subway Sarin Attack: Disaster Management, Part 1: Community Emergency Response," *Acad Emerg Med*, 1998, 5: 613-7.

Okumura T, Suzuki K, Fukuda Atsuhito, et al, "Tokyo Subway Sarin Attack: Disaster Management, Part 2: Hospital Response," *Acad Emerg Med*, 1998, 5:618-24.

Okumura T, Suzuki K, Fukuda Atsuhito, et al, "Tokyo Subway Sarin Attack: Disaster Management, Part 3: National and International Responses," *Acad Emerg Med* 1998, 5:625-8.

http://en.wikipedia.org/wiki/Sarin_gas_attack_on_the_Tokyo_subway

http://en.wikipedia.org/wiki/Matsumoto_incident

http://www.bt.cdc.gov/stockpile/index.asp

Strategic National Stockpile. April 15, 2006

http://www.bt.cdc.gov/planning/continuationguidance/docs/chempack-attachj.doc

http://www.cdc.gov/programs/php09.htm

Strategic National Stockpile

Obtained from the CDC - Center for Disease Control - www.cdc.gov/print.do?url=http://
www.cdc.gov.stockpile/

CDC's Strategic National Stockpile (SNS) has large quantities of medicine and medical supplies to protect the American public if there is a public health emergency (terrorist attack, flu outbreak, earthquake) severe enough to cause local supplies to run out. Once Federal and local authorities agree that the SNS is needed, medicines will be delivered to any state in the U.S. within 12 hours. Each state has plans to receive and distribute SNS medicine and medical supplies to local communities as quickly as possible.

Helping State and Local Jurisdictions Prepare for a National Emergency

An act of terrorism (or a large scale natural disaster) targeting the U.S. civilian population will require rapid access to large quantities of pharmaceuticals and medical supplies. Such quantities may not be readily available unless special stockpiles are created. No one can anticipate exactly where a terrorist will strike and few state or local governments have the resources to create sufficient stockpiles on their own. Therefore, a national stockpile has been created as a resource for all.

In 1999, Congress charged the Department of Health and Human Services (HHS) and the Centers for Disease Control and Prevention (CDC) with the establishment of the National Pharmaceutical Stockpile (NPS). The mission was to provide a resupply of large quantities of essential medical materiel to states and communities during an emergency within twelve hours of the federal decision to deploy.

The Homeland Security Act of 2002 tasked the Department of Homeland Security (DHS) with defining the goals and performance requirements of the SNS Program, as well as managing the actual deployment of assets. Effective on 1 March 2003, the NPS became the Strategic National Stockpile (SNS) Program managed jointly by DHS and HHS. With the signing of the BioShield legislation, the SNS Program was returned to HHS for oversight and guidance. The SNS Program works with governmental and nongovernmental partners to upgrade the nation's public health capacity to respond to a national emergency. Critical to the success of this initiative is ensuring capacity is developed at federal, state, and local levels to receive, stage, and dispense SNS assets.

A National Repository of Life-Saving Pharmaceuticals and Medical Material

The SNS is a national repository of antibiotics, chemical antidotes, antitoxins, life-support medications, IV administration, airway maintenance supplies, and medical/surgical items. The SNS is designed to supplement and resupply state and local public health agencies in the event of a national emergency anywhere and at anytime within the U.S. or its territories.

The SNS is organized for flexible response. The first line of support lies within the immediate response 12-hour Push Packages. These are caches of pharmaceuticals, antidotes, and medical supplies designed to provide rapid delivery of a broad spectrum of assets for an ill-defined threat in the early hours of an event. These Push Packages are positioned in strategically located, secure warehouses ready for immediate deployment to a designated site within 12 hours of the federal decision to deploy SNS assets.

If the incident requires additional pharmaceuticals and/or medical supplies, follow-on vendor-managed inventory (VMI) supplies will be shipped to arrive within 24 to 36 hours. If the agent is well defined, VMI can be tailored to provide pharmaceuticals, supplies and/or products specific to the suspected or confirmed agent(s). In this case, the VMI could act as the first option for immediate response from the SNS Program.

Determining and Maintaining SNS Assets

To determine and review the composition of the SNS Program assets, HHS and CDC consider many factors, such as current biological and/or chemical threats, the availability of medical material, and the ease of dissemination of pharmaceuticals. One of the most significant factors in determining SNS composition, however, is the medical vulnerability of the U.S. civilian population.

The SNS Program ensures that the medical material stock is rotated and kept within potency shelf-life limits. This involves quarterly quality assurance/quality control checks (QA/QCs) on all 12-hour Push Packages, annual 100% inventory of all 12-hour Push Package items, and inspections of environmental conditions, security, and overall package maintenance.

Supplementing State and Local Resources

During a national emergency, state, local, and private stocks of medical material will be depleted quickly. State and local first responders and health officials can use the SNS to bolster their response to a national emergency, with a 12-hour Push Package, VMI, or a combination of both, depending on the situation. The SNS is not a first-response tool.

Rapid Coordination & Transport

The SNS Program is committed to have 12-hour Push Packages delivered anywhere in the U.S. or its territories within 12 hours of a federal decision to deploy. The 12-hour Push Packages have been configured to be immediately loaded onto either trucks or commercial cargo aircraft for the most rapid transportation. Concurrent to SNS transport, the SNS Program will deploy its Technical Advisory Response Unit (TARU). The TARU staff will coordinate with state and local officials so that the SNS assets can be efficiently received and distributed upon arrival at the site.

Transfer of SNS Assets to State and/or Local Authorities

HHS will transfer authority for the SNS material to the state and local authorities once it arrives at the designated receiving and storage site. State and local authorities will then begin the breakdown of the 12-hour Push Package for distribution. SNS TARU members will remain on site in order to assist and advise state and local officials in putting the SNS assets to prompt and effective use.

When and How Is the SNS Deployed?

The decision to deploy SNS assets may be based on evidence showing the overt release of an agent that might adversely affect public health. It is more likely, however, that subtle indicators, such as unusual morbidity and/or mortality identified through the nation's disease outbreak surveillance and epidemiology network, will alert health officials to the possibility (and confirmation) of a biological or chemical incident or a national emergency. To receive SNS assets, the affected state's governor's office will directly request the deployment of the SNS assets from CDC or HHS. HHS, CDC, and other federal officials will evaluate the situation and determine a prompt course of action.

Training and Education

The SNS Program is part of a nationwide preparedness training and education program for state and local healthcare providers, first responders, and governments (to include federal officials, governors' offices, state and local health departments, and emergency management agencies). This training not only explains the SNS Program's mission and operations, it alerts state and local emergency response officials to the important issues they must plan for in order to receive, secure, and distribute SNS assets.

To conduct this outreach and training, CDC and SNS Program staff are currently working with HHS, Regional Emergency Response Coordinators at all of the U.S. Public Health Service regional offices, state and local health departments, state emergency management offices, the Metropolitan Medical Response System cities, the Department of Veterans' Affairs, and the Department of Defense.

Strategic National Stockpile (SNS) Contents

(12-Hour Push Package)

Pharmaceuticals

- Ciprofloxacin 400 mg in D_5W 200 mL bag I.V.
- Doxycycline hyclate 100-mg powder vial I.V.
- Sterile water for injection (SWFI), preservative free, 10-mL vial
- Gentamicin sulfate 40-mg/mL (20 mL) multidose vial for injection
- 0.9% NaCl flush, preservative free, 3 mL Carpuject®
- NaCl 0.9% 1000 mL I.V. solution
- NaCl 0.9% 100 mL I.V. piggyback mix
- Polymyxin B/bacitracin ointment, 0.9-g packets
- Mark 1 (600 mg pralidoxime/2 mg atropine) autoinjector
- Diazepam HCl 10-mg autoinjector
- Atropine sulfate 0.4-mg/mL × 20-mL multidose vial for injection
- Pralidoxime HCl 1-g powder vial for injection
- Diazepam HCl 10 mg (5 mg/mL) 10 mL single-dose vial for injection
- Dopamine HCl 400-mg (80 mg/mL × 5 mL) vial I.V.
- Epinephrine HCl 1:10,000 (10 mL) syringe/needle for injection
- Methylprednisolone Na succinate 125-mg (2 mL) vial for injection
- Albuterol metered dose inhaler 17 g
- Epinephrine autoinjector (0.3 mg × 1:1000)
- Epinephrine autoinjector (0.15 mg × 1:2000)
- Morphine sulfate 10 mg/mL (1 mL), 25-gauge needle, Carpuject®
- Lorazepam HCl 2 mg/mL (1 mL), 22-gauge needle, Carpuject®
- Ciprofloxacin 500 mg 20-tablet unit of use
- Ciprofloxacin HCl po 500-mg tablets (100#)
- Ciprofloxacin HCl po susp 250-mg/5-mL (100 mL) bottle
- Doxycycline hyclate 100 mg 20-tablet unit of use
- Doxycycline hyclate 100 mg (500# bottle)

Medical Supplies

- Calibrates oral dosing syringes 10 mL
- Carpuject® device
- Syringe/needle 10 mL, 20G × 1-1/2"
- Intravenous catheter & needle unit 18G × 2"
- Intravenous catheter & needle unit 18G × 1 - 1/4"
- Intravenous catheter & needle unit 20G × 1 - 1/4"
- Intravenous catheter & needle unit 24G × 5/8"
- Intravenous set, butterfly, 12" tubing, 21G × 3/4"
- I.V. admin set, 10 drops/mL, vented, y-site

- I.V. admin set, 10 drops/mL, y-site
- I.V. admin set, 60 drops/mL, vented, y-site
- I.V. admin set, 60 drops/mL, 2 y-sites
- Isopropyl alcohol pads, 70%, 1 - 1/4′ × 2 - 1/2″
- Povidone iodine swabsticks, 10% triples
- Gloves, medium, nonsterile, nonlatex, single
- Gloves, large, nonsterile, nonlatex, single
- Tape, cloth, 1″ × 10 yards (Durapore or equivalent), roll
- I.V. site transparent dressing, 2″ × 3″ (Tegaderm® or equiv)
- Tourniquet, latex-free, 3/4″ × 18″
- Intermittent I.V. injection site, long with Luer lock
- Endotracheal tube, 3 mm ID, uncuffed, Murphy, std. connect
- Endotracheal tube, 4 mm ID, uncuffed, Murphy, std. connect
- Endotracheal tube, 5 mm ID, uncuffed, Murphy, std. connect
- Endotracheal tube, 6 mm ID, HVLP cuff, Murphy, std. connect
- Endotracheal tube, 7 mm ID, HVLP cuff, Murphy, std. connect
- Endotracheal tube, 8 mm ID, HVLP cuff, Murphy, std. connect
- Nasogastric tube, adult, 14Fr
- Nasogastric tube, adult, 16Fr
- Nasogastric tube, pediatric, 8Fr
- Endotracheal tube guide (stylette), adult, 10Fr OD
- Endotracheal tube guide (stylette), ped & small adult
- Endotracheal tube guide (stylette), infant, 6Fr OD
- Yankauer suction, with control vent
- Suction catheter 18Fr sterile, flexible, w/control valve
- Suction catheter 14Fr sterile, flexible, w/control valve
- Suction unit (aspirator), portable
- Oropharyngeal (Berman) airway, neonatal, 40-mm length
- Oropharyngeal (Berman) airway, pediatric, 60-mm length
- Oropharyngeal (Berman) airway, adult, 90-mm length
- MPR, adult; bag, mask
- MPR, pediatric; bag, mask, pop-off valve
- Easy cap II CO2 detector (works w/both size MPR)
- Laryngoscope, disposable, large (equivalent of Macintosh 3)
- Laryngoscope, disposable, small (equivalent of Miller 2)
- Laryngoscope, illuminator (reusable with battery)
- Oxygen tubing, 7-ft accommodates 5-7 mm male fittings
- Oxygen mask, nonrebreather, pediatric, safety vent 7-ft tubing
- Oxygen mask, nonrebreather, adult, safety vent 7-ft tubing
- Nasal cannula, 7-ft tubing
- Conforming gauze, sterile, 4″ × 4.1 yard
- Sterile dressing, 4″ × 4″
- Sterile dressing, 8″ × 10″

Designed for 850,000 patients. Quantity not listed above.

Acetylcysteine

CAS Number 19542-74-6 (Acetylcysteine Sodium); 616-91-1

U.S. Brand Names Acetadote®

Use Adjunctive mucolytic therapy in patients with abnormal or viscid mucous secretions in acute and chronic bronchopulmonary diseases; pulmonary complications of surgery and cystic fibrosis; diagnostic bronchial studies; antidote for acute acetaminophen toxicity; phosgene exposure

Use: Unlabeled/Investigational Prevention of radiocontrast-induced renal dysfunction (oral); distal intestinal obstruction syndrome (DIOS, previously referred to as meconium ileus equivalent); phosgene inhalation

Mechanism of Action An antioxidant; exerts mucolytic action through its free sulfhydryl group which opens up the disulfide bonds in the mucoproteins thus lowering the viscosity. The exact mechanism of action in acetaminophen toxicity is unknown. It may act by maintaining or restoring glutathione levels or by acting as an alternative substrate for conjugation with the toxic metabolite.

Adverse Reactions

Inhalation:

>10%:

Stickiness on face after nebulization

Miscellaneous: Unpleasant odor during administration

1% to 10%:

Central nervous system: Drowsiness, chills, fever

Gastrointestinal: Vomiting, nausea, stomatitis

Local: Irritation

Respiratory: Bronchospasm, rhinorrhea, hemoptysis

Miscellaneous: Clamminess

Systemic:

1% to 10%:

Central nervous system: Fever, drowsiness, dizziness (10%; prevention of radiocontrast-induced renal function)

Gastrointestinal: Nausea, vomiting

<1% (Limited to important or life-threatening): Bronchospastic allergic reaction, anaphylactoid reaction, ECG changes (transient)

Pharmacodynamics/Kinetics

Onset of action: Upon inhalation, within 1 minute; direct instillation, immediate peak within 5-10 minutes

Duration: Can persist for longer than 1 hour

Absorption: Mostly acts directly on mucus in lungs; remainder is absorbed by pulmonary epithelium; after oral administration, absorbed from GI tract

Distribution: Unknown

Protein binding: 50%

Metabolism: Hepatic; undergoes rapid deacetylation *in vivo* to yield cysteine or oxidation to yield diacetylcysteine

Half-life: V_d: 0.33-0.47 L/kg

Reduced acetylcysteine: 6.25 hours

Total acetylcysteine: 2.27 hours (4.9 hours in cirrhosis)

Elimination: Unknown; renal clearance may account for 30% elimination (0.21 L/kg/hour)

Usual Dosage

Acetaminophen poisoning: Children and Adults:

Oral: 140 mg/kg; followed by 17 doses of 70 mg/kg every 4 hours; repeat dose if emesis occurs within 1 hour of administration; therapy should continue until all doses are administered even though the acetaminophen plasma level has dropped below the toxic range

I.V. (Acetadote®): Loading dose: 150 mg/kg over 60 minutes. Loading dose is followed by 2 additional infusions: Initial maintenance dose of 50 mg/kg infused over 4 hours, followed by a second maintenance dose of 100 mg/kg infused over 16 hours. To avoid fluid overload in patients <40 kg and those requiring fluid restriction, decrease volume of D_5W proportionally. Total dosage: 300 mg/kg administered over 21 hours.

> **Note:** If commercial I.V. form is unavailable, the following dose has been reported for I.V. administration using solution for oral inhalation (unlabeled): Loading dose: 140 mg/kg, followed by 70 mg/kg every 4 hours, for a total of 13 doses (loading dose and 48 hours of treatment); infuse each dose over 1 hour through a 0.2 micron Millipore filter (in-line).

Experts suggest that the duration of acetylcysteine administration may vary depending upon serial acetaminophen levels and liver function tests obtained during treatment. In general, patients without measurable acetaminophen levels and without significant LFT elevations (>3 times the ULN) can safely stop acetylcysteine after ≤24 hours of treatment. The patients who still have detectable levels of acetaminophen, and/or LFT elevations (>1000 units/L) continue to benefit from additional acetylcysteine administration.

Three-Bag Method Dosage Guide by Weight, Patients <40 kg							
Body Weight		Loading Dose		Second Dose		Third Dose	
(kg)	(lb)	Acetadote (mL)	5% Dextrose (mL)	Acetadote (mL)	5% Dextrose (mL)	Acetadote (mL)	5% Dextrose (mL)
30	66	22.5	100	7.5	250	15	500
25	55	18.75	100	6.25	250	12.5	500
20	44	15	60	5	140	10	280
15	33	11.25	45	3.75	105	7.5	210
10	22	7.5	30	2.5	70	5	140

Adjuvant therapy in respiratory conditions (such as phosgene): **Note:** Patients should receive an aerosolized bronchodilator 10-15 minutes prior to acetylcysteine.

> Inhalation, nebulization (face mask, mouth piece, tracheostomy): Acetylcysteine 10% and 20% solution (Mucomyst®) (dilute 20% solution with sodium chloride or sterile water for inhalation); 10% solution may be used undiluted
>> Infants: 1-2 mL of 20% solution or 2-4 mL 10% solution until nebulized given 3-4 times/day
>> Children and Adults: 3-5 mL of 20% solution or 6-10 mL of 10% solution until nebulized given 3-4 times/day for about 20 minutes; dosing range: 1-10 mL of 20% solution or 2-20 mL of 10% solution every 2-6 hours
> Inhalation, nebulization (tent, croupette): Children and Adults: Dose must be individualized; may require up to 300 mL solution/treatment
> Direct instillation: Adults:
>> Into tracheostomy: 1-2 mL of 10% to 20% solution every 1-4 hours
>> Through percutaneous intrathecal catheter: 1-2 mL of 20% or 2-4 mL of 10% solution every 1-4 hours via syringe attached to catheter
> Diagnostic bronchogram: Nebulization or intrathecal: Adults: 1-2 mL of 20% solution or 2-4 mL of 10% solution administered 2-3 times prior to procedure
> Prevention of radiocontrast-induced renal dysfunction (unlabeled use): Adults: Oral: 600 mg twice daily for 2 days (beginning the day before the procedure); may be given as powder in capsules, some centers use solution (diluted in cola beverage or juice). Hydrate patient with saline concurrently.
> Treatment of phosgene inhalation (investigational): Children and Adults: 10-20 mL of a 20% solution by aerosolization

Contraindications Known hypersensitivity to acetylcysteine

Dosage Forms Solution, as sodium: 10% [100 mg/mL] (4 mL, 10 mL, 30 mL); 20% [200 mg/mL] (4 mL, 10 mL, 30 mL, 100 mL)

Stability Acetadote® is compatible with 5% dextrose, 1/2 normal saline (0.45% sodium chloride injection), and water for injection.

Drug Interactions Adsorbed orally by activated charcoal; clinical significance is minimal, though, once a pure acetaminophen ingestion requiring N-acetylcysteine is established; further charcoal dosing is unnecessary once the appropriate initial charcoal dose is achieved (5-10 g:g acetaminophen); concomitant administration of acetylcysteine and carbamazepine can lead to decreased carbamazepine levels.

Pregnancy Issues B; Based on limited reports using acetylcysteine to treat acetaminophen overdose in pregnant women, acetylcysteine has been shown to cross the placenta in a limited manner and may provide protective levels in the fetus; no teratogenic effects.

Nursing Implications Assess patient for nausea, vomiting, and skin rash following oral administration for treatment of acetaminophen poisoning; intermittent aerosol treatments are commonly given when patient arises, before meals, and just before retiring at bedtime.

Reference Range Determine acetaminophen level as soon as possible, ideally at least 4 hours after ingestion; toxic concentration with possible hepatotoxicity >150 mcg/mL (probably hepatotoxicity >200 mcg/mL) at 4 hours or 50 mcg/mL at 12 hours; serum N-acetylcysteine level is ~500 mg/L 15 minutes after a loading dose of 150 mg/kg. An I.V. dose of 150 mg/kg results in an approximate serum N-acetylcysteine level of 554 mg/L. Peak plasma N-acetylcysteine level in the I.V. protocol is approximately 6-35 times higher than in the oral protocol.

Additional Information
I.V. use: Approved by FDA in February 2004 for acetaminophen overdose within 8-10 hours of ingestion; rate of adverse effects from intravenous N-acetylcysteine is 11% to 14%.

Acetadote® may turn from an essentially colorless liquid to a slight pink or purple color once the stopper is punctured. Acetadote® is a hyperosmolar solution.

Specific References
Appelboam AV, Dargan PI, Jones AL, et al, "Fatal Anaphylactoid Reaction to N-Acetylcysteine: Caution in Asthmatics," *J Toxicol Clin Toxicol*, 2002, 40(3):366-7.

Cascinu S, Catalano V, Cordella L, et al, "Neuroprotective Effect of Reduced Glutathione on Oxaliplatin-Based Chemotherapy in Advanced Colorectal Cancer: A Randomized, Double-Blind, Placebo-Controlled Trial," *J Clin Oncol*, 2002, 20(16):3478-83.

Chyka PA, Butler AY, Holliman BJ, et al, "Utility of Acetylcysteine in Treating Poisonings and Adverse Drug Reactions," *Drug Saf*, 2000, 22(2):123-48.

Clifton JC 2nd, "I'll Have My N-Acetylcysteine With an Orange Twist," *Pediatr Emerg Care*, 2001, 17(1):82.

Grissinger M, "Acetylcysteine Oral Solution Mistaken for Parenteral Use," *P&T*, 2002, 27(6):280.

Merl W, Koutsogiannis Z, Kerr D, et al, "How Safe is Intravenous N-Acetylcysteine for the Treatment of Acetaminophen Toxicity?" *Acad Emerg Med*, 2006, 13(5):S176-7.

Safirstein R, Andrade L, and Vieira JM, "Acetylcysteine and Nephrotoxic Effects of Radiographic Contrast Agents - A New Use for an Old Drug," *N Engl J Med*, 2000, 343(3):210-2.

Salhanick SD, Orlow D, Holt DE, et al, "Endothelially-Derived Nitric Oxide Affects the Severity of Early Acetaminophen-Induced Hepatic Injury in Mice," *Acad Emerg Med*, 2006, 13:479-85.

Tanen DA, LoVecchio F, and Curry SC, "Failure of Intravenous N-Acetylcysteine to Reduce Methemoglobin Produced by Sodium Nitrite in Human Volunteers: A Randomized Controlled Trial," *Ann Emerg Med*, 2000, 35(4):369-73.

Tenenbein PK, Sitar DS, and Tenenbein M, "Interaction Between N-Acetylcysteine and Activated Charcoal: Implications for the Treatment of Acetaminophen Poisoning," *Pharmacotherapy*, 2001, 21(11):1331-6.

Tepel M, van der Giet M, Schwarzfeld C, et al, "Prevention of Radiographic-Contrast-Agent-Induced Reductions in Renal Function By Acetylcysteine," *N Engl J Med*, 2000, 343(3):180-4.

Vlachos P, Vadala H, Sofidiotou V, et al, "Late Antidote Administration in 40 Paracetamol (Acetaminophen) Poisoning Cases," *J Toxicol Clin Toxicol*, 2001, 39(3):231-2.

Amyl Nitrite

CAS Number 110-46-3; 463-04-7

Use Coronary vasodilator in angina pectoris; an adjunct in treatment of cyanide poisoning; also used to produce changes in the intensity of heart murmurs

Mechanism of Action Vasodilator (vascular smooth muscle relaxant)

Adverse Reactions

Cardiovascular: Postural hypotension, cutaneous flushing of head, neck, and clavicular area, palpitations, tachycardia, sinus tachycardia, vasodilation

Central nervous system: Headache, incoherent speech

Dermatologic: Contact dermatitis

Gastrointestinal: Nausea, colitis, vomiting

Genitourinary: Penile erection is enhanced, retarded ejaculation

Hematologic: Heinz body hemolysis/hemolytic anemia, methemoglobinemia followed by hemolysis

Ocular: Increased intraocular pressure, blurred vision, yellow vision

Respiratory: Tracheobronchitis

Pharmacodynamics/Kinetics

Onset of action: Within 30 seconds

Duration: 5 minutes

Absorption:

Readily through the respiratory tract

Rapidly with oral ingestion

Metabolism: Hepatic; metabolized by liver to form inorganic nitrates (less potent)

Half-life:

Parent: <1 hour

Methemoglobin: 1 hour

Elimination: Renal: ~33%

Usual Dosage 1-6 inhalations from one capsule are usually sufficient to produce the desired effect

Contraindications Severe anemia; hypersensitivity to nitrates

Warnings Use with caution in patients with increased intracranial pressure, low systolic blood pressure

Dosage Forms Vapor for inhalation [crushable glass perles]: 0.3 mL

Stability Store in cool place, protect from light; insoluble in water; inflammable

Drug Interactions Alcohol

Pregnancy Issues X

Monitoring Parameters Monitor blood pressure during therapy or have patient lie down during inhalation; methemoglobin levels; arterial blood gas

Methemoglobin levels:

15%: Chocolate-colored blood

20%: Symptomatic

40%: Tachycardia

50%: Stupor

70%: Lethal

Nursing Implications Administer by nasal inhalation; patient should be sitting; crush ampul in woven covering between fingers and then hold under patient's nostrils

Signs and Symptoms of Acute Exposure Ataxia, bradycardia, coma, cyanosis, dyspnea, methemoglobinemia (may be followed by hemolysis), stupor, tachycardia, hypotension due to vasodilitation

Anthrax Vaccine Adsorbed

Synonyms AVA

U.S. Brand Names BioThrax™

Use Immunization against *Bacillus anthracis*. Recommended for individuals who may come in contact with animal products that come from anthrax-endemic areas and may be contaminated with *Bacillus anthracis* spores; recommended for high-risk persons such as veterinarians and others handling potentially infected animals. Routine immunization for the general population is not recommended.

The Department of Defense is implementing an anthrax vaccination program against the biological warfare agent anthrax, which will be administered to all active duty and reserve personnel.

Use: Unlabeled/Investigational Postexposure prophylaxis in combination with antibiotics

Mechanism of Action Active immunization against *Bacillus anthracis*. The vaccine is prepared from a cell-free filtrate of *B. anthracis*, but no dead or live bacteria and produced in a protein-free medium with aluminum hydroxide adjuvant and benzethonium chloride preservative.

Adverse Reactions (Includes pre- and postlicensure data; systemic reactions reported more often in women than in men)

>10%:

Central nervous system: Malaise (4% to 11%)

Local: Tenderness (58% to 71%), erythema (12% to 43%), subcutaneous nodule (4% to 39%), induration (8% to 21%), warmth (11% to 19%), local pruritus (7% to 19%)

Neuromuscular & skeletal: Arm motion limitation (7% to 12%)

1% to 10%:

Central nervous system: Headache (4% to 7%), fever (<1% to 7%)

Gastrointestinal: Anorexia (4%), vomiting (4%), nausea (<1% to 4%)

Local: Mild local reactions (edema/induration <30 mm) (9%), edema (8%)

Neuromuscular & skeletal: Myalgia (4% to 7%)

Respiratory: Respiratory difficulty (4%)

<1%: Chills, body aches, delayed hypersensitivity reaction (started approximately day 17), moderate local reactions (edema/induration >30 mm and <120 mm), severe local reactions (edema/induration >120 mm in diameter or accompanied by marked limitation of arm motion or marked axillary node tenderness)

Postmarketing and/or case reports: Anaphylaxis, angioedema, aplastic anemia, arthralgia, aseptic meningitis, asthma, atrial fibrillation, cardiomyopathy, cellulitis, cerebrovascular accident, CNS lymphoma, cysts, collagen vascular disease, dizziness, encephalitis, endocarditis, facial palsy, fatigue, glomerulonephritis, Guillain-Barré syndrome, hearing disorder, idiopathic thrombocytopenia purpura, immune deficiency, inflammatory arthritis, injection site pain/tenderness, leukemia, liver abscess, lymphoma, mental status change, multiple sclerosis, myocarditis, neutropenia, pemphigus vulgaris, peripheral swelling, polyarteritis nodosa, psychiatric disorders, renal failure, seizure, sepsis, spontaneous abortion, sudden cardiac arrest, suicide, syncope, transverse myelitis, tremor, systemic lupus erythematosus, visual disorder

Pharmacodynamics/Kinetics

Duration: Unknown; may be 1-2 years following two inoculations based on animal data

Usual Dosage SubQ:

Children <18 years: Safety and efficacy have not been established

Children ≥18 years and Adults:

Primary immunization: Three injections of 0.5 mL each given 2 weeks apart, followed by three additional injections given at 6-, 12-, and 18 months; it is not necessary to restart the series if a dose is not given on time; resume as soon as practical.

Subsequent booster injections: 0.5 mL at 1-year intervals are recommended for immunity to be maintained

43

Elderly: Safety and efficacy have not been established for patients >65 years of age.

Administration Administer SubQ; shake well before use. Do not use if discolored or contains particulate matter. Do not use the same site for more than one injection. Do not mix with other injections. After administration, massage injection site to disperse the vaccine. Federal law requires that the date of administration, the vaccine manufacturer, lot number of vaccine, and the administering person's name, title, and address be entered into the patient's permanent medical record.

Contraindications Hypersensitivity to anthrax vaccine or any component of the formulation; severe anaphylactic reaction to a previous dose of anthrax vaccine; individuals who have recovered from anthrax; history of Guillain-Barré syndrome; pregnancy

Warnings Immediate treatment for anaphylactic/anaphylactoid reaction should be available during vaccine use. Patients with a history of Guillain-Barré syndrome should not be given the vaccine unless there is a clear benefit that outweighs the potential risk of recurrence. Defer dosing during acute respiratory disease or other active infection; defer dosing during short-term corticosteroid therapy, chemotherapy, or radiation; additional dose required in patients on long-term corticosteroid therapy; discontinue immunization in patients with chills or fever associated with administration; use caution with latex allergy; immune response may be decreased with immunodeficiency. Safety and efficacy in children <18 years of age or adults >65 years of age have not been established. Administer with caution to individuals with latex sensitivity.

Dosage Forms Injection, suspension: 5 mL [vial stopper contains dry natural rubber]

Pregnancy Issues Inform prescriber if you are or intend to become pregnant; consult prescriber if breast feeding.

Monitoring Parameters Monitor for local reactions, chills, fever, anaphylaxis

Storage Instructions
Store under refrigeration at 2°C to 8°C (36°F to 46°F). Do not freeze.

Additional Information Not commercially available in the U.S.

Local reactions increase in severity by the fifth dose. Moderate local reactions (>5 cm) may be pruritic and may occur if given to a patient with a previous history of anthrax infection. Federal law requires that the date of administration, the vaccine manufacturer, lot number of vaccine, and the administering person's name, title, and address be entered into the patient's permanent medical record.

Presently, all anthrax vaccine lots are owned by the U.S. Department of Defense. The Centers for Disease Control (CDC) does not currently recommend routine vaccination of the general public.

Specific References
Advisory Committee on Immunization Practices, "Use of Anthrax Vaccine in the United States," *MMWR Recomm Rep*, 2000, 49(RR-15):1-20.

Antitoxin Botulinin Types A, B, and E

Use Prophylaxis or active treatment of individuals who have eaten food known or strongly suspected of being infected with *Clostridium botulinum*

Adverse Reactions
Cardiovascular: Tachycardia, cardiovascular collapse, flushing, cyanosis, sinus tachycardia
Dermatologic: Urticaria
Ocular: Dry eyes, lagophthalmos, ptosis, photophobia, vertical deviation
Respiratory: Wheezing, bronchospasm, apnea
Miscellaneous: Signs or symptoms of anaphylaxis (typically occur within 20-60 minutes); serum sickness may occur within 14 days of administration and is more likely to follow a repeat injection of equine serum; hypersensitivity

Pharmacodynamics/Kinetics Half-life (circulating antitoxin): 5-8 days

Administration Administer 1-2 vials slowly I.V. in a 1:10 dilution with 0.9% normal saline (may also give a dose of 1 vial I.M.), and then subsequent doses every 2-4 hours I.V. based on clinical findings. Avoid magnesium since it may enhance the action of the toxin. Trivalent (ABE) antitoxin is usually most effective if administered within 24 hours. Usually, only one vial dose is necessary.

Contraindications History of anaphylaxis to equine-derived serums is a relative contra-indication.

Warnings Always precede treatment with a skin test for sensitivity, as described in the dosage section; when administering this agent, have ready access to drugs of resuscitation, including epinephrine and ventilatory support equipment. Botulism equine antitoxin does not limit the microbial spread of *C. botulinum*. Likewise, antibiotics do not decrease adverse effects to the toxin. Botulinum toxoid is required to confer long-lasting immunity. The preparation contains 0.25% phenol and 0.005% thimerosal in antivenin, and 1:100,000 phenylmercuric nitrate in bacteriostatic water for diluent. Not effective for botulinum type C toxicity.

Stability Excessive agitation or shaking of the reconstituted vial may cause foaming, which may lead to denaturation of the antivenin

Pregnancy Issues C; Use only if the potential benefits outweigh the risks; it is not known if botulinum antitoxin antibodies cross the placenta.

Monitoring Parameters Resolution of symptoms of botulism, or emergence of symptoms of sensitivity to the antitoxin

Additional Information Trivalent botulinum antitoxin will not neutralize toxin that is already bound to the synaptic cleft, and will therefore not reverse established paralysis. Antitoxin is only available from the Immunobiologics Unit, Clinical Medicine Branch, Division of Host Factors, Center for Infectious Diseases, Centers for Disease Control, Atlanta, GA. To order this product, call (404) 639-3334 or (404) 639-3670 Monday through Friday, 8 AM to 4:30 PM EST, or (404) 639-2206 or (404) 639-3670 weekends and after working hours (emergency requests only). Additional information may be obtained through the CDC Drug Information Services at (404) 639-3356.

Specific References

Shapiro RL, Hatheway C, and Swerdlow DL, "Botulism in the United States: A Clinical and Epidemiologic Review," *Ann Intern Med*, 1998, 129(3):221-8.

Asoxime Chloride

CAS Number 34433-31-3

Use Investigational: Has been used in dichlorvos and dimethoate poisoning; rodent experiments indicate that it may be effective in tabun, sarin, VX soman or GF exposure with atropine used as a pretreatment

Mechanism of Action A cholinesterase reactivator; may be more effective for the diethoxy group of organophosphates than the dimethoxy type; may be effective in treating VX, sarin, soman, or GF-type nerve gas

Pharmacodynamics/Kinetics

Absorption: Intramuscular (rapid and complete)
Distribution: V_d: 0.4 L/kg
Bioavailability: Oral: Low
Elimination: Renal clearance: 1.7-2 mL/kg/minute

Usual Dosage Nerve gas or parathion exposure: 500 mg diluted in 3 mL of distilled water intramuscularly 4 times/day for 2-7 days; total dosage range: 4-14 g

Reference Range Therapeutic serum HI-6 level is $\sim$4 mcg/mL

Additional Information May be obtained from the Bosnalijek Drug Company in Sarajevo; does not penetrate red blood cell membrane

Atropine

CAS Number 51-55-8; 55-48-1; 5908-99-6

U.S. Brand Names AtroPen®; Atropine-Care®; Isopto® Atropine; Sal-Tropine™

Use Preoperative medication to inhibit salivation and secretions; treatment of bradycardia (sinus); management of peptic ulcer; treat exercise-induced bronchospasm; antidote for organophosphate pesticide poisoning, anticholinesterase, cholinergic medication poisoning, nerve agent poisoning, and muscarinic symptoms of mushroom poisoning (*Clitocybe, Inocybe*); used to produce mydriasis and cycloplegia for examination of the retina and optic disk and accurate measurement of refractive errors; uveitis

Mechanism of Action Blocks the action of acetylcholine at parasympathetic sites in smooth muscle, secretory glands, and the central muscarinic receptors; increases cardiac output, dries secretions, antagonizes histamine and serotonin

Adverse Reactions Severity and frequency of adverse reactions are dose-related and vary greatly; listed reactions are limited to significant and/or life-threatening.

Cardiovascular: Arrhythmia, flushing, hypotension, palpitation, tachycardia

Central nervous system: Ataxia, coma, delirium, disorientation, dizziness, drowsiness, excitement, fever, hallucinations, headache, insomnia, nervousness, weakness

Dermatologic: Anhidrosis, urticaria, rash, scarlatiniform rash

Gastrointestinal: Bloating, constipation, delayed gastric emptying, loss of taste, nausea, paralytic ileus, vomiting, xerostomia

Genitourinary: Urinary hesitancy, urinary retention

Ocular: Angle-closure glaucoma, blurred vision, cycloplegia, dry eyes, mydriasis, ocular tension increased

Respiratory: Dyspnea, laryngospasm, pulmonary edema

Miscellaneous: Anaphylaxis

Pharmacodynamics/Kinetics

Onset of action: I.V.: Effects on heart rate peak within 2-4 minutes

Duration: Oral: 4-6 hours; Parenteral: Brief

Absorption: Well absorbed from all dosage forms

Distribution: V_d: 2.3 L/kg; widely distributed; crosses the blood-brain barrier

Protein binding: Moderate 18%

Metabolism: Liver to tropic acid, atropine

Half-life: 2-3 hours (10 hours in elderly)

Time to peak serum concentration: I.M.: 30-45 minutes

Elimination: Into urine of both metabolites and unchanged drug (30% to 50%); clearance: 660 mL/minute

Usual Dosage

Children:

Preanesthetic: I.M., I.V., SubQ:

<5 kg: 0.02 mg/kg/dose 30-60 minutes preop then every 4-6 hours as needed

>5 kg: 0.01-0.02 mg/kg/dose to a maximum 0.4 mg 30-60 minutes preop; minimum dose: 0.1 mg

Bradycardia: I.V., intratracheal: 0.02 mg/kg every 5 minutes

Minimum dose: 0.1 mg (if administered via endotracheal tube, dilute to 1-2 mL with normal saline prior to endotracheal administration)

Maximum single dose: 0.5 mg (adolescents: 1 mg)

Total maximum dose: 1 mg (adolescents: 2 mg)

When using to treat bradycardia in neonates, reserve use for those patients unresponsive to improved oxygenation

Nerve agent, organophosphate or carbamate poisoning: I.V.: 0.02-0.05 mg/kg every 10-20 minutes until atropine effect (dry flushed skin, tachycardia, mydriasis, fever) is observed then every 1-4 hours for at least 24 hours

Bronchospasm: Inhalation: 0.03-0.05 mg/kg/dose 3-4 times/day; maximum: 1 mg

Ophthalmic, 0.5% solution: Instill 1-2 drops twice daily for 1-3 days before the procedure

Adults (doses <0.5 mg have been associated with paradoxical bradycardia):

Asystole: I.V.: 1 mg; may repeat every 3-5 minutes as needed; may give intratracheal in 1 mg/10 mL dilution only, intratracheal dose should be 2-2.5 times the I.V. dose

Preanesthetic: I.M., I.V., SubQ: 0.4-0.6 mg 30-60 minutes preop and repeat every 4-6 hours as needed

Bradycardia: I.V.: 0.5-1 mg every 5 minutes, not to exceed a total of 3 mg or 0.04 mg/kg; may give intratracheal in 1 mg/10 mL dilution only, intratracheal dose should be 2-2.5 times the I.V. dose

Neuromuscular blockade reversal: I.V.: 25-30 mcg/kg 30 seconds before neostigmine or 10 mcg/kg 30 seconds before edrophonium

Organophosphate or carbamate poisoning: I.V.: 1-2 mg/dose every 10-20 minutes until respiratory status improves; then every 1-4 hours for at least 24 hours; up to 50 mg in first 24 hours and 2 g over several days may be given in cases of severe intoxication; over 30 g has been given to manage this toxicity

Bronchospasm: Inhalation: 0.025-0.05 mg/kg/dose every 4-6 hours as needed (maximum: 5 mg/dose)

Ophthalmic solution: 1%: Instill 1-2 drops 1 hour before the procedure

Uveitis: 1-2 drops 4 times/day

Ophthalmic ointment: Apply a small amount in the conjunctival sac up to 3 times/day; compress the lacrimal sac by digital pressure for 1-3 minutes after instillation

Administration Administer undiluted by rapid I.V. injection; slow injection may result in paradoxical bradycardia

Contraindications Hypersensitivity to atropine sulfate or any component; narrow-angle glaucoma; tachycardia; thyrotoxicosis; obstructive disease of the GI tract; obstructive uropathy

Warnings Use with caution in children with spastic paralysis; use with caution in elderly patients. Low doses cause a paradoxical decrease in heart rates. Some commercial products contain sodium metabisulfite, which can cause allergic-type reactions; use with caution in patients with autonomic neuropathy, prostatic hypertrophy, hyperthyroidism, congestive heart failure, cardiac arrhythmias, chronic lung disease, biliary tract disease; with massive dosing in treatment of severe pesticide poisoning, hemolysis from free water dilution can occur.

Dosage Forms

Injection, solution, as sulfate: 0.1 mg/mL (5 mL, 10 mL); 0.4 mg/mL (1 mL, 20 mL); 0.5 mg/mL (1 mL); 1 mg/mL (1 mL)

Ointment, ophthalmic, as sulfate: 1% (3.5 g)

Solution, ophthalmic, as sulfate: 1% (5 mL, 15 mL)

Atropine-Care®: 1% (2 mL)

Atropisol®: 1% (1 mL)

Isopto® Atropine: 1% (5 mL, 15 mL)

Tablet, as sulfate (Sal-Tropine™): 0.4 mg

Stability Store injection below 40°C, avoid freezing

Drug Interactions Phenothiazines, amantadine, antiparkinsonian drugs, glutethimide, meperidine, tricyclic antidepressants, antiarrhythmic agents, some antihistamines; all have anticholinergic activity; not compatible with alkali solutions

Pregnancy Issues C; Crosses the placenta; trace amounts appear in breast milk

Monitoring Parameters Heart rate (EKG), respiratory status

Nursing Implications Give by rapid I.V. injection since slow infusion may cause a paradoxical bradycardia; may give intratracheal in 1 mg/10 mL dilution only

Signs and Symptoms of Acute Exposure Ataxia, blurred vision, coma, dilated unreactive pupils, diminished or absent bowel sounds, dryness of mucous membranes, flushing, foul breath, hallucinations (lilliputian), hypertension, hyperthermia, ileus, increased respiratory rate, ototoxicity (deafness), seizures, swallowing difficulty, tachycardia, tinnitus, urinary retention

Reference Range Therapeutic range: 3-25 ng/mL. Peak serum atropine level after a 1 mg I.V. dose: ~0.003 mg/L (within 30 minutes of administration); blood level (postmortem) of 0.2 mg/L and urine level of 1.5 mg/L associated with fatality

Additional Information
Expired atropine products have been shown to have significant potency and may be an option in mass casualty events.

Anticholinergic toxicity is caused by strong binding of the drug to cholinergic receptors. For anticholinergic overdose with severe life-threatening symptoms, physostigmine 1-2 mg (0.5 or 0.02 mg/kg for children) SubQ or I.V., slowly may be given to reverse these effects. Response to atropine may not be observed until after I.V. calcium administration. A single drop of ophthalmologic atropine sulfate (1%) contains 0.6 mg of atropine.

Protocol for extemporaneous preparation of atropine sulfate for injection:
- Reconstitute 30 g of atropine sulfate USP powder in 30 mL normal saline (1 g/mL)
- Draw up 1 mL into a syringe
- Attach a Millex 0.22 micron filter to the syringe
- Attach a sterile needle to Millex filter
- Add 1 mL (1 g atropine sulfate) to a 500-mL bottle of normal saline (2 mg/mL)

OR
- Add 1 mL (1 g atropine sulfate) to a 1-L bottle of normal saline (1 mg/mL)

Specific References
Barrueto F Jr, Salleng K, Sahni R, et al, "Histopathologic Effects of the Single Intramuscular Injection of Ketamine, Atropine and Midazolam in a Rat Model," *Vet Hum Toxicol*, 2002, 44(5):306-10.

Bentur Y, Layish I, Krivoy A, et al, "Civilian Adult Self Injections of Atropine-Trimedoxime (TMB4) Auto-Injectors," *Clin Toxicol*, 2006, 44:301-6.

Brady WJ and Perron AD, "Administration of Atropine in the Setting of Acute Myocardial Infarction: Potentiation of the Ischemic Process?" *Am J Emerg Med*, 2001, 19(1):81-3.

Bryant SM, Wills BK, Rhee JW et al, "Intramuscular Opthalmic Homotropine vs. Atropine to Prevent Lethality in Rates with Dichlorvos Poisoning," *J Medical Toxicol*, 2006, 2(4):156-9.

Corner B, "Intravenous Atropine Treatment in Infantile Hypertrophic Pyloric Stenosis," *Arch Dis Child*, 2003, 88(1):87; author reply 87.

Dix J, Weber RJ, Frye RF, et al, "Stability of Atropine Sulfate Prepared for Mass Chemical Terrorism," *J Toxicol Clin Toxicol*, 2003, 41(6):771-5.

Freigang SH, Krohs S, Tiefenbach B, et al, "Poisoning After Local Application of Atropine Eye Drops in Children," *J Toxicol Clin Toxicol*, 2001, 39(3):264.

Geller RJ, Lopez GP, Cutler S, et al, "Atropine Availability as an Antidote for Nerve Agent Casualties: Validated Rapid Reformulation of High-Concentration Atropine From Bulk Powder," *Ann Emerg Med*, 2003, 41(4):453-6.

Henretig FM, Mechem C, and Jew R, "Potential Use of Autoinjector-Packaged Antidotes for Treatment of Pediatric Nerve Agent Toxicity," *Ann Emerg Med*, 2002, 40(4):405-8.

Kozak RJ, Siegel S, and Kuzma J, "Rapid Atropine Synthesis for the Treatment of Massive Nerve Agent Exposure," *Ann Emerg Med*, 2003, 41(5):685-8.

Schier JG, Mehta R, Mercurio-Zappala M, et al, "Preparing for Chemical Terrorism: Stability of Expired Atropine," *J Toxicol Clin Toxicol*, 2002, 40(5):625.

Schier JG, Ravikumar PR, Nelson LS, et al, "Preparing for Chemical Terrorism: Stability of Injectable Atropine Sulfate," *Acad Emerg Med*, 2004, 11(4):329-34.

Tomassoni AJ and Simone KE, "Development and Use of a Decentralized Antidote Stockpile in a Rural Site," *J Toxicol Clin Toxicol*, 2004, 42(5):824.

Botulinum Pentavalent (ABCDE) Toxoid

Synonyms Botulinum Toxoid, Pentavalent Vaccine (Against Types A/B/C/D/E Strains of *C. botulinum*); PBT Vaccine

Use: Unlabeled/Investigational
Investigational: Prophylaxis for *C. botulinum* exposure (high-risk research laboratory personnel actively working with, or expect to work with, known cultures and purified botulinum toxin). Available for this purpose since 1967.

Adverse Reactions Moderate-to-severe effects (initial series 5.6%, booster 12.3%, systemic 4.5% initial & booster)
Central nervous system: Fever (0.4%), headache (0.5%), malaise (0.6%), dizziness
Dermatologic: Rash, pruritus (0.9%)
Gastrointestinal: Nausea, vomiting, diarrhea
Local: Pain/soreness at injection site (7.3%)
Neuromuscular & skeletal: Muscle pain
Ocular: Blurred vision

Usual Dosage Do not inject intracutaneously or into superficial structures.
Initial vaccination series: 0.5 mL deep SubQ at 0-, 2-, 12- and 24 weeks
First booster: 0.5 mL deep SubQ 12 months after first injection of the initial series
Subsequent boosters: 0.5 mL deep SubQ at 1-year intervals

Drug Interactions Might make Botox® or MyoBloc™ less effective

Additional Information
Each vial contains 0.022% formaldehyde and 1:10,000 thimerosal as a preservative; manufactured by Michigan Biologic Products Institute, Lansing, MI 48909.
Between 1970-2002, 21,320 injections had been administered.
For advice on vaccine administration and contraindications, contact the Division of Immunization, CDC, Atlanta, GA 30333 (404-639-3670, FAX 404-639-3717).

Incidence and Severity of Local Reactions Following PBT Vaccination, CDC Data 1970-2002 (% of 21,320 Injections)					
Injection Number	Local Reaction			No Record	Total
	None to Mild	Moderate	Severe		
Primary Injections					
1	5305	164	5	80	5554
2	4878	304	8	72	5262
3	4254	363	14	85	4716
Other Primary	5	1	-	2	8
Subtotal	14,442	832	27	239	15,540
	(67.7%)	(3.9%)	(0.1%)	(1.1%)	(72.9%)
Booster Injections					
1	2452	349	27	53	2881
2	964	130	11	24	1129
3	475	66	6	7	554
≥4	1078	107	1	4	1190
Unknown	23	2	0	1	26
Subtotal	4992	654	45	89	5780
	(23.4%)	(3.1%)	(0.2%)	(0.4%)	(27.1%)
Total	19,434	1486	72	328	21,320
	(91.1%)	(7.0%)	(0.3%)	(1.5%)	(100%)

None = No reactions.
Mild = Erythema only; edema or induration that is measurable but ≤30 mm in any diameter.
Moderate = Edema or induration measuring >30 mm and <120 mm in any diameter.
Severe = Any reaction measuring >120 mm in any one diameter or any reaction accompanied by marked limitation of motion of the arm or marked axillary node tenderness.

Source: CDC, 2005.

Budesonide

U.S. Brand Names Entocort® EC; Pulmicort Respules®; Pulmicort Turbuhaler®; Rhinocort® Aqua®

Use
Intranasal: Children ≥6 years of age and Adults: Management of symptoms of seasonal or perennial rhinitis
Nebulization: Children 12 months to 8 years: Maintenance and prophylactic treatment of asthma; may be useful for hydrocarbon inhalation injuries
Oral capsule: Treatment of active Crohn's disease (mild to moderate) involving the ileum and/or ascending colon
Oral inhalation: Maintenance and prophylactic treatment of asthma; includes patients who require corticosteroids and those who may benefit from systemic dose reduction/ elimination

Mechanism of Action Controls the rate of protein synthesis, depresses the migration of polymorphonuclear leukocytes, fibroblasts, reverses capillary permeability, and lysosomal stabilization at the cellular level to prevent or control inflammation

Adverse Reactions Reaction severity varies by dose and duration; not all adverse reactions have been reported with each dosage form.
>10%:
Central nervous system: Oral capsule: Headache (up to 21%)
Gastrointestinal: Oral capsule: Nausea (up to 11%)
Respiratory: Respiratory infection, rhinitis, dysphonia
Miscellaneous: Symptoms of HPA axis suppression and/or hypercorticism (acne, easy bruising, fat redistribution, striae, edema) may occur in >10% of patients following administration of dosage forms that result in higher systemic exposure (ie, oral capsule), but may be less frequent than rates observed with comparator drugs (prednisolone). These symptoms may be rare (<1%) following administration via methods that result in lower exposures (topical).
1% to 10%:
Cardiovascular: Syncope, edema, hypertension
Central nervous system: Chest pain, dysphonia, emotional lability, fatigue, fever, insomnia, migraine, nervousness, pain, dizziness, vertigo
Dermatologic: Bruising, contact dermatitis, eczema, pruritus, pustular rash, rash
Endocrine & metabolic: Hypokalemia, adrenal insufficiency
Gastrointestinal: Abdominal pain, anorexia, diarrhea, dry mouth, dyspepsia, gastroenteritis, oral candidiasis, taste perversion, vomiting, weight gain, flatulence, oral candidiasis (4% to 13%)
Hematologic: Cervical lymphadenopathy, purpura, leukocytosis
Neuromuscular & skeletal: Arthralgia, fracture, hyperkinesis, hypertonia, myalgia, neck pain, weakness, paresthesia, back pain
Ocular: Conjunctivitis, eye infection, cataracts
Otic: Earache, ear infection, external ear infection
Respiratory: Bronchitis, bronchospasm, cough, epistaxis, nasal irritation, pharyngitis, sinusitis, stridor
Miscellaneous: Allergic reaction, flu-like syndrome, herpes simplex, infection, moniliasis, viral infection, voice alteration, weight gain, 22% elevation of HDL cholesterol
<1%: Aggressive reactions, alopecia, angioedema, avascular necrosis of the femoral head, depression, dyspnea, hoarseness, hypersensitivity reactions (immediate and delayed; include rash, contact dermatitis, angioedema, bronchospasm), intermenstrual bleeding, irritability, nasal septum perforation, osteoporosis, psychosis, somnolence
Postmarketing and/or case reports: Growth suppression, benign intracranial hypertension

Usual Dosage

Nasal inhalation (Rhinocort® Aqua®): Children ≥6 years and Adults: 64 mcg/day as a single 32 mcg spray in each nostril. Some patients who do not achieve adequate control may benefit from increased dosage. A reduced dosage may be effective after initial control is achieved.

Maximum dose: Children <12 years: 128 mcg/day; Adults: 256 mcg/day

Nebulization: Children 12 months to 8 years: Pulmicort Respules®: Titrate to lowest effective dose once patient is stable; start at 0.25 mg/day or use as follows:

Hydrocarbon inhalation injuries: 0.5 mg every 12 hours for 4 days

Previous therapy of bronchodilators alone: 0.5 mg/day administered as a single dose or divided twice daily (maximum daily dose: 0.5 mg)

Previous therapy of inhaled corticosteroids: 0.5 mg/day administered as a single dose or divided twice daily (maximum daily dose: 1 mg)

Previous therapy of oral corticosteroids: 1 mg/day administered as a single dose or divided twice daily (maximum daily dose: 1 mg)

Oral inhalation:

Children ≥6 years:

Previous therapy of bronchodilators alone: 200 mcg twice initially, which may be increased up to 400 mcg twice daily

Previous therapy of inhaled corticosteroids: 200 mcg twice initially, which may be increased up to 400 mcg twice daily

Previous therapy of oral corticosteroids: The highest recommended dose in children is 400 mcg twice daily

Adults:

Previous therapy of bronchodilators alone: 200-400 mcg twice initially, which may be increased up to 400 mcg twice daily

Previous therapy of inhaled corticosteroids: 200-400 mcg twice initially, which may be increased up to 800 mcg twice daily

Previous therapy of oral corticosteroids: 400-800 mcg twice daily, which may be increased up to 800 mcg twice daily

NIH Guidelines (NIH, 1997) (give in divided doses twice daily):

Children:

"Low" dose: 100-200 mcg/day

"Medium" dose: 200-400 mcg/day (1-2 inhalations/day)

"High" dose: >400 mcg/day (>2 inhalation/day)

Adults:

"Low" dose: 200-400 mcg/day (1-2 inhalations/day)

"Medium" dose: 400-600 mcg/day (2-3 inhalations/day)

"High" dose: >600 mcg/day (>3 inhalation/day)

Oral: Adults: Crohn's disease: 9 mg once daily in the morning; safety and efficacy have not been established for therapy duration >8 weeks; recurring episodes may be treated with a repeat 8-week course of treatment

Note: Treatment may be tapered to 6 mg once daily for 2 weeks prior to complete cessation. Patients receiving CYP3A4 inhibitors should be monitored closely for signs and symptoms of hypercorticism; dosage reduction may be required.

Dosage adjustment in hepatic impairment: Monitor closely for signs and symptoms of hypercorticism; dosage reduction may be required.

Administration

Inhalation: Inhaler should be shaken well immediately prior to use; while activating inhaler, deep breathe for 3-5 seconds, hold breath for ~10 seconds and allow ≥1 minute between inhalations. Rinse mouth with water after use to reduce aftertaste and incidence of candidiasis.

Nebulization: Shake well before using. Use Pulmicort Respules® with jet nebulizer connected to an air compressor; administer with mouthpiece or facemask. Do not use ultrasonic nebulizer. Do not mix with other medications in nebulizer. Rinse mouth following treatments to decrease risk of oral candidiasis (wash face if using face mask).

Oral capsule: Capsule should be swallowed whole; do not crush or chew.

Contraindications Hypersensitivity to budesonide or any component of the formulation

Inhalation: Contraindicated in primary treatment of status asthmaticus, acute episodes of asthma; not for relief of acute bronchospasm

Warnings May cause hypercorticism and/or suppression of hypothalamic-pituitary-adrenal (HPA) axis, particularly in younger children or in patients receiving high doses for prolonged periods. Particular care is required when patients are transferred from systemic corticosteroids to products with lower systemic bioavailability (ie, inhalation). May lead to possible adrenal insufficiency or withdrawal from steroids, including an increase in allergic symptoms. Patients receiving prolonged therapy of ≥20 mg per day of prednisone (or equivalent) may be most susceptible. Aerosol steroids do not provide the systemic steroid needed to treat patients having trauma, surgery, or infections.

Controlled clinical studies have shown that orally inhaled and intranasal corticosteroids may cause a reduction in growth velocity in pediatric patients. (In studies of orally inhaled corticosteroids, the mean reduction in growth velocity was approximately 1 centimeter per year [range 0.3-1.8 cm per year] and appears to be related to dose and duration of exposure.) To minimize the systemic effects of orally inhaled and intranasal corticosteroids, each patient should be titrated to the lowest effective dose. Growth should be routinely monitored in pediatric patients.

May suppress the immune system; patients may be more susceptible to infection. Use with caution in patients with systemic infections or ocular herpes simplex. Avoid exposure to chickenpox and measles. Corticosteroids should be used with caution in patients with diabetes, hypertension, osteoporosis, peptic ulcer, glaucoma, cataracts, or tuberculosis. Use caution in hepatic impairment. Enteric-coated capsules should not be crushed or chewed.

Dosage Forms

Capsule, enteric coated (Entocort™ EC): 3 mg

Powder for oral inhalation (Pulmicort Turbuhaler®): 200 mcg/inhalation (104 g) [delivers ~160 mcg/inhalation; 200 metered doses]

Additional dosage strengths available in Canada: 100 mcg/inhalation, 400 mcg/inhalation

Suspension, nasal spray (Rhinocort® Aqua®): 32 mcg/inhalation (8.6 g) [120 metered doses]

Suspension for oral inhalation (Pulmicort Respules®): 0.25 mg/2 mL (30s), 0.5 mg/2 mL (30s)

Stability

Nebulizer: Store upright at 20°C to 25°C (68°F to 77°F) and protect from light. Do not refrigerate or freeze. Once aluminum package is opened, solution should be used within 2 weeks. Continue to protect from light.

Nasal inhaler: Store with valve up at 15°C to 30°C (59°F to 86°F). Use within 6 months after opening aluminum pouch. Protect from high humidity.

Nasal spray: Store with valve up at 20°C to 25°C (68°F to 77°F) and protect from light. Do not freeze.

Drug Interactions **Substrate** of CYP3A4 (major)

Cimetidine: Decreased clearance and increased bioavailability of budesonide

CYP3A4 inhibitors: Serum level and/or toxicity of budesonide may be increased; this effect was shown with ketoconazole, but not erythromycin. Other potential inhibitors include amiodarone, cimetidine, clarithromycin, delavirdine, diltiazem, dirithromycin, disulfiram, fluoxetine, fluvoxamine, grapefruit juice, indinavir, itraconazole, ketoconazole, nefazodone, nevirapine, propoxyphene, quinupristin-dalfopristin, ritonavir, saquinavir, verapamil, zafirlukast, zileuton.

Proton pump inhibitors (omeprazole, pantoprazole, rabeprazole): Theoretically, alteration of gastric pH may affect the rate of dissolution of enteric-coated capsules. Administration with omeprazole did not alter kinetics of budesonide capsules.

Salmeterol: The addition of salmeterol has been demonstrated to improve response to inhaled corticosteroids (as compared to increasing steroid dosage).

Pregnancy Issues There are no adequate and well-controlled studies in pregnant women; use only if potential benefit to the mother outweighs the possible risk to the fetus. Hypoadrenalism has been reported in infants.

Monitoring Parameters Monitor growth in pediatric patients.

Reference Range Plasma budesonide level of 5 nmol/L obtained within 5 hours of a 9-mg oral dose

Additional Information Effects of inhaled/intranasal steroids on growth have been observed in the absence of laboratory evidence of HPA axis suppression, suggesting that growth velocity is a more sensitive indicator of systemic corticosteroid exposure in pediatric patients than some commonly used tests of HPA axis function. The long-term effects of this reduction in growth velocity associated with orally inhaled and intranasal corticosteroids, including the impact on final adult height, are unknown. The potential for "catch up" growth following discontinuation of treatment with inhaled corticosteroids has not been adequately studied.

Specific References

Gurkan F and Bosnak M, "Use of Nebulized Budesonide in Two Critical Patients With Hydrocarbon Intoxication," *Am J Ther*, 2005, 12(4):366-7.

Calcium Chloride

CAS Number 10035-04-8; 10043-52-4; 7774-34-7

Use Cardiac resuscitation when epinephrine fails to improve myocardial contractions, cardiac disturbances of hyperkalemia, hypocalcemia or calcium channel blocking agent toxicity; also used for fluoride, hydrogen fluoride, ethylene glycol, magnesium sulfate, oxalate, and black widow spider bites

Adverse Reactions

Cardiovascular: Vasodilation, hypotension, bradycardia, cardiac arrhythmias, fibrillation (ventricular), syncope, shock, sinus bradycardia, arrhythmias (ventricular)

Central nervous system: Lethargy, coma

Dermatologic: Erythema

Endocrine & metabolic: Hypomagnesemia, hypercalcemia

Gastrointestinal: Elevated serum amylase

Local: Tissue necrosis (more irritant than calcium gluconate), extravasation injury

Neuromuscular & skeletal: Muscle weakness

Renal: Hypercalciuria

Pharmacodynamics/Kinetics

Absorption: I.V. calcium salts are absorbed directly into the bloodstream

Elimination: Mainly in feces as unabsorbed calcium with 20% eliminated by the kidneys

Usual Dosage I.V. (calcium chloride is 3 times as potent as calcium gluconate):

Cardiac arrest in the presence of hyperkalemia or hypocalcemia, magnesium toxicity, or calcium antagonist toxicity:

Infants and Children: 10-30 mg/kg; may repeat in 5-10 minutes, if necessary

Adults: 10 mL/dose every 10-20 minutes

Hypocalcemia:

Infants and Children: 10-20 mg/kg/dose, repeat every 4-6 hours, if needed

Adults: 500 mg to 1 g at 1- to 3-day intervals

Hydrofluoric acid burns: Intra-arterial infusion: 10 mL of 10% calcium chloride in 40 mL of normal saline over 4 hours

Tetany:

Neonates: 2.4 mEq/kg/day

Infants and Children: 10 mg/kg over 5-10 minutes; may repeat after 6 hours or follow with an infusion with a maximum dose of 200 mg/kg/day

Adults: 1 g over 10-30 minutes; may repeat after 6 hours

Electrical-mechanical dissociation:

Due to propranolol: 1 g

Due to disopyramide: 1-3 g

Reversal of neuromuscular blockade due to polymixin antibiotics and anesthetic agents: I.V.: 1 g

Hypocalcemia secondary to citrated blood transfusion give 0.45 mEq **elemental** calcium for each 100 mL citrated blood infused

Administration Generally, I.V. infusion rates should not exceed 0.7-1.5 mEq/minute (0.5-1 mL/minute); stop the infusion if the patient complains of pain or discomfort; do not inject calcium chloride I.M. or administer SubQ since severe necrosis and sloughing may occur. Do not use scalp vein or small hand or foot veins for I.V. administration. Warm to body temperature; administer slowly, do not exceed 1 mL/minute (inject into ventricular cavity - not myocardium).

Contraindications Ventricular fibrillation during cardiac resuscitation, and in patients with risk of digitalis toxicity, renal or cardiac disease

Warnings Avoid too rapid I.V. administration; avoid extravasation; use with caution in digitalized patients, respiratory failure or acidosis; acidifying effect; give only 2-3 days then change to another calcium salt

Dosage Forms Elemental calcium listed in brackets:

Injection: 10% = 100 mg/mL [27.2 mg/mL, 1.36 mEq/mL] (10 mL)

Stability Admixture incompatibilities: Carbonates, phosphates, sulfates, tartrates

Test Interactions Increases calcium (S); decreases magnesium

Drug Interactions Administer cautiously to a digitalized patient, may precipitate arrhythmias; calcium may antagonize the effects of verapamil; renders tetracycline and quinolone antibiotics inactive

Pregnancy Issues C; Crosses the placenta; appears in breast milk

Monitoring Parameters EKG

Nursing Implications Monitor EKG if calcium is infused faster than 2.5 mEq/minute (occasionally necessary in treating hyperkalemia)

Extravasation: Give hyaluronidase (1:10 dilution of a 150-unit vial in saline equivalent to 15 units/mL) SubQ in multiple (usually about 5) injections of 0.2 mL each to help increase absorption

Signs and Symptoms of Acute Exposure Coma, lethargy, nausea, vomiting

Treatment Strategy Supportive therapy: Can be used to treat electromechanical dissociation caused by atenolol overdose

Following withdrawal of the drug, treatment consists of bedrest, liberal intake of fluids, reduced calcium intake, and cathartic administration. Severe hypercalcemia requires I.V. hydration and forced diuresis. Urine output should be monitored and maintained at >3 mL/kg/hour. I.V. saline and natriuretic agents (eg, furosemide) can quickly and significantly increase excretion of calcium.

Reference Range Serum 9-10.4 mg/dL; due to a poor correlation between the serum ionized calcium (free) and total serum calcium, particularly in states of low albumin or acid/base imbalances, direct measurement of ionized calcium is recommended. In low albumin states, the corrected **total** serum calcium may be estimated by this equation (assuming a normal albumin of 4 g/dL); corrected total calcium = total serum calcium + 0.8 (4- measured serum albumin).

Additional Information 14 mEq/g/10 mL; 270 mg elemental calcium/g (27% elemental calcium); ingested calcium chloride solutions can result in hypercalcemia (calcium levels of 16-20 mg/dL); may convert paroxysmal supraventricular tachycardia to normal sinus rhythm (dose: 10 mL of 10% solution)

Symptoms of overdose include lethargy, nausea, vomiting, coma; following withdrawal of the drug, treatment consists of bed rest, liberal intake of fluids, reduced calcium intake, and cathartic administration. Severe hypercalcemia requires I.V. hydration and forced diuresis with I.V. furosemide (20-40 mg I.V. every 4-6 hours for adults). Urine output should be monitored and maintained at >3 mL/kg/hour. I.V. saline can quickly and significantly increase excretion of calcium into urine. Calcitonin, cholestyramine, prednisone, sodium EDTA, biphosphonates, and mithramycin have all been used successfully to treat the more resistant cases of vitamin D-induced hypercalcemia.

Specific References

Kagen MH, Bansal MG, and Grossman M, "Calcinosis Cutis Following the Administration of Intravenous Calcium Therapy," *Cutis*, 2000, 65(4):193-4.

Van Deusen SK, Birkhahn RH, and Gaeta TJ, "Treatment of Hyperkalemia in a Patient with Unrecognized Digitalis Toxicity," *J Toxicol Clin Toxicol*, 2003, 41(4):373-6.

Calcium Gluconate

CAS Number 18016-24-5; 299-28-5

Use Treatment and prevention of hypocalcemia, treatment of tetany, cardiac disturbances of hyperkalemia, cardiac resuscitation when epinephrine fails to improve myocardial contractions, hypocalcemia, calcium channel blocker toxicity, black widow spider bites

Mechanism of Action Moderates nerve and muscle performance via action potential excitation threshold regulation

Adverse Reactions

Cardiovascular: Vasodilation, hypotension, bradycardia, cardiac arrhythmias, fibrillation (ventricular), syncope, sinus bradycardia, arrhythmias (ventricular)

Central nervous system: Lethargy, coma, mental confusion, mania

Dermatologic: Erythema

Endocrine & metabolic: Hypomagnesemia, hypercalcemia

Gastrointestinal: Elevated serum amylase

Local: Tissue necrosis, extravasation injury

Neuromuscular & skeletal: Muscle weakness

Ocular: Corneal calcification

Renal: Hypercalciuria

Pharmacodynamics/Kinetics

Absorption: I.M. and I.V. calcium salts are absorbed directly into the bloodstream; absorption from the GI tract requires vitamin D

Elimination: Mainly in feces as unabsorbed calcium with 20% eliminated by the kidneys

Usual Dosage Dosage is in terms of elemental calcium

Dietary Reference Intake:

0-6 months: 210 mg/day

7-12 months: 270 mg/day

1-3 years: 500 mg/day

4-8 years: 800 mg/day

Adults, Male/Female:

9-18 years: 1300 mg/day

19-50 years: 1000 mg/day

≥51 years: 1200 mg/day

Female: Pregnancy: Same as for Adults, Male/Female

Female: Lactating: Same as for Adults, Male/Female

Dosage expressed in terms of calcium gluconate

Hypocalcemia: I.V.:

Neonates: 200-800 mg/kg/day as a continuous infusion or in 4 divided doses

Infants and Children: 200-500 mg/kg/day as a continuous infusion or in 4 divided doses

Adults: 2-15 g/24 hours as a continuous infusion or in divided doses

Hypocalcemia: Oral:

Children: 200-500 mg/kg/day divided every 6 hours

Adults: 500 mg to 2 g 2-4 times/day

Osteoporosis/bone loss: Oral: 1000-1500 mg in divided doses/day

Hypocalcemia secondary to citrated blood infusion: I.V.: Give 0.45 mEq elemental calcium for each 100 mL citrated blood infused

Hypocalcemic tetany: I.V.:

Neonates: 100-200 mg/kg/dose, may follow with 500 mg/kg/day in 3-4 divided doses or as an infusion

Infants and Children: 100-200 mg/kg/dose (0.5-0.7 mEq/kg/dose) over 5-10 minutes; may repeat every 6-8 hours **or** follow with an infusion of 500 mg/kg/day

Adults: 1-3 g (4.5-16 mEq) may be administered until therapeutic response occurs

Calcium antagonist toxicity, magnesium intoxication, or cardiac arrest in the presence of hyperkalemia or hypocalcemia: Calcium chloride is recommended calcium salt: I.V.:

Infants and Children: 60-100 mg/kg/dose (maximum: 3 g/dose)

Adults: 500-800 mg; maximum: 3 g/dose

Maintenance electrolyte requirements for total parenteral nutrition: I.V.: Daily requirements: Adults: 8-16 mEq/1000 kcal/24 hours

Dosing adjustment in renal impairment: Cl_{cr} <25 mL/minute: Dosage adjustments may be necessary depending on the serum calcium levels

Administration I.M. injections should be administered in the gluteal region in adults, usually in volumes <5 mL; avoid I.M. injections in children and adults with muscle mass wasting; do not use scalp veins or small hand or foot veins for I.V. administration; generally, I.V. infusion rates should not exceed 0.7-1.5 mEq/minute (1.5-3.3 mL/minute); stop the infusion if the patient complains of pain or discomfort. Warm to body temperature; administer slowly, usually no faster than 1.5-3.3 mL/minute, do not inject into the myocardium when using calcium during advanced cardiac life support. Topical therapy for hydrofluoric acid exposures with 2.5% gel applied to affected area every 4 hours as needed. 2.5% gel can be prepared by mixing 5 oz of K-Y® Jelly or Surgilube® with 25 mL of 10% calcium gluconate.

Contraindications Ventricular fibrillation during cardiac resuscitation and in patients with risk of digitalis toxicity, renal or cardiac disease

Warnings Avoid too rapid I.V. administration; use with caution in digitalized patients, respiratory failure or acidosis; avoid extravasation; may produce cardiac arrest. The serum calcium level should be monitored twice weekly during the early dose adjustment period. The serum calcium times phosphate product should not be allowed to exceed 66.

Dosage Forms Elemental calcium listed in brackets:

Injection: 10% = 100 mg/mL [9 mg/mL] (10 mL, 50 mL, 100 mL, 200 mL)

Tablet: 500 mg [45 mg], 650 mg [58.5 mg], 975 mg [87.75 mg], 1 g [90 mg]

Stability Admixture incompatibilities: Carbonates, phosphates, sulfates, tartrates; store at room temperature; do not use if precipitate occurs

Test Interactions ↑ calcium (S); ↓ magnesium

Drug Interactions Administer cautiously to a digitalized patient, may precipitate arrhythmias; calcium may antagonize the effects of verapamil; renders tetracycline antibiotics inactive

Pregnancy Issues C; Crosses the placenta; appears in breast milk

Monitoring Parameters Serum calcium

Nursing Implications Do not administer I.M. or SubQ

Extravasation: Give hyaluronidase (1:10 dilution of a 150-unit vial in saline equivalent to 15 units/mL) SubQ in multiple (usually about 5) injections of 0.2 mL each to help increase absorption

Signs and Symptoms of Acute Exposure Coma, lethargy, nausea, vomiting

Treatment Strategy

Acute single ingestions of calcium salts may produce mild gastrointestinal distress, but hypercalcemia or other toxic manifestations are extremely unlikely. Treatment is supportive.

Reference Range Serum: 9.0-10.4 mg/dL; due to a poor correlation between the serum ionized calcium (free) and total serum calcium, particularly in states of low albumin or acid/base imbalances, direct measurement of ionized calcium is recommended. If ionized calcium is unavailable, in low albumin states, the corrected total serum calcium may be estimated by this equation (assuming a normal albumin of 4 g/dL); corrected total calcium = total serum calcium + 0.8 (4 - measured serum albumin).

Additional Information 4.65 mEq/g; 93 mg elemental calcium/g (9.3% elemental calcium)

A commercial 2.5% calcium gluconate gel (as "H-F Antidote Gel") is available in 25 g tubes from Pharmascience, Inc, Montreal, Quebec (514-340-1114)

To make calcium gel for dermal exposure to hydrogen fluoride, 3.5 g of calcium gluconate can be added to 5 oz tube of a surgical lubricant (water soluble such as K-Y® Jelly); alternatively can be ground into a fine powder and added to 20 mL of surgical lubricant (water soluble such as K-Y® Jelly), resulting in a 32.5% slurry. Calcium chloride should not be used for these purposes.

Symptoms of overdose include lethargy, nausea, vomiting, coma; following withdrawal of the drug, treatment consists of bed rest, liberal intake of fluids, reduced calcium intake, and cathartic administration. Severe hypercalcemia requires I.V. hydration and forced diuresis with I.V. furosemide (20-40 mg I.V. every 4-6 hours for adults). Urine output should be monitored and maintained at >3 mL/kg/hour. I.V. saline can quickly and significantly increase excretion of calcium into urine. Calcitonin, cholestyramine, prednisone, sodium EDTA, biphosphonates, and mithramycin have all been used successfully to treat the more resistant cases of vitamin D-induced hypercalcemia.

Charcoal

CAS Number 16291-96-6

U.S. Brand Names Actidose-Aqua® [OTC]; Actidose® with Sorbitol [OTC]; Char-Caps [OTC]; CharcoAid G® [OTC] [DSC]; Charcoal Plus® DS [OTC]; Charcocaps® [OTC]; EZ-Char™ [OTC]; Kerr Insta-Char® [OTC]

Use Emergency treatment in poisoning by drugs and chemicals; repetitive doses for gastrointestinal dialysis for drug overdose and in uremia to adsorb various waste products; hyperbilirubinemia; has been used as an aid in the diagnosis of colouterine fistula

Mechanism of Action Adsorbs toxic substances or irritants, thus inhibiting GI absorption and for selected drugs increasing clearance by interfering with enterohepatic recycling or dialysis across intestinal vascular membranes; adsorbs intestinal gas; the addition of sorbitol results in hyperosmotic laxative action causing catharsis

Adverse Reactions

Gastrointestinal: Vomiting 6%-15% with charcoal alone, 16%-56% with sorbitol; diarrhea with sorbitol, constipation, intestinal obstruction can occur (only with multiple dosing of activated charcoal), stools will turn black

Ocular: Corneal abrasions (2 cases)

Respiratory: Aspiration usually does not cause major problems in adults, but can cause tracheal obstruction in infants; very rarely, aspiration pneumonitis (8 case reports; 3 fatalities) more likely with ingestions treated with multiple doses

Pharmacodynamics/Kinetics

Absorption: Not absorbed from GI tract

Metabolism: Not metabolized

Elimination: As charcoal in feces

Usual Dosage Oral: Ideally, achieving a 10 g charcoal:1 g toxin dose is the desired outcome

Acute poisoning: Single dose: Charcoal with sorbitol (Note: Check product label for sorbitol content):

Children and Adults: At least 5-10 times the weight of the ingested poison and other coingestants on a g:g ratio; minimum dose is probably 15-30 g; in young children sorbitol dose should not exceed 1.5 g/kg/day

Adults: 30-100 g

Charcoal in water: Same as above, but sorbitol should be added in appropriate daily doses
Single dose:
Infants and Children 1-12 years: 15-30 g
Adults: 30-100 g
Multiple dose (use only one appropriate dose of cathartic daily):
Infants <1 year: 15-30 g every 4-6 hours
Children 1-12 years: 20 g every 2 hours until clinical observations and serum drug concentration have returned to a subtherapeutic range or the development of absent bowel sounds or ileus
Adults: 20 g every 2 hours

Contraindications Not effective for cyanide, mineral acids, caustic alkalis, organic solvents, iron, ethanol, methanol poisoning, lithium; do not use charcoal with sorbitol in patients with fructose intolerance; charcoal with sorbitol is not recommended in children <1 year of age

Warnings Charcoal may cause vomiting, which is hazardous in petroleum distillate and caustic ingestions; marked variation in sorbitol content (fourfold) occurs in commercially prepackaged products; appropriate charcoal dosing may be accompanied by an overdosage of sorbitol; charcoal may adsorb maintenance medications additionally putting patients at risk for exacerbation of concomitant disorders; this is most likely with multidose regimens; food can limit charcoal's binding capacity; if charcoal in sorbitol is administered, doses should be limited to prevent excessive fluid and electrolyte losses

Dosage Forms
Capsule (Charcocaps®): 260 mg
Granules, activated (CharcoAid®-G): 15 g (120 mL)
Liquid, activated:
Actidose-Aqua®: 15 g (72 mL); 25 g (120 mL); 50 g (240 mL)
CharcoAid® 2000: 15 g (72 mL); 50 g (240 mL)
Liqui-Char®: 15 g (75 mL); 25 g (120 mL); 30 g (120 mL)
Liquid, activated [with propylene glycol]: 12.5 g (60 mL); 25 g (120 mL)
Liquid, activated [with sorbitol]:
Actidose®: 25 g (120 mL); 50 g (240 mL)
CharcoAid® 2000: 15 g (72 mL); 50 g (240 mL)
Liqui-Char®: 25 g (120 mL); 50 g (240 mL)
Powder for suspension, activated: 15 g, 30 g, 40 g, 120 g, 240 g

Stability Adsorbs gases from air, store in closed container

Pregnancy Issues C

Nursing Implications Instruct patient to drink slowly, rapid administration appears to increase frequency of vomiting; for persistent vomiting, activated charcoal can be administered as a continuous enteral infusion at doses of 10-25 g/hour; fluid volume and sorbitol dosing must be reviewed carefully; concentrated slurries may clog airway; stools will turn black; vigorous shaking of the product is suggested

Additional Information Charcoal in water with extemporaneous addition of cathartic is preferred; variation exists in absorptive surface area between commercial products, importance is not known at this time; granisetron (10 mcg/kg I.V.) can decrease nausea and emesis associated with activated charcoal/sorbitol administration; *in vivo* studies indicate that a 30- to 60-minute delay reduces efficacy by 25% to 50%. Most efficient if used within 30 minutes postingestion (84% reduction in bioavailability). Commercial charcoal products may also contain propylene glycol and can increase serum osmolarity as much as 10 mOsm.

Specific References
Alaspää AO, Kusima MJ, Hoppu K, et al, "Feasibility Study on Activated Charcoal Given Prehospital by Emergency Medical System (EMS) in Acute Intoxications," *J Toxicol Clin Toxicol*, 2002, 40(3):312-3.
Arnold TD, Cady FM, Zhang S, et al, "Low Pressure Ventilation Does Not Prevent Microvascular Permeability Lung Injury in a Rat Model of Activated Charcoal Aspiration," *J Toxicol Clin Toxicol*, 2000, 38(5):507.

Bond GR, "Activated Charcoal in the Home: Helpful and Important or Simply a Distraction?" *Pediatrics*, 2002, 109(1):145-6.

Bond GR, "The Role of Activated Charcoal and Gastric Emptying in Gastrointestinal Decontamination: A State-of-the-Art Review," *Ann Emerg Med*, 2002, 39(3):273-86.

Caravati EM, Knight HH, Linscott MS, et al, "Esophageal Laceration and Charcoal Mediastinum Complicating Gastric Lavage," *J Emerg Med*, 2001, 20(3):273-6.

Christophersen AB, Hoegberg LC, Angelo HR, et al, "Activated Charcoal in a Simulated Paracetamol Overdose: Downscaling of Dose to 10 Grams - Preliminary Results," *J Toxicol Clin Toxicol*, 2002, 40(5):696.

Cooper GM, Le Coteur DG, Richardson DB, et al, "A Randomized Clinical Trial of Activated Charcoal for the Routine Management of Oral Drug Overdose," *Acad Emerg Med*, 2002, 9(5):487.

Dorrington C, Johnson DW, Brant R, et al, "Pulmonary Aspiration and Gastrointestinal Obstruction Associated With the Use of Multiple Dose Activated Charcoal," *J Toxicol Clin Toxicol*, 2000, 38(5):507.

Dorrington CL, Johnson DW, and Brant R, "The Frequency of Complications Associated With the Use of Multiple-Dose Activated Charcoal," *Ann Emerg Med*, 2003, 41(3):370-7.

Eizember FL, Tomaszewski CA, and Kerns WP, "Acupressure for Prevention of Emesis in Patients Receiving Activated Charcoal," *J Toxicol Clin Toxicol*, 2002, 40(6):775-80.

Eroglu A, Kucuktulu U, Erciyes N, et al, "Multiple Dose-Activated Charcoal as a Cause of Acute Appendicitis," *J Toxicol Clin Toxicol*, 2003, 41(1):71-3.

Higgins T, Curry S, Brooks D, et al, "Iatrogenic Administration of Propylene Glycol With Activated Charcoal," *J Toxicol Clin Toxicol*, 2000, 38(5):529.

Hofbauer RD and Holger JS, "Cholecystokinin As an Adjunctive Measure in Charcoal Decontamination," *Acad Emerg Med*, 2002, 9(5):530.

Isbister G, Dawson A, and Whyte I, "Prehospital Activated Charcoal Who Should Be Treated?" *J Toxicol Clin Toxicol*, 2002, 40(5):613.

Justice HM, Knapp BJ, Pianalto DA, et al, "Failure of Emergency Medical Services Providers to Administer Activated Charcoal Add to Emergency Delay," *Acad Emerg Med*, 2006, 13(5):S180.

McGoodwin L and Schaeffer S, "Availability of Activated Charcoal in the Metropolitan Area of Oklahoma City," *J Toxicol Clin Toxicol*, 2000, 38(5):562-3.

McGrath JC, Klein-Schwartz W, and Coop A, "The Effectiveness of Activated Charcoal in Adsorbing GBL," *J Toxicol Clin Toxicol*, 2002, 40(5):686.

Merigian KS and Blaho KE, "Single-Dose Oral Activated Charcoal in the Treatment of the Self-Poisoned Patient: A Prospective, Randomized, Controlled Trial," *Am J Ther*, 2002, 9(4):301-8.

Rangan C, Nordt SP, Hamilton R, et al, "Treatment of Acetaminophen Ingestion With a Superactivated Charcoal-Cola Mixture," *Ann Emerg Med*, 2001, 37(1):55-8.

Roberts MS, Magnusson BM, Burczynski FJ, et al, "Enterohepatic Circulation: Physiological, Pharmacokinetic and Clinical Implications," *Clin Pharmacokinet*, 2002, 41(10):751-90.

Rose R and Waring E, "Consumer Access to Activated Charcoal Products," *J Toxicol Clin Toxicol*, 2000, 38(5):561-2.

Spiller HA and Rodgers GC Jr, "Evaluation of Administration of Activated Charcoal in the Home," *Pediatrics*, 2001, 108(6):E100.

Su M, Fong J, Howland MA, et al, "Multiple-Dose Activated Charcoal Used to Treat Valproic Acid Overdose," *J Toxicol Clin Toxicol*, 2002, 40(3):381-2.

Tenenbein PK, Sitar DS, and Tenenbein M, "Interaction Between N-Acetylcysteine and Activated Charcoal: Implications for the Treatment of Acetaminophen Poisoning," *Pharmacotherapy*, 2001, 21(11):1331-6.

Cholestyramine Resin

CAS Number 11041-12-6

U.S. Brand Names Prevalite®; Questran® Light; Questran®

Use Adjunct in the management of primary hypercholesterolemia; pruritus associated with elevated levels of bile acids; diarrhea associated with excess fecal bile acids; diarrhea due to Pfiesteria; binding toxicologic agents, such as digitoxin, possibly phenobarbital, warfarin, lindane, lorazepam, methotrexate, chlordecone; pseudomembraneous colitis (*Clostridium difficile*), oxaluria

Mechanism of Action Forms a nonabsorbable complex with bile acids in the intestine, releasing chloride ions in the process; inhibits enterohepatic reuptake of intestinal bile salts and thereby increases the fecal loss of bile salt-bound low density lipoprotein cholesterol; an anion-exchange resin (chloride form)

Adverse Reactions

Dermatologic: Rash, irritation of perianal area, skin

Endocrine & metabolic: Hyperchloremic acidosis, hypoprothrombinemia, hypernatremia from free water loss due to diarrhea

Gastrointestinal: Constipation, nausea, vomiting, abdominal distention and pain, malabsorption of fat-soluble vitamins, fecal impaction, steatorrhea, tongue irritation

Genitourinary: Increased urinary calcium excretion

Hepatic: Elevation of alkaline phosphatase and liver-function test results

Pharmacodynamics/Kinetics

Onset of action: Reduction of plasma cholesterol concentrations generally occurs within 24-48 hours after initiation of cholestyramine therapy, but may continue to fall for up to 1 year; in some patients after the initial decrease, serum cholesterol concentrations return to or exceed baseline levels with continued therapy; relief of pruritus associated with biliary stasis usually occurs within 2-3 weeks after initiation of therapy; relief of diarrhea associated with bile acids occurs within 24 hours

Duration:

Reduction of plasma cholesterol concentration: After withdrawal of cholestyramine, cholesterol concentrations return to baseline in ~2-4 weeks

Relief of pruritus associated with biliary stasis: Pruritus returns within 1-2 weeks when medication is withdrawn

Absorption: Not absorbed from the GI tract

Distribution: None

Metabolism: None

Elimination: In the feces as an insoluble complex with bile acids

Usual Dosage Oral (dosages are expressed in terms of anhydrous resin):

Children: 240 mg/kg/day in 3 divided doses; need to titrate dose depending on indication

Adults: 4 g 1-6 times/day to a maximum of 16-32 g/day

Administration Administer with water or pulpy fruit

Contraindications Avoid using in complete biliary obstruction

Warnings Use with caution in patients with constipation or phenylketonuria

Dosage Forms

Powder: 4 g of resin/9 g of powder (9 g, 378 g)

Powder for oral suspension:

With aspartame: 4 g of resin/5 g of powder (5 g, 210 g)

With phenylalanine: 4 g of resin/5.5 g of powder (60 s)

Stability Suspension may be used for up to 48 hours after refrigeration

Test Interactions ↑ prothrombin time (S); ↓ cholesterol (S), iron (B)

Drug Interactions Decreased absorption (oral) of digitalis glycosides, warfarin, thyroid hormones, thiazide diuretics, loperamide, phenylbutazone, propranolol, phenobarbital, fat soluble vitamins, amiodarone, methotrexate, NSAIDs, and other drugs by binding to the drug in the intestine

Pregnancy Issues C

Nursing Implications Do not administer the powder in its dry form; just prior to administration, mix with fluid or with applesauce; administer warfarin at least 1-2 hours prior to, or 6 hours after cholestyramine because cholestyramine may bind warfarin and decrease its total absorption. (**Note:** Cholestyramine itself may cause hypoprothrombinemia in patients with impaired enterohepatic circulation.)

Signs and Symptoms of Acute Exposure Gastrointestinal obstruction/concretion

Additional Information Questran® Light contains aspartame

Specific References

Balagani R, Wills B, and Leikin JB, "Cholestyramine Improves Tropical-Related Diarrhea," *Am J Ther*, 2006, 13(3):281-2.

Roberge RJ, Rao P, Miske GR, et al, "Diarrhea-Associated Over-Anticoagulation in a Patient Taking Warfarin: Therapeutic Role of Cholestyramine," *Vet Hum Toxicol*, 2000, 42(6):351-3.

White CM, Kalus JS, Caron MF, et al, "Cholestyramine Ointment Used on an Infant for Severe Buttocks Rash Resistant to Standard Therapeutic Modalities," *J Pharm Technol*, 2003, 19:11-3.

Cyanide Antidote Kit

U.S. Brand Names Cyanide Antidote Package

Use Treatment of cyanide poisoning

Usual Dosage For cyanide poisoning, a 0.3 mL ampul of amyl nitrite is crushed every minute and vapor inhaled for 15-30 seconds until an I.V. sodium nitrite infusion is available. Following administration of 300 mg I.V. sodium nitrite, inject 12.5 g sodium thiosulfate I.V. (over ~10 minutes), if needed; injection of both may be repeated at $^1/_2$ the original dose.

Contraindications Hypersensitivity to any component

Dosage Forms Kit: Sodium nitrite 300 mg/10 mL (#2); sodium thiosulfate 12.5 g/50 mL (#2); amyl nitrite 0.3 mL (#12); also disposable syringes, stomach tube, tourniquet, and instructions

Additional Information Shelf-life: 18 months. The amyl nitrite pearls achieve a methemoglobin level of ~3% and are primarily used when there is no intravenous access. Available from Acorn Pharmaceuticals, Inc, PO Box 3950, Chicago, IL 60674; (800) 932-5676

Deferiprone

CAS Number 30652-11-0

Use: Unlabeled/Investigational Prevention of chronic iron overload in patients with transfusion-dependent beta-thalassemia

Mechanism of Action Orally active iron chelator

Adverse Reactions

Cardiovascular: Vasculitis

Hematologic: Bone marrow suppression, thrombocytopenia, leukopenia

Neuromuscular & skeletal: Arthritis, arthralgia

Pharmacodynamics/Kinetics

Absorption: Delayed by food

Half-life: 3-4 hours

Usual Dosage Oral: 75-100 mg/kg/day in 3 divided doses

Test Interactions ↓ serum ferritin; ↑ urinary excretion of iron

Reference Range Oral dose of 25 mg/kg produces a peak plasma level of 14-17 mcg/mL

Additional Information Criteria proposed for effective chronic chelation in thalassemia major are hepatic iron concentration <80 μmol/g and serum ferritin level <2500 mcg/L; 3 molecules of deferiprone binds one molecule of iron

Specific References
Anderson LJ, Wonke B, Prescott E, et al, "Comparison of Effects of Oral Deferiprone and Subcutaneous Desferrioxamine on Myocardial Iron Concentrations and Ventricular Function in Beta-Thalassaemia," *Lancet*, 2002, 360(9332):516-20.

Deferoxamine

CAS Number 138-14-7; 1950-39-6; 70-51-9
Synonyms BA-33112; Desferrioxamine; DFM Mesylate; DFO
U.S. Brand Names Desferal®
Use Acute iron intoxication (>350 mcg/dL); chronic iron overload secondary to multiple transfusions; diagnostic test for iron overload
Use: Unlabeled/Investigational Treatment of aluminum accumulation in renal failure (serum aluminum levels >60 mcg/L); iron overload secondary to congenital anemias; hemochromatosis; removal of corneal rust rings following surgical removal of foreign bodies; has also been used for porphyria cutanea tarda and rheumatoid arthritis, aluminum toxicity
Mechanism of Action Complexes with trivalent ions (ferric ions) to form ferrioxamine, which is removed by the kidneys
Adverse Reactions
Cardiovascular: Flushing, hypotension, tachycardia, shock, edema, sinus tachycardia
Central nervous system: Fever, seizures
Dermatologic: Erythema, urticaria, pruritus, rash, cutaneous wheal formation
Endocrine & metabolic: Growth suppression
Gastrointestinal: Abdominal discomfort, diarrhea, appendicitis
Genitourinary: Dysuria, urine discoloration (pink)
Local: Pain and induration at injection site
Neuromuscular & skeletal: Leg cramps, exacerbation of myasthenia gravis
Ocular: Blurred vision, cataract, visual field defects, night blindness, macular edema, retinopathy
Otic: Hearing loss, tinnitus
Renal: Renal insufficiency
Miscellaneous: Anaphylaxis, possible risk of fungal and *Y. enterocolitica* infections (increased); *Yersinia*, Zygomycetes and *Aeromonas hydrophila* infection have been associated with deferoxamine administration; phycomycosis; thrombocytopenia, toxicity, fibrosis and edema at high doses (over 24 hours) - restrictive lung pathology has been noted
Pharmacodynamics/Kinetics
Absorption: Oral: <15%
Distribution: V_d: 2.6 L/kg in healthy individuals; 1.9 L/kg in patients with chronic iron overload; distributed widely into body
Metabolism: In the liver to ferrioxamine
Half-life:
Deferoxamine: 6.1 hours
Ferrioxamine: 5.8 hours
Hemochromatosis patients: 5.6 hours
Elimination: Renal excretion of the metabolite and unchanged drug (tubular reabsorption can occur)
Clearance:
Healthy: 0.296 L/hour/kg
Hemochromatosis patients: 0.234 L/hour/kg
Thalassemic children: 1.8-2.4 L/hour/kg

Usual Dosage Oral use is probably not effective; 100 mg DFO binds ~10 mg iron and 4.1 mg of aluminum
Children:
Acute iron intoxication:
I.M.: 90 mg/kg/dose every 8 hours; maximum: 6 g/day
I.V.: 10-15 mg/kg/hour (up to 35 mg/kg/hour with caution in severe poisoning); in nonpoisoning settings, maximum dose: 6 g/day
Chronic iron overload:
I.V.: 15 mg/kg/hour
SubQ: 20-40 mg/kg/day over 8-12 hours
Aluminum-induced bone disease: 20-40 mg/kg every hemodialysis treatment, frequency dependent on clinical status of the patient
Dialysis patients for general aluminum toxicity (levels >60 mcg/dL): 40-80 mg/kg I.V. once weekly prior to dialysis with a dose reduction to 20-60 mg/kg with a positive response
Adults:
Acute iron intoxication:
I.M.: 1 g stat, then 0.5 g every 4 hours for two doses, then 0.5 g every 4-12 hours up to 6 g/day
I.V.: 15 mg/kg/hour, up to 6-8 g/day; maximum given 16 g/day without side effects
Chronic iron overload:
I.M.: 0.5-1 g every day
SubQ: 1-2 g every day over 8-24 hours

Contraindications Patients with anuria

Warnings Use with caution in patients with severe renal disease, pyelonephritis; may increase susceptibility to *Yersinia enterocolitica*, phycomycosis, *Cunninghamella bertholletiae*, *Pneumocystis carinii*, and *Staphylococcus aureus*

Dosage Forms Injection, powder for reconstitution, as mesylate: 500 mg, 2 g

Stability Protect from light; reconstituted solutions (sterile water) may be stored at room temperature for 7 days

Test Interactions May interfere with colorimetric iron assays, along with total iron binding capacity

Drug Interactions Can cause loss of consciousness when administered with prochlorperazine

Pregnancy Issues C; Do not withhold chelation treatment for iron overdose solely due to pregnancy; has caused fetal skeletal abnormalities in animal models but is probably safe to use in the gravid patient.

Monitoring Parameters Serum iron; ophthalmologic exam and audiometry with chronic therapy

Nursing Implications Iron chelate colors urine salmon pink, which is concentration and pH dependent; I.M. is preferred route; maximum I.V. rate is 35 mg/kg/hour; incompatible with heparin

Signs and Symptoms of Acute Exposure Seizures

Additional Information One proposed endpoint of therapy after standard (15 mg/kg/hour) dose of deferoxamine is to obtain a urine iron to creatinine ratio; a value >12.5 indicates need for continued therapy of deferoxamine; hypotension (with rapid I.V. administration: >15 mg/kg/hour); acetylcysteine (I.V. at 140 mg/kg, then 70 mg/kg for 2 days) has been used to treat deferoxamine-induced pulmonary injury; high doses of deferoxamine may reduce cadmium induced liver and renal toxicity (in a rodent model).
Indications for use in iron poisoning:
• Asymptomatic patients with peak iron levels >500 mcg/dL
• Symptomatic patients with peak iron levels >350 mcg/dL
• Patients with multiple radiopaque tablets on abdominal radiograph
• Persistent minor symptoms
• Acidosis
• Hypovolemia/hypotension
• Mental status changes
• Abdominal pain
• Repeated episodes of emesis or loose stools

DEXTROSE

Endpoints of therapy for deferoxamine use in iron poisoning:
- Reversion to normal urine color from *vin rosé* urine
- Clearance of radiopacities on abdominal radiograph
- Clearance of systemic signs or symptoms relating to iron toxicity
- Serum iron concentration <100 mcg/dL
- Investigational: Evidence of continued ferruresis as demonstrated by urine iron to creatinine ratio >12.5 (at I.V. doses of deferoxamine of 15 mg/kg/hour) is an indication for continued therapy.

Specific References
Sánchez Rodríguez A, Martín Oterino JA, and Fidalgo Fernandez MA, "Unusual Toxicity of Deferoxamine," *Ann Pharmacother*, 1999, 33(4):505-6.

Dextrose

CAS Number 50-99-7; 5996-10-1
Synonyms $D_{10}W$; $D_{20}W$; $D_{50}W$; D_5W; $D_{70}W$; Glucose
Use Patients with altered mental status due to hypoglycemia
Mechanism of Action A carbohydrate substrate for aerobic metabolism
Adverse Reactions
Central nervous system: Confusion, has been associated with worsening of ischemic strokes
Endocrine & metabolic: Hypokalemia, hyperglycemia, hypomagnesemia, hypophosphatemia
Gastrointestinal: Sweet taste
Local: Vein irritation, thrombophlebitis, extravasation injury
Respiratory: Pulmonary edema
Miscellaneous: Hyperosmolar solution, which can cause a Volkmann's contraction of an extremity if extravasation occurs

Pharmacodynamics/Kinetics Metabolism: To carbon dioxide and water
Usual Dosage Dextrose 50% solution (parenteral administration)
Children: Dilute to 10% to 25% and give 2-4 mL/kg body weight

"Rule of 50" for Providing 0.5 g of Dextrose Intravenously in Pediatric Patients with Sulfonylurea-induced Hypoglycemia	
Dextrose Concentration of I.V. Fluid	**Volume to be Administered**
10%	5 mL/kg
25%	2 mL/kg
50%	1 mL/kg

Adults: 50-100 mL
Warnings Do not administer within same tubing as whole blood in that hemolysis can occur; administer thiamine (50-100 mg I.V.) to prevent Wernicke's encephalopathy
Test Interactions May cause false elevation of serum creatinine through interference with laboratory determination
Pregnancy Issues C
Monitoring Parameters Serum glucose
Additional Information Has been used as a sclerosing agent (for varicose vein treatment or to produce adhesive pleuritis); for most insulin overdoses, anticipate a need of 400-600 mg of glucose/kg/hour; continuous infusions of glucose with concentrations >20% should be given by central venous line
Specific References
Wiesli P, Spinas GA, Pfammatter T, et al, "Glucose-Induced Hypoglycaemia," *Lancet*, 2002, 360(9344):1476.
Wood S, "Is D_{50} Too Much of a Good Thing?," *JEMS*, 2007, 32(3):103-112.

Diethyldithiocarbamate Trihydrate

CAS Number 148-18-5

Synonyms DDC; DDTC; Dithiocarb Sodium; DTC

Use: Unlabeled/Investigational Chelating agent for nickel; a metabolite of disulfiram, it has also been used to treat thallium poisoning, but is not effective; shown to be active against HIV *in vitro*

Mechanism of Action Binds to nickel; the nickel-diethyldithiocarbamate complex is lipophilic thus enhancing its elimination

Adverse Reactions Gastrointestinal: Nausea

Pharmacodynamics/Kinetics
Metabolism: To carbon disulfide and diethylamine
Half-life: 15.5 hours
Elimination: Urine and feces

Usual Dosage
Mild or doubtful nickel exposure (urine nickel < 100 mcg/L): Oral: 2 g in divided doses (every 4 hours)
Severe nickel exposure (levels 100-500 mcg/L): Oral: Initial: 2 g with 1 g sodium bicarbonate, then 1 g at 4 hours, 600 mg at 8 hours, 400 mg at 16 hours, and 400 mg every 8 hours thereafter
Alternatively, 35-45 mg/kg/day on day one, then 400 mg every 8 hours until symptoms are resolved or urine nickel levels are < 100 mcg/L; I.V. dose is 12.5 mg/kg
Urine nickel levels > 500 mcg/L: Give 25-100 mg/kg/day

Warnings Do not use DDC for divalent nickel poisoning.

Additional Information May facilitate nickel GI absorption; may actually cause an increase in CNS levels of thallium; can cause disulfiram reaction with ethanol. Avoid use of ethanol; not useful to treat ocular copper nitrate toxicity; probably not useful to treat nickel dermatitis. Sodium bicarbonate dosage is used to prevent nausea; experimental use (in rodents) to remove cadmium bound to metallothionein.

Specific References
Barceloux DG, "Nickel," *J Toxicol Clin Toxicol*, 1999, 37(2):239-58.

Diethylene Triamine Penta-Acetic Acid

CAS Number 67-43-6

Synonyms Calcium Chel 330 (Ca-DTPA); DPTA: Pentetic Acid; DTPA; IND 4041 (Ca-DTPA); IND-14603 (Zn-DTPA)

Use Chelating absorbed multivalent radioisotopes of the actinide series (plutonium, neptunium, americium) as well as cesium and actinide; may be useful for uranium, curium, californium, cerium, yttrium, lanthanum, scandium, promethium, niobium, manganese, thorium, lutetium, zirconium, and zinc; on IND status

Mechanism of Action Chelation for multivalent radioisotopes

Adverse Reactions
Central nervous system: Chills, fever, headache
Dermal: Pruritus
Gastrointestinal: Transient nausea, vomiting, diarrhea, metallic taste
Neuromuscular & skeletal: Muscle cramps in first day of therapy
Renal: Nephrotoxicity
Miscellaneous: Ca-DTPA may cause zinc depletion leading to transient anosmia

Pharmacodynamics/Kinetics
Half-life, plasma: 20-60 minutes
Elimination: Renal, fecal is < 3%

Usual Dosage 1 g of the calcium salt diluted in 250 mL of 5% dextrose in water infused over 60-90 minutes; may be repeated daily for 5 days; zinc salt can be used for pregnant patients; may be aerosolized through nebulization. Most effective if used within 24 hours of exposure.

Contraindications Ca-DTPA: Do not use in pregnant females, children, patients with nephrotic syndrome or patients with bone marrow depression (Zn-DTPA may be used in above circumstances).

Warnings Both Zn-DTPA or Ca-DTPA should be used with extreme caution for uranium (due to nephrotoxic effects) or neptunium (due to increased bone deposition).

Stability Store in cool environment and away from sunlight

Pregnancy Issues C (Zn-DTPA); D (Ca-DTPA)

Monitoring Parameters Urine analysis, radioassay, renal function and blood counts during therapy, pulse and blood pressure

Additional Information Calcium salt appears to more effective than zinc salt during first hours postexposure; combination with dimercaprol has been demonstrated to be effective in rodents to enhance the elimination of cadmium; may be obtained from Radiation Emergency Assistance Center/Training Site (REAC/TS) in Oak Ridge, TN; (615) 576-3131 or (615) 576-1004 or (615) 481-1000.

Zinc replacement (with 220 mg of zinc sulfate tablets) may be required if long-term Ca-DTPA treatment is performed; IND forms need to be followed. Questions regarding use of these agents may be directed to Robert C Ricks, PhD; (615) 576-3131, Shirley A Fry, MPH; (615) 576-3480, WW Burr, MD, PhD; (615) 576-5262, Dale E Minner, MD; (615) 576-2124, or AS Garrett, MD; (615) 576-7431 at the Medical Sciences Division of Oak Ridge Associated Universities, PO Box 117, Oak Ridge, TN 37831-0117.

Dimercaprol

CAS Number 59-52-9

Synonyms 2,3-Dimercapto-1-propanol; BAL; British Anti-Lewisite; Dithioglycerol

U.S. Brand Names BAL in Oil®

Use Antidote to gold, arsenic, mercury, methyl bromide, methyl iodide, or trivalent antimony poisoning; adjunct to edetate calcium disodium in lead poisoning; possibly effective for bismuth, polonium, chromium, copper, nickel, tungsten, or zinc

Mechanism of Action Sulfhydryl group combines with ions of various heavy metals to form relatively stable, nontoxic, soluble chelates, which are excreted in bile and urine.

Adverse Reactions

>10%:

Cardiovascular: Hypertension, tachycardia (dose-related)

Central nervous system: Headache

1% to 10%: Gastrointestinal: Nausea, vomiting

<1%: Nervousness, fever, convulsions, salivation, transient neutropenia, thrombocytopenia, increased PT, pain at the injection site, abscess formation, myalgia, paresthesia, blepharospasm, burning eyes, nephrotoxicity, dysuria, burning sensation of the lips, mouth, throat, and penis

Pharmacodynamics/Kinetics

Duration of action: ~4 hours; frequent doses at 3- to 4-hour intervals over prolonged periods are necessary to maintain therapeutic effect

Absorption: Slow through the skin

Distribution: To all tissues including the brain; highest concentrations in liver and kidneys

Metabolism: Rapid to inactive products

Half-life: Short

Time to peak serum concentration: Obtained in 30-60 minutes

Elimination: In the urine within 4 hours

Usual Dosage Children and Adults:

I.M.:

Mild arsenic and gold poisoning: 2.5 mg/kg/dose every 6 hours for 2 days, then every 12 hours on the third day, and once daily thereafter for 10 days

Severe arsenic and gold poisoning: 3 mg/kg/dose every 4 hours for 2 days then every 6 hours on the third day, then every 12 hours thereafter for 10 days

Mercury poisoning: Initial: 5 mg/kg followed by 2.5 mg/kg/dose 1-2 times/day for 10 days

Lead poisoning (use with edetate calcium disodium):

Mild: 3-5 mg/kg/dose every 4 hours for 2 days, then 2.5-3 mg/kg/dose every 6 hours for 2 days, then 2.5-3 mg/kg/dose every 12 hours for 1 week

Severe: 4 mg/kg/dose every 4 hours for 5-7 days

Acute encephalopathy: Initial: 4 mg/kg/dose, then every 4 hours

Topical dermal/ocular dose for Lewisite exposure: <20% solution

Contraindications Hepatic insufficiency; do not use on iron, cadmium, or selenium poisoning; contraindicated in patients allergic to peanuts

Warnings Urinary alkalinization recommended to protect kidney; use with caution in patients with oliguria. Potentially a nephrotoxic drug, use with caution in patients with oliguria; keep urine alkaline to protect kidneys; use with caution in patients with hypertension or with glucose 6-phosphate dehydrogenase deficiency; give all injections deep I.M. at different sites. Also may contain benzl benzoate and peanut oil. (Do not use in patients allergic to peanuts.)

Dosage Forms Injection, oil: 100 mg/mL (3 mL) [contains benzyl benzoate and peanut oil]

Test Interactions May ↓ Iodine I-131 thyroid uptake values

Drug Interactions Toxic complexes with iron, cadmium, selenium, or uranium

Pregnancy Issues C

Nursing Implications Administer deep I.M. only; urine should be kept alkaline

Additional Information Preadministration with ephedrine or antihistamines may reduce side effects; nephrotoxic effects can be minimized by maintaining an alkaline urine flow; effectiveness questionable for arsenic-induced liver damage or short-chain organic mercury compounds; ineffective for thallium, tellurium, or vanadium

Information can be obtained from Acorn Pharmaceuticals, Inc, 2500 Millbrook Drive, Buffalo Grove, IL 60089; (800) 932-5676.

Specific References

Vilensky JA and Redman K, "British Anti-Lewisite (Dimercaprol): An Amazing History," *Ann Emerg Med*, 2003, 41(3):378-83.

DOBUTamine

CAS Number 49745-95-1

Use Short-term management of patients with cardiac decompensation; useful for hypotension induced by tricyclic antidepressants, beta-adrenergic blockers, doxazosin, calcium channel blockers, or sedative-hypnotic toxicity

Mechanism of Action Stimulates $beta_1$-adrenergic receptors, causing increased contractility and heart rate, with little effect on $beta_2$- or alpha-receptors

Adverse Reactions Incidence of adverse events is not always reported.

Cardiovascular: Increased heart rate, increased blood pressure, increased ventricular ectopic activity, hypotension, premature ventricular beats (5%, dose-related), anginal pain (1% to 3%), nonspecific chest pain (1% to 3%), palpitations (1% to 3%)

Central nervous system: Fever (1% to 3%), headache (1% to 3%), paresthesia

Endocrine & metabolic: Slight decrease in serum potassium

Gastrointestinal: Nausea (1% to 3%)

Hematologic: Thrombocytopenia (isolated cases)

Local: Phlebitis, local inflammatory changes and pain from infiltration, cutaneous necrosis (isolated cases)

Neuromuscular & skeletal: Mild leg cramps
Respiratory: Dyspnea (1% to 3%)

Pharmacodynamics/Kinetics
Onset of action: I.V.: Within 1-10 minutes
Duration: A few minutes
Distribution: Wide throughout the body
Metabolism: In tissues and the liver to inactive metabolites
Time to peak serum concentration: Within 10-20 minutes
Half-life: 2 minutes
Elimination: Urine

Usual Dosage I.V. infusion:
Neonates and Children: 2.5-15 mcg/kg/minute, titrate to desired response
Adults: 2.5-15 mcg/kg/minute; maximum: 40 mcg/kg/minute, titrate to desired response. See table.

DOBUTamine	
Creatinine Clearance (mL/min)	**Dosage Interval**
30-40	q8h
15-30	q12h
<15	q24h

Maximum I.V. dose: 40 mcg/kg/minute
Maximum survivable dose on record: 130 mcg/kg/minute for 30 minutes. See table.

Infusion Rates of Various Dilutions of DOBUTamine		
Desired Delivery Rate (mcg/kg/min)	**Infusion Rate (mL/kg/min)**	
	500 mcg/mL*	**1000 mcg/mL†**
2.5	0.005	0.0025
5.0	0.01	0.005
7.5	0.015	0.0075
10.0	0.02	0.01
12.5	0.025	0.0125
15.0	0.03	0.015

*500 mg per liter or 250 mg per 500 mL of diluent.
†1000 mg per liter or 250 mg per 250 mL of diluent.

Contraindications Hypersensitivity to sulfites (commercial preparation contains sodium bisulfite) and/or dobutamine; patients with idiopathic hypertrophic subaortic stenosis

Warnings Hypovolemia should be corrected prior to use; infiltration causes local inflammatory changes, extravasation may cause dermal necrosis; use with extreme caution following myocardial infarction; potent drug, must be diluted prior to use; patient's hemodynamic status should be monitored

Dosage Forms
Infusion, as hydrochloride [premixed in dextrose]: 1 mg/mL (250 mL, 500 mL); 2 mg/mL (250 mL); 4 mg/mL (250 mL)
Injection, solution, as hydrochloride: 12.5 mg/mL (20 mL, 40 mL, 100 mL) [contains sodium bisulfite]

Stability Remix solution every 24 hours; incompatible with sodium bicarbonate solutions; store reconstituted solution under refrigeration for 48 hours or 6 hours at room temperature; pink discoloration of solution indicates slight oxidation but no significant loss of potency

Test Interactions Dobutamine (and to a lesser extent dopamine) can cause profound negative interference with the enzymatic method for serum creatinine determination, producing a false depression of serum creatinine.

Drug Interactions General anesthetics (ie, halothane or cyclopropane) and usual doses of dobutamine have resulted in ventricular arrhythmias in animals. In animals, the cardiac effects of dobutamine are antagonized by beta-adrenergic blockers, resulting in predominance of alpha-adrenergic effects and increased peripheral resistance. Hypotension may develop with simultaneous administration of dopamine and dipyridamole; low dose dobutamine interaction with concomitant carvedilol leading to hypotension.

Pregnancy Issues B

Monitoring Parameters Blood pressure, EKG, heart rate, CVP, RAP, MAP, urine output; if pulmonary artery catheter is in place, monitor CI, PCWP, and SVR

Nursing Implications Alkaline solutions (sodium bicarbonate); do not give through same I.V. line as heparin, hydrocortisone sodium succinate, cefazolin, or penicillin; administer into large vein; use infusion device to control rate of flow

Extravasation: Use phentolamine as antidote; mix 5 mg with 9 mL of normal saline; inject a small amount of this dilution into extravasated area; blanching should reverse immediately. Monitor site; if blanching should recur, additional injections of phentolamine may be needed.

Signs and Symptoms of Acute Exposure Fatigue, nervousness

Additional Information

Standard diluent: 250 mg/500 mL D_5W

Minimum volume: 500 mg/250 mL D_5W

Dobutamine stress echocardiography: Incidence of 1 severe adverse reaction per every 335 examinations

Specific References

Dribben WH, Kirk MA, Trippi JA, et al, "A Pilot Study to Assess the Safety of Dobutamine Stress Echocardiography in the Emergency Department Evaluation of Cocaine-Associated Chest Pain," *Ann Emerg Med*, 2001, 38(1):42-8.

Lattanzi F, Picano E, Adamo E, et al, "Dobutamine Stress Echocardiography: Safety in Diagnosing Coronary Artery Disease," *Drug Saf*, 2000, 22(4):251-62.

DOPamine

CAS Number 51-61-6; 62-31-7

Synonyms 3-Hydroxytyramine; Dopamine Hydrochloride; Intropin

Use Adjunct in the treatment of shock, which persists after adequate fluid volume replacement; dose-related inotropic and vasopressor effects; stimulates dopaminergic, beta- and alpha-receptors

Mechanism of Action Stimulates both adrenergic and dopaminergic receptors, lower doses are mainly dopaminergic stimulating and produce renal and mesenteric vasodilation, higher doses also are both dopaminergic and beta$_1$-adrenergic and produce cardiac stimulation and renal vasodilation, large doses stimulate alpha-adrenergic receptors. A metabolic precursor to norepinephrine.

Adverse Reactions Frequency not defined.

Most frequent:

Cardiovascular: Ectopic beats, tachycardia, anginal pain, palpitations, hypotension, vasoconstriction

Central nervous system: Headache

Gastrointestinal: Nausea and vomiting

Respiratory: Dyspnea

Infrequent:

Cardiovascular: Aberrant conduction, bradycardia, widened QRS complex, ventricular arrhythmias (high dose), gangrene (high dose), hypertension

Central nervous system: Anxiety

Endocrine & metabolic: Piloerection, serum glucose increased (usually not above normal limits)

Local: Extravasation of dopamine can cause tissue necrosis and sloughing of surrounding tissues

Ocular: Intraocular pressure increased, dilated pupils

Renal: Azotemia, polyuria

Pharmacodynamics/Kinetics

Children: With medication changes, may not achieve steady-state for ~1 hour rather than 20 minutes

Adults:

Onset of action: 5 minutes

Duration: <10 minutes

Metabolism: In the plasma, kidneys, and liver 75% to inactive metabolites by monoamine oxidase and 25% to norepinephrine (active)

Half-life: 2 minutes

Elimination: Metabolites are excreted in the urine; neonatal clearance varies and appears to be age-related; clearance is more prolonged with combined hepatic and renal dysfunction. Dopamine has exhibited nonlinear kinetics in children.

Usual Dosage I.V. infusion:

Neonates: 1-20 mcg/kg/minute continuous infusion, titrate to desired response

Children: 1-20 mcg/kg/minute, maximum: 50 mcg/kg/minute continuous infusion, titrate to desired response

Adults: 1-5 mcg/kg/minute up to 50 mcg/kg/minute, titrate to desired response; infusion may be increased by 1-4 mcg/kg/minute at 10- to 30-minute intervals until optimal response is obtained

If dosages >20-30 mcg/kg/minute are needed, a more direct-acting pressor may be more beneficial (ie, epinephrine, norepinephrine).

The hemodynamic effects of dopamine are dose-dependent:

Low-dose: 1-5 mcg/kg/minute, increased renal blood flow and urine output

Intermediate-dose: 5-15 mcg/kg/minute, increased renal blood flow, heart rate, cardiac contractility, and cardiac output

High-dose: >15 mcg/kg/minute, alpha-adrenergic effects begin to predominate, vasoconstriction, increased blood pressure

Administration Administer into large vein to prevent the possibility of extravasation; use infusion device to control rate of flow; due to short half-life, withdrawal of the drug is often the only necessary treatment

Extravasation: Use phentolamine as antidote; mix 5 mg with 9 mL of normal saline; inject a small amount of this dilution into extravasated area; blanching should reverse immediately. Monitor site; if blanching should recur, additional injections of phentolamine may be needed.

Contraindications Hypersensitivity to sulfites (commercial preparation contains sodium bisulfite); pheochromocytoma, or ventricular fibrillation

Warnings Potent drug; must be diluted prior to use. Patient's hemodynamic status should be monitored; patients with peripheral vascular disease may develop gangrene. Not effective for hypotension induced by ethanol-disulfiram interaction; avoid tricyclic antidepressants; alpha$_1$-adrenergic antagonist, phenothiazine, or hydrocarbon toxicity.

Dosage Forms

Infusion, as hydrochloride [premixed in D$_5$W]: 0.8 mg/mL (250 mL, 500 mL); 1.6 mg/mL (250 mL, 500 mL); 3.2 mg/mL (250 mL)

Injection, solution, as hydrochloride: 40 mg/mL (5 mL, 10 mL); 80 mg/mL (5 mL); 160 mg/mL (5 mL)

Stability Protect from light; solutions that are darker than slightly yellow should not be used; incompatible with alkaline solutions or iron salts; compatible when coadministered with dobutamine, epinephrine, isoproterenol, and lidocaine

Test Interactions Dobutamine (and to a lesser extent dopamine) can cause profound negative interference with the enzymatic method for serum creatinine determination, producing a false depression of serum creatinine; may interfere with TSH assay resulting in falsely low TSH levels.

Drug Interactions Dopamine's effects are prolonged and intensified by MAO inhibitors, alpha- and beta-adrenergic blockers, general anesthetics, phenytoin. Paradoxical hypotension can develop when tolazoline is also given; concomitant use with ergometrine can cause gangrene.

Pregnancy Issues C; Not known if it crosses the placenta

Monitoring Parameters Potassium, glucose, blood pressure, EKG, heart rate, CVP, RAP, MAP, urine output; if pulmonary artery catheter is in place, monitor Cl, PCWP, SVR, and PVR

Nursing Implications Monitor continuously for free flow

Extravasation: Use phentolamine as antidote; mix 5 mg with 9 mL of normal saline; inject a small amount of this dilution into extravasated area; blanching should reverse immediately. Monitor site; if blanching should recur, additional injections of phentolamine may be needed. If phentolamine is not available, topical nitroglycerin paste or subcutaneous administration of terbutaline (at a 1:1 to 1:10 dilution) may be effective alternative treatment.

Signs and Symptoms of Acute Exposure Fixed and dilated pupils, tachycardia, severe hypertension

Additional Information Compatible with heparin, lidocaine, potassium chloride, and calcium chloride

Specific References

Holmes CL and Walley KR, "Bad Medicine: Low-Dose Dopamine in the ICU," *Chest*, 2003, 123(4):1266-75.

Kellum JA and M Decker J, "Use of Dopamine in Acute Renal Failure: A Meta-analysis," *Crit Care Med*, 2001, 29(8):1526-31.

Yeo V, Young K, and Tsuen CH, "Anticholinesterase-Induced Hypotension Treated With Pulmonary Artery Catheterization-Guided Vasopressors," *Vet Hum Toxicol*, 2002, 44(2): 99-100.

Edetate Calcium Disodium

CAS Number 23411-34-9; 62-33-9

Synonyms Calcium EDTA; Sodium Calcium Edetate

U.S. Brand Names Calcium Disodium Versenate®

Use Treatment of acute and chronic lead, manganese, scandium, lanthanum, plutonium, or zinc poisoning; may aid in the diagnosis of lead poisoning

Use: Unlabeled/Investigational Cholelitholytic agent (through biliary duct infusion) for pigment bile duct stones

Mechanism of Action Calcium is displaced by divalent and trivalent heavy metals, forming a nonionizing soluble complex that is excreted in urine.

Adverse Reactions

Cardiovascular: Hypotension, arrhythmias, EKG changes

Central nervous system: Fever, headache, chills

Dermatologic: Skin lesions, cheilosis

Endocrine & metabolic: Hypercalcemia, zinc deficiency

Gastrointestinal: GI upset, anorexia, nausea, vomiting

Hematologic: Transient marrow suppression, anemia

Hepatic: Mild increase in liver function test results

Local: Pain at injection site following I.M. injection, thrombophlebitis following I.V. infusion (at concentration >5 mg/mL)

Neuromuscular & skeletal: Arthralgia, tremor, numbness, paresthesia

Ocular: Lacrimation

Renal: Renal tubular necrosis, proteinuria, microscopic hematuria

Respiratory: Sneezing, nasal congestion

Miscellaneous: Zinc depletion

Pharmacodynamics/Kinetics

Distribution: Into extracellular fluid

Metabolism: Not metabolized

Elimination: Rapid in urine; urinary excretion of chelated lead begins in 1 hour and peak excretion of chelated lead occurs within 24-48 hours

Usual Dosage Several regimens have been recommended:

Diagnosis of lead poisoning: Mobilization test (not recommended by AAP guidelines): I.M., I.V.:

Children: 500 mg/m^2/dose (maximum dose: 1 g) as a single dose or divided into 2 doses

Adults: 500 mg/m^2/dose

Treatment of lead poisoning: Children and Adults (each regimen is specific for route):

Symptoms of lead encephalopathy and/or blood lead level >70 mcg/dL: Treat 5 days; give first dose after, then given in conjunction with dimercaprol; wait a minimum of 2 days with no treatment before considering a repeat course:

I.M.: 250 mg/m^2/dose every 4 hours

I.V.: 50 mg/kg/day as 24-hour continuous I.V. infusion or 1-1.5 g/m^2 I.V. as either an 8- to 24-hour infusion or divided into 2 doses every 12 hours

Symptomatic lead poisoning without encephalopathy or asymptomatic with blood lead level >70 mcg/dL: Treat 3-5 days; treatment with dimercaprol is recommended until the blood lead level concentration <50 mcg/dL:

I.M.: 167 mg/m^2 every 4 hours

I.V.: 1 g/m^2 as an 8- to 24-hour infusion or divided every 12 hours

Asymptomatic children with blood lead level 45-69 mcg/dL: I.V.: 25 mg/kg/day for 5 days as an 8- to 24-hour infusion or divided into 2 doses every 12 hours

Depending upon the blood lead level, additional courses may be necessary; repeat at least 2-4 days and preferably 2-4 weeks apart

Adults with lead nephropathy: An alternative dosing regimen reflecting the reduction in renal clearance is based upon the serum creatinine. See table.

Alternative Dosing Regimen for Adults with Lead Nephropathy	
Serum Creatinine (mg/dL)	Ca EDTA Dosage
≤2	1 g/m^2/day for 5 days
2-3	500 mg/m^2/day for 5 days
3-4	500 mg/m^2/dose every 48 hours for 3 doses
>4	500 mg/m^2/week
Repeat these regimens monthly until lead excretion is reduced toward normal.	

Administration For intermittent I.V. infusion, administer the dose I.V. over at least 1 hour in asymptomatic patients, 2 hours in symptomatic patients; for I.V. continuous infusion, dilute to 2-4 mg/mL in D$_5$W or normal saline and infuse over at least 8 hours, usually over 12-24 hours; for I.M. injection, 1 mL of 1% procaine hydrochloride may be added to each mL of EDTA calcium to minimize pain at injection site

Contraindications Severe renal disease, anuria

Warnings Potentially nephrotoxic; monitor BUN, creatinine, urinalysis, I & O during therapy; acidosis (renal tubular) and fatal nephrosis may occur, especially with high doses; EKG changes may occur during therapy; do not exceed recommended daily dose; avoid rapid I.V. infusion in the management of lead encephalopathy, may increase intracranial pressure to lethal levels. If anuria, increasing proteinuria, or hematuria occurs during therapy, discontinue calcium EDTA. Can also chelate zinc and iron.

Dosage Forms Injection, solution: 200 mg/mL (5 mL)

Stability Dilute with 0.9% sodium chloride or D_5W; physically incompatible with $D_{10}W$, lactated Ringer's, Ringer's

Test Interactions If calcium EDTA is given as a continuous I.V. infusion, stop the infusion for at least 1 hour before blood is drawn for lead concentration, to avoid a falsely elevated value.

Drug Interactions Do not use simultaneously with zinc insulin preparations, do not mix in the same syringe with dimercaprol

Pregnancy Issues B

Monitoring Parameters BUN, serum creatinine, urinalysis, fluid balance, EKG, blood and urine lead concentrations

Specific References

Morgan BW, Singleton K, and Thomas JD, "Adverse Effects in 5 Patients Receiving EDTA at an Outpatient Chelation Clinic," *Vet Hum Toxicol*, 2002, 44(5):274-6.

Edetate Disodium

CAS Number 139-33-3; 150-38-9; 6381-92-6

Synonyms Edathamil Disodium; EDTA; Sodium Edetate

U.S. Brand Names Endrate®

Use Emergency treatment of hypercalcemia; also used as contact lens cleaner; also used as a 0.05 molar topical solution for alkali ocular exposure due to calcium hydroxide (lime) from motor or cement exposures; also chelates lead, copper, manganese, zinc, lanthanum, plutonium, scandium, or yttrium; useful in treating extravasation due to mithramycin

Mechanism of Action Chelates with divalent or trivalent metals to form a soluble complex that is then eliminated in urine

Adverse Reactions

Cardiovascular: Arrhythmias, transient hypotension, shock, vasculitis, pericardial effusion/ pericarditis

Central nervous system: Seizures, fever, headache, chills

Dermatologic: Skin eruptions, exfoliative dermatitis, dermatitis, skin lesions

Endocrine & metabolic: Hypocalcemic tetany, hypomagnesemia, hypokalemia

Gastrointestinal: Nausea, vomiting, abdominal cramps, diarrhea

Genitourinary: Urinary frequency

Hematologic: Anemia

Local: Pain at the site of injection, thrombophlebitis

Neuromuscular & skeletal: Back pain, muscle cramps, paresthesia, myalgia, leg cramps

Renal: Nephrotoxicity, acute tubular necrosis

Respiratory: Death from respiratory arrest, respiratory failure, sneezing

Miscellaneous: Yawning

Pharmacodynamics/Kinetics

Metabolism: Not metabolized

Half-life: 20-60 minutes

Elimination: Following chelation, 95% excreted in urine as chelates within 24-48 hours

Usual Dosage Hypercalcemia: I.V.:

Children: 40-70 mg/kg/day slow infusion over 3-4 hours or more to a maximum of 3 g/24 hours; administer for 5 days and allow 5 days between courses of therapy

Adults: 50 mg/kg/day over 3 or more hours to a maximum of 3 g/24 hours; a suggested regimen of 5 days followed by 2 days without drug and repeated courses up to 15 total doses

Mithramycin extravasation: 150 mg of sodium edetate either through offending I.V. catheter of through multiple SubQ injections; apply ice to affected area

Administration Must be diluted before use in 500 mL D_5W or normal saline to <30 mg/mL

Contraindications Severe renal failure or anuria

Warnings Monitor cardiac function (EKG monitoring), blood pressure during infusion; renal function should be assessed before and during therapy; monitor calcium, magnesium, potassium levels; use of this drug is recommended only when the severity of the clinical condition justifies the aggressive measures associated with this type of therapy. Use with caution in patients with intracranial lesions, seizure disorders, coronary or peripheral vascular disease, or patients with tuberculosis.

Dosage Forms Injection, solution: 150 mg/mL (20 mL)

Drug Interactions May decrease blood glucose concentrations and reduce insulin requirements in diabetic patients treated with insulin

Pregnancy Issues C

Monitoring Parameters Serum calcium

Nursing Implications Avoid extravasation; patient should remain supine for a short period after infusion

Additional Information Sodium content of 1 g: 5.4 mEq; hypotension, dysrhythmias, tetany, seizures; have I.V. calcium salts ready to treat hypocalcemia-related adverse reactions; replace calcium cautiously in patients on digitalis

Epinephrine

CAS Number 51-43-4

U.S. Brand Names Adrenalin®; EpiPen® Jr; EpiPen®; Primatene® Mist [OTC]; Raphon [OTC]; S2® [OTC]; Twinject™

Use Bronchospasms, anaphylactic reactions, bradycardia, cardiac arrest, management of open-angle (chronic simple) glaucoma, hypotension, allergic reactions to antivenin, specifically improved mortality secondary to chloroquine overdose

Mechanism of Action Stimulates alpha-adrenergic, $beta_1$- and $beta_2$-adrenergic receptors; small doses can causes vasodilation via $beta_2$ vascular receptors; decreases production of aqueous humor and increases aqueous outflow; dilates the pupil by contracting the dilator muscle

Adverse Reactions Frequency not defined.

Cardiovascular: Tachycardia (parenteral), pounding heartbeat, flushing, hypertension, pallor, chest pain, increased myocardial oxygen consumption, cardiac arrhythmias, sudden death, angina, vasoconstriction

Central nervous system: Nervousness, anxiety, restlessness, headache, dizziness, light-headedness, insomnia

Gastrointestinal: Nausea, vomiting, xerostomia, dry throat

Genitourinary: Acute urinary retention in patients with bladder outflow obstruction

Neuromuscular & skeletal: Weakness, trembling

Ocular: Precipitation or exacerbation of narrow-angle glaucoma, transient stinging, burning, eye pain, allergic lid reaction, ocular irritation

Renal: Decreased renal and splanchnic blood flow

Respiratory: Wheezing, dyspnea

Miscellaneous: Diaphoresis (increased)

Pharmacodynamics/Kinetics

Onset of action: Inhalation: Within 1 minute; SubQ: Within 3-5 minutes

Following conjunctival instillation intraocular pressures fall within 1 hour with a maximal response occurring within 4-8 hours; ocular effects persist for 12-24 hours

Absorption: Absorbed well; rapid and prolonged following administration of aqueous suspension

Distribution: Does not cross blood-brain barrier; distributes throughout the body

Metabolism: Following administration, the drug is taken up into the adrenergic neuron and metabolized by monoamine oxidase and catechol-o-methyltransferase, circulating drug is metabolized in the liver; inactive metabolites (metanephrine and the sulfate and hydroxy derivatives of mandelic acid)

Elimination: Small amount of unchanged drug is excreted in urine

Usual Dosage

Neonates: Cardiac arrest: I.V.: 0.01-0.03 mg/kg (0.1-0.3 mL/kg of 1:10,000 solution) every 3-5 minutes as needed. Although I.V. route is preferred, may consider administration of doses up to 0.1 mg/kg through the endotracheal tube until I.V. access established; dilute intratracheal doses to 1-2 mL with normal saline.

Infants and Children:

Asystole/pulseless arrest, bradycardia, VT/VF (after failed defibrillations):

I.V., I.O.: 0.01 mg/kg (0.1 mL/kg of 1:10,000 solution) every 3-5 minutes as needed (maximum: 1 mg)

Intratracheal: 0.1 mg/kg (0.1 mL/kg of 1:1000 solution) every 3-5 minutes (maximum: 10 mg)

Continuous I.V. infusion: 0.1-1 mcg/kg/; doses <0.3 mcg/kg/minute generally produce β-adrenergic effects and higher doses generally produce α-adrenergic vasoconstriction; titrate dosage to desired effect

Bronchodilator: SubQ: 0.01 mg/kg (0.01 mL/kg of 1:1000) (single doses not to exceed 0.5 mg) every 20 minutes for 3 doses

Nebulization: 1-3 inhalations up to every 3 hours using solution prepared with 10 drops of 1:100

Children <4 years: S2® (racepinephrine, OTC labeling): Croup: 0.05 mL/kg (max 0.5 mL/dose); dilute in NS 3 mL. Administer over ~15 minutes; do not administer more frequently than every 2 hours.

Inhalation: Children ≥4 years: Primatene® Mist: Refer to Adults dosing.

Decongestant: Children ≥6 years: Refer to Adults dosing

Hypersensitivity reaction:

SubQ, I.V.: 0.01 mg/kg every 20 minutes; larger doses or continuous infusion may be needed for some anaphylactic reactions

SubQ, I.M.:

15-30 kg: Twinject™: 0.15 mg (for self-administration following severe allergic reactions to insect stings, food, etc)

>30 kg: Refer to Adults dosing

I.M.:

<30 kg: Epipen® Jr: 0.15 mg (for self-administration following severe allergic reactions to insect stings, food, etc)

>30 kg: Refer to Adults dosing

Adults:

Asystole/pulseless arrest, bradycardia, VT/VF:

I.V., I.O.: 1 mg every 3-5 minutes; if this approach fails, higher doses of epinephrine (up to 0.2 mg/kg) may be indicated for treatment of specific problems (eg, beta-blocker or calcium channel blocker overdose)

Intratracheal: Administer 2-2.5 mg for VF or pulseless VT if I.V./I.O. access is delayed or cannot be established; dilute in 5-10 mL NS or distilled water. Note: Absorption is greater with distilled water, but causes more adverse effects on PaO_2.

Bradycardia (symptomatic) or hypotension (not responsive to atropine or pacing): I.V. infusion: 2-10 mcg/minute; titrate to desired effect

Bronchodilator:

SubQ: 0.3-0.5 mg (1:1000) every 20 minutes for 3 doses

Nebulization: 1-3 inhalations up to every 3 hours using solution prepared with 10 drops of the 1:100 product

S2® (racepinephrine, OTC labeling): 0.5 mL (~10 drops). Dose may be repeated not more frequently than very 3-4 hours if needed. Solution should be diluted if using jet nebulizer.

Inhalation: Primatene® Mist (OTC labeling): One inhalation, wait at least 1 minute; if relieved, may use once more. Do not use again for at least 3 hours.

Decongestant: Intranasal: Apply 1:1000 locally as drops or spray or with sterile swab

Hypersensitivity reaction:

I.M., SubQ: 0.3-0.5 mg (1:1000) every 15-20 minutes if condition requires (I.M route is preferred)

>30 kg: Twinject™: 0.3 mg (for self-administration following severe allergic reactions to insect stings, food, etc)

I.M.: >30 kg: Epipen®: 0.3 mg (for self-administration following severe allergic reactions to insect stings, food, etc)

I.V.: 0.1 mg (1:10,000) over 5 minutes. May infuse at 1-4 mcg/minute to prevent the need to repeat injections frequently.

Contraindications Hypersensitivity to epinephrine or any component (particularly sulfites); cardiac arrhythmias; angle-closure glaucoma

Warnings Use with caution in elderly patients, patients with diabetes mellitus, cardiovascular diseases, thyroid disease, or cerebral arteriosclerosis; use with caution in intoxications with halogenated and aromatic hydrocarbons, digitalis derivatives, neuroleptics, propranolol, tremor, and ergot; rapid I.V. infusion may cause death from cerebrovascular hemorrhage or cardiac arrhythmias; often preserved with sulfites

Dosage Forms

Aerosol for oral inhalation (Primatene® Mist): 0.22 mg/inhalation (15 mL, 22.5 mL)

Injection, solution [prefilled auto injector]:

EpiPen®: 0.3 mg/0.3 mL [1:1000] (2 mL) [contains sodium metabisulfite; available as single unit or in double-unit pack with training unit]

EpiPen® Jr: 0.15 mg/0.3 mL [1:2000] (2 mL) [contains sodium metabisulfite; available as single unit or in double-unit pack with training unit]

Injection, solution, as hydrochloride: 0.1 mg/mL [1:10,000] (10 mL); 1 mg/mL [1:1000] (1 mL)

Adrenalin®: 1 mg/mL [1:1000] (1 mL, 30 mL)

Solution for oral inhalation, as hydrochloride: Adrenalin®: 1% [10 mg/mL, 1:100] (7.5 mL) [contains sodium bisulfite]

Solution, ophthalmic, as hydrochloride (Epifrin®): 0.5% (15 mL); 1% (15 mL); 2% (15 mL) [contains benzalkonium chloride and sodium metabisulfite]

Stability Protect from light, oxidation turns drug pink, then a brown color; solutions should not be used if they are discolored or contain a precipitate; stability of injection of parenteral admixture at room temperature and refrigeration: 24 hours; unstable in alkaline solution; do not use D_5W as a diluent; D_5W is incompatible with epinephrine

Test Interactions ↑ bilirubin (S), catecholamines (U), glucose, uric acid (S)

Drug Interactions Increased cardiac irritability if administered concurrently with halogenated inhalational anesthetics; in presence of beta-blockers, an exaggerated increase in blood pressure can occur; avoid if possible concomitant infusion with sodium bicarbonate; entacapone can potentiate the chronotropic and arrhythmogenic effects of epinephrine

Pregnancy Issues C; Crosses the placenta; appears in breast milk

Monitoring Parameters Heart rate, blood pressure

Nursing Implications

Protect from light; oxidation turns dark pink, then brown - solutions should not be used if they are discolored or contain a precipitate. Epinephrine is unstable in alkaline solution.

Extravasation: Use phentolamine as antidote; mix 5 mg with 9 mL of normal saline; inject a small amount of this dilution into extravasated area; blanching should reverse immediately. Monitor site; if blanching should recur, additional injections of phentolamine may be needed. If phentolamine is not available, topical nitroglycerin paste or subcutaneous administration of terbutaline (at a 1:1 to 1:10 dilution) may be effective alternative treatment.

Signs and Symptoms of Acute Exposure Cardiac arrhythmias, hyperglycemia, hypertension (which may result in subarachnoid hemorrhage and hemiplegia), metabolic acidosis, pulmonary edema, renal failure, unusually large pupils

Reference Range Therapeutic: 31-95 pg/mL (SI: 170-520 pmol/L); norepinephrine level of 438 pg/mL associated with myocardial ischemia in children

Additional Information There is no specific antidote for epinephrine intoxication and the bulk of the treatment is supportive. Hyperactivity and agitation usually respond to reduced sensory input; however, with extreme agitation, haloperidol (2-5 mg I.M. for adults) may be required. Hyperthermia is best treated with external cooling measures, or when severe or unresponsive, muscle paralysis with pancuronium may be needed. Hypertension is usually transient and generally does not require treatment unless severe. For diastolic blood pressures > 110 mm Hg, a nitroprusside infusion should be initiated. Seizures usually respond to diazepam I.V. and/or phenytoin maintenance regimens.

Specific References

Goldhaber SZ, "Administration of Epinephrine by Emergency Medical Technicians," *N Engl J Med*, 2000, 342(11):822.

Hahn I, Schnadower D, Dakin RJ, et al, "Electromagnetic Interference From a Cellular Phone As a Cause of Acute Epinephrine Poisoning," *J Toxicol Clin Toxicol*, 2000, 38(5):524-5.

Hatton RC, "Bismuth Subgallate-Epinephrine Paste in Adenotonsillectomies," *Ann Pharmacother*, 2000, 34(4):522-5.

Loeckinger A, Kleinsasser A, Wenzel V, et al, "Pulmonary Gas Exchange After Cardiopulmonary Resuscitation With Either Vasopressin or Epinephrine," *Crit Care Med*, 2002, 30(9):2059-62.

Plint AC, Osmond MH, and Klassen TP, "The Efficacy of Nebulized Racemic Epinephrine in Children With Acute Asthma: A Randomized Double-Blind Trial," *Acad Emerg Med*, 2000, 7(10):1097-103.

Putland M, Kerr D, and Kelly AM, "Adverse Events Associated With the Use of Intravenous Epinephrine in Emergency Department Patients Presenting With Severe Asthma," *Ann Emerg Med*, 2006, 47(6):559-63.

Raymondos K, Panning B, Leuwer M, et al, "Absorption and Hemodynamic Effects of Airway Administration of Adrenaline in Patients With Severe Cardiac Disease," *Ann Intern Med*, 2000, 132(10):800-3.

Sato Y, Tanaka M, and Nishikawa T, "Reversible Catecholamine-Induced Cardiomyopathy by Subcutaneous Injections of Epinephrine Solution in an Anesthetized Patient," *Anesthesiology*, 2000, 92(2):615-9.

Sicherer SH, Forman JA, and Noone SA, "Use Assessment of Self-Administered Epinephrine Among Food-Allergic Children and Pediatricians," *Pediatrics*, 2000, 105(2):359-62.

Von Derau K and Martin TG, "Supportive Treatment for Intradigital Epinephrine Injections," *J Toxicol Clin Toxicol*, 2002, 40(5):614-5.

Wyer PC, Perera P, Jin Z, et al, "Vasopressin or Epinephrine for Out-of-Hospital Cardiac Arrest," *Ann Emerg Med*, 2006, 48:86-97.

Filgrastim

Synonyms G-CSF; Granulocyte Colony Stimulating Factor

U.S. Brand Names Neupogen®

Use Stimulation of granulocyte production in patients with malignancies, including myeloid malignancies; receiving myelosuppressive therapy associated with a significant risk of neutropenia; severe chronic neutropenia (SCN); receiving bone marrow transplantation (BMT); undergoing peripheral blood progenitor cell (PBPC) collection

Mechanism of Action Stimulates the production, maturation, and activation of neutrophils, G-CSF activates neutrophils to increase both their migration and cytotoxicity. Natural proteins that stimulate hematopoietic stem cells to proliferate, prolong cell survival, stimulate cell differentiation, and stimulate functional activity of mature cells. CSFs are produced by a wide variety of cell types. Specific mechanisms of action are not yet fully understood, but possibly work by a second-messenger pathway with resultant protein production. Specific activity: 1×10^8 units/mg; neutrophil production rate increases 9.4-fold.

Adverse Reactions
>10%:
    Cardiovascular: Chest pain
    Central nervous system: Fever
    Dermatologic: Alopecia
    Endocrine & metabolic: Fluid retention
    Gastrointestinal: Nausea, vomiting, diarrhea, mucositis; splenomegaly - up to 33% of patients with cyclic neutropenia/congenital agranulocytosis receiving filgrastim for $\geq$14 days; rare in other patients
    Neuromuscular & skeletal: Bone pain (24%), commonly in the lower back, posterior iliac crest, and sternum
1% to 10%:
    Cardiovascular: S-T segment depression (3%)
    Central nervous system: Headache
    Dermatologic: Rash
    Gastrointestinal: Anorexia, constipation, sore throat
    Hematologic: Leukocytosis
    Local: Pain at injection site
    Neuromuscular & skeletal: Weakness
    Respiratory: Dyspnea, cough
<1%: Transient supraventricular arrhythmias, pericarditis, hypotension, thrombophlebitis, hypersensitivity reactions

Pharmacodynamics/Kinetics
Onset of action: Rapid elevation in neutrophil counts within the first 24 hours, reaching a plateau in 3-5 days; immature neutrophils (bands) may be elevated within 1 hour
Duration: ANC decreases by 50% within 2 days after discontinuing G-CSF; white counts return to the normal range in 4-7 days
Absorption: SubQ: 100%; peak plasma levels can be maintained for up to 12 hours
Distribution: V_d: 0.15 L/kg; no evidence of drug accumulation over a 11- to 20-day period
Metabolism: Systemic
Bioavailability: Oral: Not bioavailable
Half-life: 1.3-7.2 hours (longer with doses >10 mcg/kg)
Time to peak serum concentration: SubQ: Within 2-6 hours

Usual Dosage Refer to individual protocols.
Dosing, even in morbidly obese patients, should be based on actual body weight. Rounding doses to the nearest vial size often enhances patient convenience and reduces costs without compromising clinical response.
Myelosuppressive therapy: 5 mcg/kg/day - doses may be increased by 5 mcg/kg according to the duration and severity of the neutropenia.
Bone marrow transplantation: 5-10 mcg/kg/day - doses may be increased by 5 mcg/kg according to the duration and severity of neutropenia; recommended steps based on neutrophil response:
    When ANC >1000/mm^3 for 3 consecutive days: Reduce filgrastim dose to 5 mcg/kg/day
    If ANC remains >1000/mm^3 for 3 more consecutive days: Discontinue filgrastim
    If ANC decreases to <1000/mm^3: Resume at 5 mcg/kg/day
    If ANC decreases <1000/mm^3 during the 5 mcg/kg/day dose, increase filgrastim to 10 mcg/kg/day and follow the above steps
Peripheral blood progenitor cell (PBPC) collection: 10 mcg/kg/day or 5-8 mcg/kg twice daily in donors. The optimal timing and duration of growth factor stimulation has not been determined.

Severe chronic neutropenia:
  Congenital: 6 mcg/kg twice daily
  Idiopathic/cyclic: 5 mcg/kg/day
Colchicine overdose: Daily doses of 300 mg SubQ for 5 days has been used in adults

Administration The UHC Colony-Stimulating Factors Expert Panel has made the following clinical and administrative recommendations:

Clinical:

Very few data are available to support the therapeutic interchangeability of G-CSF and GM-CSF. On the basis of the panel's clinical experience and in the absence of conclusive data, interchangeability is acceptable in limited situations, as indicated in the UHC CSF guidelines.

For most indications, an ANC of $<500 \times 10^6$ cells/L can be recommended as a starting point for the initiation of therapy. Stopping CSF therapy is based on a biologic endpoint, and the panel felt that CSF therapy could be discontinued in most patients at an ANC $\geq 5000 \times 10^6$ cells/L, depending on the expected nadir of chemotherapy. Several panel members suggested an ANC endpoint to CSF therapy at 2000×10^6 cells/L.

Food and Drug Administration labeling recommends daily dosing of G-CSF at 5 mcg/kg, but panel members felt that doses as low as 2 mcg/kg might be effective for some indications. Daily doses of >10 mcg/kg were considered not appropriate for any chemotherapy regimens. Overall, CSF dosing on the basis of patient weight has not been well studied.

In general, patients should not receive CSFs with their first cycle of cancer chemotherapy. Patients who have previously demonstrated febrile neutropenia following chemotherapy or have had a previous positive response to CSF therapy following chemotherapy would be appropriate for CSF therapy.

Certain chemotherapeutic regimens may warrant CSF use for every patient. These clinical situations should be determined at each UHC member institution. Some examples cited by the panel include patients in the following categories: adult acute lymphocytic leukemia, induction therapy; previously radiated Hodgkin's disease; previous pelvic radiation; adjuvant breast cancer chemotherapy (to avoid any treatment delays caused by neutropenia); and AIDS lymphoma treatment with myelosuppressive chemotherapy.

Administrative:

CSF guidelines developed by an institution for inpatient use should apply to the outpatient setting

Several examples of strategies for CSF cost-containment were offered by the panel, including requiring blood count results prior to dispensing CSF doses; strictly enforced inpatient and outpatient use guidelines; institutional reimbursement precertification for CSF patients; accessing pharmaceutical company reimbursement assistance/indigent programs; forming biotechnology review committees; and dosing based on vial size, rather than on a mcg/kg basis, using the smallest clinically acceptable vial size

Contraindications Hypersensitivity to *E. coli* derived proteins or G-CSF

Warnings Complete blood count and platelet count should be obtained prior to chemotherapy. Do not use G-CSF in the period 24 hours before to 24 hours after administration of cytotoxic chemotherapy because of the potential sensitivity of rapidly dividing myeloid cells to cytotoxic chemotherapy. Precaution should be exercised in the usage of G-CSF in any malignancy with myeloid characteristics. G-CSF can potentially act as a growth factor for any tumor type, particularly myeloid malignancies. Tumors of nonhematopoietic origin may have surface receptors for G-CSF.

Dosage Forms

Injection, solution [preservative free, vial]: 300 mcg/mL (1 mL, 1.6 mL)

Injection, solution [preservative free, prefilled Singleject® syringe]: 600 mcg/mL (0.5 mL, 0.8 mL)

Stability Store at 2°C to 8°C (36°F to 46°F); do not expose to freezing or dry ice. Prior to administration, filgrastim may be allowed to be at room temperature for a maximum of 24 hours. It may be diluted in dextrose 5% in water to a concentration ≥ 15 mcg/mL for I.V.

infusion administration. Minimum concentration is 15 mcg/mL; concentrations <15 mcg/mL require addition of albumin (1 mL of 5%) to the bag to prevent absorption. This diluted solution is stable for 7 days under refrigeration or at room temperature. Filgrastim is incompatible with 0.9% sodium chloride (normal saline).

Standard diluent: ≥375 mcg/25 mL D_5W

Test Interactions ↑ uric acid, lactate dehydrogenase, alkaline phosphatase; may ↓ cholesterol

Drug Interactions Acts with synergy with interleukin-3 to increase megakaryocyte and platelet production.

Pregnancy Issues C; Excretion in breast milk unknown; use caution

Monitoring Parameters Complete blood cell count and platelet count should be obtained twice weekly. Leukocytosis (white blood cell counts ≥100,000/mm^3) has been observed in ~2% of patients receiving G-CSF at doses >5 mcg/kg/day. Monitor platelets and hematocrit regularly. Monitor patients with pre-existing cardiac conditions closely as cardiac events (myocardial infarctions, arrhythmias) have been reported in premarketing clinical studies.

Nursing Implications Do not mix with sodium chloride solutions

Reference Range Peak plasma level as high as 600 ng/mL achieved after I.V. doses of 3-60 mcg/kg

Additional Information Reimbursement hotline: 1-800-28-AMGEN; monitor CBC at least twice weekly when used in conjunction with chemotherapy; analgesic agents can be used to control skeletal pain

Signs and symptoms: Leukocytosis, which was not associated with any clinical adverse effects; after discontinuing the drug there is a 50% decrease in circulating levels of neutrophils within 1-2 days, return to pretreatment levels within 1-7 days. No clinical adverse effects seen with high-dose producing ANC >10,000/mm^3.

Total duration of drug-induced neutropenia decreases from ~10 days to ~7 days with colony-stimulating factor treatment

Specific References

Andres E, Kurtz JE, Martin-Hunyadi C, et al, "Nonchemotherapy Drug-Induced Agranulocytosis in Elderly Patients: The Effects of Granulocyte Colony-Stimulating Factor," *Am J Med*, 2002, 112(6):460-4.

Moleski RJ, "Comparison of G-CSF and GM-CSF Adverse Event Profiles in Office-Based Practices: Preliminary Study Results," *Pharmacotherapy*, 2000, 20(7 Pt 2):112S-7S.

Flumazenil

CAS Number 78755-81-4

U.S. Brand Names Romazicon®

Use Benzodiazepine antagonist - reverses sedative effects of benzodiazepines used in general anesthesia; for management of benzodiazepine or zolpidem overdose; flumazenil does **not** antagonize the CNS effects of other GABA agonists (such as ethanol, barbiturates, or general anesthetics), **does not** reverse narcotics; may be useful for baclofen, carbamazepine, chloral hydrate, and zopiclone overdoses; chlorzoxazone overdose; gabapentin-induced coma; may be used to reverse paradoxical reactions to benzodiazepines in children; possible antidotal effect by flumazenil in zaleplon-induced coma and bluish-green urine

Mechanism of Action Antagonizes the effect of benzodiazepines on the GABA/benzodiazepine receptor complex, including synergistic effects with the other GABA agonists.

Adverse Reactions

>10%: Gastrointestinal: Vomiting, nausea

1% to 10%:

Cardiovascular: Palpitations

Central nervous system: Headache, anxiety, nervousness, insomnia, abnormal crying, euphoria, depression, agitation, dizziness, emotional lability, ataxia, depersonalization, increased tears, dysphoria, paranoia

Endocrine & metabolic: Hot flashes

Gastrointestinal: Xerostomia

Local: Pain at injection site

Neuromuscular & skeletal: Tremor, weakness, paresthesia

Ocular: Abnormal vision, blurred vision

Respiratory: Dyspnea, hyperventilation

Miscellaneous: Diaphoresis

<1%: Abnormal hearing, altered blood pressure (increases and decreases), anxiety and sensation of coldness, bradycardia, chest pain, generalized convulsions, hiccups, hypertension, shivering, somnolence, tachycardia, thick tongue, ventricular extrasystoles, withdrawal syndrome

Pharmacodynamics/Kinetics

Onset of benzodiazepine reversal: Within 1-3 minutes

Peak effect: 6-10 minutes

Duration: Usually <1 hour; duration is related to dose given and benzodiazepine plasma concentrations; reversal effects of flumazenil may wear off before effects of benzodiazepine

Distribution: Adults:

Initial V_d: 0.5 L/kg

V_{dss}: 0.77-1.6 L/kg

Endotracheal: V_d: 1.02 L/kg

Protein binding: ~50%

Half-life: Adults:

Alpha: 7-15 minutes

Terminal: 41-79 minutes

Endotracheal: 80 minutes

Elimination: Clearance dependent upon hepatic blood flow, hepatically eliminated; <1% excreted unchanged in urine

Usual Dosage I.V. (most patients respond to doses under 1 mg):

Children:

Reversal of conscious sedation or general anesthesia: Minimal information available; an initial dose of 0.01 mg/kg (maximum dose: 0.2 mg), then 0.005 mg/kg (maximum dose: 0.2 mg) given every minute to a maximum total dose of 1 mg has been used to reverse midazolam after circumcision in 40 children 3-12 years of age (mean: 7 years); mean total dose required: 0.024 mg/kg; further studies are needed

Management of benzodiazepine overdose: Minimal information available; a few cases have reported an initial dose of 0.01 or 0.02 mg/kg (maximum dose: 0.125 mg) with repeat doses of 0.01 mg/kg or follow up continuous infusions of 0.05 mg/hour (0.004 mg/kg/hour) or 0.05 mg/kg/hour (n = 1) for 2-6 hours; further studies are needed

Adults:

Reversal of conscious sedation or general anesthesia: 0.2 mg over 15 seconds; may repeat 0.2 mg every 60 seconds up to a total of 1 mg, usual dose: 0.6-1 mg. In event of resedation, may repeat doses at 20-minute intervals with maximum 1 mg/dose given as 0.2 mg/minute, maximum of 3 mg in 1 hour.

Management of benzodiazepine overdose: 0.2 mg over 30 seconds; may give 0.3 mg dose after 30 seconds if desired level of consciousness is not obtained; additional doses of 0.5 mg can be given over 30 seconds at 1-minute intervals up to a cumulative dose of 3 mg, usual cumulative dose: 1-3 mg; rarely, patients with partial response at 3 mg may require additional titration up to total dose of 5 mg; if the patient has not responded 5 minutes after cumulative dose of 5 mg, the major cause of sedation is likely not due to benzodiazepines. In the event of resedation, may repeat doses at 20-minute intervals with maximum of 1 mg/dose (given at 0.5 mg/minute); maximum: 3 mg in 1 hour.

Continuous infusion: Investigational: 0.5 mg/hour for 5 hours

Endotracheal: 1 mg in 10 mL saline

Contraindications Known hypersensitivity to flumazenil or benzodiazepines; patients given benzodiazepines for control of potentially life-threatening conditions (eg, control of intracranial

81

pressure or status epilepticus); patients who are showing signs of serious tricyclic antidepressant overdosage

Warnings Monitor patients for return of sedation or respiratory depression. Flumazenil should be used with caution in the intensive care unit because of increased risk of unrecognized benzodiazepine dependence in such settings. Flumazenil may produce severe withdrawal reactions in patients physically dependent on benzodiazepines.

Dosage Forms Injection, solution: 0.1 mg/mL (5 mL, 10 mL)

Stability For I.V. use only; compatible with D_5W, lactated Ringer's or normal saline; once drawn up in the syringe or mixed with solution use within 24 hours; discard any unused solution after 24 hours

Drug Interactions Increased toxicity: Use with caution in overdosage involving mixed drug overdose; toxic effects may emerge (especially with tricyclic antidepressants) with the reversal of the benzodiazepine effect by flumazenil

Pregnancy Issues C

Monitoring Parameters Respiratory rate, level of sedation

Nursing Implications Flumazenil does not effectively reverse hypoventilation, even in alert patients

Reference Range Plasma flumazenil levels between 10-20 mcg/L for 1-2 hours reverse benzodiazepine-induced CNS depression; average peak flumazenil blood level of 66 ng/mL achieved within 1 minute of administering 1 mg of flumazenil endotracheally

Additional Information Reverses benzodiazepine-induced depression of ventilatory responses to hypercapnia and hypoxia. Does not decrease antiepileptic effect of diazepam when given concomitantly and may have intrinsic antiepileptic properties. At high doses, flumazenil can inhibit adenosine uptake in the neurons. Flumazenil (50 mg I.V.) has been utilized to reverse carbamazepine-induced coma (serum level 27.8 mg/L), but it should not be utilized in any patient with an underlying seizure disorder; since carbamazepine can be epileptogenic, it should not be used routinely for this overdose. Resedation does not usually occur with adult midazolam doses <10 mg. Flumazenil (0.4 mg I.V.) has been used successfully in reversing midazolam-induced laryngospasm. Flumazenil is not a substitute for airway management if benzodiazepine related respiratory depression occurs. Flumazenil has demonstrated some benifit to treating coma resulting from hepatic encephalopathy given that this entity is mediated through the benzodiazepine 1 receptor.

Specific References

Brammer G, Gilby R, Walter FG, et al, "Continuous Intravenous Flumazenil Infusion for Benzodiazepine Poisoning," *Vet Hum Toxicol*, 2000, 42(5):280-1.

Butler TC, Rosen RM, Wallace AL, et al, "Flumazenil and Dialysis for Gabapentin-Induced Coma," *Ann Pharmacother*, 2003, 37(1):74-6.

Grob U, Schildknecht M, Rentsch K, et al, "Has Flumazenil an Antagonistic Effect in GHB-Induced CNS Depression?" *J Toxicol Clin Toxicol*, 2002, 40(5):615.

Gueye PN, Lofaso F, Borron SW, et al, "Mechanism of Respiratory Insufficiency in Pure or Mixed Drug-Induced Coma Involving Benzodiazepines," *J Toxicol Clin Toxicol*, 2002, 40(1):35-47.

Hojer J, Salmonson H, and Sundin P, "Zaleplon-Induced Coma and Bluish-Green Urine: Possible Antidotal Effect by Flumazenil," *J Toxicol Clin Toxicol*, 2002, 40(5):571-2.

Mathieu-Nolf M, Babé MA, Coquelle-Couplet V, et al, "Flumazenil Use in an Emergency Department: A Survey," *J Toxicol Clin Toxicol*, 2001, 39(1):15-20.

Folic Acid

CAS Number 59-30-3; 6484-89-5

Synonyms Folacin; Folate; Pteroylglutamic Acid

Use Treatment of megaloblastic and macrocytic anemias due to folate deficiency; used in methanol toxicity, methotrexate toxicity, chronic alcoholism

Mechanism of Action Necessary for normal erythropoiesis

Adverse Reactions

Central nervous system: Irritability, difficulty sleeping, confusion, insomnia, hyperthermia, fever

Dermatologic: Pruritus

Gastrointestinal: GI upset, bitter taste, flatulence

Miscellaneous: Hypersensitivity reactions, anaphylactic shock, zinc depletion, anaphylaxis

Pharmacodynamics/Kinetics

Peak effect: Oral: Within 30-60 minutes

Absorption: In the proximal part of the small intestine

Usual Dosage Oral, I.M., I.V., SubQ: Minimal daily requirement: 50 mcg; pregnant/critically ill: 100-200 mcg

Infants: 0.1 mg/day

Children: Initial: 1 mg/day

Deficiency: 0.5-1 mg/day

Maintenance dose:

<4 years: Up to 0.3 mg/day

>4 years: 0.4 mg/day

Adults: Initial: 1 mg/day

Deficiency: 1-3 mg/day

Maintenance dose: 0.5 mg/day

Pregnant and lactating women: 0.8 mg/day

Administration Oral, but may also be administered by deep I.M., SubQ, or I.V. injection; a diluted solution for oral or for parenteral administration may be prepared by diluting 1 mL of folic acid injection (5 mg/mL), with 49 mL sterile water for injection; resulting solution is 0.1 mg folic acid/1 mL

Contraindications Pernicious, aplastic, or normocytic anemias

Warnings Doses >0.1 mg/day may obscure pernicious anemia; patients with pernicious anemia may show hematologic improvement with doses as low as 0.25 mg/day with continuing irreversible nerve damage progression; resistance to treatment may occur with depressed hematopoiesis, alcoholism, deficiencies of other vitamins; injection contains benzyl alcohol (1.5%) as preservative

Dosage Forms

Injection, solution, as sodium folate: 5 mg/mL (10 mL) [contains benzyl alcohol]

Tablet: 0.4 mg, 0.8 mg, 1 mg

Stability Incompatible with oxidizing and reducing agents and heavy metal ions

Test Interactions Falsely low serum concentrations may occur with the *Lactobacillus casei* assay method in patients on anti-infective (i.e., tetracycline) therapy.

Drug Interactions In folate-deficient patients, folic acid therapy may increase phenytoin metabolism resulting in decreased phenytoin serum concentrations. Concurrent administration of chloramphenicol and folic acid in these patients may result in antagonism of the hematopoietic response to folic acid. Oral contraceptives may impair folate metabolism and produce folate depletion; may interfere with the antimicrobial actions of pyrimethamine against toxoplasmosis.

Pregnancy Issues A (C if dose exceeds RDA recommendation); 400 mcg/day needed to prevent neural tube defects (spina bifida)

Monitoring Parameters Hemoglobin

Nursing Implications Oral, but may also be administered by deep I.M., SubQ, or I.V. injection; a diluted solution for oral or for parenteral administration may be prepared by diluting 1 mL of folic acid injection (5 mg/mL), with 49 mL sterile water for injection; resulting solution is 0.1 mg folic acid per 1 mL

Reference Range Therapeutic: 0.005-0.015 mcg/mL

Additional Information Water-soluble vitamin with a wide margin of safety

Fomepizole

CAS Number 7554-65-6
Synonyms 4-Methylpyrazole; 4-MP
U.S. Brand Names Antizol®
Use FDA approved for ethylene glycol and methanol toxicity; may be useful in propylene glycol; usefulness in disulfiram-ethanol reactions is unclear; may be useful for diethylene glycol and triethylene glycol
Mechanism of Action Complexes and inactivates alcohol dehydrogenase competitively, thus preventing formation of the toxic metabolites of the alcohols
Adverse Reactions
>10%:
Central nervous system: Headache (14%)
Gastrointestinal: Nausea (11%)
1% to 10% (≤3% unless otherwise noted):
Cardiovascular: Bradycardia, facial flush, hypotension, phlebosclerosis, shock, tachycardia
Central nervous system: Dizziness (6%), increased drowsiness (6%), agitation, anxiety, lightheadedness, seizure, vertigo
Dermatologic: Rash
Gastrointestinal: Bad/metallic taste (6%), abdominal pain, decreased appetite, diarrhea, heartburn, vomiting
Hematologic: Anemia, disseminated intravascular coagulation, eosinophilia, lymphangitis
Hepatic: Increased liver function tests
Local: Application site reaction, inflammation at the injection site, pain during injection, phlebitis
Neuromuscular & skeletal: Backache
Ocular: Nystagmus, transient blurred vision, visual disturbances
Renal: Anuria
Respiratory: Abnormal smell, hiccups, pharyngitis
Miscellaneous: Multiorgan failure, speech disturbances
Postmarketing and/or case reports: Mild allergic reactions (mild rash, eosinophilia)
Pharmacodynamics/Kinetics
Maximum effect: 1.5-2 hours
Absorption: Oral: Readily absorbed
Distribution: V_d: 0.6-1.0 L/kg; unknown distribution, probably very similar to ethanol
V_d: Healthy volunteers: 0.69 L/kg
Protein binding: Negligible
Half-life: During dialysis: 3 hours (plasma clearance during dialysis: 230 mL/min); No dialysis: 5 hours
Metabolism: Hepatic. Primary metabolite: 4-carboxypyrazole
Elimination: Nonlinear elimination; at suggested therapeutic doses of 10-20 mg/kg the apparent elimination rate is 4-5 µmol/L/hour; 4-MP is dialyzable
Usual Dosage Oral: 15 mg/kg followed by 5 mg/kg in 12 hours and then 10 mg/kg every 12 hours until levels of toxin are not present
One other protocol (from France) suggests an infusion of 10-20 mg/kg before dialysis and intravenous infusion of 1-1.5 mg/kg/hour during hemodialysis
Loading dose of 15 mg/kg I.V. followed by 10 mg/kg I.V. every 12 hours for 48 hours, then 15 mg/kg every 12 hours until methanol or ethylene glycol levels are <20 mg/dL; supplemental doses required during dialysis
Dosing in patients requiring hemodialysis:
Dose at the beginning of hemodialysis. If <6 hours since last dose, do not administer dose; if ≥6 hours since last dose, administer next scheduled dose.
Dosing during hemodialysis: every 4 hours

Dosing at the time hemodialysis is completed:
Time between last dose and the end of hemodialysis:
<1 hour: Do not administer dose at the end of hemodialysis
1-3 hours: Administer $^1/_2$ of next scheduled dose
>3 hours: Administer next scheduled dose
Maintenance dosing off hemodialysis: Give next scheduled dose 12 hours from last dose administered

Administration Antizol® solidifies at temperatures <25°C (<77°F). If the Antizol® solution has become solid in the vial, the solution should be liquefied by running the vial under warm water or by holding in the hand. Solidification does not affect the efficacy, safety, or stability of Antizol®. Using sterile technique, the appropriate dose of Antizol® should be drawn from the vial with a syringe and injected into at least 100 mL of sterile 0.9% sodium chloride injection or dextrose 5% injection. Mix well. The entire contents of the resulting solution should be infused over 30 minutes. Antizol®, like all parenteral products, should be inspected visually for particulate matter prior to administration, whenever solution and container permit.

Contraindications Documented serious hypersensitivity to pyrazoles

Dosage Forms Injection, solution [preservative free]: 1 g/mL (1.5 mL)

Stability Store vials at room temperature. Antizol® diluted in 0.9% sodium chloride injection or 5% dextrose injection is stable for at least 48 hours when stored refrigerated or at room temperature. After dilution, do not use beyond 24 hours.

Drug Interactions Inhibitory effects on alcohol dehydrogenase are increased in presence of ethanol; reduces rate of elimination of ethanol by 40%; ethanol also decreases metabolism and elimination of 4-MP by 50%; 4-MPO induces cytochrome P450 mixed function oxidases *in vitro*; 4-MP may worsen the ethanol-chloral hydrate central nervous system interaction

Pregnancy Issues C

Reference Range Serum fomepizole concentrations >15 mg/L provide complete inhibition of alcohol dehydrogenase

Additional Information Approved in December 1997 for the treatment of ethylene glycol poisoning and in December 2000 for methanol toxicity; may be useful in ethanol-disulfiram reactions; hemodialysis clearance of 4-MP: 52-80 mL/minute. Direct questions to Orphan Medical, Inc (888) 867-7426.

Specific References

Battistella M, "Fomepizole as an Antidote for Ethylene Glycol Poisoning," *Ann Pharmacother*, 2002, 36(6):1085-9.

Bekka R, Borron SW, Astier A, et al, "Treatment of Methanol and Isopropanol Poisoning With Intravenous Fomepizole," *J Toxicol Clin Toxicol*, 2001, 39(1):59-67.

Benitez JG, Swanson-Biearman B, and Krenzelok EP, "Nystagmus Secondary to Fomepizole Administration in a Pediatric Patient," *J Toxicol Clin Toxicol*, 2000, 38(7):795-8.

Brent J, McMartin K, Phillips S, et al, "Fomepizole for the Treatment of Methanol Poisoning," *N Engl J Med*, 2001, 344(6):424-9.

Casavant MJ, "Fomepizole in the Treatment of Poisoning," *Pediatrics*, 2001, 107(1):170.

Megarbane B, Fompeydie D, Garnier R, et al, "Treatment of a 1,4-Butanediol Poisoning With Fomepizole," *J Toxicol Clin Toxicol*, 2002, 40(1):77-80.

Moestue S, Akervik O, Svendsen J, et al, "Fomepizole Treatment Prevents Renal Failure in Severe Ethylene Glycol Poisoning: Report of Two Cases," *J Toxicol Clin Toxicol*, 2002, 40(3):269-70.

Mycyk MB and Leikin JB, "Antidote Review: Fomepizole for Methanol Poisoning," *Am J Ther*, 2003, 10(1):68-70.

Najafi CC, Hertko LJ, Leikin JB, et al, "Fomepizole in Ethylene Glycol Intoxication," *Ann Emerg Med*, 2001, 37(3):358-9.

Osterhoudt KC, "Fomepizole Therapy for Pediatric Butoxyethanol Intoxication," *J Toxicol Clin Toxicol*, 2002, 40(7):929-30.

Quang L, Maher T, Shannon M, et al, "Pretreatment of CD-1 Mice With 4-Methylpyrazole (4-MP) Blocks 1,4-Butanediol (BD) Toxicity," *J Toxicol Clin Toxicol*, 2000, 38(4):527.

Quang L, Maher T, Shannon M, et al, "Use of 4-Methylpyrazole (4-MP) As an Antidote for 1,4-Butanediol (BD) Toxicity in CD-1 Mice," *J Toxicol Clin Toxicol*, 2000, 38(5):581.

Velez LI, Kulstad E, Shepherd G, et al, "Inhalational Methanol Toxicity in Pregnancy Treated Twice With Fomepizole," *Vet Hum Toxicol*, 2003, 45(1):28-30.

von Krogh A, Rygnestad T, and Iversen H, "Overdose of Disulfiram and Ethanol Treated With Fomepizole," *J Toxicol Clin Toxicol*, 2002, 40(3):392-3.

Glucagon

CAS Number 16941-32-5

U.S. Brand Names GlucaGen® Diagnostic Kit; GlucaGen® HypoKit™; GlucaGen®; Glucagon Diagnostic Kit [DSC]; Glucagon Emergency Kit

Use Hypoglycemia; diagnostic aid in the radiologic examination of GI tract when a hypotonic state is needed; useful in esophageal food impaction, cardiogenic shock; used with some success as a cardiac stimulant in management of severe cases of beta-adrenergic or calcium channel blocking agent, calcium channel blocker, or oral hypoglycemia overdosage; may be useful in treating hypotension due to imipramine toxicity; noninsulin-dependent diabetes mellitus; also used in allergic reactions

Use: Unlabeled/Investigational Nebulized for asthma

Mechanism of Action Stimulates adenylate cyclase to produce increased cyclic AMP, which promotes hepatic glycogenolysis and gluconeogenesis, causing a raise in blood glucose levels

Adverse Reactions Frequency not defined.

Gastrointestinal: Nausea, vomiting (high incidence with rapid administration of high doses)

Miscellaneous: Hypersensitivity reactions (hypotension, respiratory distress, urticaria)

Pharmacodynamics/Kinetics

Onset of action: Effects on blood glucose levels occur within 5-20 minutes following parenteral administration

Duration: 60-90 minutes

Absorption: Destroyed in the GI tract

Distribution: Not fully understood

Metabolism: In the liver with some inactivation occurring in the kidneys and plasma

Half-life: 3-10 minutes

Elimination: Renal

Usual Dosage

Allergic reactions: I.V.:

Children: 0.03-0.1 mg/kg/dose up to 1 mg; repeat every 5-20 minutes

Adults: 1 mg; repeat every 5 minutes; can give up to 5 mg/dose; for infusion, can titrate at a rate of 1-5 mg/hour

Hypoglycemia or insulin shock therapy: I.M., I.V., SubQ:

Neonates: 0.3 mg/kg/dose; maximum: 1 mg/dose

Children: 0.025-0.1 mg/kg/dose, not to exceed 1 mg/dose, repeated in 20 minutes as needed

Adults: 0.5-1 mg, may repeat in 20 minutes as needed

Cardiotoxic agent toxicity: I.V.: 0.05 mg/kg (up to 10 mg) over 1 minute, then continuous infusion of 2-5 mg/hour in 5% dextrose

Diagnostic aid: Adults: I.M., I.V.: 0.25-2 mg 10 minutes prior to procedure

Facilitate passage of foreign bodies from distal esophagus into stomach: I.V.: 0.5-1 mg over 30 seconds, may be repeated

Nebulized for asthma: 2 mg diluted in 3 mL of saline

Administration Reconstitute powder for injection by adding 1 or 10 mL of sterile diluent to a vial containing 1 or 10 units of the drug, respectively, to provide solutions containing 1 mg of glucagon/mL; if dose to be administered is <2 mg of the drug →, use only the diluent provided by the manufacturer; if >2 mg →, use sterile water for injection; use immediately after reconstitution

Contraindications Hypersensitivity to glucagon or any component

Warnings Use with caution in patients with a history of insulinoma and/or pheochromocytoma

Dosage Forms Injection, powder for reconstitution, as hydrochloride:
GlucaGen®, Glucagon: 1 mg [1 unit]
GlucaGen® Diagnostic Kit: 1 mg [1 unit] [packaged with sterile water]
Glucagon Diagnostic Kit, Glucagon Emergency Kit: 1 mg [1 unit] [packaged with diluent syringe containing glycerin 12 mg/mL and water for injection]

Stability After reconstitution, use immediately; may be kept at 5°C for up to 48 hours, if necessary

Pregnancy Issues B

Monitoring Parameters Blood pressure, blood glucose, potassium, and glucose

Signs and Symptoms of Acute Exposure Allergic reactions, dizziness, hyperglycemia, hypertension, hypokalemia, hypotension, nausea, weakness, vomiting

Reference Range Normal range: 50-60 pg/mL

Additional Information 1 unit = 1 mg; the diluent used to contain 2 mg of phenol/mL as a preservative; normal basal levels 50-60 pg/mL; I.V. infusion contains phenol; 1 mg glucagon (I.M.) used to remove orogastric tube due to esophageal (distal) spasm; glucagon may be quite effective for treatment of beta-blocker induced anaphylaxis; less effective in children or malnourished patients with significant glycogen stores

Specific References

Bailey B, "Glucagon in Beta-Blocker and Calcium Channel Blocker Overdoses: A Systematic Review," *J Toxicol Clin Toxicol*, 2003, 41(5):595-602.

Mehta DI, Attia MW, Quintana EC, et al, "Glucagon Use for Esophageal Coin Dislodgment in Children: A Prospective, Double-Blind, Placebo-Controlled Trial," *Acad Emerg Med*, 2001, 8(2):200-3.

Papadopoulos J and O'Neil MG, "Utilization of a Glucagon Infusion in the Management of a Massive Nifedipine Overdose," *J Emerg Med*, 2000, 18(4):453-5.

Shrestha M and Roberts J, "Glucagon Treatment of Mild Intestinal Obstruction from Illicit Drug Body Packing," *J Toxicol Clin Toxicol*, 2002, 40(3):381.

Wilber ST, Wilson JE, Blanda M, et al, "The Bronchodilator Effect of Intravenous Glucagon in Asthma Exacerbation: A Randomized, Controlled Trial," *Ann Emerg Med*, 2000, 36(5): 427-31.

Glycopyrrolate

Synonyms Glycopyrronium Bromide

U.S. Brand Names Robinul® Forte; Robinul®

Use Vagal-mediated bradycardia, gastric secretion reduction; also used as an adjunct to relieve GI tract spasms, as an antisialagogue agent, adjunct in anesthesia, to reverse effects of nondepolarizing muscle relaxants; may be helpful in treating organophosphate poisoning; gustatory sweating (Frey syndrome)

Use: Unlabeled/Investigational Asthma

Mechanism of Action Quaternary ammonium anticholinergic agent 5 times as potent as atropine with fewer cardiovascular or CNS side effects

Adverse Reactions
>10%:
Dermatologic: Dry skin
Gastrointestinal: Constipation, dry throat, xerostomia
Local: Irritation at injection site
Respiratory: Dry nose
Miscellaneous: Diaphoresis (decreased)

1% to 10%:
  Dermatologic: Increased sensitivity to light
  Endocrine & metabolic: Decreased flow of breast milk
  Gastrointestinal: Dysphagia
<1%: Orthostatic hypotension, ventricular fibrillation, tachycardia, palpitations, confusion, drowsiness, headache, loss of memory, fatigue, ataxia, rash, bloated feeling, nausea, vomiting, dysuria, weakness, increased intraocular pain, blurred vision

Pharmacodynamics/Kinetics
  Absorption: Oral: 10% to 25%
  Distribution: V_d: 0.64 L/kg
  Half-life: I.V.: 2.22 minutes
  Peak effects: I.M.: 30-45 minutes
  Elimination: Renal (48%); biliary clearance: 1.12 L/hour/kg

Usual Dosage
  Pediatric:
    Preanesthetic medication: I.M.: 4.4-8.8 mcg/kg
    Preanesthetic medication: I.V.: 10 mcg/kg up to 500 mcg
    Intraoperative: I.V.: 4.4 mcg/kg repeat every 2-3 minutes
    Reversal of neuromuscular blockade: I.V.:
      0.2 mg for each 1 mg of neostigmine
      0.2 mg for each 5 mg of pyridostigmine
    Inhaled dose for asthma: 25-50 mcg/kg up to 1 mg
    Organophosphate poisoning: $\sim \frac{1}{2}$ the dose one would use with atropine
  Adults: Oral: 1-2 mg (2-3 times/day)
    Intraoperative: I.V.: 0.1 mg repeat as needed every 2-3 minutes
    Reversal of neuromuscular blockade: I.V.:
      0.2 mg for each 1 mg of neostigmine
      0.2 mg for each 5 mg of pyridostigmine
    GI disorders: 0.1-0.2 mg 3-4 times/day
    Inhaled dose for asthma: Up to 2 mg
    Organophosphate poisoning: $\sim \frac{1}{2}$ the dose one would use for atropine

Contraindications Hypersensitivity to glycopyrrolate or any component of the formulation; ulcerative colitis; narrow-angle glaucoma; acute hemorrhage; tachycardia; obstructive uropathy; paralytic ileus, obstructive disease of GI tract; myasthenia gravis

Dosage Forms
  Injection, solution (Robinul®): 0.2 mg/mL (1 mL, 2 mL, 5 mL, 20 mL) [contains benzyl alcohol]
  Tablet:
    Robinul®: 1 mg
    Robinul® Forte: 2 mg

Drug Interactions B; As per atropine: Ventricular arrhythmias can occur with concomitant cycloperopane anesthesia exposure; concomitant use with cannabinoids or ritodrine has resulted in supraventricular tachycardia

Pregnancy Issues B; no teratogenic effects; in normal doses (4 mcg/kg), does not appear to affect fetal heart rate

Reference Range Peak plasma glycopyrrolate level was −6.3 ng/mL 10 minutes after intramuscular injection of 6 mcg/kg

Additional Information Note: If a pregnant organophosphate patient is treated with glycopyrrolate, the fetus is less likely to be adequately treated than if atropine is used.

Specific References
  Choi PT, Quinonez LG, Cook DJ, et al, "The Use of Glycopyrrolate in a Case of Intermediate Syndrome Following Acute Organophosphate Poisoning," *Can J Anaesth*, 1998, 45(4):337-40.
  Sivilotti ML, Bird SB, Lo JC, et al, "Multiple Centrally-Acting Antidotes Protect Against Severe Organophosphate Toxicity," *Clin Toxicol (Phila)*, 2005, 43:693.

HI-6

CAS Number 34433-31-3

Synonyms Asoxime Chloride; HJ-6

Military Classification Bispyridinium-Dioxime Anticholinesterase Reactivator; Nerve Agent Antidote

Use: Unlabeled/Investigational Has been used in dichlorvos and dimethoate poisoning; rodent experiments indicate that it may be effective in sarin, VX, soman, or GF exposure with atropine used as a pretreatment

Mechanism of Action

A broad-spectrum reactivator. Compounds with an oxime moiety (RCH=NOH) attached to a quaternary nitrogen pyridinium ring dephosphorylate the enzyme active site thereby reactivating nerve agent inhibited acetylcholinesterase. Reactivation occurs through nucleophilic attack by the oxime on the phosphorus, splitting an oxime-phosphonate away from the active site. The reactivated esteratic site is thus able to bind to and cleave its normal substrate acetylcholine. May have direct central and peripheral anticholinergic effects in addition to reactivating acetylcholinesterase.

Oximes have a relatively wide index of safety. HI-6 can produce hypotension, warm-flush facial feeling and pain at the injection site. Hepatic function should be monitored; obidoxime has caused hepatic dysfunction after multiple dosing; HI-6 may cause similar effects.

Pharmacodynamics/Kinetics

Absorption: I.M.: Within 8 minutes

Distribution: V_d: 0.4 L/kg

Elimination: Renal (56%); renal clearance of 1.74-1.98 mL/min/kg

Usual Dosage Investigational: I.M.: 500 mgm dissolved in 3 mL distilled water four times daily for 2-7 days. Usual total dose: 4-14 g

Reference Range Therapeutic serum concentration: ~4 mcg/mL

Additional Information Does not penetrate red blood cell membrane. May be superior to other oximes in the treatment of soman and sarin. Probably not effective for tabun. Has a high affinity for both intact and phosphonylated acetylcholinesterase. Animal studies show protection from toxic effects and improved survival even against supralethal doses. Currently not available in the United States for the treatment of nerve agents. May be obtained from the Bosnalijek Drug Company in Sarajevo.

Specific References

American Academy of Pediatrics Committee on Environmental Health and Committee on Infectious Diseases, "Chemical-Biological Terrorism and Its Impact on Children: A Subject Review," *Pediatrics*, 2000, 105(3 Pt 1):662-70.

Arnold JF, "CBRNE - Nerve Agents, G-Series: Tabun, Sarin, Soman," http://www.emedicine.com/EMERG/topic898.htm. Last accessed October 12, 2003.

Ben Abraham R, Weinbroum AA, Rudick V, et al, "Perioperative Care of Children With Nerve Agent Intoxication," *Paediatr Anaesth*, 2001, 11(6):643-9.

Benitez FL, Velez-Daubon LI, and Keyes DC, "CBRNE - Nerve Agents, V-Series: Ve, Vg, Vm, Vx," http://www.emedicine.com/emerg/topic899.htm. Last accessed October 3, 2003.

Eyer P, "The Role of Oximes in the Management of Organophosphorus Pesticide Poisoning," *Toxicol Rev*, 2003, 22(3):165-90.

Henretig FM, Cieslak TJ, and Eitzen EM Jr, "Biological and Chemical Terrorism," *J Pediatr*, 2002, 141(3):311-26. Erratum in: *J Pediatr*, 2002, 141(5):743-6.

Karalliedde L, Wheeler H, Maclehose R, et al, "Possible Immediate and Long-Term Health Effects Following Exposure to Chemical Warfare Agents," *Public Health*, 2000, 114(4):238-48.

Kassa J, "Comparison of the Effects of BI-6, A New Asymmetric Bipyridine Oxime, With HI-6 Oxime and Obidoxime in Combination With Atropine on Soman and Fosdrine Toxicity in Mice," *Ceska Slov Farm*, 1999, 48(1):44-7.

Kassa J, "Review of Oximes in the Antidotal Treatment of Poisoning by Organophosphorus Nerve Agents," *J Toxicol Clin Toxicol*, 2002, 40(6):803-16.

Kuca K, Bielavsky J, Cabal J, et al, "Synthesis of a New Reactivator of Tabun-Inhibited Acetylcholinesterase," *Bioorg Med Chem Lett*, 2003, 13(20):3545-7.

LeJeune KE, Dravis BC, Yang F, et al, "Fighting Nerve Agent Chemical Weapons With Enzyme Technology," *Ann N Y Acad Sci*, 1998, 864:153-70.

Moreno G, Nocentini S, Guggiari M, et al, "Effects of the Lipophilic Biscation, Bis-Pyridinium Oxime BP12, on Bioenergetics and Induction of Permeability Transition in Isolated Mitochondria," *Biochem Pharmacol*, 2000, 59(3):261-6.

"Poison Warfare (Nerve) Gases," http://www.emergency.com/nervgas.htm. Last accessed October 10, 2003.

Rotenberg JS and Newmark J, "Nerve Agent Attacks on Children: Diagnosis and Management," *Pediatrics*, 2003, 112(3 Pt 1):648-58.

Rousseaux CG and Dua AK, "Pharmacology of HI-6, an H-series Oxime," *Can J Physiol Pharmacol*, 1989, 67(10):1183-9.

Salem H and Sidell FR, "Nerve Gases," *Encyclopedia of Toxicology*, Vol 1, Wexler P, ed: Academic Press, 2002, 380-5.

Shiloff JD and Clement JG, "Comparison of Serum Concentrations of the Acetylcholinesterase Oxime Reactivators HI-6, Obidoxime, and PAM to Efficacy Against Sarin (Isopropyl Methylphosphonofluoridate) Poisoning in Rats," *Toxicol Appl Pharmacol*, 1987, 89(2):278-80.

Velez-Daubon L, Benitez FL, Keyes DC, "CBRNE - Nerve Agents, Binary: GB2, VX2," *Emedicine*, 2003, http://www.emedicine.com/EMERG/topic 900.htm. Last accessed October 13, 2003.

HLo-7

CAS Number 120103-35-7

Military Classification Bispyridinium-Dioxime Anticholinesterase Reactivator; Nerve Agent Antidote

Mechanism of Action

A broad-spectrum reactivator. Compounds with an oxime moiety (RCH=NOH) attached to a quaternary nitrogen pyridinium ring dephosphorylate the enzyme active site thereby reactivating nerve agent inhibited acetylcholinesterase. Reactivation occurs through nucleophilic attack by the oxime on the phosphorus, splitting an oxime-phosphonate away from the active site. The reactivated esteratic site is thus able to bind to and cleave its normal substrate acetylcholine.

Oximes have a relatively wide index of safety. However if administered intravenously too rapidly, hypertension may ensue. In animals, death due to respiratory paralysis is related to pyridine aldoximes.

Toxicodynamics/Kinetics

Onset of action: Rapid

Half-life: 45 minutes with maximal plasma concentration at 30 minutes; first order kinetics

Usual Dosage Not established

Additional Information The dimethanesulfonate of the reactivator obidoxime. May be superior to other oximes in the treatment of soman and sarin. Has a high affinity for both intact and phosphonylated acetylcholinesterase. May not be effective for VX exposure.

Specific References

American Academy of Pediatrics Committee on Environmental Health and Committee on Infectious Diseases, "Chemical-Biological Terrorism and Its Impact on Children: A Subject Review," *Pediatrics*, 2000, 105(3 Pt 1):662-70.

Arnold JF, "CBRNE - Nerve Agents, G-Series: Tabun, Sarin, Soman," http://www.emedicine.com/EMERG/topic898.htm. Last accessed October 12, 2003.

Ben Abraham R, Weinbroum AA, Rudick V, et al, "Perioperative Care of Children With Nerve Agent Intoxication," *Paediatr Anaesth*, 2001, 11(6):643-9.

Benitez FL, Velez-Daubon, and Keyes DC, "CBRNE - Nerve Agents, V-Series: Ve, Vg, Vm, Vx," http://www.emedicine.com/emerg/topic899.htm. Last accessed October 3, 2003.

Eyer P, Hagedorn I, Klimmek R, et al, "HLo 7 Dimethanesulfonate, A Potent Bispyridinium-Dioxime Against Anticholinesterases," *Arch Toxicol*, 1992, 66(9):603-21.

Henretig FM, Cieslak TJ, and Eitzen EM Jr, "Biological and Chemical Terrorism," *J Pediatr*, 2002, 141(3):311-26. Erratum in: *J Pediatr*, 2002, 141(5):743-6.

Howland MA, "Pralidoxime - Antidotes in Depth," *Goldfrank's Toxicological Emergencies*, 6th ed, Norwalk, CT: Appleton and Lange, 1998, 1361-5.

Karalliedde L, Wheeler H, Maclehose R, et al, "Possible Immediate and Long-Term Health Effects Following Exposure to Chemical Warfare Agents," *Public Health*, 2000, 114(4):238-48.

Kassa J, "Comparison of the Effects of BI-6, A New Asymmetric Bipyridine Oxime, With HI-6 Oxime and Obidoxime in Combination With Atropine on Soman and Fosdrine Toxicity in Mice," *Ceska Slov Farm*, 1999, 48(1):44-7.

Kassa J, "Review of Oximes in the Antidotal Treatment of Poisoning by Organophosphorus Nerve Agents," *J Toxicol Clin Toxicol*, 2002, 40(6):803-16.

Kuca K, Bielavsky J, Cabal J, et al, "Synthesis of a New Reactivator of Tabun-Inhibited Acetylcholinesterase," *Bioorg Med Chem Lett*, 2003, 13(20):3545-7.

LeJeune KE, Dravis BC, Yang F, et al, "Fighting Nerve Agent Chemical Weapons With Enzyme Technology," *Ann N Y Acad Sci*, 1998, 864:153-70.

Moreno G, Nocentini S, Guggiari M, et al, "Effects of the Lipophilic Biscation, Bis-Pyridinium Oxime BP12, on Bioenergetics and Induction of Permeability Transition in Isolated Mitochondria," *Biochem Pharmacol*, 2000, 59(3):261-6.

"Poison Warfare (Nerve) Gases," http://www.emergency.com/nervgas.htm. Last accessed October 10, 2003.

Rotenberg JS and Newmark J, "Nerve Agent Attacks on Children: Diagnosis and Management," *Pediatrics*, 2003, 112(3 Pt 1):648-58.

Rousseaux CG and Dua AK, "Pharmacology of HI-6, an H-series Oxime," *Can J Physiol Pharmacol*, 1989, 67(10):1183-9.

Salem H and Sidell FR, "Nerve Gases," *Encyclopedia of Toxicology*, Vol 1, Wexler P, ed: Academic Press, 2002, 380-5.

Shiloff JD and Clement JG, "Comparison of Serum Concentrations of the Acetylcholinesterase Oxime Reactivators HI-6, Obidoxime, and PAM to Efficacy Against Sarin (Isopropyl Methylphosphonofluoridate) Poisoning in Rats," *Toxicol Appl Pharmacol*, 1987, 89(2):278-80.

Velez-Daubon L, Benitez FL, Keyes DC, "CBRNE - Nerve Agents, Binary: GB2, VX2," *Emedicine*, 2003, http://www.emedicine.com/EMERG/topic 900.htm. Last accessed October 13, 2003.

Hydroxocobalamin

CAS Number 13422-52-1

Synonyms Vitamin B_{12a}

Use Treatment of pernicious anemia, vitamin B_{12} deficiency, increased B_{12} requirements due to pregnancy, thyrotoxicosis, hemorrhage, malignancy, liver or kidney disease

Mechanism of Action Coenzyme for various metabolic functions, including fat and carbohydrate metabolism and protein synthesis, used in cell replication and hematopoiesis. Also binds to cyanide ion preferentially than iron in cytochrome oxidase.

Adverse Reactions

Cardiovascular: Transient hypertension and tachycardia

Dermatologic: Itching, urticaria, erythema, pink discoloration to skin/mucous membranes (at doses >4 g, resolves in 1-2 days)

Endocrine & metabolic: Hypokalemia

Gastrointestinal: Diarrhea, feces discoloration (red)

Genitourinary: Urine discoloration (pink/deep red)

Hematologic: Polycythemia
Local: Pain at injection site
Respiratory: Pulmonary edema
Miscellaneous: Anaphylactoid reactions

Pharmacodynamics/Kinetics
Distribution: V_d: 0.45 L/kg
Half-life: 26.2 hours
Elimination: Renal (clearance: $\sim$0.31 L/hour): Cyanide exposure: 37%; Normal subjects: 62%

Usual Dosage
Vitamin B_{12} deficiency: I.M.:
  Children: 1-5 mg given in single doses of 100 mcg over 2 or more weeks, followed by 30-50 mcg/month
  Adults: 30 mcg/day for 5-10 days, followed by 100-200 mcg/month
I.V.: Cyanide overdose: 20-50 g of hydroxocobalamin required to bind $\sim$1 g cyanide infused at doses of 4 g associated with 8 g sodium thiosulfate; 5 g hydroxocobalamin binds about 250 mg cyanide (50:1 hydroxocobalamin:cyanide molecular weight ratio) the initial dose of Cyanokit is five grams (two vials) in adults (i.v.). may repeat to a total dose of ten grams.

Contraindications Hypersensitivity to cyanocobalamin or any component, cobalt; patients with hereditary optic nerve atrophy

Warnings Some products contain benzoyl alcohol; avoid use in premature infants; an intradermal test dose should be performed for hypersensitivity; use only if oral supplementation not possible or when treating pernicious anemia; appropriate doses for cyanide poisoning requires excessive fluid administration.

Dosage Forms Injection, solution: 1000 mcg/mL (30 mL)

Test Interactions In therapeutic concentrations, hydroxocobalamin falsely elevates bilirubin and magnesium and falsely lowers AST and creatinine; may interfere (decrease) aspartate aminotransferase and creatinine assays; may cause increase in bilirubin and magnesium)

Pregnancy Issues C

Nursing Implications Administer I.M. only; may require coadministration of folic acid

Reference Range Normal vitamin B_{12} plasma level: 200-800 pg/mL; peak serum level after a 5 g dose: $\sim$300 μmol/L

Additional Information An investigational combination of hydroxocobalamin (4 g) and sodium thiosulfate (8 g) is under study in the U.S. and is possibly available from the manufacturer (EVREKA); a 5 g dose of hydroxocobalamin appears to bind all cyanide ions in patients with initial cyanide levels up to 40 μmol/L. Approved by the FDA as part of the Cyanokit in December, 2006. Reduces whole blood cyanide concentration in dogs by 55%. Information can be also obtained via Dey, L.P. at www.Dey.com (800-429-7751).

Specific References
Baud FJ, Favier C, Borron SW, et al, "Clinical Safety of High Doses of Hydroxocobalamin in Fire Victims," *J Toxicol Clin Toxicol*, 2001, 39(3):267.
Campbell A and Jones AL, "Cyanide Poisoning Managed With Hydroxocobalamin in the U.K.," *J Toxicol Clin Toxicol*, 2001, 39(3):294.
DesLouriers CA, Burda AM, and Wahl M, "Hydroxocobalamin as a Cyanide Antidote," *Am J Ther*, 2006, 13:161-5.
Geller RJ, Barthold C, Saiers JA, Hall AH, "Pediatric cyanide poisoning: Causes, Manifestations, Management and Unmet Needs," *Pediatrics*, 2006, 118 (5):2146-58.
Mannaioni G, Vannacci A, Marzocca C, et al, "Acute Cyanide Intoxication Treated With a Combination of Hydroxycobalamin, Sodium Nitrite, and Sodium Thiosulfate," *J Toxicol Clin Toxicol*, 2002, 40(2):181-3.
Mégarbane B, Borron SW, and Baud F, "Hydroxycobalamin Versus Conventional Treatment in Cyanide Poisoning," *J Toxicol Clin Toxicol*, 2002, 40(3):314.
Sauer SW and Keim ME, "Hydroxocobalamin: Improved Public Health Readiness for Cyanide Disasters," *Ann Emerg Med*, 2001, 37(6):635-41.

Insect Sting Kit

U.S. Brand Names Ana-Kit®

Use Anaphylaxis; emergency treatment of insect bites or stings (ie, yellow jacket, honeybee, hornet, wasp, deerfly, kissing bug) by the sensitive patient when self-treatment may occur within minutes of insect sting or exposure to an allergenic substance

Usual Dosage Children and Adults:
Epinephrine:
<2 years: 0.05-0.1 mL
2-6 years: 0.15 mL
6-12 years: 0.2 mL
>12 years: 0.3 mL
Chlorpheniramine:
<6 years: 1 tablet
6-12 years: 2 tablets
>12 years: 4 tablets

Contraindications Hypertensive or hyperthyroid patients

Warnings Use caution in elderly patients, patients with diabetes mellitus or cardiovascular diseases and/or history of sensitivity to sympathomimetic amines; epinephrine should be used with caution in patients also taking beta-adrenergic blockers as unopposed alpha-adrenergic effects may occur (primarily hypertension).

Dosage Forms Kit: Epinephrine hydrochloride 1:1000 [prefilled syringe, delivers two 0.3 mL doses; contains sodium bisulfite] (1 mL), chlorpheniramine maleate chewable tablet 2 mg (4), sterile alcohol pads [isopropyl alcohol 70%] (2), tourniquet (1)

Stability Protect from light, store at room temperature, prevent from freezing

Additional Information Not intended for I.V. use (I.M. or SubQ only); not intended as a substitute for professional medical treatment; hypotension due to a systemic allergic reaction does not respond well to epinephrine; not effective for skin reactions due to fire ant bites.

Isoproterenol

CAS Number 299-95-6; 51-30-9; 6700-39-6; 7683-59-2

Synonyms Isoprenaline Hydrochloride; Isoproterenol Hydrochloride; Isoproterenol Sulfate

U.S. Brand Names Isuprel®

Use Ventricular arrhythmias due to AV nodal block; hemodynamically compromised bradyarrhythmias or atropine- and dopamine-resistant bradyarrhythmias (when transcutaneous/venous pacing is not available); temporary use in third-degree AV block until pacemaker insertion

Mechanism of Action Relaxes bronchial smooth muscle by action on beta$_2$-receptors; causes increased heart rate and contractility by action on beta$_1$-receptors

Adverse Reactions
Cardiovascular: Premature ventricular beats, bradycardia, hypertension, hypotension, chest pain, palpitations, tachycardia, ventricular arrhythmias, myocardial infarction size increased
Central nervous system: Headache, nervousness, restlessness
Gastrointestinal: Nausea, vomiting
Respiratory: Dyspnea

Pharmacodynamics/Kinetics
Distribution: V_d: 0.5 L/kg
Protein binding: 65%

Metabolism: Metabolized by conjugation in many tissues including the liver and lungs by catecholamine-O-methyltransferase (COMT) to 3-O-methyl-isoproterenol (weak beta-blocker activity)

Half-life: Biphasic: First phase: 2.5-5 minutes; second phase: 3-7 hours

Time to peak serum concentration: Oral: Within 1-2 hours

Elimination: Principally in urine as sulfate conjugates

Usual Dosage I.V.: Cardiac arrhythmias:

Children: Initial: 0.1 mcg/kg/minute (usual effective dose 0.2-2 mcg/kg/minute)

Adults: Initial: 2 mcg/minute; titrate to patient response (2-10 mcg/minute)

Administration I.V. infusion administration requires the use of an infusion pump. To prepare infusion: See formula.

$\dfrac{6 \times \text{weight (kg)} \times \text{desired dose (mcg/kg/min)}}{100 \text{ mL I.V. fluid}}$ I.V. infusion rate (mL/h) = mg of drug added to

Contraindications Angina, pre-existing cardiac arrhythmias (ventricular); tachycardia or AV block caused by cardiac glycoside intoxication; allergy to sulfites or isoproterenol or other sympathomimetic amines; avoid in hydrocarbon exposure

Warnings Elderly patients, diabetics, renal or cardiovascular disease, hyperthyroidism; excessive or prolonged use may result in decreased effectiveness

Dosage Forms Injection, solution, as hydrochloride: 0.02 mg/mL (10 mL); 0.2 mg/mL (1:5000) (1 mL, 5 mL) [contains sodium metabisulfite]

Stability Do not use discolored solutions; limit exposure to heat, light or air; stability of parenteral admixture at room temperature (25°C) and at refrigeration temperature (4°C): 24 hours; incompatible when mixed with aminophylline, furosemide; incompatible with alkaline solutions

Drug Interactions Increased toxicity with sympathomimetics (increased blood pressure and headache), general anesthetics (arrhythmias); increases clearance of theophylline; entacapone can potentiate the chronotropic and arrhythmogenic effects of isoproterenol

Pregnancy Issues C

Monitoring Parameters EKG, heart rate, respiratory rate, arterial blood gas, arterial blood pressure, CVP

Nursing Implications Give around-the-clock to promote less variation in peak and trough serum levels

Reference Range Peak plasma isoproterenol level of 0.0004 mg/L achieved after an I.V. injection of 0.063 mcg/kg postmortem level of 0.1 mg/L noted following a death in an asthmatic patient relatable to isoproterenol use

K027

Synonyms 1-(4-hydroxyiminomethylpyridinium)-3-(carbamoylpyridinium) propane dibromide; Oxime K027

Use: Unlabeled/Investigational Organophosphate agent/nerve agent reactivator antidote

Mechanism of Action Reactivation of organophosphate on nerve agent-induced acetylcholinesterase

Usual Dosage Not established

Additional Information Can reactivate (*in vitro*) acetylcholinesterase inhibition due to tabun, sarin, or VX at relatively low doses. Sarin-inhibited cholinesterase requires higher dose for reactivation.

Specific References

Kuca K and Kassa J, "*In vitro* Reactivation of Acetylcholinesterase Using the Oxime K027," *Vet Hum Toxicol*, 2004, 46(1):15-8.

Mafenide

CAS Number 13009-99-9; 138-39-6

Synonyms Mafenide Acetate

U.S. Brand Names Sulfamylon®

Use Adjunct in the treatment of second- and third-degree burns to prevent septicemia caused by susceptible organisms such as *Pseudomonas aeruginosa*; vesicant exposure

Orphan drug: Prevention of graft loss of meshed autografts on excised burn wounds

Mechanism of Action Interferes with bacterial folic acid synthesis through competitive inhibition of para-aminobenzoic acid.

Adverse Reactions Frequency not defined.

Cardiovascular: Facial edema

Central nervous system: Pain

Dermatologic: Rash, erythema, pruritus

Endocrine & metabolic: Hyperchloremia, metabolic acidosis

Hematologic: Porphyria, bone marrow suppression, hemolytic anemia, bleeding, methemoglobinemia (in 2 pediatric patients)

Local: Burning sensation, excoriation

Respiratory: Hyperventilation, tachypnea, dyspnea

Miscellaneous: Hypersensitivity

Pharmacodynamics/Kinetics

Absorption: Diffuses through devascularized areas and is rapidly absorbed from burned surface

Metabolism: To para-carboxybenzene sulfonamide, a carbonic anhydrase inhibitor

Time to peak, serum: 2-4 hours

Elimination: Urine (as metabolites)

Usual Dosage Children and Adults: Topical: Apply once or twice daily with a sterile gloved hand; apply to a thickness of approximately 16 mm; the burned area should be covered with cream at all times

Contraindications Hypersensitivity to mafenide, sulfites, or any component of the formulation

Warnings Use with caution in patients with renal impairment and in patients with G6PD deficiency; prolonged use may result in superinfection

Dosage Forms

Cream, topical, as acetate: 85 mg/g (60 g, 120 g, 454 g) [contains sodium metabisulfite]

Powder, for topical solution: 5% (5s) [50 g/packet]

Stability

Mafenide 5% topical solution preparation:

Dissolve the 50 g mafenide acetate (Sulfamylon®) packet in 200 mL of either sterile water for irrigation or sterile saline for irrigation (minimum solubility of 50 g of mafenide is in 200 mL of either solution).

Sterilize this solution by pushing through a 0.22 micron filter

Further dissolve this 200 mL of sterile Sulfamylon® solution in 800 mL of the initial diluent (either sterile water for irrigation or normal saline for irrigation)

This solution is stable and sterile for a total of 48 hours at room temperature.

Note: Mafenide acetate topical solution CANNOT be mixed with nystatin due to reduced activity of mafenide.

Note: Pilot *in vitro* studies: Silvadene® and Furacin® cream combined with nystatin cream were equally effective against the microorganisms as were the individual drugs. However, Sulfamylon® cream combined with nystatin lost its antimicrobial capability. [J Burn Care Rehabil 1989, 109:508-11.]

Drug Interactions No data reported

Pregnancy Issues C; Excretion in breast milk unknown

Monitoring Parameters Acid base balance

Nursing Implications For external use only; monitor acid base balance

Reference Range After topical use, maximum serum mafenide levels range from 0.05-1.7 μmol/L after 2- to 4-hour exposure.

Methylene Blue

CAS Number 61-73-4; 7220-79-3

U.S. Brand Names Urolene Blue®

Use Antidote for drug-induced methemoglobinemia, indicator dye, chronic urolithiasis; bacteriostatic genitourinary antiseptic. Has been used topically (0.1% solutions) in conjunction with polychromatic light to photoinactivate viruses such as herpes simplex; this is an unlabeled indication; has been used alone or in combination with vitamin C for the management of chronic urolithiasis; has reversed ifosfamide (Ifex®) associated encephalopathy; also has been suggested to treat ackee fruit encephalopathy.

Mechanism of Action Weak germicide; in low concentrations, hastens the conversion of methemoglobin to hemoglobin; has opposite effect at high concentrations by converting ferrous iron of reduced hemoglobin to ferric iron to form methemoglobin; in cyanide toxicity, it combines with cyanide to form cyanmethemoglobin preventing the interference of cyanide with the cytochrome system; indicated for symptomatic methemoglobinemia (usually when methemoglobin levels are >20%)

Adverse Reactions Frequency not defined.

Cardiovascular: Hypertension, precordial pain

Central nervous system: Dizziness, mental confusion, headache, fever

Dermatologic: Staining of skin

Gastrointestinal: Fecal discoloration (blue-green), nausea, vomiting, abdominal pain

Genitourinary: Discoloration of urine (blue-green), bladder irritation

Hematologic: Anemia

Miscellaneous: Diaphoresis

Pharmacodynamics/Kinetics

Maximum effect: 30 minutes

Absorption: Oral: 53% to 97%; absorbed poorly from GI tract

Metabolism: Reduced to leukomethylene blue by the tissues

Elimination: Bile, feces, and urine as leukomethylene blue

Usual Dosage

Children: NADPH-methemoglobin reductase deficiency: Oral: 1-1.5 mg/kg/day (maximum: 4 mg/kg/day) given with 5-8 mg/kg/day of ascorbic acid

Children and Adults: Methemoglobinemia: I.V.: 1-2 mg/kg or 25-50 mg/m^2 over several minutes; may be repeated in 1 hour if necessary; use the higher dose if methemoglobin levels exceed 60%

Adults: Genitourinary antiseptic: Oral: 55-130 mg 3 times/day with a full glass of water (maximum: 300 mg/day)

Contraindications Severe renal insufficiency; intraspinal injection; hypersensitivity to methylene blue or any component

Warnings Use with caution in young patients and in patients with G6PD deficiency and in aniline-induced or dapsone methemoglobinemia; can cause cyanide release in patients with sodium nitrite overdosage; use will not reverse cyanosis due to sulfhemoglobinemia (as with dapsone); can exacerbate hemolysis in aniline or dapsone poisoning

Dosage Forms

Injection, solution: 10 mg/mL (1 mL, 10 mL)

Tablet (Urolene Blue®): 65 mg

Test Interactions Can reduce arterial lactate concentration in septic shock

Pregnancy Issues C (D if injected intra-amniotically)

Monitoring Parameters Hemoglobin, methemoglobin concentrations should drop within 1 hour

Nursing Implications Inject over several minutes to avoid high concentration; SubQ injection may cause necrotic abscess; may be diluted with normal saline

Signs and Symptoms of Acute Exposure Feces discoloration (black; blue), jaundice, urine discoloration (blue; blue-green; green; green-yellow; yellow-brown). Doses >20 mg/kg can cause hemolysis and hypotension.

Additional Information Skin stains may be removed using a hypochlorite solution; does not interfere with spectrophotometric measurement of methemoglobin; adverse effects seen in neonates exposed to methylene blue include Heinz body hemolytic anemia, hyperbilirubinemia, methemoglobinemia, respiratory distress and photosensitization; may be given by intraosseous infusion over 3-5 minutes with standard intravenous doses.

Specific References

Bradberry SM, "Occupational Methaemoglobinaemia. Mechanisms of Production, Features, Diagnosis and Management Including the Use of Methylene Blue," *Toxicol Rev*, 2003, 22(1):13-27.

Donati A, Conti G, Loggi S, et al, "Does Methylene Blue Administration to Septic Shock Patients Affect Vascular Permeability and Blood Volume?" *Crit Care Med*, 2002, 30(10):2271-7.

Gaudette NF and Lodge JW, "Determination of Methylene Blue and Leucomethylene Blue in Male and Female Fischer 344 Rat Urine and B6C3F$_1$ Mouse Urine," *J Anal Toxicol*, 2005, 29:28-33.

Liao YP, Hung DZ, and Yang DY, "Hemolytic Anemia After Methylene Blue Therapy for Aniline-Induced Methemoglobinemia," *Vet Hum Toxicol*, 2002, 44(1):19-21.

Sharma AN, Chawla S, Nelson L, et al, "Methylene Blue Safely and Effectively Treats Methemoglobinemia in a Known G6PD Deficient Patient With Hemolytic Anemia," *J Toxicol Clin Toxicol*, 2000, 38(5):541.

Naloxone

CAS Number 357-08-4; 465-65-6; 51481-60-8

U.S. Brand Names Narcan® [DSC]

Use Reverses CNS and respiratory depression in suspected narcotic overdose; neonatal opiate depression; coma of unknown etiology

Use: Unlabeled/Investigational Shock, alcohol ingestion; pruritus due to cholestasis; also useful in clonidine, camylofin, valproic acid, and captopril overdose; may be useful in reversing cardiac depression due to dextropropoxyphene

Mechanism of Action Competes and displaces narcotics at narcotic receptor sites

Adverse Reactions

Cardiovascular: Hypertension, hypotension, tachycardia, arrhythmias (ventricular), shock, sinus tachycardia

Central nervous system: Insomnia, irritability, anxiety, narcotic withdrawal, psychosis, dysphoria

Dermatologic: Rash

Gastrointestinal: Nausea, vomiting

Genitourinary: Urinary urgency

Hematologic: Coagulopathy

Ocular: Blurred vision

Respiratory: Pulmonary edema has been described

Miscellaneous: Diaphoresis

Pharmacodynamics/Kinetics

Onset of effect: Endotracheal, I.M., SubQ: Within 2-5 minutes. I.V.: Within 2 minutes

Duration: 20-60 minutes; since shorter than that of most opioids, repeated doses are usually needed

Distribution: V$_d$: 2.7 L/kg; crosses the placenta

Metabolism: Primarily by glucuronidation in the liver to naloxone 3-glucuronide

Half-life: Neonates: 1.2-3.5 hours; Adults: 1-1.5 hours

Elimination: In urine as metabolites

NALOXONE

Usual Dosage

I.M., I.V. (preferred), intratracheal, SubQ:

Postanesthesia narcotic reversal: Infants and Children: 0.01 mg/kg; may repeat every 2-3 minutes, as needed based on response

Opiate intoxication:

Children:

Birth (including premature infants) to 5 years or <20 kg: 0.1 mg/kg; repeat every 2-3 minutes if needed; may need to repeat doses every 20-60 minutes

>5 years or ≥20 kg: 2 mg/dose; if no response, repeat every 2-3 minutes; may need to repeat doses every 20-60 minutes

Children and Adults: Continuous infusion: I.V.: If continuous infusion is required, calculate dosage/hour based on effective intermittent dose used and duration of adequate response seen, titrate dose 0.04-0.16 mg/kg/hour for 2-5 days in children, adult dose typically 0.25-6.25 mg/hour (short-term infusions as high as 2.4 mg/kg/hour have been tolerated in adults during treatment for septic shock); alternatively, continuous infusion utilizes $^2/_3$ of the initial naloxone bolus on an hourly basis; add 10 times this dose to each liter of D_5W and infuse at a rate of 100 mL/hour; $^1/_2$ of the initial bolus dose should be readministered 15 minutes after initiation of the continuous infusion to prevent a drop in naloxone levels; increase infusion rate as needed to assure adequate ventilation; 2 mg naloxone in 3 mL saline can be delivered via nebulization

Narcotic overdose: Adults: I.V.: 0.4-2 mg every 2-3 minutes as needed; may need to repeat doses every 20-60 minutes, if no response is observed after 10 mg, question the diagnosis. **Note:** Use 0.1-0.2 mg increments in patients who are opioid dependent and in postoperative patients to avoid large cardiovascular changes.

Contraindications Hypersensitivity to naloxone or any component

Warnings Due to an association between naloxone and acute pulmonary edema, use with caution in patients with cardiovascular disease or in patients receiving medications with potential adverse cardiovascular effects (eg, hypotension, pulmonary edema or arrhythmias). Excessive dosages should be avoided after use of opiates in surgery, because naloxone may cause an increase in blood pressure and reversal of anesthesia; may precipitate withdrawal symptoms in patients addicted to opiates, including pain, hypertension, sweating, agitation, irritability; in neonates: shrill cry, failure to feed. Recurrence of respiratory depression is possible if the opioid involved is long-acting; observe patients until there is no reasonable risk of recurrent respiratory depression.

Dosage Forms

Injection, neonatal solution, as hydrochloride: 0.02 mg/mL (2 mL)

Injection, solution, as hydrochloride: 0.4 mg/mL (1 mL, 10 mL); 1 mg/mL (2 mL, 10 mL)

Stability Protect from light; stable in 0.9% sodium chloride and D_5W at 4 mcg/mL for 24 hours; do not mix with alkaline solutions

Test Interactions Will not give a false-positive enzymatic urine screen for opiates

Drug Interactions Decreased effect of narcotic analgesics; use with cocaine; may have caused ventricular arrhythmias and atrial fibrillation in one case report; naloxone can precipitate withdrawal syndrome (fatigue, headache, weakness) in patients who abuse anabolic steroids; animal studies have indicated that use of naloxone or metoclopramide may attenuate the hyperthermia caused by interaction of tranylcypromine and meperidine

Pregnancy Issues C; Consider benefit to the mother and the risk to the fetus before administering to a pregnant woman who is known or suspected to be opioid dependent. May precipitate withdrawal in both the mother and fetus. Excretion in breast milk is unknown; breast-feeding is not recommended.

Monitoring Parameters Respiratory rate, heart rate, blood pressure

Nursing Implications The use of neonatal naloxone is no longer recommended because unacceptable fluid volumes will result, especially to small neonates; the 0.4 mg/mL preparation is available and can be accurately dosed with appropriately sized syringes (1 mL).

Signs and Symptoms of Acute Exposure Agitation, anxiety, anorexia, bradycardia, diaphoresis (at 2-4 mg/kg), focal seizures, hypotension, irritability, laryngospasm, nausea

Reference Range Plasma naloxone levels at 2 and 5 minutes after a 0.4 mg I.V. dose: 0.01 mg/L and 0.004 mg/L, respectively

Additional Information In Talwin® Nx to prevent abuse of tablets via parenteral administration; not compatible with alkaline solutions, bisulfite, or metabisulfite. Naloxone is the drug of choice for respiratory depression that is known or suspected to be caused by an overdose of an opiate or opioid. **Caution:** Naloxone's effects are due to its action on narcotic reversal, not due to any direct effect upon opiate receptors. Therefore, adverse events occur secondarily to reversal (withdrawal) of narcotic analgesia and sedation, which can cause severe reactions. Has been given intrathecally to reverse intrathecal morphine overdose. Intrathecal use is associated with significant hypertension.

Specific References

Liu M and Wittbrodt E, "Low-Dose Oral Naloxone Reverses Opioid-Induced Constipation and Analgesia," *J Pain Symptom Manage*, 2002, 23(1):48-53.

Mycyk MB, Szyszko AL, and Aks SE, "Nebulized Naloxone Gently and Effectively Reverses Methadone Intoxication," *J Emerg Med*, 2003, 24(2):185-7.

Neostigmine

CAS Number 114-80-7; 51-60-5; 59-99-4

U.S. Brand Names Prostigmin®

Use Second-line agent for envenomation by Asian snakes or intoxication with tetradotoxin exposures if edrophonium supplies are depleted; reversal of the effects of nondepolarizing neuromuscular blocking agents after surgery

Mechanism of Action Inhibits destruction of acetylcholine by acetylcholinesterase, which facilitates transmission of impulses across myoneural junction

Adverse Reactions Frequency not defined.

Cardiovascular: Arrhythmias (especially bradycardia), hypotension, decreased carbon monoxide, tachycardia, AV block, nodal rhythm, nonspecific ECG changes, cardiac arrest, syncope, flushing

Central nervous system: Convulsions, dysarthria, dysphonia, dizziness, loss of consciousness, drowsiness, headache

Dermatologic: Skin rash, thrombophlebitis (I.V.), urticaria

Gastrointestinal: Hyperperistalsis, nausea, vomiting, salivation, diarrhea, stomach cramps, dysphagia, flatulence

Genitourinary: Urinary urgency

Neuromuscular & skeletal: Weakness, fasciculations, muscle cramps, spasms, arthralgias

Ocular: Small pupils, lacrimation

Respiratory: Increased bronchial secretions, laryngospasm, bronchiolar constriction, respiratory muscle paralysis, dyspnea, respiratory depression, respiratory arrest, bronchospasm

Miscellaneous: Diaphoresis (increased), anaphylaxis, allergic reactions

Pharmacodynamics/Kinetics

Onset of action: I.V.: 1-20 minutes

Duration: 1-2 hours

Absorption: Poor from GI tract

Distribution: V_d: ~1 L/kg

Protein binding: 15% to 25%

Metabolism: In the liver

Half-life: Oral: 52 minutes; I.M.: 51-90 minutes; I.V.: 53 minutes

Elimination: 50% excreted renally as unchanged drug

Usual Dosage

Myasthenia gravis: Diagnosis: I.M.:

Children: 0.04 mg/kg as a single dose

Adults: 0.02 mg/kg as a single dose

Myasthenia gravis: Treatment:
  Children:
    Oral: 2 mg/kg/day divided every 3-4 hours
    I.M., I.V., SubQ: 0.01-0.04 mg/kg every 2-4 hours
  Adults:
    Oral: 15 mg/dose every 3-4 hours
    I.M., I.V., SubQ: 0.5-2.5 mg every 1-3 hours
Reversal of nondepolarizing neuromuscular blockade after surgery in conjunction with atropine or glycopyrrolate: I.V.:
  Infants: 0.025-0.1 mg/kg/dose
  Children: 0.025-0.08 mg/kg/dose
  Adults: 0.5-2.5 mg; total dose not to exceed 5 mg
Bladder atony: Adults: I.M., SubQ:
  Prevention: 0.25 mg every 4-6 hours for 2-3 days
  Treatment: 0.5-1 mg every 3 hours for 5 doses after bladder has emptied

Contraindications Hypersensitivity to neostigmine, bromides or any component; GI or GU obstruction

Warnings Use with caution in patients with epilepsy, asthma, bradycardia, hyperthyroidism, cardiac arrhythmias or peptic ulcer; adequate facilities should be available for cardiopulmonary resuscitation when testing and adjusting dose for myasthenia gravis

Dosage Forms
  Injection, solution, as methylsulfate: 0.5 mg/mL (1 mL, 10 mL); 1 mg/mL (10 mL)
  Tablet, as bromide: 15 mg

Test Interactions ↑ aminotransferase [ALT (SGPT)/AST (SGOT)] (S), amylase (S)

Drug Interactions Neuromuscular blocking agents, when taken with digitalis glycosides, prolonged AV node conduction time may result

Pregnancy Issues C; Does not cross placental barrier except with large doses; not distributed into breast milk

Monitoring Parameters Respiratory status/muscle weakness

Nursing Implications In the diagnosis of myasthenia gravis, all anticholinesterase medications should be discontinued for at least 8 hours before administering neostigmine

Signs and Symptoms of Acute Exposure Blurred vision, bradycardia, excessive sweating, muscle weakness, nausea, tachypnea, tearing and salivation, vomiting

Additional Information
  Neostigmine bromide: Prostigmin® tablet
  Neostigmine methylsulfate: Prostigmin® injection

Specific References
  Isbister GK, Oakley P, Whyte I, et al, "Treatment of Anticholinergic-Induced Ileus With Neostigmine," *Ann Emerg Med*, 2001, 38(6):689-93.
  Seed MJ and Ewan PW, "Anaphylaxis Caused by Neostigmine," *Anaesthesia*, 2000, 55(6):574-5.

Nerve Agent Antidote Kit (NAAK)

Related Information Nerve Agents - Patient Information Sheet, Pralidoxime p105-107

Synonyms Combopen MC; Mark I Kit; NAAK

Warnings Elderly and pediatric populations may be particularly susceptible to adverse effects of agents; The injector needles are deployed at high speed, can penetrate plywood, and are potentially dangerous if the safety component is removed.

Additional Information Increasingly, local EMS are stocking Mark I kits; Mark I and Combopen MC autoinjector marketed by Survival Technology, Bethesda, Maryland; Mark I is not to be used in children <10 years of age.

Therapeutic plasma 2-PAM levels (over 4 mcg/mL) are achieved in 4-8 minutes following intramuscular injection

A kit that contains two sequentially numbered autoinjectors: the Atropen (see Atropine monograph) which delivers 2 mg of Atropine 2-PAM injector which contains 600 mg of pralidoxime (see pralidoxime monograph) can be purchased by authorized responder agencies and health care facilities through Meridien Medical Technologies/King Pharmaceuticals, Columbia, Maryland.

Specific References

American Academy of Pediatrics Committee on Environmental Health and Committee on Infectious Diseases, "Chemical-Biological Terrorism and Its Impact on Children: A Subject Review," *Pediatrics*, 2000, 105(3 Pt 1):662-70.

Arnold JF, "CBRNE - Nerve Agents, G-Series: Tabun, Sarin, Soman," http://www.emedicine.com/EMERG/topic898.htm. Last accessed October 12, 2003.

Ben Abraham R, Weinbroum AA, Rudick V, et al, "Perioperative Care of Children With Nerve Agent Intoxication," *Paediatr Anaesth*, 2001, 11(6):643-9.

Benitez FL, Velez-Daubon, and Keyes DC, "CBRNE - Nerve Agents, V-Series: Ve, Vg, Vm, Vx," http://www.emedicine.com/emerg/topic899.htm. Last accessed October 3, 2003.

Eyer P, Hagedorn I, Klimmek R, et al, "HLo 7 Dimethanesulfonate, A Potent Bispyridinium-Dioxime Against Anticholinesterases," *Arch Toxicol*, 1992, 66(9):603-21.

Henretig FM, Cieslak TJ, and Eitzen EM Jr, "Biological and Chemical Terrorism," *J Pediatr*, 2002, 141(3):311-26. Erratum in: *J Pediatr*, 2002, 141(5):743-6.

Karalliedde L, Wheeler H, Maclehose R, et al, "Possible Immediate and Long-Term Health Effects Following Exposure to Chemical Warfare Agents," *Public Health*, 2000, 114(4):238-48.

Kassa J, "Comparison of the Effects of BI-6, A New Asymmetric Bipyridine Oxime, With HI-6 Oxime and Obidoxime in Combination With Atropine on Soman and Fosdrine Toxicity in Mice," *Ceska Slov Farm*, 1999, 48(1):44-7.

Kassa J, "Review of Oximes in the Antidotal Treatment of Poisoning by Organophosphorus Nerve Agents," *J Toxicol Clin Toxicol*, 2002, 40(6):803-16.

Kuca K, Bielavsky J, Cabal J, et al, "Synthesis of a New Reactivator of Tabun-Inhibited Acetylcholinesterase," *Bioorg Med Chem Lett*, 2003, 13(20):3545-7.

LeJeune KE, Dravis BC, Yang F, et al, "Fighting Nerve Agent Chemical Weapons With Enzyme Technology," *Ann N Y Acad Sci*, 1998, 864:153-70.

Moreno G, Nocentini S, Guggiari M, et al, "Effects of the Lipophilic Biscation, Bis-Pyridinium Oxime BP12, on Bioenergetics and Induction of Permeability Transition in Isolated Mitochondria," *Biochem Pharmacol*, 2000, 59(3):261-6.

"Poison Warfare (Nerve) Gases," http://www.emergency.com/nervgas.htm. Last accessed October 10, 2003.

Rotenberg JS and Newmark J, "Nerve Agent Attacks on Children: Diagnosis and Management," *Pediatrics*, 2003, 112(3 Pt 1):648-58.

Rousseaux CG and Dua AK, "Pharmacology of HI-6, an H-series Oxime," *Can J Physiol Pharmacol*, 1989, 67(10):1183-9.

Salem H and Sidell FR, "Nerve Gases," *Encyclopedia of Toxicology*, Vol 1, Wexler P, ed: Academic Press, 2002, 380-5.

Scott JA, Miller GT, Issenberg SB, et al, "Skill Improvement During Emergency Response to Terrorism Training," *Prehosp Emerg Care*, 2006, 10(4):507-514.

Shiloff JD and Clement JG, "Comparison of Serum Concentrations of the Acetylcholinesterase Oxime Reactivators HI-6, Obidoxime, and PAM to Efficacy Against Sarin (Isopropyl Methylphosphonofluoridate) Poisoning in Rats," *Toxicol Appl Pharmacol*, 1987, 89(2):278-80.

Velez-Daubon L, Benitez FL, Keyes DC, "CBRNE - Nerve Agents, Binary: GB2, VX2," *Emedicine*, 2003, http://www.emedicine.com/EMERG/topic 900.htm. Last accessed October 13, 2003.

Norepinephrine

CAS Number 51-40-1; 51-41-2; 6981-49-5
Synonyms Levarterenol Bitartrate; Noradrenaline Acid Tartrate
U.S. Brand Names Levophed®
Use Treatment of shock, which persists after adequate fluid volume replacement; useful for hypotension induced by tricyclic antidepressants, disulfiram-ethanol interaction, phenothiazine antidysrhythmic agents, nicotine, cyanide agents
Mechanism of Action Stimulates beta$_1$-adrenergic receptors and alpha-adrenergic receptors causing increased contractility and heart rate as well as vasoconstriction, thereby increasing systemic blood pressure and coronary blood flow
Adverse Reactions
Cardiovascular: Cardiac arrhythmias, palpitations, bradycardia, tachycardia, hypertension, chest pain, superior mesenteric artery thrombosis, pallor, ischemic necrosis and sloughing of superficial tissue after extravasation, sinus bradycardia, cardiomyopathy, cardiomegaly, angina, sinus tachycardia, vasoconstriction
Central nervous system: Anxiety, fear, headache
Gastrointestinal: Vomiting
Genitourinary: Uterine contractions
Local: Organ ischemia (due to vasoconstriction of renal and mesenteric arteries)
Ocular: Photophobia, nystagmus
Respiratory: Respiratory distress
Miscellaneous: Diaphoresis
Pharmacodynamics/Kinetics
Onset of action: I.M.: Very rapid acting
Duration: 1-2 minutes
Absorption: SubQ: Poor
Distribution: Localizes primarily in sympathetic tissue; does not cross the blood-brain barrier
Metabolism: By catechol-o-methyltransferase (COMT) and monoamine oxidase (MAO)
Elimination: In urine (84% to 96% as inactive metabolites) and in the liver
Usual Dosage I.V. infusion (dose stated in terms of norepinephrine base):
Children: Initial: 0.05-0.1 mcg/kg/minute, titrate to desired effect
Rate (mL/hour) = dose (mcg/kg/minute) × weight (kg) × 60 minutes/hour divided by concentration (mcg/mL)
Adults: 8-12 mcg/minute as an infusion; initiate at 4 mcg/minute and titrate to desired response
Note: Dose stated in terms of norepinephrine base
Rate of infusion: 4 mg in 500 mL D$_5$W
2 mcg/minute = 15 mL/hour
4 mcg/minute = 30 mL/hour
6 mcg/minute = 45 mL/hour
8 mcg/minute = 60 mL/hour
10 mcg/minute = 75 mL/hour
12 mcg/minute = 90 mL/hour
14 mcg/minute = 105 mL/hour
16 mcg/minute = 120 mL/hour
18 mcg/minute = 135 mL/hour
20 mcg/minute = 150 mL/hour
Administration Administer into large vein to avoid the potential for extravasation; standard concentration: 4 mg/500 mL but 8 mg/500 mL has been used
Contraindications Hypersensitivity to norepinephrine or sulfites
Warnings Use with caution in elderly patients, patients with diabetes mellitus, cardiovascular diseases, thyroid disease, or cerebral arteriosclerosis; use with caution in intoxications with halogenated and aromatic hydrocarbons, digitalis derivatives, neuroleptics, propranolol, tremor, and ergot; blood/volume depletion should be corrected, if possible,

before norepinephrine therapy; extravasation may cause severe tissue necrosis, give into a large vein; the drug should not be given to patients with peripheral or mesenteric vascular thrombosis because ischemia may be increased and the area of infarct extended; use with caution during cyclopropane and halothane anesthesia; use with caution in patients with occlusive vascular disease; some products may contain sulfites

Dosage Forms Injection, solution, as bitartrate: 1 mg/mL (4 mL) [contains sodium metabisulfite]

Stability Readily oxidized; protect from light. Do not use if brown coloration. Stable in D_5NS, D_5W, LR; may dilute with D_5W or D_5NS, but not recommended to dilute in normal saline; not stable in alkaline solutions. Stability of parenteral admixture at room temperature (25°C) is 24 hours.

Y-site administration: Incompatible with insulin (regular), thiopental

Compatibility when admixed: Incompatible with aminophylline, amobarbital, chlorothiazide, chlorpheniramine, pentobarbital, phenobarbital, phenytoin, sodium bicarbonate, streptomycin, thiopental

Drug Interactions Atropine sulfate may block the reflex bradycardia caused by norepinephrine and enhances the pressor response; tricyclic antidepressants, MAO inhibitors, antihistamines (diphenhydramine, tripelennamine), guanethidine, ergot alkaloids, and methyldopa may potentiate the effect of norepinephrine

Pregnancy Issues D; Crosses the placenta

Monitoring Parameters Blood pressure, heart rate, urine output, peripheral perfusion

Nursing Implications Central line administration required; do not administer $NaHCO_3$ through an I.V. line containing norepinephrine; administer into large vein to avoid the potential for extravasation; potent drug, must be diluted prior to use

Extravasation: Use phentolamine as antidote; mix 5 mg with 9 mL of NS; inject a small amount of this dilution into extravasated area; blanching should reverse immediately. Monitor site; if blanching should recur, additional injections of phentolamine may be needed.

Signs and Symptoms of Acute Exposure Cerebral hemorrhage, hypotension, seizures, sweating

Reference Range 24-hour urine catecholamine level: <100 mcg; normal plasma basal norepinephrine level: 100-447 pg/mL

Additional Information Do not use normal saline as a diluent, normal saline is incompatible with norepinephrine; rate of infusion: 2-12 mcg/minute
Standard diluent: 4 mg/500 mL D_5W; 4 mg/250 mL D_5W
Minimum volume: 8 mg/500 mL D_5W

Obidoxime Chloride

CAS Number 114-90-9; 7683-36-5

Synonyms BH-6 Dichloride; Lu-H-6

Use A cholinesterase activator similar to pralidoxime; used primarily in Europe for organophosphate poisoning and sarin or tabun nerve gas toxicity

Mechanism of Action A quaternary oxime, which reactivates cholinesterase; used in conjunction with atropine; less toxic than pralidoxime; has weak anticholinergic activity

Adverse Reactions
Cardiovascular: Hypertension, tachycardia at doses >5 mg/kg
Gastrointestinal: Nausea, vomiting, diarrhea, xerostomia
Hepatic: Transient rise in liver enzymes
Local: Pain at injection site
Neuromuscular & skeletal: Paresthesia
Ocular: Diplopia, blurred vision
Miscellaneous: Menthol-like aftertaste, "hot-feeling"

Pharmacodynamics/Kinetics
Absorption: Oral: Poorly absorbed
Distribution: V_d (healthy volunteers): 0.17 L/kg may be as high as 2.8 L/kg in severe organophosphate-poisoned patients
Protein binding: <1%
Half-life: 83-120 minutes (may be longer in severe organophosphate or nerve gas poisoning)
Peak concentration: I.M.: 20-30 minutes
Elimination: Renal

Usual Dosage
Children: 4-8 mg/kg not to exceed 250 mg
Adults:
    I.M. or slow I.V.: Initial: 250 mg, may be repeated once or twice at 2-hour intervals; maximum daily dose: 750 mg; a 5-day course can be considered for moderate to severe exposure
Continuous infusion: About 0.5 mg/kg/hour up to a maximum dose of 750 mg daily

Reference Range Therapeutic serum level: 4 mg/L

Additional Information Ineffective for treatment of bisdimethylamide (triamphos) poisoning, or carbamate poisoning

Specific References
Bentur Y, Raikhlin-Eisenkraft B, and Singer P, "Beneficial Late Administration of Obidoxime in Malathion Poisoning," *Vet Hum Toxicol*, 2003, 45(1):33-5.

Eyer F, Haberkorn M, Felgenhauer N, et al, "Toxicity of Parathion, Cholinesterase Status and Neuromuscular Function During Antidotal Therapy in a Fatal Case of Parathione Poisoning," *J Toxicol Clin Toxicol*, 2001, 39(3):318.

Eyer P, "The Role of Oximes in the Management of Organophosphorus Pesticide Poisoning," *Toxicol Rev*, 2003, 22(3):165-90.

Haberkorn M, Eyer P, Eyer F, et al, "Effectiveness of Oxime Therapy and Clinical Data of Thirty-Four Organophosphate Intoxications," *J Toxicol Clin Toxicol*, 2002, 40(3):319.

Kassa J, "Review of Oximes in the Antidotal Treatment of Poisoning by Organophosphorus Nerve Agents," *J Toxicol Clin Toxicol*, 2002, 40(6):803-16.

Reis E, AragûÈo I, Martins H, et al, "Organophosphate Poisoned Patients: Cholinesterase Levels, Oximes Therapy and Clinical Course," *J Toxicol Clin Toxicol*, 2002, 40(3):387.

Thiermann H, Worek F, Szinicz L, et al, "Obidoxime Plasma Levels in Organophosphate Poisoned Patients," *J Toxicol Clin Toxicol*, 2002, 40(3):318-9.

Thierman H, Worek F, Szinicz L, et al, "Principles of Oxime Treatment and Its Limitations: From Case Reports to an Assessment of Oxime Effectiveness," *J Toxicol Clin Toxicol*, 2001, 39:256.

Oxygen (Hyperbaric)

Synonyms Hyperbaric Oxygen

Use Carbon monoxide, carbon tetrachloride, cyanide, hydrocarbon, hydrogen sulfide, methylene chloride, mushroom (*Amanita* toxin), brown recluse spider bite, chloroform; also decompression sickness, air emboli, and anaerobic infections; may also be useful in methemoglobinemia; treatment of helium-induced embolism; to treat ergotamine-induced peripheral ischemia; useful for necrotic arachnidism caused by *Lampona cylindrata* (spider found in Australia); radiation-related bone and soft tissue complications (osteoradionecrosis); gas gangrene

Mechanism of Action Displaces carbon monoxide from binding sites and increases elimination rate; also alleviates cerebral edema and CO-induced peroxidation

Adverse Reactions
Ocular: Temporary visual deficits, vision color changes (increased color perception), retinal vessel narrowing
Otic: Otic discomfort, ruptured tympanic membranes
Miscellaneous: Barotrauma to CNS or lung and seizures are reported with prolonged durations (greater than those used for CO poisoning)

Usual Dosage 2.5-3 atmospheres

Contraindications Absolute contraindications are pneumothorax and bowel obstruction; avoid in paraquat or bleomycin-induced pulmonary toxicity. Relative contraindications include claustrophobia, respiratory infections, seizure disorder, asthma, COPD, pneumothorax or history of spontaneous pneumothorax, thoracic or otic surgery, optic neuritis, and pulmonary lesions; also may exacerbate acetaminophen or bromobenzene-induced hepatic necrosis. Use of hyperbaric oxygen in nitrogen oxide exposure is contraindicated.

Dosage Forms Liquid system with large reservoir holding 75-100 lb of liquid oxygen; compressed gas system consisting of high-pressure tank; tank sizes are "H" (6900 L of oxygen), "E" (622 L of oxygen), and "D" (356 L of oxygen)

Pregnancy Issues Well tolerated in the treatment of carbon monoxide and indicated for the treatment of pregnant patients when symptomatic or with carboxyhemoglobin levels >20%

Reference Range Concentrations of CO >30% in nonpregnant patients or >20% in pregnant patients are indications for treatment

Additional Information Hyperbaric oxygen has a clear benefit in the treatment of direct and indirect (methylene chloride) CO poisoning; recent study demonstrated decreased incidence of delayed neurologic sequelae due to CO poisoning; its role in other toxins is not well defined, its relative unavailability and delays in initiating therapy preclude direct extension of promising experimental results; may be of some use in quinine or organophosphate poisoning; not useful in nitrogen oxides toxicity; half-life for carboxyhemoglobin elimination: −27.5 minutes with use of an inflatable portable hyperbaric chamber (modified Ganow bag) at 1.58 ATA; hyperbaric oxygen should be administered at 2.8 ATA within 6 hours of exposure to carbon monoxide for maximum benefit.

Specific References

Chou KJ, Fisher JL, and Silver EJ, "Characteristics and Outcome of Children With Carbon Monoxide Poisoning With and Without Smoke Exposure Referred for Hyperbaric Oxygen," *Pediatr Emerg Care*, 2000, 16(3):151-5.

Gilmer B, Kilkenny J, Tomaszewski C, et al, "Hyperbaric Oxygen Does Not Prevent Neurologic Sequelae After Carbon Monoxide Poisoning," *Acad Emerg Med*, 2002, 9(1):1-8.

Hampson NB and Zmaeff JL, "Outcome of Patients Experiencing Cardiac Arrest With Carbon Monoxide Poisoning Treated With Hyperbaric Oxygen," *Ann Emerg Med*, 2001, 38(1). 36-41.

Silbergleit R and Yeakley W, "Narrow Window of Opportunity for Neuroprotective Effects of Hyperbaric Oxygen in Transient Focal Cerebral Ischemia," *Acad Emerg Med*, 2000, 7(5):512.

Thom SR, "Hyperbaric-Oxygen Therapy for Acute Carbon Monoxide Poisoning," *N Engl J Med*, 2002, 347(14):1105-6.

Penicillamine

CAS Number 2219-30-9; 52-67-5; 59-53-0

Synonyms D-3-Mercaptovaline; D-Penicillamine; β,β-Dimethylcysteine

U.S. Brand Names Cuprimine®; Depen®

Use Treatment of Wilson's disease, cystinuria, adjunct in the treatment of rheumatoid arthritis; lead, mercury, cadmium, iron poisoning; primary biliary cirrhosis; has been used for polonium poisoning.

Mechanism of Action Chelates with lead, copper, mercury, iron, and other heavy metals to form stable, soluble complexes that are excreted in urine; depresses circulating IgM rheumatoid factor, depresses T-cell but not B-cell activity; combines with cystine to form a compound that is more soluble, thus cystine calculi are prevented

Adverse Reactions

>10%:

Dermatologic: Rash, urticaria, itching (44% to 50%)

Gastrointestinal: Hypogeusia (25% to 33%), diarrhea (17%)
Neuromuscular & skeletal: Arthralgia
1% to 10%:
Cardiovascular: Edema of the face, feet, or lower legs
Central nervous system: Fever, chills
Gastrointestinal: Weight gain, sore throat, anorexia, epigastric pain
Genitourinary: Bloody or cloudy urine
Hematologic: Aplastic or hemolytic anemia, leukopenia (2%), thrombocytopenia (4%)
Renal: Proteinuria (6%)
Miscellaneous: White spots on lips or mouth, positive ANA
<1%: Allergic reactions, alopecia, cholestatic jaundice, coughing, fatigue, Goodpasture's syndrome, hepatitis, increased friability of the skin, iron deficiency, lymphadenopathy, myasthenia gravis syndrome, nausea, nephrotic syndrome, pancreatitis, pemphigus, polymyositis, optic neuritis, renal vasculitis, SLE-like syndrome, spitting of blood, tinnitus, toxic epidermal necrolysis, vasculitis, vomiting, weakness, wheezing
Pharmacodynamics/Kinetics
Absorption: Oral: 40% to 70%
Protein binding: 80%
Metabolism: Small amounts of hepatic metabolism to disulfides
Half-life: 1.7-3.2 hours
Time to peak serum concentration: Within 1 hour
Elimination: Primarily (30% to 60%) in urine as unchanged drug
Usual Dosage Oral:
Rheumatoid arthritis:
Children: Initial: 3 mg/kg/day (≤250 mg/day) for 3 months, then 6 mg/kg/day (≤500 mg/day) in divided doses twice daily for 3 months to a maximum of 10 mg/kg/day in 3-4 divided doses
Adults: 125-250 mg/day, may increase dose at 1- to 3-month intervals up to 1-1.5 g/day
Wilson's disease (doses titrated to maintain urinary copper excretion >1 mg/day):
Infants <6 months: 250 mg/dose once daily
Children <12 years: 250 mg/dose 2-3 times/day
Adults: 250 mg 4 times/day
Cystinuria:
Children: 30 mg/kg/day in 4 divided doses
Adults: 1-4 g/day in divided doses every 6 hours
Lead poisoning (continue until blood lead level is <60 mcg/dL):
Children: 25-40 mg/kg/day in 3 divided doses
Adults: 250-500 mg/dose every 8-12 hours
Primary biliary cirrhosis: 250 mg/day to start, increase by 250 mg every 2 weeks up to a maintenance dose of 1 g/day, usually given 250 mg 4 times/day
Arsenic poisoning: Children: 100 mg/kg/day in divided doses every 6 hours for 5 days; maximum: 1 g/day
Mercury poisoning:
Children: 20-30 mg/kg in 4 divided doses
Adults: 250 mg 4 times/day
Dosing adjustment/comments in renal impairment: Cl$_{cr}$ <50 mL/minute: Avoid use
N-acetyl D,L-penicillamine: Chronic elemental mercury exposure:
Children: 30 mg/kg/day on an empty stomach in 4 divided doses for up to 6 days
Adults: 250 mg 4 times/day for up to 6 days; may repeat course as needed
Can be obtained through Aldrich Chemical Co, 940 West St Paul Ave, Milwaukee, WI [(414) 273-3850 or (800) 336-9719; Product No A-19008]. The D isomer can be obtained through Sigma Chemical, St. Louis, MO [(800) 325-3010; Product No A-5678]. The D isomer has not been used in humans.
Contraindications Hypersensitivity to penicillamine; rheumatoid arthritis patients with renal insufficiency; patients with previous penicillamine-related aplastic anemia or agranulocytosis

Warnings Cross-sensitivity with penicillin is possible; therefore, use cautiously in patients with a history of penicillin allergy. Patients on penicillamine for Wilson's disease or cystinuria should receive pyridoxine supplementation 25 mg/day; once instituted for Wilson's disease or cystinuria, continue treatment on a daily basis; interruptions of even a few days have been followed by hypersensitivity with reinstitution of therapy. Penicillamine has been associated with fatalities due to agranulocytosis, aplastic anemia, thrombocytopenia, Goodpasture's syndrome, and myasthenia gravis; patients should be warned to report promptly any symptoms suggesting toxicity; approximately 33% of patients will experience an allergic reaction.

Dosage Forms
Capsule (Cuprimine®): 125 mg, 250 mg
Tablet (Depen®): 250 mg

Stability Store in tight, well-closed containers

Test Interactions Positive ANA

Drug Interactions
Decreased absorption with iron and zinc salts, probenecid, antacids (magnesium, calcium, aluminum) and food; food, aluminum-based antacids, and ferrous sulfate reduce absorption by >50%
Decreased effect/levels of digoxin
Increased effect of gold, antimalarials, immunosuppressants, phenylbutazone (hematologic, renal toxicity), insulin (hypoglycemia)

Pregnancy Issues D; Correlated with cutis laxa in neonates

Monitoring Parameters Urinalysis, CBC with differential, hemoglobin, platelet count, liver function tests

Nursing Implications For patients who cannot swallow, contents of capsules may be administered in 15-30 mL of chilled puréed fruit or fruit juice; patients should be warned to report promptly any symptoms suggesting toxicity

Signs and Symptoms of Acute Exposure Acrodynia, agitation, dysphagia, hemoptysis, hypertrichosis, nausea, seizures, vomiting

Additional Information Danazol can be used to treat breast enlargement. N-acetyl form of penicillamine is less toxic but still investigational. L-isomer is the more toxic isomer. An IND process must be followed; essentially its use has been supplanted by 2,3-dimercaptosuccinic acid. Pyridoxine can be used to treat sideroblastic anemia. Following GI decontamination, treatment is supportive. Corticosteroids should be given for myositis. Neurologic symptoms may respond to zinc therapy.

Specific References
Deguti MM, Mucenic M, Cancado EL, et al, "Elastosis Perforans Serpiginosa Secondary to D-Penicillamine Treatment in a Wilson's Disease Patient," *Am J Gastroenterol*, 2002, 97(8):2153-4.
Lifshitz M and Levy J, "Efficacy of D-Penicillamine in Reducing Lead Concentrations in Children: A Prospective, Uncontrolled Study," *J Pharm Technol*, 2000, 16(3):98-101.
Shapiro M, Jimenez S, and Werth VP, "Pemphigus Vulgaris Induced by D-Penicillamine Therapy in a Patient With Systemic Sclerosis," *J Am Acad Dermatol*, 2000, 42(2 Pt 1): 297-9.
Tchebiner JZ, "Breast Enlargement Induced by D-Penicillamine," *Ann Pharmacother*, 2002, 36(3):444-5.

Phentolamine

CAS Number 65-28-1

Use Prevention or control of hypertensive episodes that may occur in a patient with pheochromocytoma, also used to treat hypertension caused by food/drug interactions with MAO inhibitors. Prevention or treatment of dermal necrosis and sloughing following intravenous administration or extravasation of dopamine or norepinephrine. Diagnosis of pheochromocytoma by the Regitine® blocking test; can be used for cocaine-induced vasospasm; also used for rebound hypertension due to clonidine withdrawal. May reverse cocaine-induced coronary artery vasoconstriction.

Mechanism of Action Competitively blocks alpha-adrenergic receptors to produce brief antagonism of circulating epinephrine and norepinephrine to reduce hypertension caused by these catecholamines; no inotropic activity

Adverse Reactions Frequency not defined.

Cardiovascular: Hypotension, tachycardia, arrhythmia, flushing, orthostatic hypotension

Central nervous system: Weakness, dizziness

Gastrointestinal: Nausea, vomiting, diarrhea

Respiratory: Nasal congestion

Case report: Pulmonary hypertension

Pharmacodynamics/Kinetics

Onset of action: I.M.: Within 15-20 minutes; I.V.: Immediate

Metabolism: In the liver

Bioavailability: Oral: 20%

Half-life: 19 minutes

Elimination: In urine (10% as unchanged drug)

Usual Dosage

Treatment of alpha-adrenergic drug extravasation: SubQ:

Children: 0.1-0.2 mg/kg diluted in 10 mL 0.9% sodium chloride infiltrated into area of extravasation within 12 hours

Adults: Infiltrate area with small amount of solution made by diluting 5-10 mg in 10 mL 0.9% sodium chloride within 12 hours of extravasation

If dose is effective, normal skin color should return to the blanched area within 1 hour.

Diagnosis of pheochromocytoma: I.M., I.V.:

Children: 0.05-0.1 mg/kg/dose, maximum single dose: 5 mg

Adults: 5 mg

Surgery for pheochromocytoma: Hypertension: I.M., I.V.:

Children: 0.05-0.1 mg/kg/dose given 1-2 hours before procedure; repeat as needed every 2-4 hours until hypertension is controlled; maximum single dose: 5 mg

Adults: 5 mg given 1-2 hours before procedure and repeated as needed every 2-4 hours

Hypertensive crisis: Adults: I.V.: 2-10 mg repeated every 5-15 minutes as necessary; infusion: 5-10 mcg/kg/minute

Reverse cocaine-induced cardiac vasoconstriction: Titrate in 1 mg doses

Administration Treatment of extravasation: Infiltrate area of extravasation with multiple small injections; use 27- or 30-gauge needles and change needle between each skin entry

Contraindications Hypersensitivity to phentolamine or any component; renal impairment; coronary or cerebral arteriosclerosis; do not use in the setting of monoamine oxidase inhibitor overdose; but can be used for hypertension due to MAO inhibitor interactions with tyramine

Warnings Myocardial infarction, cerebrovascular spasm and cerebrovascular occlusion have occurred following administration; use with caution in patients with a history of cardiac arrhythmias

Dosage Forms Injection, powder for reconstitution, as mesylate: 5 mg

Stability Reconstituted solution is stable for 48 hours at room temperature and 1 week when refrigerated.

Test Interactions ↑ LFTs, rarely

Drug Interactions Increased toxicity with ethanol (disulfiram reaction); can enhance arginine-induced gastrin secretion; tachycardia may occur when combined with tolazoline

Pregnancy Issues C

Monitoring Parameters Blood pressure, heart rate

Nursing Implications Infiltrate the area of dopamine extravasation with multiple small injections using only 27- or 30-gauge needles and changing the needle between each skin entry; take care not to cause so much swelling of the extremity or digit that a compartment syndrome occurs; monitor patient for orthostasis; assist with ambulation.

Additional Information Injection contains mannitol 25 mg/vial

Specific References

Burns MJ, Dickson EW, Sivilotti ML, et al, "Phentolamine Reduces Myocardial Injury and Mortality in a Rat Model of Phenylpropanolamine Poisoning," *J Toxicol Clin Toxicol*, 2001, 39(2):129-34.

Physostigmine

CAS Number 57-47-6; 57-64-7; 64-47-1

Use Second-line agent for envenomation of Asian snakes or intoxication with tetradotoxin exposures if edrophonium supplies are depleted; reverse toxic CNS effects caused by anticholinergic drugs; controversial role for baclofen, and tricyclic antidepressant toxicity

Mechanism of Action Inhibits destruction of acetylcholine by acetylcholinesterase, which facilitates transmission of impulses across myoneural junction

Adverse Reactions Frequency not defined.

Cardiovascular: Palpitations, bradycardia

Central nervous system: Restlessness, nervousness, hallucinations, seizures

Gastrointestinal: Nausea, salivation, diarrhea, stomach pains

Genitourinary: Frequent urge to urinate

Neuromuscular & skeletal: Muscle twitching

Ocular: Lacrimation, miosis

Respiratory: Dyspnea, bronchospasm, respiratory paralysis, pulmonary edema

Miscellaneous: Diaphoresis

Pharmacodynamics/Kinetics

Onset of action: Ophthalmic: Within 10-30 minutes; I.V.: Within 5 minutes

Duration: Ophthalmic: 12-48 hours; I.V.: 0.5-2 hours

Absorption: Readily following I.M. and SubQ administration; readily from mucous membranes, muscle and subcutaneous tissue

Distribution: Wide throughout body; readily passes blood-brain barrier

Half-life: 15 to 31 minutes; V_d: 2.4 L/kg

Elimination: Not fully understood; only small amounts found in urine; clearance: 12.43 mL/minute

Usual Dosage

Children: Anticholinergic drug overdose: Reserve for life-threatening situations only: I.V.: 0.01-0.03 mg/kg/dose, (maximum: 0.5 mg/minute); may repeat after 5-10 minutes to a maximum total dose of 2 mg or until response occurs or adverse cholinergic effects occur

Adults:

Anticholinergic drug overdose:

I.M., I.V.: 0.5-2 mg to start, repeat every 20 minutes until response occurs or adverse effect occurs

Repeat 1-4 mg every 30-60 minutes as life-threatening signs (arrhythmias, seizures, deep coma) recur; maximum I.V. rate: 1 mg/minute

Ophthalmic:

Ointment: Instill a small quantity to lower fornix up to 3 times/day

Solution: Instill 1-2 drops into eye(s) up to 4 times/day

Contraindications Hypersensitivity to physostigmine or any component; GI or GU obstruction, asthma, diabetes, gangrene, cardiovascular disease; patients receiving choline esters or depolarizing neuromuscular blocking agents

Warnings Use with caution in patients with epilepsy, bradycardia. Discontinue if excessive salivation or emesis, frequent urination, or diarrhea occur. Reduce dosage if excessive sweating or nausea occur. Administer I.V. slowly or at a controlled rate not faster than 1 mg/minute. Due to the possibility of hypersensitivity or overdose/cholinergic crisis, atropine should be readily available. Not intended as a first-line agent for anticholinergic toxicity, especially tricyclic antidepressant

Dosage Forms Injection, solution, as salicylate: 1 mg/mL (2 mL) [contains benzyl alcohol and sodium metabisulfite]

Stability Do not use solution if cloudy or dark brown

Test Interactions ↑ aminotransferase [ALT (SGPT)/AST (SGOT)] (S), amylase (S); ↑ serum cortisol, prolactin, and epinephrine levels

Drug Interactions Potentiates effects of nondepolarizing neuromuscular blockers

Pregnancy Issues C

Signs and Symptoms of Acute Exposure Blurred vision, excessive sweating, fasciculations, hallucinations, hypertension, laryngospasm, muscle weakness, nausea, seizures, tearing and salivation, vomiting

Reference Range Fifteen minutes after a 2 mg oral dose of physostigmine salicylate, plasma level was reported as 1.03 ng/mL

Additional Information Usually not stocked in quantities to treat a significant adult poisoning with anticholinergics. The drug crosses the blood-brain barrier readily and reverses both central and peripheral anticholinergic effects; I.V. rate not to exceed 1 mg/minute in adults; 0.5 mg/minute in children or 0.01 mg/kg/minute; whichever is slower. Have atropine on hand to control bradycardia or seizures; toxic effect is more severe than with physostigmine. Physostigmine therapy for drug intoxications should be used with extreme caution in patients with asthma, gangrene, severe cardiovascular disease, or mechanical obstruction of the GI or urogenital tract. In these patients, physostigmine should be used only when life-threatening. If physostigmine is to be used in the toxic patient, the patient should exhibit central and peripheral anticholinergic effects, a narrow QRS complex on EKG, be placed on a cardiac monitor, and have no exposure to toxins causing a delay in intraventricular conduction.

Specific References

Bania TC, Chu J, O'Neil M, et al, "Effects of Physostigmine Following Cessation of Chronic GHB Administration in Mice," *Acad Emerg Med*, 2003, 10:518.

Brown DV, Heller F, and Barkin R, "Anticholinergic Syndrome After Anesthesia: A Case Report and Review," *Am J Ther*, 2004, 11(2):144-53.

Burns MJ, Linden CH, Graudins A, et al, "A Comparison of Physostigmine and Benzodiazepines for the Treatment of Anticholinergic Poisoning," *Ann Emerg Med*, 2000, 35(4): 374-81.

Caldicott DG and Kuhn M, "Gamma-Hydroxybutyrate Overdose and Physostigmine: Teaching New Tricks to an Old Drug?" *Ann Emerg Med*, 2001, 37(1):99-102.

Ganetsky M, Babu KM, Lian IE, et al, "Case Series of Physostigmine for Olanzapine and Quetiapine-Induced Anticholinergic Syndrome," *Clin Toxicol (Phila)*, 2005, 43:674.

O'Donnell SJ, Burkhart KK, Donovan JW, et al, "Safety of Physostigmine Use for Anticholinergic Toxicity," *J Toxicol Clin Toxicol*, 2002, 40(5):684.

Padilla RB and Pollack ML, "The Use of Physostigmine in Diphenhydramine Overdose," *Am J Emerg Med*, 2002, 20(6):569-70.

Salen PN, Shih R, Sierzenski P, et al, "How Do Physostigmine and Gastric Lavage Affect ICU Utilization and Length of Stay in the Hospital in a *Datura stramonium*-Induced Anticholinergic Poisoning Epidemic?" *Acad Emerg Med*, 2002, 9(5):536.

Schneir AB, Offerman SR, Ly BT, et al, "Complications of Diagnostic Physostigmine Administration to Emergency Department Patients," *Ann Emerg Med*, 2003, 42(1):14-9.

Suchard JR, "Assessing Physostigmine's Contraindication in Cyclic Antidepressant Ingestions," *J Emerg Med*, 2003, 25(2):185-91.

Traub SJ, Nelson LS, and Hoffman RS, "Physostigmine as a Treatment for Gamma-Hydroxybutyrate Toxicity: A Review," *J Toxicol Clin Toxicol*, 2002, 40(6):781-7.

Watts D and Wax P, "Physostigmine Administration for Quetiapine Toxicity," *J Toxicol Clin Toxicol*, 2003, 41(5):646.

Zvosec DL, Smith SW, and Litonjua MR, "Physostigmine for Gamma Hydroxybutyrate Coma: Lack of Efficacy and Adverse Events in 5 Patients," *Clin Toxicol (Phila)*, 2005, 43:674.

Polyethylene Glycol - High Molecular Weight

CAS Number 25322-68-3

Synonyms Colon Electrolyte Lavage Preparation; Electrolyte Lavage Solution; Macrogol; PEG

U.S. Brand Names Colyte®; GoLYTELY®; NuLYTELY®; TriLyte™

Use Whole bowel irrigation in acute overdoses of iron, zinc sulfate, lead oxide, lithium, ampicillin, heavy metals, "body packers," sustained release medications; also used in bowel-cleansing procedures

Mechanism of Action Nonabsorbable agent, which increases bowel osmotic pressure

Adverse Reactions Frequency not defined.

Dermatologic: Dermatitis, rash, urticaria

Gastrointestinal: Nausea, abdominal fullness, bloating, abdominal cramps, vomiting, anal irritation, diarrhea, flatulence

Postmarketing and/or case reports: Upper GI bleeding, Mallory-Weiss tear, esophageal perforation, asystole, dyspnea (acute), pulmonary edema, vomiting with aspiration of PEG, anaphylaxis; dehydration and hypokalemia have been reported in children

Pharmacodynamics/Kinetics

Onset of effect: Oral: Within 1-2 hours

Absorption: Not absorbed

Half-life: 7.7 hours

Usual Dosage For gastrointestinal decontamination:

Oral: Children and Adults: 15-60 mL/kg/hour until clear rectal effluent appears; rate nearly universally requires a naso- or orogastric tube

Nasogastric tube: Administer through a #12 French nasogastric tube with head of bed elevated at least 45°

Children: 9 months to 6 years: 500 mL/hour

Children: 6-12 years: 1000 mL/hour

Adolescents and Adults: 1500-2000 mL/hour

If emesis occurs, decrease infusion rate by 50% for up to 1 hour and then return to the original rate. Metoclopramide can be utilized as an antiemetic and promotility agent (10 mg I.V. initial dose, with 10 mg I.V. in 8 hours with erythromycin 250 mg I.V.)

Contraindications GI obstruction, gastric retention, bowel perforation, toxic colitis, megacolon, clinically significant GI hemorrhage, unprotected compromised airway, hemodynamic instability, uncontrolled intractable vomiting

Warnings May decrease the absorptive capacity of charcoal for cocaine and/or theophylline

Dosage Forms

Powder, for oral solution:

Colyte®:

PEG 3350 240 g, sodium sulfate 22.72 g, sodium bicarbonate 6.72 g, sodium chloride 5.84 g, and potassium chloride 2.98 g (4000 mL) [available with citrus berry, lemon lime, cherry, and pineapple flavor packets]

PEG 3350 227.1 g, sodium sulfate 21.5 g, sodium bicarbonate 6.36 g, sodium chloride 5.53 g, and potassium chloride 2.82 g (4000 mL) [regular and pineapple flavor]

GoLYTELY®:

Disposable jug: PEG 3350 236 g, sodium sulfate 22.74 g, sodium bicarbonate 6.74 g, sodium chloride 5.86 g, and potassium chloride 2.97 g (4000 mL) [regular and pineapple flavor]

Packets: PEG 3350 227.1 g, sodium sulfate 21.5 g, sodium bicarbonate 6.36 g, sodium chloride 5.53 g, and potassium chloride 2.82 g (4000 mL) [regular flavor]

MiraLax™: PEG 3350 255 g (14 oz); PEG 3350 527 g (26 oz)

NuLytely®: PEG 3350 420 g, sodium bicarbonate 5.72 g, sodium chloride 11.2 g, and potassium chloride 1.48 (4000 mL) [cherry flavor, lemon-lime flavor, and orange flavor]

Solution, oral (OCL®): PEG 3350 6 g, sodium sulfate decahydrate 1.29 g, sodium bicarbonate 168 mg, potassium chloride 75 mg, and polysorbate 80 30 mg per 100 mL (1500 mL) [DSC]

Drug Interactions Reduces activity of bacitracin or penicillin

Pregnancy Issues C

Additional Information Use with caution in patients with coma and lethargy; a total dose of 3 L in adults may be sufficient for whole bowel irrigation; average amount of diethylene glycol in PEG: 4.3 mcg/mL; average amount consumed in a patient is ~11 mg; largest reported volume of polyethylene glycol electrolyte solution for use for whole bowel irrigation in a toxic ingestion: 44.3 L (2.953 mL/kg) over a 5-day period in a boy 33 months of age who ingested at least 160 mg/kg elemental iron (initial serum iron level: 367 mcg/dL); optimal bowel cleansing in pediatric patients (0.5-12 years of age) is 6-10 hours of therapy at a rate of 25-35 mL/kg/hour of polyethylene glycol

Specific References

Farmer JW, Chan SB, Beranek G, et al, "Whole Bowel Irrigation for Contraband Body-packers: A Case Series," *Ann Emerg Med*, 2000, 36(4):585.

Guzman DD, Teoh D, Velez LI, et al, "Accidental Intravenous Infusion of Golytely® in a 4-Year-Old Female," *J Toxicol Clin Toxicol*, 2002, 40(3):361-2.

Ly BT, Schneir AB, Williams SR, et al, "Effect of Whole Bowel Irrigation on the Pharmacokinetics of Tylenol Extended Relief," *Acad Emerg Med*, 2002, 9(5):529.

Narsinghani U, Chadha M, Farrar H, et al, "Life-Threatening Respiratory Failure Following Accidental Infusion of Polyethylene Glycol Electrolyte Solution Into the Lung," *J Toxicol Clin Toxicol*, 2001, 39(1):105-7.

Tuckler V, Cramm K, Martinez J, et al, "Accidental Large Intravenous Infusion of Golytely®," *J Toxicol Clin Toxicol*, 2002, 40(5):687-8.

Potassium Iodide

CAS Number 7681-11-0

U.S. Brand Names Iosat™ [OTC]; Pima®; SSKI®; ThyroSafe™ [OTC]; ThyroShield™ [OTC]

Use Block thyroidal uptake of radioactive isotopes of iodine in a radiation emergency; also used to treat *Sporothrix schenckii* (lymphocutaneous form)

Mechanism of Action Inhibits uptake of I-131 or technetium by thyroid and minimizes the risk for radioactive iodine-induced thyroid cancer

Adverse Reactions

Cardiovascular: Leukocytoclastic vasculitis

Central nervous system: Fever, headache, confusion

Dermatologic: Urticaria, acne, angioedema, cutaneous hemorrhage

Endocrine & metabolic: Goiter with hypothyroidism, parotitis

Gastrointestinal: Metallic taste, GI upset (more frequent in children), soreness of teeth and gums

Hematologic: Eosinophilia, hemorrhage (mucosal)

Neuromuscular & skeletal: Arthralgia, numbness

Respiratory: Rhinitis

Miscellaneous: Lymph node enlargement

Pharmacodynamics/Kinetics

Onset of action: 24-48 hours

Peak effect: 10-15 days after continuous therapy

Absorption: Adequate from GI tract

Elimination: If the patient is euthyroid, renal clearance rate is 2 times that of the thyroid; renal excretion is 90%

Usual Dosage

Oral:

Adults: RDA: 150 mcg (iodide)

Expectorant:

Children (Pima®):

<3 years: 162 mg 3 times day

>3 years: 325 mg 3 times/day

Adults:
 Pima®: 325-650 mg 3 times/day
 SSKI®: 300-600 mg 3-4 times/day
Preoperative thyroidectomy: Children and Adults: 50-250 mg (1-5 drops SSKI®) 3 times/day **or** 0.1-0.3 mL (3-5 drops) of strong iodine (Lugol's solution) 3 times/day; administer for 10 days before surgery
Radiation protectant to radioactive isotopes of iodine (Pima®):
 Children:
  Infants up to 1 year: 65 mg once daily for 10 days; start 24 hours prior to exposure
  >1 year: 130 mg once daily for 10 days; start 24 hours prior to exposure
 Adults: 195 mg once daily for 10 days; start 24 hours prior to exposure
To reduce risk of thyroid cancer following nuclear accident (dosing should continue until risk of exposure has passed or other measures are implemented; usual treatment should not exceed 1 or 2 weeks): Predicted exposure:
 ≥5 rad (roentgen-equivalent-man):
  Infants ≤1 month: 16 mg/day
  Children 1 month to 3 years: 32 mg/day
  Children 3-18 years (children >70 kg should be treated with adult dosage): 65 mg/day
  Pregnant or lactating women: 130 mg/day
 ≥10 rad: Adults 18-40 years: 130 mg/day
 ≥500 rad: Adults >40 years: 130 mg/day (less effective for individuals over 40 years old)
Thyrotoxic crisis:
 Infants <1 year: 150-250 mg (3-5 drops SSKI®) 3 times/day
 Children and Adults: 300-500 mg (6-10 drops SSKI®) 3 times/day or 1 mL strong iodine (Lugol's solution) 3 times/day
Sporotrichosis (cutaneous, lymphocutaneous): Adults: Oral: Initial: 5 drops (SSKI®) 3 times/day; increase to 40-50 drops (SSKI®) 3 times/day as tolerated for 3-6 months
 Note: Ineffective for disseminated sporotrichosis.

Contraindications Known hypersensitivity to iodine; hyperkalemia

Warnings Prolonged use can lead to hypothyroidism; can cause acne flare-ups, can cause dermatitis, some preparations may contain sodium bisulfite (allergy). Essentially ineffective if administered more than 6 hours after radiation exposure. Use with caution in patients with Graves' disease, multinodular goiter, and autoimmune thyroiditis.

Dosage Forms
Solution, oral:
 SSKI®: 1 g/mL (30 mL, 240 mL) [contains sodium thiosulfate]
 Lugol's solution, strong iodine: Potassium iodide 100 mg/mL and iodine 50 mg/mL (15 mL, 480 mL)
 Syrup (Pima®): 325 mg/5 mL [equivalent to iodide 249 mg/5 mL] (473 mL) [black raspberry flavor]
 Tablet: 65 mg [equivalent to iodide 50 mg]
  Iosat™: 130 mg

Stability Store in tight, light-resistant containers at temperature <40°C; freezing should be avoided

Test Interactions Thyroid function tests

Drug Interactions Lithium (can cause hypothyroidism)

Pregnancy Issues D; Crosses the placenta; abnormal thyroid function or goiter may occur in the fetus

Signs and Symptoms of Acute Exposure Angioedema, laryngeal edema

Additional Information 10 drops of SSKI® = potassium iodide 500 mg; tablets are radiopaque. Potassium iodide doses can block up to 90% of radioiodine absorption if the first dose is taken immediately postexposure. At 3-4 hours postexposure, approximately 50% of the radioiodine absorption can be blocked. These protective effects last for about 24 hours. SSKI solution can be diluted in milk, water, or formula.

Specific References

Gusev IA, Guskova AK, and Mettler FA, *Management of Radiation Accidents*, 2nd ed, Boca Raton, FL: CRC Press, 2001.

Kauffman CA, Hajjeh R, and Chapman SW, "Practice Guidelines for the Management of Patients With Sporothrichosis. For the Mycoses Study Group. Infectious Diseases Society of America," *Clin Infect Dis*, 2000, 30(4):684-7.

U.S. Food and Drug Administration, "FDA's Guidance on Protection of Children and Adults Against Thyroid Cancer in Case of Nuclear Accident," FDA Talk Paper. Available at: http://www.fda.gov/bbs/topics/answers/2001/ans01126.html, last accessed January 11, 2002.

Pralidoxime

CAS Number 51-15-0

Synonyms 2-PAM; 2-Pyridine Aldoxime Methochloride; Pralidoxime Chloride

U.S. Brand Names Protopam®

Use Reverse muscle paralysis with toxic exposure to organophosphate anticholinesterase pesticides and chemicals; control of overdose of drugs used to treat myasthenia gravis; may be effective for tacrine toxicity; nerve gas agents

Not generally indicated for carbamate ingestions, although recent reports have shown a benefit in reversing nicotinic symptoms refractory to atropine; questionable efficacy for selected organophosphates (ciodrin, dimefox, dimethoate, methyl diazinon, phorate, schaadan, weesyn)

Mechanism of Action Reactivates cholinesterase that had been inactivated by phosphorylation due to exposure to organophosphate pesticides by displacing the enzyme from its receptor sites; most effective if given within 24 hours of exposure; has greater impact on reversing nicotinic effects versus muscarinic

Adverse Reactions

Cardiovascular: Tachycardia, hypertension

Central nervous system: Dizziness, headache, drowsiness

Dermatologic: Rash

Gastrointestinal: Nausea

Hepatic: Transient increases in ALT, AST

Local: Pain at injection site after I.M. administration

Neuromuscular & skeletal: Muscle rigidity, weakness

Ocular: Accommodation impaired, blurred vision, diplopia

Renal: Renal function decreased

Respiratory: Hyperventilation, laryngospasm

Pharmacodynamics/Kinetics

Absorption: Variable and incomplete

Distribution: Throughout extracellular fluids; crosses blood-brain barrier very slowly, if at all

V_d: Intermittent dosing: 0.4-1.6 L/kg. Continuous dosing: Children: 1.7-13.8 L/kg; Adults: 1.8-2.7 L/kg. **Note:** V_d is higher in severely poisoned patients.

Protein binding: Not bound to plasma proteins

Metabolism: Hepatic

Half-life: 90-110 minutes

Time to peak serum concentration: Within 5-15 minutes

Elimination: 0.57-0.88 L/kg/hour; quickly excreted in urine, mostly as metabolites

Usual Dosage

Organic phosphorus poisoning (use in conjunction with atropine; atropine effects should be established before pralidoxime is administered): I.V. (may be given I.M. or SubQ if I.V. is not feasible):

Children: 20-50 mg/kg/dose; repeat in 1-2 hours if muscle weakness has not been relieved, then at 8- to 12-hour intervals if cholinergic signs recur

Adults: 1-2 g; repeat in 1 hour if muscle weakness has not been relieved, then at 8- to 12-hour intervals if cholinergic signs recur. When the poison has been ingested, continued absorption from the lower bowel may require additional doses; patients should be titrated as long as signs of poisoning recur; dosing may need repeated every 3-8 hours.

Organic phosphorus chemical poisoning (unlabeled use): I.M.: Adults: 600 mg, may repeat after 15 minutes; administered with atropine

Treatment of acetylcholinesterase inhibitor toxicity: Adults: I.V.: Initial: 1-2 g followed by increments of 250 mg every 5 minutes until response is observed

Elderly: Refer to Adults dosing; dosing should be cautious, considering possibility of decreased hepatic, renal, or cardiac function

Dosing adjustment in renal impairment: Dose should be reduced

Contraindications Hypersensitivity to pralidoxime or any component of the formulation; poisonings due to phosphorus, inorganic phosphates, or organic phosphates without anticholinesterase activity; poisonings due to pesticides or carbamate class (may increase toxicity of carbaryl)

Warnings Use with caution in patients with myasthenia gravis; dosage modification required in patients with impaired renal function; use with caution in patients receiving theophylline, succinylcholine, phenothiazines, respiratory depressants (eg, narcotics, barbiturates).

Dosage Forms Injection, powder for reconstitution, as chloride: 1 g

Drug Interactions

Decreased effect: Atropine, although often used concurrently with pralidoxime to offset muscarinic stimulation, these effects can occur earlier than anticipated

Increased effect: Barbiturates (potentiated)

Increased toxicity: Use with aminophylline, morphine, theophylline, and succinylcholine is contraindicated; use with reserpine and phenothiazines should be avoided in patients with organophosphate poisoning

Pregnancy Issues C

Monitoring Parameters Blood or plasma cholinesterase

Signs and Symptoms of Acute Exposure Blurred vision, dizziness, nausea, tachycardia, vertigo

Reference Range Pralidoxime concentration of 4 mcg/mL therapeutic *in vitro*; pralidoxime plasma level >14 mcg/mL is associated with side effects (ie, dizziness and blurred vision)

Additional Information Total doses as high as 92 g have been used

Specific References

Bouchard NC, Mercurio-Zappala M, Abrey EM, et al, "Expired 2-PAM Effectively Reverses Cholinergic Crisis in Humans," *J Toxicol Clin Toxicol*, 2004, 42(5):742.

Burillo-Putze G, Hoffman RS, Howland MA, et al, "Late Administration of Pralidoxime in Organophosphate (Fenitrothion) Poisoning," *Am J Emerg Med*, 2004, 22(4):327-8.

Corvino TF, Nahata MC, Angelos MG, et al, "Availability, Stability, and Sterility of Pralidoxime for Mass Casualty Use," *Ann Emerg Med*, 2006, 47(3):272-7.

Deng JF, Wu ML, Tsai WJ, et al, "Acute Hemolysis Induced by Percutaneous Trichlorfon Exposure," *J Toxicol Clin Toxicol*, 2002, 40(3):358-9.

Eyer P, "The Role of Oximes in the Management of Organophosphorus Pesticide Poisoning," *Toxicol Rev*, 2003, 22(3):165-90.

Henretig FM, Mechem C, and Jew R, "Potential Use of Autoinjector-Packaged Antidotes for Treatment of Pediatric Nerve Agent Toxicity," *Ann Emerg Med*, 2002, 40(4):405-8.

Kassa J, "Review of Oximes in the Antidotal Treatment of Poisoning by Organophosphorus Nerve Agents," *J Toxicol Clin Toxicol*, 2002, 40(6):803-16.

Pannbacker RG and Oehme FW, "Pralidoxime Hydrolysis of Thiocholine Esters," *Vet Hum Toxicol*, 2003, 45(1):39-41.

Roh HK, Um WH, and Kin JS, "Prolonged Treatment With 2-PAM in a Large Amount of Organophosphate Insecticide Intoxication," *Clin Toxicol (Phila)*, 2005, 43:738.

Sungur M and Guven M, "Intensive Care Management of Organophosphate Insecticide Poisoning," *Crit Care*, 2001, 5(4):211-5.

Wolowich WR, Weisman RS, Cacace JL, et al, "The Stabiity of Pralidoxime Solution After Discharge From a Mark-1 Autoinjector," *J Toxicol Clin Toxicol*, 2004, 42(5):715.

Wong A, Sandron CA, MagalhûÈes AS, et al, "Comparative Efficacy of Pralidoxime vs Sodium Bicarbonate in Rats and Humans Severely Poisoned With O-P Pesticide," *J Toxicol Clin Toxicol*, 2000, 38(5):554-5.

Prussian Blue

CAS Number 12240-15-2 (potassium containing); 14038-43-8

Synonyms Berlin Blue; Ferric (III) Hexacyanoferrate (II); Ferricyanide; Paris Blue; Pigment Blue 27

Use Treatment for internal contamination of cesium, thallium, and its radioisotopes

Adverse Reactions

Endocrine & metabolic: Hypokalemia (7%)

Gastrointestinal: Constipation (24%)

Miscellaneous: Blue discoloration of stools, sweat, or tears (transient)

Pharmacodynamics/Kinetics

Absorption: Prussian blue: Oral: none

Half-life elimination:

Cesium-137: Effective: Adults: 80 days, decreased by 69% with Prussian blue; adolescents: 62 days, decreased by 46% with Prussian blue; children: 42 days, decreased by 43% with Prussian blue

Nonradioactive thallium: Biological: 8-10 days; with Prussian blue: 3 days

Elimination:

Cesium-137: Without Prussian blue: Urine ($\sim$80%); feces ($\sim$20%)

Thallium: Without Prussian blue: Fecal to urine excretion ration: 2:1

Prussian blue: Feces (99%, unchanged)

Toxicodynamics/Kinetics Absorption: Minimal

Usual Dosage To be given orally only; 150-250 mg/kg daily (orally or through nasogastric tube) in 2-4 divided doses. It is often dissolved in 50 mL of 15% mannitol to prevent constipation. Therapy is continued for thallium exposure until urinary thallium concentrations fall to below 0.5 mg/day. Treatment for cesium should be for a minimum of 30 days.

Contraindications None

Warnings Prussian blue increases the rate of elimination of thallium and cesium; it does not treat complications of radiation exposure. Supportive treatment for radiation toxicity should be given concomitantly. Use caution with decreased gastric motility; constipation should be avoided to prevent increased radiation absorption from the gastrointestinal tract. Use caution with pre-existing cardiac arrhythmias or electrolyte imbalances. Patients should be instructed to minimize radiation exposure to others. Additional decontamination and/or treatment may be needed if exposure to other radioactive isotopes is known or suspected.

Dosage Forms Capsule: 500 mg

Drug Interactions May decrease absorption of orally administered drugs

Pregnancy Issues C; Prussian Blue is not absorbed from the gastrointestinal tract and reproduction studies have not been conducted. Cesium-137 crosses the placenta; in one case, reported levels were equal in the mother and the neonate. Thallium also crosses the placenta; fetal death, failure to thrive, and alopecia in the neonate have been reported. Toxicity from exposure to thallium or radioactive cesium is expected to be greater than the risk of toxicity to ferric hexacyanoferrate. Oligospermia or azoospermia has been reported following whole body radiation in doses >1 Gy of cesium-137.

Breast-feeding/lactation: Excretion in breast milk unknown/not recommended; Cesium and thallium are excreted in breast milk; internally contaminated mothers should not breast feed.

Additional Information If given within 10 minutes, can reduce gut absorption of cesium by 40%, and reduce half-life by 43% to 69%. The maximal adsorption capacity (MAC) for thallium is 72 mg of thallium per gram of Prussian blue. Half-life of thallium reduction is from 8 days (untreated) to 3 days (treated). Constipation due to Prussian blue responds to oral fiber administration.

Specific References

Hoffman RS, "Thallium Toxicity and the Role of Prussian Blue in Therapy," *Toxicol Rev*, 2003, 22(1):29-40 (review).

Pyridostigmine

CAS Number 101-26-8

U.S. Brand Names Mestinon® Timespan®; Mestinon®; Regonol®

Use Second-line agent for envenomation by Asian snakes or intoxication with tetradotoxin exposures if edrophonium supplies are depleted; reversal of the effects of nondepolarizing neuromuscular blocking agents after surgery; pretreatment for chemical warfare agents (ie, Soman nerve gas), thus making the use of atropine and pralidoxime more effective; symptomatic treatment of myasthenia gravis

Mechanism of Action Inhibits destruction of acetylcholine by acetylcholinesterase, which facilitates transmission of impulses across myoneural junction

Adverse Reactions

Cardiovascular: Arrhythmias (especially bradycardia), hypotension, tachycardia, AV block, nodal rhythm, nonspecific EKG changes, cardiac arrest, syncope, flushing

Central nervous system: Convulsions, dysarthria, dysphonia, dizziness, loss of consciousness, drowsiness, headache

Dermatologic: Skin rash, thrombophlebitis (I.V.), urticaria

Gastrointestinal: Hyperperistalsis, nausea, vomiting, salivation, diarrhea, stomach cramps, dysphagia, flatulence, abdominal pain

Genitourinary: Urinary urgency

Neuromuscular & skeletal: Weakness, fasciculations, muscle cramps, spasms, arthralgias, myalgia

Ocular: Small pupils, lacrimation, amblyopia

Respiratory: Increased bronchial secretions, laryngospasm, bronchiolar constriction, respiratory muscle paralysis, dyspnea, respiratory depression, respiratory arrest, bronchospasm

Miscellaneous: Diaphoresis (increased), anaphylaxis, allergic reactions

Pharmacodynamics/Kinetics

Onset of action: Oral, I.M.: 15-30 minutes; I.V. injection: 2-5 minutes

Duration: Oral: Up to 6-8 hours (due to slow absorption); I.V.: 2-3 hours

Absorption: Oral: Very poor

Metabolism: Hepatic

Bioavailability: 10% to 20%

Half-life elimination: 1-2 hours; Renal failure: ≤6 hours

Elimination: Urine (80% to 90% as unchanged drug)

Usual Dosage

Myasthenia gravis:

Oral:

Children: 7 mg/kg/24 hours divided into 5-6 doses

Adults: Highly individualized dosing ranges: 60-1500 mg/day, usually 600 mg/day divided into 5-6 doses, spaced to provide maximum relief

Sustained release formulation: Highly individualized dosing ranges: 180-540 mg once or twice daily (doses separated by at least 6 hours); Note: Most clinicians reserve sustained release dosage form for bedtime dose only.

I.M., slow I.V. push:

Children: 0.05-0.15 mg/kg/dose

Adults: To supplement oral dosage pre- and postoperatively during labor and postpartum, during myasthenic crisis, or when oral therapy is impractical: ∼1/30th of oral dose; observe patient closely for cholinergic reactions

or

I.V. infusion: Initial: 2 mg/hour with gradual titration in increments of 0.5-1 mg/hour, up to a maximum rate of 4 mg/hour

Pretreatment for Soman nerve gas exposure (military use): Oral: Adults: 30 mg every 8 hours beginning several hours prior to exposure; discontinue at first sign of nerve agent exposure, then begin atropine and pralidoxime

Reversal of nondepolarizing muscle relaxants: Note: Atropine sulfate (0.6-1.2 mg) I.V. immediately prior to pyridostigmine to minimize side effects: I.V.:

Children: Dosing range: 0.1-0.25 mg/kg/dose*

Adults: 0.1-0.25 mg/kg/dose; 10-20 mg is usually sufficient*

*Full recovery usually occurs ≤15 minutes, but ≥30 minutes may be required

Dosage adjustment in renal dysfunction: Lower dosages may be required due to prolonged elimination; no specific recommendations have been published

Administration Do **not** crush sustained release tablet.

Contraindications Hypersensitivity to pyridostigmine, bromides, or any component; GI or GU obstruction

Warnings Use with caution in patients with epilepsy, asthma, bradycardia, hyperthyroidism, arrhythmias, or peptic ulcer

Dosage Forms

Injection, solution, as bromide:

Mestinon®: 5 mg/mL (2 mL)

Regonol® [DSC]: 5 mg/mL (2 mL, 5 mL) [contains benzyl alcohol 1%]

Syrup, as bromide (Mestinon®): 60 mg/5 mL (480 mL) [raspberry flavor; contains alcohol 5%, sodium benzoate]

Tablet, as bromide: 30 mg [blister pack of 21]

Mestinon®: 60 mg

Tablet, sustained release, as bromide (Mestinon® Timespan®): 180 mg

Stability Protect from light

Test Interactions ↑ aminotransferase [ALT (SGPT)/AST (SGOT)] (S), amylase (S)

Drug Interactions Depolarizing neuromuscular blockers (eg, succinylcholine or decamethonium)

Pregnancy Issues B; Safety has not been established for use during pregnancy. The potential benefit to the mother should outweigh the potential risk to the fetus. When pyridostigmine is needed in myasthenic mothers, giving dose parenterally 1 hour before completion of the second stage of labor may facilitate delivery and protect the neonate during the immediate postnatal state.

Breast-feeding/lactation: Enters breast milk/compatible

Nursing Implications Observe for cholinergic reactions, particularly when administered I.V.

Signs and Symptoms of Acute Exposure Blurred vision, excessive sweating, muscle weakness, nausea, salivation, tearing, vomiting

Additional Information Not a cure; patient may develop resistance to the drug; normally, sustained release dosage form is used at bedtime for patients who complain of morning weakness; atropine counteracts cholinergic effects. Atropine is the treatment of choice for intoxications manifesting with significant muscarinic symptoms. Atropine I.V. 2-4 mg every 3-60 minutes (or 0.04-0.08 mg I.V. every 5-60 min if needed for children) should be repeated to control symptoms and then continued as needed for 1-2 days following the acute ingestion.

Pyridoxine

CAS Number 58-56-0; 65-23-6

Synonyms Vitamin B_6

U.S. Brand Names Aminoxin® [OTC]

Use Prevents and treats vitamin B_6 deficiency, pyridoxine-dependent seizures in infants, adjunct to treatment of acute toxicity from acrylamide, isoniazid, cycloserine, penicillamine, altretamine, or hydrazine overdose; optic neuritis due to isoniazid or chloramphenicol; hydrazine-containing mushrooms (*Gyromitra*); useful for primary oxaluria; questionable and unproven use in carbon disulfide toxicity; has been used to treat mitomycin C extravasation; seizures due to ginkgo seed ingestion

Mechanism of Action Precursor to pyridoxal, which functions in the metabolism of proteins, carbohydrates, and fats; pyridoxal also aids in the release of liver and muscle stored glycogen, inhibits lactation

Adverse Reactions

Central nervous system: Sensory neuropathy (after chronic administration of large doses), seizures (following I.V. administration of very large doses), headache, hypotonia, memory disturbance

Dermatologic: Photosensitivity, bullous lesions, palmar-plantar erythrodysesthesia syndrome

Endocrine & metabolic: Decreased serum folic acid concentration

Gastrointestinal: Nausea

Hepatic: Elevated AST

Local: Burning or stinging at injection site

Neuromuscular & skeletal: Paresthesia

Respiratory: Respiratory distress

Miscellaneous: Allergic reactions have been reported; may suppress lactation at doses >600 mg/day

Pharmacodynamics/Kinetics

Absorption: Enteral, parenteral: Well absorbed in the jejunum

Distribution: V_d: 0.07-0.17 L/kg

Protein binding: 0%

Metabolism: In 4-pyridoxic acid, and other metabolites

Half-life: 2-3 weeks: I.V. infusion: 107 minutes; I.V. bolus: 18 minutes; Oral: 43 minutes

Elimination: Urinary excretion

Usual Dosage Neuropathy may occur at doses >2 g/day.

Acute hydrazine toxicity (treatment): A pyridoxine dose of 25 mg/kg in divided doses I.M./I.V. has been used.

Acute isoniazid toxicity (treatment of seizures and/or coma): A dose of pyridoxine hydrochloride equal to the amount of INH ingested can be given I.M./I.V. in divided doses together with other anticonvulsants; can give as much as 1 g/kg in adults or 250 mg/kg in children

Dietary deficiency: Oral:

Children: 5-10 mg/24 hours for 3 weeks

Adults: 10-20 mg/day for 3 weeks

Drug-induced neuritis (eg, isoniazid, hydralazine, penicillamine, cycloserine): Oral:

Children: 10-50 mg/24 hours; prophylaxis: 1-2 mg/kg/24 hours

Adults: 100-200 mg/24 hours; prophylaxis: 10-100 mg/24 hours

Ginko seed ingestion (seizure treatment and prevention): I.V.: 2 mg/kg

Mitomycin C extravasation: SubQ: Inject (100 mg/mL) around affected area

Pyridoxine-dependent Infants:

Oral: 2-100 mg/day

I.M., I.V.: 10-100 mg

Contraindications Hypersensitivity to pyridoxine or any component

Warnings Dependence and withdrawal may occur with doses >200 mg/day

Dosage Forms

Capsule, as hydrochloride: 250 mg

Injection, solution, as hydrochloride: 100 mg/mL (1 mL)

Tablet, as hydrochloride: 25 mg, 50 mg, 100 mg, 250 mg, 500 mg

Tablet, enteric coated, as hydrochloride (Aminoxin®): 20 mg

Stability Protect from light

Test Interactions Urobilinogen

Drug Interactions

Decreased serum levels of levodopa, phenobarbital, and phenytoin

Drugs that deplete pyridoxine: Levodopa, cycloserine, penicillamine, hydrazine, isoniazid, dimethylhydrazine, phenylhydrazine, nialamide, acrylamide, hydralazine, procarbazine, iproniazid, monomethylhydrazine-containing mushrooms

Pregnancy Issues A (C if dose exceeds RDA recommendation); Seizures in an infant following *in utero* exposure

Nursing Implications Burning may occur at the injection site after I.M. or SubQ administration; seizures have occurred following I.V. administration of very large doses

Signs and Symptoms of Acute Exposure Ataxia, lethargy, seizures, sensory neuropathy

Reference Range A broad normal range is ~25-80 ng/mL (SI: 122-389 nmol/L)
HPLC method for pyridoxal phosphate normal range: 3.5-18 ng/mL (SI: 17-88 nmol/L)
Peak serum level following 1 g: I.V. infusion: 9.97 ng/μL; I.V. bolus: 44.6 ng/μL
Oral: 15.75 ng/μL

Additional Information Ataxia, sensory neuropathy with doses of 50 mg to 2 g daily over prolonged periods. Pyridoxine hydrochloride infusion may produce a transient worsening of acidosis due to its pH being 2-3.

Specific References
Burda A, Sigg T, Haque D, et al, "Inadequate Pyridoxine Stock and Effect on Patient Outcome," *J Toxicol Clin Toxicol*, 2002, 40(5):692.
Kajiyama Y, Fujii K, Takeuchi H, et al, "Ginko Seed Poisoning," *Pediatrics*, 2002, 109(2):325-7.
LoVecchio F, Curry SC, Graeme KA, et al, "Intravenous Pyridoxine-Induced Metabolic Acidosis," *Ann Emerg Med*, 2001 38(1):62-4.
Nagappan R and Riddell T, "Pyridoxine Therapy in a Patient With Severe Hydrazine Sulfate Toxicity," *Crit Care Med*, 2000, 28(6):2116-8.

Rifampin

U.S. Brand Names Rifadin®

Use Management of active tuberculosis in combination with other agents; eliminate meningococci from asymptomatic carriers; may be used as adjunctive therapy in brucellosis or anthrax

Usual Dosage Oral (I.V. infusion dose is the same as for the oral route):
Tuberculosis therapy: Note: A four-drug regimen (isoniazid, rifampin, pyrazinamide, and either streptomycin or ethambutol) is preferred for the initial, empiric treatment of TB. When the drug susceptibility results are available, the regimen should be altered as appropriate.
Infants and Children <12 years:
Daily therapy: 10-20 mg/kg/day usually as a single dose (maximum: 600 mg/day)
Directly observed therapy (DOT): Twice weekly: 10-20 mg/kg (maximum: 600 mg); 3 times/week: 10-20 mg/kg (maximum: 600 mg)
Adults:
Daily therapy: 10 mg/kg/day (maximum: 600 mg/day)
Directly observed therapy (DOT): Twice weekly: 10 mg/kg (maximum: 600 mg); 3 times/week: 10 mg/kg (maximum: 600 mg)
Latent tuberculosis infection (LTBI):
As an alternative to isoniazid:
Children: 10-20 mg/kg/day (maximum: 600 mg/day)
Adults: 10 mg/kg/day (maximum: 600 mg/day) for 2 months. Note: Combination with pyrazinamide should not generally be offered (*MMWR*, Aug 8, 2003).
Meningococcal meningitis prophylaxis:
Infants <1 month: 10 mg/kg/day in divided doses every 12 hours for 2 days
Infants ≥1 month and Children: 20 mg/kg/day in divided doses every 12 hours for 2 days
Adults: 600 mg every 12 hours for 2 days
***H. influenzae* prophylaxis** (unlabeled use):
Infants and Children: 20 mg/kg/day every 24 hours for 4 days, not to exceed 600 mg/dose
Adults: 600 mg every 24 hours for 4 days

Leprosy: Adults:
Multibacillary: 600 mg once monthly for 24 months in combination with ofloxacin and minocycline
Paucibacillary: 600 mg once monthly for 6 months in combination with dapsone
Single lesion: 600 mg as a single dose in combination with ofloxacin 400 mg and minocycline 100 mg
Brucellosis: Adults: 15-20 mg/kg/day may be added to doxycycline or tetracycline; maximum: 900 mg/day
Anthrax: Recommended by CDC as an additional antibiotic in treatment of inhalation, gastrointestinal, or oropharyngeal anthrax
Nasal carriers of *Staphylococcus aureus* (unlabeled use):
Children: 15 mg/kg/day divided every 12 hours for 5-10 days in combination with other antibiotics
Adults: 600 mg/day for 5-10 days in combination with other antibiotics
Synergy for *Staphylococcus aureus* infections (unlabeled use): Adults: 300-600 mg twice daily with other antibiotics
Dosing adjustment in hepatic impairment: Dose reductions may be necessary to reduce hepatotoxicity
Hemodialysis or peritoneal dialysis: Plasma rifampin concentrations are not significantly affected by hemodialysis or peritoneal dialysis.

Serpacwa

Synonyms Skin Exposure Reduction Paste Against Chemical Warfare Agents

Use Topical agent serving as protectant (physical barrier) to urushiol and methyl nicotinate (surrogates for chemical warfare agents); currently indicated only for military personnel wearing mission-oriented protective posture (MOPP) gear; it should be applied only when chemical warfare is imminent and not for training purposes

Mechanism of Action Contains a 50:50 mixture of perfluoroakylpolyether and polytetra-fluoroethylene, which acts as a physical barrier to agents

Adverse Reactions Miscellaneous: Flu-like syndrome

Pharmacodynamics/Kinetics
Absorption: Topical: Virtually none
Duration of action: Not evaluated for >5 hours

Usual Dosage Apply to skin by hand until there is a barely noticeable white film layer. One-third of the packet should be applied evenly around wrists, neck, and boot tops (of the lower leg); remaining two-thirds of the packet should be applied evenly to axillae, groin, and waist. Prior to application, use a dry towel or cloth to remove perspiration, insect repellant, camouflage paint or dirt from skin.

Contraindications Prior hypersensitivity to any component

Warnings Avoid smoking; use of camouflage paints or insect repellants may decrease effectiveness of serpacwa; avoid open wound application or mucous membrane or eye exposure

Pregnancy Issues C

Additional Information Following chemical warfare exposure, standard decontamination procedures should be followed. Store pouches at 20°C or 30°C. Manufactured for the U.S. Army by McKesson Bioservices. Animal studies have shown decreased toxicity of sulfur mustard, VX, soman, T-2 mycotoxin, and CS (lacrimator). Protection has been partial in most cases studied.

Specific References
Liu DK, Wannemacher RW, Snider TH, et al, "Efficacy of the Tropical Skin Protectant in Advanced Development," *J Anal Toxicol*, 1999(Supplement):S41-5.

Silver Sulfadiazine

CAS Number 22199-08-2

U.S. Brand Names Silvadene®; SSD® AF; SSD®; Thermazene®

Use Prevention and treatment of infection in second and third degree burns; vesicant exposure

Mechanism of Action Acts upon the bacterial cell wall and cell membrane. Bactericidal for many gram-negative and gram-positive bacteria and is effective against yeast. Active against *Pseudomonas aeruginosa*, *Pseudomonas maltophilia*, *Enterobacter* species, *Klebsiella* species, *Serratia* species, *Escherichia coli*, *Proteus mirabilis*, *Morganella morganii*, *Providencia rettgeri*, *Proteus vulgaris*, *Providencia* species, *Citrobacter* species, *Acineto-bacter calcoaceticus*, *Staphylococcus aureus*, *Staphylococcus epidermidis*, *Enterococcus* species, *Candida albicans*, *Corynebacterium diphtheriae*, and *Clostridium perfringens*

Adverse Reactions Frequency not defined.
Dermatologic: Itching, rash, erythema multiforme, discoloration of skin, photosensitivity
Hematologic: Hemolytic anemia, leukopenia (usually transient), agranulocytosis, aplastic anemia
Hepatic: Hepatitis
Renal: Interstitial nephritis, crystalluria
Miscellaneous: Allergic reactions may be related to sulfa component, hyperthermia

Pharmacodynamics/Kinetics
Absorption: Significant percutaneous absorption of silver sulfadiazine can occur especially when applied to extensive burns
Half-life elimination: 10 hours; prolonged with renal impairment (22 hours)
Time to peak, serum: 3-11 days of continuous therapy
Elimination: Urine (~60% as unchanged drug)

Usual Dosage Children and Adults: Topical: Apply once or twice daily with a sterile-gloved hand; apply to a thickness of $^1/_{16}$"; burned area should be covered with cream at all times

Administration Apply with a sterile-gloved hand. Apply to a thickness $^1/_{16}$". Burned area should be covered with cream at all times.

Contraindications Hypersensitivity to silver sulfadiazine or any component of the formulation; premature infants or neonates <2 months of age (sulfonamides may displace bilirubin and cause kernicterus); pregnancy (approaching or at term)

Warnings Use with caution in patients with G6PD deficiency, renal impairment, or history of allergy to other sulfonamides; sulfadiazine may accumulate in patients with impaired hepatic or renal function; fungal superinfection may occur; use of analgesic might be needed before application; systemic absorption is significant and adverse reactions may occur

Dosage Forms
Cream, topical: 1% [10 mg/g] (25 g, 85 g, 400 g)
Silvadene®, Thermazene®: 1% (20 g, 50 g, 85 g, 400 g, 1000 g)
SSD®: 1% (25 g, 50 g, 85 g, 400 g)
SSD AF®: 1% (50 g, 400 g)

Stability Silvadene® cream will occasionally darken either in the jar or after application to the skin. This color change results from a light catalyzed reaction, which is a common characteristic of all silver salts. A similar analogy is the oxidation of silverware. The product of this color change reaction is silver oxide, which ranges in color from gray to black. Silver oxide has rarely been associated with permanent skin discoloration. Additionally, the antimicrobial activity of the product is not substantially diminished because the color change reaction involves such a small amount of the active drug and is largely a surface phenomenon.

Drug Interactions
Decreased effect: Topical proteolytic enzymes are inactivated
Concomitant cimetidine use may increase the likelihood of development of leukopenia (73% to 77% incidence)

Pregnancy Issues B

Monitoring Parameters Serum electrolytes, urinalysis, renal function tests, CBC in patients with extensive burns on long-term treatment

Environmental Persistency No information available to require special precautions
Additional Information Contains methylparaben and propylene glycol

Sodium Bicarbonate

CAS Number 144-55-8
Synonyms Baking Soda; Monosodium Carbonate; $NaHCO_3$; Sal de Vichy; Sodium Acid Carbonate; Sodium Hydrogen Carbonate
U.S. Brand Names Brioschi® [OTC]; Neut®
Use Management of metabolic acidosis; antacid; alkalinize urine; severe diarrhea; can reverse QRS prolongation in antidepressant overdose, cocaine, and propoxyphene; cardiac conduction defects due to quinidine-like action of cardiotoxic drugs; increases protein binding of tricyclic antidepressants; nebulized sodium bicarbonate is useful to treat chlorine gas exposure; also used to prevent rhabdomyolysis-induced renal failure; useful in extravasation injury due to carmustine; metformin-induced lactic acidosis
Mechanism of Action Dissociates to provide bicarbonate ion, which neutralizes hydrogen ion concentration and raises blood and urinary pH
Adverse Reactions Frequency not defined.
  Cardiovascular: Cerebral hemorrhage, CHF (aggravated), edema
  Central nervous system: Tetany
  Endocrine & metabolic: Hypernatremia, hyperosmolality, hypocalcemia, hypokalemia, increased affinity of hemoglobin for oxygen-reduced pH in myocardial tissue necrosis when extravasated, intracranial acidosis, metabolic alkalosis, milk-alkali syndrome (especially with renal dysfunction)
  Gastrointestinal: Flatulence (with oral), gastric distension, belching
  Respiratory: Pulmonary edema
Pharmacodynamics/Kinetics
  Oral:
    Onset of action: Rapid
    Duration: 8-10 minutes
  I.V.:
    Onset of action: 15 minutes
    Duration: 1-2 hours
  Absorption: Oral: Well absorbed
  Distribution: Bicarbonate occurs naturally and is confined to systemic circulation
  Elimination: Reabsorbed by kidney and <1% is excreted by urine
Usual Dosage
  Cardiac arrest: Routine use of $NaHCO_3$ is not recommended and should be given only after adequate alveolar ventilation has been established and effective cardiac compressions are provided
    Children and Infants: I.V.: 0.5-1 mEq/kg/dose repeated every 10 minutes or as indicated by arterial blood gases
    Adults: I.V.: Initial: 1 mEq/kg/dose one time; maintenance: 0.5 mEq/kg/dose every 10 minutes or as indicated by arterial blood gases
  Metabolic acidosis: I.V.:
    Neonates: 1-3 mEq/kg/dose slowly (usually over 20-60 minutes)
    Older Children and Adults: 2-5 mEq/kg/dose over 4-8 hours infusion or calculate dose based on base deficit (re-evaluate acid-base status) mEq $NaHCO_3 = 0.3 \times$ body weight (kg) $\times$ base deficit (mEq/L); give up to 1 mEq/kg/dose over several minutes or dilute larger doses in maintenance fluids for slow infusion; re-evaluate acid-base status frequently
      Maximum daily dose: 200 mEq in adults <60 years and 100 mEq in adults >60 years
  Maintenance electrolyte requirements of sodium: Daily requirements: 3-4 mEq/kg/24 hours or 25-40 mEq/1000 kcal/24 hours

Chronic renal failure: Oral: Initiate when plasma HCO_3 <15 mEq/L
   Children: 1-3 mEq/kg/day
   Adults: Start with 20-36 mEq/day in divided doses, titrate to bicarbonate level of 18-20 mEq/L
Renal tubular acidosis: Oral:
   Distal: Children: 2-3 mEq/kg/day; Adults: 1 mEq/kg/day
   Proximal: Children: Initial: 5-10 mEq/kg/day; maintenance: Increase as required to maintain serum bicarbonate in the normal range
Urine alkalinization: Oral:
   Children: 1-10 mEq (84-840 mg)/kg/day in divided doses every 4-6 hours; dose should be titrated to desired urinary pH
   Adults: Initial: 48 mEq (4 g), then 12-24 mEq (1-2 g) every 4 hours; dose should be titrated to desired urinary pH; doses up to 17 g/day (200 mEq) in patients <60 years and 8 g (100 mEq) in patients >60 years
Antacid: Adults: Oral: 325 mg to 2 g 1-4 times/day
Carmustine extravasation: 5 mL of a 8.4% solution of sodium bicarbonate either through offending I.V. cannulas or through multiple SubQ injections after cannula removal; apply ice to affected area
Nebulized sodium bicarbonate for chlorine gas exposure: A mixture of 3 ml of 8.4% sodium bicarbonate with 2 ml of normal saline creates 5 ml of a 5% sodium bicarbonate solution. A 3.75% solution may be prepared by mixing 2 ml of a 7.5% solution with 2 ml of normal saline. Administer for 20 minutes: can repeat as necessary.

Contraindications Alkalosis, hypocalcemia; unknown abdominal pain, inadequate ventilation during cardiopulmonary resuscitation

Warnings Use of I.V. $NaHCO_3$ should be reserved for documented metabolic acidosis and for hyperkalemia-induced cardiac arrest; routine use in cardiac arrest is not recommended. Avoid extravasation, tissue necrosis can occur due to the hypertonicity of $NaHCO_3$; may cause sodium retention especially if renal function is impaired; not to be used in treatment of peptic ulcer; use with caution in patients with CHF, edema, cirrhosis, or renal failure.

Dosage Forms
Granules, effervescent (Brioschi®): 6 g, 120 g, 240 g [lemon flavor]
Injection, solution:
   4.2% [42 mg/mL = 5 mEq/10 mL] (10 mL)
   5% [50 mg/mL = 5.95 mEq/10 mL] (500 mL)
   7.5% [75 mg/mL = 8.92 mEq/10 mL] (50 mL)
   8.4% [84 mg/mL = 10 mEq/10 mL] (10 mL, 50 mL)
   Neut®: 4% [40 mg/mL = 2.4 mEq/5 mL] (5 mL)
Infusion [premixed in sterile water]: 5% (500 mL)
Powder: 120 g, 480 g
Tablet: 325 mg [3.8 mEq]; 650 mg [7.6 mEq]

Stability Store injection at room temperature; protect from heat and from freezing; use only clear solutions; do not mix $NaHCO_3$ with calcium salts, catecholamines, atropine

Drug Interactions Amphetamines, quinidine levels may increase due to urinary alkalinization reduced excretion; mixing amiodarone with 8.4% sodium bicarbonate can cause a precipitate; I.V. ciprofloxacin lactate (2 mg/mL) can cause precipitation with low concentrations (0.1 mEq/mL) of sodium bicarbonate but may not cause precipitation when administered with high (1 mEq/mL) concentrations of sodium bicarbonate

Pregnancy Issues C

Monitoring Parameters Serum electrolytes including calcium, urinary pH, arterial blood gases (if indicated)

Nursing Implications Advise patient of milk-alkali syndrome if use is long-term; observe for extravasation when giving I.V.; incompatible with acids, acidic salts, alkaloid salts, calcium salts, catecholamines, atropine

Signs and Symptoms of Acute Exposure Confusion, cyanosis, hypocalcemia, hypokalemia, hypokalemic/hypochloremic metabolic alkalosis, hypernatremia, muscle cramps, nausea, pulmonary edema, seizures, tetany, weakness

Reference Range Therapeutic (sodium): 135-145 mEq/L (SI: 135-145 mM/L)

Additional Information

May cause sodium retention especially if renal function is impaired; not to be used in treatment of peptic ulcer

Baking soda contains 35-55 mEq per teaspoon of sodium; serum sodium concentrations in children with baking soda ingestions range from 155-210 mEq/L with serum bicarbonates ranging form 29-54 mEq/L; hypocalcemia may also be present; metabolic alkalosis is usually present

Sodium content of injection 50 mL, 8.4% = 1150 mg = 50 mEq; each 6 mg of $NaHCO_3$ contains 12 mEq sodium; 1 mEq $NaHCO_3$ = 84 mg mEq $NaHCO_3$ = 0.3 × body weight (kg) × base deficit (mEq/L)

Each 84 mg of sodium bicarbonate provides 1 mEq of sodium and bicarbonate ions; each gram of sodium bicarbonate provides 12 mEq of sodium and bicarbonate ions

Specific References

Bryant SM, Feldman S, Milner MM, et al, "Baking Soda Poisoning Following Prophylactic 'Nipple Dipping' to Prevent Colic," *J Toxicol Clin Toxicol*, 2002, 40(5):611-2.

Burda AM, Kubric A, Wahl M, "Nebulized Sodium Bicarbonate for Chlorine Gas Exposure," *Chicago Med*, 2007, 109(16):30-31.

Proudfoot AT, Krenzelok EP, Vale JA, et al, "AACT/EAPCCT Position Paper on Urine Alkalinization," *J Toxicol Clin Toxicol*, 2002, 40(3):310-1.

Tanen DA, Ruha AM, Curry SC, et al, "Hypertonic Sodium Bicarbonate is Effective in the Acute Management of Verapamil Toxicity in a Swine Model," *Ann Emerg Med*, 2000, 36(6):547-53.

Wang R, Raymond R, Lawler R, et al, "The Effects of Sodium Bicarbonate in Imipramine-Induced Cardiotoxicity," *J Toxicol Clin Toxicol*, 2000, 38(5):528.

Sodium Nitrite

CAS Number 7632-00-0

Synonyms Natril; Natrium Nitrosum; Nitris; Nitrous Acid; Sodium Salt

Use Cyanide toxicity in conjunction with amyl nitrite pearls and sodium thiosulfate; may be of use in hydrogen sulfide poisoning

Mechanism of Action Vasodilation and methemoglobin producer

Adverse Reactions

Cardiovascular: Tachycardia, hypotension from vasodilation, syncope, cyanosis, flushing, sinus tachycardia

Central nervous system: Headache

Gastrointestinal: Nausea, vomiting

Miscellaneous: Forms methemoglobin

Pharmacodynamics/Kinetics

Metabolism: To ammonia

Elimination: Renal (33%)

Usual Dosage

Cyanide toxicity:

Children (without anemia): 4.5-10 mg/kg (0.15-0.33 mL/kg of a 3% solution up to 10 mL)

Adults: 300 mg (10 mL of a 3% solution)

Acceptable daily intake: 0.4 mg/kg

Monitoring Parameters Methemoglobin levels

Signs and Symptoms of Acute Exposure Cardiovascular collapse, coma, hemolysis/hemolytic anemia, seizures

SODICONTENT## SODIUM POLYSTYRENE SULFONATE

Reference Range Sodium nitrite levels associated with fatalities range from 0.5-350 mg/L
Additional Information Estimated lethal dose: Adult: 2.6 g
Used in photography, meat preservative (up to 200 ppm), metal corrosion inhibitor, fertilizers, manufacture of diazo dyes; arterial blood appears "chocolate brown"
Specific References
Kage S, Kuto K, and Ikeda N, "Simultaneous Determination of Nitrate and Nitrite in Human Plasma by Gas Chromatography-Mass Spectrometry," *J Anal Toxicol*, 2002, 26(6):320-4.
Pelclová D, Kredba V, Pokorná P, et al, "Sodium Nitrite Intoxication in a Newborn," *J Toxicol Clin Toxicol*, 2002, 40(3):352-3.
Tanen DA, LoVecchio F, and Curry SC, "Failure of Intravenous N-Acetylcysteine to Reduce Methemoglobin Produced by Sodium Nitrite in Human Volunteers: A Randomized Controlled Trial," *Ann Emerg Med*, 2000, 35(4):369-73.

Sodium Polystyrene Sulfonate

CAS Number 25704-18-1; 9003-59-2
U.S. Brand Names Kayexalate®; Kionex™; SPS®
Use Treatment of hyperkalemia; gastric decontamination for lithium
Mechanism of Action Removes potassium by exchanging sodium ions for potassium ions in the intestine before the resin is passed from the body
Adverse Reactions Frequency not defined.
Endocrine & metabolic: Hypokalemia, hypocalcemia, hypomagnesemia, sodium retention
Gastrointestinal: Fecal impaction, constipation, loss of appetite, nausea, vomiting
Pharmacodynamics/Kinetics
Onset of action: Within 2-24 hours
Absorption: Remains in GI tract
Elimination: Completely excreted in feces (primarily as potassium polystyrene sulfonate)
Usual Dosage
Children:
Oral: 1 g/kg/dose every 6 hours
Rectal: 1 g/kg/dose every 2-6 hours (In small children and infants, employ lower doses by using the practical exchange ratio of 1 mEq K$^+$/g of resin as the basis for calculation)
Adults:
Oral: 15 g (60 mL) 1-4 times/day
Rectal: 30-50 g every 6 hours
Instructions for sodium polystyrene sulfonate enema:
Insert the tubing into the rectum ~20 cm. Administer a body-temperature tap water enema (250-500 mL) to remove any feces in the rectum and sigmoid colon.
Shake the sodium polystyrene sulfonate suspension well and mix 2:1 with tap water in the enema bucket prior to administration (ie, 60 mL sodium polystyrene sulfonate and 30 mL water).
Reinsert the enema tube into the rectum ~20 cm and administer the prescribed sodium polystyrene sulfonate dose (usual range, 15-60 g), then remove the tube.
Make sure the enema is retained for several hours, to allow exchange of sodium for potassium ions. If leakage occurs, elevate the patient's hips with pillows.
Reinsert the tube to 20 cm and administer cleansing enema(s) of body-temperature tap water (250-1000 mL) to flush any remaining sodium polystyrene sulfonate enema out of the colon to remove potassium ions and prevent adverse reactions.
For additional information, call your pharmacist or the drug information center.
Contraindications Hypernatremia
Warnings Use with caution in patients with severe congestive heart failure, hypertension, or edema. Avoid using the commercially available liquid product in neonates due to the preservative content, also may not be as effective in low birth weight infants. Enema may be

prepared with powder and diluted with sorbitol 10% solution or oral solution with 25% sorbitol solution. Enema will reduce the serum potassium faster than oral administration, but the oral route will result in a greater reduction over several hours.

Dosage Forms
Powder for suspension, oral/rectal:
Kayexalate®: 480 g
Kionex™: 454 g
Suspension, oral/rectal: 15 g/60 mL (60 mL, 120 mL, 200 mL, 500 mL) [with sorbitol and alcohol]
SPS®: 15 g/60 mL (60 mL, 120 mL, 480 mL) [contains alcohol 0.3% and sorbitol; cherry flavor]

Test Interactions ↑ sodium; ↓ potassium (S), calcium (S), magnesium (S)

Drug Interactions Cation-donating antacids and saline cathartics should be avoided

Pregnancy Issues C

Monitoring Parameters Serum electrolytes, EKG

Nursing Implications Administer oral (or NG) as ~25% sorbitol solution, never mix in orange juice; enema route is less effective than oral administration; retain enema in colon for at least 30-60 minutes and for several hours, if possible

Reference Range Serum potassium: Adults: 3.5-5.2 mEq/L

Additional Information 1 g of resin binds approximately 1 mEq of potassium; chilling the oral mixture will increase palatability; sodium content of 1 g: 31 mg (1.3 mEq); may be useful in preventing lithium absorption, especially when given in multiple doses, and most likely for sustained-release lithium preparations

Specific References
Shepherd G, Klein-Schwartz W, and Burstein AH, "Efficacy of the Cation Exchange Resin, Sodium Polystyrene Sulfonate, to Decrease Iron Absorption," *J Toxicol Clin Toxicol*, 2000, 38(4):389-94.

Sodium Thiosulfate

CAS Number 10102-17-7; 7772-98-7

Synonyms Sodium Hyposulfate; Sodium Oxide Sulfide

U.S. Brand Names Versiclear™

Use Alone or with sodium nitrite or amyl nitrite (or hydroxocobalamin) in cyanide poisoning; to reduce the risk of nephrotoxicity associated with cisplatin therapy; topically in the treatment of tinea versicolor; an inorganic reducing agent used as a fixative bleaching of bone; used in manufacture of leather; for selenium dioxide burns; can reduce cisplatin nephrotoxicity; oral lavage use (1% to 5%) for use in gastric decontamination for iodine exposure, can be used for mechlorethamine extravasation along with actinomycin D and mitomycin C; may be useful in chlorate salt toxicity and bromate toxicity; may be used alone in smoke inhalations and to reduce the toxicity of sodium nitroprusside

Mechanism of Action
Cyanide toxicity: Increases the rate of detoxification of cyanide by the enzyme rhodanese by providing an extra sulfur
Cisplatin toxicity: Complexes with cisplatin to form a compound that is nontoxic to either normal or cancerous cells

Adverse Reactions
Cardiovascular: Hypotension
Central nervous system: CNS depression secondary to thiocyanate intoxication, psychosis, confusion, coma
Dermatologic: Contact dermatitis
Gastrointestinal: Nausea, vomiting, abdominal cramps, diarrhea
Local: Local irritation
Neuromuscular & skeletal: Weakness
Otic: Tinnitus

Pharmacodynamics/Kinetics
Absorption: Well with parenteral administration
Distribution: V_d: 0.15 L/kg
Half-life: 0.65 hour
Elimination: 28.5% unchanged in urine

Usual Dosage
Cyanide and nitroprusside antidote: I.V.:
Children <25 kg: 50 mg/kg after receiving 4.5-10 mg/kg sodium nitrite; a half dose of each may be repeated if necessary
Children >25 kg and Adults: 12.5 g after 300 mg of sodium nitrite; a half dose of each may be repeated if necessary
Cyanide poisoning: I.V.: Dose should be based on determination as with nitrite, at rate of 2.5-5 mL/minute to maximum of 50 mL.
See table.

Variation of Sodium Nitrite and Sodium Thiosulfate Dose With Hemoglobin Concentration*			
Hemoglobin (g/dL)	Initial Dose Sodium Nitrite (mg/kg)	Initial Dose Sodium Nitrite 3% (mL/kg)	Initial Dose Sodium Thiosulfate 25% (mL/kg)
7	5.8	0.19	0.95
8	6.6	0.22	1.10
9	7.5	0.25	1.25
10	8.3	0.27	1.35
11	9.1	0.30	1.50
12	10.0	0.33	1.65
13	10.8	0.36	1.80
14	11.6	0.39	1.95

*Adapted from Berlin DM Jr, "The Treatment of Cyanide Poisoning in Children", *Pediatrics*, 1970, 46:793.

Cisplatin rescue should be given before or during cisplatin administration: I.V. infusion (in sterile water): 12 g/m² over 6 hours or a 9 g/m² I.V. push followed by 1.2 g/m² continuous infusion for 6 hours
Arsenic poisoning: I.V. 1 mL first day, 2 mL second day, 3 mL third day, 4 mL fourth day, 5 mL on alternate days thereafter
For use with nitroprusside, give 1 g of sodium thiosulfate for every 100 mg of nitroprusside administered to prevent cyanide toxicity
Extravasation injury due to actinomycin, mechlorethamine, and mitomycin mix: 4 mL of 10% sodium thiosulfate with 6 mL of sterile water, administer through either the offending I.V. catheter or through multiple SubQ injections after catheter removal; use ice to extravasated area
Children and Adults: Topical: 20% to 25% solution: Apply a thin layer to affected areas twice daily
Contraindications Hypersensitivity to any component, hydrogen sulfide poisoning
Warnings Safety in pregnancy has not been established; discontinue if irritation or sensitivity occurs; rapid I.V. infusion has caused transient hypotension and EKG changes in dogs
Dosage Forms
Injection, solution [preservative free]: 100 mg/mL (10 mL); 250 mg/mL (50 mL)
Lotion (Versiclear™): Sodium thiosulfate 25% and salicylic acid 1% (120 mL) [contains isopropyl alcohol 10%]
Stability Explosive when titrated with chlorates, nitrates, or permanganates
Pregnancy Issues C
Monitoring Parameters Monitor for signs of thiocyanate toxicity; administer slow I.V.; chest X-ray for inhalation injuries

Nursing Implications Given I.V. as slow I.V. push only over 10 minutes

Reference Range Serum levels of 11.13 ± 1.1 mg/L are normal in adults; levels may be decreased to 5-8 mg/L in postoperative coronary artery bypass graft patients; levels may be elevated up to 22 mg/L in patients kept NPO for 1-3 weeks

Additional Information White, odorless crystals or powder with a salty taste; normal body burden: 1.5 mg/kg

Specific References

Matteucci MJ, Reed WJ, and Tanen DA, "Failure of Sodium Thiosulfate to Reduce Methemoglobin Produced *In Vitro*," *Acad Emerg Med*, 2001, 8(5):443.

Matteucci MJ, Reed WJ, and Tanen DA, "Sodium Thiosulfate Fails to Reduce Nitrite-Induced Methemoglobinemia *In Vitro*," *Acad Emerg Med*, 2003, 10(4):299-302.

Succimer

CAS Number 304-55-2

Synonyms 2,3-Dimercaptosuccinic Acid; DMSA

U.S. Brand Names Chemet®

Use: Orphan drug Treatment of lead poisoning in children with blood levels >45 mcg/dL. It is not indicated for prophylaxis of lead poisoning in a lead-containing environment. Following oral administration, succimer is generally well tolerated and produces a linear dose-dependent reduction in serum lead concentrations. This agent appears to offer advantages over existing lead chelating agents; also has been used for arsenic and mercury poisoning; also used in children with blood lead levels <45 mcg/dL; also useful for mercury and arsenic toxicity

Mechanism of Action Succimer is an analog of dimercaprol. It forms water soluble chelates with heavy metals, which are subsequently excreted renally. Initial data have shown encouraging results in the treatment of mercury and arsenic poisoning. Succimer binds heavy metals; however, the chemical form of these chelates is not known.

Adverse Reactions

>10%:

Central nervous system: Fever

Gastrointestinal: Nausea, vomiting, diarrhea, appetite loss, hemorrhoidal symptoms, metallic taste

Neuromuscular & skeletal: Back pain

1% to 10%:

Central nervous system: Drowsiness, dizziness

Dermatologic: Rash

Endocrine & metabolic: Serum cholesterol

Gastrointestinal: Sore throat

Hepatic: Elevated AST/ALT, alkaline phosphatase

Respiratory: Nasal congestion, cough

Miscellaneous: Flu-like symptoms

<1%: Arrhythmias

Pharmacodynamics/Kinetics

Absorption: Rapid but incomplete

Metabolism: Rapid and extensive to mixed succimer cysteine disulfides

Half-life, elimination: 2 days

Time to peak serum concentration: ~1-2 hours

Elimination: ~25% in urine with peak urinary excretion occurring between 2-4 hours after dosing; of the total amount of succimer eliminated in urine, 90% is eliminated as mixed succimer-cysteine disulfide conjugates; 10% is excreted unchanged; fecal excretion of succimer probably represents unabsorbed drug

Usual Dosage Children and Adults: Oral: 10 mg/kg/dose every 8 hours for 5 days followed by 10 mg/kg/dose every 12 hours for 14 days

Dosing adjustment in renal/hepatic impairment: Administer with caution and monitor closely Concomitant iron therapy has been reported in a small number of children without the formation of a toxic complex with iron (as seen with dimercaprol); courses of therapy may be repeated if indicated by weekly monitoring of blood lead levels; lead levels should be stabilized <15 mcg/dL; 2 weeks between courses is recommended unless more timely treatment is indicated by lead levels

Contraindications Known hypersensitivity to succimer

Warnings Caution in patients with renal or hepatic impairment; adequate hydration should be maintained during therapy

Dosage Forms Capsule: 100 mg

Test Interactions False-positive ketones (U) using nitroprusside methods, falsely elevated serum CPK; falsely decreased uric acid

Drug Interactions Not recommended for routine concomitant administration with edetate calcium disodium or penicillamine

Pregnancy Issues C; No evidence for mutagenicity

Monitoring Parameters Blood lead levels, serum amino transferases

Nursing Implications Adequately hydrate patients; rapid rebound of serum lead levels can occur; monitor closely

Signs and Symptoms of Acute Exposure Anorexia, GI bleeding, hepatotoxicity, nephritis, renal tubular necrosis, respiratory depression, vomiting

Additional Information This agent appears to offer advantages over existing lead chelating agents by not facilitating gastric absorption of lead, not significantly chelating other divalent minerals (iron, copper, and zinc) and safety in G6PD deficiency. Adverse drug reactions should be reported to: Bock Pharmacal Company, Attn: Customer Service Department, PO Box 419056, St Louis, MO 63141; (800) 727-2625.

Specific References

Buchwald AL, "Intentional Overdose of Dimercaptosuccinic Acid in the Course of Treatment for Arsenic Poisoning," *J Toxicol Clin Toxicol*, 2001, 39(1):113-4.

Chisolm JJ Jr, "Safety and Efficacy of Meso-2,3-Dimercaptosuccinic Acid (DMSA) in Children With Elevated Blood Lead Concentrations," *J Toxicol Clin Toxicol*, 2000, 38(4):365-75.

Thrombopoietin

Synonyms TPO

Use: Unlabeled/Investigational Thrombocytopenia relatable to cytotoxic drugs or cancer; may be useful in treating drug-induced thrombocytopenia not due to immunological destruction

Mechanism of Action Promotes the proliferation and maturation of megakaryocyte progenitors into platelet-producing megakaryocytes; a ligand for the C-Mpl receptor on megakaryocytes

Adverse Reactions

Cardiovascular: Thrombosis is a theoretical concern

Central nervous system: Mild headache has been described although not temporarily related

Pharmacodynamics/Kinetics Half-life: 20-30 hours

Usual Dosage I.V.: 0.3-1 mcg/kg/day for up to 10 days; alternatively, a single dose of 0.3-2.4 mcg/kg 3 weeks before chemotherapy has been used; all doses are intravenous

Reference Range Peak serum thrombopoietin level after a 2.4 mcg/kg dose: ~50 ng/mL; endogenous serum thrombopoietin levels range from 0.096-0.24 ng/mL

Additional Information Following a single dose (0.3-2.4 mcg/kg) the platelet rise starts at 4 days and peaks in 10-15 days; increases of 1.3-3.6 fold have been noted

Specific References

Vadhan-Raj S, Verschraegen CF, Bueso-Ramos C, et al, "Recombinant Human Thrombopoietin Attenuates Carboplatin-Induced Severe Thrombocytopenia and the Need for Platelet Transfusions in Patients With Gynecologic Cancer," *Ann Intern Med*, 2000, 132(5):364-8.

Trimedoxime Bromide

CAS Number 56-97-3

Synonyms TMB4; Dioxime

Mechanism of Action Acetylcholinesterase reactivator, which is a bispyridium oxime that appears to be particularly effective for tabun and minimally effective for soman.

Adverse Reactions

To TMB$_4$: Headache, paresthesia, sensation of warmth, respiratory depression and hypotension at doses >30 mg/kg

To autoinjectors: Local effects (pain, swelling) in 23%; systemic effects of atropinization in 18%

Usual Dosage As autoinjector (injected into thigh) combined with atropine:

Children <2 years (blue-colored): 0.5 mg atropine/20 mg TMB$_4$

Children 3-10 years and adults >60 years (rose-colored): 1 mg atropine/40 mg TMB$_4$

Children and Adults 10-60 years (yellow-colored): 2 mg atropine/80 mg TMB$_4$

Stability Stable at a pH of 3

Specific References

Kassa J, Kuca K, and Cabul J, "A Comparison of the Potency of Trimedoxime and Other Currently Available Oximes to Reactivate Tabun-Inhibited Acetylcholinesterase and Eliminate Acute Toxic Effects of Tabun," *Biomed Pap Med Fac Palacky Olomouc Czech Republic*, 2005, 149(2):419-23.

Kozar E, Mordel A, Haim SB, et al, "Pediatric Poisoning From Trimedoxime (TMB$_4$) an Atropine Autoinjectors," *J Pediatr*, 2005, 146:41-4.

Unithiol

CAS Number 4076-02-2

Synonyms 2,3-Dimercaptopropanesulphonate; DMPS; Unitiol

Use May be effective for Wilson's disease

Use: Unlabeled/Investigational (in the U.S.): Antidote for arsenic, bismuth, lead, zinc, mercury (inorganic and organic), chromium, antimony, and cobalt

Mechanism of Action A water-soluble chemical analogue of dimercaprol

Adverse Reactions

Dermatologic: Macular erythematous rash

Cardiovascular: Hypotension at high doses

Gastrointestinal: Nausea, vomiting

Hematologic: Leukopenia

Pharmacodynamics/Kinetics

Absorption: Oral: 46% to 60%

Protein binding: 70% to 90%

Half-life (intravenous): 0.9 hour (alpha); 19 hours (beta)

Usual Dosage Oral:

Antimony: Children: 50-100 mg 3 times/day

Arsenic: Mild poisoning: 200 mg orally 3 times/day; severe poisoning: 200 mg I.V. or 400 mg orally initially, then 100-200 mg I.V., or 200-400 mg orally every 2 hours; taper dose slowly

Bismuth: 250 mg every 4 hours I.V., then 250 mg orally 3 times/day for 14 days

Metal chelation: 100 mg 3 times/day for 5 days

Wilson's disease: 200 mg 2 times/day

Dosing scheme for mercury intoxication: I.V.: 250 mg every 4 hours for 48 hours, followed by 250 mg every 6 hours for the succeeding 48 hours, and then 250 mg every 8 hours thereafter; if no gastrointestinal lesions are present, conversion to oral route of administration can occur; following 96 hours of administration, 300 mg 3 times/day can be given orally; continue treatment until mercury levels in blood and urine fall below 100 mcg/L and 300 mcg/L, respectively. Neurological improvement has occurred following three 5-day courses of 30 mg/kg/day.

Contraindications Arsine poisoning or hypersensitivity to DMPS or its salts

Warnings Use with caution in patients with renal insufficiency (serum creatinine >2.5 mg/dL), in acute infections, or zinc deficiency

Pregnancy Issues Not embryotoxic

Additional Information Clearance half-life of elemental mercury decreases by 66% with use of unithiol; not useful for cadmium removal; can cause zinc elimination. Information can be obtained from the Heyltex Corporation, 10655 Richmond Ave, #170, Houston, Texas 77042; (800) 237-6793.

Specific References

Angle CR, Centeno JA, and Guha Mazumder DN, "DMSA, DMPS Treatment of Chronic Arsenicism," *J Toxicol Clin Toxicol*, 1998, 36(5):495.

Dargan PI, Giles L, House IM, et al, "A Case of Severe Mercuric Sulfate Ingestion Treated With 2,3-Dimercaptopropane-1-Sulphonate (DMPS) and Hi-Flow Hemodiafiltration," *J Toxicol Clin Toxicol*, 1999, 37(5):622-3.

Torres-Alanís O, Garza-Ocañas L, Bernal MA, et al, "Urinary Excretion of Trace Elements in Humans After Sodium 2,3-Dimercaptopropane-1-Sulfonate Challenge Test," *J Toxicol Clin Toxicol*, 2000, 38(7):697-700.

Torres-Alanís O, Garza-Ocañas L, Bernal-Hernández MA, et al, "Urinary Excretion of Trace Elements in Humans After Sodium 2,3-Dimercaptopropane-1-Sulfonate (DMPS) Therapy," *J Toxicol Clin Toxicol*, 2000, 38(2):253.

Wax PM and Thornton CA, "Recovery From Severe Arsenic-Induced Peripheral Neuropathy With 2,3-Dimercapto-1-propanesulphonic Acid," *J Toxicol Clin Toxicol*, 2000, 38(7):777-80.

Vasopressin

CAS Number 11000-17-2; 113-79-1; 50-57-7

Synonyms 8-Arginine Vasopressin; ADH; Antidiuretic Hormone; Vasopressin Tannate (discontinued in 1990)

U.S. Brand Names Pitressin®

Use Treatment of diabetes insipidus; prevention and treatment of postoperative abdominal distention; differential diagnosis of diabetes insipidus

Use: Unlabeled/Investigational Adjunct in the treatment of GI hemorrhage and esophageal varices; pulseless ventricular tachycardia (VT); ventricular fibrillation (VF); vasodilatory shock (septic shock)

Mechanism of Action Increases cyclic adenosine monophosphate (cAMP), which increases water permeability at the renal tubule resulting in decreased urine volume and increased osmolality; causes peristalsis by directly stimulating the smooth muscle in the GI tract

Adverse Reactions

Cardiovascular: Elevated blood pressure, bradycardia, arrhythmias, venous thrombosis, vasoconstriction with higher doses, angina, myocardial infarction, tachycardia (ventricular), torsade de pointes, chest pain, peripheral gangrene, superior mesenteric artery thrombosis, arrhythmias (ventricular), sinus bradycardia

Central nervous system: Pounding in the head, fever, headache, vertigo

Dermatologic: Urticaria, cutaneous gangrene, circumoral pallor, bullous lesions, skin necrosis

Endocrine & metabolic: Hyponatremia, water intoxication, hyperprolactinemia

Gastrointestinal: Flatulence, abdominal cramps, nausea, vomiting, mesenteric occlusion, ischemic colitis, diarrhea, colonic ischemia

Hepatic: Hepatic steatosis

Neuromuscular & skeletal: Tremor, rhabdomyolysis

Miscellaneous: Diaphoresis, anaphylaxis, allergic reaction, anaphylactic shock

Toxicodynamics/Kinetics
Nasal:
    Onset of action: 1 hour
    Duration: 3-8 hours
Parenteral:
    Duration of action: I.M., SubQ: 2-8 hours
    Absorption: Destroyed by trypsin in GI tract, must be administered parenterally or intranasally
Nasal:
    Metabolism: In the liver, kidneys
    Half-life: 10-20 minutes
    Elimination: In urine
Parenteral: Metabolism: Most of dose metabolized by liver and kidneys; Half-life: 10-20 minutes; Elimination: 5% of SubQ dose (aqueous) excreted unchanged in urine after 4 hours

Usual Dosage
Cyclophosphamide-induced hemorrhagic cystitis: Adults: I.V.: 0.4 units/minute
Diabetes insipidus (highly variable dosage; titrated based on serum and urine sodium and osmolality in addition to fluid balance and urine output):
    I.M., SubQ:
        Children: 2.5-5 units 2-4 times/day as needed
        Adults: 5-10 units 2-4 times/day as needed (dosage range 5-60 units/day)
    Continuous I.V. infusion: Children and Adults: 0.5 milliunit/kg/hour (0.0005 unit/kg/hour); double dosage as needed every 30 minutes to a maximum of 0.01 unit/kg/hour
    Intranasal: Administer on cotton pledget, as nasal spray, or by dropper
Abdominal distention: Adults: I.M.: 5 units stat, 10 units every 3-4 hours
GI hemorrhage (unlabeled use): I.V. infusion: Dilute in NS or D_5W to 0.1-1 unit/mL
    Children: Initial: 0.002-0.005 units/kg/minute; titrate dose as needed; maximum: 0.01 unit/kg/minute; continue at same dosage (if bleeding stops) for 12 hours, then taper off over 24-48 hours
    Adults: Initial: 0.2-0.4 unit/minute, then titrate dose as needed, if bleeding stops; continue at same dose for 12 hours, taper off over 24-48 hours
Pulseless VT/VF (ACLS protocol): I.V.: 40 units (as a single dose only); if no I.V. access, administer 40 units diluted with NS (to a total volume of 10 mL) endotracheally
Vasodilatory shock/septic shock (unlabeled use): Adults: I.V.: Vasopressin has been used in doses of 0.01-0.1 units/minute for the treatment of septic shock. Doses >0.05 units/minute may have more cardiovascular side effects. Most case reports have used 0.04 units/minute continuous infusion as a fixed dose.
Dosing adjustment in hepatic impairment: Some patients respond to much lower doses with cirrhosis

Administration
I.V.: Use extreme caution to avoid extravasation because of risk of necrosis and gangrene. In treatment of varices, infusions are often supplemented with nitroglycerin infusions to minimize cardiac effects.
    GI hemorrhage: Administration requires the use of an infusion pump and should be administered in a peripheral line.
    Vasodilatory shock: Administration through a central catheter is recommended.
        Infusion rates: 100 units in 500 mL D_5W rate
        0.1 unit/minute: 30 mL/hour
        0.2 unit/minute: 60 mL/hour
        0.3 unit/minute: 90 mL/hour
        0.4 unit/minute: 120 mL/hour
        0.5 unit/minute: 150 mL/hour
        0.6 unit/minute: 180 mL/hour
Intranasal (topical administration on nasal mucosa): Administer injectable vasopressin on cotton plugs, as nasal spray, or by dropper. Should not be inhaled.

VASOPRESSIN

Contraindications Hypersensitivity to vasopressin or any component

Warnings Use with caution in patients with seizure disorders, migraine, asthma, vascular disease, renal disease, cardiac disease, chronic nephritis with nitrogen retention, goiter with cardiac complications, arteriosclerosis, I.V. infiltration may lead to severe vasoconstriction and localized tissue necrosis; also gangrene of extremities, tongue, and ischemic colitis

Dosage Forms Injection, solution, aqueous: 20 pressor units/mL (0.5 mL, 1 mL, 10 mL)
Pitressin®: 20 pressor units/mL (1 mL)

Stability Store injection at room temperature; protect from heat and from freezing; use only clear solutions

Drug Interactions
Decreased effect: Lithium, epinephrine, demeclocycline, heparin, and alcohol block antidiuretic activity to varying degrees
Increased effect: Chlorpropamide, phenformin, urea and fludrocortisone potentiate antidiuretic response

Pregnancy Issues C; Animal reproduction studies have not been conducted. Vasopressin and desmopressin have been used safely during pregnancy and nursing based on case reports. Enters breast milk/use caution

Monitoring Parameters Serum and urine sodium, urine output, fluid input and output, urine specific gravity, urine and serum osmolality

Nursing Implications Watch for signs of I.V. infiltration and gangrene; elderly patients should be cautioned not to increase their fluid intake beyond that sufficient to satisfy their thirst in order to avoid water intoxication and hyponatremia; under experimental conditions, the elderly have shown to have a decreased responsiveness to vasopressin with respect to its effects on water homeostasis

Reference Range Plasma: 0-2 pg/mL (SI: 0-2 ng/L) if osmolality <285 mOsm/L; 2-12 pg/mL (SI: 2-12 ng/L) if osmolality >290 mOsm/L

Additional Information Due to prolongation of QT_c interval on EKG, avoid vasopressin in arsenic poisoning; not useful in treating lithium-induced diabetes insipidus

Specific References
Barry JD, Durkovich DW, and Williams SR, "Vasopressin for Hypotension in Severe Amitriptyline Poisoning," *J Toxicol Clin Toxicol*, 2002, 40(5):622.
Sharshar T, Carlier R, Blanchard A, et al, "Depletion of Neurohypophyseal Content of Vasopressin in Septic Shock," *Crit Care Med*, 2002, 30(3):497-500.
Wenzel V, Krismer AC, Arntz HR, et al, "A Comparison of Vasopressin and Epinephrine for Out-of-Hospital Cardiopulmonary Resuscitation," *N Engl J Med*, 2004, 350(2):105-13.
Wyer PC, Perera P, Jin Z, et al, "Vasopressin or Epinephrine for Out-of-Hospital Cardiac Arrest," *Ann Emerg Med*, 2006, 48:86-97.

SECTION II
BIOLOGICAL AGENTS

Avian Influenza

R.B. McFee, DO, MPH, FACPM

Background

The term "influenza" describes an acute viral disease of the respiratory tract is caused by viruses that belong to the orthomyxovirus family, which includes the genera of influenza virus A, B, and C as defined by the antigenicity of the nucleocapsid and matrix proteins.[1-6] Generally, influenza A viruses are associated with more severe human illness, epidemics, and pandemics.[7,8] Influenza A virus is a negative sense, single-stranded RNA virus, with an 8-segment genome that encodes for 10 proteins. Influenza A virus is further subtyped by 2-surface proteins – haemagglutinin (H), which attaches the viral particle to the host cell for cell entry, and neuraminidase (N), which facilitates the spread of progeny virus.[1,9-12] It is the latter that is a target for the class of antiviral therap y referred to as neuraminidase inhibitors, which will be discussed shortly.[9-11]

There are 16 H and 9 N subtypes that make up all the subtypes of influenza A by various combinations.[1,4,11,12] The term "antigenic drift" refers to the various mutations and changes in surface antigenicity of these surface proteins as a response to host immunity. This is why every year, usually in February, the World Health Organization (WHO) decides the strain of viruses to be incorporated into the annual influenza vaccine.[13] More worrisome is the potential for "antigenic shift"; an event that can lead to the creation of a novel virus against which humans have little or no immunity. This can occur because influenza has a segmented genome – shuffling of gene segments can occur if 2 different subtypes of influenza A virus infect the same cell. If a human flu virus, such as H3N2 and an avian H5N1 virus coinfect a human or pig, a new virus "H5N2" could emerge – a hybrid that could bring the high virulence and case fatality rate of H5N1 with the efficiency of human-to-human transmission found in the "parent" human virus.[1,2,8] Studies suggest this is what happened in the 1957 and 1968 influenza pandemics. A pandemic is considered an epidemic that crosses continents. Of note, while Mother Nature has demonstrated its capacity to do this, such viral reassortment also could be accomplished in a laboratory for bioterrorism purposes.[14-16] Therefore, bioterrorism preparedness, increased surveillance, and efforts to enhance physician training in unusual or emerging diseases is of significant importance.

Unlike usual patterns of mortality associated with seasonal flu, namely the very young and old, both the 1918 pandemic and what is currently occurring with H5N1 involved all age groups. Once avian influenza A H5N1 becomes a more human-like virus with the ability to efficiently cause person-to-person spread, in the absence of an effective, prepositioned, widely available vaccine, a pandemic could rapidly develop. Illnesses can traverse the country, given, we must consider, our nation is highly susceptible with little to no immunity. Most scenarios describe a significant number of people becoming quite sick over a relatively short period of time. The ability to contain such a public health crisis may well rest on the community physician who will be called upon to rapidly diagnose, treat, and, provide care for a potentially overwhelming number of patients (and protect his or her family, too)!

There are many genotypes of H5N1 – the predominant one is "Z," which is associated with high virulence for a wide range of animals from poultry to felines. It appears to be stable in the environment for up to 6 days, but can be transmitted, albeit inefficiently, from person to person.[1,4,13,14]

Influenza virus (the "seasonal" forms, which have caused significant respiratory illness for centuries), remains a major, global public health problem that continues to result in millions of cases of severe illness, as well as approximately 500,000 deaths worldwide and 36,000 deaths in the U.S. annually. Yet it is a vaccine-preventable disease.[16-19]

An influenza pandemic occurs when 1) a new flu virus emerges for which people have little or no immunity, 2) the virus spreads readily from person to person, and 3) no vaccine is available. Avian influenza virus A H5N1 is a virus that has resulted in the death of over 140 million birds, is continuing to mutate, and, unlike other strains of flu, has demonstrated an ability to cause significant pulmonary damage.[1,4,20-23] Avian flu, for the moment, passes easily from bird to bird. It infects birds via the intestinal tract – which allows the virus to be found in feces. Of note, the more ordinary strains of avian flu usually infect migratory birds and do not result in disease; the H5N1 is killing these very animals suggesting the virus has adapted and thus increased its virulence and capacity to kill. This persistence in the ground promotes contamination to people and other animals as well as a risk to water. Avian flu infects people via the respiratory tract, which can be accomplished by fomites, inhaling the virus or getting it on hands and then contacting the pulmonary mucosa. Currently, H5N1 affects domesticated poultry (chickens, ducks, turkeys) as well as migratory birds (wild ducks, geese, swans); it is the latter that do not honor borders and are not readily contained thus enhancing the global spread. Fortunately, people cannot get avian flu from eating properly cooked poultry. Eating raw eggs, poultry blood, or undercooked bird meat are modalities that one could become infected and are practices in countries where human avian flu cases have been recorded. The virus is demonstrating an ability to infect other species like cats and people.

The following must occur before avian flu can cause human pandemics the way it has caused bird pandemics: 1) avian flu must be able to infect humans, 2) it must be virulent, and 3) it must spread easily from person to person. Avian flu already has demonstrated the first two factors, but it has not quite achieved the last factor. While data suggest there have been a few cases resulting from human-to-human transmission, we have been fortunate that the infection stopped at the second person, usually a family member.[24] As of this writing, H5N1 is inefficient at person-to-person transmission unlike the highly contagious nature of the seasonal flu virus (actually viruses) in general.

Potential Magnitude of a Pandemic

However, if avian influenza virus A H5N1 mutates into a more human-like influenza (ie, develops the ability to be as contagious as seasonal flu), and is then likely to spread very rapidly in a sustained fashion across the globe and result in thousands, perhaps millions of deaths similar to the flu pandemic of 1918 (which resulted in deaths estimated at between 20 and 50 million people worldwide), some experts suggest upwards of 2.5 percent of the world's population could be affected in a matter of months while others pose more dire projections. Estimates suggest an influenza pandemic could sicken upwards of 90 million people in the United States, including over 1/3 of the healthcare workforce. Federal officials are concerned at least 10 million influenza patients could require hospitalization at least for one night including almost 1.5 million requiring intensive care and possibly 750,000 needing ventilator assistance.[25] Given the persistent problem of diminished surge capacity, hospital over-crowding from the emergency department to the ventilated beds in the critical care units is inevitable when a pandemic occurs.[26-28] Overall, poor infection control practices within healthcare facilities, in addition to healthcare workers demonstrating low vaccine rates, inconsistent handwashing and deficiencies in other respiratory hygiene practices are the building blocks to promote rapid spread of an emerging contagion.[13,18,29,30]

The exact impact of a global pandemic remains unknown in spite of numerous projection models. World Health Organization leaders are concerned that it is not a matter of if, but when, avian influenza breaks loose, not only arriving in the West, but becoming more efficient at person-to-person transmission (24). If that occurs, primary care physicians will likely be on the front lines, (perhaps even diagnosing the index case), clearly providing guidance to the worried well, treating the sick, and working with public health and other organizations in an attempt to contain the outbreak. It is a daunting task, especially in the absence of a widely available vaccine and few antivirals (of which each must be administered early in the course of illness). In a world without borders, unusual emerging pathogens from SARS to H5N1 can arrive on passenger airliners or many other entry points.

If seasonal influenza is a predictable killer, the avian flu has the potential to be a true killing machine. Despite the efforts of the international health community to contain the avian influenza epidemic that emerged in Southeast Asia in 2003/2004, sporadic cases of human influenza A H5N1 infection are still reported – and carry a high case fatality rate. Never before has an avian influenza strain killed so many hundreds of millions of birds, traveled so far so fast, and posed the potential to simultaneously affect Asia, Europe, and other continents.

Clinical

Epidemiology Unlike usual patterns of mortality associated with seasonal flu – namely the very young and old, both the 1918 pandemic and what is currently occurring with H5N1 involved all age groups, especially ages 20-40 who are confined together, such as the military and universities. The case fatality rate of H5N1 is >50% depending upon the country.

Pathogenesis of Avian Flu H5N1 induces proinflammatory cytokines (such as interferon gamma inducible protein and tumor necrosis factor (TNF) alpha) in human macrophage cells, which may lead to a cytokine storm and death without extrapulmonary viral dissemination.[20] The haemagglutinin of this influenza virus also may attach to respiratory epithelium to cause inhibition of epithelial sodium channels leading to pulmonary edema, alveolar flooding, and early acute respiratory failure – events that rarely accompany seasonal flu.[21,22]

For the moment, the most important route of acquisition for avian influenza H5N1 infection is through contact with infected birds or their excreta. However, hospital-acquired infection was revealed in a retrospective study. Healthcare workers (HCW) exposed to patients with H5N1 infection were more likely to be seropositive and this was not attributable to animal exposure. It is reasonable to assume that the route of infection for avian influenza patients, like most flu patients, will be from inhalation of infective respiratory secretions and/or contact with virus-laden secretions and subsequent contact with mucous membranes. Studies also suggest airborne transmission is possible, which would explain the occasional numerically explosive nature of influenza epidemics.[31,32]

Clinical Presentation[1-4] Avian flu (H5N1) presents with a rapid onset of severe illness including a fever spike of over 101°F and often respiratory symptoms – nearly universal is dyspnea. Significant symptoms include muscle aches and generally feeling poorly, cough, headache, shortness of breath, chest pain, and/or difficulty breathing. The latter symptoms are not common to seasonal flu (especially in younger people), which seem to be increasingly affected by avian flu. Avian flu illness is not subtle; it will not present like the common cold or the average case of seasonal flu. Of those who present for medical attention, the illness is rapidly progressive with patients often complaining of chest pain and shortness of breath. True dyspnea is rarely associated with seasonal flu, especially in young, otherwise healthy, patients. Shortness of breath under most circumstances is worrisome and should be properly evaluated, but in the context of a rapid rise in temperature should raise an alarm for avian flu or other serious infection such as Legionella pneumonia (33,34). Furthermore, avian flu can have an extrapulmonary impact, including the central nervous system resulting in encephalopathy, seizures in addition to severe headache. Of note, while gastrointestinal (GI) symptoms may be more common among children with seasonal flu, young adults and other age groups may experience abdominal pain, nausea, vomiting, or diarrhea before or during the development of respiratory symptoms from avian flu, which is not common with seasonal flu in this age group. Unlike seasonal flu deaths, which often result from secondary infections such as pneumonia, avian influenza seems to cause direct pulmonary damage that can result in noncardiogenic pulmonary edema and pneumonia.

The differential diagnosis includes *Legionella* pneumonia and severe Gram negative pneumonias. *Legionella* pneumonia causes severe community-acquired pneumonia often requiring admission to an intensive care unit.[33,34] Patients often have a nonproductive cough, pleuritic chest pain, diarrhea and gastrointestinal symptoms, not unlike the biodrome of avian influenza. A valuable clinical clue includes the finding of relative bradycardia as well as hyponatremia among Legionella patients. A recent study suggests *Legionella* is significantly underdiagnosed. Numerous clinical tests are available but each has limitations in time, sensitivity, or availability. Polymerase chain reaction (PCR) for *L. pneumophilia* is available in some reference laboratories and can detect the organism in sputum, bronchoalveolar lavage fluid, and blood. Legionella species are not detected in routine cultures; physicians should specifically request culture for *Legionella* if it is suspected.[33,34]

Tests[1-4,13] Presumptive diagnosis is clinical and based upon identifying the biodrome (symptom pattern recognition). Critical to diagnosing a potential case of avian flu include awareness in terms of most recent locations of avian influenza, asking your patient about recent travel (within 14 days) especially to countries suspected of having H5N1, an occupation history, in addition to a thorough history of present illness and careful examination. However, certain initial tests can guide the clinician, especially with early infection control and treatment while initiating contact with specialized laboratories for confirmation.

Chest X-Ray (CXR) Major findings on CXR include extensive infiltration bilaterally, lobar collapse, focal consolidation, and less commonly, interstitial lung infiltrates. Pleural effusion and widened mediastinum may be observed. Clinical deterioration associated with this is common.

Blood Work[1-4,13,14,23] Several patients had lower total peripheral white blood cell counts that are more often lymphopenic and associated with fatality. Many patients presenting with pneumonia associated with H5N1 also had abnormal liver function tests. Over thirty percent of patients exhibited impaired renal function.

RT-PCR has superior sensitivity and specificity compared to antigen detection. Commercial immunochromatographic membrane enzyme immunoassay tests are not specific for H5 and only has a 70% specificity compared with viral culture. Nasopharyngeal aspirate or bronchoalveolar lavage (BAL) followed by nasopharyngeal swab or throat swab placed in viral transport medium (VTM) should be collected with airborne precautions in patients suspected of having avian flu. A stool or rectal swab also placed in VTM should be considered. If H5N1 is suspected, the health department should be alerted for guidance as well as the laboratory to remind them to take proper precautions.

The Food and Drug Administration recently approved a laboratory test to diagnose patients suspected of being infected with avian influenza A/H5 viruses.[35] The test is referred to as Influenza A/H5 (Asian Lineage) Virus Real-time RT-PCR primer and Probe Set. This test can provide preliminary results on suspected H5 Influenza samples within four hours once sample testing begins at the lab. This is a major advance given previous technology required 2 to 3 days for similar results. If the H5 strain is identified, further testing is required to determine the specific subtype (such as N1, etc). The test will be distributed nationwide to Laboratory Response network (LRN) designated laboratories in order to enhance surveillance and diagnostic capabilities. There are approximately 140 LRN laboratories throughout the 50 United States. The CDC recommends if a clinician suspects a patient may be infected with avian influenza, it is important to contact the local or state health department for assistance in accessing the LRN capabilities. These laboratories can be accessed by calling (404) 639.2790.

Management[2,13,17] CDC recommends respiratory/airborne precautions in addition to droplet and contact and standard precautions as infection control practices for healthcare workers and healthcare facilities. The period of communicability of H5N1 remains understudied but can last for up to 3 weeks in children.

Management strategies include identifying and isolating potential respiratory contagious patients, promoting the use of seasonal flu vaccines – especially among healthcare providers, maintaining adequate stocks of oseltamivir and zanamivir, universal and respiratory precautions, ample fitted N-95 respirators and personal protective equipment (PPE) that employees have been trained when and how to safely use. New antivirals and H5N1 vaccines are being evaluated but not currently available to the public.

Current Antivirals[1-4,11,13,17] There are 2 classes of antivirals available to treat influenza virus – the neuraminidase inhibitors oseltamivir (Tamiflu®) and zanamivir (Relenza®) (refer to selected monographs). The former is approved for influenza A and B viruses, and the M blockers amantadine (Symmetrel®) and rimantadine (Flumadine®) (refer to selected monographs). Each is designed to take advantage of influenza viral structure. While both classes can treat influenza viruses, avian flu H5N1 is already resistant to amantadine and rimantadine owing to several possible factors. The present circulating H5N1 has the genotype "Z," which confers a residue on the M2 protein – making avian flu intrinsically resistant to the M blockers. Some experts suggest the Chinese practice of administering amantadine to poultry may also have contributed to the antiviral resistance. At least for the moment, the neuraminidase inhibitors remain effective against both seasonal and avian flu but should be administered early – ideally within 48 hours of onset of illness. Oseltamivir when administered for seasonal flu is usually given at a dosage of 75 mg by mouth twice daily for 5 days. A higher dose (150 mg orally given twice daily) has been recommended in clinical trials and associated with a larger reduction in viral load and shorter duration of illness. Whether a higher dose given over a longer duration would confer benefit in avian influenza remains to be further evaluated but should be considered in patients with significant pulmonary and GI symptoms. Children older than one year of age can receive twice daily oral dosing based upon weight: 30 mg per dose if 15 kg or less, 45 mg if 15–23 kg, 60 mg for 23–40 kg, and 75 mg for those >40 kg.

Unfortunately resistance to oseltamivir is emerging. Dosage may need to be adjusted in adults with renal impairment. There are insufficient human data to determine the risk to a pregnant woman or developing fetus (Pregnancy Category C). The most commonly reported adverse effects include nausea, abdominal pain, and vomiting. As with other antimicrobials, attention to early symptoms of anaphylactic/anaphylactoid risk is important.

The neuraminidase inhibitor oseltamivir can be used as a prophylactic chemotherapy for persons exposed to avian flu. WHO recommends healthcare workers exposed to H5N1 receive 75 mg orally once a day for at least 7 days and may be required for 6 weeks; duration of protection lasts for the period of dosing. Vaccination against seasonal flu should be obtained if the HCW has not been immunized.

Zanamivir (Relenza®) is an inhaled neuraminidase inhibitor, and has little systemic absorption; it may not be useful if extrapulmonary disease occurs. Data are lacking in terms of the effectiveness of Zanamivir against H5N1 either for acute treatment or as chemoprophylaxis, although experts consider it of value given the class effect of neuraminidase inhibitors.

Newer Treatments Studies are underway evaluating a new neuraminidase inhibitor – Peramivir that in early studies, when compared to other neuraminidase inhibitors, showed promise against influenza.[11] Other drugs being investigated to treat both seasonal and avian influenza include long-acting neuraminidase inhibitors, the antiviral Ribavirin, and interferon alpha.[36,37]

Other Modalities It is important to avoid aspirin-containing products as a precaution against Reyes Syndrome, in patients younger than 16 years of age. It should be noted, Reyes Syndrome has been reported, albeit rarely, in adults as well. In addition to early administration of antiviral therapy, respiratory support and intensive care are critical during the acute stage of H5N1 pneumonic illness.

Infection Control[2,13,18]

Infection control for avian flu involves a 2-tier approach: 1) Universal precautions, which apply to ALL patients at ALL times, including those who have HPAI H5N1 and 2) additional measures, which include droplet precautions, contact precautions, the use of a high-efficiency (HE) mask (n-95 or higher respirator), and negative pressure room, if possible. Place patient in single room; cohort confirmed and suspected cases in designated areas. The distance between beds should be at least 1 m and separated by a barrier. Anyone entering the room must wear appropriate PPE: mask respirator, gown, face shield or goggles, gloves.

Transportation of Patients If transportation is necessary from the isolation room, the patient should wear PPE, especially HE mask.

Waste Disposal All waste generated in the isolation area should be treated as clinical/infectious waste. Handlers should also wear appropriate PPE. Liquid waste can be safely flushed into the toilet.

Cleaning and Disinfection The survival time for the influenza virus is:
- 24-48 hours on hard, nonporous surfaces
- 8-12 hours on cloth, paper, and tissue
- 5 minutes on hands

The virus is inactivated by 70% alcohol and by chlorine. Cleaning of environmental surfaces with neutral detergent followed by a disinfectant solution is recommended. As an aside, it is very persistent in the environment, especially in farm soil in the presence of infected birds.

Disposition of the Dead[18]

Continued use of universal precautions including HE mask is necessary. The body should be fully sealed in an impermeable body bag prior to transfer to the morgue/mortuary. The outside of the bag must be clean and free of liquid. A postmortem may be performed with caution, given the lungs may still be filled with virus. Full PPE is recommended. Avoiding techniques that promote aerosolization of tissue is encouraged. Staff of the funeral home should be informed that the deceased had HPAI H5N1 and encouraged to follow universal precautions. Embalming may be conducted. Hygienic preparation of the deceased is permissible.

Vaccination Strategies[2,13,16,17] While an avian flu H5N1 vaccine is not available to the public, as of 2006, several are under clinical investigation worldwide. The U.S. government is conducting clinical tests on a vaccine that is based upon the Vietnamese H5N1 strain with early data suggesting the vaccine is both safe for human use and effective against the strain it was based upon.

Clinicians should encourage all patients to obtain seasonal flu vaccines immediately. While it is unknown if cross protection for H5N1 is possible, and probably unlikely with the current batch of flu vaccines, seasonal flu remains a consistent killer and cause of significant morbidity annually.

Postexposure Prophylaxis

Individuals exposed to avian influenza should begin Oseltamivir prophylaxis should begin immediately or within 2 days – 75 mg tablet each day for 7 days, although therapy may continue upwards of 6 weeks according to WHO.[13,18]

Travel Recommendations[38-40]

Persons planning to travel overseas for work, recreation, or humanitarian outreach should be able to do so relatively safely if precautions are taken. Vaccines against diseases endemic to the new host region may need weeks to evoke an immune response. Special precautions should be taken when contemplating visiting countries where avian flu infections are reported in birds, other animals, or people. This includes avoiding crowded places and farms, marketplaces that have poultry and/or kill chickens on demand, and changing clothes if visiting any of the above. No matter where the person visits, frequent handwashing and respiratory hygiene should be stressed. Just as referral to a cardiologist for a patient with significant cardiac history is standard of care, so should referral to a travel medicine clinic or physician who specializes in travel medicine. These would be physicians with experience in emerging infectious diseases and the appropriate background such as infectious diseases, occupational medicine, medical toxicology, tropical medicine with additional specialized training in travel-related health issues that include not only pathogens but safety, nutrition, and other relevant topics. The U.S. Department of State provides important information about most countries with timely alerts on emerging threats, political instability or other risks for U.S. citizens. The Centers for Disease Control publishes the Yellow Book – a wonderful travel health resource. Recommendations about vaccines appropriate to foreign destinations can be obtained through CDC as well. The WHO publishes an updated screening and assessment algorithm for the management of returning travelers and visitors from countries affected by H5N1 presenting with febrile respiratory illness.[13,18,41]

Human Cases of Avian Influenza A H5N1 Worldwide by National WHO Confirmed Human Cases - Updated 12 May 2006		
Location	Total # of Human Cases	Total Deaths
Azerbaijan	8	5
Cambodia	6	6
Djibouti	1	0
Thailand	22	14
Vietnam	93	42
Indonesia	33	25
China	18	12
Turkey	12	4
Egypt	13	5
Iraq	2	2
Total as of 5/18/06	208	115
Reference: World Health Organization (WHO), www.who.org		

Key Differentiating Symptoms: Avian Influenza Compared With Seasonal Flu and Influenza-Like Illnesses/Upper Respiratory Infections

	Avian Flu (H5N1)	Seasonal Influenza (Flu)	Upper Respiratory Infection	Common Cold
Elevated temperature	+++/++++	++	++	+/ −
Fever/Chills	++++	++++	++++	
Cough	+++	+++	+++	+++
Shortness of breath	++++	+/ −	+/ −	
Chest discomfort	+++	++	++	
Sore throat	+/ −	+++	+++	++
Vomiting/nausea	++	+	+	
Diarrhea	++	+ (young children)	+/ −	
CNS/Encephalopathy/ Seizures	++	−	−	
Malaise/fatigue	+	+++	+	
Runny nose/watery eyes	+/ −	+	++	+++
Headache/muscle ache	++	+++	++	+/ −
Young healthy at risk for serious illness	+++	+/ −		

Reference: Influenza Virus Vaccine, U.S. Food and Drug Administration http://www.fda.gov/cber/flu/flu.htm (last accessed 3/7/06)

References
1. Yuen KY and Wong SY, "Human Infection by Avian Influenza A H5N1," *Hong Kong Med J*, 2005, 11(3):189-99.
2. Centers for Disease Control and prevention (CDC), "Avian Influenza - Information for Physicians," www.cdc.gov/flu (Last accessed 3/7/06).
3. Gupta NE, "Everything You Should Know About Virulent Avian Flu," Cortlandt Forum, 2005, Dec 20:26-34.
4. Beigel JH, Farrar J, Han AM, et al, "Avian Influenza A (H5N1) Infection in Humans," *N Engl J Med*, 2005, 353:1374-85.
5. "Avian Flu/Pandemic Flu," National Institutes of Health, www.nih.gov (Last accessed 3/7/06).
6. Infectious Diseases Society of America, www.idsociety.org (Last accessed 3/7/06).
7. Raisuke H and Kawaoka Y, "Influenza: Lessons From Past Pandemics, Warnings From Current Incidents," *Nature Reviews Microbiology*, 2005, 3:591-600.
8. PandemicFlu.gov, AvianFlu.gov - Department of Health and Human Services (HHS), http://www.pandemicflu.gov/vaccine (Last accessed 3/7/06).
9. Osterholm MT, "Emerging Infectious Diseases - A Real Public Health Crisis?" Postgraduate Medicine online - 1996, Nov, 100(5), http://www.postgradmed.com/issues/1996/11_06/ed_nov.htm (Last accessed 2/16/06).
10. Lederberg J, Shope RE, and Oaks SC, Institute of Medicine Committee on Emerging Threats to Health, *Emerging Infections: Microbial Threats to Health in the United States*, Washington, DC: National Academy Press, 1992.
11. Sidwell RW and Smee DF, "Peramivir (BCX-1812, RWJ - 270201): Potential New Therapy for Influenza - Expert Opinion," *Investig Drug*, 2002, 11(6):859-69.
12. Fouchier RA, Munster V, Wallenstein A, et al, "Characterization of a Novel Influenza A Virus Hemagglutinin Subtype H16 Obtained From Black Headed Gulls," *J Bio*, 2005, 79:2814-22.
13. World Health Organization. "WHO Interim Guidelines on Clinical Management of Humans Infected by Influenza A (H5N1)," http://www.who.int/csr/disease/avian_influenza/guidelines/Guidelines_clinical%20management_H5N1_rev.pdf.

14. Tran TH, Nguyen TL, Nguyen TD, et al, "World Health Organization International Avian Influenza Investigative Team. Avian Flu A h5N1 in 10 Patients in Vietnam," *N Engl J Med*, 2004, 350:1179-88.
15. Krug RM, "The Potential Use of Influenza Virus as an Agent for Bioterrorism," *Antiviral Res*, 2003, 57:147-50.
16. "World Report: Nations Set Out a Global Plan for Influenza Action," *Lancet*, 2005, 23:366, www.thelancet.com (Last accessed 3/3/06).
17. Preventing the Flu. Influenza (Flu), CDC Key Facts About Flu Vaccine, www.cdc.gov/flu/protect/keyfacts.htm (Last accessed 3/7/06).
18. Influenza A (H5N1):WHO Interim Infection Control Guidelines for Health Care Facilities 2004," World Health Organization, http://www.who.int/csr/disease/avian_influenza/guidelines/infectioncontrol1/en/ (Last accessed 3/7/06).
19. National Immunization Survey 2004, *MMWR*, 2006.
20. Cheung CY, Poon LI, Lau AS, et al, "Induction of Proinflammatory Cytokines in Human Macrophages by Influenza A H5N1 Viruses: A Mechanism for the Unusual Severity of Human Disease?" *Lancet*, 2002, 360:1831-7.
21. Kunzelmann K, Beesley AH, King NH, et al, "Influenza Inhibits Amiloride-Sensitive NA+ Channels in Respiratory Epithelia," *Proc Natl Acad Sci USA*, 2000, 97:10282-7.
22. Chen XJ, Set S, Yue G, et al, "Influenza Virus Inhibits ENaC and Lung Fluid Clearance," *Am J Physiol Lung Cell Mol Physiol*, 2004, 287:L366-73.
23. de Jong MD, Bach VC, Phan TQ, et al, "Fatal Avian Influenza A (H5N1) in a Child Presenting With Diarrhea Followed by Coma," *N Engl J Med*, 2005, 17;352(7):686-91.
24. Ungchusak K, Auerwarakul P, Dowell SF, et al, "Probable Person to Person Transmission of Avian Influenza A H5N1," *N Engl J Med*, 2005, 352:333-40.
25. Avian Influenza - Pandemic Preparedness, Mass. Dept of Health Projections, March 2006.
26. Silka A, Geiderman JM, Goldberg JB, et al, "Demand on ED Resources During Periods of Widespread Influenza Activity," *Am J Emerg Med*, 2003, 21:534-9.
27. McFee RB, "Preparing for an Era of Weapons of Mass Destruction (WMD): Are We There Yet? Why We Should All Be Concerned. Part I." *Vet Hum Toxicol*, 2002, 44(4):193-9.
28. Zigmund J, "No More Room, Overcrowding Blamed for Ambulance Diversions," Modernhealthcare.com, Feb 13, 2006, www.modernhealthcare.com/printwindow.cms?articleId=38689&pageType-article (Last accessed 3/11/06).
29. "Improving Influenza Vaccination Rates in Health Care Workers: Strategies to Increase Protection for Workers and Patients," National Foundation for Infectious Diseases, 2004, Washington, DC.
30. Highly Pathogenic Avian Influenza (HPAI) Interim Infection Control Guidelines for Health Care Facilities - World Health Organization, 18 February 2004.
31. Langmuir AD, "Changing Concepts of Airborne Infection of Acute Contagious Diseases; A Reconsideration of Classic Epidemiologic Theories," *Airborne Contagion*, Kundsin RB (ed), New York, NY: Annals of the New York Academy of Sciences, 1980, 353:35-44.
32. Bridges CB, Kuehnert MJ, and Hall CB, "Transmission of Influenza: Implications for Control in Health Care Settings," *Clin Infect Dis*, 2003, 37:1094-111.
33. "Legionella Pneumophilia," *Infectious Diseases*, Lexi-Comp Online, www.online.lexi.com/crlsq/servlet/crlonline?a=doc&bc=idh&id=121387&mid=45264&mn (Last accessed 5/20/06.
34. Murdoch DR, "Diagnosis of Legionella Infection," *Clin Infec Dis*, 2003, 36(1):64-9.
35. HHS, FDA Approves New Laboratory Test to Detect Human Infections With Avian Influenza A/H5 Viruses, 2006.
36. Knight V and Gilbert BE, "Ribavirin Aerosol Treatment of Influenza," *Infect Dis Clin North Am*, 1987, 441-57.
37. Madren LK, Shipman C JR, and Hayden FG, "*In vitro* Inhibitory Effects of Combinations of Anti-influenza Agents," *Antivir Chem Chemother*, 1995, 6:109-13.
38. Smith SM, "Where Have You Been? The Potential to Overlook Imported Disease in the Acute Setting," *Eur J Emerg Med*, 2005, 12(5):230-3.

39. Stienlauf S, Segal G, Sidi Y, et al, "Epidemiology of Travel-related Hospitalization," *J Travel Med*, 2005, 12(3):136-41.
40. Ver Herck K, Van Damme P, Casttelli F, et al, "Knowledge, Attitudes and Practices in Travel-related Infectious Diseases: The European Airport Survey," *J Travel Med*, 2004, 11(1):3-8.
41. World Health Organization: Algorithm for the Management of Returning Travelers From Countries Affected by H5N1 Presenting With Febrile Respiratory Illness: Recognition, Investigation, and Initial Management," http://www.hpa.org.uk/infections/topics_az/influenza/avian/algorithm.htm (Last accessed 5/20/06).
42. Bartlett JG, "Planning for Avian Influenza," *Ann Intern Med*, 2006, 145(2):141-4.

Botulism

Britney B. Anderson, MD and Mark B. Mycyk, MD

Department of Emergency Medicine, Northwestern Memorial Hospital, Chicago, Illinois

Introduction

Botulinum toxin is the most lethal biological substance known. Botulinum toxin has a known median lethal dose (LD_{50}) of 1 nanogram of toxin per kilogram body mass, thus doses as small as 0.05 to 0.1 micrograms can cause death in humans. Even though botulinum toxin has been feared for its deadly effect on humans, this toxin (among other neurotoxins) has recently been identified as a useful treatment modality for many medical conditions in fields such as ophthalmology, neurology, and dermatology. Botulinum toxin has become so popular now that it is universally recognized by the general public as a cosmetic enhancement tool better known as Botox® in North America or Dysport® in the UK.

The development and use of botulinum toxin as a bioweapon began about 6 decades ago. In the 1930s in Manchuria, the head of the Japanese biological warfare group (Unit 731) fed cultures of *C. botulinum* to prisoners resulting in fatalities. The United States bioweapon program first produced botulinum toxin during World War II. The toxin was also investigated as a biologic agent by the British, Canadian, Japanese, and Soviet military. Some terrorist groups have been successful obtaining the toxin. During World War II, Paul Fildes, a high-ranking British specialist in bacterial weapons, alluded to the fact that by utilizing the toxin, he contributed to assassination of Reinhard Heydrick, the head of the Gestapo. After concerns that the Germans had weaponized botulinum toxin, doses of botulinum toxoid vaccine were made for Allied troops fighting Germany.

Botulinum toxin was one of several agents tested on Vozrozhdeniye Island by Soviets at their Aralsk-7 site. In the early 1990s, before the Sarin gas attack on the Tokyo subway system, the Japanese cult Aum Shinrikyo had released a *C. botulinum* preparation in Japan targeted at a U.S. military installation, but the toxin was ineffectively produced. After the 1991 Persian Gulf War, Iraq told the United Nations inspection team that they had produced 19,000 L of botulinum toxin, and 10,000 L of this was loaded into military weapons. This was 3 times the amount needed to kill the entire human population by inhalation. Iraq has chosen to weaponize more botulinum toxin than any other of its known biological agents[1].

Pathophysiology

Clostridium botulinum is a heterogeneous group of anaerobic, gram-positive, rod-shaped organisms that forms subterminal spores, which elaborate the most potent bacterial toxin known. In addition to *C. botulinum*, *Clostridium baratii* and *Clostridium butyricum* also have the capacity to produce botulinum toxin.

There are 8 known serotypes of botulism designated A-G (A, B, C1, C2, D, E, F, and G). Almost all human cases of botulism have been caused by serotypes A, B, and E. There are 6 forms of botulism that have been reported: food-borne, wound, infant-intestinal, adult-intestinal, inadvertent injection-related, and inhalational botulism.

Upon entering the body, botulinum toxin finds its way to systemic circulation and is transported to sites of acetylcholine-mediated neurotransmission such as neuromuscular junctions, postganglionic parasympathetic nerve endings, and peripheral ganglia. The active neurotoxin is a large 150-kDa molecular mass, which is incapable of crossing the blood-brain

147

barrier, therefore the central nervous system is not involved. The toxin consists of a heavy chain (100-kDa fragment responsible for neurospecific binding and translocation in the nerve cell) and a light chain (50-kDa fragment responsible for intracellular catalytic activity). The toxin binds to the neuronal cell membrane on the nerve and is taken up by endocytosis into the presynaptic nerve ending. Once inside the neuron, after reduction of a single disulfide bond, the light chain is cleaved free and acts as a zinc-dependent protease, which can attack SNARE proteins (soluble N-ethylmaleimide sensitive fusion protein attachment receptor). It is the SNARE protein complex, which is the key process that allows vesicles in the nerve terminus, which contain acetylcholine, to fuse with the neural cell wall and be released into the synaptic cleft. By preventing the release of acetylcholine, muscular contraction cannot occur and flaccid paralysis results. This process renders this neural tissue nonfunctional and recovery occurs only as new neural tissue endplates are regenerated[2].

Forms of Botulism Illness

Food-Borne Botulism In the United States, food-borne botulism results from the ingestion of preformed toxin (type A and B being most common). It usually occurs from exposure to home-canned foods, most commonly vegetables, fruit and condiments that are improperly preserved or undercooked. Type E outbreaks are frequently associated with fish products. Most cases are initially misdiagnosed, because the early GI symptom complex of nausea, vomiting, and diarrhea is similar to that of other food-borne illnesses.

Infant Botulism Infant botulism (sometimes referred to as infant-intestinal) is the most commonly reported form of the illness. Most cases occur from one week to one year of age and are caused by ingestion of spores. The immature gastrointestinal tract of infants allows the spores to germinate leading to *in vivo* production of toxin. Honey ingestion, and to a lesser extent, corn syrup and environmental exposures such as living in a rural area, or having a parent who works with soil, have been identified as important risk factors for infant botulism[3].

Wound Botulism Wound botulism occurs when a wound is contaminated with spores, which then germinate and produce toxin. Wound botulism has been documented after traumatic injury involving contamination with soil, and after cesarean delivery. Wound botulism from type A and B toxins have also recently become a frequent complication in injection drug users, particularly in those injecting black tar heroin. In 1997, California reported 99 wound botulism cases seen in IV drug users[4].

Adult-Intestinal Botulism Adult-intestinal botulism (also classified as undefined or adult-type infant botulism) is a delayed-onset neurologic syndrome. It mainly occurs in adults with abnormal gastrointestinal pathology in GI disease such as Crohn's disease, peptic ulcer disease, or GI surgery such as Billroth surgery[5].

Inadvertent Botulism Inadvertent (injection-related) botulism is an iatrogenic form associated with patients who have been treated with injections of pharmaceutical botulinum toxin[6]. In Florida, a few individuals were the result of unlicensed persons running a cosmetic clinic. They did not use medical grade, appropriately prepared botulinum toxin. Because of the increased prevalence of botulinum use in the medical and cosmetic settings, it is likely that other iatrogenic cases will occur.

Inhalational Botulism Inhalational botulism can occur through the deliberate release of the botulinum toxin as a biologic weapon.

Confirming the Threat and What to Expect

The early recognition of a botulism outbreak is based upon rapid diagnosis, and is essential to mobilizing vital resources and obtaining appropriate treatment in bioterrorist attacks. The history and motor exam are sufficient to make a presumptive diagnosis and will justify the release of botulism antitoxin from the CDC. There are confirmatory laboratory tests for botulism, but the treatment must be started before the tests are resulted, as they can take 1 to 2 days for preliminary results. Any case of botulism should raise red flags about the possibility of bioterrorism. A consensus statement published by the *Journal of the American Medical Society* in 2001 listed specific features of an outbreak that would suggest the deliberate release of botulinum toxin. These included: 1) the outbreak of a large number of cases of acute flaccid paralysis with prominent bulbar palsies, 2) an outbreak of an unusual botulinum toxin type (ie, type C, D, F, or G, or type E toxin not acquired from an aquatic food), 3) an outbreak with a common geographic factor among cases but without a common dietary exposure, and 4) multiple simultaneous outbreaks with no common source.

Bioterrorism

Aerosol Dissemination   Aerosol dissemination and food-borne botulism would be the most likely forms of botulism seen in a bioterrorist attack. Cases of botulism acquired by aerosol dissemination would not be difficult to recognize because a large number of the cases would share a common geographical and temporal exposure. Given the ease of travel and mobility in this day and age, a careful history about recent travel, locations as well as activities and occupation would be important. Not much is known about the rapidity of onset of aerosolized botulism, because so few cases have been reported. Some studies on primates showed the onset of signs and symptoms of inhaled botulism from 12-80 hours after the exposure depending on the dose. The only human cases of inhaled botulism known were in laboratory workers handling a deceased animal that had died of botulism. The onset of symptoms in this case was approximately 72 hours after exposure to an unknown amount of re-aerosolized toxin. If a deliberate dispersal of botulinum toxin were to occur, there would likely be an upper airway prodrome such as a cold (without a fever) followed by a variable onset of differing degrees of paralysis in the exposed[2].

Food-Borne Botulism   Food-borne botulism outbreaks would be slightly more difficult in that there would be a need to discern between naturally occurring food-borne disease, and deliberate food-borne disease unless numerous patients with similar symptoms presented. The onset of symptoms of food-borne botulism typically present within 10-72 hour range after the meal. The severity of the disease, and rapidity of its onset depend on the amount of toxin ingested. Botulinum toxin is colorless, tasteless, and odorless, and is readily inactivated by heat. Therefore, food-borne botulism is always transmitted by foods that are not heated (or improperly heated) before eating. On the other hand, spores can be dormant, resistant to heat and germinate in low acidity, low salinity environments. Commonly implicated foods include canned vegetables, particularly beans, corn, peppers, and other low acidic vegetables. Items preserved in garlic oil and soups have often been implicated as well. Restaurant outbreaks in the past have been from salads and dips, condiments such as garlic in oil[7], sautéed onions,[8] and other commercial foods like yogurt[9]. Food is a convenient, ubiquitous, and relatively unnoticed vehicle in which to hide the lethal botulism agent.

Water-Borne Botulism   There have been no reports of cases of water-borne botulism in the past, and water-borne botulism is speculated to be an unlikely scenario. The standard treatment of potable water using chemicals like chlorine, and aeration are thought to inactivate the toxin. However, untreated beverages may ensure stability of botulism for days, and should be suspected as a source of contamination if no other source can be identified.

Personal Protective Equipment/Decontamination There are no specific guidelines for detoxification of patients exposed to botulism except that clothing and skin should be washed with soap and water. Botulism degrades easily and decays at 1% per minute, therefore, substantial inactivation of toxin occurs by 2 days after the aerosolization. All contaminated surfaces should be cleaned with bleach solution if they cannot be avoided for the hours to days required for natural degradation[10]. Victims of botulinum toxin are not contagious, and upon admission to the hospital, patients do not need to be placed in any type of isolation environment.

Although symptomatic patients will likely arrive long after their initial exposure to botulinum toxin, activated charcoal should be considered for decontamination of any food-borne or intestinal exposure. Activated charcoal has been shown to adsorb botulinum toxin A *in vitro*. Wound debridement should be considered for patients with wound exposure.

Clinical Manifestations

Despite different routes of infection, all adult forms of botulism present similarly. Onset of symptoms after exposure is variable, but with food-borne botulism symptoms usually occur within 1 to 5 days after ingestion of the contaminated food but may be longer after inhalation exposure depending upon the dose and type of toxin. Botulism can be recognized by its classic triad: 1) symmetric, descending flaccid paralysis with prominent bulbar palsies in 2) an afebrile patient with 3) a clear sensorium[1].

The paralysis seen in botulism is an acute, symmetric descending flaccid paralysis that begins in the bulbar musculature. Usually the bulbar palsies present first as paralysis of the motor functions of the cranial nerves. Initially botulism affects the oculomotor muscles, progressing to facial muscle paralysis and to the muscles of mastication, and swallowing. Prominent neurologic findings in all forms of botulism include ptosis, dry mouth, and the "4Ds": diplopia, dysarthria, dysphonia, and dysphagia. Unlike nerve agent exposure, which usually results in miosis, botulinum can cause mydriasis in approximately 50% of cases. The mucous membranes of the mouth, tongue, and pharynx often appear erythematous and dry due to peripheral parasympathetic cholinergic blockade. The gag reflex is often depressed or absent. As paralysis extends beyond the bulbar musculature, patients experience symmetric descending muscular weakness. Neck muscles, respiratory muscles, and those in the upper and lower extremities are affected. Upper extremities are usually more affected than lower extremities, and proximal muscles weaker than distal. The abdomen may appear distended with hypoactive or absent bowel sounds; ileus may develop. Bladder distension may indicate urinary retention. Eventually the muscles of respiration, including accessory muscles and the diaphragm are affected. Respirations may become rapid and shallow. Once paralysis progresses to respiratory failure, the patient will need ventilator assisted support until the motor end-pate recovers, often for several months.

A hallmark of the disease is the absence of sensory symptoms and the absence of fever. Sensory examination is normal throughout the disease. The only sensory changes observed are infrequent circumoral and peripheral paresthesias from hyperventilation.

Differential Diagnosis

There is a vast list of metabolic and neurologic causes of motor neuropathies, and the differential diagnosis of botulism includes a wide variety of illnesses [Table from AMA BT book]. Two diagnoses in particular seem to be difficult in differentiating from botulism. Myasthenia gravis (MG) often presents with ptosis and is a disease involving bulbar palsies without sensory findings. Due to the subjectivity of the edrophonium test, it is often determined positive in cases thought to be MG, but which turn out to be botulism. However

MG is also associated with recurrent paralysis, has different electromyographic (EMG) findings, and has a sustained response to anticholinesterase therapy unlike botulism.

Another clinical mimic of botulism is the Miller-Fisher variant of the Guillain-Barre syndrome (GB). This disease also has been known to present with diplopia, ophthalmoplegia, ptosis, facial and extremity weakness. However, and this condition is usually associated with a history of antecedent infection, **ascending paralysis**, paresthesias, and early areflexia. On CSF examination, an increase in protein content of the fluid exists, which is characteristic of GB.

Other toxins must also be considered as possibilities. Anticholinergics (atropine, jimson weed, belladonna) cause papillary dilation and erythematous, dry mucous membranes. Aminoglycosides have been known to cause neuromuscular blockade. Curare toxin, which contains alkaloids that affect neuromuscular transmission, can cause paralysis without change in mental status. Tick paralysis and hypokalemic periodic paralysis can also cause symmetric myopathies, but tend to have less of an effect on the cranial nerves, and predominantly affect proximal large muscles.

Laboratory Studies

Lab testing is confirmatory. Presumptive diagnosis is based upon clinical presentation. The diagnostic laboratory tests for botulism are only available at the CDC [the CDC Director's Emergency Operation Center at 770-488-7100] and some state and municipal public health laboratories. Samples must be collected and sent to one of these specific locations. Samples include serum, stool, gastric aspirate, and suspect foods. The standard laboratory diagnostic test is the mouse bioassay. Mice, which have been pretreated with type-specific antitoxin, are subjected to the sample and the toxins therein. The bioassay can detect as little as 0.03 ng of botulinum toxin. Fecal and gastric specimens can also be cultured anaerobically and results of these cultures usually take 7-10 days to come available. Although serum samples must be obtained before therapy with antitoxin, because the antitoxin can negatively affect the mouse bioassay, this is not a reason to delay treatment. Sterile water should be used for enemas to obtain stool samples, as other solutions can confound the mouse bioassay.

Adjunct Studies

Botulism has characteristic findings on electromyographic (EMG) studies. An EMG with repetitive nerve stimulation at 20-50 Hz can show normal nerve conduction velocity, normal sensory nerve function, and a pattern of short, small-amplitude motor potentials. The most distinctive pattern is an incremental response to repetitive stimulation often seen only at 50 Hz. Additionally, other studies can be utilized to help rule out other diseases on the differential. Cerebral spinal fluid is unchanged in botulism but is abnormal in many CNS diseases. Other intracranial or spinal pathology such as hemorrhage or neoplasm can be ruled out with neuroimaging techniques such as CT scans and MRIs. Simple laboratory tests can rapidly rule out metabolic causes such as hypoglycemia, hypothyroidism, and hypokalemia.

Early Interventions

Supportive Care

The morbidity of the disease can be severe and usually results from respiratory paralysis. Therefore, treatment should begin as early as possible once botulism is suspected based on history and physical exam. The respiratory status must be closely monitored. Adequacy of gag and cough reflexes should be routinely assessed and close control of oral secretions is beneficial. Monitoring parameters such as oxygen saturation, vital capacity, and negative inspiratory force (NIF) are helpful. Reverse Trendelenburg at 20° to 25° with cervical support may improve ventilation by reducing oral secretions, which can pass into the airway, and suspending some of the weight of the abdominal viscera from the diaphragm. Controlled, early anticipatory intubation is indicated when respiratory function deteriorates. The proportion of patients who require mechanical ventilation varies depending on the outbreak. In a large outbreak of botulism, the need for mechanical ventilators, and critical-care beds could easily exceed the capacity of local resources and would require outside resource allocation.

Antibiotics may be necessary in treating complications from the therapies recommended above, such as nosocomial infections. It should be noted that aminoglycosides and clindamycin can exacerbate neuromuscular blockade and should be avoided unless the clinical benefit outweighs the risk[11]. Maintaining adequate hydration, nasogastric suctioning for ileus, bowel, and bladder care in addition to prevention of decubitus ulcers and deep venous thrombosis.

Botulism Antitoxin

Therapy includes passive immunization with equine antitoxin and supportive care. The antitoxin is available in the U.S. through the CDC via state and local health departments. Often a direct call to the CDC Assistance line will expedite the release of antitoxin. These numbers should be prominently displayed and regularly checked in case they are changed. Early administration of this passive neutralizing antibody is important, as the antitoxin can only neutralize circulating toxin; it cannot reverse the existing paralysis, but does minimize further damage to the nerves.

The antitoxin is a trivalent form, which is active against serotypes A, B, and E. It contains 7500 International Units (IU) of type A, 5500 IU of type B, and 8500 IU of type E antitoxin[12]. Monovalent type E antitoxin is available in Alaska and Canada. Current dosing recommendations for the trivalent antitoxin are one vial (10 mL) diluted 1:10 in normal saline intravenously over 30 to 60 minutes. The package insert should be reviewed with public health authorities before using the product, as dosing recommendations can change. The amount of neutralizing antibody in the antitoxins far exceeds the serum toxin levels seen in food-borne botulism patients. In an instance where a patient could have been exposed to high levels of toxin, such as a bioterrorist attack, the adequacy of neutralization can be confirmed by retesting serum for toxin after treatment. The antitoxin is a horse-derived immunoglobin and has potential to cause hypersensitivity reactions. There are few published data on the safety of the antitoxin, but a review from 1967-1977 done on 268 patients cites the relative risk of hypersensitivity reaction to be 9%[14]. Anaphylaxis occurred in 7 patients in this study (2.6%), and a delayed onset serum sickness occurred in 10 (3.7%). Notably, 228 patients (85%) received more than 1 vial of antitoxin as was the common treatment regimen at that time. To screen for hypersensitivity reactions, a skin test is recommended before receiving a full dose of the antitoxin. Patients that have a reaction to the skin test may be desensitized over 3 to 4 hours before full infusion occurs. Any decision to give the skin test must weigh the risk and benefits of delayed vs safe treatment. Neither a positive nor a negative skin test can predict or exclude an acute allergic reaction. Diphenhydramine and epinephrine should always be readily available during administration and the clinician should regularly monitor the patient for an adverse reaction. A despeciated heptavalent (A, B, C, D, E, F, and G) antitoxin is available and held by the U.S. military[13,15]. However, 4% of horse antigens remain so there is still a risk for hypersensitivity.

Special Considerations

There is limited information regarding the treatment of botulism in pregnancy, immunocompromised patients, and in children. There are case reports of both children and pregnant women receiving the antitoxin, and the risk to the fetus is unknown.

In a biological warfare event, the prophylactic immunization of designated individuals may be necessary. A pentavalent vaccine exists, and is currently used as pretreatment in laboratory workers and military personnel who are at risk for botulism exposure. The vaccine is active against serotypes A, B, C, D and E[16].

Disposition

All patients with significant botulism exposure need to be admitted to the hospital for observation. The proportion of patients who require mechanical ventilation varies depending on the outbreak. In a large outbreak of botulism, the need for mechanical ventilators, and critical-care beds could easily exceed the capacity of local resources and would require outside resource allocation. Unfortunately, patients who develop respiratory compromise and require ventilatory support may require hospitalization for months. Emotional support and behavioral health care will be a necessary part of therapy.

Forensic Issues

Whenever botulism is suspected, immediate contact with the CDC and the local health department will speed access to diagnostic and laboratory services and enhance surveillance in the early stages of a mass outbreak. If terrorism is suspected, law enforcement will become involved. Attention to the chain of evidence is a consideration.

References
1. Arnon SS, Schechter R, and Inglesby TV, "Botulism Toxin as a Biological Weapon: Medical and Public Health Management," *JAMA*, 2001, 285:1059-70.
2. Horowitz BZ, "Botulinum Toxin," *Crit Care Clin*, 2005, 21:825-39.
3. Fox CK, Keet CA, and Strober JB, "Recent Advances in Infant Botulism," *Pediatric Neurology*, 2005, 32(3):149-54.
4. Werner SB, Passaro D, McGee F, et al., "Wound Botulism in California, 1951-1998: Recent Epidemic in Heroin Injectors," *Clin Inf Dis*, 2000, 31:1018-24.
5. Shapiro RL, Hatheway C, and Swerdlow DL, "Botulism in the United States: A Clinical and Epidemiologic Review," *Ann Intern Med*, 1998, 129:221-8.
6. Bakheit AO, Ward CD, McLellan DL, "Generalized Botulism-like Syndrome After Intramuscular Injections of Botulism Type A: A Report of Two Cases," *J Neurol Neurosurg Psych*, 1997, 62:198.
7. Townes JM, Cieslak PR, Hatheway CL, et al, "An Outbreak of Type A Botulism Associated With a Commercial Cheese Sauce," *Ann Intern Med*, 1996,125:558-63.
8. MacDonald KL, Spengler RF, Hatheway CL, et al, "Type A Botulism From Sautéed Onions: Clinical and Epidemiological Observations," *JAMA*, 1985, 253:1275-8.
9. O'Mahony M, Mitchell E, Gilbert RJ, et al, "An Outbreak of Foodborne Botulism Associated With Contaminated Hazelnut Yoghurt," *Epidemiol Infect*, 1990, 104:389-95.
10. Siegel LS, "Destruction of Botulinum Toxin in Food and Water," in Hauschild AH and Dodds DL (eds), *Clostridium Botulinum: Ecology and Control in Foods*, New York, NY: Marcel Dekker Inc, 1993:323-41.
11. Santos JI, Swensen P, and Glasgow LA, "Potentiation of *Clostridium botulinum* Toxin by Aminoglycoside Antibiotics: Clinical and Laboratory Observations," *Pediatrics*, 1981, 68:50-4.

12. Tacket CO, Shandera WX, and Mann JM, "Equine Antitoxin Use and Other Factors That Predict Outcome in Type A Food Borne Botulism," *Am J Med*, 1984, 76:794-8.

13. Hibbs RG, Weber JT, Corwin A, et al, "Experience With the Use of an Investigational F(ab')2 Heptavalent Botulism Immune Globulin of Equine Origin During an Outbreak of Type E Botulism in Egypt," *Clin Infec Dis*, 1996, 23:337-40.

14. Black RE and Gunn RA, "Hypersensitivity Reactions Associated With Botulinal Antitoxin," *Am J Med*, 1980, 69:567-70.

15. Kortepeter M, *USAMRIID's Medical Management of Biological Casualties Handbook*, 4th ed, 2001, Fort Detrick, MD, 118-27.

16. Metzger JR and Lewis LE, "Human-Derived Immune Globulins for the Treatment of Botulism," *Rev Infect Dis*, 1979, 1:689-92.

Smallpox (Variola) and Poxviruses

R. B. McFee, DO, MPH, FACPM

"The single greatest threat to man's continued existence on earth is the virus." - Joshua Ledeberg, Nobel Scientist

Poxviruses

The poxvirus family is a diverse group of viruses that affect humans and animals. Their nucleosome contains double-stranded DNA; they are among the largest of all animal viruses (200-320 nm) and can be visualized with light microscopes. They preferentially infect skin epithelial cells, replicate in cell cytoplasm, and may produce eosinophilic cytoplasmic inclusion bodies, ultimately cause toxic effects on cells. Among the poxviruses known to affect humans are from the genera Orthopoxvirus, Parapoxvirus, Molluscipoxvirus, and Yatapoxvirus, of which the Orthopoxvirus variola (smallpox) is the most deadly to humans. Different poxviruses are capable of producing a localized, self-limited infection by inoculation to skin such as orf (contagious pustular dermatitis), or systemic disease such as variola (smallpox). Other poxviruses cause localized cell proliferation, that is, molluscum contagiosum. Most cases of poxvirus infection are occupational and lead to few cutaneous lesions. However, smallpox and monkeypox can cause serious illness and death. Therefore, from a bioweapon perspective, orthopoxviruses are of greatest interest because they include smallpox, monkeypox, and vaccinia cowpox viruses; the former 2 are capable of significant illness while the latter has been utilized as the basis for vaccination against smallpox. Some preparedness experts express concern that the Orthopoxvirus Camelpox poses a risk because it is genetically similar to smallpox, not uncommon and, owing to its size, may be amenable to bioweapon engineering. Infection with one Orthopoxvirus confers protection against other members of the genera.

Vaccinia

Vaccinia is an Orthopoxvirus affecting a wide variety of vertebrate hosts, is among the most widely studied virus, and, in 1796, was the first countermeasure developed against smallpox. As a laboratory isolate, this virus does not cause serious disease in immunocompetent humans. Generalized vaccinia viremia can occur 6-9 days after vaccination with full recovery expected. Other vaccine adverse events can occur; the most common is autoinoculation whereby the patient touches the vaccine site, then rubs the eyes without washing hands first. Genetically engineered and attenuated recombinant viruses are being studied as safer alternatives to vaccinia especially for immunocompromised individuals but have not been FDA-approved as of this writing.

If adverse events occur, especially potentially life-threatening disseminated vaccinia, cardiac sequela associated with vaccinia vaccine, vaccinia immune globulin (VIG) can be administered.

Monkeypox

In 2003, monkeypox appeared for the first time in the Western Hemisphere when several individuals in Midwestern United States became ill after contact with prairie dogs and Gambian giant rat. Importation of exotic animals remains a potential source of human infection with diseases not common to North America. Monkeypox is an Orthopoxvirus that usually affects primates but has infected humans, squirrels, and other animals. According to the World Health Organization, humans usually contract monkeypox through contact with an infected animal's blood, body fluids, or lesions, or by being bitten. Secondary transmission - human to human transmission after close contact with an infected human who was infected by an animal exposure - is approximately 9% based upon cases in the Congo. WHO cautions that outbreaks in one part of the world may not behave similarly to other regions. According to WHO experts, monkeypox is clinically indistinguishable from monkeypox that generally causes less serious illness. Monkeypox does produce lymphadenopathy and skin lesions that occur in crops. Vaccinia vaccination confers protection. The smallpox vaccine was about 85% effective in preventing human monkeypox during an African outbreak. Vaccination is recommended for those caring for monkeypox patients; vaccination can be given up to 14 days after exposure. Unvaccinated children are most vulnerable and deaths have resulted from infection with monkeypox. Large outbreaks have occurred and controversy as well as concern persists surrounding the actual ability for person-to-person transmission. It appears to have a lower overall secondary attack rate than smallpox. The recent appearance of monkeypox in the U.S. underscores the problem of dissemination of pathogens outside their endemic area as well as the threat of animal importation in addition to the emergence of travel-related illness and global biological threats. Moreover, the importance of early diagnosis cannot be overstated, not only to contain a potential outbreak but because of the impact upon immunosuppressed individuals. Monkeypox can be accidentally transmitted to humans and lead to small epidemics, even death. Given the relatively more common nature of monkeypox, its availability raises concern about the criminal introduction of this poxvirus, which, given its transmissibility through prairie dogs as well as person-to-person, makes it a potential bioweapon.

Clinical Monkeypox can cause a febrile illness with vesiculopustular eruptions. Presumptive diagnosis can be made based upon clinical presentation and immunohistochemical evidence of poxvirus infection in skin lesion tissues. Virus can be recovered in cell cultures. Monkeypox-specific DNA sequences can be used to identify the virus. ELISA results correlate with virologic PCR and viral culture results. If a poxvirus is suspected, respiratory precautions must be initiated immediately. In addition, local public health must be notified to initiate a CDC response, including advice and access to advanced laboratory capabilities, as well as countermeasure support that may include antivirals, immunoglobulin, and vaccines.

Presumptive diagnosis is made based upon clinical presentation, history of illness, and epidemiology.

Management Limited data exist on the effectiveness of smallpox vaccination for preventing monkeypox. Data suggest pre-exposure smallpox vaccination is highly effective at >85% in protecting persons exposed to monkeypox from developing disease. The effectiveness of postexposure vaccination requires further study. Data suggest smallpox vaccination following exposure to smallpox is effective at preventing or ameliorating disease; given the similarity among orthopoxviruses, smallpox vaccination should confer similar benefit against monkeypox. Given the mortality rate from monkeypox ranges from 1% to 10%; the risk of death from smallpox vaccine is approximately 1-2 per million vaccines. It is important to carefully screen potential vaccines for precautions and contraindications (Table). Persons without a successful vaccine "take" should be revaccinated within 2 weeks of the most recent exposure to monkeypox.

Rash illnesses suspected to be monkeypox should be confirmed by laboratory evaluation; such labs should have the capability to test varicella, vaccinia, and other similar viral infections.

No data are available concerning the benefit of VIG as treatment of monkeypox complications. It is unknown if a patient with severe monkeypox infection will thus benefit from treatment with VIG; its use, however, might still be considered. VIG can be considered for prophylactic use in persons exposed to monkeypox who have severe T-cell function immunodeficiency where smallpox vaccine would be contraindicated. VIG could be obtained under an Investigational New Drug (IND) protocol. Physicians should contact their State Health Department; subsequently referred to the CDC Clinical Information Line: 1-877-554-4625.

Cidofovir has demonstrated anti-monkeypox viral activity in animal and *in vitro* studies. The benefit to patients with severe infection is unknown. Nevertheless, one should consider using Cidofovir in such instances. Realizing the toxicity potential of Cidofovir, safety considerations suggest it should be utilized as a treatment for severe monkeypox but not as prophylaxis. It is licensed for use in the treatment of cytomegalovirus (CMV) retinitis in HIV patients. However, if it was used in the treatment of monkeypox, vaccinia vaccine, or smallpox, it would be under the restrictions of an Investigational New Drug (IND) protocol. Clinical consultation is available concerning the use of VIG and Cidofovir from the CDC at 1-877-554-4625.

Smallpox (Variola)

The threat of Variola virus (smallpox) - the leading cause of infection-related death in history - has returned, and once again smallpox has become a household word. Smallpox infections have been recognized as far back as ancient Egypt, affecting pharaohs and peasants. Unlike decades ago when smallpox was a naturally occurring infection, concerns are emerging that smallpox may become intentionally spread as a bioweapon. Intelligence sources suggest Iraq and other terrorist-friendly nations may have access to or already weaponized Variola. Rumors persist that scientists in the former Soviet Union were trying to develop an Ebola/Variola hybrid virus. Smallpox is an Orthopox virus, which consists of the infectious agents that cause camelpox, smallpox, monkeypox, and cowpox. Immunity to one member of this family confers protection against others, hence the use of vaccinia (cowpox) virus to immunize against Variola. Vaccinia is a virus with minimal pathogenicity except in immune-compromised patients. Smallpox is a painful, disabling, febrile disease.

Variola occurs as primarily two strains - variola major (smallpox) and the less pathological version variola minor - which causes a milder febrile rash, and is sometimes referred to as alastrim. Smallpox presents a particularly serious risk because of a high case-fatality rate estimated at 30 percent to 50 percent depending upon the strain, naive vaccine status of the exposed population and inconsistent availability of appropriate healthcare to prevent secondary bacterial infection. It is a highly pathogenic virus; it is estimated <10-100 smallpox virions is necessary to cause human illness! Bioterrorism experts consider smallpox to be a significant threat because it is relatively easy to produce once starting material is obtained, the aerosol infectivity - contagiousness, widespread susceptibility of nonimmunized or under-immunized populations. Contagion results from transmitting virus in airborne droplets or by contact with lesions. Primary portal of entry is the respiratory tract so population density, overcrowding, as well as immune status affect the extent of spread. No significant subclinical carrier state is believed to exist, nor are there known animal reservoirs. Infected people are contagious from the onset of illness until the last crusts of the dermal lesions are gone. The smallpox virus is relatively resistant to environmental conditions; infections have occurred from Africa to North America. The last case of smallpox occurred during the 1940s in the U.S., and the last naturally acquired case occurred in Somalia in 1977. Most clinicians have never seen this illness, so experienced diagnosticians are uncommon. Routine vaccination stopped almost 30 years ago.

Initial presentation Smallpox progresses through 3 phases: incubation, prodrome, and pox. The incubation period on average is 7-17 days postexposure. The individual is asymptomatic and not considered contagious during this period. Typically, smallpox victims will then experience a prodrome characterized by fever (100 to 105 degrees F), vomiting, headache, bodyache, or backache; this period lasts for several days (usually 2-4) and the patient is extremely ill. Some consider this stage potentially contagious.

Appearance of a rash marks the third stage; this period lasts from 14-17 days. The rash appears as small red dots in the mouth and on the tongue then develop into sores (oropharyngeal enanthem) that rupture and release virus into the upper respiratory system. This is the most contagious stage. Patients will be prostrated and may continue to complain of severe headache, chills, backache, malaise, nausea, and vomiting. At about the same time as the oral lesions rupture, the typical skin rash, the exanthem, usually appear, starting most often on the hands and face, spreading to the arms and legs then to the feet. The lesions evolve from macules to papules to vesicles to pustules and finally the lesions crust, which may take 1-2 weeks. All the lesions develop in a similar stage at any given time often referred to as "synchronous" development. The initial lesions usually appear on the palms of the hands, soles of the feet, feel firm, nodular to palpation. The distribution of smallpox lesions is centrifugal-extremity first then trunk in contrast to the most similar disease in the differential diagnosis, chickenpox, whereupon the rash starts centrally on the trunk and works toward the periphery (Figure). Smallpox lesions often are on the extensor surfaces, are deep with firm multiloculated vesicles. After the crusting phase, scarring occurs and may cause severe disfigurement. Patients who have oral lesions are contagious, as are those who develop skin rashes. Dentists and others who may be called upon to examine complaints of a febrile illness and mouth sores should be sensitized to the potential of smallpox. Patients are contagious until ALL scabs have fallen off.

Types of smallpox illness There are 4 types of smallpox illness identified by the pattern of rash.

Classic Smallpox

Smallpox lesions in the classic presentation are deep and dermal. The lesions are synchronous in their development - whatever location of the body a cluster of lesions appears on, they appear similar. The lesions progress from macula, papule, vesicle, pustule (pox), and ultimately to scabs. The time from exposure to symptoms ranges from 7 to 17 days. Chickenpox (Varicella) lesions are asynchronous in development - some are umbilicated, others are newly emerging, while the remainder may be ready to fall off. Varicella lesions are often found on flexor surfaces. Unlike smallpox, chickenpox rarely affects the soles and palms. Other illnesses that may simulate chickenpox on cursory appearance include coxsackievirus, leprosy, and syphilis. Monkeypox also resembles smallpox.

The key distinguishing features between each illness is the history and differences in the rashes. Coxsackievirus usually affects adolescents and young adults, the patients do not appear ill, and the rash is superficial not dermal. Leprosy lesions are more confluent except in the tuberous version, with possible neurologic symptoms, a clear history of the ailment. Again, especially if the patient has been treated, he/she will not appear ill.

Smallpox patients will appear very sick and prostrated. Syphilis patients will have lesions that tend to remain localized on the palms and soles, which may recollect a history of prior genital illness. Again such patients usually do not appear significantly ill. Rickettsialpox, a mite borne disease that belongs to the spotted fever group of rickettsioses is not a poxvirus. However, it can cause fever and a papulovesicular eruption. Although the associated dermal manifestation may somewhat mimic smallpox, clinically it often causes an asymptomatic vesicle that rapidly ulcerates to become an eschar. Moreover, rickettsialpox is an uncommon disease that occurs in urban populations in the Eastern U.S. as well as globally. Patients often complain of fever and malaise, including headache and fatigue. Patients generally do not appear sick, unlike smallpox patients who often are prostrated and appear quite ill. Also, unlike smallpox, which generally results in multiple lesions, localized rickettsialpox has a primary lesion developing at the bite site within 48 hours that may result in lymphadenopathy and one or two eschars that resolve within 4 weeks. Generalized cutaneous eruptions have occurred, however, and can appear on the face, trunk, and extremities with no specific sequence nor involvement of palms and sole, bearing greater similarity to chickenpox than smallpox, which classically starts at the extremities and works toward the trunk and do include those latter areas. An enanthem is possible on the tongue, tonsils, uvula, or pharynx but is uncommon.

Approximately 15 percent of smallpox victims will have delirium. Encephalitis is possible.

Hemorrhagic Smallpox

An atypical form of smallpox, the hemorrhagic variety, sometimes referred to as purpura variolosa, which is usually always fatal. It tends to occur more often in pregnant women or persons with significantly compromised immune response, and presents with epistaxis, hematemesis, hemoptysis, hematuria, subconjunctival hemorrhages, petechiae, ecchymosis, and bloody pustules on the skin and mucosa.

Flat/Confluent Smallpox

Flat (malignant) smallpox whereupon a dense confluent macular rash occurs. This version, like hemorrhagic smallpox, carries a high mortality rate. A deficiency in cellular immunity may be associated with this rare version, which occurs more often in children and is characterized by severe toxemia. The skin lesions develop slowly, become confluent, remain flat, never developing into pustules. Some have described the lesions as "velvety" to the touch. Sections of the skin may slough off.

Modified or subacute

Most often occurs in previously immunized patients. The lesions may be fewer in number and more superficial than classic smallpox. The prodrome stage may include severe headache, backache, and fever, albeit it might be shorter. Once lesions appear, they usually evolve and crust over quicker.

Medical Management

Control of Exposure Rapid recognition and isolation of an individual suspected of smallpox infection is critical. Institute universal precautions, identify all individuals in contact with the patient, contact the local health department.

Diagnosis (CDC Poster) Presumptive diagnosis is based upon the clinical picture; laboratory testing confirms the presence or absence of smallpox. Although the differential diagnosis of smallpox appears a lengthy list (Table), true variola patients appear sick - and their disease usually follow they typical biodrome. If the diagnosis of smallpox (or monkeypox) is considered, immediate isolation must be initiated. All patient contacts should be identified. While laboratory confirmation with silver impregnation or fluorescent antibody staining of smears taken from skin lesions is possible, CDC involvement, biosafety level considerations and other precautions must be taken. Contact public health, hospital laboratory about presumptive diagnosis; initiate lab precautions, obtain appropriate samples based upon laboratory capabilities, public health, and infectious disease recommendations.

Laboratory Testing

Early suspicion, collaboration with experts from the health department and CDC to identify or rule out smallpox with subsequent contact tracing, vaccination, and containment strategies are essential to contain a mass incident from this highly contagious disease. Estimates of per person transmission rates vary from 8 to 30 depending on the computer models utilized and the environment. Studies suggest that healthcare facilities are most likely to exhibit the highest transmission rates. Quarantine laws will probably be implemented under public health direction. Symptomatic and supportive care is essential. Prevention of secondary infections is critical. If such exigency occurs, antibiotics must be quickly initiated. Because smallpox is spread from airborne droplets, respiratory and fluid precautions using appropriate personal protective equipment (PPE) are necessary. Patient isolation in negative pressure rooms equipped with special filter systems is important.

Vaccines

Postexposure vaccination within 72 hours with vaccinia vaccine should confer significant protection for healthcare workers and others who have been exposed. Vaccinia immune globulin (VIG) should be readily available to treat adverse events associated with the vaccine. Vaccinia vaccine was generally considered a safe vaccine when given in the 1960s and 1970s, with relatively low adverse events recorded. Unfortunately, with the increased number of immunosuppressed persons who have HIV, transplants or chronic illnesses, and an aging society, there are subpopulations that may be at increased risk of adverse outcomes from the vaccine. Although rare, encephalitis and death can occur. The most common adverse event is autoinoculation - the patient scratches the vaccine site then rubs his or her eye, subsequently developing a lesion. Patients with eczema are more likely to experience eczema vaccinatum. Eczema vaccinatum can occur when a person with atopic dermatitis receives the vaccination.

Atopic dermatitis results from an immunological deficiency resulting in a skin abnormality. Untreated eczema vaccinatum can be fatal; treatment with VIG must be immediate. VIG is not considered effective in the treatment of postvaccination encephalitis or meningoencephalitis; rare, but life threatening events with CFR of 25%.

Generalized vaccinia results from the systemic spread of the virus from the vaccine site; usually a benign, self-limited complication of primary vaccination in healthy individuals. Realize that the risk of smallpox to these special populations is greater than the threat of vaccinia once exposure to variola occurs. As of April 2003, vaccinations have been halted in many regions due to concerns over unexpected cardiac-related deaths. Further studies reveals there was a small subset of European patients in the 1970s that also experienced cardiac reactions; usually they were minor. Some of those patients were treated with immune globulin; most of whom recovered. The patients with cardiac-related adverse events in 2003 were not treated with immune globulin; perhaps this should be considered as an intervention if mass vaccination resumes. Further studies should identify those individuals at risk or the precise mechanisms by which such deaths have occurred.

Up to date information on smallpox vaccination, contraindications, and news can be found at www.cdc.org: Smallpox Fact Sheet.

Antivirals

Cidofovir (Vistide®) [(S)-1-(3-hydroxy-2-phosphonylmethoxypropyl) cytosine] (HPMC) is an antiviral used for the treatment of cytomegalovirus (CMV) retinitis, especially in HIV patients. Currently it is under clinical trials evaluating potential systemic treatment of AIDS. Cidofovir has demonstrated significant *in vivo* and *in vitro* activity in experimental animals. Whether or not the use of Cidofovir confers benefit superior to immediate postexposure vaccination in human smallpox victims remains uncertain. Other antivirals for use against variola are under investigation. Studies suggest cidofovir antiviral has some effectiveness against smallpox. Recommended dosing in the treatment of CMV retinitis are based upon the patient's underlying kidney function creatinine clearance.

Dosing schedule - Cidofovir (Vistide®) Caveat: Use of Cidofovir for the treatment of monkeypox or smallpox is under an IND.

Cidofovir is FDA-approved for the treatment of CMV retinitis in HIV patients.

Mechanism: Cidofovir diphosphate suppresses CMV replication by selective inhibition of viral DNA synthesis. Incorporation of Cidofovir into growing viral DNA chain results in reductions in the rate of viral DNA synthesis.

Dosing must take into consideration creatinine clearance (Cl_{cr}).

Dosing in the patient with normal renal function: Induction: 5 mg/ kg I.V. over 1 hour once weekly for 2 consecutive weeks

Maintenance: 5 mg/kg over 1 hour once every other week
It is recommended to administer probenecid 2 g orally 3 hours prior to each Cidofovir dose and 1 g at 2 hours and 8 hours after completion of the infusion (total 4 g). It is also important to hydrate the patient with 1 L of 0.9% NS I.V. prior to Cidofovir infusion. A second liter may be administered over a 1-3 hour period immediately following infusion, if tolerated.

Dosing adjustment in renal impairment: If the creatinine increases by 0.3-0.4 mg/dL, reduce the cidofovir dose to 3 mg/kg; discontinue therapy for increases $\geq$0.5 mg/dL or development of $\geq$3+ proteinuria. Patients with pre-existing renal impairment: Use is contraindicated with serum creatinine >1.5 mg/dL, Cl_{cr} <55 mL/minute, or urine protein $\geq$100 mg/dL ($\geq$2+ proteinuria). However, the clinician must balance the threat of the infection against the risk and benefit associated with antiviral therapy.

Contraindications include history of severe hypersensitivity to probenecid or other sulfa-containing medications, serum creatinine >1.5 mg/dL, Cl_{cr} <55 mL/minute, or urine protein $\geq$100 mg/dL ($\geq$2+ proteinuria) or use within 7 days of nephrotoxic agents. Fanconi syndrome may occur.

Dose-dependent nephrotoxicity requires dose adjustment or discontinuation. Neutropenia and ocular hypotony have occurred. Safety and efficacy have not been established in children or the elderly. Other adverse effects include, but are not limited to, chills, fever, headache, pain, iritis, decrease in intraocular pressure, uveitis, cough, dyspnea, metabolic acidosis, cardiomyopathy/tachycardia, photosensitivity/skin discoloration, abdominal pain, and tremor. Cidofovir is a Pregnancy Category C risk. It was shown to be teratogenic and embryotoxic in animal studies, some at doses that also produced maternal toxicity.

Prepare admixtures in a class two laminar flow hood, wearing PPE, and appropriate disposal precautions.

Potential New Therapies

HDP-Cidofovir U.S. researchers reported promising results on hexadecyloxypropyl-cidofovir (HDP-cidofovir), an oral drug active against smallpox, at the 15th International Conference on Antiviral Research (Prague, Czech Republic; March 17-21). "Cidofovir itself is active against smallpox but has to be given intravenously which limits its usefulness," says researcher Karl Hostetler (Veterans Affairs Medical Center and University of California, San Diego, CA, USA). "Provided it passes all the necessary toxicity and safety tests, HDP-cidofovir could be self-administered." And, he adds, "because it is also active against cytomegalovirus and herpes, varicella zoster, and Epstein Barr viruses, the agent might also be of use in more common diseases."

Other antivirals similar to Cidofovir are being investigated. These include Adefovir (Hepsera™), which is used in the treatment of chronic hepatitis B with evidence of active viral replication including patients with lamivudine-resistant hepatitis B. A nucleotide analog, adefovir dipivoxil, has demonstrated antipoxvirus activity but is still under investigation as a potential therapy for Orthopoxvirus infection.

Unlike Cidofovir (Vistide®), Adefovir (Hepsera™) is an oral medication. Usual dose is 10 mg once daily. Patients with Cl_{cr} 20 39 mL/minute: 10 mg every 48 hours. Cl_{cr} 10-19 mL/minute: 10 mg every 72 hours.

Adefovir is a Pregnancy Category C risk. Lactic acidosis and severe hepatomegaly with steatosis (sometimes fatal) have occurred. Safety in pediatric patients has not been established. Adverse events include, but are not limited to, rash, pruritis, nausea, vomiting, diarrhea, abdominal pain, fever, and headache.

Differential Diagnosis Smallpox [Significant Differences]

- Varicella [rash distribution, general appearance of patient]
- Disseminated herpes zoster [rash distribution, history of illness, patient appearance]
- Impetigo [limited rash, appearance of patient, development of rash]
- Drug eruptions [history, appearance of rash and patient, distribution of rash]
- Contact dermatitis [history, patient appearance]
- Erythema multiforme
- Enteroviral infections
- Disseminated herpes simplex
- Scabies
- Insect bites
- Molluscum contagiosum (immunocompromised patients)
- Secondary syphilis
- Rickettsial diseases

Smallpox Vaccination is contraindicated for the following patients

- Persons who have severe immunodeficiency in T-cell function, defined as:
  - HIV - infected adults with CD4 lymphocyte count less than 200 (or age-appropriate equivalent counts for HIV-infected children)
  - Solid organ, bone marrow transplant recipients, or others currently receiving high-dose immunosuppressive therapy (ie, 2 mg/kg body weight or a total of 20 mg/day of prednisone or equivalent for persons whose weight is >10 kg, when administered for >2 weeks)
  - Persons with lymphosarcoma, hematological malignancies, or primary T-cell con-genital immunodeficiencies.
- Persons with life-threatening allergies to latex or to smallpox vaccine or any of its components (polymyxin B, streptomycin, chlortetracycline, neomycin).

These persons have a risk of severe complications from smallpox vaccination that may approach or exceed the risk of disease from monkeypox exposure.

In persons with close or intimate exposure within the past 2 weeks to a confirmed human case or probable or confirmed animal case of monkeypox, neither age, pregnancy, nor a history of eczema are contraindications to receipt of smallpox vaccination. These conditions are precautions, not contraindications. However, active eczematous disease is more concerning, but in instances when the potential vaccine has had true close or intimate exposure, the risk of monkeypox likely is greater than the risk of smallpox vaccine – related complications. Appropriate vaccine site care should be used to prevent transmission of smallpox vaccine (vaccinia virus).

Key "Look-alike" and Key Differentiators		
Rash	Patient Appearance	Patients
Varicella Zoster (Chickenpox)		
• Asynchronous development of lesions • Lesions in different stages; papules and pustules on same region • Centrifugal distribution (trunk then extremities) • Not usually involving palms of hands or soles of feet • Superficial skin involvement - soft to touch (not dermal/nodular) • Rash usually pruritic	• Patients do not usually look very sick, just uncomfortable • Febrile illness	• Usual patients are children. • Adults, especially unvaccinated and without a history of chicken pox may become infected
Shingles (Zoster)		
• Usually limited to dermatomal distribution • May appear as popular cluster; similar in appearance to chickenpox • Lesions are usually painful, itchy, or patient complains of tingling/burning • Generally present in older patients, especially under stress with history of chickenpox	• Patients do not usually look very sick, just uncomfortable • Discomfort from localized lesions may keep patients awake • Nonfebrile or low grade fever	
Molluscum contagiosum		
• Multiple pearly white nodules (2-5 mm diameter with central umbilication) • Painless • Distribution usually anogenital region, but possible anywhere on the body		• Common in AIDS patients (lesions may be larger/atypical and severe in comparison with presentation involving non-AIDS patients) • Children, sexually active adults, and sports involving close person-to-person contact, or patients with impaired cellular immunity

Preparedness Issues

1. Smallpox plan
   - Updated
   - Involves your department, facility, regional responders
   - Tested
   - Readily available to response team
2. Diagnostic chart from CDC or WHO displaying lesions with associated clinical information
3. Information to contact local public health and CDC if smallpox is suspected
   - Public health/CDC provide access to advanced lab capabilities

- Public health/CDC provide access to newer countermeasures that may only be available as IND products

4. Ability to isolate patient immediately
5. Ability to identify who was in contact with patient
   - Must be able to access these individuals for follow-up, vaccinations, and other contact-tracing that the health department will conduct
   - Your team already should have started working with public health
6. Readily available personal protective equipment (PPE)
7. Trained team to utilize PPE
8. Back-up team to assist primary team
9. Additional ventilators available
   - Sources of ventilators identified
   - Arrangements/mutual aid agreements/distributor agreements in place
   - Transportation and other logistics in place
10. Logistics
    - Cidofovir antiviral medication
    - Location to triage patients
    - Location to treat patients
    - Quarantine sites
    - Morgue
11. Resources dedications to healthcare teams and their families to ensure attendance and avoid absenteeism
    - Daycare
    - Food
    - Sleeping accommodations
    - Communications capabilities
    - Medications/healthcare
12. Healthcare facility team ready to coordinate activities with other response agencies (HEICS)
13. On speed dial
    - Director of ED
    - Directory of local health department
    - CDC office of emergency management
    - Regional LRN lab
    - Infectious disease consult

Management Considerations

One case of smallpox is a global emergency - an act of bioterrorism - a public health threat
1. Rapid/accurate diagnosis
   - History
   - Physical examination
   - Laboratory studies
     - Notify lab may be variola
     - Only tests that are approved in facility under such circumstances
2. Limit contagion - even presumptive concern over potential contagion
3. Isolate patient
4. Droplet/airborne precautions for a minimum of 17 days following exposure of all contacts. Patients are infectious until all scabs separate.
5. Contacts with fever >38°C or 101°F should be considered infected and treated accordingly with home or hospital isolation and appropriate management.
6. Initiate Preparedness Plan
   - Public health notification
   - Protect self and staff

164

○ Isolate patient
○ Obtain appropriate guidance from experts including CDC
○ Obtain access to laboratory - warn of potential variola samples
○ Obtain countermeasures
  • Vaccine
  • Immunoglobulin
  • Antivirals (eg, Cidofovir)
  • Mental health counselors for staff and ultimately for patient

Handling the Sick

• Negative pressure room
• N95 respirator
• Universal precautions/respiratory-droplet precautions

References

1. Barclay L, "Monkeypox Outbreak in the US: An Expert Interview with Cathy Roth, MD," *Medscape Medical news*, 2003. http://www.medscape.com/viewarticle/457247; last accessed 5/18/06.
2. "Bioterrorism and Weapons of Mass Destruction 2004: Physicians as First Responders," *The DO*, 2004, Mar Suppl:9-23.
3. Centers for Disease Control (CDC), Smallpox. www.cdc.gov
4. Diven DG, "An Overview of Poxviruses," *J Am Acad Derm*, 2001, 4(1):1-14.
5. Foster D, "Smallpox as a Biological Weapon: Implications for the Critical Care Clinician," *Dimens Crit Care Nurs*, 2003, 22:2-7.
6. Halloran ME, Longini IM, Nizam A, et al, "Containing Bioterrorist Smallpox," *Science*, 2002, 298:1428-32.
7. Henderson DA, *Emerging Infectious Diseases* - Special Issue - "Smallpox: Clinical and Epidemiologic Features," www.cdc.gov.ncidod/eid/vol5no4/henderson/htm.
8. Hirsch MS, *ACP Medicine Online*, 2002, 10/02 http:www.medscape.com/viewarticle/526732_print; last accessed 5/18/06.
9. Ippolito G, Nicastri E, Capobianchi M, et al, "Hospital Preparedness and Management of Patients Affected by Viral Haemorrhagic Fever or Smallpox at the Lazzaro Spallanzani Institute, Italy," *Eurosurveillance*, 2005, 10:36-9.
10. Jones SW, Dobson ME, Francesconi SC, et al, "DNA Assays for Detection, Identification, and Individualization of Select Agent Microorganisms: The Armed Forces Institute of pathology, Division of Microbiology, Washington DC. *Croatian Medical Journal*, 2005, 46(4):522-9.
11. Karem KL, Reynolds M, Braden Z, et al, "Characterization of Acute-Phase Humoral Immunity to Monkeypox: Use of Immunoglobulin M Enzyme-Linked Immunosorbent Assay (ELISA) for Detection of Monkeypox Infection During the 2003 North American Outbreak," *Clin and Diag Lab Immunol*, 2005, 12(7):867-972.
12. Kern ER, "*In vitro* Activity of Potential Anti-poxvirus Agents," *Antiviral Res*, 2003, 57(1-2):35-40.
13. Kim-Farley RJ, Celentano JT, Gunter C, et al, "Standardized Emergency Management System and Response to a Smallpox Emergency," *Prehosp Disast Med*, 2003, 18(4):313-20.
14. Krugman S, Katz SL, Gershon AA, et al, "Smallpox (Variola and Vaccinia)," *Infectious Diseases of Children*, 9th ed, St Louis, MO: Mosby, 1992, 457-62.
15. Lewis-Jones S, "Zoonotic Poxvirus Infections in Humans," *Curr Opin Infect Dis*, 2004, 17(2):81-9.
16. McFee RB, "Preparing for an Era of Weapons of Mass Destruction (WMD): Are We There Yet? Why We Should All Be Concerned. Part I," *Vet Hum Toxicol*, 2002, 44(4):193-9.

17. Neff JM, "Variola (Smallpox) and Monkeypox Viruses," *Principles and Practice of Infections Diseases*, 5th ed, Mandell G, Douglas RG, and Bennett J (eds), New York, NY: Churchill Livingstone, 2000, 1555-6.
18. O'Byrne WT, Terndrup TE, Kiefe CI, et al, "A Primer on Biological Weapons for the Clinician, Part I," *Adv Stud Med*, 2003, 3(2):75-86.
19. Reed KD, Melski JW, Graham MB, et al, "The Detection of Monkeypox in Humans in the Western Hemisphere," *N Engl J Med*, 2004, 350(4):342-50.
20. Saini R, Pui JC, and Burgin S, "Rickettsialpox: Report of Three Cases and a Review," *J Am Acad Derm*, 2004, 51:S137-9.
21. Smee DF, Sidwell RW, Kefauver D, et al, "Characterization of Wild Type and Cidofovir Resistant Strains of Camelpox, Cowpox, Monkeypox, and Vaccinia Virus," *Antimicrob Agents Chemother*, 2002, 46(5):1329-35.
22. *USAMRIID's Medical Management of Biological Casualties Handbook*, 4th ed, 2001, U.S. Army Medical Research Institute of Infectious Diseases, Fort Detrick, Frederick, MD.

Smallpox — cutaneous papules and vesicles.

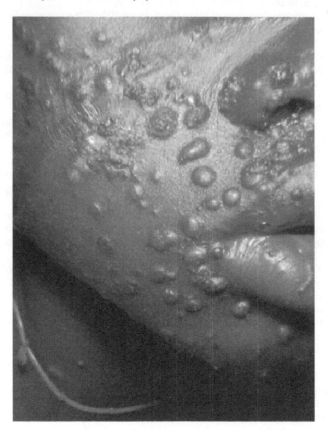

Public Health Image Library, CDC.

Smallpox immunization considerations for medicolegal death investigators*

Because the distribution of the smallpox vaccine to the civilian U.S. population was discontinued in 1983,[†] essentially all U.S. residents having contact with a smallpox case are at increased risk for infection. Although probably susceptible to smallpox, with appropriate precautions, medicolegal death investigators can reduce their risk of smallpox infection if they must examine or autopsy a decedent suspected to be infected with smallpox. Three risk- reduction activities during the postmortem period might be considered, 1) voluntary vaccination after the occurrence of smallpox has been confirmed in the community; 2) modification of autopsy procedures to limit the possible aerosolization of smallpox virus; and 3) exclusion of embalming procedures (see text).

In the event of mass fatalities resulting from a smallpox outbreak, CDC recommends that health departments consider planning for vaccinating mortuary personnel and their families.[§] This recommendation is relevant for medical examiners, coroners, and other forensic death investigators who have a high likelihood of handling smallpox-infected decedents during a mass fatality event.

In considering vaccination plans, attention should be given to the risk of adverse effects from smallpox vaccination as well as to its potential benefits. During a smallpox-associated mass fatality event, the federal government might propose that vaccinia inoculations be offered on a voluntary basis to appropriate personnel. Vaccinia inoculations have been effective in preventing smallpox infection but also pose certain risks for causing adverse reactions in the vaccinee and, less frequently, for spreading the vaccinia virus to other close contacts.

Because of the increased risk of adverse effects, the Advisory Committee on Immunization Practices (ACIP) recommends that the following persons not receive vaccinia inoculation:

- persons with immunosuppressive conditions;
- those receiving immunosuppressive medical treatments or pharmaceutical regimens;
- those with eczema or who have a close contact having eczema;
- anyone who is allergic to the vaccine or any of its components;
- women who are breastfeeding;
- anyone aged <12 months; and
- pregnant women or women expecting to become pregnant within 4 weeks.[¶]

ACIP recommends that persons be excluded from the pre-event smallpox vaccination program who have known underlying heart disease, with or without symptoms, or who have ≥3 known major cardiac risk factors (i.e., hypertension, diabetes, hypercholesterolemia, heart disease at age 50 years in a first-degree relative, and smoking).[**] Persons at increased risk for adverse reactions to the vaccine should be counseled regarding the potential risks before being vaccinated.

---

* Source: Adapted from Payne DC. Smallpox considerations for forensic professionals. National Association of Medical Examiners (NAME) News 2003;11(1):2.
† Source: CDC. Smallpox vaccine no longer available for civilians United States MMWR. 1983;32:387.
§ Source: CDC. Smallpox response plan, smallpox vaccination clinic guide. Annex 3-38. Atlanta, GA: US Department of Health and Human Services, CDC, 2002. Available at http://www.br.cdc.gov/agent/smallpox/response-plan/files/annex-3.pdf.
¶ Source: CDC. Recommendations for using smallpox vaccine in a pre-event vaccination program: supplemental recommendations of the Advisory Committee on Immunization Practices (ACIP) and the Healthcare Infection Control Practices Advisory Committee (HICPAC). MMWR 2003;52(No. RR-7):I-I6. Available at http://www.cdc.gov/mmwr/preview/mmwrhtml/rr5207a1.htm.
** Source: CDC. Supplemental recommendations on adverse events following smallpox vaccine in the pre-event vaccination program: recommendations of the Advisory Committee on Immunization Practices (Notice to readers]. MMWR 2003;52:282-4. Available at http://www.cdc.gov/mmwr/preview/mmwrhtml/mm5213a5.htm.

Smallpox — histologic section of skin with intraepidermal vesicles and ballooning degeneration of epithelial cells with viral inclusions (Guarnieri bodies [arrow])(hematoxylin and eosin stain).

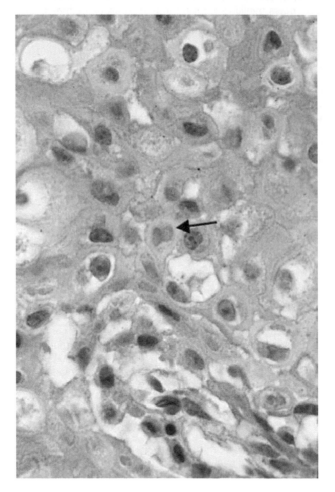

Infectious Disease Pathology Activity, CDC.

Intracellular mature variola virus particles grown in cell culture*.

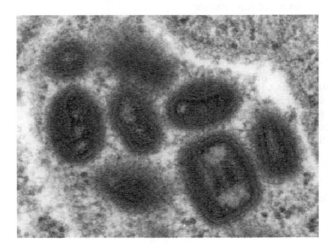

Infectious Disease Pathology Activity, CDC.
*Note: The barbell-shaped inner core and two lateral bodies are surrounded by an outer membrane. One brick-shaped particle is also illustrated (thin section electron microscopy).

Management of Healthcare Worker Exposures to HBV, HCV, and HIV

Adapted from Updated U.S. Public Health Service Guidelines for the Management of Occupational Exposures to HIV and Recommendations for Postexposure Prophylaxis, "Recommended HIV Postexposure Prophylaxis (PEP) for Percutaneous Injuries," *MMWR Recommended Rep*, 2005, 54(RR-9):3-17.

Factors to Consider in Assessing the Need for Follow-up of Occupational Exposures

- **Type of exposure**
  - Percutaneous injury
  - Mucous membrane exposure
  - Nonintact skin exposure
  - Bites resulting in blood exposure to either person involved
- **Type and amount of fluid/tissue**
  - Blood
  - Fluids containing blood
  - Potentially infectious fluid or tissue (semen; vaginal secretions; and cerebrospinal, synovial, pleural, peritoneal, pericardial, and amniotic fluids)
  - Direct contact with concentrated virus
- **Infectious status of source**
  - Presence of HB_sAg
  - Presence of HCV antibody
  - Presence of HIV antibody
- **Susceptibility of exposed person**
  - Hepatitis B vaccine and vaccine response status
  - HBV, HCV, HIV immune status

Evaluation of Occupational Exposure Sources

Known sources

- Test known sources for HB_sAg, anti-HCV, and HIV antibody
  - Direct virus assays for routine screening of source patients are not recommended
  - Consider using a rapid HIV-antibody test
  - If the source person is not infected with a blood-borne pathogen, baseline testing or further follow-up of the exposed person is not necessary
- For sources whose infection status remains unknown (eg, the source person refuses testing), consider medical diagnoses, clinical symptoms, and history of risk behaviors
- Do not test discarded needles for blood-borne pathogens

Unknown sources

- For unknown sources, evaluate the likelihood of exposure to a source at high risk for infection
  - Consider the likelihood of blood-borne pathogen infection among patients in the exposure setting

Recommended Postexposure Prophylaxis for Exposure to Hepatitis B Virus

Vaccination and Antibody Response Status of Exposed Workers*	Treatment		
	Source HB$_s$Ag[†]- Positive	Source HB$_s$Ag[†] - Negative	Source Unknown or Not Available for Testing
Unvaccinated	HBIG[‡] × 1 and initiate HB vaccine series[§]	Initiate HB vaccine series	Initiate HB vaccine series
Previously vaccinated			
Known responder[¶]	No treatment	No treatment	No treatment
Known nonresponder[#]	HBIG × 1 and initiate revaccination or HBIG × 2[•]	No treatment	If known high risk source, treat as if source was HB$_s$Ag-positive
Antibody response unknown	Test exposed person for anti-HB$_s$[°] 1. If adequate,[¶] no treatment is necessary 2. If inadequate,[#] administer HBIG × 1 and vaccine booster	No treatment	Test exposed person for anti-HB$_s$ 1. If adequate,[§] no treatment is necessary 2. If inadequate,[§] administer vaccine booster and recheck titer in 1-2 months

*Persons who have previously been infected with HBV are immune to reinfection and do not require postexposure prophylaxis.
[†]Hepatitis B surface antigen.
[‡]Hepatitis B immune globulin; dose is 0.06 mL/kg intramuscularly.
[§]Hepatitis B vaccine.
[¶]A responder is a person with adequate levels of serum antibody to HB$_s$Ag (ie, anti-HB$_s$ ≥10 mIU/mL).
[#]A nonresponder is a person with inadequate response to vaccination (ie, serum anti-HB$_s$ <10 mIU/mL).
[•]The option of giving one dose of HBIG and reinitiating the vaccine series is preferred for nonresponders who have not completed a second 3-dose vaccine series. For persons who previously completed a second vaccine series but failed to respond, two doses of HBIG are preferred.
[°]Antibody to HB$_s$Ag.

Recommended HIV Postexposure Prophylaxis (PEP) for Percutaneous Injuries

Exposure Type	Infection Status of Source				
	HIV-Positive Class 1*	HIV-Positive Class 2*	Source of Unknown HIV Status[†]	Unknown Source[‡]	HIV-Negative
Less severe[§]	Recommend basic 2-drug PEP	Recommend expanded ≥3-drug PEP	Generally, no PEP warranted; however, consider basic 2-drug PEP[¶] for source with HIV risk factors[#]	Generally, no PEP warranted; however, consider basic 2-drug PEP[¶] in settings in which exposure to HIV-infected persons is likely	No PEP warranted
More severe[•]	Recommend expanded 3-drug PEP	Recommend expanded ≥3-drug PEP	Generally, no PEP warranted; however consider basic 2-drug PEP[¶] for source with HIV risk factors[#]	Generally, no PEP warranted; however, consider basic 2-drug PEP[¶] in settings in which exposure to HIV-infected persons is likely	No PEP warranted

Table (*Continued*)

*HIV-positive, class 1 - asymptomatic HIV infection or known low viral load (eg, <1500 ribonucleic acid copies/mL). HIV-positive, class 2 - symptomatic HIV infection, AIDS, acute seroconversion, or known high viral load. If drug resistance is a concern, obtain expert consultation. Initiation of PEP should not be delayed pending expert consultation, and, because expert consultation alone cannot substitute for face-to-face counseling, resources should be available to provide immediate evaluation and follow-up care for all exposures.

†For example, deceased source person with no samples available for HIV testing.

‡For example, a needle from a sharps disposal container.

§For example, solid needle or superficial injury.

¶The recommendation "consider PEP" indicates that PEP is optional; a decision to initiate PEP should be based on a discussion between the exposed person and the treating clinician regarding the risk versus benefits of PEP.

#If PEP is offered and administered and the source is later determined to be HIV-negative, PEP should be discontinued.

*For example, large-bore hollow needle, deep puncture, visible blood on device, or needle used in patient's artery or vein.

Recommended HIV Postexposure Prophylaxis (PEP) for Mucous Membrane Exposures and Nonintact Skin* Exposures

Exposure Type	Infection Status of Source				
	HIV-Positive Class 1†	HIV-Positive Class 2†	Source of Unknown HIV Status‡	Unknown Source§	HIV-Negative
Small volume¶	Consider basic 2-drug PEP#	Recommend basic 2-drug PEP	Generally, no PEP warranted*	Generally, no PEP warranted	No PEP warranted
Large volume°	Recommend basic 2-drug PEP	Recommend expanded ≥3-drug PEP	Generally, no PEP warranted; however consider basic 2-drug PEP# for source with HIV risk factors*	Generally, no PEP warranted; how-ever, consider basic 2-drug PEP# in settings in which exposure to HIV-infected persons is likely	No PEP warranted

*For skin exposures, follow-up is indicated only if evidence exists of compromised skin integrity (eg, dermatitis, abrasion, or open wound).

†HIV-positive, class 1 - asymptomatic HIV infection or known low viral load (eg, <1500 ribonucleic acid copies/mL). HIV-positive, class 2 - symptomatic HIV infection, AIDS, acute seroconversion, or known high viral load. If drug resistance is a concern, obtain expert consultation. Initiation of PEP should not be delayed pending expert consultation, and, because expert consultation alone cannot substitute for face-to-face counseling, resources should be available to provide immediate evaluation and follow-up care for all exposures.

‡For example, deceased source person with no samples available for HIV testing.

§For example, splash from inappropriately disposed blood.

¶For example, a few drops.

#The recommendation "consider PEP" indicates that PEP is optional; a decision to initiate PEP should be based on a discussion between the exposed person and the treating clinician regarding the risks vs benefits of PEP.

*If PEP is offered and administered and the source is later determined to be HIV-negative, PEP should be discontinued.

°For example, a major blood splash.

Situations for Which Expert[1] Consultation for HIV Postexposure Prophylaxis Is Advised

- *Delayed (ie, later than 24-36 hours) exposure report*
  - ○ The interval after which there is no benefit from postexposure prophylaxis (PEP) is undefined
- *Unknown source (eg, needle in sharps disposal container or laundry)*
  - ○ Decide use of PEP on a case-by-case basis
  - ○ Consider the severity of the exposure and the epidemiologic likelihood of HIV exposure
  - ○ Do not test needles or sharp instruments for HIV
- *Known or suspected pregnancy in the exposed person*
  - ○ Does not preclude the use of optimal PEP regimens
  - ○ Do not deny PEP solely on the basis of pregnancy
- *Resistance of the source virus to antiretroviral agents*
  - ○ Influence of drug resistance on transmission risk is unknown
  - ○ Selection of drugs to which the source person's virus is unlikely to be resistant is recommended, if the source person's virus is unknown or suspected to be resistant to ≥1 of the drugs considered for the PEP regimen
  - ○ Resistance testing of the source person's virus at the time of the exposure is not recommended
- *Toxicity of the initial PEP regimen*
  - ○ Adverse symptoms, such as nausea and diarrhea, are common with PEP
  - ○ Symptoms can often be managed without changing the PEP regimen by prescribing antimotility and/or antiemetic agents
  - ○ Modification of dose intervals (ie, administering a lower dose of drug more frequently throughout the day, as recommended by the manufacturer), in other situations, might help alleviate symptoms

[1]Local experts and/or the National Clinicians' Postexposure Prophylaxis Hotline (PEPline 888-448-4911).

Occupational Exposure Management Resources	
National Clinicians' Postexposure Prophylaxis Hotline (PEPline) Run by University of California-San Francisco/San Francisco General Hospital staff; supported by the Health Resources and Services Administration Ryan White CARE Act, HIV/AIDS Bureau, AIDS Education and Training Centers, and CDC	Phone: (888) 448-4911 Internet: http://www.ucsf.edu/hivcntr
Needlestick! A website to help clinicians manage and document occupational blood and body fluid exposures. Developed and maintained by the University of California, Los Angeles (UCLA), Emergency Medicine Center, UCLA School of Medicine, and funded in part by CDC and the Agency for Healthcare Research and Quality.	Internet: http://www.needlestick.mednet.ucla.edu
Hepatitis Hotline	Phone: (888) 443-7232 Internet: http://www.cdc.gov/hepatitis
Reporting to CDC: Occupationally acquired HIV infections and failures of PEP	Phone: (800) 893-0485
HIV Antiretroviral Pregnancy Registry	Phone: (800) 258-4263 Fax: (800) 800-1052 Address: 1410 Commonwealth Drive, Suite 215 Wilmington, NC 28405 Internet: http://www.glaxowellcome.com/preg_reg/antiretroviral

Table (*Continued*)

Food and Drug Administration Report unusual or severe toxicity to antiretroviral agents	Phone: (800) 332-1088 Address: MedWatch HF-2, FDA 5600 Fishers Lane Rockville, MD 20857 Internet: http://www.fda.gov/medwatch
HIV/AIDS Treatment Information Service	Internet: http://www.aidsinfo.nih.gov

Management of Occupational Blood Exposures

Provide immediate care to the exposure site
- Wash wounds and skin with soap and water
- Flush mucous membranes with water

Determine risk associated with exposure by:
- Type of fluid (eg, blood, visibly bloody fluid, other potentially infectious fluid or tissue, and concentrated virus)
- Type of exposure (ie, percutaneous injury, mucous membrane or nonintact skin exposure, and bites resulting in blood exposure)

Evaluate exposure source
- Assess the risk of infection using available information
- Test known sources for HB_sAg, anti-HCV, and HIV antibody (consider using rapid testing)
- For unknown sources, assess risk of exposure to HBV, HCV, or HIV infection
- Do not test discarded needle or syringes for virus contamination

Evaluate the exposed person
- Assess immune status for HBV infection (ie, by history of hepatitis B vaccination and vaccine response)

Give PEP for exposures posing risk of infection transmission
- HBV: See Recommended Postexposure Prophylaxis for Exposure to Hepatitis B Virus Table
- HCV: PEP not recommended
- HIV: See Recommended HIV Postexposure Prophylaxis for Percutaneous Injuries Table and Recommended HIV Postexposure Prophylaxis for Mucous Membrane Exposures and Nonintact Skin Exposures Table
  - Initiate PEP as soon as possible, preferably within hours of exposure
  - Offer pregnancy testing to all women of childbearing age not known to be pregnant
  - Seek expert consultation if viral resistance is suspected
  - Administer PEP for 4 weeks if tolerated

Perform follow-up testing and provide counseling
- Advise exposed persons to seek medical evaluation for any acute illness occurring during follow-up

HBV exposures
- Perform follow-up anti-HB_s testing in persons who receive hepatitis B vaccine
- Test for anti-HB_s 1-2 months after last dose of vaccine
- Anti-HB_s response to vaccine cannot be ascertained if HBIG was received in the previous 3-4 months

HCV exposures
- Perform baseline and follow-up testing for anti-HCV and alanine (ALT) 4-6 months after exposures
- Perform HCV RNA at 4-6 months if earlier diagnosis of HCV infection is desired
- Confirm repeatedly reactive anti-HCV enzyme immunoassays (EIAs) with supplemental tests

HIV exposures
- Perform HIV antibody testing for at least 6 months postexposure (eg, at baseline, 6 weeks, 3 months, and 6 months)

- Perform HIV antibody testing if illness compatible with an acute retroviral syndrome occurs
- Advise exposed persons to use precautions to prevent secondary transmission during the follow-up period
- Evaluate exposed persons taking PEP within 72 hours after exposure and monitor for drug toxicity for at least 2 weeks

Basic and Expanded HIV Postexposure Prophylaxis Regimens

BASIC REGIMENS

Tenofovir DF (Viread®; TDF) + emtricitabine (Emtriva®; FTC); available as Truvada®

Preferred dosing
- TDF: 300 mg once daily
- FTC: 200 mg once daily
- As Truvada®: One tablet daily

Dosage forms
- TDF: 300 mg tablet
- FTC: See FTC
- Truvada® (TDF 300 mg plus FTC 200 mg)

Advantages
- FTC: See above
- TDF
  - Convenient dosing (single pill once daily)
  - Resistance profile activity against certain thymidine analogue mutations
  - Well tolerated

Disadvantages
- TDF
  - Same class warnings as NRTIs
  - Drug interactions
  - Increased TDF concentrations among persons taking atazanavir and lopinavir/ritonavir; need to monitor patients for TDF-associated toxicities
  - Preferred dosing of atazanavir if used with TDF: 300 mg + ritonavir 100 mg once daily + TDF 300 mg once daily

Tenofovir DF (Viread®; TDF) + lamivudine (Epivir®; 3TC)

Preferred dosing
- TDF: 300 mg once daily
- 3TC: 300 mg once daily or 150 mg twice daily

Dosage forms
- TDF: 300 mg tablet
- 3TC: See above

Advantages
- 3TC: See above
- TDF
  - Convenient dosing (single pill once daily)
  - Resistance profile activity against certain thymidine analogue mutations
  - Well tolerated

Disadvantages
- TDF
  - Same class warnings as nucleoside reverse transcriptase inhibitors (NRTIs)
  - Drug interactions
  - Increased TDF concentrations among persons taking atazanavir and lopinavir/ritonavir; need to monitor patients for TDF-associated toxicities

175

- Preferred dosage of atazanavir if used with TDF: 300 mg + ritonavir 100 mg once daily + TDF 300 mg once daily

Zidovudine (Retrovir®; ZDV; AZT) + emtricitabine (Emtriva®; FTC)

Preferred dosing
- ZDV: 300 mg twice daily or 200 mg three times daily, with food; total: 600 mg/day, in 2-3 divided doses
- FTC: 200 mg (one capsule) once daily

Dosage forms
- ZDV: See above
- FTC: 200-mg capsule, 10 mg/mL oral solution

Advantages
- ZDV: See above
- FTC
  - Convenient (once daily)
  - Well tolerated
  - Long intracellular half-life (~40 hours)

Disadvantages
- ZDV: See above
- FTC
  - Rash perhaps more frequent than with 3TC
  - No long-term experience with this drug
  - Cross resistance to 3TC
  - Hyperpigmentation among non-Caucasians with long-term use: 3%

Zidovudine (Retrovir®; ZDV; AZT) + lamivudine (Epivir®; 3TC); available as Combivir®

Preferred dosing
- ZDV: 300 mg twice daily or 200 mg three times daily, with food; total: 600 mg daily
- 3TC: 300 mg once daily or 150 mg twice daily
- Combivir®: One tablet twice daily

Dosage forms
- ZDV: 100 mg capsule, 10 mg/mL injection solution, 50 mg/5 mL oral syrup, 300 mg tablet
- 3TC: 10 mg/mL oral solution, 150 mg or 300 mg tablet
- Combivir®: Tablet, 300 mg ZDV + 150 mg 3TC

Advantages
- ZDV associated with decreased risk for HIV transmission
- ZDV used more often than other drugs for PEP for healthcare personnel (HCP)
- Serious toxicity rare when used for PEP
- Side effects predictable and manageable with antimotility and antiemetic agents
- Can be used by pregnant HCP
- Can be given as a single tablet (Combivir®) twice daily

Disadvantages
- Side effects (especially nausea and fatigue) common and might result in low adherence
- Source-patient virus resistance to this regimen possible
- Potential for delayed toxicity (oncogenic/teratogenic) unknown

ALTERNATE BASIC REGIMENS

Emtricitabine (Emtriva®; FTC) + didanosine (Videx®; ddl)

Preferred dosing
- FTC: 200 mg once daily
- ddl: See above

176

Dosage forms
- ddI: See above
- FTC: See above

Advantages
- ddI: See above
- FTC: See above

Disadvantages
- Tolerability: Diarrhea more common with buffered than with enteric-coated preparation
- Associated with toxicity: Peripheral neuropathy, pancreatitis, and lactic acidosis
- Must be taken on empty stomach except with TDF
- Drug interactions
- FTC: See above

Emtricitabine (Emtriva®; FTC) + stavudine (Zerit®; d4T)

Preferred dosing
- FTC: 200 mg daily
- d4T: 40 mg twice daily (can use lower doses of 20-30 mg twice daily if toxicity occurs; equally effective but less toxic among HIV-infected patients who developed peripheral neuropathy); if body weight is <60 kg, 30 mg twice daily

Dosage forms
- FTC: See above
- d4T: See above

Advantages
- 3TC and FTC: See above; d4T's GI side effects rare

Disadvantages
- Potential that source-patient virus is resistant to this regimen
- Unknown potential for delayed toxicity (oncogenic/teratogenic) unknown

Lamivudine (Epivir®; 3TC) + didanosine (Videx®; ddI)

Preferred dosing
- 3TC: 300 mg once daily or 150 mg twice daily
- ddI: Videx® chewable/dispersible buffered tablets can be administered on an empty stomach as either 200 mg twice daily or 400 mg once daily. Patients must take at least two of the appropriate-strength tablets at each dose to provide adequate buffering and prevent gastric acid degradation of ddI. Because of the need for adequate buffering, the 200-mg-strength tablet should be used only as a component of a once-daily regimen. The dose is either 200 mg twice daily or 400 mg once daily for patients weighing >60 kg and 125 mg twice daily or 250 mg once daily for patients weighing >60 kg.

Dosage forms
- 3TC: 150 mg or 300 mg tablets
- ddI: 125 mg, 200 mg, 250 mg, 400 mg delayed-release capsule; 25-mg , 50-mg, 100-mg, 150-mg, or 200-mg buffered white tablet

Advantages
- ddI: Once-daily dosing option
- 3TC: See above

Disadvantages
- Tolerability: Diarrhea more common with buffered preparation than with enteric-coated preparation
- Associated with toxicity: Peripheral neuropathy, pancreatitis, and lactic acidosis
- Must be taken on empty stomach except with TDF
- Drug interactions
- 3TC: See above

Lamivudine (Epivir®; 3TC) + stavudine (Zerit®; d4T)

Preferred dosing
- 3TC: 300 mg once daily or 150 mg twice daily
- d4T: 40 mg twice daily (can use lower doses of 20-30 mg twice daily if toxicity occurs; equally effective but less toxic among HIV-infected patients with peripheral neuropathy); 30 mg twice daily if body weight is <60 kg

Dosage forms
- 3TC: See above
- d4T: 1 mg/mL powder for oral solution, 15 mg, 20 mg, 30 mg, and 40 mg capsule

Advantages
- 3TC: See above
- d4T: Gastrointestinal (GI) side effects rare

Disadvantages
- Possibility that source-patient virus is resistant to this regimen
- Potential for delayed toxicity (oncogenic/teratogenic) unknown

PREFERRED EXPANDED REGIMEN BASIC REGIMEN PLUS:

Lopinavir/Ritonavir (Kaletra®; LPV/RTV)

Preferred dosing
- LPV/RTV: 400 mg/100 mg = 3 capsules twice daily with food

Dosage forms
- LPV/RTV: 133-mg/33-mg capsules

Advantages
- Potent HIV protease inhibitor
- Generally well-tolerated

Disadvantages
- Potential for serious or life-threatening drug interactions
- Might accelerate clearance of certain drugs, including oral contraceptives (requiring alternative or additional contraceptive measures for women taking these drugs)
- Can cause severe hyperlipidemia, especially hypertriglyceridemia
- GI (eg, diarrhea) events common

ALTERNATE EXPANDED REGIMENS BASIC REGIMEN PLUS ONE OF THE FOLLOWING:

Atazanavir (Reyataz®; ATV) ± ritonavir (Norvir®; RTV)

Preferred dosing
- ATV: 400 mg once daily, unless used in combination with TDF, in which case ATV should be boosted with RTV, preferred dosing of ATV 300 mg + RTV: 100 mg once daily

Dosage forms
- ATV: 100-mg, 150-mg, and 200-mg capsules
- RTV: 100-mg capsule

Advantages
- Potent HIV protease inhibitor
- Convenient dosing – once daily
- Generally well tolerated

Disadvantages
- Hyperbilirubinemia and jaundice common
- Potential for serious or life-threatening drug interactions
- Avoid coadministration with proton pump inhibitors

- Separate antacids and buffered medications by 2 hours and H$_2$-receptor antagonists by 12 hours to avoid decreasing ATV levels
- Caution should be used with ATV and products known to induce PR prolongation (eg, diltiazem)

Fosamprenavir (Lexiva™; FOSAPV) ± ritonavir (Norvir®; RTV)

Preferred dosing
- FOSAPV: 1400 mg twice daily (without RTV)
- FOSAPV: 1400 mg once daily + RTV 200 mg once daily
- FOSAPV: 700 mg twice daily + RTV 100 mg twice daily

Dosage forms
- FOSAPV: 700-mg tablet
- RTV: 100-mg capsule

Advantages
- Once daily dosing when given with ritonavir

Disadvantages
- Tolerability: GI side effects common
- Multiple drug interactions. Oral contraceptives decrease fosamprenavir concentrations
- Incidence of rash in healthy volunteers, especially when used with low doses of ritonavir. Differentiating between early drug-associated rash and acute seroconversion can be difficult and cause extraordinary concern for the exposed person.

Efavirenz (Sustiva®; EFV)

Preferred dosing
- EFV: 600 mg daily, at bedtime

Dosage forms
- EFV: 50-mg, 100-mg, 200-mg capsules
- EFV: 600-mg tablet

Advantages
- Does not require phosphorylation before activation and might be active earlier than other antiretroviral agents (a theoretic advantage of no demonstrated clinical benefit)
- Once daily dosing

Disadvantages
- Drug associated with rash (early onset) that can be severe and might rarely progress to Stevens-Johnson syndrome
- Differentiating between early drug-associated rash and acute seroconversion can be difficult and cause extraordinary concern for the exposed person
- Central nervous system side effects (eg, dizziness, somnolence, insomnia, or abnormal dreaming) common; severe psychiatric symptoms possible (dosing before bedtime might minimize these side effects)
- Teratogen; should not be used during pregnancy
- Potential for serious or life-threatening drug interactions

Indinavir (Crixivan®; IDV) ± ritonavir (Norvir®; RTV)

Preferred dosing
- IDV 800 mg + RTV 100 mg twice daily without regard to food

Alternative dosing
- IDV: 800 mg every 8 hours, on an empty stomach

Dosage forms
- IDV: 200-mg, 333-mg, and 400-mg capsule
- RTV: 100-mg capsule

Advantages
- Potent HIV inhibitor

Disadvantages
- Potential for serious or life-threatening drug interactions
- Serious toxicity (eg, nephrolithiasis) possible; consumption of 8 glasses of fluid/day required
- Hyperbilirubinemia common; must avoid this drug during late pregnancy
- Requires acid for absorption and cannot be taken simultaneously with ddI, chewable/dispersible buffered tablet formulation (doses must be separated by ≥1 hour)

Nelfinavir (Viracept®; NFV)

Preferred dosing
- NFV: 1250 mg (2 × 625 mg or 5 × 250 mg tablets), twice daily with a meal

Dosage forms
- NFV: 250-mg or 625-mg tablet

Advantages
- Generally well tolerated

Disadvantages
- Diarrhea or other GI events common
- Potential for serious and/or life-threatening drug interactions

Saquinavir (Invirase®; SQV) + ritonavir (Norvir®; RTV)

Preferred dosing
- SQV: 1000 mg (given as Invirase®) + RTV 100 mg, twice daily
- SQV : Five capsules twice daily + RTV: One capsule twice daily

Dosage forms
- SQV (Invirase®): 200-mg capsule
- RTV: 100-mg capsule

Advantages
- Generally well tolerated, although GI events common

Disadvantages
- Potential for serious or life-threatening drug interactions
- Substantial pill burden

Antiretroviral Agents Generally Not Recommended For Use as PEP

Nevirapine (Viramune®; NVP)

Disadvantages
- Associated with severe hepatotoxicity (including at least one case of liver failure requiring liver transplantation in an exposed person taking PEP)
- Associated with rash (early onset) that can be severe and progress to Stevens-Johnson syndrome
- Differentiating between early drug-associated rash and acute seroconversion can be difficult and cause extraordinary concern for the exposed person
- Drug interactions: Can lower effectiveness of certain antiretroviral agents and other commonly used medicines

Abacavir (Ziagen®; ABC)

Disadvantages
- Severe hypersensitivity reactions can occur, usually within the first 6 weeks
- Differentiating between early drug-associated rash/hypersensitivity and acute seroconversion can be difficult

Delavirdine (Rescriptor®; DLV)

Disadvantages
- Drug associated with rash (early onset) that can be severe and progress to Stevens-Johnson syndrome
- Multiple drug interactions

Zalcitabine (Hivid®; ddC)

Disadvantages
- Three times a day dosing
- Tolerability
- Weakest antiretroviral agent

Antiretroviral Agent for Use as PEP Only With Expert Consultation

Enfuvirtide (Fuzeon™; T20)

Preferred dosing
- T20: 90 mg (1 mL) twice daily by subcutaneous injection

Dosage forms
- T20: Single-dose vial, reconstituted to 90 mg/mL

Advantages
- New class
- Unique viral target; to block cell entry
- Prevalence of resistance low

Disadvantages
- Twice-daily injection
- Safety profile: Local injection site reactions
- Never studied among antiretroviral-naive or HIV-negative patients
- False-positive EIA HIV antibody tests might result from formation of anti-T20 antibodies that cross-react with anti-gp41 antibodies

Anthrax

Synonyms *Bacillus anthracis*

Clinically Resembles Respiratory influenza (no nasal symptoms)

Disease Syndromes Included Anthrax; Mad-Hatter Disease; Wool Sorter's Disease

Microbiology *Bacillus anthracis* is an aerobic, Gram-positive bacillus with a large polypeptide capsule. Some isolates may appear gram-variable rather than Gram-positive. On Gram's stain, there is a characteristic "box car" appearance to these organisms, a feature common to the genus *Bacillus*. The organism forms endospores when grown aerobically in the laboratory, but spores usually are not seen in clinical specimens (endospores in necrotic lesions). Virulence factors for this organism include anthrax toxin (made up of three components) and a capsule (which protects against host antibodies).

Postmortem Care Standard precautions, which include wearing appropriate personal protective equipment when body fluid/aerosol generation is anticipated

Epidemiology *Bacillus anthracis* causes the disease anthrax. The organism can be found in some areas of the world in soil and decaying vegetation. The endospores of the organism are hardy and survive for years under adverse conditions. These endospores can contaminate the hide of herbivores such as cattle, sheep, and goats. Animals consume the endospores and ultimately expire. In this way, anthrax is a zoonosis, with humans only incidentally exposed. Direct contact with the contaminated hides or hair of these herbivores leads to human disease sporadically. The disease has historically been more common in the Middle East than in the United States; in 1992, there was only one case of anthrax reported.

In 2001, there was a surge of bioterrorism-related anthrax in the United States. As of December 2001, there were 22 cases of confirmed or suspected cases of inhalational (11 cases) and cutaneous anthrax (11 cases) reported to the Centers for Disease Control and Prevention (CDC). These cases came from a variety of states including New York, New Jersey, Florida, the District of Columbia, Connecticut, Maryland, and Pennsylvania. Nearly all cases were acquired from exposure to the endospores of *Bacillus anthracis*, which was sent through the mail, although in several cases the source of inhalational anthrax has remained unclear. The majority of cases occurred in postal workers and mail handlers or sorters. Other cases occurred in several media workers, a hospital worker, a journalist, a bookkeeper, and an elderly woman in Connecticut.

Clinical Syndromes

• **Cutaneous anthrax:** This form of human infection results from direct inoculation of *Bacillus anthracis* endospores into the skin. Exposure results from handling insulation made from animal hair (usually commercial buildings), infected animal hides, clothing products, or wool. Less commonly, endospores come from contaminated soil. Endospores enter intact or exposed skin in an extremity such as the forearm. Disease develops 2-5 days postexposure. Starts as a nonpainful papule, which over 1-2 days becomes vesicular with localized surrounding edema. The vesicle generally ruptures in 1 week, thereby creating an ulcer, which progresses to a black eschar. The eschar falls off in ~2-3 weeks. Systemic symptoms of fever, chills, and total body edema can result from the effects of anthrax toxin. Headache and painful lymphadenopathy may also occur. Systemic manifestation may occur in 5% to 20% of patients. Fatality rate for cutaneous anthrax is 5% to 20% in untreated patients and 1% in treated patients.

• **Inhalation anthrax:** Also known as Woolsorter's disease, inhalational anthrax presents as a rapidly progressive pneumonitis. Inhaled spores are taken up by macrophages and transported to hilar and mediastinal lymph nodes where germination occurs. Incubation is 1-6 days. Low-grade fever, nonproductive cough, diaphoresis, and myalgias initially occur followed by respiratory distress, bacteremia, sepsis, shock, and death. Chest X-ray typically reveals a widened mediastinum and may be a clue to the diagnosis. Meningitis with subarachnoid hemorrhage may also occur in up to 50% of patients. Mortality is high.

• **Gastrointestinal anthrax:** This rare form of anthrax is characterized by fever, severe abdominal pain, and sepsis syndrome. This generally occurs after eating undercooked meat contaminated with anthrax endospores. The incubation period is estimated to be 1-7 days. Inflammation in the lower colon leads to hematemesis and bloody diarrhea. A form of GI anthrax involving the oropharynx has also been reported, where patients develop lesions at the base of the tongue, along with dysphagia and cervical adenopathy. The mortality of GI anthrax is approximately 50%.

• **Bioterrorism-related anthrax:** One concern has been that the presentation of bioterrorism-related anthrax may differ from the historical presentation of anthrax. The recent cases of bioterrorism-related anthrax have been studied in detail and publicly disseminated so that healthcare workers may become familiar with the presentation. For the 11 cases of inhalational anthrax, the majority presented with fever, chills, severe fatigue, and nonproductive cough. Chest discomfort or pleuritic pain was common, and some experienced nausea and vomiting with abdominal pain. Dyspnea, headache, and myalgias were also reported. Initially, WBCs were either normal or slightly elevated (7.5-13.3 $\times 10^3$/ mm^3). An increase in the number of band forms was common. The presenting chest X-ray was abnormal in all of the cases. Abnormalities on chest X-ray and CT scan of the chest included mediastinal widening, mediastinal lymphadenopathy, paratracheal/hilar fullness, hilar fullness, and pleural effusions. Two patients had no mediastinal abnormalities but presented with pleural effusions. Effusions were hemorrhagic and in some cases required chest tube placement. Pulmonary infiltrates, often involving multiple lobes were reported in 4 patients. For the 11 cases of cutaneous anthrax, the incubation period was approximately 5 days (1-10 days). The anthrax lesions were painless and found on the forearm, neck, chest, and fingers. Patients did complain of a tingling sensation or pruritus at the skin lesion site.

Infective Dose Probably significantly less than 8000-50,000 spores by inhalation. Direct exposure to vesicle secretions of cutaneous anthrax lesions may result in secondary cutaneous infection.

Diagnosis The diagnosis of cutaneous anthrax should be suspected clinically in a patient with a nontender, nonhealing ulcerative skin lesion with necrosis, which is otherwise unexplained. The skin lesions usually present as papules, followed by vesiculation and rapid development of a deep skin ulcer, often with severe surrounding edema. Necrosis is usually present. The organism can often be identified on Gram's stain of a swab or aspirate of a skin lesion. Typically, the Gram's stain of a cutaneous anthrax lesion shows few or no polymorphonuclear leukocytes and sporulating Gram-positive rods.

The diagnosis of inhalational anthrax should be suspected clinically in a patient who presents with a progressive, otherwise unexplained, respiratory illness, particularly if there is mediastinal widening on radiographs. However, the early diagnosis of inhalational anthrax may be very difficult since patients early on present with nonspecific systemic symptoms, which may mimic influenza and other common respiratory infections. The CDC has recommended the following laboratory criteria for the diagnosis of anthrax: 1) isolation of *B. anthracis* from a clinical specimen, or 2) other supportive laboratory tests, including (a) detection of *B. anthracis* DNA by polymerase chain reaction (PCR) from clinical specimens, (b) demonstration of *B. anthracis* in a clinical specimen by immunohisto-chemical staining, or (c) serologic testing.

The organism can be isolated from nasal swabs, skin lesions, cerebrospinal fluid, and induced sputum without special culture techniques. For suspected cases of anthrax, the CDC recommends at least the following specimens be obtained: 1) for inhalational anthrax, blood cultures and cerebrospinal fluid cultures if meningeal signs are present 2) cutaneous anthrax, blood cultures and vesicular fluid cultures, 3) gastrointestinal anthrax, blood cultures. In the bioterrorism-related inhalational anthrax cases, the organism was isolated in routine blood cultures in all patients who had not received prior antibiotic therapy. *B. anthracis* can also be isolated from other body fluids including pleural effusions, pleural biopsies, lymph nodes, and other tissues.

Serologic tests are available to aid in the diagnosis of anthrax. Specific serum IgG to the protective antigen component of *B. anthracis* toxin can be measured by enzyme immunoassay. Demonstration of a fourfold rise in specific antibody is confirmatory of acute anthrax. Serologic testing was used to confirm some of the bioterrorism-related cases with negative blood and tissue cultures. PCR tests to detect specific DNA sequences are also available and can be performed on blood and involved tissue specimens. (*MMWR*, 2004, 53, Figures 2–7)

Diagnostic Test/Procedure

Aerobic Culture, Appropriate Site

Aerobic Culture, Sputum

Gram's Stain

Endospore Stain

PCR

DFA 2-component

Serology

Treatment Strategy In the past, penicillin was the antibiotic of choice for *B. anthracis*. The organism has traditionally been very susceptible to a number of antibiotics including penicillin, tetracyclines, quinolones, and macrolides. However, the isolates of *B. anthracis* from the bioterrorism-related cases in 2001 demonstrated *in vitro* resistance to a number of antibiotics including vancomycin, imipenem, clarithromycin, rifampin, chloramphenicol, and clindamycin. The organisms maintained *in vitro* susceptibility to penicillin and ampicillin, but the isolates demonstrated constitutive and inducible beta-lactamases, which could lead to clinically significant penicillin and ampicillin resistance. Based on this, the CDC advised against the use of either penicillin or ampicillin as single agent therapy. Interim recommendations for treatment of inhalational anthrax and cutaneous anthrax related to bioterrorism have been published by the CDC (see tables).

Inhalational Anthrax Treatment Protocol[1,2] for Cases Associated with This Bioterrorism Attack		
Category	**Initial Therapy (I.V.)[3,4]**	**Duration**
Adults	Ciprofloxacin 400 mg every q12h[1] or Doxycycline 100 mg q12h[6] and 1 or 2 additional antimicrobials[4]	I.V. treatment initially[5]. Switch to oral antimicrobial therapy when clinically appropriate: Ciprofloxacin 500 mg po bid or Doxycycline 100 mg po bid
		Continue for 60 days (I.V. and oral combined)[7]
Children	Ciprofloxacin 10-15 mg/kg q12h[8,9] or Doxycycline:[6,10] >8 years and >45 kg: 100 mg q12h >8 years and ≤45 kg: 2.2 mg/kg q12h ≤8 years: 2.2 mg/kg q12h and 1 or 2 additional antimicrobials[4]	I.V. treatment initially.[5] Switch to oral antimicrobial therapy when clinically appropriate: Ciprofloxacin 10-15 mg/kg po q12h[9] or Doxycycline:[10] >8 years and >45 kg: 100 mg po bid >8 years and ≤45 kg: 2.2 mg/kg po bid ≤8 years: 2.2 mg/kg po bid
		Continue for 60 days (I.V. and oral combined)[7]
Pregnant women[11]	Same for nonpregnant adults (the high death rate from the infection outweighs the risk posed by the antimicrobial agent)	I.V. treatment initially. Switch to oral antimicrobial therapy when clinically appropriate.[2] Oral therapy regimens same for nonpregnant adults.
Immunocompromised persons	Same for nonimmunocompromised persons and children	Same for nonimmunocompromised persons and children
Reference: U.S. Department of Health and Human Services, "Update: Investigation of Bioterrorism-Related Anthrax and Interim Guidelines for Exposure Management and Antimicrobial Therapy, October 2001," *MMWR Morb Mortal Wkly Rep*, 2001, 50(42):909-40.		

Table (*Continued*)

[1]For gastrointestinal and oropharyngeal anthrax, use regimens recommended for inhalational anthrax.
[2]Ciprofloxacin or doxycycline should be considered an essential part of first-line therapy for inhalational anthrax.
[3]Steroids may be considered as an adjunct therapy for patients with severe edema and for meningitis based on experience with bacterial meningitis of other etiologies.
[4]Other agents with *in vitro* activity include rifampin, vancomycin, penicillin, ampicillin, chloramphenicol, imipenem, clindamycin, and clarithromycin. Because of concerns of constitutive and inducible beta-lactamases in *Bacillus anthracis*, penicillin and ampicillin should not be used alone. Consultation with an infectious disease specialist is advised.
[5]Initial therapy may be altered based on clinical course of patient; one or two antimicrobial agents (eg, ciprofloxacin or doxycycline) may be adequate as the patient improves.
[6]If meningitis is suspected, doxycycline may be less optimal because of poor central nervous system penetration.
[7]Because of the potential persistence of spores after an aerosol exposure, antimicrobial therapy should be continued for 60 days.
[8]If I.V. ciprofloxacin is not available, oral ciprofloxacin may be acceptable because it is rapidly and well absorbed from the GI tract with no substantial loss by first-pass metabolism. Maximum serum concentrations are attained 1-2 hours after oral dosing but may not be achieved if vomiting or ileus are present.
[9]In children, ciprofloxacin dosage should not exceed 1 g/day.
[10]The American Academy of pediatrics recommends treatment of young children with tetracyclines for serious infections (eg, Rocky Mountain spotted fever).
[11]Although tetracyclines are not recommended during pregnancy, their use may be indicated for life-threatening illness. Adverse effects on developing teeth and bones are dose related; therefore, doxycycline might be used for a short time (7-14 days) before 6 months of gestation.

Cutaneous Anthrax Treatment Protocol[1] for Cases Associated with This Bioterrorism Attack

Category	Initial Therapy (Oral)[2]	Duration
Adults[1]	Ciprofloxacin 500 mg bid or Doxycycline 100 mg bid	60 days[3]
Children[1]	Ciprofloxacin 10-15 mg/kg every 12 hours (not to exceed 1 g/day)[2] or Doxycycline:[4] >8 years and >45 kg: 100 mg q12h >8 years and ≤45 kg: 2.2 mg/kg q12h ≤8 years: 2.2 mg/kg q12h	60 days[3]
Pregnant women[1,5]	Ciprofloxacin 500 mg bid or Doxycycline 100 mg bid	60 days[3]
Immunocompromised persons[1]	Same for nonimmunocompromised persons and children	60 days[3]

Reference: U.S. Department of Health and Human Services, "Update: Investigation of Bioterrorism-Related Anthrax and Interim Guidelines for Exposure Management and Antimicrobial Therapy, October 2001," *MMWR Morb Mortal Wkly Rep*, 2001, 50(42):909-40.

[1]Cutaneous anthrax with signs of systemic involvement, extensive edema, or lesions on the head or neck require I.V. therapy, and a multidrug approach is recommended.
[2]Ciprofloxacin or doxycycline should be considered first-line therapy. Amoxicillin 500 mg po tid for adults or 80 mg/kg/day divided q8h for children is an option for completion of therapy after clinical improvement. Oral amoxicillin dose is based on the need to achieve appropriate minimum inhibitory concentration levels.

Table (*Continued*)
[3]Previous guidelines have suggested treating cutaneous anthrax for 7-10 days, but 60 days is recommended in the setting of this attack, given the likelihood of exposure to aerosolized *B. anthracis*. [4]The American Academy of Pediatrics recommends treatment of young children with tetracyclines for serious infections (eg, Rocky Mountain spotted fever). [5]Although tetracyclines or ciprofloxacin are not recommended during pregnancy, their use may be indicated for life-threatening illness. Adverse effects on developing teeth and bones are dose related; therefore, doxycycline might be used for a short time (7-14 days) before 6 months gestation.

The treatment of choice for inhalational anthrax is either ciprofloxacin or doxycycline, in combination with a second antibiotic. The initial treatment is with intravenous antibiotics, which may later be changed to oral antibiotics when clinically stable. The above recommendation is for empirical treatment; the antibiotic regimen may need to be altered depending on the results of antimicrobial susceptibility testing. The total recommended length of treatment (oral and intravenous) is 60 days. The treatment of choice for cutaneous anthrax is ciprofloxacin or doxycycline. In cases of dermal exposure, thoroughly wash exposed skin and clothing with soap and water.

One important issue is the role of postexposure prophylaxis (ie, antibiotic treatment of individuals who have had a potential aerosol exposure to anthrax but who have no signs or symptoms of disease). The nature of the exposure is the key to determining if a person should receive postexposure prophylaxis, since there is not a completely reliable laboratory method for determining if an exposure has occurred. Nasal swabs were obtained in a large number of individuals who had potential exposure to anthrax in 2001, but the CDC has stressed that the sensitivity of nasal swabs may be limited and thus the use of nasal swabs may only be suitable for epidemiologic purposes at this time. The CDC has recommended ciprofloxacin or doxycycline for adults and children who have had a suspected exposure to anthrax. The duration of antibiotic prophylaxis is 60 days. Because of the potential adverse effects of both of these drugs for children, the risks and benefits must be weighed on an individual basis.

A vaccine against anthrax is available and has been licensed in the U.S. This vaccine is not currently available to the general public but has been used to immunize the U.S. military. The vaccine is recommended only in certain groups such as laboratory workers who work directly with the organism, some persons who work with animal hides and furs, and veterinarians who work in foreign countries where anthrax is more common. The CDC has also offered the vaccine to those who are receiving anthrax postexposure prophylaxis, in addition to the antibiotic regimen.

Quarantine Standard precautions; private rooms are not necessary

Rescuer Contamination

Interim Guidelines for Evacuation and Personal Decontamination of Workers After a Positive Autonomous Detection System (ADS) Signal Indicates Presence of a Biologic Agent	
Worker Category	**Evacuation/Decontamination Procedures**
Group I. Workers who did not enter the production area containing the ADS device during the sampling and testing period (eg, 1.5 hours) before the positive ADS signal and who were not in an area that shares a heating, ventilating, and air conditioning (HVAC) system with the production area experiencing the positive signal	Evacuate; no special decontamination steps are needed

Worker Category	Evacuation/Decontamination Procedures
Group 2. All workers who were present in the production area containing the ADS device during the sampling and testing period before the positive ADS signal or who were in an area that shares an HVAC system with the production area experiencing the positive signal	• Evacuate immediately • Remove potentially contaminated outer garments at the site • Wash all areas of skin (eg, face, arms, hands, and legs) exposed at the time of the positive ADS signal with mild soap and copious amounts of warm water • Use replacement outer garments and shoes
Group 3. Workers identified in advance as particularly at risk of exposure to a higher concentration of deposited spores as a result of direct physical contact with aerosol-generating equipment	• Evacuate immediately • Remove potentially contaminated garments at the site • Take a shower at the site to wash all areas of exposed and unexposed skin with mild soap and warm water • Use replacement outer garments, underwear, and shoes

Adapted from Meehan PJ, Rosenstein NE, Gillen M, et al, "Responding to Detection of Aerosolized *Bacillus anthracis* by Autonomous Detection Systems in the Workplace," *MMWR*, 2004, 53(RR07):1-12.

Additional Information Can be cleared to work in research laboratory 3 weeks after 3rd anthrax vaccine dose (0.5 mL SubQ) with annual booster dose after 18 months

Presumptive *Bacillus anthracis* Identification and Similar Organisms

Test	B. anthracis	B. cereus	B. cereus var. mycoides	B. turingiensis	B. megaterium	B. subtilis
β-hemolysis on SBA	−	+	−	+	−	V (64%)
Motility	− (0%)	+ (99%)	V (63%)	+	V (42%)	V (84%)
Capsule	+	−	−	−	− (+ in rare variants)	− (+ in rare variants)
Urea hydrolysis	− (0%)	V (26%)	V (25%)	+ (weak)	V (47%)	V (47%)
Nitrate reduction	+ (100%)	V (89%)	V (89%)	+	− (10%)	V (89%)
Oxidase	V (87%)	V (63%)	V (25%)	V	V (42%)	V (61%)
Indole	− (0%)	− (0%)	− (0%)	−	− (0%)	− (0%)
Cell width >1.0 micron	+	+	+	+	+	−
Gamma phage lysis	+	−	−	−	−	−

Reference: http://www.bt.cdc.gov/documents/PPTResponse/table6anthraxid.pdf.

Specific References

Bell DM, Kozarsky PE, and Stephens DS, "Clinical Issues in the Prophylaxis, Diagnosis, and Treatment of Anthrax," *Emerg Infect Dis*, 2002, 8(2):222-5.

Coker PR, Smith KL, and Hugh-Jones ME, "Antimicrobial Susceptibilities of Diverse *Bacillus anthracis* Isolates," *Antimicrob Agents Chemother*, 2002, 46(12):3843-5.

De BK, Bragg SL, Sanden GN, et al, "A Two-Component Direct Fluorescent-Antibody Assay for Rapid Identification of *Bacillus anthracis*," *Emerg Infect Dis*, 2002, 8(10):1060-5.

Dewan PK, Fry AM, Laserson K, et al, "Inhalational Anthrax Outbreak Among Postal Workers, Washington, D.C., 2001," *Emerg Infect Dis*, 2002, 8(10):1066-72.

Leitman S, "Plasmapheresis of Anthrax Vaccinees for Production of An," *Crisp Data Base National Institutes of Health*, 2002.

Moye PK, Pesik N, Terndrup T, et al, "Bioterrorism Training in U.S. Emergency Medicine Residencies: Has It Changed Since 9/11?," *Acad Emerg Med* 2007, 14:221-227.

Quinn CP, Semenova VA, Elie CM, et al, "Specific, Sensitive, and Qualitative Enzyme-Linked Immunosorbent Assay for Human Immunoglobulin G Antibodies to Anthrax Toxin Protective Antigen," *Emerg Infect Dis*, 2002, 8(10):1103-10.

Rusnak J, Boudreau E, Bozue J, et al, "An Unusual Inhalational Exposure to *Bacillus anthracis* in a Research Laboratory," *J Toxicol Clin Toxicol*, 2004, 46(4):313-4.

Rusnak JM, Kortepeter MG, Aldis J, et al, "Experience in the Medical Management of Potential Laboratory Exposures to Agents of Bioterrorism on the Basis of Risk Assessment at the United States Army Medical Research Institute of Infectious Diseases (USAMRIID)," *J Occup Environ Med*, 2004, 46(8):801-11.

Rusnak JM, Kortepeter MG, Hawley RG, et al, "Management Guidelines for Laboratory Exposures to Agents of Bioterrorism," *J Occup Environ Med*, 2004, 46(8):791-800.

Schultz CH, "Chinese Curses, Anthrax, and the Risk of Bioterrorism," *Ann Emerg Med*, 2004, 43(3):329-32.

Wein LM, Craft DL, and Kaplan EH, "Emergency Response to an Anthrax Attack," *Proc Natl Acad Sci*, 2003, 100(7):4346-51.

Cutaneous anthrax — eschar lesion.

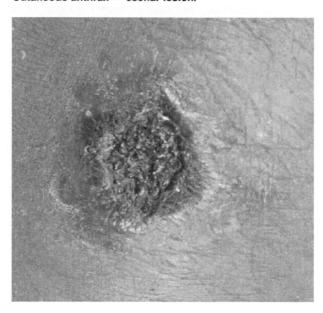

Public Health Image Library, CDC.

Inhalational anthrax — hemorrhagic mediastinal lymphadenitis surrounding trachea; inset, cross-section of trachea surrounded by hemorrhagic soft tissue and lymph nodes.

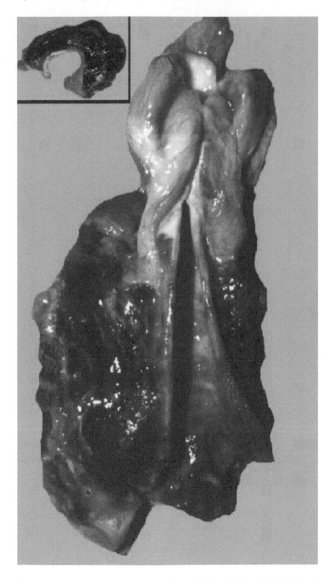

Reprinted courtesy of New York City Office of the Chief Medical Examiner. Public Health Image Library, CDC.

Inhalational anthrax — histologic section of mediastinal lymph node with hemorrhage, necrosis, and sparse inflammatory cell infiltrate (hematoxylin and eosin stain).

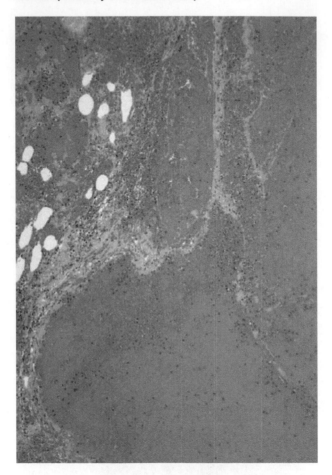

Infectious Disease Pathology Activity, CDC.

Anthrax - hemorrhagic meningitis.

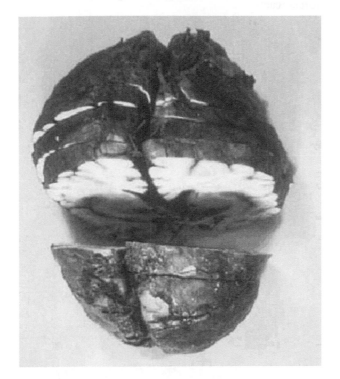

Public Health Image Library, CDC.

Anthrax — *Bacillus anthracis* rods in mediastinal lymph node (Gram stain).

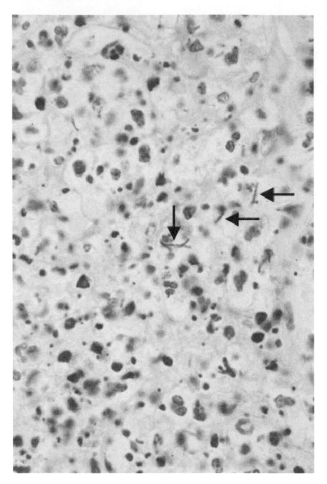

Infectious Disease Pathology Activity, CDC.

ANTHRAX

Anthrax — *Bacillus anthracis* rods, bacillary fragments and granular bacterial fragments in spleen (immunohistochemistry).

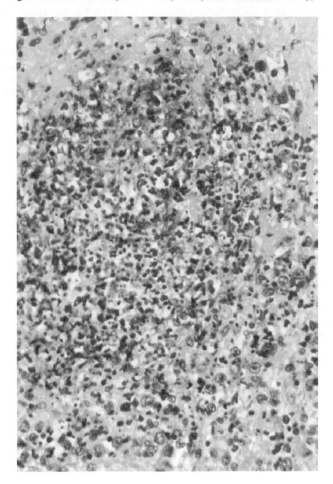

Infectious Disease Pathology Activity, CDC.

193

Arboviruses

Disease Syndromes Included Colorado Tick Fever; Congo-Crimean Hemorrhagic Fever; Dengue; Eastern Equine Encephalitis; LaCrosse Encephalitis; Powassan Encephalitis; Rift Valley Fever; St. Louis Encephalitis; Tick-Borne Encephalitis; West Nile Viral Syndrome; Western Equine Encephalitis; Yellow Fever

Microbiology The group of viruses commonly known as arboviruses (arthropod-borne viruses) is composed of approximately 500 viruses, which have in common the fact that the vectors that transmit these viruses to humans are arthropods (mosquitoes, ticks, and certain blood-sucking flies). Mosquitoes are the most common vectors. Arboviruses are morphologically diverse and can be either enveloped or nonenveloped, and either icosahedral or helical; however, most arboviruses are 40-110 nm and contain RNA. Some of the more commonly discussed and encountered arbovirus infections in the U.S. are shown in the table.

Arboviruses					
Disease	**Virus**	**Vector**	**Clinical Syndrome**	**Geographical Location**	**Mortality (%)**
Colorado tick fever*	Orbivirus	Tick	F, M, H	West U.S.	<1
Congo-Crimean hemorrhagic fever	Nariovirus	Tick, infected blood	H	Africa, Europe, Asia	10-50
Dengue	Flavivirus	Mosquito	F, H	Tropical, world-wide	5 (H)
Eastern equine encephalitis*	Alphavirus	Mosquito	M	East U.S., Central & South America	30
LaCrosse encephalitis*	Bunyavirus	Mosquito	M	East U.S.	<1
Powassan encephalitis*	Flavivirus	Tick, mosquito	M	U.S., Russia, China, Canada	
Rift Valley fever	Phlebovirus	Mosquito, infected blood	F, M, H	Africa, Middle East	<1
St Louis encephalitis*	Flavivirus	Mosquito	M	North, Central & South America	7
Tick-borne encephalitis	Flavivirus	Tick	F, M, H	Europe, Asia	1-10
Western equine encephalitis*	Alphavirus	Mosquito	M	West U.S., South America, Canada, Mexico	5
West Nile	Flavivirus	Mosquito	F, M	Europe, North America	3-15
Yellow fever	Flavivirus	Mosquito	F, H	South America, Africa	15

*One of the six significant arbovirus diseases in the U.S.

F: Febrile illness; M: Meningoencephalitis; H: Hemorrhagic fever.

Information in this table is adapted from the publications by Tsai TF. See Specific References.

Epidemiology Arboviruses are found worldwide but are more common in tropical than in temperate climates. Only about 150 of the ~500 arboviruses cause disease in humans. Generally, arboviruses are transmitted between small mammals and birds. The aforementioned arthropod vectors transmit the arboviruses from these small animals to humans who are dead-end hosts for these viruses.

Clinical Syndromes Most arbovirus infections in humans are mild viral infections, which often are indistinguishable from many other viral infections. Acute arbovirus infections often are characterized by sudden onset of headache, fever, muscle and joint pain, and other constitutional symptoms. If disease progresses beyond these symptoms, the disease can be extremely serious and usually manifests as a severe febrile illness (with or without rash, with or without arthritis), meningoencephalitis, or hemorrhagic fever. In more severe cases, more than one of these three conditions can be present (see table).

Diagnosis The basis of a definitive diagnosis of an arbovirus infection is a thorough physical examination and an extremely close examination of the exposure, travel, work, and socio-economic history of the patient. Febrile patients may exhibit relative bradycardia, sometimes referred to as pulse-temperature dissociation. This clinical sign can also be seen with typhoid fever as well as other bioweapon illnesses including certain viral hemorrhagic fevers and plague. Enzyme immunoassay, immunofluorescence, and nucleic acid hybridization with or without polymerase chain reaction amplification methods can be used to detect virus antigens in throat and blood specimens and paraffin-embedded tissues. However, these tests are not widely available and are applicable to only a few arboviruses. The test of choice in the laboratory diagnosis of an arbovirus infection is serology. Acute and convalescent sera must be collected if an accurate diagnosis is sought. Immunofluorescence, enzyme immunoassay, hemagglutination inhibition, complement fixation, and antibody neutralization tests are generally available at reference laboratories and the Centers for Disease Control and Prevention. Contact the local laboratory for details on specimen collection, availability of tests, turnaround time, and discussions regarding the most likely virus to cause the disease of a particular infection.

Diagnostic Test/Procedure Encephalitis Viral Serology

Treatment Strategy Treatment for specific arbovirus infections usually is supportive and directed toward making the patient comfortable and reducing symptoms.

Drug Therapy Comment There is no specific antiviral agent(s) for arboviruses. Supportive therapy is the mainstay.

Specific References

Beaty BJ, "Control of Arbovirus Diseases: Is the Vector the Weak Link?" *Arch Virol Suppl*, 2005, 19:73-88.

Sidwell RW and Smee DF, "Viruses of the Bunya- and Togaviridae Families: Potential as Bioterrorism Agents and Means of Control," *Antiviral Res*, 2003, 57:101-11.

Whitley RJ and Gnann JW, "Viral Encephalitis: Familiar Infections and Emerging Pathogens," *Lancet*, 2002, 359:507-13.

Brucella Species

Clinically Resembles Flu-like symptoms, appendicitis

Disease Syndromes Included Brucellosis

Microbiology *Brucella* species are small, aerobic, gram-negative coccobacilli, which primarily cause disease in domestic animals. There are four species that cause human disease:
- *B. abortus* (cattle)
- *B. melitensis* (goats and sheep)
- *B. suis* (swine)
- *B. canis* (dogs)

The organism is nonmotile and does not form spores. All species tend to be slow-growing and fastidious. Supplemental carbon dioxide is required for optimal growth of *B. abortus*. The cell walls of *B. abortus*, *B. suis*, and *B. melitensis* contain endotoxin and two major surface antigens (A and M). The endotoxin is structurally and biologically different from endotoxins produced by many other enteric gram-negative bacilli. *Brucella* produces urease and catalyzes nitrite to nitrate.

Epidemiology Human brucellosis occurs worldwide. Approximately 500,000 cases are reported annually. The incidence has declined following mandatory pasteurization of dairy products and immunization of cattle with live-attenuated *Brucella abortus* vaccine. Endemic areas include the Mediterranean basin, South America, and Mexico. In the United States, more than half the cases of human brucellosis occur in Texas, California, Virginia, and Florida. U.S. incidence is 200 per year or 0.04 per 100,000 population. A small number of cases (1% to 2%) result from exposure to *Brucella* in the microbiology or research laboratory.

Transmission to humans occurs by one of three routes:
- Direct contact of infected tissue, blood, or lymph with broken skin or conjunctivae
- Ingestion of contaminated meat or dairy products
- Inhalation of infected aerosols

Clinical Syndromes *Brucella* species enter the human via breaks in the skin or mucous membranes or via inhalation. Incubation period is 3 days to 2 months. Normal human serum has good bactericidal activity against *B. abortus* but not *B. melitensis*. Organisms not killed by polymorphonuclear leukocytes travel to regional lymph nodes, then enter the circulation, and localize in organs of the reticuloendothelial system where they are ingested by macrophages. Some will survive intracellularly and multiply, but when the macrophage is activated, the intracellular organisms are killed and release endotoxin. The host response to endotoxin can result in the signs and symptoms of acute brucellosis.

The clinical manifestations depend on both the immune status of the patient and the species of *Brucella* involved.
- **Asymptomatic infection**
- **Acute brucellosis, or "Malta fever":** Acute symptoms appear 1 week to several months following exposure. Symptoms include fever, sweats, chills, weakness, malaise, headache, and anorexia. In 25% to 50% of cases, there is weight loss, myalgias, arthralgias, and back pain. Epididymis is common in some series. Splenomegaly occurs in 20% to 30% of cases and lymphadenopathy in 10% to 20%. Pulse-temperature dissociation is possible with this illnesss.

  A number of complications of acute brucellosis have been reported. Behavioral changes have been associated with untreated brucellosis. Neurotoxicity resulting in a variety of psychiatric manifestations including paranoia have been identified and attributed to brucellosis infection, although the exact mechanism is still under study. When infection is present for more than 2 months without therapy, the complication rate can approach 30%.
- **Musculoskeletal:** Sacroiliitis, arthritis, osteomyelitis, paraspinal abscess
- **Neurologic:** Meningoencephalitis, myelitis, others; Behavioral changes are possible
- **Genitourinary:** Epididymo-orchitis, prostatitis
- **Endocarditis:** Occurs in 2% of cases
- **Granulomatous hepatitis**
- **Caseating or noncaseating granulomatous disease**
- **Splenic abscess**
- **Nodular lung lesions; lung abscess**
- **Erythema nodosum; other skin manifestations**
- **Ocular lesions**
- **Localized disease:** Can occur in any of the sites listed above
- **Subclinical infection**

Infective Dose 10-100 organisms

Diagnosis The clinician should suspect brucellosis in any individual presenting with fever and who has a history of travel to endemic areas; exposure to livestock; a history of consuming unpasteurized milk, cheese, or other dairy products; or suggestive occupational risk factors (veterinarians, laboratory workers, etc). Laboratory confirmation is essential since the signs and symptoms of human brucellosis may be nonspecific. The diagnosis is established by recovery of the organism from blood, fluid, or other tissues (bone, abscesses, etc). In patients with *B. melitensis* infections, cultures of blood will be positive in 70% of cases. Bone marrow biopsy for histopathology and culture may be positive when other specimens are negative. In many laboratories, blood cultures for *Brucella* sp. are processed using the Castañeda technique, which utilizes biphasic media in a bottle. The microbiology laboratory should always be informed if *Brucella* sp are highly suspected.

Serologic techniques for diagnosing brucellosis include a standard tube agglutination test. A *Brucella* titer ≥1:160 or a fourfold rise in titer is considered presumptive evidence of recent infection with *Brucella* sp.

Diagnostic Test/Procedure
Aerobic Culture, Appropriate Site
Blood Culture, *Brucella*
Gram's Stain - immunofluorescence

Treatment Strategy
Laboratory workers should be notified if *Brucella* is suspected, given high risk of occupational transmission in clinical laboratories.

Decontamination: Dermal: Wash with soap and water. Abscess should be drained promptly. Ceftriaxone is not effective drug therapy.

Watch for relapse as incompletely or inadequately treated Brucellosis can occur.

Uncomplicated brucellosis:

Children <8 years: Oral: Trimethoprim/sulfamethoxazole (TMP/SMX) 10 mg/kg/day **or**

Children >8 years: Oral: Doxycycline 1-2 mg/kg twice daily

Adults: Oral: Doxycycline 100 mg twice daily **or** trimethoprim/sulfamethoxazole (TMP/SMX) component twice daily for 6 weeks

Plus:

Children: I.V.: Gentamicin 2.5 mg/kg every 8 hours for 2-3 weeks **or** rifampin 15-20 mg/kg orally every day for 6 weeks

Adults: I.V.: Gentamicin 5 mg/kg/ every day for 2-3 weeks **or** rifampin 600-900 mg orally every day for 6 weeks

Complicated neurobrucellosis, spondylitis, endocarditis:

Children: I.V.: Gentamicin 2.5 mg/kg every 8 hours for 2-4 weeks

Adults: I.V.: Gentamicin 5 mg/kg/ every day for 2-4 weeks

Plus:

Children <8 years: Oral: Trimethoprim/sulfamethoxazole (TMP/SMX) 10 mg/kg/day for 8-10 weeks **or**

Children >8 years: Oral: Doxycycline 1-2 mg/kg twice daily for 8-10 weeks

Adults: Oral: Doxycycline 100 mg twice daily **or** trimethoprim/sulfamethoxazole (TMP/SMX) component twice daily for 8-12 weeks

Plus:

Children: Oral: Rifampin 20 mg/kg every day for 8-12 weeks

Adults: Oral: Rifampin 600-900 mg every day for 8-12 weeks

Steroids are indicated for: Fever, anorexia, septic shock, DIC, neurobrucellosis with neurologic signs, iritis from chronic brucellosis, or prevention of Herxheimer-like reaction.

Methylprednisolone: I.V.: 10-40 mg up to 6 times daily **or**

Prednisone: Oral: 20 mg two or three times daily; maximum: 5 days

Septic shock: I.V.: Methylprednisolone 30 mg/kg every 4-6 hours; maximum: 72 hours

Neurobrucellosis: I.V.: Dexamethasone 0.15 mg/kg every 8 hours

Additional Information

Presumptive *Brucella* spp. Identification and Similar Organisms						
Test	*Brucella* spp.	*Bordetella bronchiseptica*	*Acinetobacter* spp.	*Psychrobacter phenylpyruvicus*[1]	*Oligella ureolytica*	*Haemophilus influenzae*
Oxidase	+[2,3]	+	−	+	+	V
Motility	−	+	−	−	V	−
Urea hydrolysis	+	+	V	+	+	V
Nitrate reduction	+	+	−	V	+	+
Growth BAP[4]	+	+	+	+	+	−
Cellular morphology	tiny ccb[5] stains faint	small ccb, rods	broad ccb	broad ccb	tiny ccb	small ccb

Table   (*Continued*)

Test	Brucella spp.	Bordetella bronchiseptica	Acinetobacter spp.	Psychrobacter phenylpyruvicus[1]	Oligella ureolytica	Haemophilus influenzae
Specimen source	blood, bone marrow	respiratory tract	various sites	various sites	urinary tract	mucous membranes
X or V factor requirement	−	−	−	−	−	+

[1]Formerly *Moraxella phenylpyruvica*
[2]Oxidase: *B. abortus*, *B. melitensis*, and *B. suis* are all ≥95% +; *B. canis* is 72% +
[3](+) = ≥90% positive; (−) = ≤10% positive; V, varible = 11-89% positive
[4]BAP = Blood Agar Plate
[5]ccb = coccobacillus
Reference: http://www.bt.cdc.gov/documents/PPTResponse/table5brucellaid.pdf

Specific References

Centers for Disease Control and Prevention, "Suspected Brucellosis Case Prompts Investigation of Possible Bioterrorism-Related Activity - New Hampshire and Massachusetts, 1999," *JAMA*, 2000, 284:300-2.

Hadjinikolaou L, Triposkiadis F, Zairis M, et al, "Successful Management of *Brucella Melitensis* Endocarditis With Combined Medical and Surgical Approach," *Eur J Cardiothorac Surg*, 2001, 19:806-10.

Ko J and Splitter GA, "Molecular Host-Pathogen Interaction in Brucellosis: Current Understanding and Future Approaches to Vaccine Development for Mice and Humans," *Clin Microbiol Rev*, 2003, 16:65-78.

Pappas G, Akritidis N, Bosilkovski M, et al, "Brucellosis," *N Engl J Med*, 2005, 2;352(22):2325-36 (review).

Yagupsky P and Baron EJ, "Laboratory Exposures to Brucellae and Implications for Bioterrorism," *Emerg Infect Dis*, 2005, 11(8):1180-5.

Burkholderia mallei

Clinically Resembles Bronchopneumonia

Disease Syndromes Included Glanders; Melioidosis (Whitmore's Disease) (Burkholeria pseudo-mallei)

Microbiology The genus *Burkholderia* (formerly *Pseudomonas*) is composed of twenty-two species including *B. mallei* and *B. pseudomallei*. *Burkholderia* spp. are motile, gram-negative, nonspore-forming rods. *B. mallei* is not difficult to grow and isolate in a clinical laboratory because the bacterium grows readily on most standard media. However, physicians should notify the laboratory whenever *B. mallei* is suspected.

Epidemiology *Burkholderia* spp. are found worldwide in water and soil, and on many plants, including vegetables and fruit. This high efficiency of transmission is especially important because it occurs not only nosocomially but could be used as an agent for bioterrorism. Humans and animals (primarily horses, mules, and donkeys) are believed to acquire the infection by inhalation of dust, ingestion of contaminated water, and contact with contaminated soil especially through skin abrasions, and for military troops, by contamination of war wounds. Person-to-person transmission can occur.

Clinical Syndromes Burkholder mallei is the etiologic agent for glander, which can manifest as a localized, suppurative dermal infection within five days of exposure. It is usually found in Africa and South America. Generalized muscle aches, diarrhea, and pneumonia can occur. Septicemia usually is fatal within one week. *Burkholderia pseudomallei* is the etiological agent of melioidosis, a disease with protean manifestations, which range from asymptomatic infection to soft tissue abscesses to fulminant septicemia. Melioidosis is mostly in Southeast Asia and Australia. Incubation period is 10-14 days by inhalation. Death can occur within 10 days. Symptoms include: Fever, rigors, myalgia, headache, chest pain, cervical adenopathy, urticaria, splenomegaly, generalized papular/pustular eruptions. Leukocyte count and liver transaminases may be mildly elevated. Chest x-ray may demonstrate miliary lesions (1 cm in diameter), small lung abscess, or bronchopneumonia. Overall mortality is 40% (90% if untreated); septicemia 50%; localized disease 20%.

Diagnosis In Gram-stained specimens, *Pseudomonas* spp. and *Burkholderia* species usually are indistinguishable. The suggestion of *B. mallei* in a Gram-stained specimen or the growth of *B. mallei* in culture is not automatically indicative of *B. mallei* being an etiological agent of an infectious process. The Gram's stain and culture information must be interpreted in the context of an appropriate clinical presentation (eg, cystic fibrosis or an association with a nosocomial setting). Gram's stain or methylene blue stain of exudates revealing small bacilli. **Note:** Blood cultures may be positive in 48 hours. Automated bacterial analysis will often confuse B. mallei with *Pseudomonas fluorescens* or *Pseudomonas putida*. Mild leukocytosis may be noted.

Treatment Strategy

Localized disease: Oral:

Amoxicillin/clavulanate 60 mg/kg/day divided three times daily, with sulfamethoxazole and trimethoprim (4 mg TMP/kg/day and 20 mg SMX/kg/day) divided twice daily combined for 30 days

Then switch to monotherapy with amoxicillin and clavulanate or TMP/SMX for 60-150 days total course.

Tetracycline: Children >8 years and Adults: Oral: 40 mg/kg/day divided three times daily can also be added

Severe disease:

Ceftazidime: 120 mg/kg/day in 3 divided doses combined with TMP/SMX (8 mg/kg/day of TMX; 40 mg/kg/day in four divided doses); I.V. for 2 weeks; oral therapy (ciprofloxacin, doxycycline, TMP/SMX, or amoxicillin and clavulanate) for 6 months

Extrapulmonary signs should be treated for 6-12 months

Drain abscesses surgically

May be resistent to Chloramphenicol.

Quarantine Contact and respiratory precautions - standard isolation precautions: Person to person transfer capability is low.

Additional Information The organism is susceptible to heat (over 55°C), and disinfectants (eg, benzalkonium chloride, 1% sodium hypochlorite, 70% ethanol and iodine).

Environmental persistency - stable

There is no vaccine for glanders.

Specific References

Chetchotisakd P, Porramatikul S, Mootsikapun P, et al, "Randomized, Double-Blind, Controlled Study of Cefoperazone-Sulbactam Plus Cotrimoxazole Versus Ceftazidime Plus Cotrimoxazole for the Treatment of Severe Melioidosis," *Clin Infect Dis*, 2001, 33:29-34.

White NJ, "Melioidosis," *Lancet*, 2003, 361:1715-22.

Clostridium botulinum

See Botulism (p 147)

Clinically Resembles Myasthenia gravis, Guillain-Barré syndrome, diphtheria, cerebrovascular accident, tick paralysis, poliomyelitis, meningitis, Reye's syndrome

Disease Syndromes Included Botulism

Microbiology The etiologic agent of botulism is *Clostridium botulinum*. The organism is a gram-positive, anaerobic bacillus. The natural reservoir is soil and sediment where the organism survives by forming spores. Most foodborne epidemics are associated with home-canned foods. Restaurants and commercially canned products are also the source of outbreaks.

Deliberate contamination of food or beverages with botulinum toxin is the most likely route of dissemination for a bioterroristic attack. As the bacteria grow and undergo autolysis, a powerful polypeptide neurotoxin is released. The toxin is heat labile and thus, is destroyed by appropriate heating in commercial canning processes. An acidic pH inhibits spore germination. This is the basis for adding ascorbic acid to home-canned products. Ingestion of even a small amount of contaminated food may cause severe clinical symptoms. Dispersion of aerosolized toxin is also possible. Aerosolized particles of botulinum toxin are approximately 0.1-0.3 mcm in size and experts have estimated that 1 gram of aerosolized botulinum toxin could kill up to 1.5 million people.

Postmortem Care Standard precautions

Epidemiology *Clostridium botulinum* is found in soil worldwide. Incidence of reported botulism is increasing. Most cases are individual or occur as a small cluster. If the outbreak is the result of contaminated commercially distributed food, the cases may occur in a widespread area. Early consultation with the State Health Department in conjunction with the Centers for Disease Control is imperative for case management and epidemic control. Botulism is almost always caused by *C. botulinum* types A, B, and E in the western U.S., eastern U.S., and Alaska, respectively. Not transmitted person-to-person.

Clinical Syndromes

• **Infant botulism:** The most commonly reported form of botulism in the United States is infant botulism. *C. botulinum* toxin is released from organisms colonizing the intestinal tract. The toxin binds to synaptic membranes of cholinergic nerves and prevents the release of acetylcholine. Transmission of nerve fiber impulses to muscle is interrupted with resultant autonomic nerve dysfunction, which progresses to motor weakness (flaccid paralysis). Subsequently, cranial and peripheral nerve weakness, constipation, and autonomic instability is the result of toxin effect on the ganglionic and postganglionic parasympathetic synapses. Typically, infants 1 week to 1 year of age present with progressive descending weakness following a period of constipation. Weak suck and poor feeding are frequent complaints. The pathogenesis of intestinal botulism in adults is similar to that of infant botulism.

• **Inhalational botulism:** Disease is caused by inhalation of aerosolized preformed botulinum toxin with subsequent absorption through the lungs into the circulation.

• **Foodborne botulism (see the following monograph):** Symptoms of botulism begin as descending paralysis or weakness within a few hours to 1-2 days following ingestion of contaminated food. Symptoms may commence as late as 1 week. Dizziness, lassitude, and weakness are common. Nausea and vomiting are less common. Blockage of autonomic nerve impulse transmission causes dry tongue, mouth, and pharynx, which is unrelieved by fluids. Urinary retention, ileus, and constipation frequently follow. Paralysis descends from cranial nerves to extremities and the respiratory muscles. Speech and vision are often impaired. There are no sensory deficits. Key clinical observations include the fact that the patients are oriented, alert, or easily roused and afebrile. They may demonstrate postural hypotension and dilated unreactive pupils.

• **Wound botulism:** Botulism may result from germination of spores contaminating traumatic wounds. Wounds caused by intravenous drug and nasal cocaine abuse have also been reported as sources. The clinical syndrome resembles that of foodborne botulism. Toxin can be recovered from the serum and wound drainage of affected patients. *C. botulinum* may be recovered from debrided tissue and drainage.

Infective Dose 0.001 mcg/kg

Diagnosis Outbreaks of botulism may be documented by identifying toxin in the stool or serum of symptomatic patients or from suspected food. Isolation of *C. botulinum* from stool or demonstration of the toxin by the mouse neutralization test establishes the diagnosis. Toxin is difficult to recover from serum; therefore, toxin recovery from stool may be more productive, but acquiring a stool specimen may be difficult because of the constipation. A low volume, normal saline, colonic irrigation may yield a useful specimen. Electromyography and repetitive nerve stimulation may yield corroborative evidence. The electromyographer should be alerted to the suspicion of botulism so that the study can be directed toward the diagnosis. A cerebrospinal fluid analysis to exclude meningitis, poliomyelitis, and Fisher Guillain-Barré syndrome is a useful part of the work-up.

Diagnostic Test/Procedure
Samples collected for botulism testing should be safely handled, using Standard Precautions.
C. botulinum toxin detection should be performed only by trained individuals at level C or higher LRN laboratories.
Sodium hypochlorite (0.1%) or sodium hydroxide (0.1 N) inactivate the toxin and are recommended by CDC for decontaminating work surfaces and spills of cultures or toxin.
The mouse bioassay is currently the only diagnostic method used for detection and identification of botulinum toxin.
Culture for *C. botulinum* in addition to toxin testing for stool or gastric specimens
Polymerase chain reaction assays (PCR)
Botulism, Diagnostic Procedure
Cerebrospinal Fluid Analysis

Signs and Symptoms of Acute Exposure Symptoms may be delayed 12-36 hours. GI symptoms are usually absent in wound botulism. Adynamic ileus, aspiration, blurred vision, constipation followed by descending paralysis and respiratory failure, cranial nerve palsies, dry mouth, dry skin, dysphagia, fasciculations, hyporeflexia, mydriasis, ptosis, slurred speech

Treatment Strategy
Decontamination: **Do not** induce vomiting, lavage (within 1 hour) with activated charcoal (sorbitol as a cathartic) may be useful to remove spores if there is a known ingestion; activated charcoal avidly bind to *C. botulinum*, type A neurotoxin. Avoid magnesium-based cathartics since magnesium may enhance neurotoxicity.
Supportive therapy: Primarily respiratory: Aggressive respiratory and ventilatory support is critical and may be required for months. Mechanical ventilation may be necessary; "tensilon test" up to 10 mg of edrophonium I.V. over 5 minutes; may show improvement in muscle strength in botulism; guanidine (15-50 mg/kg/day orally) may be given (although its use is not proven). Trivalent (A, B, E) (Liosiero) antitoxin should precede administration; give 2 mL of antitoxin diluted in 100 mL of 0.9% saline over 30 minutes; 10 mL of antitoxin may be given in another 2-4 hours, then at 12- to 24-hour intervals as necessary. Sensitivity skin testing should proceed antitoxin administration. Give 20 mL of antitoxin diluted in 100 mL of 0.9% saline over 30 minutes intravenously. 10 mL of antitoxin may be given in another 2-4 hours, then at 12- to 24-hour intervals as necessary. 4-Aminopyridine (0.5 mg/kg I.V. bolus; 0.25 mg/kg/hour continuous infusion I.V.) may improve peripheral strength but not respiratory parameters. Avoid magnesium since it may enhance the action of the toxin. Trivalent (ABE) antitoxin is usually most effective if administered within 24 hours. Usually, only one vial dose is necessary. The antitoxin does not reverse the current clinical situation; it scavenges unbound toxin, once the toxin has bound to receptors, the antitoxin cannot unbind the toxin. Since all antitoxins are equine derived, there is a risk for reaction. Even the "despeciated" version poses a risk for anaphylaxis.

Quarantine Standard precautions; no isolation needed

Environmental Persistency Aerosolized toxin estimated atmospheric decay rate is 1% to 4% per minute or about 2 days total. Stable for weeks in food or water.

Additional Information Bioterrorism emergency number at CDC Emergency Response Office: 770-488-7100
Can be cleared to work in research laboratory after postvaccination (4 weeks after 3rd dose), titer ≥ 0.02 Int. units/mL and ≥ 0.25 Int. units/mL after 1 year booster dose.

Specific References
Arnon SS, Schechter R, Inglesby TV, et al, "Botulinum Toxin as a Biological Weapon: Medical and Public Health Management," *JAMA*, 2001, 285(8):1059-81.
"Biological and Chemical Terrorism: Strategic Plan for Preparedness and Response. Recommendations of the CDC Strategic Planning Workgroup," *MMWR*, 2000, 21;49(RR-4):1-14.

Craven KE, Ferreira JL, Harrison MA, et al, "Specific Detection of *Clostridium Botulinum* Types A, B, E, and F Using the Polymerase Chain Reaction," *J AOAC Int*, 2002, 85(5):1025-8.

Shukla HD and Sharma SK, "Clostridium botulinum: A Bug With Beauty and Weapon," *Crit Rev Microbiol*, 2005, 31(1):11-8.

Clostridium botulinum Food Poisoning

Synonyms Botulism; Oculinum® (Botulinum A Toxin); Sysport

Mechanism of Toxic Action An anaerobic, spore-forming, gram-positive bacteria; ingestion of the toxin or wound contamination are likely routes of exposure; a heat labile neurotoxin, which causes irreversible neuromuscular blockade and prevents acetylcholine release; the spores are heat-resistant

Pregnancy Issues Therapeutically used for blepharospasm or strabismus

Diagnostic Test/Procedure
Botulism Diagnostic Procedure
Stool Culture

Signs and Symptoms of Acute Exposure Symptoms may be delayed 12-36 hours. GI symptoms are usually absent in wound botulism. Adynamic ileus, aspiration, blurred vision, constipation followed by descending paralysis and respiratory failure, cranial nerve palsies, dry mouth, dry skin, dysphagia, fasciculations, hyporeflexia, mydriasis, ptosis, slurred speech

Admission Criteria/Prognosis Any suspected exposure or symptomatic patient should be admitted; patients with neurologic toxicity or respiratory failure should be admitted into intensive care unit

Treatment Strategy
Decontamination: **Do not** induce vomiting, lavage (within 1 hour) with activated charcoal (sorbitol as a cathartic) may be useful to remove spores if there is a known ingestion; activated charcoal avidly bind to *C. botulinum*, type A neurotoxin. Avoid magnesium-based cathartics since magnesium may enhance neurotoxicity.

Supportive therapy: Primarily respiratory: Mechanical ventilation may be necessary; "tensilon test" up to 10 mg of edrophonium I.V. over 5 minutes; may show improvement in muscle strength in botulism; guanidine (15-50 mg/kg/day orally) may be given (although its use is not proven). Trivalent (A, B, E) (Liosiero) antitoxin should precede administration; give 2 mL of antitoxin diluted in 100 mL of 0.9% saline over 30 minutes; 10 mL of antitoxin may be given in another 2-4 hours, then at 12- to 24-hour intervals as necessary. Sensitivity skin testing should proceed antitoxin administration. Give 20 mL of antitoxin diluted in 100 mL of 0.9% saline over 30 minutes intravenously. 10 mL of antitoxin may be given in another 2-4 hours, then at 12- to 24-hour intervals as necessary. 4-Aminopyridine (0.5 mg/kg I.V. bolus; 0.25 mg/kg/hour continuous infusion I.V.) may improve peripheral strength but not respiratory parameters. Avoid magnesium since it may enhance the action of the toxin. Trivalent (ABE) antitoxin is usually most effective if administered within 24 hours. Usually, only one vial dose is necessary.

Additional Information Lethal dose: Humans: 1 pg/kg
Not transmitted person-to-person
Stools or gastric contents may be contagious; infantile botulism has been related to honey ingestion; symptoms may be delayed for 1 week. Antitoxin can be obtained from the CDC (404-639-3753 [days], 404-639-2888 [nights]).
Electromyogram shows decreased evoked action potential at 2 Hz/sec and an increased evoked action potential at 50 Hz/sec; up to 15% of affected individuals will have normal electromyograms
Thirty-four cases of foodborne botulism have been reported to the CDC in 1994. Boiling food for 10 minutes (>100°C) destroys the toxin; cooked foods should not be kept at room temperature (4°C to 60°C) for longer than 4 hours; recent cases have been associated with

commercial pot pies, home-canned foods, asparagus, green beans, peppers, and onions sauteed in margarine.

Twenty-five cases of wound botulism reported in 1995 (23 of them in California, almost all due to intravenous drug injection).

Injection of "black tar" heroin (which has been the predominant form of heroin in the western United States during the 1990's either intramuscularly or subcutaneously) is the primary risk factor for development of wound botulism (46 cases in California from 1988 to 1995).

Type A toxin has the highest case fatality rate; overall fatality rate for foodborne botulism is 5% to 10%; for wound botulism it is 15%

Epidemiology, diagnosis, treatment, prevention, and reporting of infant (intestinal) botulism (*MMWR,* 2003, 52(2):23)

Epidemiology

• Intestinal botulism is the most common form of human botulism reported in the United States; approximately 100 cases are reported annually.

• The majority of cases occur among infants aged ≤6 months; intestinal botulism is seen rarely in adults.

• The majority of cases are caused by botulinum toxin types A and B.

• The case-fatality rate of hospitalized patients is <1%

• Although ingesting honey is a known risk factor, the source of spores for the majority of cases is unknown.

• Ingestion of *Clostridium botulinum* spores, which exist worldwide in the soil and dust, is believed to be the principal route of exposure.

Clinical findings

• Reporting symptoms range from constipation and mild lethargy to hypotonia and respiratory insufficiency.

• Symptoms in infants aged ≤12 months include constipation, lethargy, poor feeding, weak cry, bulbar palsies (eg, ptosis, expressionless face, and difficulty swallowing), and failure to thrive.

• Presenting symptoms might be followed by progressive weakness, impaired respiration, and sometimes death.

• Differential diagnosis includes sepsis, dehydration, Werdnig-Hoffman disease, Guillain-Barré syndrome, myasthenia gravis, drug or toxin ingestions, metabolic disorders, and meningoencephalitis or myelitis.

Laboratory testing

• Laboratory confirmation requires detection of 1) botulinum toxin in stool or serum by using mouse neutralization assay or 2) isolation of toxigenic *C. botulinum* (or related clostridia) in the feces by using stool enrichment culture techniques.

• To avoid delay, treatment should be administered without awaiting laboratory confirmation.

Recommended treatment

• Primary therapy is supportive care with mechanically assisted ventilation when necessary.

• Prompt clinical diagnosis and treatment with Botulism Immune Globulin Intravenous (human) (BIG-IV) might reduce the recovery time. BIG-IV should be requested without awaiting laboratory confirmation.

• BIG-IV can be obtained from the California Department of Health Services, Infant Botulism Treatment and Prevention Program, telephone (510) 540-2646.

Prevention and reporting

• Avoid feeding honey to infants aged ≤12 months.

• Report all cases to local and state health departments.

Epidemiology, diagnosis, treatment, and prevention of foodborne botulism (*MMWR,* 2003, 52(2):25)

Epidemiology

• Caused by eating foods contaminated with preformed toxins of *Clostridium botulinum*

• Home-canned foods and raw or fermented Alaska Native dishes commonly associated with illness

- During 1973-1998, a total of 814 cases and an annual median of 24 cases (range: 14-94 cases) of foodborne botulism reported in the United States; 236 (29%) in Alaska.
- Humans affected by toxin types A, B, E, and rarely F; type E intoxication associated exclusively with eating marine animals
- Classified as a category A terrorism agent

Clinical findings
- Cranial nerve palsies
- Symmetrically descending flaccid voluntary muscle weakness possibly progressing to respiratory compromise
- Normal body temperature
- Normal sensory nerve examination findings
- Intact mental status despite groggy appearance
- Differential diagnosis includes Guillain-Barré syndrome, myasthenia gravis, stroke, drug overdose, and other entities

Laboratory findings
- Normal cerebrospinal fluid values
- Specific electromyography (EMG) findings including:
  - normal motor conduction velocities
  - normal sensory nerve amplitudes and latencies
  - decreased evoked muscle action potential
  - facilitation following rapid repetitive nerve stimulation
- Standard mouse bioassay positive for toxin from clinical specimens and/or suspect food; requires up to 4 days for final results

Recommended treatment
- Prompt administration of polyvalent equine-source antitoxin:
  - can decrease the progression of paralysis and severity of illness
  - will not reverse existing paralysis
  - available in the United States only through the public health system
- Place suspect cases in an intensive care setting
- Monitor for respiratory function deterioration every 4 hours using forced vital capacity testing.
- Provide mechanical ventilation if necessary

Prevention and control
- Boil raw or fermented Alaska Native dishes and home-canned foods ≥10 minutes before eating.
- Follow recommended home-canning procedures.
- Notify state health department immediately of suspected cases.

Commonly found in food poisoning: Poorly preserved meat, sausage, fruit, or vegetables (type A or B), marine products (type E), liver pate, venison jerky (type F); black tar heroin-intravenous drug abusers (wound botulism)

Percentage of Patients* With Symptoms of Food Poisoning Caused by Botulism by Symptom - Nan Province, Thailand, March 2006

Abdominal pain: 77%
Dry mouth: 51%
Nausea: 51%
Dysphagia: 37%
Colicky pain: 36%
Vomiting: 35%
Diarrhea: 26%
Headache: 18%
Dyspnea: 15%
Sweating: 12%
Ptosis: 11%
Weakness of extremities: 9%

* Includes 141 hospitalized patients and 10 persons treated as outpatients who were queried about their symptoms (n = 151)

Symptoms and Physical Findings in Patients With Types A and B Foodborne Botulism			
	Percentage of Patients Developing Symptoms or Signs		Significant Difference (p<0.05)
	Type A	Type B	
Symptoms			
Neurologic Symptoms			
Dysphagia	96	97	0
Dry mouth	9683	97,100	00
Diplopia	90	92	0
Dysarthria	100	69	+
Upper extremity weakness	86	64	0
Lower extremity weakness	76	64	0
Blurred vision	100	42	+
Dyspnea	91	34	+
Paresthesia	20	12	0
Gastrointestinal Symptoms			
Constipation	73	73	0
Nausea	73	57	0
Vomiting	70	50	0
Abdominal cramps	33	45	0
Diarrhea	35	8	+
Miscellaneous Symptoms			
Fatigue	92	69	0
Sore throat	75	39	+
Dizziness	86	30	+
Physical Findings			
Cranial Nerve Examination			
Ptosis	96	55	+
Hypoactive gag	81	54	0
Ophthalmoplegia	87	46	+
Facial palsy	84	48	+
Tongue weakness	91	31	+
Pupils fixed or dilated	33	56	0
Nystagmus	44	4	+
Ataxia	24	13	0
Extremity Power			
Upper	91	62	+
Lower	82	59	0
Deep Tendon Reflexes			
Hypoactive or absent	54	29	0
Hyperactive	12	0	0
Altered Sensorium	12	7	0

Adapted from Hughes, et al, "Clinical Features of Types A and B Foodborne Botulism," *Ann Intern Med*, 1981, 95:442-5.

Quarantine Standard precautions.

Environmental Persistency Stable in food or water for several weeks.

Specific References

Centers for Disease Control and Prevention (CDC), "Botulism From Home Canned Bamboo Shoots - Nan Province, Thailand, March 2006," *MMWR*, 2006, 55(14):389-418.

Cohen MA and Hern G, "Clinicopathological Conference: Sore Throat and Weakness in an Injection Drug User," *Acad Emerg Med*, 2000, 7(6):679-86.

"Infant Botulism - New York City, 2001-2002," *MMWR Morb Mortal Wkly Rep*, 2003, 52(2):21-4.

"Outbreak of Botulism Type E Associated With Eating a Beached Whale - Western Alaska, July 2002," *MMWR Morb Mortal Wkly Rep*, 2003, 52(2):24-6.

Robinson RF and Nahata MC, "Management of Botulism," *Ann Pharmacother*, 2003, 37(1):127-31.

Weimersheimer P, "Botulism," *Clin Tox Rev*, 2000, 22(7):1-2.

Clostridium perfringens

Pregnancy Issues None

Disease Syndromes Included Gangrene; Gas Gangrene; Pigbel

Microbiology *Clostridium perfringens* is an anaerobic gram-positive rod; occasionally it can appear gram-negative or gram-variable. It is a spore-forming organism, but the spores are not usually seen on Gram's stain. It has been termed "aerotolerant" because of its ability to survive when exposed to oxygen for limited periods of time.

The organism produces 12 toxins active in tissues and several enterotoxins which cause severe diarrhea. Four toxins can be lethal. The toxins separate the species into five types, A-E.

• Alpha toxin: A lecithinase that damages cell membranes. It is produced by *C. perfringens* type A. It is the major factor causing tissue damage in *C. perfringens*-induced gas gangrene (myonecrosis). The toxin is a phospholipase that hydrolyzes phosphatidylcholine and sphingomyelin and leads to increased vascular permeability, myocardial depression, hypotension, bradycardia, and shock.

• Enterotoxin: Produced mainly by *C. perfringens* type A but also by types C and D. This toxin is responsible for the diarrheal syndromes classically ascribed to this organism. The enterotoxin binds to intestinal epithelial cells after the human ingests food contaminated with *C. perfringens*. The small bowel (ileum) is primarily involved. The toxin inhibits glucose transport and causes protein loss.

• Beta toxin: Produced by *C. perfringens* type B and C. This toxin causes enteritis necroticans or pigbel. This disease is seen in New Guinea where some natives ingest massive amounts of pork at feasts after first gorging on sweet potatoes. The sweet potatoes have protease inhibitors, which prevent the person from degrading the beta toxin that is ingested in the contaminated pork.

• Epsilon toxin is a permease enzyme produced by *C. perfringens* types B and D. It can cause disruption in intestinal and vascular permeability and potassium efflux from cells. Pancytopenia can occur. The lethal dose (inhalation) is 1 mcg/kg with illness onset within 12 hours. No known person-to-person transmission.

Epidemiology *C. perfringens* is ubiquitous in the environment, being found in soil and decaying vegetation. The organism has been isolated from nearly every soil sample ever examined except in the sand of the Sahara desert. In the human, *C. perfringens* is common in the human gastrointestinal tract. In one study, it was found in 28 of 40 adults. It can also be commonly recovered from many mammals including cats, dogs, whales, and others.

Clinical Syndromes The organism can be pathogenic, commensal, or symbiotic. Important diseases caused by *C. perfringens* include the following.

• **Food poisoning:** *C. perfringens* is one of the most common causes of food poisoning in the United States. Foods commonly contaminated with *C. perfringens* are meat, poultry, and meat products such as gravies, hash, and stew. Human disease is caused by ingestion of heat-labile toxin. The highest risk comes from meats that are partially cooked, cooled, then reheated. Spores present in the food germinate during the reheating process. Symptoms include watery diarrhea, headache (42%), and abdominal cramps, which occur about 10-12 hours after the meal. Fever is generally not part of this illness. Vomiting is unusual, occurring in 13% to 30% of patients. Duration of symptoms is about 24 hours. The diagnosis is made by culturing the stools and, for epidemiological purposes, the food.

• **Pigbel:** See description under Microbiology above. This follows massive ingestion of contaminated pork and sweet potatoes. Incubation period is 24 hours. There is intense abdominal pain, bloody diarrhea, vomiting, shock, and intestinal perforation in some cases. A vaccine is available for travelers.

• **Gynecologic infections:** *C. perfringens* has been isolated in association with septic abortions, tubo-ovarian abscesses, and uterine gas gangrene. Interestingly, *C. perfringens* has been isolated from the blood of healthy females in the immediate postnatal period; the organism appeared to be a commensal in the blood since the patients remained well without antibiotic therapy. This underscores the often enigmatic nature of *C. perfringens*.

• **Skin and soft tissue infection colonization:** Many wounds can be contaminated with *C. perfringens*, particularly when there is an open wound exposed to soil. The organism may or may not be causing disease, and its relevance should be based on clinical findings; a superficial wound growing *C. perfringens* can often be treated using local care alone.

• **Anaerobic cellulitis:** *C. perfringens* alone or in a mixed infection causes a local tissue invasion with some necrosis. Patients are afebrile, and there is little pain or swelling. Gas may be quite noticeable in the infected tissues. Anaerobic cellulitis is seen in patients with infected diabetic foot ulcers and patients with perirectal abscesses. Localized infection can spread if not appropriately treated.

• **Fasciitis:** Patients with *C. perfringens* fasciitis present with rapidly progressive infection through soft tissue planes. **This is a medical emergency.** Anaerobic organisms cause pain, swelling, and gas formation; and patients frequently present in a florid sepsis syndrome. The classic setting for this is the patient with known colonic cancer with presumed abdominal fascial metastases; this is the nidus for colonic contamination, followed by rapid movement along the fascial planes. Although the muscle is not involved, the mortality of anaerobic cellulitis remains very high even with early surgical intervention.

• **Clostridial myonecrosis:** This entity has been described after traumatic wounds, especially during war time when wounds are grossly contaminated with soil. In World War II, 30% of battlefield wounds were associated with this, but in Vietnam only 0.02%. Clostridial myonecrosis can also be seen with crush injury, colon resections, septic abortions, and other conditions. Clinically, there is systemic toxicity, with tissue hypoxia and vascular insufficiency. The involved muscles appear black and gangrenous. Often there is abundant gas felt as crepitance in the wound. Treatment is surgical removal of devitalized tissue and, often, amputation.

Diagnosis The diagnosis of infection by *C. perfringens* begins with a high index of suspicion when a patient presents with one of the clinical syndromes described above. One important caveat is that the laboratory isolation of *C. perfringens* from a necrotic wound does not necessarily imply disease from this organism; many wounds can be colonized. In the right clinical setting, however, a positive *C. perfringens* culture and a compatible clinical presentation are highly suggestive of disease caused by this agent. Gram's stain preparations of wound material, uterine tissue, cervical discharge, muscle tissues, and other relevant materials should be made; a predominance of gram-positive rods should bring anaerobic infection, including *C. perfringens*, to mind.

In patients with clostridial septicemia, accompanying abnormalities may be a clue to clostridial disease before culture confirmation is made. Patients may have disseminated intravascular coagulation with a brisk hemolysis, hemoglobinuria, and proteinuria. X-rays of diseased areas may reveal the presence of gas in muscle or soft tissue; this is suggestive of clostridial infection, although other anaerobes (and aerobes) can cause gas formation.

If clostridial myonecrosis is suspected, a muscle biopsy usually performed at the time of tissue debridement, can be diagnostic.

Diagnostic Test/Procedure
Anaerobic Culture
Gram's Stain
Skin Biopsy

Treatment Strategy In general, the treatment of clostridial infection is high-dose penicillin G, to which the organism has remained susceptible.
Penicillin G: I.V.:
Children: 100,000-250,000/kg/day divided every 4 hours
Adults: 10-40 million units daily divided every 4 hours
Alternative dosing: Clindamycin: I.V.:
Children: 20 mg/kg every 8 hours
Adults: 600 mg every 8 hours
Chloramphenical: I.V.:
Children: 50-75 mg/kg/day divided every 6 hours
Adults: 500-1000 mg every 6 hours

For skin and soft tissue infections, the extent of infection determines the need for surgical debridement. When wounds are simply colonized with clostridia, neither antibiotics nor surgery are indicated. Localized soft tissue infections can often be managed by surgical debridement alone, without antibiotics. When systemic symptoms are present and there is extension of infection into deeper tissues, antibiotics and surgical intervention are required. In fulminant cases of gas gangrene with myonecrosis, immediate surgical intervention (debridement, amputation) is the primary treatment of choice, and antibiotics have little effect. Hyperbaric oxygen with surgery and antibiotic therapy may reduce mortality, although its effect on epsilon toxin is not known.

Additional Information Equipment can be decontaminated with soap and water.

Specific References
Greenfield RA, Brown BR, Hutchins JB, et al, "Microbiological, Biological, and Chemical Weapons of Warfare and Terrorism," *Am J Med Sci*, 2002, 323:326-40.
McClane BA, "The Complex Interactions Between *Clostridium perfringens* Enterotoxin and Epithelial Tight Junctions," *Toxicon*, 2001, 39:1781-91.
Tweten RK, "*Clostridium perfringens* Beta Toxin and *Clostridium septicum* Alpha Toxin: Their Mechanisms and Possible Role in Pathogenesis," *Vet Microbiol*, 2001, 82:1-9.

Coxiella burnetii (Q Fever)

Clinically Resembles Viral illness, influenza; chronic syndrome may resemble chronic fatigue syndrome

Disease Syndromes Included Q Fever

Microbiology *Coxiella burnetii* is an obligately intracellular, gram-negative coccobacillus and a member of the group of organisms known as rickettsiae (*Rickettsia*, *Ehrlichia*, *Rochalimaea*, and *Coxiella*). *C. burnetii* survives extracellularly probably in spore form and can survive weeks to years in adverse environmental conditions. *C. burnetii* exists in two antigenic forms or "phases." Phase I organisms are avirulent, exist in nature and in laboratory animals, and react serologically with convalescent sera. Phase II organisms are virulent and react with acute sera. If phase I organisms are passed through chicken eggs, the phase I organisms become phase II organisms. *C. burnetii* is extremely infectious to humans; a single organism can cause infection.

Postmortem Care Droplet precautions/respiratory isolation

Epidemiology Q fever was first described in Australia in 1937 (Q: "query"), occurs worldwide, and is not uncommon. Cattle, goats, sheep, and ticks are natural reservoirs. Persons at high risk for Q fever are persons associated with slaughter houses, persons who work with the aforementioned animals, persons who work with hides and wool, farmers, veterinarians, and researchers. The most common causes of infection are inhalation of organisms, handling of infected birth products (especially cat and cattle placentas), skinning infected animals, and transport of infected animals. *C. burnetii* is not transmitted person-to-person and only rarely by blood products.

Clinical Syndromes The most common form of Q fever is a self-limited febrile illness. Humans are the only known animals that regularly develop infection with *C. burnetii*. Incubation period is 2-4 weeks. Pneumonia, hepatitis, and endocarditis are the most common manifestations. Q fever usually presents as an atypical pneumonia; Q fever can also present as a rapidly progressive disease or as a secondary finding in someone with fever of unknown origin. Duration of acute illness is about two weeks. The hepatitis form can present as infectious hepatitis. Endocarditis is the primary manifestation of chronic Q fever. Complete blood count is usually normal, while liver function tests demonstrate a 2- to 3-fold elevation in transaminase levels with normal serum bilirubin. The chronic syndrome occurs in about 5% of individuals and can result in osteoarthritis, vascular aneurysm, and can last over 6 months.

Infective Dose 1 to 10 organisms

Diagnosis Typical signs and symptoms include severe headache, unusually high fever (up to 105°F) associated with slow pulse rate, chills, sore throat, fatigue, and myalgia. Pulse-temperature dissociation can occur (high fever accompanied by an inappropriately low heart rate). Patients taking chronotropic (heart rate) suppressors such as beta adrenergic blockers may confound this picture. Other agents causing this important clinical manifestation include Coxiella, certain VHF, Brucellosis, and Tularemia. Other symptoms depend on which organ is affected. A rash can occur in the chronic endocarditis form of Q fever; a rash occurs only rarely in acute Q fever. Leukocytosis is seen in about 25% of cases with elevated serum creatinine levels and creatinine phosphokinase levels noted in over 20% of cases. Isolation of *C. burnetii* is an extremely dangerous laboratory procedure and is not practical. The most practical and clinically relevant method is serology. Complement fixation (CF) and immunofluorescence (IF) are the two most commonly used serological methods. CF is widely available and can demonstrate a fourfold rise in titer; however, CF does not separate IgG from IgM. Immunofluorescence uses separate phase I and II antigens and is more clinically useful. Phase II and I antibody titers are usually high in acute and convalescent disease, respectively. IgM antibody can persist for more than a year in some cases, but this is not a common finding. Some laboratories offer enzyme immunoassay (EIA) serology. EIA probably is the most sensitive of the three tests. Mortality rate: 1% to 2%

Diagnostic Test/Procedure Q Fever Serology

Treatment Strategy
  Decontamination:
    Dermal: Remove contaminated clothing and wash skin thoroughly with soap and water
    Inhalation: Administer 100% humidified oxygen
  Acute Q fever: Oral:
    Children >8 years: Tetracycline 25 mg/kg/day in divided doses for 2-3 weeks
    Adults: Doxycycline (100 mg twice daily) **or** tetracycline (500 mg 4 times/day) for 2-3 weeks or until patient is afebrile for one week
    Alternative adult therapy: Oral:
      Ofloxacin 200 mg every 12 hours for 2-3 weeks or until patient is afebrile for one week **or** Pefloxacin 400 mg I.V. or orally every 12 hours for 2-3 weeks or until patient is afebrile for one week
  Chronic Q fever: Oral:
    Doxycycline 100 mg twice daily or ofloxacin 400-600 mg/day for 3 years-lifetime
    Hydroxychloroquine 200 mg 3 times/day can be added to alkalinize fluids and thus help in eradicating organisms in phagocytes
  Granulomatous hepatitis can respond to oral prednisone (0.5 mg/kg/day) if fever persists following antibiotic therapy. Steroid dose can be tapered over 1 month.

Quarantine Droplet precautions/respiratory isolation

Environmental Persistency Very stable.

Additional Information A Q fever vaccine is available in Australia and Eastern Europe (dose is 0.5 mL SubQ and is given to individuals at high risk for exposure). Prophylactic regimen (given 8-12 days postexposure): Doxycycline for 14 days. May prolong illness if doxycycline is started before 8 doses postexposure. Recovery of Q fever without antibiotics is 99%, but antibiotics may shorten the febrile course in half.

Can be cleared to work in research laboratory 3 weeks after vaccination (0.5 mL SubQ); no booster doses required.

Specific References

Alarcon A, Villanueva JL, Viciana P, et al, "Q Fever: Epidemiology, Clinical Features and Prognosis. A Study From 1983 to 1999 in the South of Spain," *J Infect*, 2003, 47:110-6.

Houpikian P, Habib G, Mesana T, et al, "Changing Clinical Presentation of Q Fever Endocarditis," *Clin Infect Dis*, 2002, 34:E28-31.

Marrie TJ and Raoult D, "Update on Q fever, Including Q Fever Endocarditis," *Curr Clin Top Infect Dis*, 2002, 22:97-124.

Nicholson WL, McQuiston J, Vannieuwenhoven TJ, et al, "Rapid Deployment and Operation of a Q Fever Field Laboratory in Bosnia and Herzegovina," *Ann N Y Acad Sci*, 2003, 990:320-6.

Rusnak JM, Kortepeter MG, Aldis J, et al, "Experience in the Medical Management of Potential Laboratory Exposures to Agents of Bioterrorism on the Basis of Risk Assessment at the United States Army Medical Research Institute of Infectious Diseases (USAMRIID)," *J Occup Environ Med*, 2004, 46(8):801-11.

Rusnak JM, Kortepeter MG, Hawley RG, et al, "Management Guidelines for Laboratory Exposures to Agents of Bioterrorism," *J Occup Environ Med*, 2004, 46(8):791-800.

Scola BL, "Current Laboratory Diagnosis of Q Fever," *Semin Pediatr Infect Dis*, 2002, 13:257-62.

Cryptosporidium

Pregnancy Issues None known

Disease Syndromes Included Cryptosporidiosis

Microbiology Although *Cryptosporidium* was identified as early as 1907, it was not recognized as a human pathogen until 1976. *Cryptosporidium* is classified as a protozoan and is structurally related to *Toxoplasma* and *Plasmodium* species. Several species of *Cryptosporidium* have been found; the species most associated with human disease is *C. parvum*. The organism is approximately 2.5 μm in diameter and is about the same size and shape as many yeasts. *Cryptosporidium* can be identified by use of an acid-fast stain or an immunofluorescence test of a fresh stool specimen.

The parasite has four spores (sporozoites) within an oocyst. Its life cycle occurs within a single host. Human infection begins when the mature oocyst form of *Cryptosporidium* is ingested or perhaps inhaled. The four sporozoites excyst and divide asexually into meronts, which release merozoites that can either reinvade the human or develop sexually into meronts. The sexual cycle ends in the formation of oocysts. The oocysts then can exit the body of the host through the feces to begin the infectious cycle in a new human or in the same host (autoinfection).

Epidemiology Cryptosporidial infections occur worldwide. In the United States, the organism is most prevalent in persons with AIDS, with infection rates ranging from 3% to 20%. In such countries as Africa and Haiti, *Cryptosporidium* can infect over 50% of the AIDS population. The organism is also a cause of sporadic diarrheal illness in normal individuals. Outbreaks of cryptosporidiosis have been well-described in the U.S. and include a recent outbreak in Wisconsin related to a contaminated metropolitan water supply.

The modes of transmission are as follows:

• Person-to-person: Accounts for spread of infection within day care centers, households, and hospitals. Transmission is usually by a fecal-oral route.

• Environmental contamination: The oocyst can be found in rivers and other natural water supplies, and thus, the traveler and camper is at risk. *Cryptosporidium* resists standard chlorination procedures in many areas and thus, can initiate major outbreaks of diarrheal illness when water supplies are contaminated.

• Farm animal-to-human

• Foodborne illness

Clinical Syndromes

• **Acute diarrhea in the normal host:** The most common manifestation of human cryptosporidiosis is a syndrome characterized by profuse, watery diarrhea (often with mucus), malaise, and abdominal cramping or pain. Fevers to the 39°C to 40°C range have occurred even in immunologically competent individuals. The incubation period is estimated to be 2-14 days following exposure to the parasite. The diarrhea often has an acute onset and symptoms can last 2 weeks or more.

• **Acute diarrhea in the immunocompromised host:** Cryptosporidiosis can be more subacute in onset. However, it is still characterized by voluminous diarrhea and weight loss over a several-week period can occur.

• **Cryptosporidial cholecystitis:** This has been described in immunocompromised patients. Symptoms include right upper quadrant abdominal pain, nausea, vomiting, and fever. There can be a concomitant diarrheal illness.

Diagnosis The symptoms of human cryptosporidiosis are nonspecific and cannot be distinguished from other causes of acute diarrheal illnesses. A high level of suspicion must be maintained because the organism cannot be readily identified using standard fecal smears. Standard laboratory studies are not helpful in establishing the diagnosis. The leukocyte count may be normal, and there is usually not a peripheral blood eosinophilia to suggest a parasitic disease. Abnormalities in radiographic studies such as barium enemas and small bowel series have been described but again are not diagnostic of the organism. The only reliable means of identifying *Cryptosporidium* is to perform a direct exam (acid-fast stain) of a stool sample. This allows rapid identification of the organism. Because excretion of the parasite in the feces may be intermittent, it is recommended that two or more separate samples be submitted to the laboratory. Other techniques for identification of *Cryptosporidium* have been developed, including a direct immunofluorescence assay (DFA) and an enzyme immunoassay (EIA).

Diagnostic Test/Procedure

Acid-Fast Stain

Cryptosporidium Diagnostic Procedures, Stool

Treatment Strategy Supportive care and fluid replacement are treatment of choice

Usual Dosage

Nitazoxanide (100 mg orally) every 12 hours for 3 days in ages 1-4 years or 500 mg orally every 6 to 12 hours for 3 days (ages 4-11 years) may be effective. Paromomycin (25 mg/kg to 35 mg/kg per day orally) in children in 2-4 divided doses or 500 mg 3 times/day for 7 days in adults may also be effective.

Quarantine Enteric precautions

Drug Therapy Comment Antidiarrheal agents are not recommended in children or infants for routine care. Most cases are self-limiting.

Specific References

Chen XM, Keithly JS, Paya CV, et al, "Cryptosporidiosis," *N Engl J Med*, 2002, 346:1723-31.

Gradus MS, "*Cryptosporidium* and Public Health: From Watershed to Water Glass," *Clin Microbiol Newslett*, 2000, 22(4):25-32.

Hlavsa MC, Watson JC, and Beach MJ, "Cryptosporidiosis Surveillance – United States 1999-2002." *MMWR Surveill Summ*, 2005, 28;54(1):1-8.

Hoveyda F, Davies WA, and Hunter PR, "A Case of Cryptosporidiosis in Pregnancy," *Eur J Clin Microbiol*, 2002, 21:637-8.

Leav BA, Mackay M, and Ward HD, "*Cryptosporidium* Species: New Insights and Old Challenges," *Clin Infect Dis*, 2003, 36:903-8.

Encephalitis, Viral

Synonyms Viral Encephalitis

Clinical Presentation The term encephalitis refers to inflammation of the brain parenchyma, as opposed to the term meningitis, which refers to inflammation mainly confined to the meninges. There is some overlap between these two important syndromes and a number of organisms can cause either encephalitis or meningitis. Encephalitis can have both infectious and noninfectious etiologies. The most common etiology of encephalitis is from viral infection, with about 20,000 cases of viral encephalitis in the United States per year. Important epidemiologic clues to evaluate include the seasonality and exposure history (animal bites, rodent exposure, tick bites). Patients with viral encephalitis have variable degrees of fever and the presentation is usually acute or subacute. The hallmark of encephalitis is an altered level of consciousness, which can vary from mild lethargy to coma. Typically, patients are confused and delirious, and bizarre behavior is not uncommon such as hallucinations, psychosis, and personality change. On examination, common focal neurologic findings include ataxia and aphasia, hemiparesis (often with increased deep tendon reflexes and extensor plantar responses), and cranial nerve deficits such as ocular and facial palsies. Because of the disruption of the hypothalamic-pituitary axis in viral encephalitis, fever, diabetes insipidus, and the syndrome of inappropriate ADH (SIADH) can accompany the mental status changes. In general, it is not possible to distinguish one type of viral encephalitis from other forms of encephalitis on clinical grounds alone, despite the pathologic evidence that different viruses injure different areas of the brain. The CSF profile in viral encephalitis is similar to viral meningitis and consists of a modest elevation in the CSF WBCs, mainly lymphocytic. In over 95% of cases, the CSF contains >5 cells/μL. More marked CSF pleocytosis is unusual and in only 10% of the time, the CSF WBCs will be >500 cells/μL. The CSF WBCs are usually lymphocytes. If the CSF WBCs are mainly polymorphonuclear leukocytes, this is more suggestive of bacterial etiologies, leptospirosis, and noninfectious causes (acute hemorrhagic leukoencephalitis). Occasionally, viral encephalitis from eastern equine encephalitis or enteroviruses can give polymorphonuclear leukocytes in the CSF. There is usually a mild increase in the CSF protein, and the CSF glucose is usually normal.

Differential Diagnosis Infectious causes of encephalitis (nonviral): *Listeria* rhombencephalitis; *Mycoplasma pneumoniae* (especially if a pulmonary infiltrate is present); *Legionella*; abscess and subdural empyema; *Mycobacterium* tuberculosis; fungal (*Cryptococcus*, others); *Rickettsia*; *Bartonella* (cat scratch encephalitis)
Noninfectious "mimics" of encephalitis: Vascular diseases; toxic encephalopathy; subdural hematoma; Reye's syndrome; systemic lupus erythematosus

Likely Pathogens

The most common cause of community-acquired encephalitis is herpes simplex virus encephalitis (HSV-1). Other common causes of viral encephalitis include arboviruses, which are a diverse group of viruses that cause encephalitis via an arthropod vector, and enteroviruses. Mumps virus is a potential cause of encephalitis, but the number of cases has decreased as the disease has become less common due to routine vaccination. Less common causes of viral encephalitis are HIV, Epstein-Barr virus, cytomegalovirus, varicella-zoster virus, measles virus, and adenoviruses. There are a number of other viruses that can cause encephalitis, but these are rare. One important rare cause is rabies.

Herpes Simplex Virus
Arboviruses
Enterovirus
Human Immunodeficiency Virus
Measles Virus
Epstein-Barr Virus
Cytomegalovirus
Varicella-Zoster Virus
West Nile Virus

Rabies Virus

Influenza Virus

Diagnostic Test/Procedure

Since it is nearly impossible to establish a specific etiology of viral encephalitis on the basis of clinical presentation alone, laboratory diagnosis remains essential. In many cases, however, a specific viral (or nonviral) etiology of encephalitis cannot be determined, due in part to limitations in laboratory diagnostics. Cultures of cerebrospinal fluid in viral encephalitis are invariably negative, due to poor sensitivity. Polymerase chain reaction (PCR) is considered the diagnostic procedure of choice for CSF analysis. The most commonly ordered diagnostic test on CSF in cases of encephalitis is the herpes simplex virus PCR since (1) it is the most common cause of encephalitis, (2) HSV culture and antigen are insensitive, (3) findings on MRI and EEG can be highly suggestive of HSV encephalitis but are not always present, and (4) the sensitivity and specificity of CSF PCR is equivalent to brain biopsy for the diagnosis of HSV encephalitis. For other viruses, CSF PCR is available but less studied than HSV PCR. There is growing literature on the utility of enteroviral CSF PCR. Acute and convalescent serum titers for arboviruses are necessary to establish the diagnosis of arboviral encephalitis.

CAT scan or MRI of the brain should be performed in patients with encephalitis to rule out a structural brain lesion. The finding of a focal area of encephalitis in the temporal-parietal area suggests HSV encephalitis, although this finding is not entirely specific. EEG may show focal spikes on a background of slow activity in the temporal area in HSV encephalitis. The need for brain biopsy in the diagnosis of viral encephalitis has declined in recent years due to the sensitivity and specificity of HSV CSF PCR.

Cerebrospinal Fluid Analysis

Polymerase Chain Reaction

Encephalitis Viral Serology (Arbovirus serology)

Computed Transaxial Tomography, Head Studies

Magnetic Resonance Scan, Brain

Lumbar Puncture

Electroencephalography

Drug Therapy Comment Of the many potential etiologies of viral encephalitis, few are treatable. Establishing a diagnosis of HSV encephalitis is important since this is potentially treatable. Acyclovir is the drug of choice for HSV encephalitis given at a dose of 10 mg/kg every 8 hours I.V. for at least 14 days. HIV encephalitis is usually a manifestation of advanced AIDS, although occasionally has been reported in individuals with early disease; this may or may not respond to antiretroviral agents. An investigational antiviral agent (pleconaril) is being studied for enteroviral meningitis and encephalitis, but no agents are currently approved for enterovirus. There is no therapy available for arboviral encephalitis at this time. Oseltamivir may be effective in treatment of influenza B infection.

Specific References

Hinson VK and Tyor WR, "Update on Viral Encephalitis," *Curr Opin Neurol*, 2001, 14(3):369-74.

Marfin AA and Gubler DJ, "West Nile Encephalitis: An Emerging Disease in the United States," *Clin Infect Dis*, 2001, 33(10):1713-9.

Straumanis JP, Tapia MD, and King JC, "Influenza B Infection Associated With Encephalitis: Treatment With Oseltamivir," *Pediatr Infect Dis J*, 2002, 21(2):173-5.

Escherichia coli, Enterohemorrhagic

Synonyms *E. coli* O157:H7; EHEC

Microbiology Enterohemorrhagic *E. coli* causes a distinct form of hemorrhagic colitis in humans. Like other *E. coli* strains, enterohemorrhagic *E. coli* is a facultative, gram-negative bacillus. The most common serotype is O157:H7. Essentially all published information regarding enterohemorrhagic *E. coli* refers only to this serotype.

Epidemiology In 1982, the first large-scale outbreak of *E. coli* O157:H7 colitis was described. Multiple cases of severe bloody diarrhea were found to be epidemiologically linked to ingestion of contaminated hamburger meat. Since then, the organism has been recognized as an important cause of bloody diarrhea and the hemolytic uremic syndrome. Over 12 major outbreaks have been reported, along with numerous sporadic cases. The majority of cases have been traced to contaminated ground beef, although other potential sources have been cited, including unpasteurized milk, apple cider, municipal water, and roast beef. The organism inhabits the gastrointestinal tract of some healthy cattle and is thought to contaminate meat during slaughter and the processing of ground beef ("internal contamination"). If the ground beef is undercooked, the organism remains viable; undercooking of hamburger patties has proven important in several outbreaks.

In 1993, a well-publicized multistate outbreak of *E. coli* O157:H7 took place in the western United States (Washington, California, Idaho, and Nevada). Over 500 infections and four deaths were documented. The vast majority of cases were ultimately linked to contaminated hamburger meat from a particular restaurant chain. Further investigation by the Centers for Disease Control identified several slaughter plants in the United States and one in Canada as the probable source. Thousands of contaminated patties not yet consumed were discovered. In March, 1994, the USDA Food Safety and Inspection Service recommended that all raw meat should be cooked thoroughly, with an increase in the internal temperature for cooked hamburgers to 155°F.

A 2-year nationwide surveillance study by the Centers for Disease Control has found *E. coli* O157:H7 to be the most commonly identified pathogen associated with bloody diarrhea. In many parts of the U.S., *E. coli* O157:H7 is the second most common cause of bacterial diarrhea.

Acquisition of disease is usually by ingestion of contaminated food, but person-to-person transmission has been documented, especially in day care centers. Children and elderly individuals are at highest risk for severe infections. Simple and careful hand washing essentially eliminates the probability of person-to-person transmission.

Clinical Syndromes

• **Hemorrhagic colitis:** *E. coli* O157:H7 causes a bloody diarrhea associated with abdominal cramps. Pathologically, there is no invasion or inflammation of the intestinal mucosa, and thus fever is often absent. The diarrhea is caused by shiga-like toxins. In most cases, the illness resolves within 7 days, but death can occur in the elderly.

• **Hemolytic uremic syndrome:** Approximately 5% to 10% of patients with *E. coli* O157:H7 diarrhea develop a syndrome characterized by acute renal failure, thrombocytopenia, and evidence of hemolysis on a peripheral blood smear. Children are at high risk for this syndrome. The patient may be toxic-appearing, and the presentation may be confused with a variety of diseases including sepsis with disseminated intravascular coagulation, vasculitis, thrombotic thrombocytopenia purpura, and others. A good history and early index of suspicion are essential to early diagnosis, which may be life-saving. The estimated mortality is 3% to 5%.

Diagnosis Enterohemorrhagic *E. coli* should be strongly considered in any patient presenting with bloody diarrhea, whether or not hemolytic uremic syndrome is present. It is likely that many sporadic cases of *E. coli* O157:H7 diarrhea occur in the community and go unrecognized for two reasons: many clinicians do not order stool cultures for stable patients with diarrhea; many microbiology laboratories do not routinely culture stools for *E. coli* O157:H7 unless there is a specific order from the physician.

Diagnosis is confirmed by isolation of *E. coli* O157:H7 from stool specimens and subsequent serological confirmation. This requires special media in the Microbiology Laboratory (sorbitol-MacConkey medium). Other methods for the rapid detection of this organism are currently under study.

Diagnostic Test/Procedure

Shiga Toxin Test, Direct

Stool Culture, Diarrheagenic *E. coli*

Drug Therapy Comment No antibiotics proven effective; in fact, antibiotics may increase the risk of developing HUS.

Specific References

Bender JB, Hedberg CW, Besser JM, et al, "Surveillance by Molecular Subtype for *Escherichia coli* O157:H7 Infections in Minnesota by Molecular Subtyping," *N Engl J Med*, 1997, 337(6):388-94.

Boyce TG, Pemberton AG, Wells JG, et al, "Screening for *Escherichia coli* O157:H7-A Nationwide Survey of Clinical Laboratories," *J Clin Microbiol*, 1995, 33:3275-7.

Mahon BE, Griffin PM, Mead PS, et al, "Hemolytic Uremic Syndrome Surveillance to Monitor Trends in Infection With *Escherichia coli* O157:H7 and Other Shiga Toxin-Producing *E. coli*," *Emerg Infect Dis*, 1997, 3:409-12.

Mead PS and Griffin PM, "*Escherichia coli* O157:H7," *Lancet*, 1998, 352:1207-12.

Slutsker L, Ries AA, Greene KD, et al, "*Escherichia coli* O157:H7 Diarrhea in the United States: Clinical and Epidemiologic Features," *Ann Intern Med*, 1997, 126:505-13.

Francisella tularensis

Disease Syndromes Included Tularemia

Microbiology *Francisella tularensis* is a nonmotile, pleomorphic, strictly aerobic, gram-negative rod. The bacterium possesses a lipid capsule, which is a virulence factor and that might be responsible, at least in part, for the ability of the bacterium to survive weeks to months in adverse environmental conditions such as water, mud, and decaying animal carcasses. The bacterium is biochemically inert when grown (for identification purposes) *in vitro* on sugars or other substrates. The bacterium exists as two clinically significant biogroups. Biogroup *F. tularensis* (type A, mortality rate 5%) is found in North America and produces the most severe form of tularemia. Biogroup *Palearctica* (type B, extremely low mortality) is found only in the northern hemisphere (particularly Asia and Europe) and produces a milder form of tularemia. Biogroup *Francisella tularensis* causes tularemia (rabbit fever, deer fly fever), a zoonosis of wild animals and the third most common human tick-borne illness in the United States.

Postmortem Care Standard precautions; avoid procedures likely to cause aerosolization of body fluids

Epidemiology *Francisella tularensis* is found throughout the United States (except the Southeast, the Northeast, and the Great Lakes areas). Tularemia is endemic in Missouri, Arkansas, and Oklahoma (together, 50% of all United States cases). The bacterium is strikingly absent from the United Kingdom, Africa, South America, and Australia. Hundreds of wild animal species and common house pets are hosts of the bacterium, which is perpetuated freely and often in nature as it is passed from wild animal to wild animal by ectoparasites, poor environmental conditions, and less-than-respectable eating and culinary habits. The most common vectors that transmit *F. tularensis* to humans (incidental and dead-end hosts) are ticks and biting flies. In addition, humans who handle hides, woodland water, and animal carcasses can acquire the bacterium by the respiratory route. Occupations associated with a higher risk for tularemia are laboratory worker, veterinarian, sheep worker, hunter, trapper, and meat handler. Since 1965, the number of cases of tularemia in the United States has remained between 0.05 and 0.15 per 100,000. Person-to-person transmission of tularemia has not been documented.

Clinical Syndromes The severity of tularemia depends on the virulence of the biotype, portal of entry, inoculation dose, extent of dissemination, and immunocompetence of the host. After an incubation period of 2-10 days, flu-like symptoms usually occur. These symptoms can be chronic and debilitating. A nonhealing skin ulcer or lesion can develop at the cutaneous portal of entry and last for months. Febrile patients may exhibit relative bradycardia, sometimes referred to as pulse-temperature dissociation. This clinical sign can also be seen with viral hemorrhagic fevers and plague. Tularemia usually presents in one or more of the following forms.

- **Ulceroglandular:** 21% to 87% of cases; obvious nonhealing, erythematous, eroding ulcers
- **Glandular:** 3% to 20% of cases; cutaneous ulcers are not found
- **Oculoglandular:** 0% to 5% of cases; severely painful, yellow, pinpoint conjunctival ulcers
- **Esophageal:** Severely painful sore throat; enlarged tonsils; white pseudomembrane
- **Systemic:** 5% to 30% of cases; "typhoidal form"; acute septicemia; classic ulcers and lymphadenopathy usually not present
- **Gastrointestinal:** Consumption of contaminated food and water; persistent diarrhea; fulminating and often fatal
- **Pulmonary:** 7% to 20% of cases; usually presents as a nonproductive pneumonia, which is observed radiographically but not clinically

The most common complaints of these forms of tularemia are lymphadenopathy and necrosis of infected lymph nodes (even with appropriate treatment). Severe cases of tularemia are complicated - dissemination of the bacterium, toxemia, DIC, renal failure, and hepatitis. The most likely form of intentional release for *F. tularensis* organisms would be via infectious aerosols. In 1969, the World Health Organization estimated that an aerosol dispersal of 50 kg of virulent *F. tularensis* over a metropolitan area with 5 million inhabitants in a developed country would result in 250,000 illnesses, including 19,000 deaths. Most of the cases would be primary pneumonic tularemia; however, some would present with either a nonspecific febrile illness of varying severity or oculoglandular tularemia from eye contamination or glandular/ulceroglandular disease through exposure of broken skin to infectious aerosol or esophageal disease through inhalation of organisms. The incubation period in this case could be as little as one day to 14 days after exposure.

Infective Dose 1-10 organisms

Diagnosis Physicians must take a complete physical, occupational, recreational, and travel history, and, preferably, be suspicious of tularemia. Febrile patients may exhibit relative bradycardia, sometimes referred to as pulse-temperature dissociation. A pulse-temperature disassociation has been noted in 42% of patients. This clinical sign can also be seen with typhoid fever as well as other bioweapon illness including certain viral hemorrhagic fevers and plague. Cultures of blood, tissue, gastric washings, and sputum are possible and can yield the bacterium; however, culture is extremely nonproductive. Physicians must notify laboratory personnel when culture for *F. tularensis* is ordered because the bacterium is an extreme health hazard to laboratory workers. The most recommended, useful, and productive tests to help diagnose tularemia are serological methods (standard tube agglutination, haemagglu-tination, and enzyme immunoassay [the most sensitive tests]). Antibodies in sera from infected persons are detectable and are highest 2-5 weeks postinfection, respectively. A single titer ≥160 and a fourfold rise in titer are presumptive and diagnostic for tularemia, respectively. Titers ≥1024 are common late in the acute stage of disease. Both IgG and IgM titers of 20-80 can persist for years. (*MMWR*, 2004, Figure 11)

Diagnostic Test/Procedure
Tularemia Serology
Gram Stain
Culture
DFA
PCR

Treatment Strategy See tables

Working Group Consensus Recommendations for Treatment of Patients With Tularemia in a Contained Casualty Setting*	
Contained Casualty Recommended Therapy	
Adults	
Preferred choices	
	Streptomycin, 1 g I.M. twice daily
	Gentamicin, 5 mg/kg I.M. or I.V. once daily[†]
Alternative choices	
	Doxycycline, 100 mg I.V. twice daily
	Chloramphenicol, 15 mg/kg I.V. 4 times/day[†]
	Ciprofloxacin, 400 mg I.V. twice daily[†]
Children	
Preferred choices	
	Streptomycin, 15 mg/kg I.M. twice daily (should not exceed 2 g/day)
	Gentamicin, 2.5 mg/kg I.M. or I.V. 3 times/day[†]
Alternative choices	
	Doxycycline; if weight ≥45 kg, 100 mg I.V. twice daily; if weight <45 kg, 2.2 mg/kg I.V. twice daily
	Chloramphenicol, 15 mg/kg I.V. 4 times/day[†]
	Ciprofloxacin, 15 mg/kg I.V. twice daily[†‡]
Pregnant Women	
Preferred choices	
	Gentamicin, 5 mg/kg I.M. or I.V. once daily[†]
	Streptomycin, 1 g I.M. twice daily
Alternative choices	
	Doxycycline, 100 mg I.V. twice daily
	Ciprofloxacin, 400 mg I.V. twice daily[†]

*Treatment with streptomycin, gentamicin, or ciprofloxacin should be continued for 10 days; treatment with doxycycline or chloramphenicol should be continued for 14-21 days. Persons beginning treatment with intramuscular (I.M.) or intravenous (I.V.) doxycycline, ciprofloxacin, or chloramphenicol can switch to oral antibiotic administration when clinically indicated.
[†]Not a U.S. Food and Drug Administration-approved use.
[‡]Ciprofloxacin dosage should not exceed 1 g/day in children
Adapted from Dennis DT, Inglesby TV, Henderson DA, et al, "Tularemia As a Biological Weapon: Medical and Public Health Management," *JAMA*, 2001, 285(21):2763-73.

Working Group Consensus Recommendations for Treatment of Patients with Tularemia in a Mass Casualty Setting and for Postexposure Prophylaxis*	
Mass Casualty Recommended Therapy	
Adults	
Preferred choices	
	Doxycycline, 100 mg orally twice daily
	Ciprofloxacin, 500 mg orally twice daily[†]
Children	
Preferred choices	
	Doxycycline; if ≥45 kg, give 100 mg orally twice daily; if <45 kg, give 2.2 mg/kg orally twice daily
	Ciprofloxacin, 15 mg/kg orally twice daily[†‡]

Table (*Continued*)	
Pregnant Women	
Preferred choices	
	Ciprofloxacin, 500 mg orally twice daily[†]
	Doxycycline, 100 mg orally twice daily

[*]One antibiotic appropriate for patient age, should be chosen from among alternatives. The duration of all recommended therapies in this table is 14 days.
[†]Not a U.S. Food and Drug Administration-approved use.
[‡]Ciprofloxacin dosage should not exceed 1 g/day in children
Adapted from Dennis DT, Inglesby TV, Henderson DA, et al, "Tularemia As a Biological Weapon: Medical and Public Health Management," *JAMA*, 2001, 285(21):2763-73.

Postexposure vaccination is under investigation. Because of the long time to develop immunity postvaccination and the relatively short incubation period, vaccination does not appear to be effective.

Quarantine Standard precautions; isolation is not recommended; clothing and linen can be decontaminated using standard protocols. There is no postexposure prophylactic antibiotic treatment regimen.

Drug Therapy Comment Antimicrobial susceptibility testing cannot be performed with *F. tularensis* because the bacterium is too fastidious to be tested by standardized, reliable methods. Aminoglycosides (especially streptomycin) are the antimicrobial agents of choice. Doxycycline, fluoroquinolones and chloramphenicol are alternatives. Generally tetracycline, beta-lactams (except for imipenem), sulfonamides, and macrolides are not as effective as aminoglycosides. Relapses are more common with both tetracycline and chloramphenicol than the aminoglycosides probably due to its bacteriostatic action rather than cidal activity. Tetracycline should be administered at a minimum dose of 2 g/day to be effective. Chloramphenicol should be added to the aminoglycosides in the treatment of meningitis secondary to this organism. Tetracycline has also been used with some success in the treatment of tularemia. Erythromycin is also active but little clinical experience is available.

Environmental Persistency May survive in a cold, moist environment for several months

Additional Information Can be cleared to work in research lab after any positive skin reaction at site following live vaccine administration with seroconversion.

Specific References

Dennis DT, Inglesby TV, Henderson DA, et al, "Tularemia as a Biological Weapon: Medical and Public Health Management," *JAMA*, 2001, 285(21):2763-73.

Feldman KA, Enscore RE, Lathrop SL, et al, "An Outbreak of Primary Pneumonic Tularemia on Martha's Vineyard," *N Engl J Med*, 2001, 345(22):1601-6.

Grunow R, Splettstoesser W, McDonald S, et al, "Detection of *Francisella tularensis* in Biological Specimens Using a Capture Enzyme-Linked Immunosorbent Assay, an Immunochromatographic Handheld Assay, and a PCR," *Clin Diagn Lab Immunol*, 2000, 7(1):86-90.

Johansson A, Berglund L, Gothefors L, et al, "Ciprofloxacin for Treatment of Tularemia in Children," *Pediatr Infect Dis J*, 2000, 19(5):449-53.

Rusnak JM, Kortepeter MG, Aldis J, et al, "Experience in the Medical Management of Potential Laboratory Exposures to Agents of Bioterrorism on the Basis of Risk Assessment at the United States Army Medical Research Institute of Infectious Diseases (USAMRIID)," *J Occup Environ Med*, 2004, 46(8):801-11.

Rusnak JM, Kortepeter MG, Hawley RG, et al, "Management Guidelines for Laboratory Exposures to Agents of Bioterrorism," *J Occup Environ Med*, 2004, 46(8):791-800.

Primary pneumonic tularemia — histologic sections of lung with (A, left) neutrophilic infiltrate in alveolar space (hematoxylin and eosin stain) and (B, right) *Francisella tularensis* bacteria (immunohistochemistry).

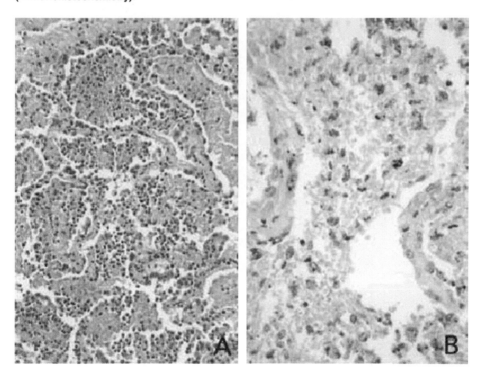

Infectious Disease Pathology Activity, CDC.

Tularemia — blood smear demonstrating *Francisella tularensis* bacteria (Giemsa stain).

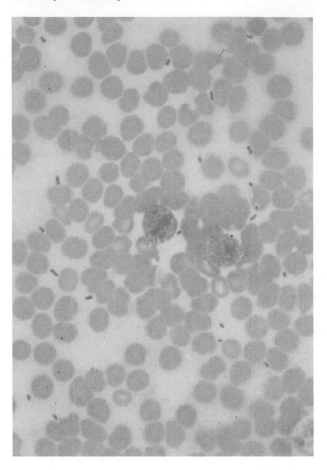

Infectious Disease Pathology Activity, CDC.

Hantavirus

Clinically Resembles Appendicitis, viral pneumonia, influenza A, and sepsis

Disease Syndromes Included Hantavirus Pulmonary Syndrome (HPS); Hemorrhagic Fever With Renal Syndrome (HFRS)

Microbiology The virus family Bunyaviridae is composed of more than 200 mostly arthropod-borne RNA viruses (arboviruses) such as California encephalitis virus, LaCrosse virus, Rift Valley fever virus, Congo-Crimean hemorrhagic fever virus, Hantaan virus, and the genus Hantavirus. Hantavirus is composed of several (hantaviruses) that usually cause one of two severe clinical syndromes: hemorrhagic fever with renal syndrome (HFRS) and hantavirus pulmonary syndrome (HPS). Hantavirus, as the term is being used in the United States, most commonly refers to Muerto Canyon/Sin Nombre virus and to Black Creek Canal virus, both of which cause HPS. Muerto Canyon virus is the virus that was responsible for the infamous outbreak of HPS in the Four Corners area (the intersection of New Mexico, Utah, Arizona, and Colorado) of the southwest United States in May, 1993. Hantaviruses are single-strand

220

RNA, 80-120 nm, spherical, pleomorphic, enveloped viruses, which are susceptible to most disinfectants.

Epidemiology Hantaviruses occur almost worldwide (see table). Human infection by hantavirus became a public health concern in the United States in the 1950s when U.S. soldiers who served in the Korean War developed Korean hemorrhagic fever (Hantaan virus). Muerto Canyon virus has been found only in North America. As of July 27, 1994, 83 cases of HPS (54% mortality) had been reported to the CDC, including the cases in the 1993 Four Corners outbreak, during which the natural history of Muerto Canyon virus was elucidated. The reservoir of the virus is *Peromyscus maniculatus*, the deer mouse, which always remains asymptomatic. If high populations of *Peromyscus* experience times of drought, scarce food, and reduced natural cover, the mice will seek food and shelter in human dwellings and outbuildings and drastically increase their contact with humans. Muerto Canyon virus is easily transmitted to humans by inhalation of aerosolized droplets of mouse urine and feces and by saliva from the bites of mice. The virus can be acquired in a laboratory setting; however, the virus is not transmitted human to human. From 1993 to early 2006, the CDC has confirmed 438 cases of HPS with 154 (35%) fatalities.

Hantavirus			
Hantavirus	**Geographical Location**	**Disease**	**Mortality (%)**
Hantaan (prototype) and Seoul	Asia	HFRS	1-15
Puumala	Scandinavia, Western Europe	HFRS	Rare
Belgrade/Dobrava	Central and Eastern Europe	HFRS	5-35
Prospect Hill	Eastern and Midwest United States	None	?
Muerto Canyon and Black Creek Canal	North America	HPS	50-70
Reference: http://www.bt.cdc.gov/documents/PPTResponse/table6anthraxid.pdf.			

Clinical Syndromes HPS usually occurs in healthy young adults, and the presentation of HPS can be considered to be similar to that of acute respiratory distress syndrome (ARDS). The incubation period of HPS is 12-16 days. HPS usually begins with a short period of general myalgia and fever, which is followed by 1-10 days of fever, cough, tachycardia, and tachypnea. Mild pulmonary edema can follow. In some cases, fulminating severe pulmonary edema manifested as ARDS can develop and can lead to cardiogenic shock and death in only a few hours. The mortality rate of HPS is extremely high (56% to 70%); however, successful respiratory therapy is possible, and complete recovery can occur in a few days. Histopathologically, lung tissue from patients with HPS show interstitial infiltration of lymphocytes and severe alveolar edema. Necrosis and polymorphonuclear infiltration is not present. The complete pathogenesis of HPS has not been completely established. HPS should be considered in any healthy adult who presents with unexplained ARDS.

Diagnosis The diagnosis of HPS is clinical and depends on careful examination of the patient's travel, work, and social history, on the patients living conditions, and on timely recognition of ARDS. The laboratory diagnosis of HPS caused by Muerto Canyon virus can be accomplished by viral culture and serology. Culture is not practical and not widely available. Almost all patients with HPS will have anti-Muerto Canyon IgG and IgM antibodies at the time of presentation and acute disease. Therefore, serological methods (enzyme immunoassay, hemagglutination inhibition, indirect immunofluorescence, complement fixation, and antibody neutralization) are the methods of choice. Both acute and convalescent sera must be tested if an accurate diagnosis of HPS is sought. Polymerase chain reaction and nucleic acid hybridization have been used to amplify and detect, respectively, Muerto Canyon virus DNA in fixed and sectioned tissue from infected patients. Thrombocytopenia, leukocytosis, elevated lactate levels, and hemoconcentration may occur.

Diagnostic Test/Procedure Hantavirus Serology

Treatment Strategy Treatment is supportive and should be given in an intensive care unit. Ventilator support with high levels of positive end-expiratory pressure is usually required within 6 hours. Volume resuscitation and dopamine may be required to treat hypotension. Supportive care: Ribavirin has been used to treat HPS; however, reproducible success has not been documented. Ribavirin used for this purpose can be obtained from the CDC.

Additional Information Spring/summer are the seasons when a majority of cases of HPS occur. Hantavirus hotline (CDC): (404) 639-1510; website: www.cdc.gov//hantavirus; special pathogens branch: Division of Viral Center for Infectious Diseases, Mailstop G-14, 1600 Clinton Road, NE Atlanta, GA 30333

Specific References

Centers for Disease Control and Prevention (CDC), "Hantavirus Pulmonary Syndrome - Five States, 2006," *MMWR Morb Mortal Wkly Rep*, 2006, 55(22):627-9.

Peters CJ and Khan AS, "Hantavirus Pulmonary Syndrome: The New American Hemorrhagic Fever," *Clin Infect Dis*, 2002, 34(9):1224-31.

Japanese Encephalitis Virus Vaccine (Inactivated)

U.S. Brand Names JE-VAX®

Use Active immunization against Japanese encephalitis for persons 1 year of age and older who plan to spend 1 month or more in endemic areas in Asia, especially persons traveling during the transmission season or visiting rural areas; consider vaccination for shorter trips to epidemic areas or extensive outdoor activities in rural endemic areas; elderly (>55 years of age) individuals should be considered for vaccination, since they have increased risk of developing symptomatic illness after infection; those planning travel to or residence in endemic areas should consult the Travel Advisory Service (Central Campus) for specific advice.

Adverse Reactions Report allergic or unusual adverse reactions to the Vaccine Adverse Event Reporting System (VAERS) (800) 822-7967.

Frequency not defined, common:
Cardiovascular: Hypotension
Central nervous system: Fever, headache, malaise, chills, dizziness
Dermatologic: Rash, urticaria, itching with or without accompanying rash
Gastrointestinal: Nausea, vomiting, abdominal pain
Local: Tenderness, redness, and swelling at injection site
Neuromuscular & skeletal: Myalgia
Frequency not defined, rare:
Cardiovascular: Angioedema
Central nervous system: Seizure, encephalitis, encephalopathy
Dermatologic: Erythema multiforme, erythema nodosum
Neuromuscular & skeletal: Peripheral neuropathy, joint swelling
Respiratory: Dyspnea
Miscellaneous: Anaphylactic reaction

Usual Dosage U.S. recommended primary immunization schedule:
Children 1-3 years: SubQ: Three 0.5 mL doses given on days 0, 7, and 30; abbreviated schedules should be used only when necessary due to time constraints
Children >3 years and Adults: SubQ: Three 1 mL doses given on days 0, 7, and 30. Give third dose on day 14 when time does not permit waiting; 2 doses a week apart produce immunity in about 80% of recipients; the longest regimen yields highest titers after 6 months.
Booster dose: Give after 2 years, or according to current recommendation
Note: Travel should not commence for at least 10 days after the last dose of vaccine, to allow adequate antibody formation and recognition of any delayed adverse reaction

Advise concurrent use of other means to reduce the risk of mosquito exposure when possible, including bed nets, insect repellents, protective clothing, avoidance of travel in endemic areas, and avoidance of outdoor activity during twilight and evening periods

Administration The single-dose vial should only be reconstituted with the full 1.3 mL of diluent supplied; administer 1 mL of the resulting liquid as one standard adult dose; discard the unused portion

Contraindications
Serious adverse reaction (generalized urticaria or angioedema) to a prior dose of this vaccine; proven or suspected hypersensitivity to proteins or rodent or neural origin; hypersensitivity to thimerosal (used as a preservative). CDC recommends that the following should not generally receive the vaccine, unless benefit to the individual clearly outweighs the risk:
- those acutely ill or with active infections
- persons with heart, kidney, or liver disorders
- persons with generalized malignancies such as leukemia or lymphoma
- persons with a history of multiple allergies or hypersensitivity to components of the vaccine
- pregnant women, unless there is a very high risk of Japanese encephalitis during the woman's stay in Asia

Warnings Severe adverse reactions manifesting as generalized urticaria or angioedema may occur within minutes following vaccination, or up to 17 days later; most reactions occur within 10 days, with the majority within 48 hours; observe vaccinees for 30 minutes after vaccination; warn them of the possibility of delayed generalized urticaria and to remain where medical care is readily available for 10 days following any dose of the vaccine; because of the potential for severe adverse reactions, Japanese encephalitis vaccine is not recommended for all persons traveling to or residing in Asia; safety and efficacy in infants <1 year of age have not been established; therefore, immunization of infants should be deferred whenever possible; it is not known whether the vaccine is excreted in breast milk

Dosage Forms Injection, powder for reconstitution: 1 mL, 10 mL

Stability Refrigerate, discard 8 hours after reconstitution

Additional Information Japanese encephalitis vaccine is currently available only from the Centers for Disease Control. Contact Centers for Disease Control at (404) 639-6370 (Mon-Fri) or (404) 639-2888 (nights, weekends, or holidays). Federal law requires that the date of administration, the vaccine manufacturer, lot number of vaccine, and the administering person's name, title, and address be entered into the patient's permanent medical record.

Junin Virus

Disease Syndromes Included Argentine Hemorrhagic Fever; New World Hemorrhagic Fever

Microbiology An RNA virus of the *Arenaviridae* family, it was first isolated in 1958. The natural host is the field mouse endemic to South America. As an intracellular virus, Junin Virus resides in macrophages and lymphatic tissue causing microvascular damage by inducing secretion of inflammatory mediators from the macrophages. Bone marrow toxicity to the megakaryocytes can occur. Transmission occurs via aerosolization of the virus from infectious blood or body fluids. Mortality is 15% to 50% (untreated).

Postmortem Care Do not embalm; prompt burial or cremation is recommended. Personnel should utilize all PPE precautions with HEPA filtered respirator and negative pressure room.

Epidemiology Found in Central Argentina, usually active in late summer to early winter. Incubation period is 1-2 weeks.

Clinical Syndromes
Initially, a nonspecific influenza-like prodrome can occur usually lasting less than 1 week. Symptoms include fever (lasting 6-8 days), headache, flushing, nausea, vomiting, abdominal pain, myalgia, tachypnea, hypotension, relative bradycardia, conjunctivitis, and pharyngitis. Hemorrhagic and neurologic (tremors, encephalopathy) occurs at 6-10 days of illness, whereupon 80% of individuals improve.

223

The remainder may exhibit petechia, bleeding, generalized seizures, and further central nervous system dysfunction. Myocarditis may also be present along with acute lung injury and elevated transaminases and thrombocytopenia (mild).

Infective Dose 1-10 organisms

Diagnosis Antigen-capture ELISA or reverse transcriptase polymerase chain reaction (RT-PCR) methodology. A fourfold elevation of IgG antibody titer between acute and convalescent phase is diagnostic. A WBC $<2500/mm^3$, a platelet count $<100,000/mm^3$, and urinary protein excretion >1 g/L are associated with Argentine hemorrhagic fever.

Treatment Strategy Decontamination:

Ocular: Irrigate with large amounts of saline or water for at least 20 minutes

Dermal: Remove contaminated clothing; wash skin with soap and water. Equipment can be washed with dilute household bleach (1:100) or glutaraldehyde and phenolic disinfectants (3% to 5%)

Inhalation: Ribavirin is most effective if given within 6 days of illness. Loading dose is 30 mg/kg I.V. (maximum of 2 g) once followed by 16 mg/kg I.V. (maximum 1 g/dose) every 6 hours for 4 days, followed by 8 mg/kg (maximum 500 mg/dose) every 8 hours for 6 days. Hypotension responds poorly to crystalloid solution resuscitation. Dopamine appears to be the vasoconstrictor of choice. Seizures can be treated with benzodiazepines. Phenobarbital can be used as a second-line agent.

Recommendations for Ribavirin Therapy in Patients With Clinically-Evident Viral Hemorrhagic Fever of Unknown Etiology or Secondary to Arenaviruses or Bunyaviruses[1]

	Contained Casualty Setting	Mass Casualty Setting[2]
Adults	Loading dose: 30 mg/kg I.V. (maximum: 2 g) once, followed by 16 mg/kg I.V. (maximum: 1 g/dose) every 6 hours for 4 days, followed by 8 mg/kg I.V. (maximum: 500 mg/dose) every 8 hours for 6 days	Loading dose: 2000 mg orally once, followed by 1200 mg/day orally in 2 divided doses (if weight >75 kg), or 1000 mg/day orally in 2 doses (400 mg in morning and 600 mg in evening) (if weight ≤75 kg) for 10 days[3]
Pregnant women	Same as for adults	Same as for adults
Children	Same as for adults, dosed according to weight	Loading dose: 30 mg/kg orally once, followed by 15 mg/kg/day orally in 2 divided doses for 10 days

[1]Recommendations are not approved by the U.S. Food and Drug Administration for any of these indications and should always be administered under an investigational new drug protocol. However, in a mass casualty setting, these requirements may need to be modified to permit timely administration of the drug.
[2]The threshold number of cases at which parenteral therapy becomes impossible depends on a variety of factors, including local healthcare resources.
[3]Although a similar dosage (1000 mg/day in 3 divided doses) has been used in a small number of patients with Lassa fever, this regimen would be impractical because the current formulation of oral ribavirin in the United States consists of 200 mg capsules, and ribavirin capsules may not be broken open.

Quarantine Strict isolation with restricted access/negative air pressure room with 6-12 air changes per hour; dedicated medical equipment

Environmental Persistency Not persistent in the environment.

Personal Protection Equipment Strict adherence to hard hygiene with double gloves, impermeable gowns, N-95 respirators or powered air-purified respirators, leg/shoe coverings, face shields/goggles

Additional Information Laboratory clearance to work in research laboratory can be obtained 4 weeks following vaccination with Junin live attenuated investigational vaccine (0.5 mL I.M.) with no booster dose needed. High-risk contacts, after viral exposure, can be given chemoprophylaxis with ribavirin at doses of 500 mg orally every 6 hours for 7-10 days in adults or 400 mg every 6 hours for 7-10 days in children ages 6-9 years.

Specific References

Borio L, Inglesby T, and Peters CJ, "Hemorrhagic Fever Viruses as Biological Weapons," *JAMA*, 2002, 287:2391-405.

Rusnak JM, Kortepeter MG, Aldis J, et al, "Experience in the Medical Management of Potential Laboratory Exposures to Agents of Bioterrorism on the Basis of Risk Assessment at the United States Army Medical Research Institute of Infectious Diseases (USAMRIID)," *J Occup Environ Med*, 2004, 46(8):801-11.

Rusnak JM, Kortepeter MG, Hawley RG, et al, "Management Guidelines for Laboratory Exposures to Agents of Bioterrorism," *J Occup Environ Med*, 2004, 46(8):791-800.

Meningitis, Community-Acquired, Adult

Synonyms Community-Acquired Meningitis, Adult

Clinical Presentation Patients present with fever, headache, nausea, vomiting, and nuchal rigidity. Focal neurologic abnormalities and papilledema are uncommon. In a prospective study by Thomas KE, et al, the 3 classic meningeal signs (Kernig's sign, Brudzinski's sign, and nuchal rigidity) were of limited clinical diagnostic value for adults with suspected meningitis. None of these meningeal signs were able to accurately discriminate patients with meningitis from those without it.

Differential Diagnosis Bacterial, viral, or other infectious agents; drug-induced; neoplastic; granulomatous; infectious endocarditis; collagen-vascular diseases; subdural hemorrhage; subarachnoid hemorrhage

Likely Pathogens

Herpes simplex
Haemophilus influenzae
Neisseria meningitidis
Streptococcus pneumoniae, Drug-Susceptible
Listeria monocytogenes (elderly, immunocompromised)

Diagnostic Test/Procedure

Blood Culture, Aerobic and Anaerobic
Cerebrospinal Fluid Analysis
Gram's Stain
Lumbar Puncture
Computed Transaxial Tomography, Head Studies
Cryptococcal Antigen Serology, Serum or Cerebrospinal Fluid
Enterovirus Culture
Magnetic Resonance Scan, Brain

Treatment Strategy Hydrate with crystalloid fluids. Acetaminophen can be used as an antipyretic. Treat with antibiotics and steroids (ie, dexamethasone, if indicated). Give first steroid dose 15-20 minutes before or concomitant with first dose of antibiotic.

Bacterial meningitis: I.V. antibiotic:

Neonates (7 days to 1 month): Ampicillin 50 mg/kg every 6 hours plus cefotaxime 50 mg/kg every 8 hours **or** gentamicin 2.5 mg/kg every 12 hours

Children >1 month: Cefotaxime 75 mg/kg every 6 hours or ceftriaxone 100 mg/kg divided twice daily **plus** vancomycin 15 mg/kg every 6 hours

Adults*: Cefotaxime 2 g every 4-6 hours or ceftriaxone 1-2 g every 12 hours plus vancomycin 500-750 mg every 6 hours in conjunction with I.V. steroid: Dexamethasone (if indicated):

Children >2 months (should be given as sodium phosphate): 0.6 mg/kg/day in 4 divided doses every 6 hours for the first 4 days of antibiotic treatment; start dexamethasone at the time of the first dose of antibiotic

Adults (may be useful for cerebral edema): 10 mg every 6 hours for 4 days

*Patients aged >50 years, immunosuppressed, or chronic alcohol abusers: Add ampicillin 2 g every 4-6 hours
Antiviral (suspected herpes simplex virus): I.V.:
  Children: Acyclovir 20 mg/kg every 8 hours for 2-3 weeks
  Adults: Acyclovir 10 mg/kg every 8 hours for 2-3 weeks

Specific References
Thomas KE, Hasbun R, Jekel J, et al, "The Diagnostic Accuracy of Kernig's Sign, Brudzinski's Sign, and Nuchal Rigidity in Adults With Suspected Meningitis," *Clin Infect Dis*, 2002, 35(1):46-52.

Ricin

CAS Number 9009-86-3; 9067-26-9

UN Number 3172

Synonyms Ricinus Lectin; Ricinus Toxin

CDC Classification for Diagnosis
Biologic: CDC can assess selected specimens on a provisional basis for urinary ricinine, an alkaloid in the castor bean plant. Only urinary ricinine testing is available at CDC for clinical specimens.
Environmental: Detection of ricin in environmental samples, as determined by CDC or FDA. Ricin can be detected qualitatively by time-resolved fluoroimmunoassay (TRFIA) and polymerase chain reaction (PCR) in environmental specimens (eg, filters, swabs, or wipes).

CDC Case Classification
Suspected: A case in which a potentially exposed person is being evaluated by healthcare workers or public health officials for poisoning by a particular chemical agent, but no specific credible threat exists.
Probable: A clinically-compatible case in which a high index of suspicion (credible threat or patient history regarding location and time) exists for ricin exposure, or an epidemiologic link exists between this case and a laboratory-confirmed case.
Confirmed: A clinically-compatible case in which laboratory tests have confirmed exposure. The case can be confirmed if laboratory testing was not performed because either a predominant amount of clinical and nonspecific laboratory evidence of a particular chemical was present or a 100% certainty of the etiology of the agent is known.

Use Agroterrorism agent; can be found in powder, mist, or pellet

Mechanism of Action Protein toxin (toxal bumen) derived from the bean of the castor plant (*Ricinus communis*); inhibition of protein synthesis

Pharmacodynamics/Kinetics Half-life: 8 days

Pregnancy Issues May speed up onset of labor; no effect on fertility

Clinically Resembles Sepsis syndrome; gastroenteritis toxic mushroom ingestion, arsenic, colchicine

Postmortem Care As per usual routine

Infective Dose Oral: 3-5 mcg/kg

Signs and Symptoms of Acute Exposure While the incubation period may be 18-24 hours, gastrointestinal symptoms of nausea, vomiting, and abdominal pain may occur within a few hours. Other symptoms following oral ingestion include diarrhea, anuria, mydriasis, fever, thirst, headache, sore throat followed by vascular collapse and death. Aerosol exposure can result in pulmonary irritation, weakness, fever, cough, pulmonary edema, respiratory distress, and death due to hypoxemia within 36-72 hours. Leukocytosis (2-5 fold increase in WBC) may also be noted, as can hematuria. Erythema of skin and eyes also noted. Seizures are uncommon. Hepatic and renal function tests may be abnormal.

Treatment Strategy
Decontamination:
Dermal: Wash with soap and water; remove clothing
Inhalation: Administer 100% humidified oxygen
Ocular: Irrigate copiously with water or normal saline
Oral: Activated charcoal within 1 hour of ingestion or lavage within 1 hour
Supportive therapy: Volume replacement for GI fluid losses; no antibiotic therapy

Quarantine No quarantine required; normal hospital routine

Rescuer Contamination Person-to-person transmission: None

Additional Information Dissolves in water or weak acid. Lethal dose is about 500 mcg (~ the size of the head of a pin). Detoxified in 10 minutes at 176°F (80°C) and in 1 hour at 122°F (50°C) The CDC has developed an assay for measuring ricinine in urine through the Laboratory Response Network (LRN) - (404) 639-2790

Specific References
Belson MG, Schier JG, and Patel MM, "Case Definitions for Chemical Poisoning," *MMWR Recomm Rep*, 2005, 54(RR-1):1-24.

Bradberry SM, Dickers KJ, Rice P, et al, "Ricin Poisoning," *Toxicol Rev*, 2003, 22(1):65-70.

Centers for Disease Control and Prevention (CDC), "Investigation of a Ricin-Containing Envelope at a Postal Facility - South Carolina, 2003," *MMWR Morb Mortal Wkly Rep*, 2003, 52(46):1129-31.

Dickers KJ, Bradberry SM, Rice P, et al, "Abrin Poisoning," *Toxicol Rev*, 2003, 22(3):137-42.

Doan LG, "Ricin: Mechanism of Toxicity, Clinical Manifestations, and Vaccine Development. A Review," *J Toxicol Clin Toxicol*, 2004, 42(2):201-8.

Johnson RC, Lemire SW, Woolfitt AR, et al, "Quantification of Ricinine in Rat and Human Urine: A Biomarker for Ricin Exposure," *J Anal Toxicol*, 2005, 29:149-55.

Kiefer M, Schier J, Patel M, et al, "Environmental Sampling and Laboratory Analysis for Ricin," *J Toxicol Clin Toxicol*, 2004, 42(5):809.

Lord MJ, Jolliffe NA, Marsden CJ, et al, "Ricin - Mechanisms of Cytotoxicity," *Toxicol Rev*, 2003, 22(1):53-64.

Olsnes S and Kozlov JV, "Ricin," *Toxicon*, 2001, 39(11):1723-8.

Salmonella Species

Clinically Resembles Brucellosis, tuberculosis, Dengue fever, endocarditis, ricin, shigella, hemorrhagic fevers, plague

Disease Syndromes Included Salmonellosis; Typhoid Fever

Microbiology *Salmonella* species are gram-negative bacilli, which are important causes of bacterial gastroenteritis, septicemia, and a nonspecific febrile illness called typhoid fever. Unfortunately, the classification system for the different *Salmonella* species is complex and confusing to most clinicians. Over 2000 separate serotypes have been identified, and, in the past and currently, each is named as if it was a species. Most experts agree that there are seven distinct subgroups of *Salmonella* (1, 2, 3a, 3b, 4, 5, and 6) each of which contain many serotypes ("species"). The main pathogens in humans are serotypes, *S. choleraesuis*, *S. typhi*, and *S. paratyphi* (all in subgroup 1).
Salmonella is an aerobic gram-negative bacillus in the family Enterobacteriaceae. It is almost always associated with disease when isolated from humans and is not considered part of the normal human flora. As with other members of this family, *Salmonella* carries the endotoxin lipopolysaccharide on its outer membrane, which is released upon cell lysis.

Epidemiology It is estimated that about 3 million new cases of salmonellosis occur each year. *Salmonella* species are found worldwide. Many are easily recovered from animals such as chickens, birds, livestock, rodents, and reptiles (turtles). Some serotypes cause disease almost exclusively in man (*S. typhi*) while others are primarily seen in animals but can cause severe disease when infecting humans (*S. choleraesuis*). Transmission is via ingestion of contaminated materials, particularly raw fruits and vegetables, oysters and other shellfish, and contaminated water. Eggs, poultry, and other dairy products are important sources. Outbreaks have been described in the summer months where children consume contaminated egg salad. Other outbreaks have been associated with pet turtles, other pets, marijuana, and, rarely, food handlers. The incubation period is about 1-3 weeks. The period of communicability lasts until all *Salmonella* have been eradicated from the stool or urine.

Clinical infection is favored when there is a high inoculum of bacteria in the ingested food, since experimental models suggest that 10^6 bacteria are needed for clinical disease. Contaminated food improperly refrigerated will allow such multiplication. Host factors are important and disease is more likely in immunocompromised individuals, sickle cell disease, or achlorhydria (gastric acid decreases the viable bacterial inoculum). Although salmonellosis can occur at any age, children are most commonly infected.

Clinical Syndromes The following are the major syndromes associated with salmonellosis. It should be emphasized that these syndromes are often overlapping.

• **Gastroenteritis:** The most common manifestation of *Salmonella* infection. After ingestion of contaminated food, the bacteria are absorbed in the terminal portion of the small intestine. The organisms then penetrate into the lamina propria of the ileocecal area. Following this, there is reticuloendothelial hypertrophy with usually a brisk host immune response. As the organisms multiply in the lymphoid follicles, polymorphonuclear leukocytes attempt to limit the infection. There is release of prostaglandins and other mediators, which stimulates cyclic AMP. This results in intestinal fluid secretion, which is nonbloody. Clinically, the patient complains of nausea, vomiting, and diarrhea from several hours to several days after consumption of contaminated food. Other symptoms include fever (over 37.7°C), malaise, muscle aches, and abdominal pain. Symptoms usually resolve from several days to 1 week, even without antibiotics.

• **Sepsis syndrome:** Patients may present in flora sepsis, indistinguishable from other forms of gram-negative sepsis. Fever, confusion, hypotension, end-organ damage, and poor perfusion may all be seen. *Salmonella* bacteremia may lead to multiple metastatic foci, such as liver abscess, osteomyelitis (particularly with sickle cell disease), septic arthritis, and endocarditis. Mycotic aneurysms may develop following bacteremia, and *Salmonella* is a leading cause of infected aortic aneurysms. Bacteremia is also common in AIDS and repeated relapses with *Salmonella* are common, despite prolonged antibiotics.

• **Typhoid fever:** Also known as enteric fever, this febrile illness is caused classically by *S. typhi*. Following ingestion of the bacteria, the organisms pass into the ileocecal area where intraluminal multiplication occurs. There is a mononuclear host cell response, but the organisms remain viable within the macrophages. The bacteria are carried to the organs of the reticuloendothelial system (spleen, liver, bone marrow) by the macrophages and clinical signs of infection become apparent. Patients complain of insidious onset of fever, myalgias, headache, malaise, and constipation, corresponding to this phase of bacteremia. Febrile patients may exhibit relative bradycardia, sometimes referred to as pulse-temperature dissociation. This clinical sign can also be seen with typhoid fever as well as other bioweapon illness including certain viral hemorrhagic fevers and plague. A characteristic rash may be seen in about 50% of patients, called "rose spots," which are 2-4 mm pink maculopapular lesions that blanch with pressure, usually on the trunk. Symptoms last for 1 or more weeks. Leukopenia, mild hyponatremia, and mild liver transaminase elevation may occur. During this time, bacteria multiply in the mesenteric lymphoid tissue, and these areas eventually exhibit necrosis and bleeding. There are microperforations of the abdominal wall. *Salmonella* spreads from the liver through the gallbladder and eventually back into the intestines. This phase of intestinal reinfection is characterized by prominent gastrointestinal symptoms including diarrhea. Overall, fatality with treatment is <2%. A similar, but milder syndrome, can occur with *S. paratyphi*, called paratyphoid fever.

• **Chronic carrier state:** Following infection with *S. typhi*, up to 5% of patients will excrete the bacteria for over 1 year. Such patients are termed chronic carriers and are asymptomatic. Millions of viable bacteria are present in the biliary tree and are shed into the bile and into the feces. Urinary carriage can also occur, particularly in patients who are coinfected with

Schistosoma haematobium. The chronic carrier state (which is more common in females and the elderly) is less important for other *Salmonella* species, where the carriage rate is <1%.

Diagnosis Laboratory confirmation is generally required, since the major syndromes are seldom distinctive enough to be diagnosed solely on clinical criteria. *Salmonella* grows readily on most media under standard aerobic conditions. Cultures from blood, joint aspirations, and cerebrospinal fluid can be plated on routine media. Specimens, which are likely to contain other organisms such as stool or sputum, require selective media, and the laboratory should be appropriately notified. Recovery of *Salmonella* from the stool is the most common means of establishing the diagnosis, and enrichment media are available to maximize the yield. Other laboratory findings may suggest salmonellosis, including a profound leukopenia often seen with typhoid fever.

Recent antimicrobial therapy may render blood and stool cultures negative. In such cases, proctoscopy with biopsy and culture of ulcerations may establish the diagnosis in the enterocolitis syndrome. Serologic tests are not particularly useful in this instance. When typhoid fever is suspected but the patient has already received antimicrobial agents, bone marrow biopsies, as well as skin biopsies of any rose spots, may yield *Salmonella typhi* in culture. Serologic studies are more helpful in diagnosing typhoid fever, but >50% of patients will fail to show the expected rise in agglutinins against the typhoid O antigen.

Diagnostic Test/Procedure
Blood Culture, Aerobic and Anaerobic
Bone Marrow Culture, Routine
Stool Culture

Treatment Strategy Note: *Salmonella* species resistant to multiple antimicrobials are increasing in frequency. *In vitro* susceptibility studies should be performed, particularly in severe cases. Treatment guidelines vary with the type of syndrome, as follows:

• **Enterocolitis:** The majority of cases are self-resolving within 3-5 days and do not need antibiotics. Clinical trials have demonstrated that a variety of antibiotics fail to influence the course of mild infections and may prolong excretion of the organisms. For severe cases, or in the immunosuppressed host, a number of antibiotics are usually effective including:
  Adults: Oral, I.V., I.M.: Ampicillin 150 mg/kg/day; children: 50 mg/kg/day in 3 divided doses
  Older Children and Adults: Oral: Co-trimoxazole 160/800 mg every 12 hours
  Ciprofloxacin: Oral: 750 mg twice daily for 14 days
  Third-generation cephalosporins for 5-7 days

• **Typhoid fever:** Cases should be treated promptly. Chloramphenicol and ampicillin are effective and have been the most extensively studied.
  Recent studies show that ciprofloxacin is highly active: Oral: 750 mg twice daily for 14 days **or** ofloxacin 400-800 mg/day in divided doses.
  Third-generation cephalosporins and co-trimoxazole are useful in organisms that are resistant to standard agents.

• **Bacteremia:** Ampicillin, chloramphenicol, co-trimoxazole, and third-generation cephalosporins are all effective.
  Children: Oral: Cefixime 20 mg/kg/day in divided doses every 12 hours for at least 12 days **or** 50-60 mg/kg/day I.V. in 2 divided doses.
  Adults: I.M.: Ceftriaxone 2 g daily for 5-8 days
  Chloramphenicol should be avoided in endocarditis or mycotic aneurysms. Ciprofloxacin is effective in treating recurrent *Salmonella* bacteremia in AIDS.

• **Chronic carriage:** Ampicillin for 6 weeks, or amoxicillin (2 g orally 3 times/day for 28 days), although relapses are common; if there is underlying gallbladder disease, cholecystectomy may be an option with repeated relapses; aztreonam 2 g I.V. every 6 hours for 16 days may be effective, as well as ciprofloxacin.

Quarantine Enteric precautions

Specific References
Frenzen PD, "Deaths Due to Unknown Foodborne Agents," *Emerg Infect Dis*, 2004 [9/07/04]. Available from http://www.cdc.gov/ncidod/EID/vol10no9/03-0403.htm.

Santos RL, Tsolis RM, Baumler AJ, et al, "Pathogenesis of *Salmonella*-Induced Enteritis," *Braz J Med Biol Res*, 2003, 36:3-12.

Zhang S, Kingsley RA, Santos RL, et al, "Molecular Pathogenesis of *Salmonella enterica* Serotype Typhimurium-Induced Diarrhea," *Infect Immun*, 2003, 71:1-12.

Severe Acute Respiratory Syndrome (SARS)

Microbiology

Coronavirus particles contain RNA, are irregularly-shaped, are approximately 60-220 nm in diameter, and have an outer envelope bearing distinctive, "club-shaped" peplomers approximately 20 nm long $\times$ 10 nm wide. The name Coronavirus (Corona = crown in Latin) refers to the projection of these peplomers around the virus. When viewed with electron microscopy, the projections give the appearance of a "corona" around the virus. First isolated from chickens in 1937, Coronaviruses cause a large proportion of cold-like illnesses in humans. More recently, a severe, acute respiratory syndrome (SARS) with significant mortality has been attributed to a member of this family.

The Coronavirus envelope contains two glycoproteins, a spike glycoprotein (S) which participates in receptor binding and cell fusion, and a membrane glycoprotein (M) which participates in budding and envelope formation. Genetic material of Coronaviruses consists of a single-stranded (+) sense RNA, approximately 27-31 kb in length. The entire 29,736 nucleotide sequence has been identified. The genome is associated with a basic phosphoprotein (N). The polymerase gene is the most highly conserved portion of the Coronavirus genome. Analysis of the genome by PCR and sequencing has demonstrated that the recently recognized (in 2003) severe acute respiratory syndrome (SARS) is caused by a novel Coronavirus, which has not been previously identified in humans.

Clinically, most Coronavirus infections cause a mild, self-limited disease (classical "cold" symptoms) in which growth appears to be localized to the epithelium of the upper respiratory tract. However, SARS infection encompasses the lower respiratory tract, and results in a much higher severity of symptoms.

The prototype strains HCoV 229E and HCoV OC43 are primarily associated with common cold syndrome. Reinfections with Coronaviruses appear to occur throughout life, implying multiple serotypes (at least four are known) and/or antigenic variation, which may limit the possibility of successful vaccine development.

Epidemiology

Coronaviruses are transmitted by aerosols of respiratory secretions, by the fecal-oral route, and by mechanical transmission. Viral replication occurs primarily in epithelial cells; however, infection of other cell types, including macrophage, liver, kidneys, and heart have been described. Coronavirus infection is very common and occurs worldwide. A strong seasonal relationship (greatest incidence in children during the winter months) is present for most Coronavirus infections.

In March 2003, the World Health Organization (WHO) reported a multicountry outbreak of an atypical pneumonia referred to as severe acute respiratory syndrome (SARS), which has since caused significant morbidity and mortality. SARS has been described in patients in Asia, North America, and Europe. As of March 21, 2003, the majority of patients identified as having SARS have been previously healthy adults between 25 and 70 years of age. Few suspected cases of SARS have been reported among children. The SARS outbreak is believed to have originated in February 2003 in the Guangdong province of China, where 300 people became ill, and at least five died. The Coronavirus responsible for SARS has several unusual properties, including growth in cell culture (most Coronaviruses cannot be cultivated). Global case counts are available at http://www.who.int. The most recent human cases of SARS-CoV infection were reported in China in April 2004 in an outbreak resulting from laboratory-acquired infections. There have not been any reported cases anywhere in the world since.

Clinical Syndromes

In humans, Coronaviruses may cause a variety of clinical syndromes, including respiratory infections (primarily cold-type illnesses), enteric infections (primarily in infants), and rare neurological syndromes. These are generally self-limiting. New attention has focused on this family following the description of severe lower respiratory tract infections (SARS).

• **Severe Acute Respiratory Syndrome (SARS):** The incubation period for SARS is typically 2-7 days and may be as long as 10 days. A prodrome has been described, consisting of high fever [>100.4°F (>38.0°C)], possibly associated with chills, rigors, and other flu-like symptoms (headache, myalgia, malaise). Mild respiratory symptoms may be present at the onset of illness. Rash, neurologic symptoms, or gastrointestinal disturbances are typically absent, although diarrhea during the febrile prodrome has been reported.

After 3-7 days, a second phase involving the lower respiratory tract begins. Symptoms include a dry, nonproductive cough and/or dyspnea, which may progress to hypoxemia. The severity of illness appears to be highly variable, ranging from mild illness to death. Respiratory compromise requiring intubation and mechanical ventilation may occur in 10% to 20% of cases. The case-fatality rate among persons with illness meeting the current WHO case definition of SARS is approximately 3%. Some close contacts, including healthcare workers, have developed similar illnesses.

In a substantial proportion of patients, the respiratory phase is characterized by early focal interstitial infiltrates. These frequently progress to generalized, patchy, interstitial infiltrates. Consolidation may be observed in some patients during the late stages of SARS. In some cases, chest radiographs remain normal during the prodrome and throughout the course of the illness. Laboratory findings may include thrombocytopenia and leukopenia. Early in the respiratory phase, elevated creatine phosphokinase levels (as high as 3000 IU/L) and elevated hepatic transaminases have been noted.

Diagnosis The CDC has provided the following SARS case definition criteria (as of December 2003):

Revised CST SARS Surveillance Case Definition (Summary of Criteria) (available at http://www.cdc.gov/mmwr/preview/mmwrhtml/mm5249a2.htm)

Clinical Criteria:

Early illness:

• Two or more of the following: fever, chills, rigors, myalgia, headache, sore throat, rhinorrhea

Mild to moderate respiratory illness:

• Temperature >100.4°F, *and*

• One or more clinical findings of LRT illness (eg, cough, SOB, difficulty breathing)

Severe respiratory illness:

• Meets clinical criteria of mild-to-moderate respiratory illness, *and*
  • radiographic evidence of pneumonia
  • acute respiratory distress syndrome
  • autopsy findings consistent with pneumonia or acute respiratory distress syndrome

Epidemiological evidence:

Possible exposure to SARS-associated coronavirus (SARS-CoV):

• One of more of the following in the 10 days before symptoms:
  • Travel to a location with recent transmissions of SARS, *or*
  • Close contact with a person with LRT illness and the aforementioned travel history

Likely exposure to SARS-CoV:

• One or more of the following in the 10 days before symptoms:
  • Close contact with a confirmed case of SARS-CoV disease, *or*
  • Close contact with a person with LRT illness for whom a chain of transmission can be linked to a case of SARS-CoV in the 10 days prior to symptoms

Laboratory criteria (not completely established):

• Detection of antibody to SARS-CoV, *or*
• Isolation of SARS-CoV in cell culture, *or*
• Detection of SARS-CoV RNA molecular methods and subsequent confirmation

SEVERE ACUTE RESPIRATORY SYNDROME (SARS)

Diagnostic Test/Procedure Consult the clinical microbiology laboratory for advice and information regarding specimen selection, collection, and transport before selecting, collecting, and transporting specimens for the laboratory diagnosis of SARS.

CDC-Recommended Specimens for Evaluation of Potential Cases of SARS			
Specimen	**Outpatient**	**Inpatient**	**Fatal**
Blood	• Serum (acute and convalescent >28 days past onset) • Plasma	• Serum (acute and convalescent >28 days postonset)	• Serum • Plasma
Upper respiratory	• N/P wash/aspirate • N/P or O/P swabs	• N/P wash/aspirate • N/P or O/P swabs	• N/P wash/aspirate • N/P or O/P swabs
Lower respiratory	• Sputum	• Bronchoalveolar lavage, tracheal aspirate, pleural fluid • Sputum	• Bronchoalveolar lavage, tracheal aspirate, pleural fluid
Stool	Yes	Yes	Yes
Tissue			• Fixed from all major organs • Frozen from lung and upper airway
Source: www.cdc.gov/ncidod/sars/guidance/f/app4.htm.			

Environmental Persistency Several days depending on temperature or humidity.

Treatment Strategy Generally involves supportive treatment only, including hemodynamic and ventilatory support. A variety of antiviral agents are currently being evaluated.

In some cases, antiviral agents such as oseltamivir or ribavirin have been given empirically without known efficacy. In addition, steroids have been administered orally or intravenously in combination with ribavirin. The efficacy of these agents has not been confirmed.

In the United States, clinicians who suspect cases of SARS are requested to report such cases to their state health departments. CDC requests that reports of suspected cases from state health departments, international airlines, cruise ships, or cargo carriers be directed to the SARS Investigative Team at the CDC Emergency Operations Center, telephone (770) 488-7100. Outside the United States, clinicians who suspect cases of SARS are requested to report such cases to their local public health authorities. Additional information about SARS (eg, infection control guidance and procedures for reporting suspected cases) is available at http://www.cdc.gov/ncidod/sars.

Personal Protection Equipment Standard precautions with contact precautions (gown/gloves), eye protection for all patient contact and NIOSH-approved respirator (ie, N-95); respirator may have an exhalation valve.

Additional Information Initial neutrophil count >7000/uL, initial C-Reactive protein concentration >47.5 mg/L and lactic acid dehydrogenase (LDH) >593.5 IU/L are important predictors or mortality from SARS. Other predictive factors include dyspnea, red blood cell count under 4.1×10^6/uL and serum aspartate aminotransferase over 57 IU/L.

Specific References

Centers for Disease Control, "Guidelines for Collection of Specimens From Potential Cases of SARS," www.cdc.gov/ncidod/sars/specimen_collection_sars2.htm

Centers for Disease Control, "Update: Outbreak of Severe Acute Respiratory Syndrome - Worldwide, 2003," *MMWR*, 2003, 52(13).

Ksiazek TG, Erdman D, Goldsmith C, et al, "A Novel Coronavirus Associated With Severe Acute Respiratory Syndrome," *N Engl J Med*, 2003, 348(20):1953-66.

Oxford JS, Balasingam S, Chan C, et al, "New Antiviral Drugs, Vaccines, and Classic Public Health Interventions Against SARS Coronavirus," *Antivir Chem Chemother*, 2005, 16(1):13-21.

Skowronski DM, Astell C, Brunham RC, et al, "Severe Acute Respiratory Syndrome (SARS): A Year in Review," *Annu Rev Med*, 2005, 56:357-81.

Srikantiah P, Charles MD, Reagan S, et al, "SARS Clinical Features, United States, 2003," *Emerg Infect Dis*, 2005, 11(1):135-8.

Shigella Species

Clinically Resembles Salmonella, ricin, Dengue fever, brucellosis

Disease Syndromes Included Dysentery; Shigellosis

Microbiology *Shigella* species are gram-negative rods, which cause a severe diarrheal syndrome, called shigellosis or bacillary dysentery. *Shigella* species belong to the family Enterobacteriaceae and is, for all practical purposes, biochemically and genetically identical to *E. coli*. Four species of *Shigella* have been identified: *S. dysenteriae, S. flexneri, S. boydii*, and *S. sonnei*. There are approximately 40 serotypes. *Shigella sonnei* is most common in the industrial world and accounts for about 64% of the cases in the United States. *S. flexneri* is seen primarily in underdeveloped countries. Shigellosis can be seen following ingestion of as few as 200 organisms.

Epidemiology Infection with *Shigella* sp. is primarily a problem in the pediatric population, with most infections in the 1- to 4-year age group. Outbreaks of epidemic proportions have been described in daycare centers and nurseries. The reservoir for the bacteria is in humans. Transmission is by direct or indirect fecal-oral transmission from patient or carrier. Hand transmission and kitchen hygiene are important. Less commonly, transmission occurs by consumption of contaminated water, milk, and food. The organism is able to produce outbreaks in areas of poor sanitation, in part due to the low number of organisms required to produce disease. Shigellosis is the most communicable of the bacterial diarrheas.

Clinical Syndromes *Shigella* species invade the intestinal mucosa wherein they multiply and cause local tissue damage. The organisms rarely penetrate beyond the mucosa, and thus the isolation in blood cultures is unusual, even with the toxic patient. Mucosal ulcerations are common. Some strains are known to elaborate a toxin (the shiga-toxin), which contributes to mucosal destruction and probably causes the watery diarrhea seen initially.

• **Dysentery:** Initially, the patient complains of acute onset of fever, abdominal cramping, and large volumes of very watery diarrhea. This phase is enterotoxin-mediated and reflects small bowel involvement. Within 24-48 hours, the fever resolves but the diarrhea turns frankly bloody, with mucous and pus in the stools as well. Fecal urgency and tenesmus are common. This phase reflects direct colonic invasion. This two-phased "descending infection" is suggestive of dysentery. Diarrhea and abdominal pain are almost universally present, but the other symptoms may be absent. Physical examination is variable, and patients may be comfortable or frankly toxic. Rectal examination is often painful due to friable and inflamed rectal mucosa. The course may be complicated by dehydration from diarrhea and vomiting, particularly in the elderly and in infants. Normally, the infection is self limited and resolves within about 1 week even without antibiotics. Complications are unusual and include febrile seizures (particularly in infants), septicemia, and the hemolytic uremic syndrome (usually from the shiga-toxin from *S. dysenteriae* 1).

• **Reactive arthritis:** Following dysentery from *Shigella*, a postinfectious arthropathy resembling Reiter's syndrome has been described, particularly in patients who are HLA-B27 positive.

Infective Dose 10-200 organisms

Diagnosis Dysentery from *Shigella* should be suspected in any patient presenting with fever and bloody diarrhea. A history of a "descending infection" as described above is further suggestive. However, the differential diagnosis of fever with bloody diarrhea is broad and includes salmonellosis, *Campylobacter* enteritis, infection with *E. coli* O157:H7, and inflammatory bowel disease. The WBC count may show either a leukocytosis, leukopenia, or be normal. There are two important laboratory tests indicated in suspected cases.

1. Stool exam for fecal leukocytes: Numerous white blood cells will be present during the colonic phase of the infection. Note that this is not diagnostic for *Shigella* infections; it

indicates that the colonic mucosa is inflamed, from whatever cause. The finding of sheets of fecal leukocytes on smear narrows the differential diagnosis of infectious diarrheas considerably.

2. Stool culture for *Shigella*: Recovery of the organism from stool is more easily performed early in the illness when the concentration in the stool is highest. Samples should be brought to the laboratory as soon as possible to maximize viability, and specific culture for *Shigella* sp. should be requested.

Diagnostic Test/Procedure
Fecal Leukocyte Stain
Stool Culture

Treatment Strategy Most cases of *Shigella* dysentery are self-resolving. Some have suggested that antibiotics be reserved for severe cases, but this does not eliminate the reservoir for infection in the community. Antibiotics have been shown to shorten the period of excretion of the organism in the feces as well as decreasing morbidity. The antibiotic of choice is co-trimoxazole for both children and adults.

Sulfamethoxazole and trimethoprim: Oral:
Children: 8-10 mg/kg TMP/50 mg/kg SMX in two divided doses for 5 days; in severe cases, give same dose in 3-4 divided doses
Adults: 160 mg TMP/800 mg SMX twice daily for 5 days
Ciprofloxacin:
Children: 10 mg/kg every 12 hours for 5 days
Adults: 500 mg twice daily for 5 days or a single dose of 750 mg
Azithromycin: 500 mg orally on day 1 then, 250 mg once daily for 4 days is also effective
Levofloxacin: 200-300 mg/day in divided doses for 5-7 days
Ofloxacin: Oral: 400 mg initially then, 5 additional 200 mg doses twice daily
Colistin: Oral: 5-15 mg/kg/day in 3 divided doses
Alternative dosing:
Ampicillin: Oral:
Children: 50-100 mg/kg/day in 4-6 divided doses for 7 days
Adults: 500 mg 4 times/day for 7 days
I.M./I.V.: 100-200 mg/kg/day in 4-6 divided doses up to 6 g/day (Note: Amoxicillin is not effective)
Some strains are resistant to trimethoprim, particularly in Africa and Southeast Asia, and *in vitro* susceptibility testing should be performed on all isolates. The quinolones have been effective for shigellosis in clinical trials and are useful alternatives. Antimotility agents such as opiates, paregoric, and diphenoxylate (Lomotil®) should be avoided, because of the potential for worsening the dysentery and for predisposing to toxic megacolon.

Quarantine Benzodiazepines can be used for seizure control; fevers can be treated with antipyretics and ice baths. Intestinal antipyretic agents should only be used with antibiotic therapy; mass antibiotic prophylaxis is not helpful; enteric precautions

Specific References
Fernandez MI and Sansonetti PJ, "*Shigella* Interaction With Intestinal Epithelial Cells Determines the Innate Immune Response in Shigellosis," *Int J Med Microbiol*, 2003, 293(1):55-67.

Smallpox

Synonyms Variola
Clinically Resembles Chicken pox, herpes simplex
Microbiology A member of the family Poxviridae, subfamily Chordopoxvirinae, genus orthopoxvirus, variola is enveloped, and has a characteristic brick-shaped structure, measuring approximately 250-300 nm by 200 nm. The virus has a dumbbell-shaped core,

and its genetic material consists of linear, double-stranded DNA. One of the largest and structurally most complex of viruses, it is possible to image the refractile virion using a high-quality UV microscope. Variola has a complex interaction with the immune system, and replicates within the host cell cytoplasm. It is closely related to four other viruses - camelpox, monkeypox, vaccinia, and cowpox; the last three viruses may infect humans; however, smallpox is the only disease from among this group, which may be readily transmitted from human to human. Humans are the only known host, and the virus does not survive for more than a few days outside of the host. The virus is usually contracted via the respiratory route (aerosolization or fomites), but may also be spread through direct contact.

Postmortem Care Airborne and contact precautions; cremation is the preferred modality

Epidemiology Smallpox has officially been eradicated with no known animal reservoir. The last naturally occurring case was reported in 1977 in Zaire, and the World Health Organization (WHO) declared formal eradication in 1980. However, variola virus has been discussed as a possible biological weapon, which may be used in aerosol form or by inoculation onto fomites. Widespread smallpox vaccinations were discontinued in the 1980s; therefore, the use of this virus as a weapon constitutes a significant threat. In the United States, virtually no person under the age of 30 has been previously vaccinated. Exceptions may include military and primary healthcare workers.

Secondary transmission rates among unvaccinated contacts have been estimated to range from 37% to 88%. Patients are most infective from the onset of exanthema through the first 7-10 days of rash.

Clinical Syndromes Smallpox is an acute exanthematous disease caused by infection with the poxvirus variola. Following inhalation, mucous membranes and local lymph nodes are rapidly infected. The virus replicates within the reticuloendothelial system during the latent period. Overall, the incubation period for smallpox is between 7 and 17 days. A 2-4 day prodromal period, characterized by pharyngitis, high fever, malaise, rigors, headache, backache, and vomiting are followed by a generalized vesicular or pustular eruption. Enanthema of the oral mucosa normally precedes the development of rash by approximately 24 hours. The rash is generalized and centrifugal, following a rapid succession from papule to vesicle to pustules, which crust over during a period of 7-14 days. Development of lesions is synchronous across multiple body sites. Smallpox eruption may be more prevalent on the face and distal extremities, with fewer lesions on the abdomen and back. The presence of lesions on the palms and soles of the patient are important diagnostic features. Fever may reappear 7 days after the onset of the rash. Mortality has been estimated between 20% to 50%.

The classic symptoms pattern is termed variola major, and accounts for approximately 90% of cases. However, less common presentations have also been described. Modified smallpox cases have been reported to occur in approximately 2% of unvaccinated contacts and up to 25% of vaccinated contacts. These cases are associated with milder symptomatology, superficial lesions, and a low mortality rate. Flat-type smallpox, which involves a slow evolution of flat, soft, focal skin lesions along with severe systemic toxicity, has been noted in 2% to 5% of patients. Mortality for flat-type smallpox was 66% in vaccinated patients and up to 97% in unvaccinated contacts. In addition, 3% of patients may have a hemorrhagic form of smallpox characterized by extensive petechiae, mucosal hemorrhage, and intense toxemia. Mortality associated with this presentation is extremely high, and death typically occurs prior to development of typical pox lesions. Finally, a fifth form of smallpox has been described in previously vaccinated patients (and in infants with maternal antibodies). Patients may be asymptomatic, or symptoms may be mild, including headache, fever and influenza-like symptoms, and mortality is <1%. (*MMWR*, 2004, Figures 13–15)

Infective Dose 10-100 organisms with human-to-human transmission within three meters

Diagnostic Test/Procedure

Possible cases of smallpox constitute a public health emergency. State health officials should be immediately contacted, who should contact the CDC and WHO. Airborne and contact precautions should be used. Scrapings from papules, vesicular fluid, pus, or scabs may be collected for rapid identification by electron microscopy (EM). Skin samples may also be used for agar gel immunoprecipitation, immunofluorescence, or polymerase chain reaction (PCR) assay. PCR allows differentiation from other pox viruses. In the event of known exposures, early postexposure (0-24 hours) nasal swabs and induced respiratory secretions may be collected for viral culture, fluorescent antibody assay, and PCR assay. After 2 days, blood or serum may be collected for viral culture. Serological tests may be useful for confirmation or early presumptive diagnosis.

Immunofluorescent Studies, Biopsy

Polymerase Chain Reaction accurate to discriminate between variola and other orthopox viruses

Treatment Strategy Supportive therapy is the mainstay of treatment. Antivirals have not been demonstrated to be useful. Vaccination administered within 3 days of exposure will prevent or significantly lessen severity of the disease. Vaccination within four to seven days may modify the severity of the illness. Cidofovir is available for treatment of serious effects related to the vaccine, although it may not prove to be beneficial. Topical ocular agents (antivirals) such as trifluridine or idoxuridine may be efficacious in treating ocular exposure. Vaccine can be obtained by calling the Centers for Disease Control (877) 554-4625 or (888) USA-RIID.

Quarantine Respiratory isolation (negative pressure rooms) 6-12 exchanges per hour. Droplet/airborne precautions for a minimum of 17 days.

Environmental Persistency Aerosol: 6-24 hours

Specimen

Label all tubes, vials, and microscope slide holders with patient's name, unique identifier, date of collection, source of specimen (vesicle, pustule, scab, or fluid), and name of person collecting the specimen.

Wear appropriate personal protective equipment.

Vesicular material:

Sanitize the patient's skin with an alcohol wipe and allow skin to dry.

Open the top of a vesicle or pustule with a scalpel, sterile 26-gauge needle, or slide. Collect the skin of the vesicle top in a dry, sterile 1.5- to 2-mL screw-capped tube. Label the tube.

Scrape the base of the vesicle or pustule with the wooden end of an applicator stick or swab and smear the scrapings onto a glass or plastic light microscope slide. Allow slide to dry for 10 minutes.

Label the slide and place it in a slide holder. To prevent cross-contamination, do not place slides from more than one patient in the same slide holder.

Take another slide, and touch it repetitively to the opened lesion using progressive movements of the slide in order to make a touch prep. Allow slide to dry for 10 minutes.

Label the slides as touch preps and place in the same slide holder. To prevent cross-contamination, do not place slides from more than one patient in the same slide holder.

If plastic-coated electron microscopic (EM) grids are available, lightly touch the shiny side of 3 EM grids to the base of the open lesion, allow EM grids to air-dry for 10 minutes, and place grids in an appropriately labeled grid box. Use varying degree of pressure (minimal, light, and moderately firm) in application of the 3 grids to the unroofed lesion. EM grids and collection materials will soon be available at Laboratory Response Network (LRN) sites.

If a slide or EM grid is not available, swab the base of the lesion with a polyester or cotton swab, place in screw-capped plastic vial, break off applicator handle, and seal

Repeat this procedure for 2 or more lesions

Scab specimens:

Sanitize the patient's skin with an alcohol wipe and allow skin to dry.

Use a 26-gauge needle to remove 2 to 4 scabs.

Place 1 or 2 scabs in each of 2 dry, sterile screw-capped plastic tubes.

Wrap parafilm around the juncture of the cap and vial.

Label the tube.

Biopsy lesions: (At least 2 specimens obtained by using a 3.5- or 4-mm punch biopsy kit.)

Use sterile technique and appropriate anesthetic

Place 1 sample in formalin for immunohistochemical or histopathologic evaluation and store at room temperature.

The second specimen should be placed dry (do not add transport medium) in a sterile 1.5- to 2-mL screw-capped container (do not add transport medium).

Refrigerate if shipment occurs within 24 hours; otherwise, the specimen should be frozen.

Serum specimens:

Draw 10 mL of blood for serum separation and collection.

Send serum, stored refrigerated.

Additional Information Standard hospital disinfectants (hypochlorite and quaternary ammonia) are effective for cleaning surfaces. Autoclave or launder in hot water (with bleach) all linens and clothing.

Can be cleared to work in research labs after evidence of a vesiculopapular response from smallpox vaccine; scab resolution (day 21-28); booster doses are given every year if working with variola (at the Centers for Disease Control) or every 3 years if working with variola.

Specific References

Babinchak T, "Smallpox Vaccine," *P&T*, 2003, 28(7):457-60.

Breman JG and Henderson DA, "Diagnosis and Management of Smallpox," *N Engl J Med*, 2002, 346(17):1300-8.

Centers for Disease Control and Prevention, "Update: Adverse Events Following Civilian Smallpox Vaccination - United States, 2003," *MMWR Morb Mortal Wkly Rep*, 2003, 52(18):419-20.

Centers for Disease Control and Prevention, "Update: Cardiac and Other Adverse Events Following Civilian Smallpox Vaccination - United States, 2003," *MMWR Morb Mortal Wkly Rep*, 2003, 52(27):639-42.

Everett WW, Coffin SE, Zaoutis T, et al, "Smallpox Vaccination: A National Survey of Emergency Health Care Providers," *Acad Emerg Med*, 2003, 10(6):606-11.

Frey SE, Newman FK, Yan L, et al, "Response to Smallpox Vaccine in Persons Immunized in the Distant Past," *JAMA*, 2003, 289(24):3295-9. Erratum in: *JAMA*, 2003, 290(3):334.

Gibson WA, "Mass Smallpox Immunization Program in a Deployed Military Setting," *Am J Emerg Med*, 2004, 22(4):267-9.

Guharoy R, Panzik R, Noviasky JA, et al, "Smallpox: Clinical Features, Prevention, and Management," *Ann Pharmacother*, 2004, 38(3):440-7.

Hupert N, Wattson D, Cuomo J, Benson S, "Anticipating Demand for Emergency Health Services Due to Medication - Related Adverse Events after Rapid Mass Prophylaxis, Campaigns," *Acad Emerg Med*, 2007, 14:268-274.

LeDuc JW and Jahrling PB, "Strengthening National Preparedness for Smallpox: An Update," *Emerg Infect Dis*, 2001, 7(1):155-7.

Moran GJ, Everett WW, Karras DJ, et al, "Smallpox Vaccination for Emergency Physicians: Joint Statement of the AAEM and the SAEM," *J Emerg Med*, 2003, 24(3):351-2.

Noeller TP, "Biological and Chemical Terrorism: Recognition and Management," *Cleve Clin J Med*, 2001, 68(12):1001-2, 1004-9, 1013-6.

Rusnak JM, Kortepeter MG, Aldis J, et al, "Experience in the Medical Management of Potential Laboratory Exposures to Agents of Bioterrorism on the Basis of Risk Assessment at the United States Army Medical Research Institute of Infectious Diseases (USAMRIID)," *J Occup Environ Med*, 2004, 46(8):801-11.

Rusnak JM, Kortepeter MG, Hawley RG, et al, "Management Guidelines for Laboratory Exposures to Agents of Bioterrorism," *J Occup Environ Med*, 2004, 46(8):791-800.

Thorne CD, Hirshon JM, Himes CD, et al, "Emergency Medicine Tools to Manage Smallpox (Vaccinia) Vaccination Complications: Clinical Practice Guideline and Policies and Procedures," *Ann Emerg Med*, 2003, 42(5):665-80.

Staphylococcal Enterotoxin B

Clinically Resembles Q fever, tularemia, plague

Microbiology Staphylococci derives its name from the Greek word staphyle meaning "bunch of grapes." On Gram's stain, staphylococci are gram-positive cocci, 0.7-1.2 mcm, nonspore-forming, occurring singly, in pairs, in short 4-5 cocci chains or clusters. Staphylococci grow rapidly both as an aerobe and anaerobe on blood agar. The colonies are sharply defined, smooth, and 1-4 mm in diameter. Staphylococci are catalase-positive and differ from micrococci by the following: anaerobic acid production from glucose, sensitivity <200 mg/mL lysostaphin, and production of acid from glycerol in the presence of 0.4 mg/mL erythromycin. *Staphylococcus aureus* may have a golden pigmentation secondary to carotenoid and produce beta-hemolysis on horse, sheep, or human blood agar after an incubation of 24-48 hours. *Staphylococcus epidermidis* and coagulase-negative staph (CNS) are often used interchangeably but recognize that there are over 30 species of CNS of which *S. epidermidis* is the most common.

Epidemiology Almost 50% of all strains of *S. aureus* produce enterotoxins. There are five distinct toxins with B accounting for less than 10% of the "naturally" occurring toxins associated with food poisoning outbreaks.

Clinical Syndromes

Food poisoning: Usually caused by ingestion of heat stable enterotoxin A. It is also resistant to freezing. Second most common cause of acute food poisoning. May occur by person-to-person transmission. Incubation 4-10 hours after ingestion (or 3-12 hours by inhalation) of toxin contained in custard filled bakery goods, canned foods, processed meats, potato salads, and ice cream. Patients present with acute salivation, fever (which may last for 5 days and range from 103°F to 106°F), chills, myalgia, nausea, vomiting progressing to abdominal cramps, and watery, nonbloody diarrhea (risk of dehydration); cough may occur and last up to 4 weeks. Severe intoxication may lead to septic shock and death.

Inhalation Poisoning: Incubation time to symptoms is 3 to 12 hours; Symptoms are initially characterized by headache, mild fever, cough, dyspaca and pleuritic chest pain. Conjunctivitis may develop within six hours. Pulmonary symptoms can progress to respiratory failure.

Infective Dose 30 mcg

Diagnosis Diagnosis is made by Gram's stain and culture of lesions on hands suspected of handling contaminated food. Phage typing is a useful epidemiological tool. Suspect isolates can be tested for the production of enterotoxins by immunodiffusion, ELISA and RIA techniques (can decontaminate equipment with soap and water). Clinically Resembles, Gastroenteris, Legionella pneumonia (when inhaled).

Diagnostic Test/Procedure

Aerobic Culture, Appropriate Site

Enterotoxin assays

Treatment Strategy Supportive care with attention to oxygenation and hydration is the mainstay of therapy; vaccine (prophylaxis) is investigational

Quarantine No person-to-person spread

Additional Information Weaponized by aerosolization in the 1960s; pulmonary signs dominate (along with fever and headache) when inhaled. Inhaled lethal dose (50%) is about 0.2 mcg/kg, while incapacitation dose (50%) is about 0.0004 mcg/kg. Resistent to freezing.

Specific References

Beno DW, Uhing MR, Goto M, et al, "Chronic Staphylococcal Enterotoxin B and Lipopolysaccharide Induce a Bimodal Pattern of Hepatic Dysfunction and Injury," *Crit Care Med*, 2003, 31:1154-9.

Greenfield RA, Brown BR, Hutchins JB, et al, "Microbiological, Biological, and Chemical Weapons of Warfare and Terrorism," *Am J Med Sci*, 2002, 323:326-40.

Hoffman RJ, Shea JM, Sandoval M, et al, "In Vitro Charcoal Binding of Staphylococcal Enterotoxin B," *Clin Toxicol (Phila)*, 2005, 43:675.

Khan AS, Cao CJ, Thompson RG, et al, "A Simple and Rapid Fluorescence-Based Immunoassay for the Detection of Staphylococcal Enterotoxin B," *Mol Cell Probes*, 2003, 17:125-6.

T-2 Mycotoxin

Mechanism of Toxic Action A trichothecene mycotoxin which is a potent inhibitor of protein and DNA synthesis; additionally inhibits mitochondrial electron transport; produced by the Fusarium fungal species

Toxicodynamics/Kinetics
Absorption: Oral: poor; dermal: poor
Half-life: 30 minutes (animals: 1.6 hours)
Metabolism: De-expoxidation and glucuronidation to less toxic metabolites
Elimination: Primarily through feces

Pregnancy Issues No human studies performed

Clinically Resembles Septic shock, nitrogen mustard, radiation

Signs and Symptoms of Acute Exposure Vomiting, conjunctivitis, hypotension, dyspnea, seizures, headache, diarrhea (bloody), leukopenia, dermal irritation progressing to skin necrosis and blisters. Pancytopenia (with lymphocytosis) may develop in 10-14 days with elevated erythrocyte sedimentation rate noted.

Treatment Strategy
Decontamination:
Dermal: Remove all clothing; wash with copious amounts of soap and water. Dermal irrigation with polyethylene glycol 300 may also be effective.
Oral ingestion: Activated charcoal (2 g/kg) within 4 hours of ingestion may be useful; binding capacity is 0.48 mg of charcoal (dissociation constant is 0.078). Multiple dosing may enhance elimination.
Supportive therapy:
Neutropenia can be treated with filgrastim (5 mcg/kg daily subcutaneously or intravenously). Hypotension can be treated with isotonic fluids (10-20 mL/kg) and Trendelenburg placement. Dopamine or norepinephrine are vasopressors of choice.

Additional Information Leukopenia can be caused by ingestion of 0.1 mg/kg daily (for 2 weeks); fatal dose: About 0.5 mg/kg
WHO recommendation for maximum tolerable dietary daily intake is 60 ng/kg body weight per day; no bioaccumulation
Water soluble/stable in water; resistant to chlorine and heat

Specific References
Fung E and Clark RF, "Health Effects of Mycotoxins: A Toxicological Overview," *J Tox Clin Tox*, 2004, 42:1-18.

Vibrio cholerae

Disease Syndromes Included Cholera

Microbiology *Vibrio cholerae* is an oxidase-positive, fermentative, gram-negative rod that can have a comma-shaped appearance on initial isolation. *V. cholerae* can be subdivided by the production of endotoxin and agglutination in 0-1 antisera with the nomenclature of *V. cholerae* 0-1, atypical or nontoxigenic 0-1, or non-0-1. The serogroup 0-1 can further be subdivided into the El Tor and Classic cholera biotypes, which can be further subdivided into a variety of serotypes. The major virulence factor of *V. cholerae* is the extracellular enterotoxin produced by the 0-1 strain, although nontoxin-producing organisms have been implicated in some outbreaks.
V. cholerae is a facultative anaerobe, which grows best at a pH of 7.0 and at a wide temperature range (18°C to 37°C). *Vibrio* species can be differentiated from other gram-negative bacilli by their sensitivity to 0129 and may be speciated by a variety of biochemical tests.

Epidemiology Cholera is usually spread by contamination of water and food by infected feces, with fecal contamination of water being the principal vehicle of transmission. Person-to-person transmission is less common due to the large organism load necessary for infection. Asymptomatic carriers play a minor role in cholera outbreaks. Outbreaks may be seasonally dependent based on either temperature or rainfall. Transmission by food can be eliminated by thorough cooking. Adequate sanitation is the best means of cholera prevention.

There have been several pandemics of cholera reported since 1817, originating in Bengal and subsequently spreading to a variety of geographic locations, responsible for hundreds of thousands of deaths. The latest pandemic originated in Indonesia in 1961 and moved to the Western hemisphere. In 1991, a cholera outbreak in Peru and 20 other countries in the Western hemisphere accounted for over 600,000 cases with 5000 deaths caused by El Tor 0-1.

Clinical Syndromes After a 2- to 5-day incubation period, classic cholera is characterized by an abrupt onset of vomiting and profuse, painless, watery-grayish diarrhea with flecks of mucus (rice water stool). Fluid losses can be significant (up to 20 L/day). Hypovolemic shock, hypokalemia, and metabolic acidosis can cause death within a few hours of onset, especially in children. Mortality, in untreated cases, is as high as 60%. Milder forms of the disease also occur, especially with the non-0-1 cholera. Fever is unusual and occurs in <5% of cases.

Infective Dose 10-500 organisms

Diagnosis Organisms can be identified by darkfield microscopy showing large numbers of comma-shaped organisms with significant motility. However, this test is relatively insensitive and is nonspecific. Thiosulfate-citrate-bile salt-sucrose agar (TCBS) or alkaline peptone broth are used to facilitate growth and identification. Positive identification depends on serologic and biochemical testing.

Diagnostic Test/Procedure
Fecal Leukocyte Stain
Stool Culture, Uncommon Organisms

Treatment Strategy Early and rapid replacement of fluid and electrolytes can decrease the mortality to <1%. Oral rehydration is usually successful, but in severe cases, intravenous replacement is required. When fluid and electrolyte imbalances are corrected, cholera is a short, self-limiting disease lasting a few days. Doxycycline, tetracycline, co-trimoxazole, erythromycin, and furazolidone have all demonstrated effectiveness in decreasing the diarrhea and bacterial shedding in this disease.

Usual recommendation: Doxycycline: Adults: 300 mg as a single dose or 100 mg twice daily for 3 days. (oral)

Alternative dosing:
Single dose tetracycline 1-2 g or 500 mg every 6 hours for 4 doses
Co-trimoxazole 5 mg/kg as trimethoprim, twice daily for 3 days
Single dose azithromycin 20 mg/kg P.O.; maximum dose: 1 g
Single dose ciprofloxacin: Adults: 1 g or 250 mg per day for 3 days
Norfloxacin: 400 mg twice daily for 3 days
Supportive therapy: Azithromycin at a single oral 1 g dose can shorten the duration of diarrhea

Quarantine Enteric precautions

Additional Information Survives up to 24 hours in sewage; 6 weeks in impure water containing organic matter. Boiling or chlorination of water can eliminate vibrio cholera. Does not survive in carbonated or acidic beverages.

Specific References

Bhattacharya SK, "An Evaluation of Current Cholera Treatment," *Expert Opin Pharmacother*, 2003, 4:141-6.

Centers for Disease Control and Prevention (CDC), "Cholera Epidemic Associated With Raw Vegetables - Lusaka, Zambia, 2003-2004," *MMWR Morb Mortal Wkly Rep*, 2004, 53(34):783-6.

Guerrant RL, "Cholera - Still Teaching Hard Lessons," *N Engl J Med*, 2006, 354(23):2500-2.

Guerrant RL, Carneiro-Filho BA, and Dillingham RA, "Cholera, Diarrhea, and Oral Rehydration Therapy: Triumph and Indictment," *Clin Infect Dis*, 2003, 37:398-405.

Ryan ET and Calderwood SB, "Cholera Vaccines," *Clin Infect Dis*, 2000, 31:561-5.

Saha D, Karim MM, Khan WA, et al, "Single-Dose Azithromycin for the Treatment of Cholera in Adults," *N Engl J Med*, 2006, 354(23):2452-62.

Shears P, "Recent Developments in Cholera," *Curr Opin Infect Dis*, 2001, 14:553-8.

Viral Hemorrhagic Fever (VHF)

Applies to Ebola Virus; Marburg Virus

Microbiology Members of the Filoviridae, these RNA viruses are stable, highly infective particles, approximately 80-100 nm in diameter. They are filamentous, elongated, flexible, and enveloped, with genetic material composed of a single nonsense, nonsegmented RNA strand. Four subtypes of Ebola virus have been identified, in addition to two forms of the Marburg virus. Although some cross reactivity has been demonstrated between Ebola subtypes, no cross reactivity have been demonstrated between the Marburg and Ebola viruses. Viruses may be antigenically differentiated by their transmembrane spike protein. The spike protein is highly glycosylated, which has been speculated to inhibit immune recognition and response.

Postmortem Care Personnel should utilize all PPE precautions including HEPA-filtered respirators and a negative pressure room. Body should be minimally handled; wrap the corpse in sealed/leakproof material; do **not** embalm; cremate or bury promptly in a sealed casket.

Epidemiology Initially isolated in primates (monkeys and macaques), epidemics in humans have occurred in several African countries including Zaire, Sudan, and the Ivory Coast. In addition, isolated cases have been associated with monkey importation (including the original Marburg, Germany importation). Blood exposure is a consistent factor in the development of epidemics.

The mode of transmission appears to involve close contact with blood and/or tissue from an infected individual or host. Airborne transmission does not appear to be a significant route of human infection, although experimental evidence suggests that this route is a possible source of infection. The reservoir for this virus has not been adequately characterized. During African epidemics, it has been noted that interhuman spread among hospital workers may be greatly reduced with the implementation of standard barrier infection control procedures (glove-gown-mask).

The risk for person-to-person transmission of VHF is greatest during the latter stages of disease when the viral burden is highest. Transmission of VHF during the incubation phase has not been reported. The largest Marburg VHF outbreak occurred October 2004 to July 2005 in Angola, which was attributed to poor medical hygiene when standard precautions were ignored.

Clinical Syndromes

• **Hemorrhagic fever:** Following an incubation period of 5-10 days (may be as long as 3 weeks), the onset of hemorrhagic fever begins with an abrupt fever, headache, and myalgia (particularly in the lumbar area). After 1-3 days, additional gastrointestinal symptoms develop, which may include nausea, vomiting, watery diarrhea, abdominal cramping and pain. Chest pain, cough, and pharyngitis may also be present. Conjunctivitis is present in nearly 50% of cases. Photophobia, conjunctival injection, lymphadenopathy, jaundice, pancreatitis, as well as central nervous system involvement (somnolence, delirium, and coma) may occur. Febrile patients may exhibit relative bradycardia, sometimes referred to as pulse-temperature dissociation. This clinical sign can also be seen with typhoid fever as well as other bioweapon illness including certain viral hemorrhagic fevers and plague.

The disease progresses rapidly over a 5- to 7-day interval, as hemorrhagic manifestations assume higher degree of prominence. Petechiae, ecchymoses, and mucous membrane hemorrhages occur in approximately 50% of cases. A prominent maculopapular rash frequently develops on the trunk. Fine desquamation of the palms and soles may be apparent.

The second week of infection may be accompanied by defervescence; however, this is frequently followed by a second febrile interval. Splenomegaly and hepatomegaly, as well as complications including orchitis, myocarditis, and pancreatitis commonly accompany this phase. A small number of patients will not progress and a protracted convalescence may begin. However, mortality associated with Ebola has been as high as 90% and is 100% when acquired through infected needles (percutaneously)

Infective Dose 400 plaque-forming units by inhalation is fatal in 5 days

Diagnosis Due to its characteristic course and epidemiology, a history of exposure in an endemic area (ie, sub-Saharan Africa) may prompt clinical suspicion. Culture is positive during the acute stages, and laboratory confirmation via polymerase chain amplification and/or antigen detection may be used.

Treatment Strategy

Decontamination:

Dermal: Remove contaminated clothing and wash skin with soap and water

Ocular: Irrigate with copious amounts of saline or water for at least 15 minutes

Supportive therapy: Monitor fluids and electrolytes; replace fluids with crystalloid solutions: I.V.:

Children: 10-20 mL/kg over 1 hour

Adults: 1-2 liters over 1 hour

Transfuse with fresh frozen plasma or red blood cells as necessary

Supportive treatment is the mainstay. No antiviral therapy has been utilized successfully; vaccines are in development. Vasopressors of choice include dopamine or norepinephrine.

Quarantine Strict isolation/barrier precautions; private negative air pressure room with 6-12 air exchanges per hour; restrict visitor access; dedicated medical equipment

Environmental Persistency Not persistent

Personal Protection Equipment

Strict adherence to hard hygiene with double gloves, impermeable gowns, N-95 respirators or powered air-purified respirators, leg/shoe coverings, face shields/goggles

Additional Information Contact CDC: (404) 639-6425 or (404) 639-2888

In 2000, 24 people died in Uganda during an ebola outbreak. Disinfect equipment with household bleach (1:100 dilution)

Specific References

Casillas AM, Nyamathi AM, Sosa A, et al, "A Current Review of Ebola Virus: Pathogenesis, Clinical Presentation, and Diagnostic Assessment," *Biol Res Nurs*, 2003, 4:268-75.

Polesky A and Bhatia G, "Ebola Hemorrhagic Fever in the Era of Bioterrorism," *Semin Respir Infect*, 2003, 18:206-15.

Sullivan N, Yang ZY, and Nabel GJ, "Ebola Virus Pathogenesis: Implications for Vaccines and Therapies," *J Virol*, 2003, 77:9733-7.

Sullivan NJ, Geisbert TW, Geisbert JB, et al, "Accelerated Vaccination for Ebola Virus Haemorrhagic Fever in Non-Human Primates," *Nature*, 2003, 424:681-4.

Towner JS, Khristova ML, Sealy TK, et al, "Marburgvirus Genomics and Association With a Large Hemorrhagic Fever Outbreak in Angola," *J Virol*, 2006, 80(13):6497-516.

Ebola hemorrhagic fever — necrotic hepatocytes with filamentous intracytoplasmic inclusions (arrows) (hematoxylin and eosin stain).

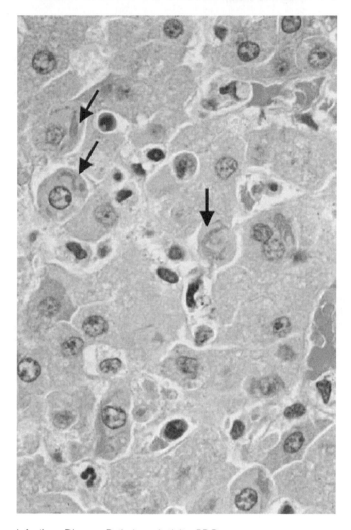

Infectious Disease Pathology Activity, CDC.

Ultrastructural appearance of Ebola virus (electron microscopy negative stain).

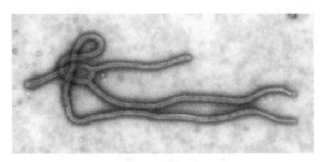

Infectious Disease Pathology Activity, CDC.

West Nile Virus

Related Information West Nile Virus Diagnostic Procedures

Microbiology A member of the flaviridae virus family, West Nile virions are 40-60 nm in size, enveloped, with an icosahedral nucleocapsid. Genetic material consists of a positive-sense, single stranded RNA approximately 10,000-11,000 bases.

Epidemiology First isolated from a febrile adult woman in the West Nile District of Uganda in 1937, West Nile virus is recognized as a cause of severe human meningoencephalitis. West Nile virus is maintained in a primary enzootic cycle between birds and mosquitoes. Domestic and wild birds are known to be infected, and mortality in these species is high. Human and equine infections are transmitted from infected mosquitoes. West Nile virus has emerged in recent years in temperate regions of Europe and North America, presenting a threat to the public, equine, and animal health. The first appearance in North America occurred in 1999, and cases were reported in both humans and horses. Originally reported on the East Coast, viral activity and human infection spread to wider regions each ensuing year. In 2002, West Nile virus cases extended across the United States. In addition to typical transmission, a small number of cases have been attributed to blood transfusions from infected individuals.

Clinical Syndromes The incubation period of West Nile virus is between 3 and 14 days. Peak incidence is in August. Most people who are infected with the West Nile virus either have no symptoms or experience mild illness such as fever (90%), headache (45%), diarrhea (27%), and body aches before fully recovering. Some persons may also develop mild rash or lymphadenopathy. Encephalitis is the most serious manifestation of viral infection. Fewer than 1% of people infected with West Nile virus develop encephalitis, and among those hospitalized with West Nile encephalitis, the case fatality rate changes from 3% to 15%. Therefore, fewer than 1 in 1000 people infected with West Nile virus die. The elderly are at increased risk for the development of encephalitis, which may cause permanent neurologic sequelae, paralysis, or death. In addition, a polio-like syndrome has been reported. Genetic factors may influence the potential for severe encephalitis. Symptoms of encephalitis (inflammation of the brain) include the abrupt onset of severe headache, high fever (90%), nuchal rigidity, muscle weakness, confusion, or coma.

Diagnostic Test/Procedure

West Nile Virus Diagnostic Procedures

IgM antibody capture testing by ELISA on serum or cerebrospinal fluid specimens. Recommended clinical and laboratory case definitions are available at www.cdc.gov/ncidod/dvbid/westnile/resources/wnv-guidelines-apr-2001.pdf.

Treatment Strategy There is no specific treatment for West Nile virus encephalitis. Treatment is supportive in nature. *In vitro* evidence suggests ribavirin and interferon alfa-2b may have activity against the virus, but clinical trials have not been reported.

Additional Information West Nile virus vaccine (with killed virus) has been effective in animal studies.

Specific References

Centers for Disease Control and Prevention, "West Nile Virus Activity - United States, June 9-15, 2004," *MMWR Morb Mortal Wkly Rep*, 2004, 53(23):511-2.

Charatan F, "Organ Transplants and Blood Transfusions May Transmit West Nile Virus," *BMJ*, 2002, 325(7364):566.

Hayes EB and O'Leary DR, "West Nile Virus Infection: A Pediatric Perspective," *Pediatrics*, 2004, 113(5):1375-81.

Lawrence D, "Susceptibility to West Nile Virus Could Be Genetically Determined," *Lancet*, 2002, 360(9333):624.

Nash D, Mostashari F, Fine A, et al, "1999 West Nile Outbreak Response Working Group. The Outbreak of West Nile Virus Infection in the New York City Area in 1999," *N Engl J Med*, 2001, 344(24):1807-14.

Petersen LR and Marfin AA, "West Nile Virus: A Primer for the Clinician," *Ann Intern Med*, 2002, 137(3):173-9.

Samuel CE, "Host Genetic Variability and West Nile Virus Susceptibility," *Proc Natl Acad Sci USA*, 2002, 99(18):11555-7.

"West Nile Virus Infection in Organ Donor and Transplant Recipients - Georgia and Florida, 2002," *MMWR Morb Mortal Wkly Rep*, 2002, 51(35):790.

Yersinia pestis

Related Information *Yersinia pestis* Diagnostic Procedures

Disease Syndromes Included Bubonic Plague; Plague; Pneumonic Plague

Microbiology *Yersinia pestis* is the etiological agent of plague. All of the 11 species of *Yersinia* are aerobic, gram-negative rods, are well established members of the family Enterobacteriaceae, and have been isolated from human clinical specimens. At least three species of *Yersinia* are unequivocal human pathogens: *Y. pestis*, *Y. enterocolitica*, and *Y. pseudotuberculosis*.

Postmortem Care Standard precautions and droplet precautions; avoid aerosol-generating procedures (such as bone sawing). If these procedures are needed, use high- efficiency particulate air filtered masks and negative pressure rooms.

Epidemiology *Yersinia pestis* infections are rare in the United States. From 1970 to 1991, 295 cases were reported to the Centers for Disease Control and Prevention. Plague occurs worldwide; most cases occur in Asia and Africa. In the United States, plague occurs (rarely) in New Mexico, Arizona, California, Utah, and Colorado. Yersinioses are zoonotic infections that usually affect rodents, small animals, and birds. Rats are the natural reservoir of *Y. pestis* in areas of urban ("city") plague; small animals such as ground squirrels, wood rats, rabbits, and cats are the natural reservoirs of *Y. pestis* in areas of sylvatic ("country") plague. Humans are accidental hosts of *Yersinia* species. Humans become infected with the bacterium after being bitten by fleas of the aforementioned animals, or, much less commonly, by handling infected animals or inhaling aerosolized *Y. pestis* generated by a person with pulmonary plague.

Clinical Syndromes

Plague: Plague presents in many different and protean clinical forms: bubonic, septicemic, pneumonic, cutaneous, and meningitis. After an incubation period of 2-8 days, most patients with bubonic plague, the most common form, experience fever, chills, headache, aches, extreme exhaustion, and lymphadenitis. Bubos are actually enlarged, infected lymph nodes which are exquisitely painful. These extremely painful bubos develop most commonly in the groin, axilla, or neck. These patients have overwhelming numbers of bacteria in their blood. Patients with septicemic plague also have tremendous numbers of bacteria in their blood; however, these patients have a more septic presentation, and bubos usually are not present.

Pneumonic plague is a complication of bubonic plague and is characterized by cough, chest pain, difficulty breathing, and hemoptysis; bubos might or might not be present. Febrile patients may exhibit relative bradycardia, sometimes referred to as pulse-temperature dissociation. Other bioweapons that can cause this include certain viral hemorrhagic fevers and tularemia. Pneumonic plague, if not treated immediately, has a very high mortality rate and can be rapidly fatal within 1 day. The natural progression of plague is from the bubonic, to the septicemic, to the pneumonic forms. The overall mortality rate for untreated cases of plague is 50% to 60%.

Infective Dose <100 organisms

Diagnosis Diagnosis of plague is mainly a clinical diagnosis. Laboratories must be notified if *Y. pestis* is suspected as an etiological agent because many laboratories do not routinely suspect *Y. pestis* or culture clinical specimens for *Y. pestis* because special media and techniques easily can be used to enhance isolation of the bacterium and because the colonial morphology of *Y. pestis* is not typical of many other gram-negative rods. Culture is the most productive test for the laboratory diagnosis of plague. *Yersinia pestis* is not fastidious and grows well on blood agar media and many enteric media. Many commercial bacterial identification systems do not include *Y. pestis* in their databases. The most appropriate clinical specimens for culture include blood, biopsy or aspirate of bubo, sputum, cerebrospinal fluid, and cutaneous biopsy. Fine needle drainage of the bubo(s) is not only diagnostic, it is therapeutic. Relieving the pressure within the tense, suppurative lymph node will yield fluid for testing and reduce the pain.

Diagnostic Test/Procedure
Aerobic Culture, Appropriate Site
Blood Culture, Aerobic and Anaerobic

Treatment Strategy Any person with a fever of 38.5°C or higher should begin prophylactic antibiotic therapy during a pneumonic plague epidemic; any infected patient should be isolated for the first 2 days of antibiotic therapy until clinical improvement occurs; respiratory droplet precautions should be utilized; exudates or discharge should be handled with rubber gloves; seizures can be treated with a benzodiazepine or phenobarbital.
Supportive therapy: See table (*MMWR*, 2004, Figures 9-10)

Recommendations for Treatment of Patients with Pneumonic Plague		
	Recommended Therapy	
Patient Category	**Contained Casualty Setting***	
Adults	Preferred choices	Streptomycin 1 g I.M. twice daily Gentamicin, 5 mg/kg I.M. or I.V. once daily or 2 mg/kg loading dose followed by 1.7 mg/kg I.M. or I.V. 3 times/day[†]
	Alternative choices	Doxycycline, 100 mg I.V. twice daily or 200 mg I.V. once daily Ciprofloxacin, 400 mg I.V. twice daily[‡] Chloramphenicol, 25 mg/kg I.V. 4 times/day[§]
Children¶	Preferred choices	Streptomycin, 15 mg/kg I.M. twice daily (maximum daily dose, 2 g) Gentamicin, 2.5 mg/kg I.M. or I.V. 3 times/day[†]
	Alternative choices	Doxycycline: If ≥45 kg, give adult dosage; if <45 kg, give 2.2 mg/kg I.V. twice daily (maximum, 200 mg/day) Ciprofloxacin, 15 mg/kg I.V. twice daily[‡] Chloramphenicol, 25 mg/kg I.V. 4 times/day[§]

Table   (*Continued*)

Patient Category		Recommended Therapy
		Contained Casualty Setting*
Pregnant women	Preferred choice	Gentamicin, 5 mg/kg I.M. or I.V. once daily or 2 mg/kg loading dose followed by 1.7 mg/kg I.M. or I.V. 3 times/day[†]
	Alternative choices	Doxycycline, 100 mg I.V. twice daily or 200 mg I.V. once daily Ciprofloxacin, 400 mg I.V. twice daily[‡]
Patient Category		**Mass Casualty Setting and Postexposure Prophylaxis***
Adults	Preferred choices	Doxycycline, 100 mg orally twice daily** Ciprofloxacin, 500 mg orally twice daily[‡]
	Alternative choice	Chloramphenicol, 25 mg/kg orally 4 times/day[§°]
Children[¶]	Preferred choices	Doxycycline**: If ≥45 kg, give adult dosage; if <45 kg, then give 2.2 mg/kg orally twice daily Ciprofloxacin, 20 mg/kg orally twice daily
	Alternative choice	Chloramphenicol, 25 mg/kg orally 4 times/day[§°]
Pregnant women#	Preferred choices	Doxycycline, 100 mg orally twice daily** Ciprofloxacin, 500 mg orally twice daily
	Alternative choice	Chloramphenicol, 25 mg/kg orally 4 times/day[§°]

*These are consensus recommendations of the Working Group on Civilian Biodefense and are not necessarily approved by the FDA. One antimicrobial agent should be selected; therapy should be continued for 10 days; oral therapy should be substituted when the patient's condition improves.
I.M. = intramuscular; I.V. = intravenous.

[†]Aminoglycosides must be adjusted according to renal function. Evidence suggests that gentamicin 5 mg/kg I.M. or I.V. once daily, would be efficacious in children, although this is not yet widely accepted in clinical practice. Neonates up to 1 week of age and premature infants should receive gentamicin, 2.5 mg/kg I.V. twice daily.
[‡]Other fluoroquinolones can be substituted at doses appropriate for age. Ciprofloxacin dosage should not exceed 1 g/day in children.
[§]Concentration should be maintained between 5 and 20 mcg/mL. Concentrations >25 mcg/mL can cause reversible bone marrow suppression.
[¶]In children, ciprofloxacin dose should not exceed 1 g/day, chloramphenicol should not exceed 4 g/day. Children younger than 2 years of age should not receive chloramphenicol.
[#]In neonates, gentamicin loading dose of 4 mg/kg should be given initially.
*Duration of treatment of plague in mass casualty setting is 10 days. Duration of postexposure prophylaxis to prevent plague infection is 7 days.
[°]Children younger than 2 years of age should not receive chloramphenicol. Oral formulation available only outside the U.S.
**Tetracycline could be substituted for doxycycline.
Adapted from Inglesby TV, Dennis DT, Henderson DA, et al, "Plague as a Biological Weapon: Medical and Public Health Management," *JAMA*, 2000, 283:2281-2290.

Quarantine Droplet precautions should be implemented until the patient has completed 48 hours of antimicrobial therapy. Surgical-type mask should be worn when within 3 feet of infected patient. Place infected patient in a private room. Special air handling is not necessary; doors may remain open.

Environmental Persistency Aerosol - ~one hour; sensitive to sunlight and heat: Soil - up to one year

Additional Information Environmental persistency when released as an aerosol: ~1 hour (sensitive to sunlight and heat)

Specific References

Drancourt M, Roux V, Dang LV, et al, "Genotyping, Orientalis-like *Yersinia pestis*, and Plague Pandemics," *Emerg Infect Dis*, [serial on the Internet], 2004, [9/7/04]. Available from http://www.cdc.gov/ncidod/EID/vol10no9/03-0933.htm.

Inglesby TV, Dennis DT, Henderson DA, et al, "Plague As a Biological Weapon: Medical and Public Health Management. Working Group on Civilian Biodefense," *JAMA*, 2000, 283(17):2281-90.

Rusnak JM, Kortepeter MG, Aldis J, et al, "Experience in the Medical Management of Potential Laboratory Exposures to Agents of Bioterrorism on the Basis of Risk Assessment at the United States Army Medical Research Institute of Infectious Diseases (USAMRIID)," *J Occup Environ Med*, 2004, 46(8):801-11.

Rusnak JM, Kortepeter MG, Hawley RG, et al, "Management Guidelines for Laboratory Exposures to Agents of Bioterrorism," *J Occup Environ Med*, 2004, 46(8):791-800.

Bubonic plague — lymphadenitis.

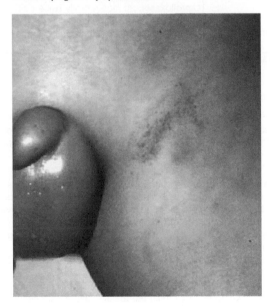

New Mexico Office of the Medical Investigator, CDC.

Secondary pneumonic plague — histologic sections of lung with (A, left) neutrophilic infiltrate in alveolar space (hematoxylin and eosin stain) and (B, right) *Yersinia pestis* bacteria (immunohistochemistry).

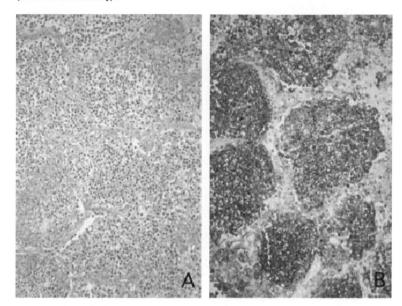

Infectious Disease Pathology Activity, CDC.

SECTION III
CHEMICAL AGENTS

Bioterrorism and the Skin

Rhonda M. Brand, PhD

Feinberg School of Medicine, Northwestern University, Chicago, Illinois

The skin is a major target site for bioterrorism agents due to both its size and contact with the environment. The primary function of the skin is to maintain fluid homeostasis while preventing compounds from entering from the external environment. Even with an intact barrier, the penetration through the skin is still the major source of absorption for numerous classes of chemicals.

Bioterrorism agents are chemically similar to the organophosphate pesticides; therefore, many conclusions drawn about these chemicals can be extended to bioterrorism agents. The greatest route of occupational exposure to pesticides is through the skin, with transdermal absorption rates during pesticide spraying being higher than from respiratory exposure. Furthermore, approximately 60% to 70% of all cases of unintentional acute pesticide poisoning are due to dermal occupational exposure. Changes to the dermal barrier could result in greater absorption of potentially toxic chemicals, thereby increasing the risk of chemically induced diseases.

Why Is the Skin Such an Effective Barrier?

To be an effective barrier, the skin has developed two layers – the epidermis and the dermis. The epidermis can be subdivided into the viable epidermis and the stratum corneum. The outermost layer of the skin, the stratum corneum, serves as the skin's principle barrier. It is approximately 15-20 μm thick in human skin and consists of terminally differentiated keratinocytes (corneocytes) that are embedded in a matrix of lipid bilayers. It can be thought of as a brick wall, with corneocytes serving as the bricks and lipids surrounding them as mortar. The viable epidermis is located just below the stratum corneum and is responsible for the production of the stratum corneum. The dermis is the innermost layer, comprised mainly of collagen fibers in an aqueous gel matrix that imparts elastic properties to the skin, provides the physiological support mechanism for the epidermis, as well as containing blood vessels, lymphatics, and nerve endings.

Keratinocytes are formed at the epidermal/dermal junction. They then begin the process of terminal differentiation by moving away from the epidermal/dermal barrier toward the stratum corneum as they are replaced with newer cells. By the time they reach the stratum corneum, they have elongated and flattened, lost their nuclei and other organelles, and are surrounded by a thick band of protein forming a cornified envelope. After approximately two weeks in the stratum corneum, the corneocytes reach the outermost layer and are sloughed off in a process called desquamation.

The lipids associated with the keratinocytes also change as the cells differentiate. The stratum corneum contains ceramides, cholesterol, fatty acids, sterol esters, and cholesteryl sulfate unlike most biological tissues, which rely on phospholipids. The lipids form a bilayer that becomes the major barrier to water and water-soluble chemicals. A chemical must, therefore, pass through the stratum corneum's brick wall to successfully penetrate through the skin. For a chemical to transverse this barrier it must rely on partitioning into the lipid mortar surrounding the keratinocytes. It then remains in the lipids and diffuses around the keratinocytes until it has successfully passed through the entire stratum corneum. The chemical must then diffuse into the more hydrophilic viable epidermis before entering the blood vessels in the dermis. The need to partition into stratum corneum lipids and then into the hydrophilic bloodstream favors molecules that are moderately lipophilic.

Movement through the skin is, therefore, different from many other tissues because it relies on passive diffusion instead of active transport. Penetration can be mathematically described using Fick's Law and demonstrates a linear relationship between transdermal absorption, exposure concentration, and size of the exposed area. Dermal absorption can be modeled based on physical properties of the compound with the combination of molecular size and lipophilicity being the best predictor of dermal absorption. Smaller and uncharged chemicals tend to penetrate better than comparable larger and charged molecules. Hair follicles and sebaceous glands can also act as shunts for chemical absorption into the skin. The surface area of these appendages, however, is quite small when compared to the total surface area of skin and, therefore, they have a minor role in transdermal penetration.

Summary of Compounds

There are two types of chemical agents that are likely to be used in bioterrorism attacks – vesicants and nerve agents. The vesicants are mustard gas and lewisite and the nerve agents are tabun, sarin, soman, and VX. The neurotoxins, VX, sarin, and soman, are organophosphates, as are the pesticides parathion, malathion, fenitrothion, chlorpyrifos, and diazinon. The transdermal absorption of the bioterrorism nerve agents are frequently examined using the safer pesticides.

Vesicants

Sulfur Mustard   Sulfur mustard (bis-2-chloroethyl sulfide) is an oily, fat soluble, light yellow-brown substance, which smells like garlic. It is heavier than water and becomes liquid at room temperature. The liquid is rapidly absorbed through the skin. The severity of the exposure correlates with the speed at which the clinical effects begin. This can range from 2-48 hours after exposure, but usually occurs within 4-8 hours. The skin lesions begin with erythema that may be combined with itching and progress to vesicles, blisters, and finally bullae. Deep burning to full skin loss may occur especially in the face, armpits, groin, and neck. Dark coloration of the skin may also be seen at these sites as will as the flexor regions of the elbow and knee joints.

The transdermal penetration of sulfur mustard has been measured to be 71-294 mg/cm^2-hr *in vitro* from a starting dose of 100 mg/cm^2. These data are in agreement with *in vivo* results of 60-240 mg/cm^2-hr. Vapor penetration was found to be 100-160 mg/cm^2-hr. A substantial reservoir of sulfur 14% to 36% has been measured in heat-separated epidermal membranes of the applied dose for up to 24 hours.

Clinically, it is interesting to note that as little as 20 mg/cm^2 of liquid sulfur mustard and 4 mg/cm^2 of vapor sulfur mustard can cause injury to human skin. The difference is due to the fact that up to 80% of the liquid dose will evaporate before penetrating the epidermis.

Lewisite   Lewisite (2-chlorovinyl dichloroarsine) is an oily, clear liquid that smells like geraniums. It is absorbed more rapidly and is more potent than sulfur mustard, resulting in deeper lesions with a thicker blister and more severe inflammatory reaction. The onset of symptoms can occur within 1 minute of exposure with a deep aching pain. After 5 minutes, the skin may develop the gray appearance of dead epithelium and by 30 minutes, redness and itching occurs. The skin heals more rapidly than after sulfur mustard exposure. Lewisite absorption can also induce signs of systemic arsenic toxicity.

Nerve Agents

Nerve agents overstimulate the muscarinic and nicotinic receptors. These agents readily penetrate through skin, but there are no physical signs on the skin as with the vesicants. Symptoms may be delayed up to 18 hours after topical exposure.

VX The transdermal absorption of VX (o-ethyl-S-(2[di-isopropylamino]ethyl) was found to be 0.6% $\pm$ 0.19% in human volunteers after a 6-hour exposure. A dose of 5.1, 11.2, 31.8, and 40.0 µg/kg placed on the cheek, forehead, abdomen, and forearm, respectively, produced a 70% depression in CHE levels. This inhibition occurred without any obvious effects on the skin.

Sarin The *in vitro* penetration of sarin (isopropylmethylphosphofluoridate) through human stratum corneum was found to range from 0.21-0.49 mg/cm^2-hr depending on the hydration state of the membrane from a donor solution of 7.8 mM.

Soman When 50 µL of soman (1,2,2-trimethylpropylmethylphosphonofluoridate) were applied over 3.1 cm^2 of the skin, the animal died approximately 15 minutes later. This corresponded to a transdermal absorption rate of 0.27 µg/min-cm^2.

Diethylmalonate is a simulant for the nerve agent soman, based on its similar physical properties including volatility and solubility. Diethylmalonate penetrates rapidly through human skin with maximum flux at 5 hours after exposure. Approximately 16% of the dose penetrates in 24 hours, 45% to 50% turns to vapor, and the remainder persists in the lipophilic stratum corneum.

Tabun Very few studies have been conducted examining the transdermal absorption of tabun (ethyl N,N-dimethylphosphoramideocyanidate). Tabun was applied to cat skin at a dose of 109 µL over a 4.5 cm^2 area and the animal died approximately 50 minutes later making it less toxic than soman. This corresponded to a transdermal delivery rate of 0.78 µg/min-cm^2.

A number of factors can dramatically influence the transdermal absorption of these chemicals. As it is frequently difficult to work with chemical warfare agents, researchers often use substitute chemicals with similar molecular structures.

Factors Influencing Transdermal Penetration

Anatomic Site Within the body, there are striking regional variations in the levels transdermal absorption for a given compound. Comparing the percutaneous penetration of parathion across body site demonstrates that relative to the forearm (1.0), the palm of the hand (1.3), and ball of the foot (1.6) have slightly elevated absorption levels. Penetration more than doubles through both the abdomen (2.1) and the back of the hand (2.4). The enhancement is even greater in the head, where penetration in 3.7 times greater for the scalp, 3.9 times greater for the jaw angle, 4.2 times greater for the forehead, and an amazing 7.4 times greater for the axilla. However, the region at greatest risk for absorption is the scrotal area, which is 11.8 times more permeable than the forearm. Therefore, men who are doused with a toxicant on the front of their pants will be at greater risk than someone who is exposed on a sleeve.

Temperature & Humidity The environmental conditions (temperature and humidity) can greatly influence the transdermal absorption of chemicals. For example, the percent of absorption of VX ranged from 4% at 18°C to 32% at 46°C when applied to the cheek and from 0.4% at 18°C to 2.9% at 46°C when given at the forearm. *In vitro* studies demonstrated that parathion penetration significantly increased both when the temperature was raised from 37°C to 42°C and when the humidity was raised for 20% to 90%. Furthermore, the increase in heat and humidity will lead to more sweating. The transdermal absorption of parathion through a uniform saturated with sweat is significantly greater than if the uniform is dry (0.65% vs 0.29%).

Clothing Clothing can be a major source of re-exposure for bioterrorism agents, even after laundering. In one case, three pesticide formulators from a single manufacturing plant developed organophosphate poisoning due to residual parathion in clothing that had been washed repeatedly. The first patient spilled a 76% parathion solution on his legs and his inguinal and scrotal areas. He immediately removed his clothes, showered, and changed his coveralls. Two days later he went to the emergency room with a plasma cholinesterase level of 340 U/L (normal range, 700 to 19,000). He was treated with atropine and pralidoxime, admitted for observation and was released the next morning. Eight days later, a second worker from the same plant was then sent to the emergency room and had a plasma cholinesterase level of 410 U/L. This formulator, however, had not been working with any organophosphates, so the method of his poisoning was unclear. A third worker at the plant became ill three days after the second while packing pesticides. His plasma cholinesterase level was 500 U/L. Again he was treated with pralidoxime and sent home.

Upon investigating the cause of these poisonings, the plant safety officer discovered that after initial spill patient one place his coveralls in a plastic bag to be burned. Instead, they were laundered and the worker wore them again until he became ill. The coveralls were laundered again and were subsequently worn by both of the other poisoning victims. This was confirmed by analyzing portions of the uniform where the spill had occurred for parathion using gas chromatography, which demonstrated that 70,000 ppm (7%) ethyl parathion remained in the coveralls. The clothes were then washed twice more, this time with soda ash, bleach and detergent and 2% still remained.

Parathion has a half-life of 120 days at 25°C when it is not applied to crops. This has led to a number of poisonings due to persistence in contaminated equipment, containers, and clothing. Poisonings have been reported in children from clothing or sheets that had been contaminated with organophosphates during shipping, and from using a contaminated burlap sack as a swing. Health hazards have been reported by leaving "empty" metal parathion containers in fields for up to a year.

The World Health Organization recommends daily washing of all clothing used by pesticide workers that have not been obviously contaminated. If pesticides have been spilled on the clothing or skin, the following has been suggested. First, contaminated clothing should be removed and the affected areas should be quickly washed with soap and water. The area should then be scrubbed again thoroughly, including nails, toenails, and hair, if appropriate. The area should then be washed again with soap and water and then rinsed with clear water and then with rubbing alcohol. Clean clothing should then be put on and the contaminated clothing should be burned.

Decontamination

The key to decontamination is to remove the source of contamination as soon as possible. Clothing should be removed immediately. One recommendation is that the skin should be washed with soap and water or a 0.5% sodium hypochlorite solution. Fuller's earth was found to be effective as a decontaminant for sulfur mustard, reducing transdermal penetration by 91% to 95%.

Furthermore, it is essential that the skin decontamination process does not itself enhance transdermal penetration. This can be tricky because both decontamination methods and barrier creams can actually increase the absorption of bioterrorism agents. For example, a number of different decontamination solutions were used to remove the soman agonist diethylmalonate. These solvents actually increased the penetration during the first 2 hours after decontamination, if they were not applied within the first 15 minutes after exposure. This can be explained by the fact that once a chemical has been absorbed into the stratum corneum, it cannot easily be removed by washing.

A number of barrier creams have been evaluated for their protective effects against bioterrorism agents. Pretreatment of skin with barrier creams may either increase or reduce lesions from sulfur mustard. Pefluorinated barrier creams were effective at blocking sulfur mustard absorption under occluded conditions, but increased penetration when used under unoccluded conditions. Furthermore, some barrier creams reduced the effectiveness of Fuller's earth as a decontaminant. It is hypothesized that the enhancement is due to prevention of sulfur mustard evaporation, which can account for up to 80% of the applied dose. Other barrier creams have been developed, which have been shown to effectively reduce the transdermal absorption of sulfur mustard, soman, and VX.

Conclusions

This review has demonstrated that the skin is a major target for bioterrorism agents even if there are no direct symptoms. Environmental factors and anatomical site can significantly alter the quantity of chemicals that will be absorbed. Decontamination should be performed quickly after exposure, but care should be taken or penetration can actually increase after the decontamination procedure.

References
al-Saleh I, "A. Pesticides: A Review Article," *J Environ Pathol Toxicol Oncol*, 1994, 13: 151-61.
Chilcott RP, Brown RF, and Rice P, "Noninvasive Quantification of Skin Injury Resulting From Exposure to Sulphur Mustard and Lewisite Vapours," *Burns*, 2000, 26:245-50.
Chilcott RP, Jenner J, Carrick W, et al, "Human Skin Absorption of Bis-2-(chloroethyl)sulphide (Sulphur Mustard) *in vitro*," *J Appl Toxicol*, 2000, 20:349-55.
Chilcott RP, Jenner J, Hotchkiss SA, et al, "Evaluation of Barrier Creams Against Sulphur Mustard. I. *In vitro* Studies Using Human Skin," *Skin Pharmacol Appl Skin Physiol*, 2002, 15:225-35.
Clifford NJ and Nies AS, "Organophosphate Poisoning From Wearing a Laundered Uniform Previously Contaminated With Parathion," *JAMA*, 1989, 262:3035-6.
Craig FN, Cummings EG, and Sim VM, "Environmental Temperature and the Percutaneous Absorption of a Cholinesterase Inhibitor, VX," *J Invest Dermatol*, 1977, 68:357-61.
Cullander C and Guy RH, "(D) Routes of Delivery: Case Studies (6). Transdermal Delivery of Peptides and Proteins," *Adv Drug Delivery Rev*, 1992, 8:291-329.
Devereaux A, Amundson DE, Parrish JS, et al, "Vesicants and Nerve Agents in Chemical Warfare. Decontamination and Treatment Strategies for a Changed World," *Postgrad Med*, 2002, 112:90-6, quiz 4.
Elias PM, "Epidermal Lipids, Barrier Function, and Desquamation," *J Invest Dermatol*, 1983, 80(Suppl):44s-9s.
Fartasch M, "The Nature of the Epidermal Barrier: Structural Aspects," *Adv Drug Delivery Rev*, 1996, 18:273-82.
Fenske RA and Elkner KP, "Multiroute Exposure Assessment and Biological Monitoring of Urban Pesticide Applicators During Structural Control Treatments With Chlorpyrifos," *Toxicol Ind Health*, 1990, 6:349-71.
Fredriksson T, "Influence of Solvents and Surface Active Agents on the Barrier Function of the Skin Towards Sarin. 3. Restoration of the Barrier Function," *Acta Derm Venereol*, 1969, 49:481-3.
Fredriksson T, "Percutaneous Absorption of Soman and Tabun, Two Organophosphorus Cholinesterase Inhibitors," *Acta Derm Venereol*, 1969, 49:484-9.
Junginger HE, Bodde HE, and de Haan FH, "Visualization of Drug Transport Across Human Skin and the Influence of Penetration Enhancers," *Drug Permeation Enhancement - Theory and Applications*, Hsieh DS (ed), Malvern, PA: Marcel Dekker, Inc, 1990, 59-89.

Kadar T, Fishbine E, Meshulam J, et al, "A Topical Skin Protectant Against Chemical Warfare Agents," *Isr Med Assoc J*, 2003, 5:717-9.

Lindsay CD, Hambrook JL, Brown RF, et al, "Examination of Changes in Connective Tissue Macromolecular Components of Large White Pig Skin Following Application of Lewisite Vapour," *J Appl Toxicol*, 2004, 24:37-46.

Liu DK, Wannemacher RW, Snider TH, et al, "Efficacy of the Topical Skin Protectant in Advanced Development," *J Appl Toxicol*, 1999, 19(Suppl 1):S40-5.

Loke WK, U, Lau SK, Lim JS, et al, "Wet Decontamination-induced Stratum Corneum Hydration - Effects on the Skin Barrier Function to Diethylmalonate," *J Appl Toxicol*, 1999, 19:285-90.

Maibach HI, Feldman RJ, Milby TH, et al, "Regional Variation in Percutaneous Penetration in Man. Pesticides," *Arch Environ Health*, 1971, 23:208-11.

Monteiro-Riviere NA, "Anatomical Factors Affecting Barrier Function," *Dermatotoxicology*, Marzulli FN and Maibach HI (eds), Washington, DC: Taylor and Francis, 1996, 3-19.

Potts RO and Guy RH, "Predicting Skin Permeability," *Pharm Res*, 1991, 9:663-9.

Walters KA, "Penetration Enhancers and Their Use in Transdermal Therapeutic Systems," *Transdermal Drug Delivery*, Guy HJ and Guy RH (eds), New York, NY: Marcel Dekker, Inc, 1989, 197-247.

Wertz PW and Downing DT, "Stratum Corneum: Biological and Biochemical Considerations," *Transdermal Drug Delivery, Developmental Issues and Research Initiatives*, Hadgraft J and Guy RH (eds), New York, NY: Marcel Dekker, Inc, 1989, 1-22.

Wester RC and Maibach H, " *In vivo* Percutaneous Absorption and Decontamination of Pesticides in Humans," *J Toxicol Environ Health*, 1985, 16:25-37.

Wester RM, Tanojo H, Maibach HI, et al, "Predicted Chemical Warfare Agent VX Toxicity to Uniformed Soldier Using Parathion *in vitro* Human Skin Exposure and Absorption," *Toxicol Appl Pharmacol*, 2000, 168:149-52.

Disaster and Mass Casualty Incidents

J. Marc Liu, MD

Department of Emergency Medicine, Medical College of Wisconsin
Milwaukee, Wisconsin

D. Robert Rodgers, MD

Department of Emergency Medicine
Northwestern University Feinberg School of Medicine, Northwestern Memorial Hospital,
Chicago, Illinois

Recent world events have increased the salience of mass casualty and disaster medicine in the consciousness of healthcare providers and the general public. Rare and always unexpected, mass casualty events can take many forms. Almost everyone can recall the events of September 11, 2001 or the Tokyo sarin gas attack. Just as deadly, and more common, are disasters such as a hurricane, earthquake, or building collapse. What events as the Rhode Island nightclub fire and the Oklahoma City bombing have in common is the presentation of numerous casualties to the local medical system. By virtue of their position, emergency physicians are often among the first healthcare workers to encounter victims of a disaster. Being trained in various fields of medicine, the emergency physician is often also the best qualified to handle these patients. Yet few emergency physicians have ever been exposed to a real disaster. Although many receive some formal training in mass casualty events, such education is often limited in scope (the authors themselves can recall attending a dozen lectures on anthrax in the fall of 2001, yet none on mass casualty triage). Much of the formal didactics and rehearsal are abbreviated by lack of time, funding, the need to focus on "day-to-day" issues, and general apathy.[1,2] The result is that many incorrect assumptions and misconceptions about disaster medicine are perpetuated.[3]

This article is intended to provide a concise overview for the physicians who find themselves in the emergency department (ED) when a mass casualty incident occurs. Rather than focus on specific threats and treatments, it will highlight the major concepts in disaster and mass casualty medicine to provide general guidelines applicable to many situations.

Disaster Terminology

Webster's Dictionary defines a disaster as "an event causing great loss, hardship, or suffering to many people."[4] In medical terms, a disaster is any event that results in injuries that cannot be adequately cared for by the medical resources of the community involved.[5] A disaster can be further classified by origin (natural or man-made), as well as by location, predictability, time (onset, duration, frequency), and magnitude.[6]

A *mass casualty* event is an event that results in a number of patients that exceed a medical facility's ability to care for them, which should be differentiated from a *multiple casualty event*, where the number of patients does not exceed a facility's resources.[7] It is important to note that no specific number of patients defines either a disaster or mass casualty incident. Instead, a disaster or mass casualty incident is defined by the fact that the number and severity of injuries overwhelm the medical services available. This article, for ease of discussion, will use the terms *disaster, mass casualty, incident, event, emergency*, and *crisis* to refer generally to any situation where the ED or hospital is faced with a number of patients that cannot be handled through normal operations.

Physicians should be acquainted with several other terms used frequently in disaster medicine. Man-made disasters can include nuclear, biological, or chemical (NBC) events.[8]

Hazardous materials (hazmat) are chemical, radioactive, or biological substances that can cause injury, whether natural or man-made.[9,10] Weapons of mass destruction (WMD) are man-made radioactive, biological, chemical, or explosive devices intended to cause multiple injuries and disruption of normal activities.[1] Disaster planning (or emergency preparedness) is the phrase used to describe the preparations and predetermined procedures made for use in the event of a mass casualty incident or disaster. Disaster management (or disaster response) refers to the actual actions taken when such an event occurs.[5]

Preparing for the Unexpected - Disaster Planning

In order to have an effective response to a mass casualty incident, it is important to have an organized approach. Organization requires thought and preplanning. In this light, it can be said that disaster response actually begins long before any damage is incurred. Emergency preparedness, unfortunately, is repeatedly placed low in the list of priorities. This general lack of interest is due to a mix of factors, including frequency (mass casualty incidents are rare), denial (the belief is "That will never happen here!"), and limits in time and money. This apathy usually only changes immediately after a large crisis does occur, and even then only for a short period of time. For the emergency physician, however, disaster planning is part of an everyday routine. It is part of the job. An emergency physician must remain, as the United States (U.S.) Coast Guard says, "Semper Paratus" ["Always Ready"], for anything that can happen. This is a difficult, although not insurmountable, task.

Around 350 B.C., the Chinese military strategist Sun Tzu wrote, "Know the enemy and know yourself; in a hundred battles you will never be in peril."[11] From this statement can be derived the main elements of emergency preparedness. A proper approach to disaster planning consists of knowledge of what can happen ("knowing the enemy") and being familiar with the resources one has available to respond ("knowing yourself"). Following Sun Tzu's advice can aid the physician both in planning for and responding to a disaster.

"Knowing your enemy" first involves the fund of medical knowledge required to treat patients and their ailments. Physicians must take care to maintain their knowledge base and procedural skills. One should also anticipate common problems in managing a large number of patients. These will occur regardless of the specific nature of the incident. Whether in a multiple vehicle crash or a nerve gas attack, triage, patient flow, security, and communications become issues that need organized resolution.[2,3,8,12,13,14] A prepared physician and a good disaster plan looks for these commonalities, takes them into account, and has a method ready to deal with them. Analysis of the many wildfires of the 1970's and 80's led California hospitals and emergency medical services (EMS) systems to develop HICS, the Hospital Incident Command System.[6,15] HICS is an administrative framework designed to manage hospital resources during any disruption of normal functions. It has become widely studied in the U.S., and incorporated into many a disaster plan. HICS uses a multidisciplinary/multidepartmental strategy to avoid the pitfall of delineating a separate response for every conceivable specific crisis.

Physicians and their institutions should additionally identify possible specific threats in the environment. For example, an urban ED will anticipate receiving multiple penetrating trauma victims, while one in a rural area may think more about possible pesticide poisonings and farm equipment accidents. A wise physician who "knows the enemy" will expect certain events to be more likely encountered, but will have a mental schema that can apply to any situation.

The other major component, "knowing yourself," entails the physician being aware of the resources that can be applied to a disaster. An effective emergency response requires efficient use of the finite amount of resources. To use resources efficiently, one must have a good knowledge of what resources are present, and what those resources can do. Resources include, first and foremost, people. One needs to know how many personnel are available, the types and levels of training those personnel have, and how they normally operate. People

260

work most efficiently when they are given tasks for which they have been trained. A stressful situation like a mass casualty incident is the worst time to attempt to perform something new. Whenever possible, personnel should be given roles utilizing skills that are known to them.[6] In addition to ED personnel, it is important to have a familiarity with the personnel in other areas of the hospital that can come into play in an emergency - operating rooms, intensive care units, pharmacy, central supply, engineering, security, hospital administration, and other departments are integral parts of a disaster response. Supplies and physical space are also resources. Even on routine days, the emergency physician needs to keep track of how many open beds are in-house, or whether pharmacy has run out of a certain medication. This type of awareness is even more critical in an emergency. By knowing where extra beds are, or how many laceration trays or ventilators are available, the physician can make informed decisions as to how best to distribute them.

"Knowing yourself" also includes a formal disaster plan. A disaster plan is developed with input from all departments and services in the hospital.[5] It defines what is considered a disaster, and how the facility would respond. In addition, each individual department or service must have its own protocols for an emergency. A written plan alone, though, is not sufficient. Even a thorough disaster plan is of no value if no one has read and practiced it. Indeed, one can develop the "paper plan syndrome" - the false sense of security engendered by the mere existence of a written document.[2] Although hospitals are required by law to have a plan, the levels of familiarity with them vary widely from hospital to hospital and person to person. Emergency physicians are both required and assumed to be well-versed in the disaster protocols of the hospital.

Training and rehearsal are essential for true emergency preparedness. They help combat the loss of knowledge and skills that occur with infrequency of use. ED physicians should consider training for disasters as part of their occupation, and convey that attitude to everyone they work with. Physicians involved with teaching should include in the curriculum lectures on the general concepts of disaster medicine, as well as on specific threats (eg, burns or bioweapons). The individual physician must maintain his or her own procedural skills, medical knowledge, and familiarity with the disaster plan. Regular refresher sessions will aid in keeping this knowledge from fading.

The last step in disaster planning is the disaster drill, which is the only controlled way to gain experience. Drills serve a valuable purpose in testing the adequacy of a disaster plan, as well as personnel's knowledge of the plan. Rehearsals acclimate people to the mass casualty situation, and can help decrease the stress of an actual event.[1] They also help in identifying deficiencies in planning and training. Unfortunately, most institutions rarely, if ever perform drills.[2,6] Emergency physicians should actively encourage, organize, and participate in disaster drills. Drills can take many forms, from "tabletop" paper exercises to large, full-scale drills involving the entire hospital. A comprehensive schedule of drills would include a mix of the different types. Tabletop drills, which are inexpensive, easy to organize, and involve no direct disruption of normal ED activities, can be held more frequently. Partial drills, involving a small portion of the department, should be held regularly, but are slightly more disruptive. Department and hospital-wide drills, because of cost and disruptions to patient care, will occur less frequently, but must be run regularly to maintain a level of proficiency. A useful tactic that the authors advocate to improve readiness is to have different services or sections (eg, nursing, housekeeping, patient transport, etc.) to hold separate drills. Service-specific rehearsals help personnel learn and practice emergency protocols without the distractions of a full-scale drill. This method also has the benefit of being less costly and time-consuming, allowing it to be done more frequently. Once a unit has drilled itself to a level of proficiency, it can be integrated into larger drills.

By thinking about and practicing for mass casualty incidents, a physician can improve response. Having a solid comprehension of potential threats, the disaster plan, and the resources available allows an organized approach to a real event. It also allows one to tailor a response to the particulars of a situation, what Auf der Heide calls "planned improvisation."[2]

Every situation has unique challenges that may not be foreseen in even the best of disaster plans. Only through an understanding of what is and is not feasible can a physician skillfully counter the inevitable unexpected problem.

Onto the Breach - Disaster Response

One of the most dreaded calls by emergency room staff is one that heralds the impending arrival of multiple injured patients. This section will discuss general steps for the emergency physician who becomes responsible for coordinating a mass casualty response. It will focus on organizing the ED to treat patients from such an event.

Activating the Plan   When one receives word of a possible disaster, the first step is to obtain as much information as possible. Ideally, this would come from on-scene first responders, but this may not be the case. Of course, there is the caveat that, in the initial confusion of an event, early information can be misleading or even false. Based on the best information available, the on-duty physician (or the senior physician) must make a determination on whether the predicted number and severity of injuries is beyond the current capabilities of the ED and hospital. If the physician believes this is the case, then the mass casualty/disaster plan should be activated in order to mobilize additional resources. Activation is also considered if there is a high level of suspicion of a hazardous materials incident, or if there has been an internal emergency (within the hospital). Different hospitals have defined different levels of activation based on predicted numbers and/or the nature of the event. Physicians have to be aware of the specific plans used by their facility. Activating the plan entails informing a predetermined list of people, which usually includes the hospital telephone operator, on-call hospital administrator, on-duty security chief, as well as other key personnel. In some hospitals, the ED need only call and inform the operator, who will then contact the other necessary persons. It is also important for the physician to notify the rest of the ED.

Appointing the Command Staff and Allocating Personnel   Once it is clear that the department is responding to a mass casualty incident, the next action is to assign personnel. First, one must form a command staff to oversee ED operations during the incident. A unified and clear chain of command is necessary for efficient functioning.[6,8] The first and most critical action is to identify the physician in charge of the ED's entire response. This ED commander/ supervisor is responsible for ensuring the rapid and appropriate treatment of the patients, and coordinates requests for resources with the other departments and services in the hospital.[15] The medical supervisor is typically a senior ED physician who is experienced with mass casualty incidents. The supervisor may decide to designate an assistant who would serve to help communicate the decisions to the rest of the staff. This assistant may also have the authority to act when the supervisor is not present.

The supervisor is responsible for appointing the rest of the ED command staff and aiding them in selecting their teams. The triage officer is the most important of these positions. This triage leader should be the most experienced medically trained person available (triage is discussed in more detail below). The ED supervisor also designates skilled and experienced physicians or nurses to act as treatment unit leaders (immediate, delayed, minor, and morgue). If there is a hazardous materials incident, one must also appoint a decontamination unit leader. These unit leaders would be responsible for the patients in their areas. The command staff also is responsible for assigning the other members of the ED to the various treatment units. The supervisor then briefs the unit leaders on their roles and responsibilities. A popular way to speed this process is to have preprinted cards or sheets listing each person's responsibilities. All members of the command staff have to be easily identifiable. Brightly colored vests, hats, and/or large badges with the specific position printed on the item are often used. To minimize confusion, frequent switching of persons in and out of designated roles should be avoided.

When an officer is relieved, any identification clothing is handed over to the new person, and other members of the command staff are informed of the change.

It is important to consider the skills and experience of each person when assigning them particular duties. People work best when given tasks familiar to them. It may be more appropriate to assign a first-year resident to the delayed or minor treatment team rather than the immediate treatment area. An experienced critical care nurse would be better utilized on the triage or immediate treatment unit, rather than tending to minor injuries. Unit leaders are the most medically proficient personnel available, who also have skill in administration and management. In the end, though, circumstances will dictate who is available, and so it is up to the ED supervisor to allocate personnel in the best way possible.

Communications   For an organized response, all teams, departments, and services need an effective method to communicate with each other. The various parties will frequently convey their status, along with requests for additional help and supplies. However, in almost every disaster, communications become a problem.[3,6,12,16,17]

Telephones and cellular phones are the most widely available means of communication. They have the advantage of being universally familiar. However, in a disaster, phone lines can be cut or otherwise rendered inoperable. Numerous calls placed due to the emergency can overload the local phone network. Also, it may be difficult for personnel to keep track of the phone numbers needed. Another option is to use computers and the Internet, which is a skill becoming more and more widespread. Computers require electricity, and, unless special software is installed, do not allow for real-time conversations. Two-way radios provide portable, real-time voice communications and are the preferred system. Nonetheless, even radios have drawbacks. They require charged batteries, require personnel to know the devices and frequencies, and are subject to interference. If no other system is available, one can use people as runners/couriers to physically deliver written or verbal messages (this is an excellent way to utilize nonmedical personnel).

Everyone should be reminded to use as plain language as possible, minimizing confusion by avoiding slang or jargon that may be specific to their own department or service. Another common problem is the overuse of communication. Too many messages can be as bad as too few.[6] Messages should be as concise as possible, but convey the necessary information. Too much information can be confusing and time consuming. Unit leaders must regularly update the supervisor on their team's general condition.

Security and Access   Although maintaining security is not directly performed by ED personnel, it is a major portion of any mass casualty incident.[9,13] Part of the physician's role is to coordinate with security personnel to maintain a safe and efficient environment for patient care. Any large group of people pose security concern, especially when emotions run high, as is likely in a crisis. Victims, who are in distress and unaccustomed to the procedures of mass casualty medicine, may become impatient or even hostile. The same can occur with the multitudes of family and friends attempting to locate loved ones. There are also the scores of unsolicited volunteers who will arrive. In hazmat events, contaminated patients, desperate for aid, may attempt to enter a facility without proper decontamination. The ED supervisor assists the security officer in ensuring that there are no disruptions to patient flow and no dangers to patients or hospital personnel. All access to the hospital and ED must either be locked or guarded. Security personnel should be stationed at triage and every treatment area. All hospital workers must be required to prominently display their identification badges on their person. They should be encouraged to report any potential security problems to their unit leader or to security.

Patient Flow and Treatment Areas   A functional hospital plan has predesignated spaces as main treatment areas. The trauma or main resuscitation room is generally used as the immediate treatment area (for the most critical victims). Alternatives include beds closer to the nursing station with ample room for critical care equipment and providers. The delayed treatment area must also have space for monitors and crash carts. Any large space, such as the waiting room or hospital lobby, can be used as the minor treatment area. Having quantities of commonly used supplies (intravenous catheters, gauze, gloves, tape, etc.) moved from the central storeroom and placed in each treatment area will improve efficiency. Consider also moving special equipment to the area most likely to need it. For example, moving airway equipment, ventilators, and chest tube trays to the immediate treatment area would be a wise precaution, as is placing large numbers of laceration trays in the minor treatment area.

All personnel should be aware of the designated areas and the designated routes for moving patients between these areas and the rest of the hospital. Routes should be as short and intersect as little as possible to avoid both "traffic jams" and confusion between the acuity level of patients. A critical aspect of ensuring overall efficiency is to remember to maintain forward patient flow. A patient never moves backwards in the system.[18] This means, once patients leave the triage area, they proceed directly to a treatment area. Once they leave a treatment area, they do not return, and proceed to the end destination (for instance, a patient needing a computerized tomography scan leaves the delayed treatment area, goes to radiology, and then straight to an in-patient bed, not back to the delayed treatment area). Having patients return back to triage or treatment areas will only overwhelm the teams and slow the entire system.

As the hospital prepares to receive casualties, the ED must attempt to open as many beds as possible within minutes. The physicians and nurses determine who can be discharged and who is stable enough to transport. Any patient who is already admitted is sent immediately to the hospital bed. All other patients are sent to a temporary holding area to complete their work-up. Hospital plans must contain policies for expedited admissions and transfers. If there is time, a brief report can be given, relaying room number, patient gender, age, and chief complaint or diagnosis. Remember that the main goal is to clear the emergency room in order to make space for the expected patients.

The Patient Population While preparations are being made to receive victims, the wise physician is already anticipating what types and numbers of injuries will present. A common misconception concerning mass casualty incidents is that a hospital will quickly receive a sudden influx of ambulances arriving with a large number of critically wounded victims. Analysis of multiple disasters has shown this is usually not the case. It is important for the physician to understand the general patterns of patient presentation.

A mass casualty incident can range anywhere from a handful of patients to thousands. There have only been seven U.S. civilian disasters with more than 1000 fatalities.[2,3,6] On average, there are 10 to 15 incidents a year with more than 40 injuries.[2] Most large disasters will produce 100-200 casualties.[6]

In terms of severity, the overwhelming majority of patients have minor injuries.[2,14,19] A review of over two dozen mass casualty incidents by the Disaster Research Center determined that fewer than 10% of victims would meet criteria for hospital admission in normal circumstances. Many patients were admitted mainly because they were involved in the incident, and stayed less than 24 hours.[2] After Hurricane Hugo in 1989, almost half of patients seen in one area of North Carolina were for wounds and insect stings.[19] A study found that the most common diagnosis in patients from the Oklahoma City bombing in 1995 were lacerations and soft tissue injury, followed by fractures.[14] Therefore, most patients presenting to the ED will be the so-called "walking wounded."

In most cases, the initial wave of patients arriving at the ED will be ambulatory with injuries of low acuity. These patients are able to move relatively quickly to the hospital. More critical patients, who are often unable to refer themselves, usually begin arriving later. This "second wave" phenomenon has been observed in numerous incidents.[14] This pattern has important implications for response. The physician in charge cannot allow all ED space to be taken by the first wave. Minor injuries must be quickly triaged and moved to a definitive care area in order to keep beds available for more serious injuries.

The majority of patients will not arrive via EMS transport.[2,14,17,20] In the Oklahoma City bombing, 56% of the patients seen in area hospitals arrived by privately owned vehicles.[14] 80% of the patients treated after the Tokyo subway nerve gas attack referred themselves to the hospital.[20] Patients and bystanders will go to the closest hospital as quickly as possible, often bypassing on-scene triage and aid stations.[2,14] Again, this pattern of presentation can affect a facility's response. The physician can expect a large number of victims who will not have had any evaluation, triage, or intervention by EMS. There can also be a risk in hazmat incidents, where patients may arrive at the hospital without having received any on-scene decontamination. Security and medical personnel must cooperate to route all patients through the triage process. Self referral also implies that one cannot expect EMS to equally distribute casualties between local hospitals. This problem is often compounded by the natural tendency for rescue workers to transport to the nearest hospital. The earlier quoted Disaster Research Center study of 29 mass casualty events found that two-thirds of patients in each event were seen by one hospital.[3] The other hospitals in the area had an average of 20% of their beds empty.[2] Workers in a hospital geographically closest to an incident can expect to bear the brunt of the response.

Triage Triage is the process of prioritizing medical care by severity of injury. The word triage comes from the French word trier, "to sort."[21,22] Triage was born in military medicine, where the goal was to treat and return soldiers to battle as rapidly as possible. The father of modern triage is considered to be Baron Dominique Jean Larre, a surgeon in the French army under Napoleon. Larre developed a system where the most severely injured were treated first, during combat. Before Larre, casualties were only attended to after a battle, and priority was based on rank. Triage in the United States was introduced mainly by John Morgan in the Revolutionary War, and John Wilson in the Civil War. Both attempted to sort out casualties by severity and organized ambulance corps for transport to field hospitals.[22]

The single most important medical role that the physician can undertake is that of the triage officer. Rapid and accurate triage is essential to a successful emergency response. For this reason, selection of the triage officer is one of the supervisor's most important duties. A triage officer must be able to apply medical knowledge and the concepts of disaster triage to make sound decisions with little time and information. The triage officer is traditionally the most experienced physician available. Triage requires knowledge of multiple aspects of medicine. This requirement makes the emergency physician, who is cross-trained in various fields, ideal for assignment as a triage leader. Burkle lists ten characteristics in the ideal triage leader:

1. surgical experience
2. easily recognized and respected
3. good judgment and leadership
4. handles stress well
5. decisive
6. familiar with personnel and resources
7. sense of humor
8. creative (able to improvise solutions)
9. able to anticipate problems
10. presence in or very near the hospital[18]

The key concept of mass casualty triage is the maximal benefit to the largest number of people using the limited resources available. Stated in other words, triage aims to provide the greatest good for the greatest number.[18,21,22] In a mass-casualty incident, one must alter the fundamental approach to delivering medical care. Instead of the usual focus on the individual patient, the focus is on the whole patient population. These triage principles conflict with the normal medical principle to do everything possible in order to save the patient's life. Triage requires delaying the care of certain patients, and declaring others beyond recovery. Anyone who is involved in mass-casualty triage must be able to shift the mindset away from the patient and toward the group as a whole, and must be willing to accept greater levels of morbidity and mortality than in usual practice.[8,11] Physicians who cannot do this would be poor candidates for triage officer.[22]

There are numerous scoring systems that have been developed for triage. Each system uses its own combination of mechanism of injury, vital signs, and physical exam criteria to triage patients. Although many of these systems are able to accurately predict mortality, they are less accurate at differentiating between levels of acuity. Almost all of the systems were tested primarily on trauma patients.[22] This, along with the need to remember the criteria and obtain such information, limit their practical applicability. To date, there have been no data to show that any one scoring system is superior to the others.[21,22,23]

Undertriage is defined as inappropriately classifying a patient's injuries as less critical than they actually are. An acceptable rate of undertriage is considered less than 5%. Overtriage is classifying a patient's injuries as more serious than they are. Because it is better to overtriage, thus minimizing the risk of missing serious conditions, an acceptable overtriage rate can be as high as 50%.[21,22]

Many disaster plans triage patients into one of four categories based on severity. These categories are often given a color or numeric designation. "Immediate" (or "emergency," "priority 1," "red") are patients that require immediate intervention to save their lives. "Delayed" ("urgent," "priority 2," or "yellow") patients require actions, but intervention can be delayed for a short period of time without significantly reducing their chance of survival. "Minor" ("nonurgent," "priority 3," or "green") patients are ones with injuries that do not need to be attended to until the first two groups have been treated. Such patients are expected to survive even without medical treatment. "Expectant" ("priority 4," "black") category patients are dead or expected to die.[21,22] This last group of patients are those where the chance of survival does not justify the expenditure of resources to attempt successful intervention. There should be a predetermined method for patients to be labeled once they have been triaged. This can be as simple as verbally relaying information. Colored tags or cards can also be used. A popular system available commercially is the Medical Emergency Triage Tag, which has perforated black, red, yellow, and green portions that can be torn off as needed.[21]

A triage system that is common in North America is the Simple Triage and Rapid Treatment (START) system.[21,23] Versions of START have been used in several U.S. disasters.[21] The modified START algorithm is as follows:

1. If a patient is not breathing after simple airway maneuvers, he is dead.
2. If a patient is walking, then he is triaged as minor (nonurgent).
3. If a patient is not walking, but can obey commands, with a respiratory rate less than 30, and a palpable radial pulse, then he is triaged as delayed (urgent).
4. Anyone else is triaged as immediate (emergency).[24]

Triage scoring systems can serve as guidelines to help classify victims. However, "There is no substitute for good judgment, sound experience, and attention to the principles of triage."[18] The triage officer first makes a general inspection of the patient while rapidly assessing airway, breathing, and circulation (the ABCs). Penetrating wounds to the chest or abdomen, or severe head or neck injuries usually warrant a higher triage status.[25] Patients who are ambulatory, alert, and oriented will probably warrant a less critical classification. A resting tachypnea can indicate a pulmonary process, airway problem, or early shock, and may also indicate a higher triage status.

The triage team's priority is to classify, and not to treat, patients. Interventions at triage are kept to a minimum.[18,21,22] A good mnemonic device is the acronym "BASIC." BASIC is an abbreviation for Bleeding, Airway control, Shock prevention, Immobilization, and Classification.[21] This means that simple airway maneuvers (head tilt/chin lift, jaw thrust) can be performed. Obvious hemorrhaging can be tamponaded with direct pressure. Extremity deformities and/or the spine can be immobilized as needed. Actions are limited to minimal, lifesaving measures. Any interventions are performed by members of the triage team, and not the triage officer. The triage officer always continues triaging, and never stops to treat patients. Patients should not be moved before triage unless there is a danger at that location or a new triage area has been designated. Once triaged, a patient is moved to the appropriate treatment area.[18] Keeping patients in the triage area, allowing patients to return to triage, or having the triage officer tend to patients will only slow the entire system and endanger the lives of following patients.

All healthcare providers need to remember that triage is a dynamic process.[21,22] Patients should be reassessed frequently to look for changes in their status. If necessary, patients can be reassigned to another category and transferred to the appropriate treatment area.

Patient Assessment and Treatment The goal of the treatment areas in a mass casualty incident is to perform an assessment and interventions necessary to stabilize the patient until definitive management can be delivered. To ensure the most efficient care possible for a large number of patients, patient flow out of the ED to end destinations must be maintained at a maximum. The ED is to be kept as open as possible in order to accommodate incoming patients. Extensive diagnostic testing and definitive care is performed in other areas of the hospital, and not in the emergency room. Such a mass casualty approach is employed regularly in Israeli hospitals, where disaster victims are stabilized in the ED, and, as quickly as possible, moved up to the operating room, intensive care unit, or general bed. There, the patients receive diagnostic tests and definitive treatment. The possible exception is in the minor treatment area, where lacerations, minor fractures, and other minor injuries can be treated and sent to the designated hospital discharge area for discharge.

The physician assigned to a treatment area performs a more thorough evaluation than is done at triage. Such an assessment focuses on finding conditions that can threaten the patient's life. An excellent guideline is to follow the primary and secondary surveys used in the ATLS and ACLS (Advanced Trauma and Advanced Cardiac Life Support) protocols. The primary survey focuses exclusively on vital signs and the ABCs. It also includes starting intravenous access, oxygen, and cardiac monitoring as needed. The secondary survey includes a head-to-toe examination, including the extremities and a neurologic exam.[7] As in routine patient care, vital signs can be important indications of problems. Tachycardia, narrowed pulse pressure, or a weak pulse can be signs of hypovolemia. Difficulty obtaining a blood pressure can be an indicator of shock, as can cold extremities with weak pulses.[25] One should be especially alert for life-threatening conditions such as airway edema/injury, pulmonary edema, pneumothorax, hemothorax, flail chest, cardiac tamponade, cardiac arrhythmias, and hemorrhage. If the patient is awake and oriented enough, a brief history is obtained. An AMPLE history (Allergies, Medications, Past medical history, Last meal, pre- and postinjury Events) is usually adequate. More details can be elicited as time permits. Any immediate life threats must be dealt with. When the patient is stabilized, the physician can determine whether the patient is to be admitted, and, if so, to where. Only minimal diagnostic testing (such as portable chest X-rays or electrocardiograms) is to be done in the ED. Other testing can be done on the way to the patient's end destination. Patients should be reassessed often, and the physician can consider increasing their triage acuity if there are reasons to do so. Once patients leave a treatment area, they do not return to that area (no backwards flow!). This will help maintain patient flow and reduce delays.

Neuropsychiatric Triage Psychiatric issues play a major role in any disaster or mass-casualty incident. Victims, bystanders, and responders alike will have suffered emotional trauma. Only relatively recently has disaster medicine begun considering psychiatric injuries.[21,26] When faced with a patient exhibiting bizarre behavior or affect, the physician's first priority is to rule out a medical cause. Hypoxia, shock, and head injuries are common causes of altered mental status in disaster victims.[21] If there is reasonable evidence to indicate a psychiatric cause, the patient can be referred for further psychiatric evaluation. In addition, physicians should remain alert for signs of psychological stress in responders and medical personnel. Treatment-unit leaders should make use of available psychiatric, social work, and chaplain personnel to help patients and rescuers with the stresses of an emergency incident. However, it is important to consider that there are mixed data on the use of mandatory psychological debriefings within 72 hours of traumatic events. Some studies have demonstrated no benefit, while others have found a worsening of symptoms after such debriefings.[26] In general, all persons demonstrating abnormal stress responses should be given information on how to seek psychiatric assistance.

When faced with a patient exhibiting bizarre behavior or affect, the physician's first priority is to rule-out a medical cause. Hypoxia, shock, and head injuries are common causes of altered mental status in disaster victims.[21] If there is reasonable evidence to indicate a true psychiatric cause, the patient can be given a psychiatric triage category in addition to the medical category. The Navy/Marine Corps system divides patients into immediate, delayed, minor, and expectant categories. The immediate group includes patients with severe injuries, functional losses, and/or cosmetic losses (eg, loss of a limb). These patients receive a full psychiatric evaluation as soon as possible. Patients in the delayed group do not have severe injuries, but may have injuries that can affect function or cosmesis. They also require a psychiatric evaluation, but that can occur at a later time. Minor category patients can usually be treated on-scene by untrained personnel with referral to professional assistance as needed. This group includes patients with little to no injuries, but who are suffering emotional distress from having been involved in or witnessing a traumatic event firsthand. The expectant category includes patients who are fatally wounded. A chaplain may be ideal for these patients, though often untrained personnel can help.[21,26]

Hazardous Materials and Decontamination   Hazardous materials incidents present special challenges to the emergency provider. Every hospital is required to have procedures in place to handle patients contaminated with biologic, chemical, or radioactive substances.[16] While a complete discussion of NBC threats is beyond the scope of this article, it is important to mention some general principles.

With hazmat incidents, the main issue is to ensure an adequate level of protection for healthcare personnel in order to prevent secondary contamination. Secondary contamination is the exposure of providers to substances on a patient's body or clothing.[9] Secondary contamination is a real risk.[10,16,17] Horton et al., in a review of U.S. hazmat incidents from 1995-2001, found 32 healthcare providers who required medical care due to secondary contamination.[10] In 1995, 110 of 472 hospital workers in Tokyo reported symptoms possibly due to secondary exposure to sarin gas.[17] The heavy majority of these providers were not using any sort of protective equipment.[10,17] There are various levels of protection. An emergency physician must be familiar with the type of protective equipment available at the hospital, and the limitations of such equipment. He or she should also be aware of who has been fitted and trained to use the equipment.[16] All personnel who may come in contact with contaminated patients must wear protective gear. There are little data to form recommendations on the levels of protection used by hospitals.[9,16,27] In general, in biological and radiation incidents, personnel must use at least disposable liquid-proof gowns, shoe covers/boots, caps, eye protection, N-95 masks, and gloves.[27,28] Chemical incidents often require chemical-resistant suits and respiratory protection.[16,29] Normal gowns, covers, and latex gloves provide little protection against most chemical agents. Nitrile gloves are more resistant to chemicals than latex, but are still only provide limited resistance.[9] All personnel, once finished, have to undergo decontamination themselves, and remove all protective coverings and clothing and place them in labeled, sealed bags or containers that are kept in one location. They should then shower.[9,27,28]

A hazmat event requires the formation of a decontamination (decon) team.[30] Depending on the circumstances, there may be considerable overlap with the triage team. A decon team leader is responsible for overall management of patient decontamination. A decon safety officer is responsible for maintaining personnel safety by monitoring decontamination and coordinating with security to maintain separate contaminated ("warm") and decontaminated ("cold") zones. The decon safety officer also is responsible for rotating personnel within the team, as decontamination can be a physically grueling task.[9,30,31] Whenever possible, the decon team leader and safety officers should not be given other duties or titles, since proper decontamination requires careful attention to detail. The team leader selects properly trained people to perform the actual decontamination. The team should also have persons without protective gear to handle cleaned patients and to bring supplies.

In order to avoid secondary contamination, it is crucial to have well-defined warm and cold zones and routes for patient movement. All patients potentially involved in the hazmat incidents are moved to one place. All other entry points to the ED and hospital must be locked and/or guarded. Any area where a contaminated patient has entered should be considered warm and sealed off until properly decontaminated. Any equipment or supplies used in a warm zone are also to be considered contaminated. Patients can only be allowed to enter the treatment area after undergoing proper decontamination.[9,27]

The goal of decontamination is to remove enough of the substance to eliminate any risk of secondary contamination and any risk of further injury to the patient.[9] The triage officer (who is wearing protective equipment) quickly assesses the patient, and instructs the triage team to perform minimal lifesaving maneuvers as needed. The triage officer determines whether the patient will require assistance with decontamination. Patients who can do so are instructed to remove their clothes (with every effort to maintain privacy) and place them in sealed bags or containers. The patient then is instructed to wash the entire body with soap and water.[27] There remains controversy as to what decontamination solution is ideal, but current data suggest that removing clothing and copious washing with soap and copious amounts of water eliminates most contaminants (with the notable exception of reactive metals, in which case water should not be used).[9,27] Patients who cannot wash themselves should have their clothing removed by the decon team. Team members then wash the entire body including the back, with proper immobilization and spinal precautions.[9]

Once decontaminated, the patient is dried and covered (again, with every effort to maintain privacy). The patient is then re-triaged and sent to the appropriate area.[27] Patients suspected of being exposed to infectious agents may need to be isolated.[32] After decontamination of all patients is complete, any decon personnel working in a warm zone then undergo decontamination. The decon safety officer examines members of the team to check for any signs or symptoms of exposure that may require medical attention.

Documentation and Record-Keeping An aspect of emergency response that is easily overlooked is that of maintaining documentation.[3,6] During the frenzied process of treating large numbers of patients, record-keeping often becomes a low priority. There is a fine line between inadequate documentation that provides no useful information, and excessive documentation, which can slow care.[22] Good records allow caregivers to keep track of a patient's condition and what tests and interventions have been performed. Documentation also allows for ease of identifying and locating victims of mass casualty incidents. It can also be used for legal and scientific purposes.[3] Members of the hospital's medical records/registration staff should be part of disaster drills, and be assigned to all triage and treatment units. At triage, patients must receive some form of identifier (usually a number) on a card, tag, or wristband. If known, patients' names can be added. Personnel need to keep a list of when patients enter a unit, when they leave, and to where they were sent. There also needs to be a simple chart to record important details of medical history, physical examination, and care received. The HICS model includes a suggested medical record format.

Volunteers   Hospitals involved with disasters have to be prepared to deal with the large number of well-meaning volunteers that inevitably arrive. Contrary to popular wisdom, many people do not remain stunned and confused after an incident, but will instead rush to see what assistance they can provide.[2,3,6] The result is a large quantity of unsolicited volunteers, both with and without medical training. In fact, in many events, hospitals found that there were too many, not too few, people available.[2] Often resources must be diverted to manage these volunteers. Since many are inexperienced with disaster medicine (and many have no medical knowledge at all), they can slow the system's response and create disorganization.[2,3,6]

It may be necessary to designate a person to coordinate the use and assignment of volunteers. It is important that volunteers who claim to have a healthcare background be required to provide proof of such training. They should be assigned, based on their skills and experience, to augment existing teams/units. Nonmedical volunteers can be used to deliver supplies or messages, or in other roles that do not require any clinical knowledge. All volunteers are given some sort of prominent identification to indicate that they are volunteers and not regular staff. Also, all volunteers should understand what their duties are, and to whom they report. Volunteers are to follow the incident's chain of command. If, in the labor pool leader's judgment, the use of volunteers would hinder the incident response, then the volunteers should be politely and firmly turned away.

The Media and Public Information   The physician may also find him or herself having to deal with members of the media, as well as concerned relatives and friends. Any questions should be directed to a member of the hospital's media relations staff. Members of the press cannot be allowed into triage or treatment areas, where they may serve as a distraction to personnel and may violate the privacy of patients. The physician who is unfortunate enough to be selected to speak to the media needs to be cautious about what is said. Care should be taken to avoid violating patient privacy, providing inaccurate information, or stating anything that could be misinterpreted. It is a wise precaution to always have a member of the media relations staff present if one must speak to the press.

Internal Disasters   An internal disaster is an incident occurring within a medical facility that prevents or affects routine operations.[33] This can take many forms – a chemical spill, fire, loss of electricity, loss of water, etc. An ED physician should be aware of what hazardous materials are used in the hospital and how to treat exposures. There must be a plan in case of loss of power, water, medical gases, suction, or other services. If necessary, a command staff can be appointed, just like with an external event. If evacuation is necessary, ventilated and critical patients should be moved first, followed by stable nonambulatory patients, and finally, ambulatory patients.

After the Disaster

Once the incident commander and ED supervisor determines that all casualties have been triaged, assessed, and have arrived at their end destinations for definitive care, the incident can enter the recovery phase. This, though, does not mean that operations can instantly return to normal.[2,3,5,6] The hospital must still provide services for the large number of patients and the personnel mobilized to care for them. The ED must remain ready to receive patients that come in after the event. These patients may be injured in the course of cleaning up afterwards. Also, in a large disaster, people may not be able to access their usual medical care, and may have to go to the emergency room.[2]

All involved personnel should participate in an incident debriefing. Incident debriefing serves two main functions. First, it is a way to identify successful and unsuccessful techniques and strategies in the emergency response. Second, it provides an opportunity for all to receive formal psychological and stress management counseling. Past incidents have shown that responders and victims are at risk for post-traumatic stress disorder, depression, and substance abuse.[6,34,35] Stress incident teams are a part of any disaster recovery. They can provide immediate interventions, as well as assist in arranging for follow-up counseling. Emergency physicians should have a high index of suspicion for psychiatric issues and encourage patients and coworkers (and themselves) to utilize psychiatric resources.

Conclusion

Disaster Planning - Key Concepts

1. Know the hospital plan. The best plan is useless if no one is familiar with it.
2. Know the enemy. Be prepared with one's medical knowledge and knowledge of mass casualty medicine. Know what is more likely to happen.
3. Know yourself. Be familiar with personnel and equipment. Know what they can and cannot do.
4. Practice, practice, practice! Hold frequent drills of different types, including regular full-scale drills. Individual services should also drill separately and more frequently to maintain proficiency. It is said in the military, "The more you sweat in training, the less you bleed in war."
5. Be prepared to improvise. This can only be done if one has fulfilled the preceding steps.

Disaster Response - Key Concepts

1. Consider activating a disaster plan for any internal or external event that may overwhelm the normal ED capacity.
2. Appoint an experienced command staff. Such persons must be easily identifiable.
3. Clear the ED before the patients arrive.
4. Most patients will have relatively minor injuries and will be self-referred. They will not have had any on-scene evaluation or decontamination.
5. Triage is the most important part of a mass casualty response. Therefore, the triage officer must be a proficient medical professional who can remember that the needs of the many outweigh the needs of the one.
6. Keep the ED clear. Maintaining forward patient flow is vital. In a mass-casualty response, only the lifesaving/stabilizing care is to be provided in the ED. Patients are moved as rapidly as possible to other areas of the hospital for further testing and definitive management.
7. Efficient documentation and communications will greatly enhance emergency response. Security and crowd control are important to maintain a safe and efficient environment.
8. Once again, the goal is to provide the greatest good for the greatest number.

Disaster Recovery - Key Concepts

1. Stay alert for psychological and stress-related problems in patients, employees, and oneself. Seek help when needed.
2. Be prepared for above-average patient volume for days or even weeks after an incident.
3. Analyze what went right and what went wrong in the response. Apply this experience to the next event.

Disasters and mass-casualty incidents are among the most intimidating occurrences an emergency physician can face. Adding to their inherent difficult nature is the fact that they are always unexpected. In such an event, the emergency department becomes the center of patient care as well as community support.[8] The emergency physician will frequently be placed in a leadership role.[5] The multidisciplinary training of emergency medicine makes the emergency physician well-suited to assuming such a responsibility. One must always be prepared to deal with the unexpected. Proper handling of a large-scale event requires a fluency in the principles of disaster and mass-casualty medicine, as well as a familiarity with available assets. With knowledge, preparation, and good judgment, one can deliver quality medical care even in the most trying times.

Acknowledgments

The authors would like to express their gratitude to Dr. Mark Courtney for his advice. JML authored this chapter during his appointment as the National Association of EMS Physicians-Zoll EMS Resuscitation Fellow.

References

1. Waeckerle JF, Seamans S, Whiteside M, et al, "Executive Summary: Developing Objectives, Content, and Competencies for the Training of Emergency Medical Technicians, Emergency Physicians, and Emergency Nurses to Care for Casualties Resulting From Nuclear, Biological, or Chemical (NBC) Incidents," *Ann Emerg Med*, 2001, 37:587-601.
2. Auf der Heide E, "Disaster Planning, Part II: Disaster Problems, Issues, and Challenges Identified in the Research Literature," *Emerg Med Clin North Am*, 1996, 14:453-80.
3. Kaji AH and Waeckerle JF, "Disaster Medicine and the Emergency Medicine Resident," *Ann Emerg Med*, 2003, 41:865-70.
4. Cayne B, ed, *New Webster's Dictionary and Thesaurus of the English Language*, Danbury, CT: Lexicon Publishers, 1993.
5. Gans L, "Disaster Planning and Management," Harwood-Nuss A, ed, *The Clinical Practice of Emergency Medicine*, Philadelphia, PA: Lippincott Williams & Wilkins, 2001.
6. Waeckerle JF, "Disaster Planning and Response," *N Engl J Med*, 1991, 324:815-21.
7. Committee of Trauma, American College of Surgeons, *Advanced Trauma Life Support for Doctors*, Chicago, IL: American College of Surgeons, 1997.
8. Lovejoy JC, "Initial Approach to Patient Management After Large-Scale Disasters," *Clin Ped Emerg Med*, 2002, 3:217-23.
9. Cox RD, "Decontamination and Management of Hazardous Materials Exposure Victims in the Emergency Department," *Ann Emerg Med*, 1994, 23:761-70.
10. Horton DK, Berkowitz Z, and Kaye WE, "Secondary Contamination of ED Personnel From Hazardous Materials Event, 1995-2001," *Am J Emerg Med*, 2003, 21:199-204.
11. Tzu S, *The Art of War*, Griffith SB, ed, London, UK: Oxford University Press, 1963.
12. Lai TI, Shih FY, Chiang WC, et al, "Strategies of Disaster Response in the Health Care System for Tropical Cyclones: Experiences Following Typhoon Nari in Taipei City," *Acad Emerg Med*, 2003, 10:1109-12.
13. Dacey MJ, "Tragedy and Response - the Rhode Island Nightclub Fire," *N Engl J Med*, 2003, 349:1990-2.
14. Hogan DE, Waeckerle JF, Dire DJ, et al, "Emergency Department Impact of the Oklahoma City Terrorist Bombing," *Ann Emerg Med*, 1999, 34:160-7.
15. San Mateo County Health Services Agency, *HEICS The Hospital Emergency Incident Command System*, June 1998, San Mateo, CA: San Mateo County Health Services Agency. Available at http://www.heics.com. Accessed January 19, 2004.
16. Hick JL, Hanfling D, Burstein JL, et al, "Protective Equipment for Healthcare Facility Decontamination Personnel: Regulations, Risks, and Recommendations," *Ann Emerg Med*, 2003, 42:370-80.
17. Okumura T, Suzuki K, Fukada A, et al, "The Tokyo Subway Sarin Attack: Disaster Management, Part 2: Healthcare Facility Response," *Acad Emerg Med*, 1998, 5:618-24.

18. Burkle FM, Sanner PH, Wolcott BW, ed, *Disaster Medicine: Application for the Immediate Management and Triage of Civilian and Military Disaster Victims*, New Hyde Park, NY: Medical Examination Publishing Co, 1984.
19. Brewer RD, Morris PD, and Cole TB, "Hurricane-Related Emergency Department Visits in an Inland Area: An Analysis of the Public Health Impact of Hurricane Hugo in North Carolina," *Ann Emerg Med*, 1994, 23:731-6.
20. Okumura T, Suzuki K, Atsuihito F, et al, "The Tokyo Subway Sarin Attack: Disaster Management, Part 1: Community Emergency Response," *Acad Emerg Med*, 1998, 5:613-7.
21. Lanoix R, Wiener DE, and Zayas VD, "Concepts in Disaster Triage in the Wake of the World Trade Center Terrorist Attack," *Top Emerg Med*, 2002, 24:60-71.
22. Kennedy K, Aghababian RV, Gans L, et al, "Triage: Techniques and Applications in Decision-making," *Ann Emerg Med*, 1996, 28:136-44.
23. Garner A, Lee A, Harrison K, et al, "Comparative Analysis of Multiple-Casualty Incident Triage Algorithms," *Ann Emerg Med*, 2001, 38:541-8.
24. Benson M, Koenig KL, and Schultz CH, "Disaster Triage: START, then SAVE - A New Method of Dynamic Triage for Victims of a Catastrophic Earthquake," *Prehosp Disaster Med*, 1996, 11:117-24.
25. Burkle FM, Newland C, and Orebaugh S, "Emergency Medicine in the Persian Gulf War - Part 2: Preparations for Triage and Combat Casualty Care," *Ann Emerg Med*, 1994, 23:748-54.
26. Katz CL, Pellegrino L, Pandya A, et al. "Research on Psychiatric Outcomes and Interventions Subsequent to Disasters: A Review of the Literature," *Psychiatry Res*, 2002, 110:201-17.
27. Macintyre AG, Christopher GW, Eitzen E, et al, "Weapons of Mass Destruction Events With Contaminated Casualties: Effective Planning for Healthcare Facilities," *JAMA*, 2000, 283:242-9.
28. Heifetz IN, "Radiation Accidents," Harwood-Nuss A, ed, *The Clinical Practice of Emergency Medicine*, Philadelphia, PA: Lippincott Williams & Wilkins, 2001.
29. Schultz M, Cisek J, and Wabeke R, "Simulated Exposure of Hospital Emergency Personnel to Solvent Vapors and Respirable Dust During Decontamination of Chemically Exposed Patients," *Ann Emerg Med*, 1995, 26:324-9.
30. Hick JL, Penn P, Hanfling D, et al, "Establishing and Training Healthcare Facility Decontamination Teams," *Ann Emerg Med*, 2003, 42:381-90.
31. Carter BJ and Cammermeyer M, "Emergence of Real Casualties During Simulated Chemical Warfare Training Under High Heat Conditions," *Mil Med*, 1985, 150:657-63.
32. Flowers LK, Mothershead JL, and Blackwell TH, "Bioterrorism Preparedness: II: The Community and Emergency Medical Services System," *Emerg Med Clin North Am*, 2002, 20:457-76.
33. Aghababian R, Lewis CP, Gans L, et al, "Disasters Within Hospitals," *Ann Emerg Med*, 1994, 23:771-7.
34. North CS, Tivis L, McMillen JC, et al, "Coping Functioning and Adjustment of Rescue Workers After the Oklahoma City Bombing," *J Trauma Stress*, 2002, 15:171-5.
35. North CS, "The Course of Post-traumatic Stress Disorder After the Oklahoma City Bombing," *Mil Med*, 2001, 166:51-2.

Examples of Mass Exposures Involving the Pediatric Population

Carl R. Baum, MD, FAAP, FACMT

Yale-New Haven Children's Hospital, New Haven, Connecticut

The Case of the Hapless Hyperventilating Hockey Players

In Massachusetts, 12 school-age children and some of their family members went to an away peewee league hockey tournament in the western part of the state, where they stayed in a local hotel, swam in the hotel pool, and had meals within the town. Over the course of the next 2-5 weeks, 16 of the children, 12 boys and 4 girls, some hockey players, and some siblings of the players, presented to a regional toxicology clinic on referral from their pediatricians with a similar constellation of symptoms, most noting fatigue as well as skin and respiratory problems, and most of the symptoms had persisted since the hockey weekend. The children, whose ages ranged from 7 to 14 years, had some or all of the following: fatigue with significantly decreased tolerance for exercise; a subtle white or ashy-appearing rash; a dry, nonproductive hacking cough; chest tightness; wheezing; irritated mucous membranes, mostly involving the eyes with some nasal irritation; diffuse abdominal pain; nausea; and some emesis. The patients were afebrile, with vital signs within the normal ranges, and their physical exams were otherwise unremarkable, other than the aforementioned rash and irritated mucous membranes. With the exception of one child who had asthma, none of the others had significant medical histories, and all were healthy and active prior to the hockey weekend. Some of the children experienced onset of symptoms during the hockey weekend, and some within 1-2 days after the weekend. None of the adults who attended the hockey tournament were known to have any of these symptoms.[1]

Consider the following:
1. How would you plan and organize the assessment and evaluation of multiple patients with multiple symptoms?
2. What further information and what laboratory evaluations, if any, would you need to obtain?
3. What would you tell the patients, their parents, and their pediatricians?

This article describes a series of incidents in this century during which large numbers of people have become ill or even died because of the exposures to a variety of toxins. This "world tour" will bring us to a number of sites, beginning in the U.S. (1937); Minamata Bay, Japan (1956); Seveso, Italy (1976); Port Pirie, S. Australia (1979-82); New Orleans, LA (1981); Bhopal, India (1984); Chernobyl, Ukraine (1986); Tokyo, Japan (1995); and Haiti (1996).

What can be learned from mass exposures? While these events may not reflect the day-to-day experiences of physicians in the clinic or emergency room, much can be learned from the way the incidents are reported and studied, which helps us describe and understand the effect of these toxins over the very long term, the importance of regulatory agencies, and the elements of an environmental history. Finally, mass exposures remind us that there are no accidents; rather, series of errors occur, sometimes over many years, that go unnoticed and culminate in catastrophic events.

United States (1937)

This was an important year in infectious diseases because the antibiotic sulfanilamide was introduced. Pharmaceutical companies rushed to produce this miracle drug. Among them was the Massengill Company, which anticipated a need, especially among children, for a liquid preparation. The company knew that the drug could not simply be dissolved without the use of a solvent. In this case, they chose diethylene glycol (DEG) to help develop the elixir of sulfanilamide. DEG, like ethylene glycol, is an antifreeze and solvent and was an inexpensive and effective solvent for the elixir. The elixir was subjected to and passed Massengill's tests for appearance, flavor, and fragrance. The company did not test the safety of the new product.

DEG, discovered in 1869, is a condensation product of ethylene glycol manufacture. It did not become commercially available until 1928. By 1937, the entire world literature on DEG consisted of a few rodent studies that suggested probable toxicity. These data were ignored, and production of 228 gallons of the DEG-based elixir proceeded. Over the subsequent 4 weeks, 353 patients received the elixir for a variety of indications, such as gonorrhea, tonsillitis, otitis, and soft-tissue infections. The instructions read as follows:

... begin with 2-3 teaspoonfuls in water every 4 hours. Decrease in 24-48 hours to 1-2 teaspoonfuls and continue at this dose until recovery.

Many did not recover. In fact, a number of patients taking the elixir developed new symptoms, including nausea and emesis, followed by flank pain, anuria, coma, and seizures. Gastrointestinal disturbances may have limited absorption of the elixir. Some died. It was estimated that a fatal cumulative dose was 10 teaspoonfuls for children or 20 teaspoonfuls for adults. Mean survival after the first dose was 9 days (range 2-22 days). There were clusters of cases around the country, in East St Louis, Tulsa, and Charleston. Only 6 gallons of elixir were eventually distributed, but the death toll was 34 children and 71 adults. Postmortem examination revealed "hydropic tubular nephrosis," or vacuolar nephropathy. Unlike ethylene glycol poisoning, no calcium oxalate deposition was observed in the renal tubules. A centrilobular hepatic degeneration was seen.

The disaster did have some beneficial effects. Until this time, federal regulation of pharmaceuticals was limited to the 1906 Pure Food and Drug Act, which prohibited misbranding and adulteration of food, beverages, and drugs. There were no safety requirements. Planned revision of the Act was stalled in Congress, but the emotion aroused by this disaster led to the 1938 Food, Drug, and Cosmetic Act, which led to the modern Food and Drug Administration.[2]

Minamata Bay, Japan (1956)

Like the events described above, this story actually begins many years before and represents a typical pattern of errors, which culminate in disaster. In 1932, the Chisso Corporation began using mercury as a catalyst in the synthesis of acetaldehyde. This reaction produced a toxic organic form of mercury, methyl mercury chloride, which was discharged into Minamata Bay. Fish living in the contaminated waters of the Bay were consumed by larger fish. Over the years, through a process known as bioaccumulation, methyl mercury was incorporated progressively into animal tissues and the food chain. By the 1950s, small animals, such as cats, were dying mysteriously, and in 1956, humans began presenting to hospitals with a variety of neurologic symptoms. Within 5 months, methyl mercury was implicated as the cause for what became known as "Minamata Disease." In the 1960s, methyl mercury levels in fish peaked at 60 times normal background levels. Cases of intrauterine intoxication became known as "fetal Minamata disease."[3] Masumoto[4] described this toxic encephalopathy, which was acquired prenatally and resulted in cerebral palsy. Harada[5] examined EEGs in 32 children; in 19 with congenital disease, 10 had abnormal findings; among 13 with acquired disease, 8 had abnormal studies. No focal abnormalities were seen on the EEGs. Methyl mercury is postulated to interfere with migration of neural cells from the neural tubule during embryogenesis. A syndrome of microcephaly, cerebral palsy, and contractures resulted, portrayed in the famous Life magazine photograph depicting a Japanese mother cradling an affected child.

An expert panel convened to assess the net impact of fish consumption on cognitive development concluded that prenatal exposure to methyl mercury might account for a loss of up to 1.5 intelligence-quotient (IQ) points.[6]

Seveso, Italy (1976)

In this story, a disinfectant, trichlorophenol was produced from a variety of starting materials at a Roche plant. On July 9, a Friday, the night shift arrived at the plant and went to work around 10:00 in the evening. On Saturday at 4:45 AM, a foreman shut down the heat in a reaction vessel in an attempt to interrupt the distillation. His shift ended at 6:00 AM, but the reaction did not. The last recorded temperature in the vessel was 158°C, and pressure continued to build until just after noon that day. With only the cleaning staff on hand, the excessive pressure burst a safety valve, and an aerosol cloud escaped toward the southeast. Inhabitants of the surrounding area were warned not to touch fruits and vegetables, although the warning efforts were delayed because it was now the weekend. Within 5 days, obvious skin inflammation was noted among some inhabitants, particularly in children. Ultimately, dioxin was found in the plant and from the surrounding areas. Children began presenting with a late-occurring acne, known as chloracne.[7] Later studies revealed indirect markers of hepatic microsomal enzyme activity, as well as modest elevations of liver function tests (GGT and ALT), which persisted for up to 6 years after the accident.[8]

Dioxin is actually a family of chemicals. The particular species was identified as tetra-chlorodibenzo-p-dioxin, a contaminant of herbicide and germicide production. Most of the childhood exposures around Seveso resulted from consumption of contaminated foods or contact with soil. The U.S. Environmental Protection Agency estimates that a lifetime exposure of 1 ng/kg/day could result in 1560 additional cases of cancer per 10,000 population. Children, therefore, may bear out this legacy in years to come.[9] A study conducted in Seveso, approximately 20 years after the accident, revealed that dioxin toxicity was confined to acute dermatological effects, and that chloracne occurrence was related to younger age and light hair color.[10]

Port Pirie, S. Australia (1979-82)

This exposure occurred over many years. This industrial town, with a population of 16,000, was situated immediately downwind of a large lead-smelting factory. There was extensive lead contamination in the town's topsoil and yard and house dust. McMichael et al,[11] undertook a large, prospective study of inhabitants. He followed over 500 children who were born in the period 1979-82, and measured blood lead levels in both the birth mothers, antenatally and at delivery, and in the children at birth, at 6, 15, and 24 months, and then annually. The investigators used one of the well-known instruments for assessing childhood development, the McCarthy Scales of Children's Abilities. They found that mean lead levels in midpregnancy among the mothers was 9 g/dL; the peak at age 2 years among the children was 21 g/dL. At that time, the CDC's "cut-off" level for lead in the United States was 30 g/dL. McMichael and coworkers found that the blood lead was inversely related to development at age 4 years. This study was one of a series that suggested even low levels of lead had adverse effects on development and helped provide impetus to further lower "acceptable" lead levels subsequently, to 25 g/dL in 1985 and to 10 g/dL in 1991.

New Orleans, Louisiana (1981)

In the spring of 1981, a neonatologist named Juan Gershanik joined the staff of a small Louisiana hospital, where he cared for a 750 g infant who suddenly died. This was a child who otherwise had been doing quite well and had every reason to survive. Gershanik thought that a poison might have been responsible. His colleagues did not agree with this theory, but Gershanik persisted and noticed a pattern with other deaths: the victims were small and ill, but many had been improving before a rapid demise. He described a syndrome of multisystem disease, characterized by severe metabolic acidosis and gasps that signaled impending death. This became known as the gasping syndrome, or "gasping baby syndrome."[12] A postmortem urinalysis for organic acids on a GC-MS revealed large amounts of benzoic acid and hippuric acid. That same evening Gershanik returned to the nursery and discovered, to his horror, a vial of bacteriostatic water atop one of the isolettes contained 0.9% benzyl alcohol, a preservative. He realized that benzyl alcohol was converted via alcohol dehydrogenase to benzoic acid and hippuric acid. He immediately removed these vials (and others) containing benzyl alcohol from the nursery, and abruptly, there were no new cases of the syndrome. Gershanik corresponded with the manufacturer of the bacteriostatic water and with the FDA. Both largely ignored his concerns about its safety. In January of 1982, he reported his findings, and other institutions began to corroborate the pattern of unexpected deaths. Similar urinary organic acid profiles were found among these victims. Nearly 1 year after Gershanik removed the offending vials from his nursery, the FDA issued a "Dear Doctor" letter warning of the association between benzyl alcohol and the syndrome, and in September 1983, the AAP issued a position statement warning of the dangers of benzyl alcohol as a preservative in very sick children. After knowledge of the association became widespread, there were dramatic falls in mortality and intraventricular hemorrhage; overall, benzyl alcohol was blamed for about 3000 deaths and countless permanent disabilities. During the period in which its use was limited, there were dramatic falls in the incidence of cerebral palsy, kernicterus, especially significant among small (<1000 g) and premature (<27 weeks gestation) infants.

Why did these events occur in 1981? This was a period of aggressive respiratory therapy. With limited transcutaneous monitoring available, therapists performed frequent blood gas testing, which required frequent flushing of arterial lines. Ironically, as an infant became more ill, more blood gases were obtained, leading to greater benzyl alcohol exposure and risk for the syndrome.

Bhopal, India (1984)

Once again, a large corporation was involved in this story. Years before, in 1969, Union Carbide built a plant in Bhopal to package U.S.-produced pesticides. In 1980, expansion allowed for production of carbamate insecticides, requiring a chemical known as methyl isocyanate, or MIC. By 1984, squatters' rights had been granted to inhabitants of the area, and 100,000 of the city's population of 900,000 lived within 1 km of the plant. Shortly after midnight on December 3, water inadvertently entered a tank containing 41,000 kg of liquid MIC. This led to an exothermic reaction, and a cloud of MIC gas, an irritant that causes an obstructive pattern in the lung, was explosively released. A total of 1000 were dead by morning. A number of studies were done on victims in the area. In one, the death rate among 1337 affected children was 8.9% (119 dead);[13] another study of 211 victims found that respiratory symptoms at 100 days were persistent among over 50% of those within 2 km of the plant compared to 8.5% of those beyond 8 km of the plant. Apprehension and depression was noted among older children, some of whom awoke after the accident to find themselves on a pile of dead bodies.[14] A follow-up study 3 years later revealed persistent eye complaints among those exposed to MIC.[15]

Chernobyl, Ukraine (1986)

In the former Soviet Union, one of the nuclear power reactors in use was known as an RBMK type. This particular design, which was used to produce both electrical power and weapon-grade plutonium, required a tall structure that precluded the usual containment shell. This resulted in a somewhat unforgiving design, because production of steam increased plant reactivity, unlike other reactors, in which steam inhibited the nuclear reaction. Xenon, another reactor product, tended to shut down the reactor if control rods were in place. In a test designed to prevent plant shutdown, the control rods were removed. But the control rods were also required as a safety feature to prevent a runaway nuclear reaction. Needless to say, steam accumulated, causing an enormous increase in the nuclear reaction. This caused a steam explosion that blew the top off the reactor. The story might have ended at this point, because the reaction stopped, but another design feature, a graphite moderator, caught fire and burned for 9 days. This helped spread radionuclides, resulting in the largest known release of radioactive material. The reactor was subsequently encased in a concrete tomb, but not before over 21,000 km² of soil were contaminated. In the first few months, the radioisotope iodine-131 (half-life 8 days) was most responsible for exposures and is blamed for causing thyroid cancers. Cesium-137 was responsible for longer-term exposures. Much of the radiation was spread to the northwest, as far as Scandinavia. Overall, an estimated 17 million people were exposed; 2.5 million were younger than 5 years of age. Delayed acknowledgment of the disaster complicated the aftermath and resulted in needless exposures. Most at risk were the so-called liquidators, who went to the scene to clean up the site and encase the reactor. Also at risk were inhabitants of the area living within 30 km, and children, whose rapidly growing bodies and long lives leave them especially susceptible to both immediate and latent effects of radiation.[16]

Most of the research done in the first decade following the accident looked at short- and medium-term effects. Language barriers and the fact that few studies appeared in western journals hampered interpretation of these studies. Furthermore, the breakup of the Soviet Union led to many small republics that guarded their data, which did not facilitate progress. An overdiagnosis bias resulted from poor monitoring, especially among liquidators. Baseline levels of radiation and rates of cancers were not well known for the region. Weinberg et al,[16] examined the plight of numerous groups who have emigrated from the Soviet Union, including 250,000 Soviet Jews. These investigators note that many èmigrès are seeking, to this day, healthcare for perceived long-term illness. How do we assess these problems? The Japanese model of radiation released at Hiroshima and Nagasaki cannot be applied because different radionuclides were released at Chernobyl. Furthermore, the above-noted fire in the graphite moderator sustained the release of radionuclides over a 9-day period, leading to more significant exposures via inhalation and ingestion. A U.S. National Chernobyl Registry has been established to sort out the details of the exposure and has recommended regular physical examinations in exposed children, with attention to the thyroid gland and overall body growth. If clinically indicated, complete blood counts, thyroid function tests, and neuropsychiatric evaluations are recommended. Another study by Dubrova[17] found a higher-than-normal mutation rate in the genome of children born in the region near the reactor after the accident compared to a control group in the United Kingdom. This difference could not be attributed to either parent or to genetic differences. A follow-up study of thyroid cancer incidence revealed significant radiation risk only for those exposed at an age of 0-9 years, with higher risk associated with younger age.[18]

Tokyo, Japan (1995)

In the mid-1990s, this populous city of 12 million had 3 million commuters per day. During the rush hour on the morning of March 20, 1995, the group Aum Shinrikyo planted a toxic substance on 5 subway cars on 3 different lines that converged beneath the government offices. Okamura[19] reported the events: 11 people died in this exposure, and 5000 required emergency evaluation. At St Luke's Hospital, 640 patients were seen in the Emergency Department within a few hours. Five patients were pregnant, and 3 presented in cardiac arrest. The gaseous chemical agent was quickly identified as sarin, developed in the 1930's as a potent organophosphate insecticide, which blocks the effect of the enzyme acetylcholinesterase at the myoneural junction, leading to a cholinergic crisis. Lethal effects result from respiratory insufficiency, with a lethal dose estimated at 1 mg. Treatment strategy revolves around blocking cholinergic activity with atropine, as well as regeneration of acetylcholinesterase with 2-pyridine aldoxime methiodide (2-PAM). Decontamination and supportive care

are essential, along with prevention of secondary contamination among healthcare providers. In the post-9/11 era, terrorism preparedness has included a focus on sarin and other "nerve" agents.

Haiti (1996)

During an epidemic of renal failure, Haiti's Ministry of Health called upon the U.S. Centers for Disease Control to investigate the situation. Many children were found to have had emesis, abdominal pain, lethargy, and malaise. Some children were flown to the centers in the U.S. for renal dialysis. Parents of two of the deceased children provided samples of antipyretic elixirs, sold under the brand names "Afebril" and "Valodon." These elixirs were found to contain acetaminophen as an active ingredient, with DEG as an excipient. One 7-year-old transferred to the U.S. had a renal ultrasound, which revealed enlarged, swollen kidneys consistent with classic DEG changes. In this case, investigators learned DEG had been inappropriately substituted for propylene glycol as a diluent. Ultimately, 30 children died.[20]

In another occurrence, 33 children died following treatment with a DEG-containing cough expectorant in Gurgaon, India.[21]

There have been other fatal DEG contaminations of pharmaceutical products since the 1937 elixir of sulfanilamide disaster; sedative mixtures in South Africa (1969), glycerine in India (1986), and acetaminophen in Nigeria (1990) and Bangladesh (1990-2). In each case, deliberate or accidental substitution of a diluent with the less expensive DEG was responsible.[2]

The Case of the Hapless Hyperventilating Hockey Players, Revisited (1996)

The children presented in the first case had played in a hotel swimming pool that, in fact, had been excessively brominated. Bromine is used as a pool disinfectant and is generally less irritating than chlorine to mucous membranes. Adults exposed to excessive bromine have developed a constellation of findings known as reactive airways dysfunction syndrome (RADS), which includes asthma-like symptoms and rash. In order to sort out these hockey players, data obtained on each patient included a complete history, a standardized patient inventory of symptoms consistent with RADS, a physical examination, and a laboratory evaluation. The latter included a complete blood count, tests of renal function, quantitative immunoglobulins, and pulmonary function testing.

On physical examination, a few had dry, excoriated skin; examinations were otherwise unremarkable. One boy had an IgE that was marginally elevated, and in 3 (19%) children, PFTs were abnormal, with bronchial hyperactivity to cold-air challenge. This hyperactivity was consistent with RADS. All of the children had follow-up visits at least 3 months later. The one boy with past medical history significant for asthma remained asymptomatic, while the patient with elevated IgE at the first visit demonstrated persistent RADS, despite use of inhaled β_2 agonists and steroids.[1]

The Environmental History

What are the elements of an environmental history? Questions about home, occupation, tobacco smoke, food, and lead should be included when interviewing parents, other caregivers, and older children themselves. The timing of questions varies; in the prenatal months, ask about renovations, smoking, and feeding. When the child is 6 months of age, ask about possible poison exposures; in preschool, arts, and crafts. In the teenage years, ask about hobbies and occupational exposures. Other timing issues concern the season of the year: lawn, garden, and other chemical applications in the spring and summer, wood and gas stoves and fireplaces in the fall and winter.[22]

Conclusion

In summary, much can be learned from mass exposures. We learn how to report these events and conduct research in the aftermath. Regulatory agencies may play a role in investigating these disasters and preventing their recurrence. A properly obtained environmental history may help sort out the details of an exposure. Finally, we learn something about the nature of "accidents"; most of these events occur not because of an isolated catastrophic failure but because of a series of less significant failures that may go unnoticed.

Footnotes

1. Perry HE, Shannon MW, Baum CR, et al, "Persistent Symptoms in Sixteen Children Following Bromine Exposure From A Pool," Presented at the Ambulatory Pediatric Association Annual Meeting, Washington, DC, May, 1997.
2. Wax PM, "It's Happening Again – Another Diethylene Glycol Mass Poisoning," *Clin Toxicol*, 1996, 34(5):517-20.
3. Powell PP, "Minamata Disease: A Story of Mercury's Malevolence," *South Med J*, 1991, 84(11):1352-8.
4. Matsumoto H, Koya G, and Takeuchi T, "Fetal Minamata Disease. A Neuropathological Study of Two Cases of Intrauterine Intoxication by a Methyl Mercury Compound," *J Neuropathol Exp Neurol*, 1965, 24(4):563-74.
5. Harada Y, Miyamoto Y, Nonaka I, et al, "Electroencephalographic Studies on Minamata Disease in Children," *Dev Med Child Neurol*, 1968, 10(2):257-8.
6. Cohen JT, Bellinger DC, and Shaywitz BA, "A Quantitative Analysis of Prenatal Methyl Mercury Exposure and Cognitive Development," *Am J Prev Med*, 2005, 29(4):353-65.
7. Caputo R, Monti M, Ermacora E, et al, "Cutaneous Manifestations of Tetrachlorodibenzo-p-Dioxin in Children and Adolescents. Follow-Up 10 Years After the Seveso, Italy Accident," *J Am Acad Dermatol*, 1988, 19 (5 Pt 1):812-9.
8. Ideo G, Bellati G, Bellobuono A, et al, "Increased Urinary D-Glucaric Acid Excretion by Children Living in an Area Polluted With Tetrachlorodibenzoparadioxin (TCDD)," *Clin Chim Acta*, 1982, 120(3):273-83.
9. "TCDD," Agency for Toxic Substances and Disease Registry, Atlanta, GA, June, 1989.
10. Baccarelli A, Pesatori AC, Consonni D, et al, "Health Status and Plasma Dioxin Levels in Chloracne Cases 20 Years After the Seveso, Italy Accident," *Br J Derm*, 2005, 152:459-65.
11. McMichael AJ, Baghurst PA, Wigg NR, et al, "Port Pirie Cohort Study: Environmental Exposure to Lead and Children's Abilities at the Age of Four Years," *N Engl J Med*, 1988, 319:468-75.
12. Gershanik J, Boecler B, Ensley H, et al, "The Gasping Syndrome and Benzyl Alcohol Poisoning," *N Engl J Med*, 1982, 307(22):1384-8.
13. Sutcliffe M, "An Eyewitness in Bhopal," *BMJ*, 1985, 290:1883-4.
14. Irani SF and Mahashur AA, "A Survey of Bhopal Children Affected by Methyl Isocyanate Gas," *J Postgrad Med*, 1986, 32(4):195-8.
15. Andersoon N, Ajwani MK, Mahashabde S, et al, "Delayed Eye and Other Consequences From Exposure to Methyl Isocyanate: 93% Follow Up of Exposed and Unexposed Cohorts in Bhopal," *Br J Ind Med*, 1990, 47(8):553-8.
16. Weinberg AD, Kripalani S, McCarthy PL, et al, "Caring for Survivors of the Chernobyl Disaster: What the Clinician Should Know," *JAMA*, 1995, 274(5):408-12.
17. Dubrova YE, Nesterov VN, Krouchinsky NG, et al, "Human Minisatellite Mutation Rate After the Chernobyl Accident," *Nature*, 1996, 380:686-6.
18. Ivanov VK, Gorski AI, Tsyb AF, et al, "Radiation-Epidemiological Studies of Thyroid Cancer Incidence Among Children and Adolescents in the Bryansk Oblast of Russia After the Chernobyl Accident (1991-2001 Follow-up Period)," *Radiat Environ Biophys*, 2006, Mar 17[Epub ahead of print].

19. Okumura T, Takasu N, Ishimatsu S, et al, "Report on 640 Victims of the Tokyo Subway Sarin Attack," *Ann Emerg Med*, 1996, 28(2):129-35.
20. Scalzo AJ, "Diethylene Glycol Toxicity Revisited: The 1996 Haitian Epidemic," *Clin Toxicol*, 1996, 34(5):513-6.
21. Singh J, Dutta AK, and Khare S, "Diethylene Glycol Poisoning in Gurgaon, India, 1998," *Bull World Health Organ*, 2001, 79(2):88-95.
22. Balk SJ, "The Environmental History: Asking the Right Questions," *Contemp Ped*, 1996, 13(2):19-36.

Organophosphate Poisoning: Case Presentation

Multiple Exposure and Disaster Management in Parathion Poisoning

Yona Amitai, MD, MPH

Ministry of Health, Jerusalem, Israel

Introduction

Organophosphates (OP) are the largest class of chemicals used in agriculture with over 25,000 brand names registered in the U.S. Their acute toxic effects include muscarinic, nicotinic, and central nerve system manifestations, whereas, chronic exposure may cause peripheral neuropathy and neurobehavioral alterations. Environmental hazards of OP often involve multiple exposures. Their use as chemical warfare has resulted in mass casualties. The following description is a simulation of multiple-victim exposure to parathion, a highly toxic and frequently used OP with emphasis on decontamination, triage, and organizational management of chemical disaster with multiple victims.

Case Reports

A fire broke out in a farm in Florida; a ten-gallon can of parathion, technical grade 95%, caught fire and exploded at 10 AM. The farm owner alerted the fire control and called for help. Within 20 minutes, he lost consciousness and his son summoned rescue teams and additional help. At 10:15, fire fighters and emergency medical services (EMS) arrived. CHEMTREC (1-800-424-9300) was notified. Since structural firefighters' protective clothing are not effective for OP, an additional rescue team with positive pressure self-contained breathing apparatus (SCBA) and chemical protective clothing was called. Firefighters controlled the fire with water sprayed from a distance upwind. At 10:40 AM, the rescue team with SCBA and chemical protective clothing arrived at the scene. At 10:50 AM, decontamination was started. By this time, the farm owner was dead and 6 other farm workers and firefighters were injured.

Decontamination and initial management were organized. Decontamination was carried out at the scene, whereas the triage and treatment area were located upwind. The contaminated area was separated from the "clean area" (150 feet upwind from the disaster site) by a red plastic ribbon.

Within the contaminated area, there was a physician with emergency medicine training who directed operations, two rescuers, and two paramedics, all equipped with chemical protective clothing and SCBA. Two EMS teams were located at the "clean area" and a third team was expected to join in 20 minutes. These teams prepared their equipment for advance life support. Police outlined the disaster area at a range of 150 feet from the explosion site, and denied access to the area to noninvolved people. Police maintained phone contact with CHEMTREC for disaster management guidelines. Patients disposition upon triage (11:10 AM) and doses of antidotal therapy are listed in the table. At triage, two subjects (nos. 1, 6) were moderately injured and two (nos. 4, 5) were mildly affected. All symptoms were related to OP toxicity. There were no burns or any evidence of trauma.

The rescuers undressed the patients and splashed them with ample amounts of water, the paramedics started I.V. lines, and the physician assigned treatment priorities and ordered individual doses of antidotes.

At 11:25 AM, while starting I.V. lines and injecting antidotes in the patients, patient no. 2 elapsed into a coma and had a generalized seizure with prolonged apnea. At the same time, patient no. 3 was drowsy with severe respiratory distress, cyanosis, frothy oral secretions, and muscle fasciculation. Both patients required tracheal intubation and assisted ventilation. Since patient decontamination had not been completed and the EMTs in the clean area were not equipped to enter into the contaminated area, the physician decided to intubate patient no. 3 with the assistance of one paramedic, and ordered the other paramedic to continue with the insertion of I.V. lines and initial antidotal therapy to the other patients. Within 5 minutes (11:30 AM), patient no. 2 had cardiorespiratory arrest. At 11:45 AM, the decontamination process, I.V. lines, and first dose of antidotes were completed in all patients. Patient no. 3 was stabilized with assisted ventilation, oxygen supply, and manual suction. All patients were moved to the clean area for further treatment. Patients no. 1, 3, and 6 were evacuated via stretchers, whereas, patients no. 4 and 5 walked out on their own.

The patients underwent field stabilization before being transported to the local hospital. Two patients, nos. 1 and 6, deteriorated and underwent endotracheal intubation with assisted ventilation and suction of tracheal secretions. During the process of field stabilization, one of the paramedics was assigned the responsibility of recording patients' names, vital signs, time and dose of medication, I.V. fluids, and other therapeutic procedures.

Between 12:15-12:45 PM, the patients were transported to the local hospital 15 miles away for definitive care. The final outcome of the disaster was three casualties; the farm owner and patient no. 2 died at the scene, and patient no. 3 deteriorated and died at the hospital on the seventh hospital day from polymorphic ventricular arrhythmia and ventricular fibrillation. The other 3 patients had transient aggravation of symptoms but were discharged after 2-5 days hospitalization in good health without sequelae.

Discussion

Disaster from environmental toxins, which result in multiple victims, present several unique problems to rescue teams:
1. The hazard of continuous poison exposure to victims and rescuers.
2. Shortage of medical staff and facilities needed for prompt response in large disasters.
3. The low state of preparedness of most medical personnel for medical emergencies resulting from hazardous materials

This complex situation may create panic reactions and chaos among victims and inexperienced medical personnel, which may further reduce efficacy of medical care. Under these circumstances, there is a tendency to evacuate and transport victims to hospitals before completion of field stabilization. In anticipation of these problems, a specific strategy should be employed to cope with chemical disasters. The main principles of this strategy are:
1. Organization and coordination of all rescue teams under the supervision or remote control of a professional authority (eg, CHEMTREC, or a regional poison center).
2. Decontamination by well-equipped rescuers (chemical protective clothing and SCBA) while controlling the area to minimize further poison exposure.
3. Triage and prioritization of treatment. When the demands for medical care in a large scale disaster far exceed the ability of the personnel to provide medical care, triage should be aimed at rescuing those with reasonable chances of survival and not necessarily toward those who are more severely affected. In the present case, the lengthy and probably unsuccessful resuscitation of patient no. 2 could have resulted in critical delays in providing care to several victims with better potential for survival.
4. Periodic evaluation and strict recording of patient status. This is particularly important in chemical disasters, because poisoning by substances such as OP often result in delayed symptoms with further deterioration of victims after the initial triage.
5. Field stabilization of patients before transfer to hospital. Life-saving procedures such as insertion of central lines, endotracheal intubation, and tracheostomy are much easier and safer to perform on the ground rather than in an ambulance or helicopter.

6. Preparedness of medical staff to such disasters, including recognition of the major potential chemical hazards and adequate equipment, including antidotes and protective clothing, may improve the efficacy of such a complicated operation.

7. In areas with particular risk for OP exposure (ie, military personnel who anticipate confrontation with chemical warfare), the supplying of individuals with SCBA and protective clothing for chemical hazards and personal automatic autoinjectors of antidotes, such as atropine and oximes with adequate training in operation of these means, can save critical time and enable hundreds of individuals to protect themselves, and use effective antidotes simultaneously, independent of the feasibility of medical assistance. Pretreatment with 10 mg of pyridostigmine may provide some protection in the event of OP poisoning.

Patient No.	Age	CNS Status	HR*	BP	Respiratory Symptoms	Pupils	Other	Triage State (category)	Antidotal Therapy	
									Atropine†	Pralidoxime
1	23	Anxiety	110	130/90	Dyspnea, cyanosis	Constricted	Diaphoresis, vomiting, abdominal cramps	Moderate	2 mg/min	1 g q6h
2	50	Coma, seizures	45	60/0	Cyanotic, gasping	Dilated	Vomiting, diarrhea	Severe/critical		2 g
3	34	Drowsy, fasciculations	60	100/60	Dyspnea, cyanosis, frothy secretions	Constricted	Vomiting, diarrhea, diaphoresis	Severe	5 mg/min	2 g q6h
4	40	Alert	80	125/80	Tachypnea	Mid-position	Abdominal pain	Mild	2 mg/min	1 g
5	36	Alert	90	120/70	No distress	Mid-position	Diaphoresis, abdominal pain	Mild	2 mg/min	1 g
6	52	Alert	72	150/90	Tachypnea, chest† tightness	Constricted	Diaphoresis, vomiting, abdominal cramps	Moderate	2 mg/min	1 g q6h

*Blood pressure was recorded only after completion of the first dose of antidotes.

†Repeated doses of atropine, until achieving atropinization (ie, drying of respiratory secretions, increase in heart rate).

Toxicology Basics of Nonmedicinal Agent Exposures

Edward P. Krenzelok, PharmD, FAACT, DABAT

Childrens Hospital of Pittsburgh, Pittsburgh, Pennsylvania

Introduction

New chemical compounds, mixtures, and products are produced throughout the world at an astounding rate. Production of a substance often precedes adequate toxicological testing to fully assess its hazards to man and the environment. Even for chemicals used in commerce for decades, there may be surprisingly little data to help clinicians plan immediate treatment and answer exposed patients' questions about long-term effects. Following a chemical-exposure incident, this lack of information may produce confusion, anxiety, and anger in patients, caregivers, and the public. Rarely are there absolute answers to all the health-related questions that arise from such incidents. Healthcare providers, however, can obtain valuable assistance from poison information centers, toxicologists, and occupational medicine experts, and through the use of toxicology references and databases. Physicians charged with acute treatment and subsequent management need to understand certain toxicological principles as well as the strengths and limitations of the toxicology literature. This chapter is intended to aid clinicians in applying this knowledge to patient care.

What Is Toxicity?

Stedman's Medical Dictionary defines toxicity as "the state of being poisonous" and toxin as a "noxious or poisonous substance..."[1] These definitions are nonspecific and reflect the ambiguity and art that exists within the field of toxicology. It is important to remember that everything is potentially toxic. Dose is the factor that principally determines whether a chemical exposure produces obvious harm to health, no demonstrable effect, or even beneficial effects. Exposure level, duration of exposure, and rate of absorption determine dose; tissue and plasma binding, excretion, and metabolism of an agent all modify its toxicity. Other potential modifying factors are listed in Table 1. Toxins also come in a variety of physical forms (Table 2).

Table 1. Factors That Modify Toxic Effect of a Given Dose
Host factors (age, sex, genomics, body fat, nutritional status)
Health behaviors (EtOH, smoking)
Other medical conditions, medications
Reproductive status
Repair
Biotransformation and excretion
Incident factors
Route of entry
Environmental factors

Table 2. Physical Forms of Toxins	
Solid	Gas
Liquid	Mist
Aerosol	Vapor
Dust	Fumes
Smoke	Fog

Every individual has a unique threshold that must be attained after exposure to most agents for an adverse effect to become manifest. Adverse effects can be overly clinical, damaging but not apparent clinically, or cumulative. This dose-response relationship for a chemical can be altered by many factors (Table 1); subclinical, latent, and idiosyncratic nondose-related effects must also be considered. Published acceptable exposure limits are not absolute; even when they have safety margins added, they are often designed for average groups of healthy workers. Furthermore, data may be based upon *in vitro* or animal testing models. These models are frequently controversial and their validity must be considered when the data are extrapolated to human exposures.

In addition, these exposure limits are designed for workplace exposures by adults. Brief, acute, high-dose exposures, exposures to children, women, or the elderly, and continuous low-level residential exposures are less well-understood. Limits for these variables are less common and less applicable.

Dose-response relationships are based upon bell-shaped curves with intense and minimal responses at the tails of the curves. While the majority of individuals in a population respond within two standard deviations of the mean dose of an agent, it is imperative that clinicians use reported toxic doses and air concentrations merely as guidelines. They are but one more factor to be added to the history, physical examination, and other laboratory data in planning patient evaluation and treatment.

The word hazard in the context of a chemical exposure incident is used commonly, but often inappropriately. Whereas toxicity reflects the ability of a substance to cause injury in an exposed individual, hazard to responders and the public depends on many factors beyond its simple toxicity (Table 3). If one can be confident that there will be no exposure to a highly toxic substance, then there is little hazard. Conversely, an agent of low toxicity can be very harmful, given sufficient exposure - consider prolonged painting in the confined space of an unventilated room. Conveying these concepts is especially important when communicating with the "worried well," and the media.

Table 3. Contributions to the Hazard Factor
Agent and Release Factors
• Amount, concentration, physical state, and density
• Volatility and reactivity under accident conditions
• Potential for explosion, thermal and corrosive injury
• Presence of ignition sources or fire
• Weather, terrain, and ventilation conditions
Exposure Factors
• Number, age, and health of potentially exposed populations
• Robustness of storage and containment structures
• Adequacy of identification, monitoring, and warning
• Adequacy of shelter and evacuation
• Threat to water supplies and environment
Human and Planning Factors
• Maintenance of equipment and facilities for the agent and response
• Adequacy of preplanning and coordination
• Adequacy of emergency response and medical services
• Training, number, and quality of operational and response personnel
• Adequacy of actions for containment, mitigation, and medical care

Finally, it is useful for healthcare providers to have some knowledge of the physicochemical properties of a chemical involved in an incident. Vapor density indicates the weight of a gas or vapor relative to dry air and provides clues as to the degree of patient exposure in inhalation injuries. For example, chlorine is heavier than air and anhydrous ammonia is lighter. This has obvious, though imperfect, exposure implications for a victim who remained flat on the ground until rescued. A liquid with a high vapor pressure will volatilize easily into the air from a contaminated patient; inadequate decontamination of the patient may pose a threat to healthcare providers in a poorly ventilated ambulance or decontamination room. Lipid and water solubility bear on absorption and decontamination methods. Other physicochemical properties such as boiling point, ignition point, and upper and lower explosive limits help define hazard, particularly to workers, first responders, and the public near an incident scene. Fire services, and especially their hazardous materials units, are usually very knowledgeable in this area.

Acute vs Chronic Toxicity

There are 3 types of exposures that can lead to toxicity: acute, subchronic, and chronic. Acute exposure is a single exposure to a chemical that may occur momentarily, briefly over several minutes, or for up to approximately 8 hours. Inhalations or dermal exposures occurring over the course of a workday or a single exposure episode can be classified in this manner. Toxic manifestations that develop secondary to short, nonrepeated exposures are classified as acute toxicity. Subchronic exposures are intermittent acute exposures that do not exceed 90 days. The terms subchronic or subacute are applied rarely to toxicity in the clinical setting, and such exposures are usually labeled inappropriately as chronic exposures. It is important to properly characterize the temporal nature of a toxic exposure, since the literature differentiates carefully between the toxic manifestations that may ensue from acute, subchronic, and chronic exposures to poisons and toxins. Misinterpretation of these classifications and the information regarding them can lead to inaccurate diagnosis and prognosis. For example, an acute exposure to toluene may produce narcosis but not the renal and hepatic pathology that is associated with chronic toluene exposure. There are significant medical, psychological, and financial consequences to such errors. Chronic exposures are repeated acute or continuous exposures that occur over an extended temporal period, which may be days, weeks, or years. Lead poisoning in bridge painters, silicosis in miners, and neuropathies secondary to prolonged solvent exposure are examples of chronic toxicity.

Toxin Identification and Toxicity Rating Systems

The toxicology and industrial hygiene literature is replete with abbreviations, numeric, and alpha-numeric codes that serve as identifiers and indicators of safety as well as toxicity. What is a TLV? How are CAS numbers used? Can an LD_{50} be used prognostically in humans? Awareness of the most clinically relevant terms will help the clinician identify toxins, assess toxicity, and perform hazard risk assessments.

Toxin Descriptors   Community and worker right-to-know legislation as well as SARA Title III regulations have spawned the wide distribution of Material Safety Data Sheets (MSDS). These documents contain limited information about an individual product's composition, adverse health effects and their treatment, physicochemical properties, and provide guidance on spill containment and cleanup. MSDS may be inaccurate and are usually inadequate for comprehensive medical care. However, they are useful in product identification and frequently accompany exposed patients to the emergency department. Chemical ingredients are listed usually by name and by the Chemical Abstracts Service (CAS) Registry number, the most common numeric code used to identify toxins in the U.S. The CAS number is a unique identifier that enables immediate identification of the substance, and helps avoid the confusion inherent in the many synonyms, closely related compounds, and "sound-alikes" present in chemical nomenclature. For instance, the CAS number of nitric acid is 7697-37-2. A call to a regional poison information center with this number allows their personnel to identify immediately the toxin and to provide toxicity and hazard information. Another commonly used identifier is the Registry of Toxic Effects of Chemical Substances (RTECS) number, which also uniquely identifies chemicals. It is found rarely on Material Safety Data Sheets; its primary use is to search databases for the toxicity literature. Field information such as data from truck placards and shipping labels may provide general hazard classifications (eg, poisonous vs corrosive chemicals) and tentative identification of a substance. However, these numbers are often not specific and must be used with caution. Again, using the example of nitric acid, its placard (for some forms and concentrations) is United Nations number 1760. This number is shared by some 30 other product groups, including aluminum sulfate solution, some weed killers, and some cosmetics!

When possible, it is judicious to use one of the specific numbers (especially the CAS number) to identify what chemicals may have been implicated in an exposure. Synonyms should be avoided since they may be regional or colloquial terms that are inadequate and possibly dangerous in decision-making and treatment. For example, an exposure to "alcohol" could involve ethanol, methanol, or isopropanol, all of which have unique chemical and toxicological properties.

Toxicity Rating   A variety of descriptive terms are used to describe dose-response relationships in the literature. These terms usually reflect animal mortality data and may be based upon acute, subchronic, or chronic exposures. The term used most commonly is the Lethal Dose 50 (LD_{50}). This is the amount of chemical required to kill 50% of an experimental animal population and is commonly referred to simply as the "lethal dose." The limitation of this value lies in the uncertainty of extrapolating it to a human exposure. However, if the value is either very low or very high, an assumption is generally made that a given chemical is either relatively safe or toxic to humans. This assumption must be tempered by existing human data, usually in the form of case reports, and even then, caution must be exercised. For example, the LD_{50}s of the organophosphate insecticides, parathion and malathion, in rats are 5 mg/kg and 1000 mg/kg, respectively. This means that considerably smaller amounts of parathion caused 50% of the rats to die. However, exposure to either of these can produce significant human cholinergic toxicity, though it may require a larger amount of malathion to do so. Similarly, the Lethal Dose Low (LD_{LO}) is the lowest amount known to be fatal when dose-response testing is conducted. Human toxicity data is reported as the Toxic Dose Low (TD_{LO}). This is the lowest amount that has actually been reported to produce toxicity in a human. Unfortunately, there is a conspicuous absence of this information in the literature. Inhalation toxicity is described by similar terminology - Lethal Concentration-$_{50}$, Lethal Concentration-$_{Low}$, Toxic Concentration-$_{Low}$, etc.

Modifications of the Hodge-Sterner Toxicity Table is published in many toxicology references and rates the degree of toxicity from "Dangerously Toxic" (LD_{50} <1 mg/kg) in 6 gradations to "Practically Nontoxic" (LD_{50} >15 g/kg). These data are supposed to be applicable to a 70-kg male. This type of rating system provides limited insight into the potential toxicity of an agent, but it does not consider the acute or chronic nature of the exposure, has rather broad ranges, and is supported by neither scientific nor clinical data. Reliance solely upon this system may result in either complacency or inappropriate therapy.

Exposure Standards   What is a permissible exposure level? Ideally, there should be no exposures to potentially noxious chemicals. However, given the ubiquity of noxious chemicals in industrialized society, reality dictates that a risk-to-benefit ratio exists for most products. Levels of acceptable exposure have been established by both the Occupational Safety and Health Administration (OSHA) and the American Conference of Governmental Industrial Hygienists (ACGIH). It is of paramount importance to understand that these values are arbitrary, and are designed to protect nearly all workers. Some individuals are more sensitive than others, that is, dose-response relationships vary within similar populations. The values, unless otherwise specified, reflect the air concentrations to which an individual can be exposed over a specific period of time and are customarily reported in parts per million (ppm). Caution must be exercised when interpreting OSHA and ACGIH exposure limits, since these values do not take into consideration factors such as underlying health problems or performance of demanding work (which may increase respiration and accordingly increase the amount of inhaled toxin). The most commonly used exposure standard is referred to as the Threshold Limit Value (TLV), which reflects the maximum allowable exposure over a specific period of time. The TLV-Time-Weighted Average (TWA) (also referred to as the Permissible Exposure Limit (PEL)) is the average concentration to which an individual can be exposed in the workplace over a standard 8-hour day or 40-hour work week without developing any adverse effects. For example, the TLV-TWA for formaldehyde is 3 ppm. However, sensitive individuals may develop mucosal irritation at air concentrations of <0.5 ppm. The TLV-Short-Term Exposure Limit (STEL) is the maximum concentration to which a worker can be exposed for 15 continuous minutes without developing intolerable irritation, chronic or irreversible tissue damage, or narcosis that may impair work efficiency or endanger the worker. This standard allows an individual to have 4 such exposures separated by at least 60 minutes per day. The TLV-Ceiling (TLV-C) is the concentration that should never be exceeded, even instantaneously. Special attention should be paid to the Immediately Dangerous to Life or Health (IDLH) Level. The IDLH is the maximum concentration from which one could escape without impairment of judgment or the development of irreversible health effects. These values are rarely available when patients with toxic effects are being treated in an emergency department or when first responders are at a hazardous materials incident. However, if the values are known, they can assist the healthcare provider in making both diagnostic and prognostic decisions regarding the patient.

Toxicokinetics

The absorption, distribution, metabolism, and elimination of drugs is called pharmacokinetics. The application of those principles to toxicology is referred to as toxicokinetics. An appreciation of the basic principles of toxicokinetics can facilitate patient assessment and management. Only factors that affect the route of absorption are addressed here.

Routes of Exposure
- Oral
- Ocular
- Dermal
- Inhalation
- Injection

Due to the nature of hazardous materials incidents, many of the exposures to noxious agents are inhalational and can result in both local and systemic toxicity. A variety of factors dictate how well a chemical is absorbed via the inhalational route; of considerable importance is particle size. Particulate matter with particle diameters >10 microns tend to lodge in the pharynx or nasal cavity; they generally do not produce effects in the lower respiratory tract. Particles smaller than 10 microns are respirable and, depending upon their size and shape, tend to gain access to the lower respiratory tract. If the diameter of a particle is in the range of 1-5 microns, it may be deposited in the bronchioles; those particles <1 micron may reach the alveoli.

Solubility also plays an important role in the inhalational absorption of chemicals. Respirable particles that are water-soluble may be scrubbed out partially by the mucous membranes of the upper respiratory tract. This tends to limit an injury to the upper respiratory tract. In contrast, respirable substances that have lower water solubility, such as nitrogen dioxide (Silo Filler's Disease) can be inspired into the alveoli. There they may be converted to nitrous acid and lead to the development of pulmonary edema.

The skin is not impervious to the absorption of toxins. Lipid-soluble chemicals such as solvents can penetrate the dermis and cause systemic toxicity. It is important to know the solubility and corrosive characteristics of a chemical as well as the duration of contact, to determine its potential for toxicity. Skin damaged by abrasions, burns, or wounds permits greater absorption and requires special attention during decontamination.

Summary

As stated previously, toxicology is more of an art than a science. However, some of the mystery of toxicology can be eliminated by understanding terminology and a few basic concepts. A considerable amount of knowledge exists about the toxicology of drug overdoses but there is a paucity of human data regarding hazardous materials. Common sense, basic life-support considerations, and the incorporation of toxicological principles allows healthcare providers to make informed decisions in patient management and to contribute to important community decisions such as evacuation.

Footnote
1. *Stedman's Medical Dictionary*, Philadelphia, PA: Lippincott Williams and Wilkins, 2006.

2,4,6-Trinitrotoluene

CAS Number 118-96-7

UN Number 0209 (<30% water); 1356 (>30% water)

Synonyms Alpha TNT; TNT; Tolit; Trilit; Tritol; Triton

Use High explosive agent used in military or civilian blasting procedures; chemical intermediate in production of dyes and photographic chemicals

Mechanism of Toxic Action Cause of methemoglobin; dermal irritant

Adverse Reactions
    Dermatologic: Contact dermatitis; yellowish orange staining of hands, arms, face
    Hematologic: Glucose-6-phosphate dehydrogenase deficiency-induced hemolytic anemia
    Ocular: Retrobulbar neuritis

Toxicodynamics/Kinetics
    Absorption: Inhalation, oral, and dermal exposure
    Metabolism: Hepatic to nitro reduction to 2,6-dinitro-4-aminotoluene (DNAT)
    Elimination: Renal

Signs and Symptoms of Acute Exposure Altered taste, anemia (aplastic), cataract, constipation, dermal burns, elevated liver enzymes, hemolytic anemia (especially in individuals deficient in G6PD enzyme), methemoglobinemia, paresthesia, red color to urine, toxic hepatitis

Admission Criteria/Prognosis Any patient with change in mental status, cardiopulmonary complaints, or methemoglobin levels >30% should be admitted. Asymptomatic patients with methemoglobin levels <30% may be considered for discharge after 6 hours of observation and if methemoglobin levels fall to <15%.

Treatment Strategy
    Decontamination:
        Dermal: Wash with soap and water
        Ocular: Irrigate with saline
    Supportive therapy: Methylene blue for symptomatic methemoglobinemia; treat burns with standard burn therapy; monitor for carbon monoxide

Additional Information Combustible, pale yellow solid. Atmospheric half-life: 18-184 days. Water half-life: 14-84 hours. PEL-TWA (skin designation): 0.5 mg/m^3. TLV-TWA (skin designation): 0.5 mg/m^3

Specific References
    Vorisek V, Pour M, Ubik K, et al, "Analytical Monitoring of Trinitrotoluene Metabolites in Urine by GC-MS. Part I. Semiquantitative Determination of 4-Amino-2,6-dinitrotoluene in Human Urine," *J Anal Toxicol*, 2005, 29:62-5.

2,4-Dinitrotoluene

CAS Number 121-14-2

UN Number 1600 (molten); 2038 (solid or liquid)

Synonyms 2,4-DNT

Use In the production of polyurethane polymers, explosives, and automobile air bags; an isomer of DNT

Mechanism of Toxic Action Producer of methemoglobin in rodent studies; not reported in humans; cytotoxic to hepatocytes

Adverse Reactions
    Cardiovascular: Cyanosis
    Central nervous system: Headache, vertigo, dizziness
    Dermatologic: Dermatitis
    Gastrointestinal: Nausea, vomiting

Hematologic: Anemia, hemolysis, although not described in humans, methemoglobinemia may occur (onset may be delayed for 4 hours)

Hepatic: Hepatitis

Neuromuscular & skeletal: Arthralgia (especially in the knees)

Miscellaneous: Cutaneous T-cell lymphoma

Toxicodynamics/Kinetics

Absorption: By oral (rapid and complete) and inhalation routes

Elimination: Renal

Drug Interactions Reduced tolerance to ethanol; ethanol many intensify symptoms of 2,4-dinitrotoluene

Monitoring Parameters CBC, methemoglobin

Treatment Strategy

Decontamination:

Oral: Do not induce emesis. Lavage within 4 hours. Activated charcoal may be useful.

Dermal: Wash with soap and water.

Ocular: Irrigate with water and saline.

Supportive therapy: Symptomatic methemoglobinemia (or methemoglobin levels >30%) can be treated with methylene blue.

Reference Range Urinary 2,4 DNT levels in ammunition plant workers range from 2-9 mcg/L

Additional Information Water soluble; mean concentration in U.S. waters is <10 mcg/L. TWA (skin): 1.5 mg/m^3; IDLH: 200 ng/m^3

Boiling point: 300°C

Specific References

Bruning T, Thier R, Mann H, et al, "Pathological Excretion Patterns of Urinary Proteins in Miners Highly Exposed to Dinitrotoluene," *J Occup Environ Med*, 2001, 43(7):610-5.

Adamsite

CAS Number 578-94-9

UN Number 1698

Synonyms Diphenylaminochloroarsine; DM; Phenarsazine Chloride

Military Classification Riot Control Agent; Vomiting Agent

CDC Classification for Diagnosis

Biologic. No biologic marker is available for adamsite exposure

Environmental. No method is available to detect adamsite in environmental samples

CDC Case Classification

Suspected: A case in which a potentially exposed person is being evaluated by healthcare workers or public health officials for poisoning by a particular chemical agent, but no specific credible threat exists.

Probable: A clinically compatible case in which a high index of suspicion (credible threat or patient history regarding location and time) exists for adamsite exposure, or an epidemiologic link exists between this case and a laboratory-confirmed case.

Confirmed: A clinically compatible case in which laboratory tests (not available for adamsite) have confirmed exposure. The case can be confirmed if laboratory testing was not performed because either a predominant amount of clinical and nonspecific laboratory evidence of a particular chemical was present or a 100% certainty of the etiology of the agent is known.

Mechanism of Toxic Action Inhibition of sulfhydryl group containing enzymes; irritation at high concentrations.

Signs and Symptoms of Acute Exposure May be delayed for several minutes: ocular burning, lacrimation, sinus pain, throat burning, chest pain, nasal secretions, salivation, sneezing; systemic effects include headache, ataxia, paresthesias, malaise, nausea, and vomiting. Sneezing and coughing may be violent. Pulmonary edema can occur at high concentrations. Dermal exposure results in pruritus, erythema, and burning sensation, which can progress to vesicle formation. Symptoms usually resolve within two hours.

Treatment Strategy

Decontamination:

Dermal: Irrigate with copious amounts of water or dilute bleach or dilute bicarbonate of soda solution

Ocular: Irrigate with copious amounts of water or normal saline

Respiratory: Administer 100% humidified oxygen

Supportive therapy: Oral antihistamines can be given for pruritus. Ophthalmic corticosteroids may be used for persistent ocular irritation. Monitor oxygenation. Aggressive airway management and pain management if necessary. Metoclopramide or ondansetron may be effective to control vomiting.

Environmental Persistency Not persistent; loss of effects within $^1/_2$ hour

Additional Information Odorless, yellow to white to invisible gas; found in yellow to green granules; can corrode iron and bronze; rate of hydrolysis is rapid to diphenylarsenious oxide and hydrogen chloride; usually dispersed by heat; does not result in arsenic toxicity (organoarsenicals)

Incapacitation dose: Airbourne concentration of 22 mg/min/m^3 at 1 minute

Median lethal dose: 11,000 mg/min/m^3

Bronchodilators may be necessary if bronchoconstriction or pulmonary edema occur. Airway protection is critical.

Specific References

Belson MG, Schier JG, and Patel MM, "Case Definitions for Chemical Poisoning," *MMWR Recomm Rep*, 2005, 54(RR-1):1-24.

Ammonia

CAS Number 7664-41-7

UN Number 1005; 2073; 2672; 3318

Synonym Spirit of Hartshorn

Commonly Found In Household cleaners (5% to 10%) and bleach

Use Primarily in fertilizers; manufacture of nitrous oxide; petroleum refining

Mechanism of Toxic Action Tissue injury of moist mucosal membranes caused by reaction with water to form ammonia hydroxide; can cause burns by liquefaction necrosis

Adverse Reactions

Cardiovascular: Chest pain

Dermatologic: Immunologic contact urticaria

Gastrointestinal: Salivation

Respiratory: Reactive airways disease syndrome

Miscellaneous: Mucosal irritation

Personal Protection Equipment Self-contained breathing apparatus with chemical-protective clothing

Signs and Symptoms of Overdose Burns, chest pain, coma, conjunctivitis, corneal defects, cough, dyspnea, GI irritation, headache, lacrimation, nausea, pulmonary edema, salivation, swelling, upper airway irritation, urticaria, wheezing, vomiting. Long-term sequelae include bronchiolitis obliterans and peribronchial fibrosis.

Admission Criteria/Prognosis Patient may be discharged if asymptomatic 6-8 hours postexposure

Toxicodynamics/Kinetics

Absorption: Not well absorbed

Metabolism: Hepatic to urea and glutamine

Environmental persistency: Lasts about 1week in air

Overdosage/Treatment

Decontamination:

Oral: **Do not** induce emesis or perform gastric lavage; dilute with water or milk (4 oz in children, 8 oz in adults).

Ocular: Irrigate eyes copiously with normal saline.

Inhalation: Administer 100% humidified oxygen.

Supportive therapy: Flush injured surfaces with water. Treat for pulmonary edema. Use steroids for third degree esophageal burns. Inverse-ratio ventilation using lower tidal volume mechanical ventilation (6 mL/kg) and plateau pressures of 30 cm water may decrease mortality. While positive end expiratory pressures should be considered, oxygen toxicity should be monitored. Use of perfluorocarbon partial liquid ventilation and exogenous surfactant in chemical-induced ARDS is investigational.

Additional Information Irritation can occur at 400 ppm. Stomatitis can occur at ammonia concentrations of 50 ppm. The mixture of ammonia with hypochlorite bleach can result in chloramine, which can produce pulmonary edema. Highly water soluble

Colorless liquid; penetrating pungent odor; stable, colorless gas; highly water soluble; alkali (pH 11.6). Odor threshold: 25-48 ppm (air); 1.5 ppm (water). Atmospheric half-life: 2-3 days. TLV-TWA: 25 ppm; IDLH: 300 ppm

Specific References

Cavender FL, Millner GC, and Goad PT, "Use of Toxicity Data in Determining In-House Concentrations Following a Catastrophic Release of Ammonia in a Derailment of Tankcars," *Clin Toxicol (Phila)*, 2005, 43:750.

Haroz R and Greenberg MI, "Bowel Necrosis Following the Intentional Administration of an Ammonia-Containing Enema," *Clin Toxicol (Phila)*, 2005, 43:745.

Lee JH, Farley CL, Brodrick CD, et al, "Anhydrous Ammonia Eye Injuries Associated With Illicit Methamphetamine Production," *Ann Emerg Med*, 2003, 41(1):157.

Weisskopf MG, Drew JM, Hanrahan LP, et al, "Hazardous Ammonia Releases: Public Health Consequences and Risk Factors for Evacuation and Injury, United States, 1993-1998," *J Occup Environ Med*, 2003, 45(2):197-204.

References Prior to 2003

Albrecht J, "Roles of Neuroactive Amino Acids In Ammonia Neurotoxicity," *J Neurosci Res*, 1998, 51(2):133-8.

Balkissoon R, "Occupational Upper Airway Disease," *Clin Chest Med*, 2002, 23(4):717-25.

Campagnolo ER, Kasten S, and Banerjee M, "Accidental Ammonia Exposure to County Fair Show Livestock Due to Contaminated Drinking Water," *Vet Hum Toxicol*, 2002, 44(5): 282-5.

Sjoblom E, Hojer J, Kulling PE, et al, "A Placebo-Controlled Experimental Study of Steroid Inhalation Therapy in Ammonia-Induced Lung Injury," *J Toxicol Clin Toxicol*, 1999, 37(1): 59-67.

Sotiropoulos G, Kilaghbian T, Dougherty W, et al, "Cold Injury From Pressurized Liquid Ammonia: A Report of Two Cases," *J Emerg Med*, 1998, 16(3):409-12.

Tanen DA, Graeme KA, and Raschke R, "Severe Lung Injury After Exposure to Chloramine Gas From Household Cleaners," *N Engl J Med*, 1999, 341(11):848-9.

Arsenic

CAS Number 7440-38-2

UN Number 1554; 1558; 1573

Synonyms Arsenate; Arsenite

CDC Classification for Diagnosis

Biologic: A case in which elevated urinary arsenic levels (>50 mcg/L for a spot or >50 mcg total for a 24-hour urine) exist, as determined by commercial laboratory tests. Speciation is required in all cases where total urine arsenic is elevated to differentiate the amount of organic and inorganic arsenic.

Environmental: Detection of arsenic in environmental samples above typical background levels, as determined by NIOSH or FDA.

CDC Case Classification

Suspected: A case in which a potentially exposed person is being evaluated by healthcare workers or public health officials for poisoning by a particular chemical agent, but no specific credible threat exists.

Probable: A clinically compatible case in which a high index of suspicion (credible threat or patient history regarding location and time) exists for arsenic exposure, or an epidemiologic link exists between this case and a laboratory-confirmed case.

Confirmed: A clinically compatible case in which laboratory tests have confirmed exposure. The case can be confirmed if laboratory testing was not performed because either a predominant amount of clinical and nonspecific laboratory evidence of a particular chemical was present or a 100% certainty of the etiology of the agent is known.

Mechanism of Toxic Action Multisystem disease secondary to inhibition of oxidative phosphorylation; Trivalent arsenic can bind to sulfhydryl groups of dihydrolipoamide thus preventing lipoamide regeneration. Hypoglycemia can occur due to inhibition of gloconeogenesis.

Adverse Reactions

Cardiovascular: Cardiotoxicity, tachycardia, acrocyanosis, Raynaud's phenomenon, congestive heart failure, myocardial depression, myocarditis, pericardial effusion/pericarditis, sinus tachycardia, tachycardia (supraventricular), vasodilation

Central nervous system: Neurotoxicity, axonopathy (peripheral), fever, hyperthermia, memory disturbance, Jarisch-Herxheimer reaction, delirium, psychosis, leukoencephalopathy

Dermatologic: Desquamation (scaling), hyperpigmentation, alopecia; airborne contact dermatitis, exanthem; dermal effects are usually not seen with inhalation exposure, lichen planus

Gastrointestinal: Gastrointestinal pathology, nausea, vomiting, metallic taste, salivation, feces discoloration (black), bloody diarrhea, abdominal pain, garlic odor

Hematologic: Bone marrow suppression, neutropenia, pancytopenia (leukopenia, thrombocytopenia, aplastic anemia)

Renal: Tubular necrosis (acute), hematuria

Miscellaneous: Basal cell carcinoma, cutaneous T-cell lymphoma

Toxicodynamics/Kinetics

Absorption: Oral, inhalation, and dermal (low)

Distribution: V_d: 0.2 L/kg

Metabolism: Reduction/oxidation reactions that interconvert arsenate and arsenite and methylation reactions to monomethyl arsonic acid and cacodylic acid

Half-life: 4-5 days

Elimination: Renal, degree of metabolism is ingestant dependent

Pregnancy Issues Inorganic arsenic is teratogenic in rats

Diagnostic Test/Procedure

Arsenic, Blood

Arsenic, Hair, Nails

Arsenic, Urine

Electrolytes, Blood

Heavy Metal Screen, Blood

Signs and Symptoms of Acute Exposure Agranulocytosis, alopecia, blindness, cough, encephalopathy, fasciculations, fever, garlic-like breath, hematuria, hemolytic anemia, hypotension, lacrimation, leukopenia, Mees' lines (on nail beds; forms at 4-6 weeks postexposure), myoglobinuria, neuritis, nystagmus, pancytopenia, paresthesia, radiopacity, seizures, stocking-glove sensory neuropathy, sweating, tachycardia, torsade de pointes, tremor

Admission Criteria/Prognosis All suspected patients should be admitted; survival >1 week after an acute exposure is usually associated with recovery, although peripheral neuropathy may be long term

Treatment Strategy

Decontamination:

Oral: Lavage (within 1 hour)/activated charcoal. Aluminum hydroxide may prevent absorption of pentavalent arsenic compounds (due to its phosphate-binding abilities), although this has not been investigated in humans. Whole bowel irrigation is effective.

Dermal: Remove contaminated clothing; wash with soap and water.

Ocular: Copious irrigation with water or saline

Supportive therapy: Succimer (at standard doses to treat lead poisoning) is probably the treatment of choice. 2,3-Dimercaptopropane sulphonate (DMPS or unithiol) is also useful to enhance this metal's elimination at a dose of 100 mg 3 times/day orally for 5 days. 2,3-Dimercaptopropane-sulphonate (DMPS), a water-soluble derivative of dimercaprol, has been recently demonstrated to prevent polyneuropathy when started within 48 hours of exposure at a dose of 5 mg/kg I.V. every 4 hours for 24 hours, and then 400 mg orally every 4 hours for 5-7 days. BAL should be utilized for severe acute exposures except for arsine gas. Penicillamine provides long-term therapy or alternatives to BAL.

Enhanced elimination: Hemodialysis clearance of arsenic ranges from 76-87 mL/minute; the clinical utility of this modality is unknown. No studies have been performed regarding hemodialysis with metal-chelate compound. Clearance of arsenic through hemodialysis is estimated to range from 76-87 mL/minute. Due to tissue binding of arsenic, this is not felt to be a useful modality for removal of arsenic. No current studies exist on the use of chelation with extracorporeal removal methods.

Reference Range Urine concentrations in nonexposed individuals ≤50 mcg/L; hair concentrations detectable 30 hours postingestion; urine ≥100 mcg/L is suggestive for chronic exposure; blood not usually helpful, although blood arsenic levels >1000 mcg/L are usually associated with fatality; background blood arsenic level is usually <1 mcg/L

Additional Information Toxic oral dose: 120-200 mg

Causes garlic odor; seafood contains arsenobetaine and arsenocholine; dietary history is important.

A radiopaque compound, but rapid absorption makes observation unlikely. Arsine gas is the most toxic, then trivalent arsenite, then pentavalent arsenate. Acute exposure can cause chronic symptoms. Bone marrow, skin, and peripheral nervous system are usual targets of chronic exposure.

Found in some homeopathic medications; Chinese herbal balls may contain from 7.8-621.3 mg of mercury and from 0.1-36.6 mg of arsenic.

TLV-TWA: 0.2 mg/m^3

Urban arsenic levels of ambient air: 20-30 ng/m^3. Arsenic levels in groundwater: 1-2 ppb; found in the earth's crust at an arsenic level of 2 ppm.

Arsenic levels in selected media: Grains: 0.22 ppm; Meat: 0.14 ppm; Seafood: 4-5 ppm; Cigarette: 1.5 ppm

Estimated daily intake of arsenic (adult): Nonsmoker: 51.5 mcg; smoker (2 packs/day): 63.5 mcg; Drinking water: 10 mcg; Food: 45 mcg

Pesticides, rodenticides, ant poisons, wood preservative, microchips, well water, seafood.

Specific References

Ahsan H, Perrin M, Rahman A, et al, "Associations Between Drinking Water and Urinary Arsenic Levels and Skin Lesions in Bangladesh," *J Occup Environ Med*, 2000, 42(12):1195-201.

Bae M, Watanabe C, Inaoka T, et al, "Arsenic in Cooked Rice in Bangladesh," *Lancet*, 2002, 360(9348):1839-40.

Belson M, Holmes A, Funk A, et al, "Cross-Sectional Exposure Assessment of Environmental Contaminants in Churchill County, Nevada," *J Toxicol Clin Toxicol*, 2003, 41(5):722.

Bourgeais AM, Avenel-Audran M, Le Bouil A, et al, "Chronic Arsenicism: 1 Case Report," *Vet Hum Toxicol*, 2001, 43(4):244.

Burgess JL, Josyula AB, Montenegro M, et al, "Low Dose Arsenic Exposure and Urinary 8-OHdG in Arizona and Sonora," *Clin Toxicol (Phila)*, 2005, 43:748.

Burgess JL, Rowland H, Josyula A, et al, "Reduction in Total Inorganic Urinary Arsenic and Toenail Arsenic With Provision of Bottled Water," *J Toxicol Clin Toxicol*, 2004, 42(5):807.

Cantrell FL, "Look What I Found! - Poison Hunting on eBay®," *Clin Toxicol*, 2005, 43(5):375-9.

Caraccio TR, McGuigan M, and Mofenson HC, "Chronic Arsenic (As) Toxicity From Chitosan® Supplement," *J Toxicol Clin Toxicol*, 2002, 40(5):644.

Chakraborti D, Mukherjee SC, Saha KC, et al, "Arsenic Toxicity From Homeopathic Treatment," *J Toxicol Clin Toxicol*, 2003, 41(7):963-7.

Emile H, Hashmonai D, Yosef E, et al, "Wood Preservative Poisoning," *Ann Emerg Med*, 2000, 35(5):S26.

Goulle JP, Mahieu L, and Kintz P, "The Murder Weapon Was Found in the Hair!" *Annale de Toxicologie Analytique*, 2005, 17(4):243-6.

Guha Mazumder DN, "Arsenic Exposure and Health Effects," *J Toxicol Clin Toxicol*, 2002, 40(4):527-8.

Hahn I, Kline SA, Howland MA, et al, "Chronic Pediatric Arsenic Poisoning From Pressure-Treated Wood Burned in a Fireplace," *J Toxicol Clin Toxicol*, 2000, 38(5):550.

Hantson P, Haufroid V, Buchet JP, et al, "Acute Arsenic Poisoning Treated by Intravenous Dimercaptosuccinic Acid (DMSA) and Combined Extrarenal Epuration Techniques," *J Toxicol Clin Toxicol*, 2003, 41(1):1-6.

Hay E, Derazon H, Eisenberg Y, et al, "Suicide by Ingestion of a CCA Wood Preservative," *J Emerg Med*, 2000, 19(2):159-63.

Hopehayn C, Bush HM, Bingcang A, et al, "Association Between Arsenic Exposure From Drinking Water and Anemia During Pregnancy," *J Occup Environ Med*, 2006, 48(6):635-43.

Horn HJ, Eicher H, Melberg W, et al, "Acute Arsenic Trioxide Poisoning: Nonserious Course of Illness Because of High Dosage Chelation Therapy," *J Toxicol Clin Toxicol*, 2002, 40(3):384-5.

Josyula AB, Rowland H, Kurzius-Spencer M, et al, "Environmental Arsenic Exposure and Sputum Metalloproteinase Concentrations," *Clin Toxicol (Phila)*, 2005, 43:753.

Kales SN, Huyck KL, and Goldman RH, "Elevated Urine Arsenic: Unspeciated Results Lead to Unnecessary Concern and Further Evaluations," *J Anal Toxicol*, 2006, 30:80-5.

Kinoshita H, Hirose Y, Tanaka T, et al, "Oral Arsenic Trioxide Poisoning and Secondary Hazard From Gastric Content," *Ann Emerg Med*, 2004, 44(6):625-7.

Kintz P, Goulle JP, Fornes P, et al, "A New Series of Hair Analyses From Napoleon Confirms Chronic Exposure to Arsenic," *J Anal Toxicol*, 2002, 26(8):584-5.

Krause E, Nussle P, Santana D, et al, "Arsenic and Lead Soil Contamination Near a Heavy Metal Refinery in the Andes Mountains," *J Toxicol Clin Toxicol*, 2003, 41(5):739.

Lamm SH, Engel A, Kruse MB, et al, "Arsenic in Drinking Water and Bladder Cancer Mortality in the United States: An Analysis Based on 133 U.S. Counties and 30 Years of Observation," *J Occup Environ Med*, 2004, 46:298-306.

Michaux I, Haufroid V, Dive A, et al, "Repetitive Endoscopy and Continuous Alkaline Gastric Irrigation in a Case of Arsenic Poisoning," *J Toxicol Clin Toxicol*, 2000, 38(5):471-6.

Midtdal K, Solberg K, Stenehjem A, et al, "Chronic Arsenic Poisoning: Probably No Effect From Chelating Therapy Using DMSA," *J Toxicol Clin Toxicol*, 2002, 40(3):385.

Morton J and Mason H, "Speciation of Arsenic Compounds in Urine From Occupationally Unexposed and Exposed Persons in the U.K. Using a Routine LC-ICP-MS Method," *J Anal Toxicol*, 2006, 30:293-301.

Mukherjee SC, Saha KC, Pati S, "Murshidabad – One of the Nine Groundwater Arsenic-Affected Districts of West Bengal, India. Part II: Dermatological, Neurological, and Obstetric Findings," *Clin Toxicol (Phila)*, 2005, 43(7):835-48.

Mycyk MB, Crulcich M, Mucha A, et al, "Exposure Assessment of Children Exposed to Arsenic in an Urban Playlot," *J Toxicol Clin Toxicol*, 2003, 41(5):737.

Nelson LS, "Toxicologic Myocardial Sensitization," *J Toxicol Clin Toxicol*, 2002, 40(7):867-79.

Ohnishi K, Yoshida H, Shigeno K, et al, "Prolongation of the QT Interval and Ventricular Tachycardia in Patients Treated With Arsenic Trioxide for Acute Promyelocytic Leukemia," *Ann Intern Med*, 2000, 133(11):881-5.

Poklis A, "A Case of Homicide by Chronic Arsenic Poisoning With Apparent Radiographic Evidence of Arsenic Administration," *J Anal Toxicol*, 2003, 27:192.

Rahman MM, Sengupta MK, Ahamed S, et al, "Murshidabad – One of the Nine Groundwater Arsenic-Affected Districts of West Bengal, India. Part I: Magnitude of Contamination and Population at Risk," *Clin Toxicol (Phila)*. 2005, 43(7):823-34.

Rusyniak DE, Furbee RB, and Kirk MA, "Thallium and Arsenic Poisoning in a Small Midwestern Town," *Ann Emerg Med*, 2002, 39(3):307-11.

Santra A, Maiti A, Das S, et al, "Hepatic Damage Caused by Chronic Arsenic Toxicity in Experimental Animals," *J Toxicol Clin Toxicol*, 2000, 38(4):395-405.

Sawyer TS, Moran D, and Lowry JA, "Accidental Ingestion of Sodium Sulfur Arsenate Proves Rapidly Fatal," *Clin Toxicol (Phila)*, 2005, 43:747.

Szinicz L, Mueckter H, Felgenhauer N, et al, "Toxicodynamic and Toxicokinetic Aspects of the Treatment of Arsenical Poisoning," *J Toxicol Clin Toxicol*, 2000, 38(2):214-6.

Vantroyen B, Heilier JF, Meulemans A, et al, "Survival After a Lethal Dose of Arsenic Trioxide," *J Toxicol Clin Toxicol*, 2004, 42(6)889-95.

Verret WJ, Chen Y, Ahmed A, et al, "A Randomized, Double-Blind Placebo-Controlled Trial Evaluating the Effects of Vitamin E and Selenium on Arsenic-Induced Skin Lesions in Bangladesh," *J Occup Environ Med*, 2005, 47:1026-35.

Wang RY, Paschal DC, Osterloh J, et al, "Urinary-Speciated Arsenic Levels From Selected U.S. Regions," *J Toxicol Clin Toxicol*, 2004, 42(5):770-1.

Wax PM, "Features and Management of Arsenic Intoxications," *J Toxicol Clin Toxicol*, 2001, 39(3):235-7.

Wax PM and Thornton CA, "Recovery From Severe Arsenic-Induced Peripheral Neuropathy With 2,3-Dimercapto-1-propanesulphonic Acid," *J Toxicol Clin Toxicol*, 2000, 38(7):777-80.

Zilker T, Lin X, and Henkelmann R, "Napoleon's Death - Was It Arsenic Poisoning?" *J Toxicol Clin Toxicol*, 2002, 40(3):277-8.

Arsine

Related Information Arsine - Patient Information Sheet

CAS Number 7784-42-1

UN Number 2188

Synonyms As_sH_3; Arsenic Trihydride; Arsenous Hydride; Arthur; Hydrogen Arsenide; SA (NATO code)

CDC Classification for Diagnosis

Biologic: No specific test is available for arsine exposure; however, exposure to arsine might be indicated by detection of elevated arsenic levels in urine (>50 mcg/L for a spot of >50 mcg for a 24-hour urine) and signs of hemolysis (eg, hemoglobinuria, anemia, or low haptoglobin).

Environmental: Detection of arsine in environmental samples, as determined by NIOSH.

CDC Case Classification

Suspected: A case in which a potentially exposed person is being evaluated by healthcare workers or public health officials for poisoning by a particular chemical agent, but no specific credible threat exists.

Probable: A clinically compatible case in which a high index of suspicion (credible threat or patient history regarding location and time) exists for arsine exposure, or an epidemiologic link exists between this case and a laboratory-confirmed case.

Confirmed: A clinically compatible case in which laboratory tests have confirmed exposure. The case can be confirmed if laboratory testing was not performed because either a predominant amount of clinical and nonspecific laboratory evidence of a particular chemical was present or a 100% certainty of the etiology of the agent is known.

Use Produced when water comes into contact with molten arsenic; used in processing of gallium chips in the semiconductor industry; also found in jewelry, lead burners, etchers, silicone chips, fertilizer makers, aniline workers

Mechanism of Toxic Action Decline of erythrocyte glutathione concentrations after arsine binds to hemoglobin

Toxicodynamics/Kinetics
Absorption: Inhalation: Well absorbed
Metabolism: To arsenic and trimethylarsine
Elimination: Excreted as arsenic in urine, feces, hair, fingernails, and by the lungs

Monitoring Parameters Renal function

Clinically Resembles Colchicine, copper sulfate, thallium phosphine

Signs and Symptoms of Acute Exposure Abdominal cramping, abdominal tenderness, asthenia, dizziness, headache, hemolysis (potent), hypotension, jaundice (2-24 hours postexposure), pulmonary edema, nausea, painless hemoglobinuria, seizures, shivering, vomiting, oliguria may occur later (urine may be colored red or green); skin may be bronzed; bone marrow depression has been reported. Monitor renal function as insufficiency or failure may occur.

Admission Criteria/Prognosis Any symptomatic patient or evidence of hemolysis should be admitted; patients may be discharged after 24 hours if the patient is asymptomatic without signs of hemolysis

Treatment Strategy
Decontamination: Remove victim immediately from source of contamination
  Dermal: Wash with soap and water
  Inhalation: Administer 100% humidified oxygen
Supportive therapy: Alkalinization of urine to prevent renal failure; exchange transfusions helpful in severe hemolysis with a free plasma hemoglobin over 1.5 g/dL; heavy metal chelators are ineffective; hypotension can be treated with intravenous crystalloid fluids (10-20 mL/kg of isotonic fluid bolus) and placement in Trendelenburg position. Dopamine or norepinephrine are vasopressors of choice.
Enhancement of elimination: Hemodialysis may be required for renal failure but will not remove the arsine-hemoglobin complex.

Environmental Persistency Up to 30 minutes

Reference Range Toxic: Blood (arsenic): >200 mcg/dL; urine: >1000 mcg/L. A plasma-free hemoglobin level >1.5 g/dL indicates the need for definitive treatment (possibly exchange transfusion).

Additional Information One of the most potent hemolytic toxins known; colorless, nonirritating at low concentrations (<2 ppm); mild garlic-like odor at higher concentrations. Immediate lethal air concentration: 250 ppm. Case fatality rate: 25%. Potential human carcinogen. Hemolysis is blocked by carbon monoxide and partially blocked by methemoglobin. One hour exposure: 5 ppm TLV-TWA: 0.05 ppm; IDLH: 6 ppm; PEL-TWA: 0.05 ppm; high volatility
If the patient is close enough to detect the odor, the ensuing exposure is significant. Some individuals may be unable to detect the mild garlic odor.
Lethal dermal dose: 2 mg of surface area/kg

Specific References
Caravati EM and Grover J, "Arsine Poisoning From Recycling of Computer Chips," *J Toxicol Clin Toxicol*, 2000, 38(5):543-4.

Goetz R, Sweeney R, Snook CP, et al, "Case Series: A New Source of Arsine," *Clin Toxicol (Phila)*, 2005, 43:746.

O'Connor AD, Kao LW, and Furbee RB, "Arsine Gas Poisoning After Occupational Exposure," *Clin Toxicol (Phila)*, 2005, 43:746.

Skinner CG, Kneewicz AA, Coon TP, et al, "Arsine Gas Exposure Presenting as Back Pain and Hematuria: A Case Series," *Clin Toxicol (Phila)*, 2005, 43:747.

Benzene

Related Information Benzene - Patient Information Sheet
CAS Number 71-43-2
UN Number 1114

Synonyms (6)-Annulene; Carbon Oil; Mineral Naphtha; Phene; Phenyl Hydride

U.S. Brand Names Polystream®

Use In lacquers, manufacture of dyes, oil cloths, varnishes, gasoline additive, natural rubber

Mechanism of Toxic Action Hematotoxicity, CNS depression, leukemogenic

Adverse Reactions

Cardiovascular: Cardiac toxicity

Central nervous system: CNS depression, euphoria, dizziness, headache

Gastrointestinal: Feces discoloration (black), burning sensation of mucous membranes

Hematologic: Leukemia (specifically acute myeloid leukemia) usually after chronic exposure, aplastic anemia, neutropenia, megaloblastic anemia, agranulocytosis, malignant lymphoma

Hepatic: Hepatotoxicity

Respiratory: Bronchial irritation, pulmonary edema

Miscellaneous: Aneuploidy induction, cutaneous T-cell lymphoma

Toxicodynamics/Kinetics

Absorption: Inhalation: Rapid (70% to 80% within first 5 minutes). Oral: 90% to 97% of dose in rodent models. Dermal: <1%; absorption rate: 0.4 mg/cm^2/hour

Metabolism: Hepatic via cytochrome P450-dependent mixed function oxidase through two detoxification pathways: 1) via glutathione to mercapturic acid, and 2) through sulfate/glucuronide conjugation. Hematotoxic metabolites include hydroquinone, phenol, muconic dialdehyde, and catechol.

Half-life: 8 hours

Elimination: Renal (<1%) and by inhalation (16% to 42%). Through dermal absorption as much as 30% can be excreted through the kidneys as phenol. Oral ingestion (rabbit model): 43% was eliminated through the lungs (1.5% as carbon dioxide), with urinary excretion accounting for 33%.

Drug Interactions Ethyl alcohol can potentiate the severity of benzene-induced hematological abnormalities

Pregnancy Issues Weak genotoxicity; can cross the placenta

Monitoring Parameters Obtain baseline complete blood count; if abnormal, perioxidase, leukocyte alkaline phosphatase, an iron level, and a reticulocyte count should be performed. Urinary levels of muconic and S-phenylmercapturic acid samples at the end of work shift are better exposure indicators for benzene than phenol. The BEI for S-phenylmercapturic acid muconic acid are 25 mcg/g creatinine and 500 mcg/g creatinine at the end of shift, respectively.

Applies to Dyes; Lacquer; Oil Cloths; Varnishes

Signs and Symptoms of Acute Exposure Aspiration, ataxia, blistering, coma (3000 ppm), cough, dementia, dizziness, lethargy, headache, hematuria, hoarseness, leukopenia, mydriasis, ototoxicity, paresthesia, seizures, tinnitus, tremors, tachycardia, erythema, vomiting (ingestion of 9 g)

Admission Criteria/Prognosis Symptomatic patients 12 hours postinhalation exposure should be admitted; patients who have ingested a significant amount of benzene should be admitted

Treatment Strategy

Decontamination:

Oral: Basic poison management. Lavage within 2 hours of ingestion with respiratory protection. Activated charcoal may be useful. Diazepam and/or phenytoin may be helpful in controlling seizures. Indomethacin has been demonstrated to decrease myelotoxicity in rodent models, but human data are lacking. Due to the possibility of arrhythmias, avoid epinephrine.

Inhalation: Move patient to fresh air; give 100% humidified oxygen

Ocular: Irrigate with copious amounts of water or saline

Dermal: Remove contaminated clothing and wash area thoroughly with soap and water

Enhancement of elimination: Although there is no experience in its use, acetylcysteine has a theoretical role in increasing glutathione stores and enhancing elimination. Hemodialysis is not effective.

Asymptomatic patients may be discharged after 6- to 12-hour observation. Patients with urinary phenol levels >75 mg/L require periodic CBC (at least monthly for 3 months) to monitor bone marrow toxicity

Environmental Persistency Half-life in air: 2-21 days
Half life in groundwater: 10 days to 2 years
Half-life in soil: 5-16 days
Bioconcentration in aquatic organisms is low

Personal Protection Equipment
Positive pressure, self-contained breathing apparatus at air concentrations >100 ppm
Chemical-protective clothing is recommended with liquid contact but not recommended for vapor contact
Wear safety goggles and face shield

Reference Range
Normal urine phenol level <10 mg/L, blood benzene level of 0.2 mg/L is consistent with exposure of 25 ppm as is a urinary phenol level of 200 mg/L

Additional Information Estimated oral lethal dose: 1 mL/kg and lethal (rapidly) air concentration over 20,000 ppm. Serious health effects can occur with one hour exposure at air concentrations >150 ppm

Daily benzene intake of a smoker (32 cigarettes daily) is $\sim$10 times (1.8 mg/day) that of a nonsmoker. Median benzene levels in homes with smokers is $\sim$50% greater than that of nonsmokers (3.3 ppb vs 2.2 ppb). Atmospheric half-life of benzene is $\sim$5.6 days. In water, it is $\sim$17 days, although the half-life of its hydroxyl radicals in water may be 8-9 months.

Benzene was substantially reduced in automotive gasoline after 1995 (see Gasoline monograph). It has been estimated that benzene exposure to pumping gasoline at service stations is $\sim$1 ppm. Ambient median benzene levels in urban areas is $\sim$12.6 ppb, while California's median benzene levels are much lower (3.3 ppb) due to stricter gasoline requirements.

Biomarkers indicative for significant benzene ($<4000/mm^3$), low erythrocyte counts ($<4,000,000/m^3$), or elevated leukocyte alkaline phosphatase levels.

TLV-TWA: 0.5 ppm; IDLH: 500 ppm; STEL: 1 ppm

Gas density: 2.8

Specific gravity: 0.88

Slightly water soluble

Odor threshold: 1.5-5 ppm

Median level of benzene found in snow samples (near Denver, Colorado): 0.02 mcg/L

Blood testing: Use lavender top (EDTA) tube and refrigerate (minimum volume 7 mL)

Benzene is found in the air from emissions from burning coal and oil, gasoline service stations, and motor vehicle exhaust. Average ambient atmospheric benzene levels in urban areas range from 1 to 5.4 parts per billion (ppb). Benzene is a carcinogen and has caused cancer in workers exposed to high levels from workplace air. Benzene can form at the ppb level in some beverages that contain both benzoate salts and ascorbic acid (vitamin C) or erythorbic acid (a closely related substance (isomer) also known as d-ascorbic acid). Elevated temperatures and light can stimulate benzene formation in the presence of benzoate salts and vitamin C, while sugar and EDTA salts inhibit benzene formation.

Specific References

Buffler PA, Kelsh M, Chapman P, et al, "Primary Brain Tumor Mortality at a Petroleum Exploration and Extraction Research Facility," *J Occup Environ Med*, 2004, 46:257-270.

Olmos V, Lenzken SC, Lopez CM, et al, "High Performance Liquid Chromatography Method for Urinary Trans, Trans-Muconic Acid. Application to Environmental Exposure to Benzene," *J Anal Toxicol*, 2006, 30:258-61.

Wennborg H, Magnusson LL, Bonde JP, et al, "Congenital Malformations Related to Maternal Exposure to Specific Agents in Biomedical Research Laboratories," *J Occup Environ Med*, 2005, 47(1):11-9.

Wong O, "Is There a Causal Relationship Between Exposure to Diesel Exhaust and Multiple Myeloma?" *Toxicol Rev*, 2003, 22(2):91-102.

Bromobenzylcyanide

CAS Number 5798-79-8, 31938-07-05
UN Number 1694
Synonyms Bromobenzylnitrile; CA
Military Classification Lacrimator, developed near end of WWI - rarely used today
Mechanism of Toxic Action Lacrimator - irritation due to the active halogen group reacting transiently with sulfhydryl groups
Toxicodynamics/Kinetics
Onset of effect: Rapid
Duration: Usually resolves in 30 minutes, skin effects may last for 2-3 hours
Absorption: Not absorbed
Pregnancy Issues No teratogenic effects
Clinically Resembles Hydrogen chloride, ammonia, chlorine gas
Signs and Symptoms of Acute Exposure Agitation, bronchospasm, cough, erythema, eye pain, lacrimation, laryngospasm (may be delayed for 2 days; such patients may require PEEP, in addition to anticipating the need for airway protection), nausea, pharyngitis, rhinorrhea, skin irritation, sneezing, tearing, vomiting
Treatment Strategy
Decontamination: Move patient rapidly to fresh air and monitor for wheezing. Administer 100% humidified oxygen. Personnel should avoid contaminating themselves. Wash skin with soap and at least two liters of water. Remove contaminated clothing; do not pull clothing over head (cut clothing instead).
Ocular: Avoid rubbing eyes as it may prolong effect. Copious irrigation with water or saline is needed.
Supportive therapy: Solution of antacid (magnesium hydroxide - aluminum hydroxide - simethicone) gently blotted topically over exposed facial areas (avoiding eyes) may help resolve symptoms. Bronchospasm can be treated with inhaled beta$_2$ agonists.
Environmental Persistency Liquid can persist for 2 days
Additional Information Yellowish white crystals or brown, oily liquid, poorly soluble in water
Action on metal: Corrosive (except to lead); reaction with iron may be explosive
Vapor density (air): 6.7
Liquid density: 1.47 at 20°C
Decomposes at 242°C to hydrobromic acid and dicyanostilbene
Median incapacitating dose: 30 mg/min/m^3
Median lethal dose (in enclosed areas): 8000-11,000 mg/min/m^3
Odor of sour fruit

Carbon Monoxide

CAS Number 630-08-0
UN Number 1016
Synonyms Carbon Oxide; Carbonic Oxide; CO; Exhaust Gas; Flue Gas
Commonly Includes Auto exhaust, by-product of methylene chloride; produced in a closed space fire due to incomplete combustion of materials
Mechanism of Toxic Action Causes tissue hypoxia and inhibition of cellular respiration; binds to myoglobin, cytochrome a, a$_3$ and also increases xanthine oxidase and nitric oxide production resulting in oxidative stress; may cause cerebral glutamate elevation along with lipid peroxidation and cerebral edema
Adverse Reactions
Cardiovascular: Cardiac abnormalities most pronounced, syncope, sinus bradycardia, cardiomyopathy, cardiomegaly, chest pain, palpitations, sinus tachycardia

Central nervous system: CNS effects, acute neurological abnormalities including dementia, drowsiness, coma, seizures, headache, confabulation, Parkinson-like syndrome, memory disturbance, depression, psychosis, bilateral putaminal involvement, leukoencephalopathy Delayed neurological sequelae observed within 40 days of significant carbon monoxide poisoning (12% to 43% incidence) including disorientation, bradykinesia, chorea (extrapyramidal), equilibrium disturbances, apathy, cogwheel rigidity, aphasia, incontinence, personality changes, short-term memory deficit, seizure disorders, and chronic headaches

Dermatologic: Bullous eruptions

Neuromuscular & skeletal: Chorea, apraxia, neuropathy (peripheral), rhabdomyolysis

Ocular: Cortical blindness, visual evoked potential abnormalities

Otic: Deafness

Toxicodynamics/Kinetics

Absorption: Readily through the lungs; does not accumulate over time

Half-life: 5-6 hours in room air; 30-90 minutes in 100% oxygen; 30 minutes in hyperbaric oxygen; 100% oxygen in children: 44 minutes

Elimination: Through the lungs.

Monitoring Parameters Arterial blood gases, carboxyhemoglobin

Clinically Resembles Cyanide, hydrogen sulfide, sodium azide

Signs and Symptoms of Acute Exposure

Acute effects of carbon monoxide poisoning (adapted from Ernst A and Zibrak JD, "Carbon Monoxide Poisoning," *N Engl J Med*, 1998, 339(22):1603-8): Abdominal pain and muscle cramping (5%); chest pain (9%); concentration difficulties, confusion, or disorientation (43%); loss of consciousness (6%); nausea and vomiting (47%); shortness of breath (usually associated with hyperventilation) (40%); throbbing, bitemporal headache (91%); dizziness or vertigo (77%); visual changes (25%); weakness primarily in the legs (53%)

Other (<5%): Arrhythmias, AV block, bradycardia, cortical blindness in children, fever, hallucinations, hearing loss (sensorineural), hot flashes, hypotension, hypothermia, muscle necrosis, myoglobinuria, tinnitus

Other effects commonly seen are lethargy, mydriasis, nystagmus, tachycardia

Carbon monoxide presentation in 140 children (from Mathieu D and Mathieu-Nolf M, *J Toxicol Clin Toxicol*, 2001, 39:266): Abnormal extensor plantar response (8%); cerebellar impairment (8%); flaccidity (17%); headache (37%); hyperreflexia (42%); lethargy/coma (19%); loss of consciousness (37%); seizures (5%); vomiting (22%)

Admission Criteria/Prognosis Acute neurologic symptoms, abnormal ECG, metabolic acidosis, chest pain, rhabdomyolysis, carboxyhemoglobin levels >25%, or if patient remains symptomatic after 4 hours on 100% oxygen, should be considered for admission; $\sim^2/_3$ of comatose patients will recover

Treatment Strategy

Decontamination: 100% humidified oxygen by nonrebreather face mask until asymptomatic if patient is not pregnant. If pregnant, the patient should be on 100% nonrebreather face mask for 5 times the length of time needed for the carboxyhemoglobin level to be <5%. Continue therapy until carboxyhemoglobin level is <5%. Hyperbaric oxygen for patients with acute neurotoxicity, patients with angina, maternal carboxyhemoglobin of 15% in pregnant patients, or asymptomatic carboxyhemoglobin of 25%. Do not consider hypothermia or exchange transfusion. Hyperbaric oxygen should be administered at 2.8 ATA within 6 hours of exposure to carbon monoxide for maximum benefit.

Supportive therapy: Dopamine (17 mcg/kg/minute) has been shown in a case report to reverse carbon monoxide-induced blindness.

Rescuer Contamination None

Environmental Persistency Short in atmosphere and water

Personal Protection Equipment

Exposure to CO at air concentrations >350 ppm require use of a supplied air respirator or self-contained breathing apparatus (SCBA)

Air concentrations >875 ppm require use of supplied air respirator or SCBA operated in continuous flow mode. If CO air concentrations exceed 1500 ppm, a supplied air respirator or SCBA with a full face mask operated in continuous flow mode with an auxiliary SCBA should be utilized.

Reference Range Carboxyhemoglobin level by CO-Oximeter™: Endogenous level: ≤0.65; smokers may have from 3% to 8%. Severe symptoms may start at 10%; >35% is associated with fatalities from acute exposure; BEI is <8%; venous and arterial samples are equivalent

Additional Information Colorless, odorless gas. Delayed neurotoxicity consisting of short-term memory deficit, ataxia, cognitive impairment may occur 2-3 weeks after an acute exposure. Parkinson syndrome may occur. Basal ganglia/potamen volume reduction can occur after CO exposure resulting in impairment in verbal memory and mental processing speed.

Carbon monoxide detectors are usually set to alarm at 85 decibels at indoor carbon monoxide air concentrations of 70 ppm (within 189 minutes), 150 ppm (within 50 minutes) or 400 ppm (within 15 minutes). Essentially, these concentrations, with corresponding time exposure, will result in a carboxyhemoglobin level of 10% during periods of heavy exertion. A carbon monoxide detector shall operate at or below the plotted limits for the 10% COHb curve as shown in the figure. If the detector employs a variable sensitivity setting, test measurements are to be made at maximum and minimum settings. For this test, three carbon monoxide concentrations (100, 200, 400 ppm) are to be used as specified in the figure.

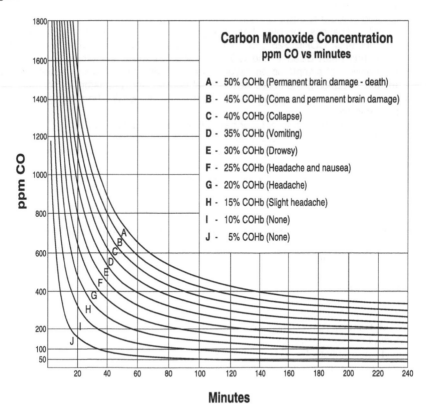

Carbon Monoxide Concentration
ppm CO vs minutes

A - 50% COHb (Permanent brain damage - death)
B - 45% COHb (Coma and permanent brain damage)
C - 40% COHb (Collapse)
D - 35% COHb (Vomiting)
E - 30% COHb (Drowsy)
F - 25% COHb (Headache and nausea)
G - 20% COHb (Headache)
H - 15% COHb (Slight headache)
I - 10% COHb (None)
J - 5% COHb (None)

Minutes

According to estimates from the Bureau of Labor statistics, carbon monoxide accounted for 867 nonfatal work-related carbon monoxide exposures in private industry and 32 work-related deaths in 1992.

TLV-TWA: 50 ppm; IDLH: 1500 ppm; PEL-TWA: 35 ppm

Carbon monoxide intoxication may also result from methylene chloride exposure.

Half-life for carboxyhemoglobin elimination is ~27.5 minutes with use of an inflatable portable hyperbaric chamber (modified Ganow bag) at 1.58 ATA.

Automobile exhaust may contain as high as 7% carbon monoxide.

Occupations particularly at risk from exposure to carbon monoxide: Blast furnace operators; bus drivers; carbide manufacturers; cooks/bakers; fire fighters; formaldehyde manufacturers; garage mechanics; iron/steel foundry (cupola); kraft paper pulp mills (Kraft recovery furnaces); lead molders; liners; operators of snow melting machines; petroleum-refining plants (catalytic cracking units); pulmonary function test practitioners; skating rinks (from Zamboni ice cleaners); tollway booth collectors; traffic policeman; warehouse storage and loading facilities (forklift operators); welders

Xenobiotics in which carbon monoxide is a metabolic by-product:
- Methylene chloride (peak of 50% COHb described in humans)
- Dibromomethane (peak of 27% COHb in rodents)*
- Diiodomethane (peak of 14.2% COHb described postingestion in a girl, 20 months of age)
- Bromochloromethane (peak of 11% COHb in rodents)*
- 1,3-Benzodioxoles*
- Tetrahalomethane (?)*
- Dibromochlormethane*

  (?) = Possible

  *Noted in rodents only

Suggested indications for hyperbaric oxygen include: Coma or any period of loss of consciousness; COHb >25% or >15% if pregnant; signs of cardiac abnormality; history of ischemic heart disease and COHb >20%; recurrent symptoms for up to 3 weeks; nonresolution of symptoms on oxygen after 6 hours; neurologic symptoms other than a transient headache; metabolic acidosis; abnormal and persistent psychometric testing; muscle necrosis, pulmonary edema, dysrhythmias, EKG changes, brain changes on neuroimaging

Risk of CO poisoning due to outdoor generators can be minimized by:

Placement of outdoor generator over 8 feet from building

Locate generators on the opposite side of building away from window air conditioners

Avoid connecting generator to a central electric panel

Celebrities in whom carbon monoxide was involved in their deaths: Ron Luciano (baseball umpire) 1995; Kevin Carter (photojournalist) 1994; Vitas Gerulaitis (tennis player) 1994; Jug McSpadden (golfer); Thelma Todd (actress) 1935; Dan White (murderer) 1985; Don Wilson (baseball pitcher) 1975; Emile Zola (novelist) 1902; Flora Disney (1938); Jimmy Campbell (bluegrass fiddle player) 2003; Rosey Nix Adams (daughter of June Carter Cash) 2003; Steve Neal (political writer, Chicago) 2004; Nick and Mary Yankovic (parents of "Weird Al" Yankovic) 2004; Zurab Zhvania (Prime Minister of Georgia) 2005; William Inge (playwright), 1973.

Specific References

Aslan S, Karcioglu O, Bilge F, et al, "Post-interval Syndrome After Carbon Monoxide Poisoning," *Vet Hum Toxicol*, 2004, 46(4):183-5.

Bayer M, Hanoian A, and Caperino CL, "A Documented Association of Poison Control Center Media Interactions and Carbon Monoxide Calls," *J Toxicol Clin Toxicol*, 2003, 41(5):708.

Below E and Lignitz E, "Cases of Fatal Poisoning in Post-mortem Examinations at the Institute of Forensic Medicine in Greifswald – Analysis of Five Decades of Post-mortems," *Forensic Sci Int*, 2003, 23;133(1-2):125-31.

Bianchini KJ, Houston RJ, Greve KW, et al, "Malingered Neurocognitive Dysfunction in Neurotoxic Exposure: An Application of the Slick Criteria," *J Occup Environ Med*, 2003, 45(10):1087-99.

Blumenthal I, "Carbon Monoxide Poisoning," *J Rog Soc Med*, 2001, 94:270-2.

Brent RL, "Environmental Causes of Human Congenital Malformations: The Pediatrician's Role in Dealing With These Complex Clinical Problems Caused by a Multiplicity of Environmental and Genetic Factors," *Pediatrics*, 2004, 113(4 Suppl):957-68.

Centers for Disease Control and Prevention (CDC), "Carbon Monoxide Poisoning After Hurricane Katrina - Alabama, Louisiana, and Missisippi, August-September 2005," *MMWR Morb Mortal Wkly Rep*, 2005, 54(39):996-8.

Centers for Disease Control and Prevention (CDC), "Carbon Monoxide Poisonings After Two Major Hurricanes - Alabama and Texas - August-October 2005," *MMWR Morb Mortal Wkly Rep*, 2006, 55(9):236-9.

Centers for Disease Control and Prevention (CDC), "Carbon Monoxide Poisonings Resulting From Open Air Exposures to Operating Motorboats - Lake Havasu City, Arizona, 2003," *MMWR Morb Mortal Wkly Rep*, 2004, 53(15):314-18.

Centers for Disease Control and Prevention (CDC), "Carbon Monoxide Release and Poisonings Attributed to Underground Utility Cable Fires - New York, January 2000-June 2004," *MMWR Morb Mortal Wkly Rep*, 2004, 53(39):920-2.

Centers for Disease Control and Prevention (CDC), "Use of Carbon Monoxide Alarms to Prevent Poisonings During a Power Outage - North Carolina, December 2002," *MMWR Morb Mortal Wkly Rep*, 2004, 53(9):189-92.

Centers for Disease Control and Prevention (CDC), "Unintentional Non-fire-related Carbon Monoxide Exposures – United States, 2001-2003," *MMWR Morb Mortal Wkly Rep*, 2005, 21;54(2):36-9.

Chang SJ, Shih TS, Chou TC, et al, "Electrocardiographic Abnormality for Workers Exposed to Carbon Disulfide at a Viscose Rayon Plant," *J Occup Environ Med*, 2006, 48:394-9.

Czogala J and Goniewicz ML, "The Complex Analytical Method for Assessment of Passive Smokers' Exposure to Carbon Monoxide," *J Anal Toxicol*, 2005, 29(8):830-4.

Dales R, "Ambient Carbon Monoxide May Influence Heart Rate Variability in Subjects With Coronary Artery Disease," *J Occup Environ Med*, 2004, 46(12):1217-21.

Delgado J, McKay C, Frankel J, et al, "Potential Carbon Monoxide Exposures: The Relationship Between Call Origin and Carboxyhemoglobin Levels," *Clin Toxicol (Phila)*, 2005, 43:718.

Domachevsky L, Adir Y, Grupper M, et al, "Hyperbaric Oxygen in the Treatment of Carbon Monoxide Poisoning," *Clin Toxicol (Phila)*, 2005, 43(3):181-8 (review).

Dueñas-Laita A, Pèrez-Castrillòn JL, Martin-Escudero JC, et al, "Study of Some Controversies in Carbon Monoxide Poisoning," *Clin Toxicol (Phila)*, 2005, 43:755.

Erdogan MS, Islam SS, Chaudhari A, et al, "Occupational Carbon Monoxide Poisoning Among West Virginia Workers' Compensation Claims: Diagnosis, Treatment Duration, and Utilization," *J Occup Environ Med*, 2004, 46(6):577-83.

Frankel J, Delgado J, Adamcewicz M, et al, "Carbon Monoxide Poisoning in Connectivut: An Analysis of Three Databases," *Clin Toxicol (Phila)*, 2005, 43:718.

Gupta A, Pasquale-Styles MA, Hepler BR, et al, "Apparent Suicidal Carbon Monoxide Poisonings With Concomitant Prescription Drug Overdoses," *J Anal Toxicol*, 2005, 29:744-9.

Hampson NB, "Trends in the Incidence of Carbon Monoxide Poisoning in the United States," *Am J Emerg Med*, 2005, 23(7):838-41.

Hampson NB and Zmaeff JL, "Carbon Monoxide Poisoning From Portable Electric Generators," *J Am Coll Cardiol*, 2005, 28:123-5.

Henry CR, Satran D, Adkinson CD, et al, "Myocardial Injury Associated With Carbon Monoxide Poisoning Predicts Long-Term Mortality," *Acad Emerg Med*, 2004, 11(5):469.

Hoffman RJ and Hoffman RS, "Effect of Mandated Residential Carbon Monoxide Detector Use on the Morbidity of Reported Cases," *Clin Toxicol (Phila)*, 2005, 43:703.

Hoffman RJ and Hoffman RS, "Mandatory Reporting of Carbon Monoxide Poisoning by Health Care Providers: Effect on Poison Center Calls," *Clin Toxicol (Phila)*, 2005, 43:719.

Holstege CP, Baer AB, Eldridge DL, et al, "Case Series of Elevated Troponin I Following Carbon Monoxide Poisoning," *J Toxicol Clin Toxicol*, 2004, 42(5):742.

Horton DK, Berkowitz Z, and Kaye WE, "Secondary Contamination of ED Personnel From Hazardous Materials Events, 1995-2001," *Am J Emerg Med*, 2003, 21(3):199-204.

Jenkins AJ, Homer CD, Engelhart DA, et al, "Carbon Monoxide-Related Deaths in Cuyahoga County, Ohio, 1988 to 1998," *J Anal Toxicol*, 2003, 27(3):178-9.

Johnson-Arbor KK and McKay CA, "Prolonged Increases in Troponin T After Carbon Monoxide Poisoning," *Clin Toxicol (Phila)*, 2005, 43:757.

Kales SN and Christiani DC, "Acute Chemical Emergencies," *N Engl J Med*, 2004, 350(8): 800-8.

Kao LW and Nanagas KA, "Carbon Monoxide Poisoning," *Emerg Med Clin N Am*, 2004, 22:985-1018.

Kao LW and Nanagas KA, "Carbon Monoxide Poisoning," *Med Clin N Am*, 2005, 89(6):1161-94 (review).

Klein KR, White SR, Herzog P, et al, "Demand for PCC Services "Surged" During the 2003 Blackout," *J Toxicol Clin Toxicol*, 2004, 42(5):814-5.

Laakso O, Haapla M, Kuitunen T, et al, "Screening of Exhaled Breath by Low-Resolution Multicomponent FT-IR Spectrometry in Patients Attending Emergency Departments," *J Anal Toxicol*, 2004, 28(2):111-7.

Lavonas E, Tomaszewski C, Kerns W, et al, "Epidemic Carbon Monoxide Poisoning Despite a CO Alarm Law," *J Toxicol Clin Toxicol*, 2003, 41(5):711-2.

Lewis RJ, Johnson RD, and Canfield DV, "An Accurate Method for the Determination of Carboxyhemoglobin in Postmortem Blood Using GC-TCD," *J Anal Toxicol*, 2003, 27(3):182.

Maready E Jr, Holstege C, Brady W, et al, "Electrocardiographic Abnormality in Carbon Monoxide-Poisoned Patients," *Ann Emerg Med*, 2004, 44:S92.

Ong JR, Hou SW, Shu HT, et al, "Diagnostic Pitfall: Carbon Monoxide Poisoning Mimicking Hyperventilation Syndrome," *Am J Emerg Med*, 2005, 23(7):903-4.

Petersen HW and Windberg CN, "Comparison of Measurements of Carboxymyoglobin in Muscles in Relation to Carbon Monoxide in Blood," *J Anal Toxicol*, 2006, 30:143.

Powell A, Eberhardt M, Bonfante G, et al, "Noninvasive Measurement of Carbon Monoxide Levels in Patients With Headaches," *Ann Emerg Med*, 2004, 44:S90.

Pulsipher DT, Hopkins RO, Weaver LK, "Basal Ganglia Volumes Following CO Poisoning: A Prospective Longitudinal Study," *Undersea Hyperb Med*, 2006, 33;245-56.

Sam-Lai NF, Saviuc P, and Danel V, "Carbon Monoxide Poisoning Monitoring Network: A Five-Year Experience of Household Poisonings in Two French Regions," *J Toxicol Clin Toxicol*, 2003, 41(4):349-53.

Sarmanaev SKh, Samolova RG, and Aidarova LF, "Epidemiology of Carbon Monoxide (CM) Poisonings in UFA," *J Toxicol Clin Toxicol*, 2003, 41(5):713.

Satran D, Henry CR, Adkinson C, et al, "Cardiovascular Manifestations of Moderate to Severe Carbon Monoxide Poisoning," *J Am Coll Cardiol*, 2005 45(9):1513-6.

Schwartz L, Martinez L, Louie J, et al, "An Evaluation of a Carbon Monoxide Poisoning Education Program," *Clin Toxicol (Phila)*, 2005, 43:703.

Stefanidou M and Athanaselis S, "Toxicological Aspects of Fire," *Vet Hum Toxicol*, 2004, 46(4):196-8.

Thomassen O, Brattebo G, and Rostrup M, "Carbon Monoxide Poisoning While Using a Small Cooking Stove in a Tent," *Am J Emerg Med*, 2004, 22(3):204-6.

Tomaszewski C, Lavonas E, Kerns R, et al, "Effect of a Carbon Monoxide Alarm Regulation on CO Poisoning," *J Toxicol Clin Toxicol*, 2003, 41(5):710.

"Use of Carbon Monoxide Alarms to Prevent Poisonings During a Power Outage - North Carolina, December 2002," *MMWR Morb Mortal Wkly Rep*, 2004, 53(9):189-92.

Wills B and Erickson T, "Drug- and Toxin-Associated Seizures," *Med Clin North Am*, 2005, 89(6):1297-321 (review).

Wolowich WR, Hadley CM, Kelley MT, et al, "Plasma Salicylate from Methyl Salicylate Cream Compared to Oil of Wintergreen," *J Toxicol Clin Toxicol*, 2003, 41(4):355-8.

Wood DM, Dargan PI, and Jones AL, "Poisoned Patients as Potential Organ Donors, A Postal Survey of Transplant Centres and Intensive Care Units," *J Toxicol Clin Toxicol*, 2003, 41(5):651.

Yeoh MJ and Braitberg G, "Carbon Monoxide and Cyanide Poisoning in Fire Related Deaths in Victoria, Australia," *J Toxicol Clin Toxicol*, 2004, 42(6):855-63.

Carfentanil Citrate

CAS Number 59708-52-0; 61380-27-6

Use Highly potent opioid analgesic usually used in veterinary medicine to immobilize large animals; used to study opiate receptors in brain imaging (research)

Mechanism of Toxic Action Respiratory and central nervous system; depression through opiate receptors

Toxicodynamics/Kinetics Duration of action: 2-24 hours

Stability 59°F to 86°F; protect from prolonged excessive heat

Signs and Symptoms of Acute Exposure Apnea, chest pain, coma, confusion, depression, dyspnea, exfoliative dermatitis, flatulence, hiccups, hypertension, hypertonia, hypotension, laryngospasm, pseudotumor cerebri, respiratory depression (especially with doses >200 mcg), seizures

Treatment Strategy Supportive therapy: Administer 100% oxygen; naloxone (0.1 mg/kg in children up to 20 kg body weight) or 2 mg in adults should be given continuous. Naloxone drip at $^2/_3$ of awakening dose may need to be given.

Additional Information 8000-10,000 times more potent than morphine and 100 times as potent as fentanyl; may be a chemical submissive agent when inhaled. Similar to opiate toxidrome.

Specific References

Bencherif B, Fuchs PN, Sheth R, et al, "Pain Activation of Human Supraspinal Opioid Pathways as Demonstrated by [11C]-Carfentanil and Positron Emission Tomography (PET)," *Pain*, 2002, 99(3):589-98.

Brown D and Baker P, "Moscow Gas Likely A Potent Narcotic: Drug Normally Used to Subdue Big Game," *Washington Post*, Saturday, November 9, 2002; Page A12. Available at: http/www.washingtonpost.com (archives). Accessed December 27, 2002.

Wang L and Bernert JT, "Analysis of 13 Fentanils, Including Sufentanil and Carfentanil, in Human Urine by Liquid Chromatography-Atmospheric-Pressure-Ionization-Tandem Mass Spectrometry," *J Anal Toxicol*, 2006, 30:335.

Wax PM, Becker CE, and Curry SC, "Unexpected 'Gas' Casualties in Moscow: A Medical Toxicology Perspective," *Ann Emerg Med*, 2003, 41(5):700-5.

Zilker TH, Pfab R, Eyer F, et al, "The Mystery About the Gas Used for the Release of the Hostages in the Moscow Musical Theater," *J Toxicol Clin Toxicol*, 2003, 41(5):661.

Chemical Warfare Agents

Synonyms Chemical Terrorism; Warfare Agents; Chemicals

Abstract Chemical warfare agents have been known for centuries. Their use in World War I, Iraqi use against the Kurds, and the Tokyo subway terrorists' attack is well documented. Although military personnel are relatively well aware of chemical warfare agents, most civilian medical communities are not adequately aware or prepared. Due to the September 2001 World Trade Center catastrophe, various anthrax attacks in the United States, and repeated terrorists threats, public and healthcare awareness to chemical and biological weapons has increased.

A United Nations report from 1969 defines chemical warfare agents as "chemical substances, whether gaseous, liquid, or solid, which might be employed because of their direct toxic effects on man, animals, and plants." The Chemical Weapons Convention defines chemical weapons as including not only toxic chemicals but also ammunition and equipment for their dispersal. Toxic chemicals are defined as "any chemical which, through its chemical effect on living processes, may cause death, temporary loss of performance, or permanent injury to people and animals." Plants are not mentioned in this context.

Additional Information Chemical warfare agents are categorized into four major classes. The table lists various chemical agents, their codes, persistence, physical state, and site of action.

309

1. **Nerve agents:** Their action is similar to organophosphates (ie, they inhibit acetylcholinesterase, resulting in accumulation of acetylcholine and overstimulation of muscarinic and nicotinic receptors). They inhibit both acetylcholinesterase and pseudocholinesterase. See Acetylcholinesterase, Red Cell, and Serum. Examples of these agents include tabun (GA, ethyl N-dimethylphophoroamidocyanidate), sarin (GB or isopropyl methylphosphonofluoridate) and soman (GD or pinacolyl methylphosphonofluoridate), and VX (methylphosphonothioic acid S-(2-(bis(1-methyl-ethyl-amino-ethyl) O-ethyl ester or V).

The clinical symptoms are associated with multiorgan effects. Symptoms include miosis, rhinorrhea, bronchial secretion, diarrhea, bronchospasm, bradycardia, muscle twitching, weakness, and paralysis.

Treatment includes injection of atropine, possibly in combination with pralidoxime chloride (protopam chloride; 2-PAMCl) an oxime. Oximes attach to the nerve agent that is inhibiting the cholinesterase and break the agent-enzyme bond to restore the normal activity of the enzyme.

Routine toxicology laboratory testing does not identify nerve agents. Measurement of RBC or serum cholinesterase has been used as an index of severity of toxicity of nerve agents, but with poor correlation. Erythrocyte enzyme activity is more sensitive to acute nerve agent exposure than is plasma enzyme activity. However, this approach is not always reliable. In the Tokyo subway sarin attack, miosis and RBC acetylcholinesterase activity were compared; miosis was a more sensitive index to sarin exposure than was RBC acetylcholinesterase activity.

Cholinesterase assay is available on most automated chemistry instruments.

2. **Vesicants or blistering agents:** These agents, on contact with body surfaces, cause burns and blisters. The most susceptible organs are eyes, mucous membranes, lungs, and skin. Examples include HD - sulphur mustard (Yperite), HN - nitrogen mustard, and L - Lewisite (arsenical vesicants may be used in a mixture with HD). Sulphur mustard was used as a weapon in World War I. Mechanism of action includes irreversibly alkylating DNA, RNA, and proteins and, thus, disrupting cell function, causing cell death. Actively dividing cells, such as epithelium and hematopoietic cells, are more severely affected. In fact, due to the alkylating properties of these agents, nitrogen mustard was one of the first chemotherapeutic agents. These agents also cause glutathione depletion, resulting in diminished protective capacity in cells from oxygen-free radicals.

Clinical symptoms include burning skin and eyes, damage of upper respiratory airway mucosa, and gastrointestinal irritation. There may be a latent period of 4-6 hours. Diagnosis is based on history and clinical symptoms. No specific treatment is available. Treatment is supportive and most patients recover. Several agents including antioxidants, anti-inflammatory agents, and sulfur-group donors (glutathione and N-acetylcysteine) are under investigation.

No laboratory tests are available to detect these agents. Leukocytosis correlates roughly with the extent of tissue injury, primarily to skin or pulmonary tissue. A leukocyte count ≤ 500 is a sign of an unfavorable prognosis.

Organisms commonly invade the damaged airway tissue. Sputum Gram stain and culture should be done for identification of the specific organism. Although gastrointestinal bleeding is unusual, declining hematocrit values should prompt serial analyses of stool for occult blood.

There is no clinical laboratory test for mustard in blood or tissue. A method for analysis of urine for thiodiglycol, a metabolite of mustard, has been described.

3. **Pulmonary intoxicants and irritants:** These agents include perfluroisobutylene (PFIB), phosgene, diphosgene, and chlorine. They cause pulmonary edema and irritation of nose, larynx, pharynx, trachea, and bronchi. Protein concentration in bronchoalveolar lavage is increased and may be useful as an indicator of the extent of edema.

4. **Cyanides - cellular asphyxiants:** These include cyanide, hydrogen cyanide, and cyanogen chloride. The mechanism of action includes binding of cyanide to iron in cytochrome oxidase, thus inhibiting intracellular oxygen utilization. After exposure to high concentrations, signs and symptoms include seizures, and respiratory and cardiac arrest.

Sodium nitrite and sodium thiosulfate are effective antidotes. Due to high LC50 and high volatility, cyanide is not considered a major threat for mass casualties. The French used 4000 tons of cyanide in World War I without significant success. Activated charcoal in a chemical protective mask effectively absorbs cyanide.

Laboratory findings include increased blood cyanide, metabolic acidosis with high concentration of lactic acid, increased anion gap, and increased oxygen content. Blood cyanide levels are generally not available and other laboratory findings are nonspecific.

Chemical Warfare Agents				
Chemical Agent	Code	Persistence (Hours) at 20°C to 32°C	State at 20°C	Site and Mechanism of Action
Nerve agents				
Tabun	GA	24-48	Liquid	Cholinesterase inhibition
Sarin	GB	1-24	Liquid	Cholinesterase inhibition
Soman	GD	24-48	Liquid	Cholinesterase inhibition
VX	VX	240-720	Liquid	Cholinesterase inhibition
Vesicants				
Distilled mustard	HD	24-48	Liquid	Blistering, alkylating
Nitrogen mustard	HN-1,2,3	24-72	Liquid	Blistering, alkylating
Lewisite	L	18-36	Liquid	Blistering, irritating
Mustard Lewisite	HL	24-36	Liquid	Blistering, irritating
Pulmonary agents				
Phosgene	CG	0.5-1	Gas	Choking, lung damaging
Diphosgene	DP	0.5-3	Liquid	Choking, lung damaging
Chlorine	CL	0.5-2	Gas	Choking, lung damaging
Cyanides				
Hydrogen cyanide	AC	0.25-0.5	Gas	Cytochrome oxidase inhibition
Cyanogen chloride	CK	0.25-0.5	Gas	Cytochrome oxidase inhibition

Specific References

Goozner B, Lutwick LI, and Bourke E, "Chemical Terrorism: A Primer for 2002," *J Assoc Acad Minor Phys*, 2002, 13(1):14-8.

Jortani SA, Snyder JW, and Valdes R Jr, "The Role of the Clinical Laboratory in Managing Chemical or Biological Terrorism," *Clin Chem*, 2000, 46(12):1883-93.

Knudson GB, Elliott TB, and Brook I, "Nuclear, Biological, and Chemical Combined Injuries and Countermeasures on the Battlefield," *Mil Med*, 2002, 167(2 Suppl):95-7.

Lawler A, "Antiterrorism Programs: The Unthinkable Becomes Real for a Horrified World," *Science*, 2001, 293(5538):2182-5.

McCauley LA, Rischitelli G, Lambert WE, et al, "Symptoms of Gulf War Veterans Possibly Exposed to Organophosphate Chemical Warfare Agents at Khamisiyah, Iraq," *Int J Occup Environ Health*, 2001, 7(2):79-89.

Smith JR, Shih ML, Price EO, et al, "Army Medical Laboratory Telemedicine: Role of Mass Spectrometry in Telediagnosis for Chemical and Biological Defense," *J Appl Toxicol*, 2001, 21(Suppl 1):S35-41.

Chlorine

Related Information Chlorine - Patient Information Sheet
CAS Number 7782-50-5
UN Number 1017
Synonyms Bertholite

CHLORINE

Antidote Sodium Bicarbonate
Military Classification Choking Agent
CDC Classification for Diagnosis
   Biologic: No biologic marker for chlorine exposure is available.
   Environmental: Detection of chlorine in environmental samples, as determined by NIOSH.
CDC Case Classification
   Suspected: A case in which a potentially exposed person is being evaluated by healthcare workers or public health officials for poisoning by a particular chemical agent, but no specific credible threat exists.
   Probable: A clinically compatible case in which a high index of suspicion (credible threat or patient history regarding location and time) exists for chlorine exposure, or an epidemiologic link exists between this case and a laboratory-confirmed case.
   Confirmed: A clinically compatible case in which laboratory tests on environmental samples are confirmatory. The case can be confirmed if laboratory testing was not performed because either a predominant amount of clinical and nonspecific laboratory evidence of a particular chemical was present or a 100% certainty of the etiology of the agent is known.
Use In pulp mills and for swimming pools; also used in bleaching; used as a water disinfectant; first used as a chemical weapon in Ypres, France in 1915
Mechanism of Toxic Action Converted to hydrogen chloride in lung parenchyma, strong irritant and corrosive agent; free radical generation may be present
Adverse Reactions
   Cardiovascular: Cardiovascular collapse, chest pain (34%), angina, tachycardia
   Central nervous system: CNS depression, headache (63%), agitation
   Ocular: Lacrimation, conjunctivitis, blepharospasm at 3-6 ppm
   Respiratory: Pulmonary edema, bronchoconstriction, reactive airways dysfunction syndrome, bronchiectasis, tachypnea, laryngeal edema
   Miscellaneous: Increased risk for lymphoma
Toxicodynamics/Kinetics Absorption: Not well absorbed, low solubility
Diagnostic Test/Procedure Lung Scan, Ventilation
Signs and Symptoms of Acute Exposure Airway irritation, chest pain, cough (52%), dermal burns, dyspnea (54%), eye irritation (4% incidence), headache, hyperchloremic metabolic acidosis, hypertension, hypotension, stomatitis, vomiting (24% to 29%), wheezing
   Severe exposures can appear like noncardiogenic pulmonary edema. Closely monitor for signs/symptoms of infection after the initial crisis if moderate to severe pulmonary involvement is suggested.
Admission Criteria/Prognosis Ongoing pulmonary symptoms 6 hours postexposure requires hospital admission
Treatment Strategy
   Decontamination:
      Dermal: Remove contaminated clothing and wash exposed area with soap and water
      Ocular: Irrigate with copious amounts of water or normal saline for at least 15 minutes
      Oral: Do not attempt to dilute ingestion as it may cause an exothermic reaction
   Supportive therapy: Administer 100% humidified oxygen. Consider airway protection and intubation in light of the risk for airway compromise. A 3.5% to 5% sodium bicarbonate solution by nebulization may be helpful for acute respiratory symptoms. Wheezing can be treated with beta-adrenergic agonists. Steroids are of no proven benefit. Avoid use of intravenous sodium bicarbonate. Inverse-ratio ventilation using lower tidal volume mechanical ventilation (6 mL/kg) and plateau pressures of 30 cm water may decrease mortality. While positive end expiratory pressures should be considered, oxygen toxicity should be monitored. Use of perfluorocarbon partial liquid ventilation and exogenous surfactant in chemical-induced ARDS is investigational.
   Endoscope is not necessary when the subject accidentally swallows a small quantity (<40 mL) of diluted bleach, the clinical examination is normal, the bleach contains <3% of

available chlorine, and bleach pH is <12. Otherwise, endoscopy should be performed within 6-12 hours after ingestion.

Consider cardiac monitoring and chest X-rays to detect delayed changes that might evolve after initial clinical presentation. Changes may appear as ARDS/NCPE. Secondary infections are of concern.

Environmental Persistency Low

Reference Range Not measurable in biological specimens

Additional Information Green-yellow gas. Taste threshold: 5 ppm. Odor threshold: 3.5 ppm. TLV-TWA: 0.5 ppm; IDLH: 10 ppm; PEL-TWA: 0.5 ppm

A few breaths at 1000 ppm can be fatal; 430 ppm: Lethal in 30 minutes; 40-60 ppm: Pulmonary edema can develop

Mixing hypochlorite (bleach) with an acid can release chlorine gas; mixing hypochlorite (bleach) with ammonia can release chloramine gas.

As a water disinfectant, chlorine is bactericidal at levels of 0.2 mg/L and cysticidal at 1.5-2 residual free chlorine (pH 7).

War gas used in World War I

In 2005, the AAPCC reported 7,643 human exposures to chlorine gas.

Specific References

Bania TC and Chu J, "Effect of Intravenous NaHCO₃ on Toxicity From Inhaled Chlorine Gas in an Animal Model," *Acad Emerg Med*, 2006, 13(5):S178.

Bania TC and Chu J, "Sucrose as a Potential Therapy for Chlorine-Induced Pulmonary Edema," *Acad Emerg Med*, 2006, 13(5):S177-8.

Belson MG, Schier JG, and Patel MM, "Case Definitions for Chemical Poisoning," *MMWR Recomm Rep*, 2005, 54(RR-1):1-24.

Burda AM, Kubic A, and Wahl M, "Nebulized Sodium Bicarbonate for Chlorine Gas Exposure," *Chicago Med*, 2007:109(16):30-1.

Cohle SD, Thompson W, Eisenga BH, et al, "Unexpected Death Due to Chloramine Toxicity in a Woman With a Brain Tumor," *Forensic Sci Int*, 2001, 124(2-3):137-9.

Eldridge DL, Richardson W, Michels JE, et al, "The Role of Poison Centers in a Mass Chlorine Exposure," *Clin Toxicol (Phila)*, 2005, 43:766.

Horton DK, Berkowitz Z, and Kaye WE, "Secondary Contamination of ED Personnel From Hazardous Materials Events, 1995-2001," *Am J Emerg Med*, 2003, 21(3):199-204.

Horton DK, Berkowitz Z, and Kaye WE, "The Public Health Consequences From Acute Chlorine Releases, 1993-2000," *J Occup Environ Med*, 2002, 44(10):906-13.

Kales SN and Christiani DC, "Acute Chemical Emergencies," *N Engl J Med*, 2004, 350(8): 800-8.

LoVecchio F, Blackwell S, Stevens D, et al, "Outcomes of Chlorine Exposure: A Five-Year Poison Center Experience," *J Toxicol Clin Toxicol*, 2002, 40(5):611.

Sarmanaev SK, Aidarova LF, Yamanaeva IE, et al, "The Structure of Poisonings by Bleaches (PB)," *J Toxicol Clin Toxicol*, 2002, 40(5):646-7.

Tobback CH and Mostin M, "Poison Centre and Product Safety Surveillance: Chlorine Exposure Due to Household Chlorine Tablets," *J Toxicol Clin Toxicol*, 2000, 38(2):242.

Traub SJ, Hoffman RS, and Nelson LS, "Case Report and Literature Review of Chlorine Gas Toxicity," *Vet Hum Toxicol*, 2002, 44(4):235-9.

Chloroacetophenone

CAS Number 532-27-4

UN Number 1697

Synonyms CAF; Chemical Mace; CN; MACE; Phenacyl Chloride; Tear Gas

Military Classification Lacrimator

Use Active ingredient in tear gas; riot control

CHLOROACETOPHENONE

Mechanism of Toxic Action Lacrimator - irritation due to the active halogen group reacting transiently with sulfhydryl groups

Adverse Reactions

Dermal: Allergic contact dermatitis, burning sensation of the skin, dermal erythema, bullous dermatitis, burns

Gastrointestinal: Metallic taste, vomiting

Ocular: Ocular burning, blepharospasms, lacrimation (eye symptoms may last for 30 minutes)

Respiratory: Bronchospasm (may be delayed for 36 hours), pulmonary edema (may be delayed for 24 hours), laryngospasm (may be delayed for 1-2 days), cough, sneezing

Toxicodynamics/Kinetics

Onset of effect: Rapid

Duration: Usually resolves in 30 minutes, skin effects may last for 2-3 hours

Absorption: Not absorbed

Pregnancy Issues No teratogenic effects

Clinically Resembles Chlorine ammonia; hydrogen chloride

Signs and Symptoms of Acute Exposure Agitation, bronchospasm, cough, erythema, eye pain, lacrimation, laryngospasm (may be delayed for 2 days), leukocytosis, nausea, pharyngitis, rhinorrhea, skin irritation, sneezing, tearing, vomiting

Treatment Strategy

Decontamination: Move patient rapidly to fresh air and monitor for wheezing. Administer 100% humidified oxygen. Personnel should avoid contaminating themselves.

Dermal: Remove all contaminated clothing, but try not to pull clothing over head (cut clothing instead). Try to leave blisters intact unless they are constricting bloodflow. Wash skin with soap and at least two liters of water.

Ocular: Avoid rubbing eyes as it may prolong effect. Copious irrigation with water or saline is needed.

Supportive therapy: Solution of antacid (magnesium hydroxide - aluminum hydroxide - simethicone) gently blotted topically over exposed facial areas (avoiding eyes) may help resolve symptoms. Bronchospasm can be treated with inhaled beta$_2$ agonists. Protect airway and consider endotracheal intubation for potential airway compromise. Inverse-ratio ventilation using lower tidal volume mechanical ventilation (6 mL/kg) and plateau pressures of 30 cm water may decrease mortality. While positive end expiratory pressures should be considered, oxygen toxicity should be monitored. Use of perfluorocarbon partial liquid ventilation and exogenous surfactant in chemical-induced ARDS is investigational.

Environmental Persistency Short

Additional Information

Colorless to white vapor with a fragrant odor of apple blossoms; insoluble in water. Can be detected using gas chromatography/mass spectrophotometry (GC/MS). TLV-TWA: 0.05 ppm; IDLH: 100 mg/m^3; PEL-TWA: 0.05 ppm. Vapor density: 5.3 (with air). Atmospheric half-life: 9.2 days

Threshold for eye irritation: 0.3 mg/m^3 (permanent eye injury may occur)

Incapacitating dose: 5-20 mg/m^3

Median incapacitating dose: 80 mg/min/m^3

Median lethal dose: 7000 mg/min/m^3 (dispersed from solvent); 14,000 mg/min/m^3 (from thermal grenade)

Decomposes to chlorine when heated; no bioconcentration

Action on metal: Tarnishes steel, mild corrosion

Specific References

Blain PG, "Tear Gases and Irritant Incapacitants," 1-chloroacetophenone, 2-chlorobenzylidene Malononitrile and Dibenz[b,f]-1,4-oxazepine," *Toxicol Rev*, 2003, 22(2):103-10 (review).

Centers for Disease Control and Prevention (CDC), "Exposure to Tear Gas From a Theft-Deterrent Device on a Safe," *MMWR Morb Mortal Wkly Rep*, 2004, 53(8):176-7.

Smith J and Greaves I, "The Use of Chemical Incapacitant Sprays: A Review," *J Trauma*, 2002, 52(3):595-600.

Cyanide

CAS Number 57-12-5; 74-90-8
UN Number 1051; 1588; 1613
Synonyms Carbon Nitride Ion
CDC Classification for Diagnosis
  Biologic: A case in which cyanide concentration is higher than the normal reference range (0.02-0.05 mcg/mL) in whole blood, as determined by a commercial laboratory.
  Environmental: Detection of cyanide in environmental samples, as determined by NIOSH or FDA.
CDC Case Classification
  Suspected: A case in which a potentially exposed person is being evaluated by healthcare workers or public health officials for poisoning by a particular chemical agent, but no specific credible threat exists.
  Probable: A clinically compatible case in which a high index of suspicion (credible threat or patient history regarding location and time) exists for cyanide exposure, or an epidemiologic link exists between this case and a laboratory-confirmed case.
  Confirmed: A clinically compatible case in which laboratory tests have confirmed exposure. The case can be confirmed if laboratory testing was not performed because either a predominant amount of clinical and nonspecific laboratory evidence of a particular chemical was present or a 100% certainty of the etiology of the agent is known.
Mechanism of Toxic Action Forms a stable complex with ferric ion of cytochrome oxidase enzymes inhibiting cellular respiration (at cytochrome aa_3), then converted to thiocyanate (less toxic form) by rhodanase enzyme. Cyanide can also cause lipid peroxidation in the brain due to inhibition of antioxidant enzymes.
Adverse Reactions
  Cardiovascular: Myocardial depression, hypotension, sinus bradycardia, angina, chest pain, congestive heart failure, sinus tachycardia, vasodilation
  Central nervous system: CNS stimulation followed by CNS depression, extrapyramidal reactions, bilateral putaminal involvement
  Endocrine & metabolic: Goiter, hypothyroidism
  Gastrointestinal: Bitter almond, burning taste
  Neuromuscular & skeletal: Rhabdomyolysis
  Ocular: Blurred vision
  Respiratory: Respiratory depression
Toxicodynamics/Kinetics
  Absorption: By inhalation (58% to 77%), oral (-50%), dermally, and ocular
  Distribution: V_d: 0.4 L/kg
  Metabolism: Hepatic by rhodanese or 3-mercaptopyruvate to thiocyanate; also combines with hydroxocobalamin to form cyanocobalamin (vitamin B_{12})
  Half-life: 0.7-2.1 hours
  Elimination: Excreted through the lungs and the kidneys
Pregnancy Issues Teratogenicity not noted
Diagnostic Test/Procedure
  Anion Gap, Blood
  Cyanide, Blood
  Electrolytes, Blood
  Methemoglobin, Blood
  Thiocyanate, Blood or Urine
Signs and Symptoms of Acute Exposure Agitation, apnea, bitter almond breath, coma, cyanosis, dizziness, flushing, headache, hyperventilation, hypothermia, hypotension, lateral gaze nystagmus, metabolic acidosis, mydriasis, myoglobinuria, nausea, pruritus, pulmonary edema, ototoxicity, seizures, skin irritation, tachycardia followed by bradycardia, tachypnea, tinnitus, vomiting

Admission Criteria/Prognosis All symptomatic patients should probably be admitted to an intensive care unit for 1-2 days; asymptomatic patients should be observed for at least 2 hours and then can be discharged; survival after 4 hours (in an acute exposure) is usually associated with recovery

Treatment Strategy

Decontamination: Basic poison management (lavage within 1 hour/activated charcoal). Emesis is contraindicated due to the rapid course of neurologic symptoms. Give 100% oxygen.

Supportive therapy: Give sodium bicarbonate for acidosis. Cyanide antidote kit (amyl nitrite, sodium nitrite followed by sodium thiosulfate) is the current primary therapeutic modality. Do not administer thiosulfate by rapid I.V. Consider aggressive airway management if necessary. Methemoglobin formation due to nitrites occurs with a target methemoglobin level of ~20%, although clinical responses do occur at levels of 5%.

Enhancement of elimination: Hemodialysis and charcoal hemoperfusion have been utilized. Hyperbaric oxygen may also be utilized, although its efficacy is uncertain. Hydroxocobalamin, dicobalt-EDTA, and 4-dimethylaminophenol are used for chelation in Europe. Typically ~4 g of hydroxocobalamin (which can bind 200 mg of cyanide) is administered with 8 g of thiosulfate.

Environmental Persistency Residence time in atmosphere of hydrogen cyanide: 2 years. In natural rivers, half-life of cyanide is 10-24 days. Normal persistency of hydrogen cyanide is up to 1 hour; not persistent in soil.

Personal Protection Equipment Pressure demand self-contained breathing apparatus with air concentrations >4.7 ppm; butyl rubber gloves, teflon or Tychem protective clothing.

Reference Range

Whole blood levels: Smoker: ≤0.5 mg/L

Flushing and tachycardia seen at 0.5-1.0 mg/L; obtundation at 1.0-2.5 mg/L

Coma and death occur at >2.5 mg/L

Plasma cyanide: Normal: 4-5 mcg/L; Asymptomatic (with metabolic acidosis): 80 mcg/L; Death: >260 mcg/L

Red blood cell cyanide: Normal: <26 mcg/L; Metabolic acidosis: 1040 mcg/L; Symptomatic: 5200 mcg/L; Fatal: >10,400 mcg/L

Note: A plasma lactate concentration >8 mM/L is 94% sensitive and 70% specific for a blood cyanide level >1 mg/L (40 μmol/L).

Additional Information Fatal dose of cyanide: 200-300 mg (adult)

Fatal dose (hydrogen cyanide): Dermal: 100 mg/kg. Inhalation for <1 hour: 110-135 ppm. **Oral:** 0.6-1.5 mg/kg

Bitter almond odor undetectable by 30% to 50% of the population; zinc cyanide is odorless. Parkinsonism has been noted after ingestion of sodium or potassium cyanide. Hydroxocobalamin and dicobalt-EDTA are used for chelation in Europe.

The ratio of red blood cell to plasma cyanide concentrations is 60:1.

Daily intake of hydrogen cyanide by inhalation: ~3.8 mcg; through drinking water: 0.4-0.7 mcg. Allowable daily intake: ~0.6 mg

Fruit juice typically contains from 1.9-5.3 mg/L of hydrogen cyanide while apricot pits contain 89-2,170 mg/kg (wet weight). Inhaled smoke from cigarettes contains from 10-400 mcg/cigarette, while sidestream smoke usually runs <27% of mainstream smoke concentrations. Average cyanide emission rate in automobiles varies from 11-14 mg/mile in cars; catalytic converters can reduce cyanide emissions by 90%.

Dermal exposure of 10% sodium cyanide to a large body surface area will cause symptoms within 20 minutes

A 5-g dose of hydroxocobalamin appears to bind all cyanide ions in patients with initial cyanide levels up to 40 μmol/L

Plasma lactate (half-life ~4 hours in treated cyanide poisoning) may be useful in determining treatment efficacy in cyanide poisoning

ACGIH-TLV-TWA: 5 mg/m^3; IDLH: 50 mg/m^3; PEL-TWA: 5 mg/m^3

Commonly found in: Gold and silver ore extraction, electroplating, fumigant, stainless steel manufacture, petroleum refining, rodenticide; plant sources include amygdalin glycodes (such as peach pits), cassava, linium, prunus, sorghum, and bamboo sprouts; also a by-product of nitroprusside and succinonitrile along with laetrile

Hydrogen cyanide (AC) was also known as Zyklon B in Germany; has a lethal dose as low as 150 mg-min/m^3.

Specific References

Aslani MR, Maleki M, Sharifi K, et al, "Mass Cyanide Intoxication in Sheep," *Vet Hum Toxicol*, 46(4):186-7.

Baud FJ, Borron SW, Megarbane B, et al, "Lactic Acidosis in Cyanide Poisoning: Pathophysiology and Clinical Considerations," *J Toxicol, Clin Toxicol*, 2001, 39:244.

Baud FJ, Borron SW, Megarbane B, et al, "Value of Lactic Acidosis in the Assessment of the Severity of Acute Cyanide Poisoning," *Crit Care Med*, 2002, 30(9):2044-50.

Bromley J, Hughes BG, Leong DC, et al, "Life-threatening Interaction Between Complementary Medicines: Cyanide Toxicity Following Ingestion of Amygdalin and Vitamin C," *Ann Pharmacother*, 2005, 39(9):1566-9.

Chin RG and Calderon Y, "Acute Cyanide Poisoning: A Case Report," *J Emerg Med*, 2000, 18(4):441-5.

Dalefield RR, "Rapid Method for the Detection of Cyanide Gas Release From Plant Material Using CYANTESMO Paper," *Vet Hum Toxicol*, 2000, 42(6):356-7.

DesLouriers CA, Burda AM, and Wahl M, "Hydroxocobalamin as a Cyanide Antidote," *KeePosted*, 2004, 23-7.

Dumas P, Gingras G, and LeBlanc A, "Isotope Dilution-Mass Spectrometry Determination of Blood Cyanide by Headspace Gas Chromatography," *J Anal Toxicol*, 2005, 29:71-3.

Eyer P, "Therapeutic Implications of the Toxicokinetics and Toxicodynamics in Cyanide Poisoning," *J Toxicol Clin Toxicol*, 2000, 38(2):212-4.

Horton DK, Berkowitz Z, and Kaye WE, "Secondary Contamination of ED Personnel From Hazardous Materials Events, 1995-2001," *Am J Emerg Med*, 2003, 21(3):199-204.

Hung YM, Hung SY, Chou KJ, et al, "Acute Poisoning of Yam Bean Seeds: Clinical Manifestation Mimicking Acute Cyanide Intoxication," *J Toxicol Clin Toxicol*, 2004, 42(5):725.

Kales SN and Christiani DC, "Acute Chemical Emergencies," *N Engl J Med*, 2004, 350(8):800-8.

Knudsen K, "Combustion Toxicology in a Disastrous Discotheque Fire," *J Toxicol Clin Toxicol*, 2000, 38(2):218.

Mannaioni G, Vannacci A, Marzocca C, et al, "Acute Cyanide Intoxication Treated With a Combination of Hydroxocobalamin, Sodium Nitrite, and Sodium Thiosulfate," *J Toxicol Clin Toxicol*, 2002, 40(2):181-3.

McFee RB, Leikin JB, and Kiernan K, "Preparing for an Era of Weapons of Mass Destruction (WMD) – Are We There Yet? Why We Should All Be Concerned. Part II," *Vet Hum Toxicol*, 2004, 46(6):347-51.

McGeorge F, Pham Cuong J, Huang D, et al, "Hemodyamic Effects of Cyanide Toxicity," *Acad Emerg Med*, 2003, 10:519b-20b.

Megarbane B, Delahaye A, Goldgran-Toledano D, et al, "Antidotal Treatment of Cyanide Poisoning," *J Chin Med Assoc*, 2003, 66(4):193-203 (review).

Musshoff F, Schmidt P, Daldrup T, et al, "Cyanide Fatalities: Case Studies of Four Suicides and One Homicide," *Am J Forensic Med Pathol*, 2002, 23(4):315-20.

Mutlu GM, Leikin JB, Oh K, et al, "An Unresponsive Biochemistry Professor in the Bathtub," *Chest*, 2002, 122(3):1073-6.

Phan P, Rosenblatt Y, Desandre P, et al, "Diagnostic Odor Recognition," *Ann Emerg Med*, 2000, 35(5):S67.

Rella JG, Marcus S, and Wagner BJ, "Rapid Cyanide Detection Using the Cyantesmo® Kit," *J Toxicol Clin Toxicol*, 2004, 42(6):897-900.

Rella JG, Marcus S, and Wagner BJ, "Rapid Cyanide Detection Using the Cyantesmo® Kit," *Clin Toxicol (Phila)*, 2005, 43:687.

Renard C, Borron SW, Renandeau C, et al, "The Diagnostic Value of Biological Markers of Cellular Hypoxia in Cyanide Poisoning in a Rat Model," *Clin Toxicol (Phila)*, 2005, 43:686.

Sauer SW and Keim ME, "Hydroxocobalamin: Improved Public Health Readiness for Cyanide Disasters," *Ann Emerg Med*, 2001, 37(6):635-41.

Soto-Blanco B, Sousa AB, Manzano H, et al, "Does Prolonged Cyanide Exposure Have a Diabetogenic Effect?" *Vet Hum Toxicol*, 2001, 43(2):106-8.

Stanford CF, Ries NL, Bogdan GM, et al, "Do Hospitals in the 50 Largest Cities of the United States Carry Sufficient Supply of the Cyanide Antidote Kit?" *Clin Toxicol (Phila)*, 2005, 43:713.

Stefanidou M and Athanaselis S, "Toxicological Aspects of Fire," *Vet Hum Toxicol*, 2004, 46(4):196-8.

Tsuge K, Kataoka M, and Seto Y, "Rapid Determination of Cyanide and Azide in Beverages by Microdiffusion Spectrophotometric Method," *J Anal Toxicol*, 2001, 25(4):228-36.

Tweyongyere R and Katongole I, "Cyanogenic Potential of Cassava Peels and Their Detoxification for Utilization as Livestock Feed," *Vet Hum Toxicol*, 2002, 44(6):366-9.

vonLandenberg F, Stonerook M, Judge K, et al, "Efficacy of Hydroxocobalamin in a Canine Model of Cyanide Poisoning: A Pilot Study," *Clin Toxicol (Phila)*, 2005, 43:692.

Yeoh MJ and Braitberg G, "Carbon Monoxide and Cyanide Poisoning in Fire Related Deaths in Victoria, Australia," *J Toxicol Clin Toxicol*, 2004, 42(6):855-63.

Zilker TH and Felgenhauer N, "4-DMAP as Cyanide Antidote: Its Efficacy and Side Effects in Human Poisoning," *J Toxicol Clin Toxicol*, 2000, 38(2):217.

Cyanogen Bromide

CAS Number 506-68-3

UN Number 1889

Synonyms Bromine Cyanide; Bromocyan; CNBr

Military Classification Blood Agent

Mechanism of Toxic Action Forms a stable complex with ferric ion of cytochrome oxidase enzymes inhibiting cellular respiration (at cytochrome aa_3), then converted to thiocyanate (less toxic form) by rhodanase enzyme, which is overwhelmed in overdose, thus necessitating the use of the cyanide antidote kit. Cyanide can also cause lipid peroxidation in the brain due to inhibition of antioxidant enzymes. Also exhibits pulmonary irritant effects along with ocular irritation. Releases hydrogen cyanide when released in water, acid, or when heated.

Signs and Symptoms of Acute Exposure Agitation, apnea, bitter almond breath, coma, cyanosis, dizziness, flushing, headache, hyperventilation, hypothermia, hypotension, metabolic acidosis, mydriasis, myoglobinuria, nausea, lateral gaze nystagmus, pruritus, pulmonary edema, ototoxicity, seizures, skin irritation, tachycardia followed by bradycardia, tachypnea, tinnitus, vomiting, ocular irritation, dermal burns

Treatment Strategy

Decontamination: Basic poison management (lavage within 1 hour/activated charcoal). Emesis is contraindicated due to the rapid course of neurologic symptoms. Give 100% oxygen.

Supportive therapy: Give sodium bicarbonate for acidosis. Cyanide antidote kit (amyl nitrite, sodium nitrite followed by sodium thiosulfate) is the current primary therapeutic modality. Do not administer thiosulfate by rapid I.V. Consider aggressive airway management. Methemoglobin formation due to nitrites occurs with a target methemoglobin level of ~20%, although clinical responses do occur at levels of 5%.

Enhancement of elimination: Hemodialysis and charcoal hemoperfusion have been utilized. Hyperbaric oxygen may also be utilized, although its efficacy is uncertain. Hydroxocobalamin, dicobalt-EDTA, and 4-dimethylaminophenol are used for chelation in Europe. Typically ~4 g of hydroxocobalamin (which can bind 200 mg of cyanide) is administered with 8 g of thiosulfate.

Environmental Persistency Low
Additional Information Introduced by Austria
Air density: 2.02
Colorless needles; does **not** bioconcentrate in soil or food chain
Colorless/white solid; volatile at room temperature
After 10 minutes exposure:
Threshold for irritation: 1.4 ppm
Lacrimation/delayed pulmonary edema: 8-10 ppm
Lethal: 92 ppm
Bitter almond odor undetectable by 30% to 50% of the population

Cyanogen Chloride

CAS Number 506-77-4
UN Number 1589
Synonyms CCIN; CK; CI-CEN
Military Classification Blood Agent
Mechanism of Toxic Action Forms a stable complex with ferric ion of cytochrome oxidase enzymes inhibiting cellular respiration (at cytochrome aa_3), then converted to thiocyanate (less toxic form) by rhodanase enzyme. Cyanide can also cause lipid peroxidation in the brain due to inhibition of antioxidant enzymes. Exhibits mucosal irritant effects.
Signs and Symptoms of Acute Exposure Agitation, apnea, bitter almond breath, coma, cyanosis, dizziness, flushing, headache, hyperventilation, hypothermia, hypotension, metabolic acidosis, mydriasis, myoglobinuria, nausea, nystagmus, pruritus, pulmonary edema (delayed), ototoxicity, seizures, skin irritation, tachycardia followed by bradycardia, tachypnea, tinnitus, vomiting, mucosal irritation, cough, ocular irritation, lacrimation; lactic acidosis is an early indicator
Treatment Strategy
Decontamination: Basic poison management (lavage within 1 hour/activated charcoal). Emesis is contraindicated due to the rapid course of neurologic symptoms. Give 100% oxygen.
Supportive therapy: Give sodium bicarbonate for acidosis. Cyanide antidote kit (amyl nitrite, sodium nitrite followed by sodium thiosulfate) is the current primary therapeutic modality. Do not administer thiosulfate by rapid I.V. Consider aggressive airway management with possible intubation. Methemoglobin formation due to nitrites occurs with a target methemoglobin level of $\sim$20%, although clinical responses do occur at levels of 5%.
Enhancement of elimination: Hemodialysis and charcoal hemoperfusion have been utilized. Hyperbaric oxygen may also be utilized, although its efficacy is uncertain. Hydroxocobalamin, dicobalt-EDTA, and 4-dimethylaminophenol are used for chelation in Europe. Typically $\sim$4 g of hydroxocobalamin (which can bind 200 mg of cyanide) is administered with 8 g of thiosulfate.
Environmental Persistency Short (about 30 minutes), although vapor may persist on plants persists longer in a hot, humid environment
Personal Protection Equipment Full NBC protection; pressure demand, self-contained breathing apparatus for air levels >0.3 ppm; use cold-insulating gloves, butyl rubber gloves, teflon or Tychem protective clothing
Additional Information Highly volatile, colorless liquid or gas with an irritating odor (nondetectable by 30% to 50% of the population); no action on metal (if cyanogen chloride is dry); used during World War I by the French
Lacrimating effect: 12 mg/m^3 (air concentration)
Median incapacitating dose: 7000 mg/m^3
Median lethal dose: 11,000 mg/m^3

Rate of hydrolysis is slow (hydrolysis products are hydrogen chloride and CNOH)
Vapor density: 1.98 (heavier than air)
Liquid density: 1.18 (at 20°C)

Cyanogen Iodide

CAS Number 506-78-5

Synonyms CIN; I-CN; Iodine Cyanide; Iodocyan; Iodocyanide

Military Classification Blood Agent

Mechanism of Toxic Action Forms a stable complex with ferric ion of cytochrome oxidase enzymes inhibiting cellular respiration (at cytochrome aa_3), then converted to thiocyanate (less toxic form) by rhodanase enzyme, which is overwhelmed in overdose, thus necessitating the use of the cyanide antidote kit. Cyanide can also cause lipid peroxidation in the brain due to inhibition of antioxidant enzymes; strong mucosal irritant.

Signs and Symptoms of Acute Exposure Agitation, apnea, bitter almond breath, coma, cyanosis, dizziness, flushing, headache, hyperventilation, hypothermia, hypotension, metabolic acidosis, mydriasis, myoglobinuria, nausea, nystagmus, pruritus, pulmonary edema, ototoxicity, seizures, skin irritation, tachycardia followed by bradycardia, tachypnea, tinnitus, vomiting, conjunctivitis, blepharospasm, lacrimation, dermal erythema

Monitoring Parameters Monitor cardiac, oxygen, and ABG

Treatment Strategy

Decontamination: Basic poison management (lavage within 1 hour/activated charcoal); strict attention to ACLS/BLS "ABCs." Emesis is contraindicated due to the rapid course of neurologic symptoms. Give 100% oxygen.

Supportive therapy: Give sodium bicarbonate for acidosis. Cyanide antidote kit (amyl nitrite, sodium nitrite followed by sodium thiosulfate) is the current primary therapeutic modality. Do not administer thiosulfate by rapid I.V. Consider aggressive airway management with possible intubation. Methemoglobin formation due to nitrites occurs with a target methemoglobin level of ~20%, although clinical responses do occur at levels of 5%.

Enhancement of elimination: Hemodialysis and charcoal hemoperfusion have been utilized. Hyperbaric oxygen may also be utilized, although its efficacy is uncertain. Hydroxocobalamin, dicobalt-EDTA, and 4-dimethylaminophenol are used for chelation in Europe. Typically ~4 g of hydroxocobalamin (which can bind 200 mg of cyanide) is administered with 8 g of thiosulfate.

Additional Information A colorless to whitish needle-like solid with a pungent odor (undetectable by 30% to 50% of the population). When heated, can form vapors of nitrogen oxides, hydrogen cyanide, and iodide. Soluble in water, alcohol, ethanol, and ether.

Eye irritation, lacrimation, blepharospasm noted at air levels of 100 mg/m^3 of cyanogen iodide.

Cyclohexyl Sarin

CAS Number 329-99-7

Synonyms GF

Use First synthesized in WWII, it may be combined with sarin to increase persistence. Iraq is the only known country to produce large amounts.

Mechanism of Toxic Action Inhibits hydrolysis of acetylcholine by acetylcholinesterase binding with this enzyme. Irreversible binding (aging) takes place. This results in acetylcholine excess at the neuronal synapse. May penetrate blood brain barrier and effect GABA transmission. Similar to organophosphate agent; inhibits the enzyme acetylcholinesterase thus resulting in acetylcholine excess at the neuronal synapse; may penetrate blood brain barrier and thus affect GABA transmission

Adverse Reactions Onset is rapid

Cardiovascular: Sinus bradycardia, tachycardia, heart rhythm disturbances

Central nervous system: Seizures, ataxia, coma, headache, fatigue, memory loss, nightmares, excess cholinergic toxicity, muscarinic activity, nicotinic activity

Endocrine & metabolic: Hypokalemia has been reported in GF intoxication but the mechanism is unclear

Gastrointestinal: Diarrhea, nausea, salivation, vomiting, abdominal pain

Genitourinary: Involuntary urination

Ocular: Lacrimation, early mydriasis rapidly followed by persistent miosis (may take 1-2 months for pupillary response to normalize), dim or blurred vision, eye pain

Respiratory: Rhinorrhea, bronchospasm, cough, tachypnea, dyspnea, wheezing, bronchorrhea; death usually results from respiratory failure

Clinically Resembles Jatropha multifida plant poisoning, organophosphate insecticides, nicotine poisoning

Treatment Strategy

Adjunctive treatment: Diazepam for seizures, when intubation and mechanical ventilation is used and when >6 mg atropine is given (or 3 Mark I kits)

Pretreatment/prophylaxis: Pyridostigmine

Decontamination:

Dermal: Remove all contaminated clothing. Wash with copious amounts of water. A 1% to 5% hypochlorite solution (household bleach diluted) with alkaline soap may be used. Immediate field/outdoor decontamination is suggested. Avoid decontamination inside the healthcare facility. Do not delay NAAK or antidotal intervention in lieu of decontamination.

Inhalation: Remove from exposure site and remove contaminated clothes. Administer 100% humidified oxygen. Protect airway. If unconscious, consider intubation with mechanical ventilation.

Ocular: Copious irrigation with water or saline

Supportive therapy: Atropine is the mainstay of therapy. Depending upon severity of exposure, 10-20 mg cumulatively over the first 2-3 hours. Atropine should be titrated to bronchial secretions, **not** ocular signs. Pralidoxime should be administered 1-2 g I.V. over 10 minutes, with repeat dose in 1 hour if weakness persists, then every 4-12 hours afterwards as needed. Pralidoxime is most effective if given 3 hours postcyclohexyl sarin exposure. If Mark I kits are available, acute treatment suggestions: mild exposure 1 kit, mild to moderate exposure 2 kits (2nd kit 5 minutes after the first), moderate to severe exposure, 3 kits (each kit separated by 5 minutes). Long term I.V. atropine and 2-PAM may be required.

Special populations: Children are especially vulnerable to these compounds owing to their smaller mass, smaller airway diameter, higher respiratory rate, and minute volumes. Due to their short stature, they are exposed to chemicals that settle close to the ground.

Pediatric dose: Atropine .05 mg/kg I.V./I.M., 2-PAM 25-50 mg/kg I.V./I.M., diazepam 0.2 to 0.5 mg/kg I.V.

Environmental Persistency Intermediate; evaporation rate about 5% of water

Personal Protection Equipment

Respiratory protection: Pressure-demand, self-contained breathing apparatus (SCBA) is recommended

Skin protection: Chemical protective clothing and butyl rubber (M3 and M4 Norton) with chemical goggles and face shield

Additional Information Colorless liquid; liquid density: 1.133; vapor density: 6.2; slightly water soluble

Median lethal dose (LD_{50}):

Mice: Oral: 17 mg/kg

Monkey: I.M.: 46.6 mcg/kg

Miosis dose: <1 mg/min/m^3 (atmospheric)

Specific References
Lynch EL and Thomas TL, "Pediatric Considerations in Chemical Exposures: Are We Prepared?" *Pediatr Emerg Care*, 2004, 20(3):198-208.
Rotenberg JS and Newmark J, "Nerve Agent Attacks on Children: Diagnosis and Management," *Pediatrics*, 2003, 112(3 Pt 1):648-58.

Cyclotrimethylenetrinitramine

CAS Number 121-82-4
Synonyms Cyclonite; RDX
Military Classification Explosive Agent
Use Major component in plastic explosives (91%), a base charge for detonation; also used in smokeless powder, torpedoes, mines, and as rat poison. Semtex contains this agent.
Mechanism of Toxic Action Neurotoxin with unknown mechanism; can cause fibrous degeneration in the CNS
Toxicodynamics/Kinetics
Absorption:
Dermal: Poorly absorbed
Inhalation/gastrointestinal tract: Well absorbed
Oral: Rapidly absorbed
Half-life: 15 hours (by ingestion)
Signs and Symptoms of Acute Exposure Headache, confusion, nausea, vomiting, malaise, seizures (onset 1-12 hours postexposure, up to 60 hours), stupor, myoclonus, lethargy, hyperreflexia, fever, tachycardia, oliguria, myalgia, leukocytosis, petechial rash
Treatment Strategy
Decontamination:
Dermal: Remove contaminated clothing; wash skin with soap and water
Inhalation: Administer 100% humidified oxygen
Ocular: Irrigate with copious amounts of water
Supportive therapy: Maintain adequate hydration; seizures can be treated with benzodiazepines and phenobarbital
Reference Range Peak serum and urine cyclotrimethylenetrinitramine levels of 10.74 mg/L and 38.41 mg/L, respectively, noted in a child with seizures
Additional Information White or colorless crystals; insoluble in water, carbon tetrachloride, or ethanol. Can be detonated by heat, shock, or mercury fulminate. Heating to decomposition results in nitrogen oxides.
Toxic oral dose: 85 mg/kg
Electroencephalogram (EEG) can show bilateral synchronous spike and wave complexes at 2-3 per second
Specific References
Arnold JL, Halpern P, Tsai MC, et al, "Mass Casualty Terrorist Bombings: A Comparison of Outcomes by Bombing Type," *Ann Emerg Med*, 2004, 43(2):263-73.

Dinitrotoluene

CAS Number 25321-14-6
Synonyms DNT
Military Classification Explosive Agent
Use Explosive manufacturing (formed by the nitration of toluene)
Mechanism of Toxic Action Causes methemoglobinemia by oxidizing hemoglobin
Adverse Reactions
Cardiovascular: Tachycardia, chest pain, hypotension
Hematologic: Methemoglobinemia, hemolysis

Hepatic: Hepatitis
Neuromuscular & skeletal: Neuropathy, tremor
Ocular: Conjunctivitis
Toxicodynamics/Kinetics
Absorption: Dermal, ingestion, inhalation
Metabolism: Hepatic
Half-life: 1-2.7 hours
Elimination: Urine
Treatment Strategy
Decontamination:
    Dermal: Remove contaminated clothing; wash skin with soap and water
    Ocular: Irrigate with copious amounts of water
Supportive therapy: Treat symptomatic methemoglobinemia (or methemoglobin levels over
    30%) with oxygen and methylene blue; monitor for carbon monoxide exposure when
    explosion occurs. Burns should be treated with standard therapy.
Additional Information
Orange-yellow crystalline solid; isomeric mixture is an oily liquid
Reacts with strong oxidizers to cause fire or explosion
Slightly soluble in water
Increased toxicity with ethanol ingestion
PEL-TWA: 0.2 ppm
Specific References
Arnold JL, Halpern P, Tsai MC, et al, "Mass Casualty Terrorist Bombings: A Comparison of
    Outcomes by Bombing Type," *Ann Emerg Med*, 2004, 43(2):263-73.
Haut MW,, Kuwabara H, Ducatman AM, et al, "Corpus Callosum Volume in Railroad Workers
    With Chronic Exposure to Solvents," *J Occup Environ Med*, 2006, 48(6):615-24.

Diphenylchloroarsine

CAS Number 712-48-1
UN Number 1699
Synonyms Clark I; DA; Sneezing Gas
Military Classification Riot Control Agent; Vomiting Agent
Mechanism of Toxic Action Inhibition of sulfhydryl group containing enzymes; irritation at
high concentrations.
Monitoring Parameters Oxygen, cardiac function, especially in elderly where pre-existing
cardiopulmonary comorbidity may predispose to more severe injury
Signs and Symptoms of Acute Exposure May be delayed for several minutes: ocular
burning, lacrimation, sinus pain, throat burning, chest pain, nasal secretions, salivation;
systemic effects include headache, ataxia, paresthesias, malaise, nausea, and vomiting.
Sneezing and coughing may be violent. Pulmonary edema can occur at high concentrations.
Dermal exposure results in pruritus, erythema, and burning sensation, which can progress to
vesicle formation. Symptoms usually resolve within two hours.
Treatment Strategy
Decontamination:
    Dermal: Irrigate with copious amounts of water or dilute bleach
    Ocular: Irrigate with copious amounts of water or normal saline for at least 20 minutes
    Respiratory: Administer 100% humidified oxygen
Supportive therapy: Oral antihistamines can be given for pruritus. Ophthalmic corticosteroids
    may be used for persistent ocular irritation. Protect airway with aggressive airway
    management and bronchodilator; chest X-ray
Environmental Persistency Not persistent; loss of effects within $^1/_2$ hour

Additional Information Usual form is aerosol or smoke (white); odorless; rate of hydrolysis is relatively slow to diphenylarsenious oxide and hydrogen chloride: Arsenic trichloride (CAS: 7784-34-1) is a precurson chemical.
Median incapacitating dose: 12 mg/min/m^3 over 10 minutes
Median lethal dose: 15,000 mg/min/m^3

Diphenylcyanoarsine

CAS Number 23525-22-6

Synonyms Clark II; DC; Diphenylarsinous Cyanide

Military Classification Riot Control Agent; Vomiting Agent

Mechanism of Toxic Action Inhibition of sulfhydryl group containing enzymes; irritation at high concentrations. More irritating than adamsite or diphenylchloroarsine; hydrolysis produce is hydrogen cyanide

Signs and Symptoms of Acute Exposure May be delayed for several minutes: ocular burning, lacrimation, sinus pain, throat burning, chest pain, nasal secretions, salivation; systemic effects include headache, ataxia, paresthesias, malaise, nausea, and vomiting. Sneezing and coughing may be violent. Pulmonary edema can occur at high concentrations. Dermal exposure results in pruritus, erythema, and burning sensation, which can progress to vesicle formation. Symptoms usually resolve within two hours.

Treatment Strategy
Dermal: Irrigate with copious amounts of water or dilute bleach
Ocular: Irrigate with copious amounts of water or normal saline for at least 20 minutes
Respiratory: Administer 100% humidified oxygen
Supportive therapy: Oral antihistamines can be given for pruritus. Ophthalmic corticosteroids may be used for persistent ocular irritation. Treat with cyanide antidote kit if signs of cyanide toxicity occur. Do not administer cyanide kit by I.V. bolus. Do not administer thiosulfate by rapid I.V. Consider aggressive airway protection and management.

Additional Information Aerosol (white in color) with an odor similar to garlic or bitter almonds (30% to 50% of the population cannot detect odor)
Hydrolyzed to hydrogen cyanide and diphenylarsenious oxide
Action on metal: None
Median incapacitating dose: 30 mg/min/m^3 (30 second exposure); 20 mg/min/m^3 (5 minute exposure)
Median lethal dose: 10,000 mg/min/m^3

Diphosgene

CAS Number 503-38-8

UN Number 1076 (phosgene)

Synonyms Carbonic Dichloride; DP; Perchloromethyl Formate; Superpalite

Military Classification Choking Agent

Clinically Resembles Phosgene

Signs and Symptoms of Acute Exposure Asthenia, chest pain, cough, dermal burns, dyspnea, headache, hemoptysis, lacrimation (strong), nausea, ocular irritation, throat burning, vomiting, pulmonary edema; symptoms can be delayed for 3 hours

Treatment Strategy See Phosgene for specific treatment
Supportive therapy: Aggressive airway management and bronchodilators; chest X-ray

Environmental Persistency More persistent than chlorine or phosgene; decomposes rapidly but may produce phosgene when degraded by water condensates (such as fog or rain)

Additional Information
Odor: Newly mown hay
Low volatility; surface decontamination with steam or ammonia; boiling point: 127°C

Vapor density (as compared to air): 6.8
Liquid density (colorless liquid): 1.65 g/mL at 20°C
Freezing point: ~57°C
Median incapacitating dose: 1600 mg/min/m^3
Median lethal dose: 3200 mg/min/m^3

Ethyldichloroarsine

CAS Number 598-14-1
Synonyms DICK; ED; TI214
Military Classification Blister Agent/Choking Agent
Mechanism of Toxic Action An organic arsenical with the odor of rotting fruit; immediately irritating blister agent used in World War I
Adverse Reactions
Dermal: Fluid-filled blisters (about 5% of the blistering activity of lewisite)
Respiratory: Cough, dyspnea, pulmonary edema
Clinically Resembles Poison ivy
Treatment Strategy
Decontamination:
Dermal: Remove contaminated clothing; wash area copiously with soap and water
Inhalation: Administer 100% humidified oxygen and remove from site
Ocular: Irrigate with water or saline for at least 20 minutes. Arsenic poisoning is unlikely with eye exposure exclusively.
Ingestion: Dilute with milk or water
Supportive therapy: Obtain arsenic level; if blood arsenic level is >7 mcg/100 mL or urine is >0.7 mg/L, consider treating as per arsenic exposure. Treat dermal burns with standard therapy; debridement of devitalized skin should occur with topical antibiotics for burns. Monitor and treat for arsenic exposure. Hypotension can be treated with intravenous crystalloid solutions and placement of the patient in the Trendelenburg position. Dopamine or norepinephrine are the vasopressor agents of choice. Use aggressive pain management.
Environmental Persistency Less persistent than mustard gas; not persistent in rain
Personal Protection Equipment Chemical protective clothing with positive-pressure self-contained breathing apparatus (SCBA)
Additional Information
Related to mustard ga, lewisite
Lethal dose: 3-5 g - minute/m^3 (air concentration)

Ethylene Glycol Dinitrate

CAS Number 628-96-6
Use Dynamite manufacturing (20% to 90% concentration) to lower freezing point of nitroglycerin; a component of ammonium nitrate/fuel oil explosive
Mechanism of Toxic Action Relaxes smooth muscle resulting in vasodilation; can cause methemoglobinemia
Adverse Reactions
Cardiovascular: Tachycardia, hypotension, syncope
Central nervous system: Headache
Gastrointestinal: Nausea, vomiting
Hematologic: Methemoglobinemia, hemolysis
Neuromuscular & skeletal: Weakness

Toxicodynamics/Kinetics

Absorption: Dermal ($\sim$3%)

Metabolism: Hepatic to ethylene glycol mononitrate, nitrites/nitrates, and ethylene glycol

Half-life: 2-20 hours

Elimination: Renal

Treatment Strategy

Decontamination:

   Dermal: Remove contaminated clothing; wash skin with soap and water

   Ocular: Irrigate with copious amounts of water

Supportive therapy: Treat symptomatic methemoglobinemia (or methemoglobin levels over 30%) with oxygen and methylene blue; monitor for carbon monoxide exposure when explosion occurs. Burns should be treated with standard therapy.

Additional Information

Odorless, yellow oily liquid insoluble in water and soluble in ethyl alcohol; reactive with acids and alkalis

Air concentrations of 5 mg/m^3 can cause symptoms

Hyperthyroidism and ethanol can increase toxicity

May be absorbed through rubber gloves

Specific References

Arnold JL, Halpern P, Tsai MC, et al, "Mass Casualty Terrorist Bombings: A Comparison of Outcomes by Bombing Type," *Ann Emerg Med*, 2004, 43(2):263-73.

Loh CH, Shih TS, Hsieh AT, et al, "Hepatic Effects in Workers Exposed to 2-Methoxy Ethanol," *J Occup Environ Med*, 2004, 46(7):707-13.

Gasoline

CAS Number 8006-61-9

UN Number 1203; 1257

Synonyms Mogas; Motor Fuel; Motor Spirit; Natural Gasoline; Petrol

Use Fuel for internal combustion engines; can be used as incendiary device

Mechanism of Toxic Action A volatile hydrocarbon with central nervous system depressant and arrhythmogenic effects

Adverse Reactions

Cardiovascular: Sinus bradycardia

Central nervous system: Panic attacks with inhalation

Dermatologic: Dermatitis or significant burn (with prolonged exposure)

Hepatic: Degenerative changes may occur after moderate to severe exposure

Neuromuscular & skeletal: Rhabdomyolysis

Ocular: Conjunctivitis

Otic: Tinnitus

Renal: Alpha 2u-globulin nephropathy, degenerative changes may occur after moderate to severe exposure

Respiratory: Burning sensation to the chest, congestion, noncardiogenic pulmonary edema, asthma-like symptoms

Toxicodynamics/Kinetics

Protein binding: None

Half-life: 17 hours

Elimination: Pulmonary/renal

Usual Dosage Lethal dose: 10 mL/kg average adult

Signs and Symptoms of Acute Exposure

Death from ingestion is usually due to aspiration. Symptoms occur at 1000 ppm after 1 hour; ataxia, confusion, delirium dysarthria, euphoria, hallucinations (visual, auditory, and tactile), headache, mania, tremor, vertigo

Inhalation: Arrhythmia, ataxia, dizziness, drowsiness, hallucinations, insomnia, intra-alveolar hemorrhage, muscle cramps, myoclonus, nausea, paresthesias, pulmonary edema, vomiting

Ingestion: Aspiration, belching, disseminated intravascular coagulation can occur, elevated hepatic enzymes, hematuria, hemolysis, hypotension, oliguria, pulmonary congestion

Admission Criteria/Prognosis Patients who develop tachypnea, cyanosis, dyspnea, CNS depression, tachycardia, or fever within 8 hours of exposure should be admitted; asymptomatic patients can be discharged after 8 hours observation

Treatment Strategy

Decontamination: **Do not** induce emesis or lavage due to risk of aspiration pneumonia. Activated charcoal can adsorb benzene.

Dermal: Wash with soap and water

Supportive therapy: Benzodiazepines for seizure control; avoid use of catecholamines due to risk of ventricular arrhythmia; treat burns with standard burn therapy; monitor for carbon monoxide

Aggressive airway management; intubation if pulmonary response is worrisome and/or loss of consciousness/CNS symptoms suggest potential loss of airway protection

For inhalation injuries, nebulized budesonide (0.5 mg every 12 hours for 4 days - pediatric dose) may be helpful

Personal Protection Equipment

Splash-proof chemical; safety goggles (or 8-inch face shield) with chemically impervious gloves, aprons, coveralls, boots

Respiratory protection includes chemical organic vapor cartridge respirator or supplied air respirator

Reference Range Urine phenol level (for benzene measurement) >40 mg/L is consistent with gasoline exposure in gasoline pump workers; blood 2 methylpentane level >50 mg/L has been associated with fatality

Additional Information Highly lipid soluble; gasoline contains a mixture of benzene (0.5%-2.5%), toluene, xylene, ethyl benzene, and possible lead (tetraethyl lead); other additives include ethylene dichloride and ethylene dibromide; lead poisoning is unusual from inhalation of gasoline containing tetraethyl lead

Odor threshold of gasoline: 0.25 ppm. Lethal inhalation concentration: 5000 ppm. Lethal ingestion concentration: 5 g/kg (12 oz)

Ambient level of gasoline at service stations is usually <100 ppm

OSHA - PEL-TWA: 300 ppm; TLV-TWA: 300 ppm

Specific gravity for gasoline is 0.75-0.8

A 1% solution of gasoline (inhaled) can produce symptoms within 5 minutes; 15-20 breaths of gasoline can result in intoxication for as long as 6 hours; Canada has replaced lead with methylcyclopentadrenyl manganese tricarbonyl (MMT), which can produce manganese toxicity

Specific References

Arnold JL, Halpern P, Tsai MC, et al, "Mass Casualty Terrorist Bombings: A Comparison of Outcomes by Bombing Type," *Ann Emerg Med*, 2004, 43(2):263-73.

Byard RW, Chivell WC, and Gilbert JD, "Unusual Facial Markings and Lethal Mechanisms in a Series of Gasoline Inhalation Deaths," *Am J Forensic Med Pathol*, 2003, 24(3): 298-302.

Gurkan F and Bosnak M, "Use of Nebulized Budesonide in Two Critical Patients With Hydrocarbon Intoxication," *Am J Ther*, 2005, 12(4):366-7.

Joseph PM and Weiner MG, "Visits to Physicians After the Oxygenation of Gasoline in Philadelphia," *Arch Environ Health*, 2002, 57(2):137-54.

Klein KR, White SR, Herzog P, et al, "Demand for PCC Services Surged During the 2003 Blackout," *J Toxicol Clin Toxicol*, 2004, 42(5):814.

Takamiya M, Niitsu H, Saigusa K, et al, "A Case of Acute Gasoline Intoxication at the Scene of Washing a Petrol Tank," *Leg Med*, 2003, 5(3):165-9.

Hydrogen Chloride

CAS Number 7647-01-0

UN Number 1050; 1789

Synonyms Chlorohydric Acid; Hydrochloride; Muriatic Acid; Spirits of Salt

Use In the manufacture of vinyl chloride and rubber; a combustible product of vinyl chloride (PVC)

Mechanism of Toxic Action Corrosive through strong acidity to the skin, eyes, nose, respiratory and GI tract; primarily local effects through denatured enzymes, minimal systemic effects; high water solubility leading to upper airway edema

Adverse Reactions Inhalation:
Cardiovascular: Chest pain, angina, tachycardia
Central nervous system: Chills, hyperthermia, fever
Nasal: Irritation, ulceration, epistaxis
Ocular: Corneal injury, conjunctivitis
Respiratory: Laryngeal spasms, noncardiogenic pulmonary edema (may be delayed for 24-72 hours), reactive airways disease syndrome, cough

Toxicodynamics/Kinetics
Absorption: Not absorbed
Metabolism: Ionized to hydronium and chloride ions

Pregnancy Issues Fetotoxic in animals

Signs and Symptoms of Acute Exposure By inhalation: tachycardia, laryngospasm, hyperchloremic metabolic acidosis, chest tightness, choking sensation, conjunctivitis, cough, dermal burns, dizziness, dyspnea, headache, hemoptysis, nasal ulcerations, nausea

Admission Criteria/Prognosis Patients with cardiopulmonary symptoms should be admitted

Treatment Strategy
Decontamination:
Dermal: Remove contaminated clothing. Wash area with copious amounts of soap and water.
Ocular: Remove contact lenses; irrigate with saline or water for at least 20 minutes until runoff pH is ~7.
Supportive therapy: Give 100% humidified oxygen. Treat wheezing with inhaled beta agonists. Corticosteroids may be helpful. Consider aggressive airway management, bronchodilators, and PEEP for bronchospasm. I.V. fluids should be administered cautiously to avoid a net positive fluid balance. Dopamine or norepinephrine are vasopressors of choice. Partial liquid ventilation may be useful. Dermal burns can be treated with standard therapy. Monitor for secondary infection. Inverse-ratio ventilation using lower tidal volume mechanical ventilation (6 mL/kg) and plateau pressures of 30 cm water may decrease mortality. While positive end expiratory pressures should be considered, oxygen toxicity should be monitored. Use of perfluorocarbon partial liquid ventilation and exogenous surfactant in chemical-induced ARDS is investigational.

Personal Protection Equipment Skin: 8 hour butyl, teflon, Tychem; 4 hour neoprene

Additional Information
Pungent odor; clear, colorless. Odor threshold: 1-5 ppm. TLV-ceiling: 5 ppm; IDLH: 100 ppm
Throat irritation at 5 ppm. Amounts over 1000 ppm can be fatal within a few minutes. Plant toxicity at 100 ppm.
Metal: Incompatible with zinc
pH of 1 N solution is 0.1

Specific References
Hsieh CH and Lin GT, "Corrosive Injury From Arterial Injection of Hydrochloric Acid," *Am J Emerg Med*, 2005, 23(3):394-6.

Kratom

Scientific Name Mitragyna Speciosa

Use As an opium substitute; the tree is native to Thailand and Malaysia. Used as a tea, or the leaves can be chewed or smoked. Typical tea dose is 10-15 g.

Mechanism of Toxic Action Contains mitragynine, an indole alkaloid that is similar to psilocybin and is a µ- and Δ-subtype receptor agonist (about 10% of the action of morphine)

Pregnancy Issues Effects not known

Signs and Symptoms of Acute Exposure Coma, vertigo, lethargy, tremors, nausea, vomiting, can be a stimulant (at doses of mitragynine >50 mg), miosis, constipation, doses >25 g of kraton leaves can be toxic.

Treatment Strategy Supportive care: Supportive care is mainstay of therapy. Mild withdrawal syndrome with chronic use may occur.

Additional Information Alkaloids resemble yohimbine; illegal in Australia, Cambodia, Malaysia, Myanmar, and Thailand. Exhibits psychoactive properties; typical effects last for 2-6 hours

Specific References

Matsumoto K, Horie S, Shikawa H, et al, "Antinociceptive Effect of 7-Hydroxymitragynine in Mice: Discovery of an Orally Active Opioid Analgesic From the Thai Medicinal Herb *Mitragyna speciosa*," *Life Sci*, 2004, 74:2143–55.

Lewisite

Related Information Lewisite and Mustard-Lewisite Mixture - Patient Information Sheet

CAS Number 541-25-3

Synonyms Chlorovinyl Arsine Dichloride

Military Classification Blister Agent

CDC Classification for Diagnosis

Biologic: A case in which sulfur mustard in biologic samples is detected, as determined by CDC or one of five IRN laboratories that have this capacity, and a case in which nitrogen mustard and lewisite are detected in biologic samples, as determined by CDC.

Environmental: Confirmation of the detection of vesicants in environmental samples is not available.

CDC Case Classification

Suspected: A case in which a potentially exposed person is being evaluated by healthcare workers or public health officials for poisoning by a particular chemical agent, but no specific credible threat exists.

Probable: A clinically compatible case in which a high index of suspicion (credible threat or patient history regarding location and time) exists for vesicant exposure, or an epidemiologic link exists between this case and a laboratory-confirmed case.

Confirmed: A clinically compatible case in which laboratory tests on biologic samples have confirmed exposure. The case can be confirmed if laboratory testing was not performed because either a predominant amount of clinical and nonspecific laboratory evidence of a particular chemical was present or a 100% certainty of the etiology of the agent is known.

Use First synthesized in U.S. in 1918 for vesicant warfare. Ten times more volatile than mustard gas; often mixed with sulfur mustard (U.S. stockpiles are scheduled to be destroyed by April, 2007). Sometimes mixed with mustard to lower the freezing point of mustard.

Pregnancy Issues Not teratogenic

Signs and Symptoms of Acute Exposure Hypotension, tachycardia, hypothermia, ocular burns, conjunctivitis, miosis, lacrimation, noncardiogenic pulmonary edema (usually associated with plural effusion), cough, dyspnea, tachypnea, weakness, nausea, vomiting, salivation, hypovolemia, hemolytic anemia (**no** bone marrow depression), dermal burns, blisters, erythema (within 30 minutes), skin edema

Treatment Strategy

Decontamination:

Dermal: Decontaminate immediately with 5% solution of sodium hypochlorite; use soap and water if bleach is not available. A 5% topical dimercaprol ointment or solution may prevent vesicant effects.

Inhalation: Administer 100% humidified oxygen; remove from source.

Ocular: Remove contact lenses, immediately irrigate copiously with tepid water or saline for at least 20 minutes. A 5% ophthalmologic compounded ointment or solution may be helpful if given within minutes of exposure.

Supportive therapy: Monitor serum/24-hour urine arsenic levels. Indications for chelation include: cough with dyspnea or signs of pulmonary edema, >1% total body surface area of dermal burn, or ≥5% of body surface in which there is skin damage or erythema noted within 30 minutes of exposure. Chelation should be continued until 24-hour urinary arsenic excretion falls under 50 mcg/L. Chelators include 2,3-dimercapto-1-propanesulfonic acid, succimer, or dimercaprol.

DMPS: Deep I.M.: 5% solution 3-5 mg/kg every 4-6 hours

DMSA: Oral: 10 mg/kg every 8 hours for 5 days, then every 12 hours for 14 days

Dimercaprol: Deep I.M.:

Mild toxicity: 2.5 mg/kg 4 times/day for 2 days, twice daily on third day, and once daily for 10 days or until complete recovery

Severe toxicity: 3 mg/kg every 4 hours for 2 days, 4 times/day on third day, and twice daily for 10 days or until complete recovery

Environmental Persistency

Shorter in hot climates; more persistent in cold areas

Soil: Intermediate

Atmospheric half-life: 1.2 days; water half-life: 5-10 days

All plants die at levels over 50 mg/L

Personal Protection Equipment Positive pressure, full-face piece, NIOSH-approved self-contained breathing apparatus with butyl rubber gloves, chemical goggles/face shield

Additional Information

Colorless, oily liquid with a geranium odor

150 mg/min/m^3 causes 1% mortality

Dermal absorption: As little as 0.5 mL can cause systemic effects (2 mL can be lethal). Vesication can occur with 14 mcg exposure and 30 mcg on skin can be lethal.

Specific References

Belson MG, Schier JG, and Patel MM, "Case Definitions for Chemical Poisoning," *MMWR Recomm Rep*, 2005, 54(RR-1):1-24.

Lindsay CD, Hambrook JL, Brown RF, et al, "Examination of Changes in Connective Tissue Macromolecular Components of Large White Pig Skin Following Application of Lewisite Vapour," *J Appl Toxicol*, 2004, 24(1):37-46.

Malathion

CAS Number 121-75-5

UN Number 2783

Synonyms O,O-Dimethyldithiophosphate Diethylmercaptosuccinate

U.S. Brand Names Ovide®

Use Marketed as insecticide granules, powder or dusting agent, or spray liquid concentrate with or without petroleum derivative as a solvent

Mechanism of Toxic Action Irreversible inhibition of acetylcholinesterase and plasma cholinesterase, resulting in excess accumulation of acetylcholine at muscarinic and nicotinic receptors, and in the central nervous system

Adverse Reactions
Cardiovascular: Hyperdynamic or hypodynamic states, QT prolongation, Raynaud's phenomenon, sinus bradycardia, sinus tachycardia, torsade de pointes
Central nervous system: Depression, seizures, hyperactivity, cognitive dysfunction, hypothermia, extrapyramidal reactions
Dermatologic: Scleroderma
Endocrine & metabolic: Diabetes insipidus, hypernatremia
Gastrointestinal: Pancreatitis
Genitourinary: Urinary incontinence
Neuromuscular & skeletal: Weakness (delayed), paralysis, sclerodermatous changes accompanied by Raynaud's phenomena, delayed paresthesia
Respiratory: Pulmonary edema, respiratory depression
Miscellaneous: Flu-like symptoms (especially with chronic exposure), intermediate syndrome

Toxicodynamics/Kinetics
Absorption: Readily through oral, dermal, or respiratory exposure
Metabolism: Rapid to weakly active compounds through hepatic hydrolysis and other pathways
Half-life: 2.9 hours
Elimination: Metabolites excreted in urine

Drug Interactions Paralysis is potentiated by neuromuscular blockade (ie, pancuronium, vecuronium, succinylcholine, atracurium, doxacurium, mivacurium); inhibition of serum esterase prolongs the half-life of succinylcholine, cocaine, and other ester anesthetics; cholinergic toxicity is potentiated by cholinesterase inhibitors such as physostigmine

Pregnancy Issues Human data regarding organophosphate exposure in pregnancy is limited and anecdotal; teratogenesis and fetal death have been produced in animal models at exposure levels that cause obvious toxicity to the pregnant animal

Diagnostic Test/Procedure
Creatinine, Serum
Pseudocholinesterase, Serum

Signs and Symptoms of Acute Exposure Abdominal pain, agitation, asystole, AV block, bradycardia, bronchorrhea, coma, confusion, cranial nerve palsies, decreased hemoglobin, decreased platelet count, decreased red blood cell count, dementia, diaphragmatic paralysis, dysarthria, excessive sweating, fecal and urinary incontinence, garlic-like breath, generalized asthenia, headache, heart block, hypertension, hypotension, lacrimation, metabolic acidosis and hyperglycemia (severe intoxication), miosis (unreactive to light), mydriasis (rarely), nausea, pallor, pulmonary edema, QT prolongation, respiratory depression, salivation, seizures, skeletal muscle fasciculation and flaccid paralysis, tachycardia, tachypnea, vomiting An "intermediate syndrome" of limb asthenia and respiratory paralysis has been reported to occur between 24 and 96 hours postorganophosphate exposure, and is independent of the acute cholinergic crisis. Late paresthesia characterized by stocking and glove paresthesia, anesthesia, and asthenia is infrequently observed weeks to months following acute exposure to certain organophosphates.

Admission Criteria/Prognosis Any symptomatic patient or an asymptomatic patient following a severe exposure should be admitted into an intensive care unit; asymptomatic patients following a moderate exposure can be discharged after 8 hours of observation.

Treatment Strategy
Decontamination: Isolation, bagging, and disposal of all contaminated clothing and other articles. All emergency medical workers and hospital staff should follow appropriate precautions regarding exposure to hazardous material including the use of protective clothing, masks, goggles, and respiratory equipment.
Oral: Activated charcoal can be administered either orally or via a nasogastric tube. Do not induce emesis because of danger of sudden respiratory compromise, alterations in mental status, seizures, coma, and possible aspiration of hydrocarbon vehicles. Do not give a cathartic.

Dermal: Prompt thorough scrubbing of all affected areas with soap and water, including hair and nails; 5% bleach can be used

Ocular: Irrigation with copious tepid sterile water or saline

Supportive therapy: Airway management, ventilatory assistance, humidified oxygen administration, and close monitoring for sudden respiratory failure

Enhancement of elimination: Dialysis and hemoperfusion are not indicated due to effectiveness of the prescribed treatment and large volumes of distribution of organophosphates.

Antidote:

Atropine: Administration should be guided by respiratory status, starting at 2-5 mg I.V. every 5-10 minutes as needed, and should be titrated to the resolution of excess pulmonary secretions. Frequent administration of large doses (cumulative doses >100 mg) may be necessary in massive exposures.

Glycopyrrolate: May be administered if atropine is unavailable (200-400 mcg I.M. or I.V. initially. or $\sim^1/_2$ the dose of atropine).

2-PAM: For more significant exposures (ie, exposures requiring large doses of atropine, or with recurring symptoms, or exposures to more lipid soluble agents), administration should follow: 1-2 g I.V. over 10-30 minutes, repeated in 1 hour if asthenia recurs, then every 4-12 hours for recurring symptoms.

Reference Range A 1.8 g ingestion with undetectable pseudocholinesterase resulted in a serum malathion level of 0.35 mg/L 2 hours postingestion.

Mild poisoning: Serum cholinesterase is 20% to 50% of normal;

Moderate poisoning: Serum cholinesterase is 10% to 20% of normal

Severe poisoning (respiratory distress and coma): Serum cholinesterase is <10%

Fatal malathion blood levels range from 0.3-1880 mg/L

Additional Information Red blood cell cholinesterase and serum pseudocholinesterase may be depressed following acute or chronic organophosphate exposure. RBC cholinesterase is typically not analyzed by in-house laboratories, and is usually not available for consideration during acute management. Pseudocholinesterase levels may be rapidly available from some in-house laboratories, but are not as reliable a marker of organophosphate exposure because of variability secondary to variant genotypes, hepatic disease, oral estrogen use, or malnutrition. Because of this variability, true indication of suppression of either of these enzymes can only be estimated through comparison to pre-exposure values; these enzymes may be useful in measuring a patient's recovery postexposure, especially if the recovery is not progressing as expected.

QT_c prolongation on EKG in the setting of organophosphate poisoning is associated with a high incidence of respiratory failure and mortality.

The intermediate syndrome is not related to delayed neuropathy.

Colorless to brown liquid with foul, skunk-like odor; thermal degradation products include sulfur and phosphorus oxides

Water solubility: 143 ppm. Vapor pressure: 0.00004 mm Hg at 20°C.

ACGIH TLV: 10 mg/m³; PEL-TWA: 10 mg/m³ (total dust); 5 mg/m³ (respirable fraction of dust); IDLH: 5000 mg/m³

Other information concerning pesticide exposures is available through the EPA-funded National Pesticide Telecommunications Network: (800) 858-7378

Specific References

Aksenov IV and Tonkopi DV, "Actoprotector Etomersol Is Effective for Acceleration of Recovery After Malathion Poisoning," *J Toxicol Clin Toxicol*, 2000, 38(5):555.

Benslama A, Moutaouakkil S, Mjahed K, et al, "Intermediary Syndrome in Acute Malathion Poisoning," *Presse Med*, 1998, 27(15):713-5.

Bentur Y, Raikhlin-Eisenkraft B, and Singer P, "Beneficial Late Administration of Obidoxime in Malathion Poisoning," *Vet Hum Toxicol*, 2003, 45(1):33-5.

Butera R, Locatelli C, Barretta S, et al, "Secondary Exposure to Malathion in Emergency Department Health-Care Workers," *J Toxicol Clin Toxicol*, 2002, 40(3):386-7.

Centers for Disease Control and Prevention, "Surveillance for Acute Insecticide-Related Illness Associated With Mosquito-Control Efforts - Nine States, 1999-2002," *MMWR Morb Mortal Wkly Rep*, 2003, 52(27):629-34.

Chambers-Emerson J, Normann SA, and Speranza VC, "Poison Center's Role During County-Wide Aerial Malathion Spraying," *J Toxicol Clin Toxicol*, 1998, 36(5):463.

Dribben W, Capiello H, and Kirk M, "Successful Organ Procurement and Transplantation From a Victim of Malathion Poisoning," *J Toxicol Clin Toxicol*, 1999, 37(5):665.

Horton DK, Berkowitz Z, and Kaye WE, "Secondary Contamination of ED Personnel From Hazardous Materials Events, 1995-2001," *Am J Emerg Med*, 2003, 21(3):199-204.

Kamijo Y, Soma K, Uchimiya H, et al, "A Case of Serious Organophosphate Poisoning Treated by Percutaneous Cardiopulmonary Support," *Vet Hum Toxicol*, 1999, 41(5): 326-8.

O'Sullivan BC, Lafleur J, Fridal K, et al, "The Effect of Pesticide Spraying on the Rate and Severity of ED Asthma," *Am J Emerg Med*, 2005, 23(4):463-7.

Sudakin DL, Mullins ME, Horowitz BZ, et al, "Intermediate Syndrome After Malathion Ingestion Despite Continuous Infusion of Pralidoxime," *J Toxicol Clin Toxicol*, 2000, 38(1):47-50.

Methyl Isocyanate

CAS Number 624-83-9

UN Number 2480

Synonyms Isocyanate; Isocyanatomethane; MIC

Use Intermediate agent in the production of carbaryl (Sevin®, a carbamate pesticide) and herbicides. Etiologic agent of inadvertent release in 1984 in Bhopal, India.

Mechanism of Toxic Action Mucosal irritant; reacts with water to form exothermic reaction; hydrolyzes to form methylamine and CO_2

Adverse Reactions

Cardiovascular: Chest pain, angina

Central nervous system: Fatigue, vertigo, ataxia, anxiety

Dermal: Erythema, edema

Gastrointestinal: Vomiting, throat irritation

Ocular: Burning eyes, lacrimation, blepharospasm, photophobia, corneal ulceration, punctate keratopathy, lid edema, corneal irritation may persist for one year postexposure

Neuromuscular & skeletal: Muscle weakness, tremor

Renal: Acidosis (renal tubular)

Respiratory: Cough, dyspnea, bronchospasm (persists for 3-7 days), wheezing, interstitial edema, bronchiolitis

Toxicodynamics/Kinetics Absorption: By inhalation and dermal routes

Pregnancy Issues

Experience after Bhopal, India 1984 exposure of 865 pregnancies revealed that 43% did not result in a live birth and an infant death rate of 14%. MIC can cross placental barrier; skeletal malformations noted in rodent studies.

Clinically Resembles Asthma

Treatment Strategy

Decontamination:

Inhalation: Give 100% humidified oxygen.

Ocular: Remove contact lenses; irrigation with copious amount of water or saline

Supportive therapy: Beta agonist agents with theophylline may be helpful; steroids (prednisone or high-dose inhaled beclomethasone) may be useful. Thiosulfate does not appear to be helpful and cyanide antidotal therapy should not be utilized. Inverse-ratio ventilation using lower tidal volume mechanical ventilation (6 mL/kg) and plateau pressures of 30 cm water may decrease mortality. While positive end expiratory pressures should be considered, oxygen toxicity should be monitored. Use of perfluorocarbon partial liquid ventilation and exogenous surfactant in chemical-induced ARDS is investigational.

Reference Range Leukocytosis, lymphocytosis and elevated erythrocyte sedimentation rate noted; blood carboxyhemoglobin, methemoglobin, and thiocyanate levels may be high, but this may be due to contaminants

Additional Information Mucous membrane irritant threshold: 0.2 ppm. Unbearable exposure: 21 ppm; respiratory response is likely at an air concentration of >0.5 ppm

Very flammable. TLV-TWA: 0.02 ppm. Odor threshold: 2 ppm

Toxic agent responsible for 3828 deaths in December 1984 in Bhopal, India

Pyrolysis decomposition products include hydrogen cyanide and carbon dioxide; MDI is more toxic when heated

Involved in the recall of >8 billion cigarettes by the Philip Morris Co in May 1995 due to contaminated plasticizer compound in the cigarette filter

Specific References

Ranjan N, Sarangi S, Padmanabhan VT, et al, "Methyl Isocyanate Exposure and Growth Patterns of Adolescents in Bhopal," *JAMA*, 2003, 290(14):1856-7.

Methyldichloroarsine

CAS Number 593-89-5
UN Number 1556
Synonyms MD
Military Classification Vesicant
Clinically Resembles Poison ivy

Signs and Symptoms of Acute Exposure Instantaneous pain and irritation of eyes, skin, and nasal pharynx; dermal burns with vesicle formation. Blepharospasm and photophobia can occur. Coughing and shortness of breath with pulmonary damage usually delayed for 3-5 days; hemolysis may occur.

Treatment Strategy
Decontamination:
Dermal: Remove liquids by any means; dilute hypochlorite (0.5% solution) can be used as an irrigating solution. Copious irrigation with water or normal saline is critical. Do not delay irrigation with water to await hypochlorite.
Inhalation: 100% humidified oxygen
Ocular: Copious irrigation with water or normal saline is critical. Do not delay irrigation with water to await hypochlorite. Irrigate until runoff pH is ~7
Supportive therapy: Small vesicles (<2 cm) can be covered with topical antibiotics while larger vesicles (>2 cm) should be denuded and irrigated with sterile saline; silver sulfadiazine can then be applied topically. Consider aggressive airway management. If laryngospasm is present, use PEEP. Systemic analgesic agents can be used for pain control; use aggressive pain management. Monitor for arsenic poisoning and treat if elevated arsenic levels are noted. Avoid salicylates when managing hyperthermia. If significant hemolysis occurs, exchange transfusion may be useful.

Personal Protection Equipment Activated charcoal cartridge with protective mask; standard mission-oriented protective posture (MOPP)

Additional Information Odor of rotting fruit

Mustard Gas

CAS Number 505-60-2
UN Number 2810
Synonyms Bis(2-Chloroethyl) Sulfide; Distilled Mustard; HD; HS; MG; S-Mustard; Yperite

Military Classification Blister Agent

Use Primarily of historical use during World War I as a vesicant chemical warfare agent; most recently used for this purpose during the Iran-Iraq War in the 1980s; derivative of this chemical has been used as an antineoplastic agent (an alkylating agent - nitrogen mustard)

Mechanism of Toxic Action Dermal, ocular, and respiratory corrosive agent; induction of structural changes to DNA/RNA (through alkylating of nucleic acids), low doses may inhibit cell division. Produces free radical-mediated oxidative stress, lipid peroxidation, and possibly glutathione depletion. DNA alkylation and cytotoxicity occur.

Adverse Reactions Children may have a shorter time for symptom onset with more severe skin lesions.

Central nervous system: Fever

Dermatologic: Erythema and pruritus; may be delayed for up to 8 hours, then ulceration, dermal burns, and blisters can result

Gastrointestinal: Vomiting, nausea, diarrhea, anorexia

Hematologic: Bone marrow suppression, pancytopenia; leukopenia usually occurs 7-10 days postexposure; eosinophilia can be seen in children

Ocular: Lacrimation, photophobia, blepharospasm, ocular irritation, opacification, blindness

Respiratory: Dyspnea, cough (usual onset is 1-12 hours postexposure)

Toxicodynamics/Kinetics

Absorption: Oral, dermal ($\sim$20%), pulmonary

Metabolism: Hepatic through hydrolysis and glutathione pathways; thiodiglycol constituents

Half-life: $\sim$1.2 days (thiodiglycol)

Elimination: Renal

Pregnancy Issues Possible teratogen (skeletal malformations/cleft palate)

Clinically Resembles Radiation injury, chemotherapeutic agent toxicity, toxic-shock syndrome, bullous pemphigoid, chemical burns, sunburn

Treatment Strategy

Decontamination: Note: Penetrates wood, leather, rubber, and paints; decontamination must occur within 2 minutes.

Dermal: Remove all contaminated clothing. Towels soaked in 0.2% chloramine-T in water (Dakin solution) placed over wounds for the first 2 hours may be helpful. Wash with soap and water (not hot). Can wash with dilute (0.5%) hypochlorite, then neutralize with 2.5% sodium thiosulfate. If no water is available, dry decontamination with Fuller's earth can be utilized. Treat wounds as burns with use of $^1/_8$ inch of silver sulfadiazine (twice daily). Leave small blisters (<1 cm) intact; unroof larger blisters and irrigate.

Ocular: Remove contact lenses; copious irrigation with saline (neutralize with 2.5% solution of sodium thiosulfate)

Respiratory: Administer 100% oxygen. A nebulized mist of 2.5% sodium thiosulfate may be helpful as a neutralizing agent if given within 15 minutes of exposure.

Oral: Activated charcoal or 150 mL of 2% sodium thiosulfate orally

Supportive therapy: N-acetylcysteine orally 4 times/day (up to 150 mg/kg), along with ascorbic acid (3 g), thienamycin (2 g twice daily), L-carnitine (3 g), and sodium thiosulfate (3-12.5 g/day), has been used to ameliorate effects, but it is unproven in its efficacy. For ocular exposure, topical antibiotics, mydriatics, and possibly corticosteroids can be utilized. Vaseline® placed on eyelid edges may prevent lids from adhering. For dermal burns, topical antibiotics with systemic analgesics are the mainstay of therapy. Do not overhydrate the patient. Do not fluid resuscitate as in thermal burns. Colony stimulating factor may be helpful in treating leukopenia. Prednisone (60-125 mg/day) orally can be helpful in treating pulmonary toxicity. Bronchospasm should be treated with β_2 adrenergic agonist agents.

Environmental Persistency Atmospheric half-life: $\sim$2.1 days

Higher wind speeds, higher temperature, and humidity increase the atmospheric vaporization rate; evaporates on surface soil within 30-50 hours at 25°C; may persist in soil for weeks.

Personal Protection Equipment Pressure demand, self-contained breathing apparatus (SCBA) with butyl rubber, neoprene, nitrile or polyvinyl chloride gloves, Responder® CSM protective clothing, PVC boots, chemical goggles, and face shield

Reference Range Urinary thiodiglycol detection limit is 1 ppb (1 µg/L). Urine thiodiglycol levels >30 ng/mL 12 days postexposure associated with severe ocular and skin lesions.

Additional Information

A combustible yellow, oily liquid with a garlic, horseradish, or mustard odor.

Median lethal dose: Inhalation: 1500 mg/min/m^3; skin absorption: 10,000 mg/min/m^3 (wet skin absorbs more mustard than dry skin)

Toxic dermal dose: 0.1% solution; fair-skinned individuals are more at risk for adverse dermal effects than dark-skinned individuals.

Total white blood cell count <200 is a harbinger for fatality. No mustard can be isolated in blister fluid.

Combat zones atmospheric concentration of mustard gas during WWI was estimated to be from 3-5 ppm; case fatality rate: 2% to 4%.

Since mustard gas binds rapidly and avidly to tissue proteins, decontamination must begin immediately, and increasing elimination of absorbed chemical is difficult.

Rate of hydrolysis (half-life) is 8.5 minutes in distilled water (25°C) and 1 hour in salt water (25°C). Hydrolyzed to hydrogen chloride and thiodiglycol.

Ocular injury occurs at 10 mg/min/m^3

Doses up to 50 mcg/cm^2 can cause skin erythema and vesicles. Exposures at 50-150 mcg/cm^2 can cause bullous-type vesicles, and doses over 150 mcg/cm^2 can cause skin necrosis and ulceration.

Liquid density: 1.27 at 25°C; vapor density: 5.4

Precursor chemical to sulfur mustard (HD), sesqui mustard, and nitrogen mustard (HN-1) is 2-chloroethanol (CAS: 107-07-3): Precursor chemical to nitrogen mustard is Triethanolamine hydrochloride. Other precursors to sulfur mustard includes sulfur dichloride (CAS: 10545-99-0); sodium sulfide (CAS: 1313-82-2) sulfur monochloride, sulfur chloride (CAS: 10025-67-9) thiorylchloride (CAS: 7719-09-7) and thiodialvcol (CAS: 111-48-8).

Typical Composition of Sulfur Mustard (H) from an Old Chemical Munition		
Compound	CAS No.	GC/MS Peak Area Percent
Sulfur mustard	505-60-2	62.2
Bis(2-chloroethyl) disulfide	1002-41-1	10.9
1,4-Dithiane	505-29-3	3.2
Bis(2-chloroethyl) trisulfide	19149-77-0	9.6
1,2-Bis(2-chloroethylthio)ethane	3563-36-8	2.6
1,2,3-Trithiolane	–	2.4
1,4-Thioxane	15980-15-1	0.1
1,2,5-Trithiepane	6576-93-8	0.9
1,2,3,4-Tetrathiane	–	1.4
1,2-Dichloroethane	107-06-2	3.2
HD Tetrasulfide	–	0.6
Tetrachloroethene	127-18-4	0.3
Sulfur	7704-34-9	0.5
Other	–	1.3

GC/MS = gas chromatography/mass spectrometry.

From *Toxicological Profile for Sulfur Mustard (Update)*, Syracuse Research Corporation, U.S. Department of Health and Human Services, 2003, 121.

Typical Composition of Sulfur Mustard (HD) in 1-Ton Storage Containers (Aberdeen, Maryland)

Compound	CAS No.	Mole Percent
Sulfur mustard	505-60-2	91.38
Q sulfonium	30843-67-5	6.08
2-Chloroethyl 4-chlorobutyl sulfide	114811-35-7	0.86
1,4-Dithiane	505-29-3	0.81
1,2-Dichloroethane	107-06-2	0.35
Bis-3-chloropropyl sulfide	22535-54-2	0.18
2-Chloropropyl 3'-chloropropyl sulfide	–	0.18
2-Chloroethyl 3-chloropropyl sulfide	71784-01-5	0.14
1-Chloropropyl 2-chloroethyl sulfide	–	0.02
1,4-Thioxane	15980-15-1	<0.01

From *Toxicological Profile for Sulfur Mustard (Update)*, Syracuse Research Corporation, U.S. Department of Health and Human Services, 2003, 122.

Physical and Chemical Properties of Sulfur Mustard

Property	Information	Reference
Molecular weight	159.08	Budavari et al, 1996
Color	Clear	Budavari et al, 1996
	Pale yellow, black if impure	Munro et al, 1999
Physical state	Oily liquid	Budavari et al, 1996
Melting point	13°C to 14°C	Budavari et al, 1996
Boiling point	217.5°C	Budavari et al, 1996
Density	1.338 at 13°C	Budavari et al, 1996
	1.2685 at 25°C	Rosenblatt et al, 1996
Odor	Weak, sweet, agreeable odor	Budavari et al, 1996
Odor threshold:		
Water	No data	
Air	0.6 mg/m^3	Bowden 1943
Solubility:		
Water	920 mg/L at 22°C	Rosenblatt et al, 1996
	684 mg/L at 25°C	Seidell 1941
Organic solvent(s)	Soluble in alcohol, ether, acetone, and benzene; miscible with petroleum ether	HSDB 2002
	Soluble in fat solvents and other common organic solvents	IARC 1975
Partition coefficients:		
Log K_{ow}	2.41	HSDB 2002
	1.37	Rosenblatt et al, 1996
Log K_{oc}	2.43	HSDB 2002

Table   (*Continued*)

Property	Information	Reference
Vapor pressure:		
at 22°C	0.082 mm Hg	Rosenblatt et al, 1996
at 25°C	0.1059 mm Hg	Rosenblatt et al, 1996
Henry's law constant	2.4×10^{-5} atm-m^3/mol	Opresko et al, 1998
	1.87×10^{-5} atm-m^3/mol	Rosenblatt et al, 1996
Autoignition temperature	No data	
Flashpoint	221°F	Sax 1989
Conversion factors	No data	
Explosive limits	No data	

From *Toxicological Profile for Sulfur Mustard (Update)*, Syracuse Research Corporation, U.S. Department of Health and Human Services, 2003, 123.

U.S. Army Toxicity Values for Sulfur Mustard		
Parameter	**Value**	**Units**
Oral reference dose	0.000007	mg/kg/day
Inhalation reference dose	0.00003	mg/kg/day
Cancer potency oral slope factor	7.7	(mg/kg/day)$^{-1}$
Cancer potency inhalation unit risk	0.085	(mcg/m^3)$^{-1}$
Cancer potency inhalation slope factor	300	(mg/kg/day)$^{-1}$

From *Toxicological Profile for Sulfur Mustard (Update)*, Syracuse Research Corporation, U.S. Department of Health and Human Services, 2003, 165.

Primary Hydrolysis Pathways of Sulfur Mustard in the Environment

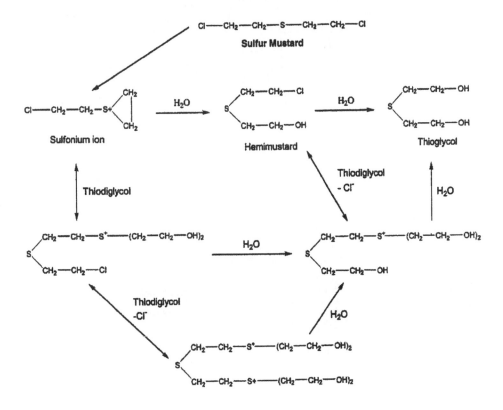

From *Toxicological Profile for Sulfur Mustard (Update)*, Syracuse Research Corporation, U.S. Department of Health and Human Services, 2003, 138.

Locations of Sulfur Mustard Stockpile and Nonstockpile Sites in the United States[1]

Source: OA 2000, 2002   (★) = Stockpile sites; (O) = Non-stockpile sites

Johnston Atoll (South Pacific) ★   o U.S. Virgin Is.

AL
  Anniston Army Depot (★)
  Ft McClellan (O)
  Camp Sibert (O)
  Huntsville Arsenal (O)
  Redstone Arsenal (O)
  Theodore Naval Ammunition
  Magazine (O) – not shown on map
AK
  Adak (O)
  Chicagoff Harbor (O)
  Gerstle River Test Center (O)
AZ
  Navajo Depot Activity (O)
  Yuma Proving Ground (O)
AR
  Pine Bluff Arsenal (★)
CA
  Ft Ord (O)
  Edwards AFB (O)
CO
  Rocky Mountain Arsenal (O)
  Pueblo Army Depot Activity (★)
FL
  Brooksville Army Air Base (O)
  Drew Field (O)
  MacDill AFB (O)
  Withlacoochee (O)
  Dry Tortuga Keys (O)
  Zephyr Hills Gunner Range (O)

GA
  Ft. Gillem (O)
  Manchester (O)
HI
  Kipapa Ammunition Storage Site (O)
  Schofield Barracks (O)
IL
  Savanna Army Depot Activity (O)
IN
  Camp Atterbury (O)
  Naval Weapons Support Center (O)
KS
  Marysville (O)
KY
  Blue Grass Army Depot (★)
LA
  Ft Polk (O)
  Concord Spur (O)
MD
  Edgewood Area- Aberdeen Proving
  Grounds (★)
MS
  Camp Shelby (O)
  Columbus Army Airfield (O)
  Horne Island (O)
NC
  Laurinburg-Maxton Army Air Base
  (O)

NE
  Nebraska Ordnance Plant (O)
NV
  Hawthorne Army Ammunition Plant (O)
NJ
  Raritan Arsenal (O)
OH
  Ravenna Army Ammunition Plant (O)
OR
  Umatilla Depot Activity (★)
SC
  Charleston Army Depot (O)
  Naval Weapons Center (O)–not shown on map
SD
  Black Hills Ordnance Depot (O)
TN
  Defense Depot Memphis (O)
TX
  Ft Hood (O)
  Camp Stanley Storage Activity (O)
  Camp Bullis (O)
UT
  Dugway Proving Grounds (O)
  Defense Depot Ogden (O)
  Tooele Army Depot (★)
VI
  Ft. Segarra (St. Thomas) (O)
Other
  Johnston Atoll (S. Pacific)(★)

[1]Post office state abbreviations used.

From *Toxicological Profile for Sulfur Mustard (Update)*, Syracuse Research Corporation, U.S. Department of Health and Human Services, 2003, 126.

Specific References

Amitai G, Adani R, Hershkovitz M, et al, "Degradation of VX and Sulfur Mustard by Enzymatic Haloperoxidation," *J Appl Toxicol*, 2003, 23(4):225-33.

Barr JR, Driskell WJ, Aston LS, et al, "Quantitation of Metabolites of the Nerve Agents Sarin, Soman, Cyclohexylsarin, VX, and Russian VX in Human Urine Using Isotope-Dilution Gas Chromatography-Tandem Mass Spectrometry," *J Anal Toxicol*, 2004, 28(5):372-8.

Boyer AE, Ash D, Barr DB, et al, "Quantitation of the Sulfur Mustard Metabolites 1,1'-Sulfonylbis[2-(methylthio)ethane] and Thiodiglycol in Urine Using Isotope-dilution Gas Chromatography-Tandem Mass Spectrometry," *J Anal Toxicol*, 2004, 28(5):327-32.

Byers CE, Holloway ER, Korte WD, et al, "Gas Chromatographic-Mass Spectrometric Determination of British Anti-Lewisite in Plasma," *J Anal Toxicol*, 2004, 28(5):384-9.

Capacio BR, Smith JR, DeLion MT, et al, "Monitoring Sulfur Mustard Exposure by Gas Chromatography-Mass Spectrometry Analysis of Thiodiglycol Cleaved From Blood Proteins," *J Anal Toxicol*, 2004, 28(5):306-10.

Dalton CH, Maidment, MP, Jenner J, et al, "Closed Cup Vapor Systems in Percutaneous Exposure Studies: What Is the Dose? *J Anal Toxicol*, 2006, 30:165-70.

Degenhardt CE, Pleijsier K, van der Schans MJ, et al, "Improvements of the Fluoride Reactivation Method for the Verification of Nerve Agent Exposure," *J Anal Toxicol*, 2004, 28(5):364-71.

Devereaux A, Amundson DE, Parrish JS, et al, "Vesicants and Nerve Agents in Chemical Warfare. Decontamination and Treatment Strategies for a Changed World," *Postgrad Med*, 2002, 112(4):90-6; quiz 4.

Dompeling E, Jobsis Q, Vandevijver NM, et al, "Chronic Bronchiolitis in a 5-yr-old Child After Exposure to Sulphur Mustard Gas," *Eur Respir J*, 2004, 23(2):343-6.

Garner JP, "Some Recollections of Portion in World War 1. Commentary," *J R Army Med Corps*, 2003, 149(2):138-41.

Han S, Espinoza LA, Liao H, et al, "Protection by Antioxidants Against Toxicity and Apoptosis Induced by the Sulphur Mustard Analog 2-Chloroethylethyl Sulphide (CEES) in Jurkat T Cells and Normal Human Lymphocytes," *Br J Pharmacol*, 2004, 141(5):795-802. Epub 2004 Feb 09.

Jakubowski EM, McGuire JM, Evans RA, et al, "Quantitation of Fluoride Ion Released Sarin in Red Blood Cell Samples by Gas Chromatography-Chemical Ionization Mass Spectrometry Using Isotope Dilution and Large-Volume Injection," *J Anal Toxicol*, 2004, 28(5):357-63.

Kadar T, Fishbine E, Meshulam J, et al, "A Topical Skin Protectant Against Chemical Warfare Agents," *Isr Med Assoc J*, 2003, 5(10):717-9.

Kales SN and Christiani DC, "Acute Chemical Emergencies," *N Engl J Med*, 2004, 350(8):800-8.

Khateri S, Ghanei M, Keshavarz S, et al, "Incidence of Lung, Eye, and Skin Lesions as Late Complications in 34,000 Iranians With Wartime Exposure to Mustard Agent," *J Occup Environ Med*, 2003, 45(11):1136-43.

Khateri S, Ghanei M, Soroush MR, et al, "Effects of Mustard Gas Exposure in Pediatric Patients (Long-Term Health Status of Mustard-Exposed Children, 14 Years After Chemical Bombardment of Sardasht)," *J Toxicol Clin Toxicol*, 2003, 41(5):733.

Lemire SW, Barr JR, Ashley DL, et al, "Quantitation of Biomarkers of Exposure to Nitrogen Mustards in Urine From Rats Dosed With Nitrogen Mustards and From an Unexposed Human Population," *J Anal Toxicol*, 2004, 28(5):320-6.

McClintock SD, Till GO, Smith MG, et al, "Protection From Half-Mustard-Gas-Induced Acute Lung Injury in the Rat," *J Appl Toxicol*, 2002, 22(4):257-62.

Mi L, Gong W, Nelson P, et al, "Hypothermia Reduces Sulphur Mustard Toxicity," *Toxicol Appl Pharmacol*, 2003, 193(1):73-83.

Noort D, Fidder A, Benschop HP, et al, "Procedure for Monitoring Exposure to Sulfur Mustard Based on Modified Edman Degradation of Globin," *J Anal Toxicol*, 2004, 28(5):311-5.

Noort D, Fidder A, Hulst AG, et al, "Retrospective Detection of Exposure to Sulfur Mustard: Improvements on an Assay for Liquid Chromatography-Tandem Mass Spectrometry Analysis of Albumin-Sulfur Mustard Adducts," *J Anal Toxicol*, 2004, 28(5):333-8.

Read RW and Black RM, "Analysis of Beta-lyase Metabolites of Sulfur Mustard in Urine by Electrospray Liquid Chromatography-Tandem Mass Spectrometry," *J Anal Toxicol*, 2004, 28(5):346-51.

Read RW and Black RM, "Analysis of the Sulfur Mustard Metabolite 1,1'-Sulfonylbis[2-S-(N-acetylcysteinyl)ethane] in Urine by Negative Ion Electrospray Liquid Chromatography-Tandem Mass Spectrometry," *J Anal Toxicol*, 2004, 28(5):352-6.

Rice P, "Sulphur Mustard Injuries of the Skin. Pathophysiology and Management," *Toxicol Rev*, 2003, 22(2):111-8.

Smith JR, "Analysis of the Enantiomers of VX Using Normal-Phase Chiral Liquid Chromatography With Atmospheric Pressure Chemical Ionization-Mass Spectrometry," *J Anal Toxicol*, 2004, 28(5):390-2.

Stuart JA, Ursano RJ, Fullerton CS, et al, "Belief in Exposure to Terrorist Agents: Reported Exposure to Nerve or Mustard Gas by Gulf War Veterans," *J Nerv Ment Dis*, 2003, 191(7):431-6.

Thomason JW, Rice TW, and Milstone AP, "Bronchiolitis Obliterans in a Survivor of a Chemical Weapons Attack," *JAMA*, 2003, 290(5):598-9.

Toxicological Profile for Sulfur Mustard (Update), Syracuse Research Corporation, U.S. Department of Health and Human Services, 2003.

van der Schans GP, Mars-Groenendijk R, de Jong LPA, et al, "Standard Operating Procedure for Immunuslotblot Assay for Analysis of DNA/Sulfur Mustard Adducts in Human Blood and Skin," *J Anal Toxicol*, 2004, 28(5):316-9.

Williems JL, "Mustard Gas: Signs, Symptoms and Treatment," *J Toxicol Clin Toxicol*, 2002, 40(3):250-1.

Wormser U, Sintov A, Brodsky B, et al, "Protective Effect of Topical Iodine Containing Anti-Inflammatory Drugs Against Sulfur Mustard-Induced Skin Lesions," *Arch Toxicol*, 2004, 78(3):156-66. Epub 2003 Nov 15.

Young CL, Ash D, Driskell WJ, et al, "A Rapid, Sensitive Method for the Quantitation of Specific Metabolites of Sulfur Mustard in Human Urine Using Isotope-Dilution Gas Chromatography-Tandem Mass Spectrometry," *J Anal Toxicol*, 2004, 28(5):339-45.

Zilker T and Felgenhauer N, "S-Mustard Gas Poisoning - Experience With 12 Victims," *J Toxicol Clin Toxicol*, 2002, 40(3):251.

Napalm B

Military Classification Incendiary Agent

Use Firebombs, landmines, and flame throwers since World War II

Mechanism of Toxic Action White phosphorus was the igniting agent (replaced with thermite). Increased viscosity as opposed to gasoline; primarily composed of napthenate and palmitate (aluminum salt); also contains oleic acid (5% to 12%)

Adverse Reactions Dermal burns, ocular burns, dyspnea, respiratory distress

Clinically Resembles Burns

Treatment Strategy

Decontamination: Dermal: Flush with copious amounts of water; administer 100% humidified oxygen; aggressive airway management/intubation may be necessary

Supportive therapy: Napalm B itself has no inherent systemic toxicity. Remove all contaminated clothing. Evaluate for carbon monoxide exposure and treat with standard burn protocol to maintain urine output at 1-2 mL/kg/hour. Topical antibiotics such as mafenide can be used in burn wound dressing. Take precautions against infection. Monitor for associated traumatic injury.

Rescuer Contamination Take precautions against flammables that have not been completely ignited or extinguished

Personal Protection Equipment Level A suits **not** protective of thermal injury

Additional Information A viscous, cloudy white sticky gelatin substance. Stable at ambient temperatures; no outgassing of hydrocarbons; no systemic toxicity. Can be adhesive while burning.

Specific References

Arnold JL, Halpern P, Tsai MC, et al, "Mass Casualty Terrorist Bombings: A Comparison of Outcomes by Bombing Type," *Ann Emerg Med*, 2004, 43(2):263-73.

Reich P and Sidel VW, "Current Concepts. Napalm," *N Engl J Med*, 1967, 277(2):86-8.

Nicotine

CAS Number 54-11-5

UN Number 1654; 1655; 1656; 1657; 1658; 1659; 3144

Commonly Includes Cigarettes (13-19 mg); nicotine gum (2-4 mg); cigars (15-40 mg); cigarette butt (5-7 mg); nicotine patch (8.3-114 mg); chewing tobacco (6-8 mg); also associated with epibatidine ("poison dart" frogs) and anatoxin-a (found in cyanobacteria, formally known as blue-green algae)

CDC Classification for Diagnosis

Biologic: A case in which increased nicotine or cotinine (the nicotine metabolite) is detected in urine, or increased serum nicotine levels occur, as determined by a commercial laboratory or CDC.

Environmental: Detection of nicotine in environmental samples, as determined by NIOSH or FDA.

CDC Case Classification

Suspected: A case in which a potentially exposed person is being evaluated by healthcare workers or public health officials for poisoning by a particular chemical agent, but no specific credible threat exists.

Probable: A clinically compatible case in which a high index of suspicion (credible threat or patient history regarding location and time) exists for nicotine exposure, or an epidemiologic link exists between this case and a laboratory-confirmed case.

Confirmed: A clinically compatible case in which laboratory tests have confirmed exposure. The case can be confirmed if laboratory testing was not performed because either a predominant amount of clinical and nonspecific laboratory evidence of a particular chemical was present or a 100% certainty of the etiology of the agent is known.

U.S. Brand Names Commit® [OTC]; NicoDerm® CQ® [OTC]; Nicorette® [OTC]; Nicotrol® Inhaler; Nicotrol® NS; Nicotrol® Patch [OTC]

Use

Treatment to aid smoking cessation for the relief of nicotine withdrawal symptoms (including nicotine craving)

Insecticide (rarely), found in tobacco leaf (1% to 6% nicotine by weight) products.

Adverse Reactions

Chewing gum/lozenge:

>10%:

Cardiovascular: Tachycardia

Central nervous system: Headache (mild)

Gastrointestinal: Nausea, vomiting, indigestion, excessive salivation, belching, increased appetite

Miscellaneous: Mouth or throat soreness, jaw muscle ache, hiccups

1% to 10%:

Central nervous system: Insomnia, dizziness, nervousness

Endocrine & metabolic: Dysmenorrhea

Gastrointestinal: GI distress, eructation

Neuromuscular & skeletal: Muscle pain

Respiratory: Hoarseness

Miscellaneous: Hiccups

<1%: Atrial fibrillation, erythema, hypersensitivity reactions, itching

Transdermal systems:

>10%:

Central nervous system: Insomnia, abnormal dreams

Dermatologic: Pruritus, erythema

Local: Application site reaction

Respiratory: Rhinitis, cough, pharyngitis, sinusitis

1% to 10%:

Cardiovascular: Chest pain

Central nervous system: Dysphoria, anxiety, difficulty concentrating, dizziness, somnolence

Dermatologic: Rash

Gastrointestinal: Diarrhea, dyspepsia, nausea, xerostomia, constipation, anorexia, abdominal pain

Neuromuscular & skeletal: Arthralgia, myalgia

<1%: Atrial fibrillation, hypersensitivity reactions, itching, nervousness, taste perversion, thirst, tremor

Toxicodynamics/Kinetics

Absorption: Through skin and oral mucosa, GI tract (except in stomach), respiratory tract; increased gastrointestinal, buccal, and dermal absorption in alkali medium

Distribution: V_d: 1-3 L/kg; higher in nonsmokers

Protein binding: 5% to 20%

Metabolism: Hepatic (cytochrome P450) to cotinine and nicotine-N-oxide; large first-pass effect

Half-life: Smokers: 0.8 hours; Nonsmokers: 1.3 hours; Cotinine: 10-20 hours; Gum/cigarette: 1-2 hours; Transdermal systems: 3-6 hours or higher

Elimination: 5% to 10% excreted unchanged, urine acidity with pH <5 may increase % excreted to 30%

Usual Dosage Smoking deterrent: Patients should be advised to completely stop smoking upon initiation of therapy.

Oral:

Gum: Chew 1 piece of gum when urge to smoke, up to 24 pieces/day. Patients who smoke <25 cigarettes/day should start with 2-mg strength; patients smoking ≥25 cigarettes/day should start with the 4-mg strength. Use according to the following 12-week dosing schedule:

Weeks 1-6: Chew 1 piece of gum every 1-2 hours; to increase chances of quitting, chew at least 9 pieces/day during the first 6 weeks

Weeks 7-9: Chew 1 piece of gum every 2-4 hours

Weeks 10-12: Chew 1 piece of gum every 4-8 hours

Inhaler: Usually 6 to 16 cartridges per day; best effect was achieved by frequent continuous puffing (20 minutes); recommended duration of treatment is 3 months, after which patients may be weaned from the inhaler by gradual reduction of the daily dose over 6-12 weeks

Lozenge: Patients who smoke their first cigarette within 30 minutes of waking should use the 4-mg strength; otherwise the 2-mg strength is recommended. Use according to the following 12-week dosing schedule:

Weeks 1-6: One lozenge every 1-2 hours

Weeks 7-9: One lozenge every 2-4 hours

Weeks 10-12: One lozenge every 4-8 hours

Note: Use at least 9 lozenges/day during first 6 weeks to improve chances of quitting; do not use more than one lozenge at a time (maximum: 5 lozenges every 6 hours, 20 lozenges/day)

Topical:

Transdermal patch: Apply new patch every 24 hours to nonhairy, clean, dry skin on the upper body or upper outer arm; each patch should be applied to a different site. Note: Adjustment may be required during initial treatment (move to higher dose if experiencing withdrawal symptoms; lower dose if side effects are experienced).

NicoDerm CQ®:

    Patients smoking ≥10 cigarettes/day: Begin with step 1 (21 mg/day) for 4-6 weeks, followed by step 2 (14 mg/day) for 2 weeks; finish with step 3 (7 mg/day) for 2 weeks

    Patients smoking <10 cigarettes/day: Begin with step 2 (14 mg/day) for 6 weeks, followed by step 3 (7 mg/day) for 2 weeks

    Note: Initial starting dose for patients <100 pounds, history of cardiovascular disease: 14 mg/day for 4-6 weeks, followed by 7 mg/day for 2-4 weeks

    Note: Patients receiving >600 mg/day of cimetidine: Decrease to the next lower patch size

Nicotrol®: One patch daily for 6 weeks

    Note: Benefits of use of nicotine transdermal patches beyond 3 months have not been demonstrated.

Ulcerative colitis (unlabeled use): Transdermal: Titrated to 22-25 mg/day

Nasal: Spray: 1-2 sprays/hour; do not exceed more than 5 doses (10 sprays) per hour [maximum: 40 doses/day (80 sprays); each dose (2 sprays) contains 1 mg of nicotine

Administration

Gum: Should be chewed slowly to avoid jaw ache and to maximize benefit.

Lozenge: Allow to dissolve slowly in the mouth. Do not chew or swallow lozenge whole. Acidic foods/beverages decrease absorption of nicotine. Avoid coffee, orange juice, or soft drinks 15 minutes prior to, during, or after lozenge.

Dosage Forms

Gum, chewing, as polacrilex (Nicorette®): 2 mg/square (48s, 108s, 168s); 4 mg/square (48s, 108s, 168s) [mint, orange, and original flavors]

Lozenge, as polacrilex (Commit™): 2 mg, 4 mg [contains phenylalanine 3.4 mg/lozenge; mint flavor]

Oral inhalation system (Nicotrol® Inhaler): 10-mg cartridge [delivering 4 mg nicotine] (42s) [each unit consists of 1 mouthpiece, 7 storage trays each containing 6 cartridges, and 1 storage case]

Patch, transdermal: 7 mg/24 (7s, 30s); 14 mg/24 hours (7s, 14s, 30s); 21 mg/24 hours (7s, 14s, 30s)

    Kit: Step 1: 21 mg/24 hours (28s); Step 2: 14 mg/24 hours (14s); Step 3: 7 mg/24 hours (14s) [kit also contains support material]

    NicoDerm® CQ® [clear patch]: 7 mg/24 hours (14s); 14 mg/24 hours (14s); 21 mg/24 hours (14s)

    NicoDerm® CQ® [tan patch]: 7 mg/24 hours (14s); 14 mg/24 hours (14s); 21 mg/24 hours (7s, 14s)

    Nicotrol®: 15 mg/16 hours (7s)

Solution, intranasal spray (Nicotrol® NS): 10 mg/mL (10 mL) [delivers 0.5 mg/spray; 200 sprays]

Pregnancy Issues D (transdermal), X (chewing gum); Pregnant smokers are almost twice as likely to have a spontaneous abortion or a low birth weight neonate (<2500 g); prematurity, placenta previa, and abruption rates are also increased in the smoker; nicotine is present in breast milk

Nursing Implications Patients should be instructed to chew slowly to avoid jaw ache and to maximize benefit

Clinically Resembles Organophosphate agents or nerve agents

Diagnostic Test/Procedure Nicotine Level

Signs and Symptoms of Acute Exposure Abdominal pain, apnea, AV block, blurred vision, cyanosis, dementia, diarrhea (may be delayed up to 24 hours in pediatric ingestion), dry mouth, dysosmia, dyspepsia, headache (early sign), hiccups, hyperglycemia, hypertension then bradycardia, hyperreflexia, hyperthermia, hyperventilation, hyponatremia, hypotension, hypotonia, increased bronchial secretions, insomnia, lacrimation, lightheadedness, mental confusion, muscle fasciculations/paralysis, myalgia, myasthenia gravis (exacerbation or precipitation), mydriasis, myoclonus, nausea, nystagmus, ototoxicity, paresthesia, respiratory depression, salivation, seizures, tachycardia, tinnitus, vomiting

Admission Criteria/Prognosis Admit pediatric ingestions >2 whole cigarettes, 6 cigarette butts or a total of >0.5 mg/kg of nicotine; symptomatic patients or patients with tachycardia or hypertension 4 hours postexposure should be admitted; survival after 4 hours is usually associated with complete recovery; any patient with change in mental status or cardiopulmonary complaints should be admitted

Treatment Strategy

Decontamination:

 Dermal: Wash area well with cool water and dry; soap (especially alkaline soaps) may increase absorption; remove any remaining transdermal systems; nicotine will continue to be absorbed several hours after removal due to depot in skin. Remove all clothing, shoes, belt, jewelry, after full body dermal exposure

 Oral: Emesis is not recommended due to seizure potential. For ingestions, lavage (within 1 hour) with 1:10,000 potassium permanganate (100 mg/L) is recommended after control of seizures. Use of activated charcoal for acute ingestions is not well established.

 Ocular: Irrigate with water or saline.

Supportive therapy: Control seizures with benzodiazepines; if continuous, use phenobarbital; atropine can be utilized for cholinergic toxicity while phentolamine can be used for hypertension; avoid antacids due to increased nicotine absorption in alkali medium. Hypotension can be treated with crystalloid fluid boluses and norepinephrine (as a vasoconstrictor agent) as necessary. Do not administer pralidoxime (can cause cholinergic excess).

Enhancement of elimination: Hemodialysis/hemoperfusion of unknown value. While acidifying the urine may enhance elimination, this modality is not recommended due to inherent dangers.

Reference Range A serum nicotine level of 13,600 ng/mL has been associated with fatality. Mean plasma level after smoking one cigarette: 5-30 ng/mL. Mean plasma level after $6^1/_2$ hours of smoking: 12-44 ng/mL. Pipe smoking can result in nicotine levels of 4-6 ng/mL. Plasma levels of cotinine averaged 0.001 mg/L in children from nonsmoking homes and 0.004 mg/L in children from homes with smoking cohabitants. Arterial levels are ~6-8 times that of venous levels, thus leading to rapidly elevated brain levels. Cotinine serum levels as high as 800 ng/mL are associated with nausea, vomiting, and severe symptomatology. Steady state plasma nicotine level from transdermal patches: 12-17 ng/mL; steady state plasma nicotine level from chewing 4 mg nicotine gum: 23 ng/mL; urine cotinine levels in nonsmokers and smokers: <20 and >50 ng/mL, respectively. Therapeutic nicotine serum level for smoking cessation: 2-17 ng/mL.

Additional Information Lethal adult dose: 40 mg

Symptoms usually do not occur in pediatric ingestions <1 mg/kg; not useful as maintenance therapy for ulcerative colitis

Nicotine withdrawal is characterized by psychological distress, difficulty concentrating, tobacco craving, and hunger; average weight gain during smoking cessation is ~4 kg; 3% annual quit rate from smoking; contact the National Capital Poison Center (1-800-498-8666) with all cases of misuse, overdose or abuse of nicotine nasal spray; each actuation of Nicotrol® NS delivers a metered 50-µL spray containing 0.5 mg of nicotine

Tobacco harvesters absorbed ~0.8 mg nicotine per day

Anatoxin-a is a light brown solid that is soluble in water; and it degrades in direct sunlight (half-life about 1 hour); avoid mouth-to-mouth resuscitation following inhalation or GI exposure due to rescuer risk. Epibatidine is an off-white powder that is soluble in alcohol.

Specific References

Arcury TA, Quandt SA, Preisser JS, et al, "The Incidence of Green Tobacco Sickness Among Latino Farmworkers," *J Occup Environ Med*, 2001, 43(7):601-9.

Belson MG, Schier JG, and Patel MM, "Case Definitions for Chemical Poisoning," *MMWR Recomm Rep*, 2005, 54(RR-1):1-24.

Ciolino LA, McCauley HA, Fraser DB, et al, "The Relative Buffering Capacities of Saliva and Moist Snuff: Implications for Nicotine Absorption," *J Anal Toxicol*, 2001, 25(1):15-25.

D'Alessandro A, Benowitz NL, Muzi G, et al, "Systemic Nicotine Exposure in Tobacco Harvesters," *Arch Environ Health*, 2001, 56(3):257-63.

Dempsey DA and Benowitz NL, "Risks and Benefits of Nicotine to Aid Smoking Cessation in Pregnancy," *Drug Saf*, 2001, 24(4):277-322.

Fukumoto M, Shirai M, Akahori F, et al, "The Acute Toxicity of Tobacco Extracts in Rats," *J Toxicol Clin Toxicol*, 2000, 38(5):517-8.

Herman EH, Vick JA, Strong JM, et al, "Cardiovascular Effects of Buccal Exposure to Dermal Nicotine Patches in the Dog: A Comparative Evaluation," *J Toxicol Clin Toxicol*, 2001, 39(2):135-42.

Nitrogen Mustard (HN-1)

Related Information Nitrogen Mustard (HN-1, HN-2, and HN-3) - Patient Information Sheet

CAS Number 538-07-8

Synonyms 2,2'-Dichlorotriethylamine

Military Classification Blister Agent

Use Primarily of historical use during World War I as a vesicant chemical warfare agent; most recently used for this purpose during the Iran-Iraq War in the 1980s; derivative of this chemical has been used as an antineoplastic agent (an alkylating agent - nitrogen mustard)

Mechanism of Toxic Action Dermal, ocular, and respiratory corrosive agent; this agent combines with DNA thus preventing cell replication

Adverse Reactions Children may have a shorter time for symptom onset with more severe skin lesions. Symptoms may be delayed for $1/_2$ day.

Central nervous system: Fever

Dermatologic: Erythema and pruritus; may be delayed for up to 8 hours, then ulceration, dermal burns, and blisters (bullae) can result

Gastrointestinal: Vomiting, nausea, diarrhea, anorexia

Hematologic: Bone marrow suppression, pancytopenia; leukopenia usually occurs 7-10 days postexposure (leukocytosis initially); eosinophilia can be seen in children

Ocular: Lacrimation, photophobia, blepharospasm, mild ocular irritation (75%), opacification, blindness

Respiratory: Dyspnea, cough (usual onset is 1-12 hours postexposure)

Pregnancy Issues Possible teratogen (skeletal malformations/cleft palate)

Clinically Resembles Radiation injury, chemotherapeutic agent toxicity, toxic-shock syndrome, bullous pemphigoid, chemical burns, sunburn

Treatment Strategy

Decontamination: **Note:** Penetrates wood, leather, rubber, and paints; decontamination must occur within 2 minutes.

Oral: Activated charcoal or 150 mL of 2% sodium thiosulfate orally

Dermal: Remove all contaminated clothing. Towels soaked in 0.2% chloramine-T in water (Dakin solution) placed over wounds for the first 2 hours may be helpful. Wash with soap and water (not hot). Can wash with dilute (0.5%) hypochlorite, then neutralize with 2.5% sodium thiosulfate. If no water is available, dry decontamination with Fuller's earth can be utilized. Treat wounds as burns with use of $1/_8$ inch of silver sulfadiazine or mafenide acetate (twice daily). Leave small blisters (<1 cm) intact; unroof larger blisters and irrigate.

Ocular: Remove contact lenses; copious irrigation with saline (neutralize with 2.5% solution of sodium thiosulfate); use cycloplegic eye drops (atropine or homatropine) three times daily if photophobia develops.

Respiratory: Administer 100% oxygen. A nebulized mist of 2.5% sodium thiosulfate may be helpful as a neutralizing agent if given within 15 minutes of exposure. While time is critical, some benefit may be derived even beyond 15 minutes postexposure.

Supportive therapy: N-acetylcysteine 4 times/day (up to 150 mg/kg), along with ascorbic acid (3 g), thienamycin (2 g twice daily), L-carnitine (3 g), and sodium thiosulfate (3-12.5 g/day), has been used to ameliorate effects, but it is unproven in its efficacy. For ocular exposure, topical antibiotics, mydriatics, and possibly corticosteroids can be utilized. Vaseline® placed on eyelid edges may prevent lids from adhering. For dermal burns, topical antibiotics with systemic analgesics are the mainstay of therapy. Do not overhydrate the patient. Do not fluid resuscitate as in thermal burns. Colony stimulating factor may be helpful in treating leukopenia. Prednisone (60-125 mg/day) orally can be helpful in treating pulmonary toxicity. Bronchospasm should be treated with β_2 adrenergic agonist agents. Aggressive airway protection and management is critical.

Environmental Persistency Persistent under cold/temperate climates. In hot climates, evaporation can lead to increased air concentrations.

Personal Protection Equipment Level A; pressure demand, self-contained breathing apparatus (SCBA) with butyl rubber, neoprene, nitrile or polyvinyl chloride gloves, Responder® CSM protective clothing, PVC boots, chemical goggles, and face shield

Additional Information

Very slightly flammable with a faint fishy or musty odor

Vesicle/blister fluid does not contain mustard and is not hazardous.

Rate of hydrolysis is slow. Hydrolysis products are hydroxyl derivatives, which are also toxic.

Median lethal dosage: Inhalation: 1500 mg/min/m³; dermal exposure (masked personnel): 20,000 mg/min/m³

Ocular injury (incapacitating median dose): 200 mg/min/m³

Liquid density: 1.09 at 25°C; vapor density: 5.9

Action on metal: Slightly corrosive of steel at 65°C.

Specific References

Carroll LS, "Sulfur Mustard: Cutaneous Exposure," *Clin Toxicol (Phila)*, 2005, 43(1):55.

Kales SN and Christiani DC, "Acute Chemical Emergencies," *N Engl J Med*, 2004, 350(8):800-8.

Nitrogen Mustard (HN-2)

Related Information Nitrogen Mustard (HN-1, HN-2, and HN-3) - Patient Information Sheet

CAS Number 51-75-2

Synonyms 2,2′-Dichloro-N-methyldiethylamine; HN₂; Mechlorethamine Hydrochloride; Mechlorethamine; Mustine

Military Classification Blister Agent

U.S. Brand Names Mustargen®

Use Primarily of historical use during World War I as a vesicant chemical warfare agent; most recently used for this purpose during the Iran-Iraq War in the 1980s; derivative of this chemical has been used as an antineoplastic agent (an alkylating agent - nitrogen mustard) Highly unstable and no longer seriously considered for use as a military chemical warfare agent. More toxic than HN-1.

Mechanism of Toxic Action Dermal, ocular, and respiratory corrosive agent; this agent combines with DNA thus preventing cell replication

Adverse Reactions Primarily dermal/ocular toxicity. Greatest blistering skin effects can be delayed for 12 hours. Children may have a shorter time for symptom onset with more severe skin lesions.

Central nervous system: Fever

Dermatologic: Erythema and pruritus; may be delayed for up to 8 hours, then ulceration, dermal burns, and blisters can result; alopecia

Gastrointestinal: Vomiting, nausea, diarrhea, anorexia

Hematologic: Bone marrow suppression, pancytopenia; leukopenia usually occurs 7-10 days postexposure; eosinophilia can be seen in children

Ocular: Lacrimation, photophobia, blepharospasm, ocular irritation, opacification, blindness

Otic: Tinnitus, ototoxicity

Respiratory: Dyspnea, cough (usual onset is 1-12 hours postexposure)

Pregnancy Issues Possible teratogen (skeletal malformations/cleft palate)

Clinically Resembles Radiation injury, chemotherapeutic agent toxicity, toxic-shock syndrome, bullous pemphigoid, chemical burns, sunburn

Treatment Strategy

Decontamination: **Note:** Penetrates wood, leather, rubber, and paints; decontamination must occur within 2 minutes.

Oral: Activated charcoal or 150 mL of 2% sodium thiosulfate orally

Dermal: Remove all contaminated clothing. Towels soaked in 0.2% chloramine-T in water (Dakin solution) placed over wounds for the first 2 hours may be helpful. Wash with soap and water (not hot). Can wash with dilute (0.5%) hypochlorite, then neutralize with 2.5% sodium thiosulfate. If no water is available, dry decontamination with Fuller's earth can be utilized. Treat wounds as burns with use of $1/8$ inch of silver sulfadiazine (twice daily). Leave small blisters (<1 cm) intact; unroof larger blisters and irrigate.

Ocular: Remove contact lenses; copious irrigation with saline (neutralize with 2.5% solution of sodium thiosulfate) or water for at least 20 minutes; use cycloplegic eye drops (atropine or homatropine) three times daily if photophobia develops.

Respiratory: Administer 100% oxygen. A nebulized mist of 2.5% sodium thiosulfate may be helpful as a neutralizing agent if given within 15 minutes of exposure.

Supportive therapy: N-acetylcysteine 4 times/day (up to 150 mg/kg), along with ascorbic acid (3 g), thienamycin (2 g twice daily), L-carnitine (3 g), and sodium thiosulfate (3-12.5 g/day), has been used to ameliorate effects, but it is unproven in its efficacy. For ocular exposure, topical antibiotics, mydriatics, and possibly corticosteroids can be utilized. Vaseline® placed on eyelid edges may prevent lids from adhering. For dermal burns, topical antibiotics with systemic analgesics are the mainstay of therapy. Do not overhydrate the patient. Do not fluid resuscitate as in thermal burns. Colony stimulating factor may be helpful in treating leukopenia. Prednisone (60-125 mg/day) orally can be helpful in treating pulmonary toxicity. Bronchospasm should be treated with $beta_2$ adrenergic agonist agents. Use aggressive airway protection and pain management. Avoid salicylates for fever.

Environmental Persistency Persistent under cold/temperate climates. In hot climates, evaporation can lead to increased air concentrations.

Personal Protection Equipment Level A; pressure demand, self-contained breathing apparatus (SCBA) with butyl rubber, neoprene, nitrile or polyvinyl chloride gloves, Responder® CSM protective clothing, PVC boots, chemical goggles, and face shield

Additional Information

A dark liquid with a soapy odor in dilute form and a fruity odor in concentrated form

Median lethal dose: Inhalation: 3000 mg/min/m^3

Ocular injury (incapacitating median dose): 100 mg/min/m^3

Dermal absorption (masked personnel): 2500-9000 mg/min/m^3

Liquid density: 1.15 at 20°C; vapor density: 5.9

Capability in vapor form: Ocular effects: Immediate

Action on metal: None

Specific References

Carroll LS, "Sulfur Mustard: Cutaneous Exposure," *Clin Toxicol (Phila)*, 2005, 43(1):55.

Kales SN and Christiani DC, "Acute Chemical Emergencies," *N Engl J Med*, 2004, 350(8):800-8.

Lemire SW, Ashley DL, and Calafat AM, "Quantitative Determination of the Hydrolysis Products of Nitrogen Mustards in Human Urine by Liquid Chromatography-Electrospray Ionization Tandem Mass Spectrometry," *J Anal Toxicol*, 2003, 27(1):1-6.

Nitrogen Mustard (HN-3)

Related Information Nitrogen Mustard (HN-1, HN-2, and HN-3) - Patient Information Sheet
CAS Number 555-77-1
Synonyms 2,2′ 2″-Trichlorotriethylamine
Use Primarily of historical use during World War I as a vesicant chemical warfare agent; most recently used for this purpose during the Iran-Iraq War in the 1980s; derivative of this chemical has been used as an antineoplastic agent (an alkylating agent - nitrogen mustard)
Mechanism of Toxic Action Dermal, ocular, and respiratory corrosive agent; this agent combines with DNA thus preventing cell replication
Adverse Reactions Children and the elderly may have a shorter time for symptom onset with more severe skin lesions.
Central nervous system: Fever
Dermatologic: Erythema and pruritus; may be delayed for up to 8 hours, then ulceration, dermal burns, and blisters can result
Gastrointestinal: Vomiting, nausea, diarrhea, anorexia
Hematologic: Bone marrow suppression, pancytopenia; leukopenia usually occurs 7-10 days postexposure; eosinophilia can be seen in children
Ocular: Lacrimation, photophobia, blepharospasm, ocular irritation, opacification, blindness
Respiratory: Dyspnea, cough (usual onset is 1-12 hours postexposure)
Pregnancy Issues Possible teratogen (skeletal malformations/cleft palate)
Clinically Resembles
Radiation injury, chemotherapeutic agent toxicity, toxic-shock syndrome, bullous pemphigoid, chemical burns, sunburn
Treatment Strategy
Decontamination: **Note:** Penetrates wood, leather, rubber, and paints; decontamination must occur within 2 minutes.
Oral: Activated charcoal or 150 mL of 2% sodium thiosulfate orally
Dermal: Remove all contaminated clothing. Towels soaked in 0.2% chloramine-T in water (Dakin solution) placed over wounds for the first 2 hours may be helpful. Wash with soap and water (not hot). Can wash with dilute (0.5%) hypochlorite, then neutralize with 2.5% sodium thiosulfate. If no water is available, dry decontamination with Fuller's earth can be utilized. Treat wounds as burns with use of 1% silver sulfadiazine (twice daily). Leave small blisters (<1 cm) intact; unroof larger blisters and irrigate.
Ocular: Remove contact lenses; copious irrigation with saline (neutralize with 2.5% solution of sodium thiosulfate); use cycloplegic eye drops (atropine or homatropine) three times daily if photophobia develops.
Respiratory: Administer 100% oxygen. A nebulized mist of 2.5% sodium thiosulfate may be helpful as a neutralizing agent if given within 15 minutes of exposure.
Supportive therapy: N-acetylcysteine 4 times/day (up to 150 mg/kg), along with ascorbic acid (3 g), thienamycin (2 g twice daily), L-carnitine (3 g), and sodium thiosulfate (3-12.5 g/day), has been used to ameliorate effects, but it is unproven in its efficacy. For ocular exposure, topical antibiotics, mydriatics, and possibly corticosteroids can be utilized. Vaseline® placed on eyelid edges may prevent lids from adhering. For dermal burns, topical antibiotics with systemic analgesics are the mainstay of therapy. Do not overhydrate the patient. Do not fluid resuscitate as in thermal burns. Colony stimulating factor may be helpful in treating leukopenia. Prednisone (60-125 mg/day) orally can be helpful in treating pulmonary toxicity. Bronchospasm should be treated with beta$_2$ adrenergic agonist agents.
Personal Protection Equipment Level A; pressure demand, self-contained breathing apparatus (SCBA) with butyl rubber, neoprene, nitrile or polyvinyl chloride gloves, Responder® CSM protective clothing, PVC boots, chemical goggles, and face shield
Additional Information Most stable and most toxic of the nitrogen mustards.
Median incapacitating dose: Ocular injury: 200 mg/min/m^3; dermal absorption (masked personnel): 2500 mg/min/m^3

Median lethal dose: Inhalation: 1500 mg/min/m^3; dermal absorption (masked personnel): 10,000 mg/min/m^3

Liquid density: 1.24 at 25°C; insoluble in water

Vapor density: 7.1; decomposes at atmospheric pressure below boiling point (256°C).

Specific References

Carroll LS, "Sulfur Mustard: Cutaneous Exposure," *Clin Toxicol (Phila)*, 2005, 43(1):55.

Kales SN and Christiani DC, "Acute Chemical Emergencies," *N Engl J Med*, 2004, 350(8):800-8.

Nitroglycerin

CAS Number 55-63-0

UN Number 1204; 3064; 3319; 3343; 3357

Synonyms Glyceryl Trinitrate

U.S. Brand Names Minitran™; Nitrek®; Nitro-Bid®; Nitro-Dur®; Nitrolingual®; NitroQuick®; Nitrostat®; NitroTime®

Use As an explosive agent; often used in oil well drilling, combating oil well fires, and in rocket propellants

Mechanism of Toxic Action An ester of glycerol with explosive action; resultant gases do not appear to be toxic; vasodilator as therapeutic agent through nitric oxide generation; also has local dermal irritant effect

Toxicodynamics/Kinetics

Absorption: Skin, lungs, mucosa

Distribution: V$_d$: 3 L/kg

Protein binding: 11% to 60%

Metabolism: Denitration to slightly active metabolites

Half-life: 1-4.4 minutes (parent compound); 40 minutes (metabolites)

Elimination: Renal

Drug Interactions Hypotensive action exacerbated by ethanol

Signs and Symptoms of Acute Exposure In addition to thermal or traumatic injury from explosion: Nausea, vomiting, headache, confusion, seizures, delirium, bradycardia, hypotension, cyanosis, methemoglobinemia, syncope, diaphoresis, leukopenia

Treatment Strategy

Decontamination:

Dermal: Wash with soap and water

Ocular: Irrigate with saline

Supportive therapy: Methylene blue for symptomatic methemoglobinemia; treat burns with standard burn therapy; monitor for carbon monoxide

Environmental Persistency Half-life: air: 1.7-17.6 hours; surface water: 2-7 days; ground water: 4-14 days; soil: 2-7 days

Personal Protection Equipment Positive pressure self-contained breathing apparatus; structural firefighters protective clothing offers only limited protection. Do not wear contact lenses when working with nitroglycerin. Appropriate gloves include materials made of nylon, neoprene, or polyethylene with a sweat-absorbent cotton liner

Additional Information Pale yellow or brown oily liquid at room temperature; volatile (explosive) at 100°C. Can be exploded by percussion or heat (highly flammable).

Modern dynamite contains 60% ammonium and sodium nitrate, 20% to 25% ethylene glycol dinitrate, 0% to 5% nitroglycerin, 10% dinitrotoluene, 3% nitrocellulose, sawdust or chalk.

IDLH: 75 ppm; vapors are heavier than air (specific gravity: 1.6); headaches can occur at air levels >0.1 mg/m^3

When heated to 135°C, yellow vapors of nitric oxide can be produced

Does not bioconcentrate in aquatic organisms. Soluble in alcohol/slightly soluble in water.

EPA drinking water guideline: 5 mcg/L

PEL ceiling: 0.2 ppm

Specific References
Fung HL, "Biochemical Mechanism of Nitroglycerin Action and Tolerance: Is This Old Mystery Solved?" *Annu Rev Pharmacol Toxicol*, 2004, 44:67-85.
Meyjohann D, Zell L, and Buchter A, "Nitrate Headache in Blasting Work," *Med Klin* (Munich), 2001, 96(5):295-7.

O-Chlorobenzylidene Malononitrile

CAS Number 2698-41-1
UN Number 2810
Synonyms 2-ChloroBMN; CS Gas; CS
Military Classification Lacrimator
Use Riot control; used by law enforcement for training purposes since the late 1950's. Ten times more potent than chloroacetophenone.
Mechanism of Toxic Action Lacrimator - irritation due to the active halogen group reacting with sulfhydryl groups
Adverse Reactions
Ocular: Ocular burning, blepharospasms, cataract. Corneal edema can occur for 2-6 hours postexposure.
Respiratory: Bronchospasm (may be delayed for 36 hours), pulmonary edema (may be delayed for 24 hours)
Toxicodynamics/Kinetics
Onset of action: Rapid
Metabolism: Hydrolysis rapidly to O-chlorobenzaldehyde and malononitrile
Pregnancy Issues No teratogenic effects
Clinically Resembles Chlorine, ammonia, hydrogen chloride
Treatment Strategy
Decontamination: Move patient rapidly to fresh air and monitor for wheezing. Treat burns in typical manner. Administer 100% humidified oxygen. Personnel should avoid contaminating themselves. Wash skin with soap and at least two liters of water. Remove contaminated clothing, but do not pull clothing over patient's head (cut clothing instead).
Ocular: Avoid rubbing eyes as it may prolong effect. Blow dry air into eyes to help the agent vaporize; copious irrigation with water or saline is needed.
Supportive therapy: Solution of antacid (magnesium hydroxide - aluminum hydroxide - simethicone) gently blotted topically over exposed facial areas (avoiding eyes) may help resolve symptoms. Bronchospasm can be treated with inhaled beta$_2$ agonists. Use aggressive airway protection and PEEP for bronchospasm. Inverse-ratio ventilation using lower tidal volume mechanical ventilation (6 mL/kg) and plateau pressures of 30 cm water may decrease mortality. While positive end expiratory pressures should be considered, oxygen toxicity should be monitored. Use of perfluorocarbon partial liquid ventilation and exogenous surfactant in chemical-induced ARDS is investigational.
Environmental Persistency Varies: Water half-life is 15 minutes at 15°C. Incapacitating dosages lose effect in ~10 minutes.
Additional Information White crystalline solid (gas is white) with usual purity of 96%. It is insoluble in water and ethanol but soluble in methylene chloride. Exhibits a pepper-like odor.
CS1 is a mixture consisting of 95% crystalline CS blended with 5% silica aerogel to reduce agglomeration and micropulverized to 3-10 micron size to achieve the desired respiratory effects when it is dispersed as a solid aerosol.
CS2 is a mixture of 94% CS, 5% Cab-O-Sil colloidal silica, and 1% hexamethyldisilazane. This compound improves the physical characteristics of CS by reducing agglomeration and hydrolysis.
Boiling point: 310°C to 315°C; density: 1.04 g/mL; OSHA: PEL 8 hour TWA: 0.05 ppm
Action on metal: Very slight

See table

Effects of O-Chlorobenzylidene Malononitrile		
Effect	Dose	Time
Eye irritation	5 mg/m^3	20 seconds
Headache/lacrimation	1.5 mg/m^3	90 minutes
Median incapacitation	10-20 mg/min/m^3	—
Median lethal (M7A3 grenade)	61,000 mg/min/m^3	—
Intolerant	1-3 mg/m^3	4 minutes
Minimal irritation	0.1-1 mg/m^3	—
IDLH	2 mg/m^3	—

Specific References

Blain PG, "Tear Gases and Irritant Incapacitants, 1-chloroacetophenone, 2-chlorobenzylidene Malononitrile and Dibenz[b,f]-1,4-oxazepine," *Toxicol Rev*, 2003, 22(2):103-10 (review).

Horton DK, Berkowitz Z, and Kaye WE, "Secondary Contamination of ED Personnel From Hazardous Materials Events, 1995-2001," *Am J Emerg Med*, 2003, 21(3):199-204.

Solomon I, Kochba I, Eizenkraft E, et al, "Report of Accidental CS Ingestion Among Seven Patients in Central Israel and Review of the Current Literature," *Arch Toxicol*, 2003, 77(10):601-4.

Yih JP, "CS Gas Injury to the Eye," *BMJ*, 1995, 311(7000):276.

Osmium Oxide

CAS Number 20816-12-0
UN Number 2471
Synonyms Milas' Reagent; OS IV Oxide
Military Classification Choking Agent
Use May be added to explosive devices; used as catalyst for certain reactions in photography and tissue staining
Mechanism of Toxic Action Corrosive agent; respiratory irritant
Adverse Reactions
Onset is usually rapid
Central nervous system: Headache (about 0.6 mg/m^3)
Dermal: Rash (black)
Gastrointestinal: Abdominal cramps
Ocular: Irritation, lacrimation, eye pain, blurred vision, conjunctivitis
Respiratory: Dyspnea, cough, pulmonary edema (may be delayed)
Pregnancy Issues Has not been tested
Treatment Strategy
Decontamination:
Ocular: Lavage eyes with saline for at least 20 minutes
Dermal: Remove all clothing and wash with copious amounts of water
Supportive therapy: Excise necrotic tissue; support inhalation injuries with oxygen and beta$_2$ agonist bronchodilator
Personal Protection Equipment Supplied air respirators for environments containing >0.01 ppm. Corn oil or warm water can be used for surface decontamination
Additional Information
Crystalline solid (colorless or pale yellow) with a chlorine-like odor
Odor threshold: 0.0019 ppm
Liquid density: 4.9

Vapor density: 8.8
IDLH: 0.1 ppm
OSHA-PEL: 0.0002 ppm
Agent is not flammable; water soluble (5-7 g/100 mL at 25°C); boiling point: 130°C
Odor threshold in water: 0.012 ppm

Pentaerythritol Tetranitrate

CAS Number 78-11-5
Synonyms PETN
Military Classification Explosive Agent
Use Manufacturing of detonating fuses, demolition explosives, and blasting caps; mixed with trinitrotoluene in grenades; combined with Sentex in plastic explosives. Also used in medicine as an oral vasodilator for angina pectoris.
Mechanism of Toxic Action Venous and arterial vasodilation
Adverse Reactions
Cardiovascular: Hypotension, tachycardia
Central nervous system: Headache, vertigo
Dermal: Erythematous rash
Ocular: Lens opacification
Toxicodynamics/Kinetics
Onset of hemodynamic effect: 20-60 minutes
Absorption: Oral/inhalation; not dermally
Half-life: 10-11 hours
Elimination: Feces/urine
Usual Dosage Antianginal: 10-40 mg 3-4 times daily; maximum: 240 mg/day
Treatment Strategy
Decontamination:
Dermal: Remove contaminated clothing. Wash skin with soap and water.
Inhalation: Administer 100% humidified oxygen
Supportive therapy: Treat burns with standard therapy; monitor for carbon monoxide
Additional Information A white, crystal powder insoluble in water; explodes on physical shock or when heated to 210°C. Decomposition products include nitrogen oxides and carbon monoxide.
Specific References
Arnold JL, Halpern P, Tsai MC, et al, "Mass Casualty Terrorist Bombings: A Comparison of Outcomes by Bombing Type," *Ann Emerg Med*, 2004, 43(2):263-73.

Pepper Spray

CAS Number 404-86-4 (capsaicin)
Military Classification Lacrimator
Use Riot control; active ingredient in tear gas
Mechanism of Toxic Action Lacrimator - irritation due to the active halogen group reacting transiently with sulfhydryl groups; contains the mucosal irritant oleoresin of capsicum (which stimulates pain fibers) in a 1% to 10% concentration.
Adverse Reactions
Dermal: Allergic contact dermatitis, burning sensation of the skin, dermal erythema, bullous dermatitis, vesiculation of skin (rare)
Gastrointestinal: Metallic taste, vomiting
Ocular: Burning (56%), conjunctival irritation (44%), lacrimation (16%), corneal abrasion (9%), blepharospasms, lacrimation (eye symptoms may last for 30 minutes)

Respiratory: Bronchospasm (may be delayed for 36 hours), pulmonary edema (may be delayed for 24 hours), cough, sneezing

Pregnancy Issues No direct effects on the fetus

Treatment Strategy

Decontamination: Move patient rapidly to fresh air and monitor for wheezing. Administer 100% humidified oxygen. Personnel should avoid contaminating themselves. Wash skin with soap and at least two liters of water.

Ocular: Avoid rubbing eyes as it may prolong effect. Copious irrigation with water or saline is needed.

Supportive therapy: Solution of antacid (magnesium hydroxide - aluminum hydroxide - simethicone) gently blotted topically over exposed facial areas (avoiding eyes) may help resolve symptoms. Bronchospasm can be treated with inhaled beta$_2$ agonists. Use antihistamines for allergic reaction. Consider aggressive airway management and PEEP for bronchospasm. There may be delayed, serious respiratory compromise.

Additional Information Exists as a white solid

Specific References

Forrester MB and Stanley SK, "The Epidemiology of Pepper Spray Exposures Reported in Texas in 1998-2002," *Vet Hum Toxicol*, 2003, 45(6):327-30.

Perfluroisobutylene

CAS Number 382-21-8

Synonyms Octafluoroisobutylene; PFIB

Mechanism of Toxic Action A fluoroalkene gas, which is produced through thermal decomposition (over 315°C) of polytetrafluoroethylene (Teflon) and perfluoroethylpropylene (Halon 1301) that can cause respiratory tract mucosal irritation resulting in rapid onset pulmonary edema, probably due to damage to the integrity of the alveolar capillary membrane and resultant plasma leakage

Monitoring Parameters Coagulation parameters, DIC panel, renal function

Signs and Symptoms of Acute Exposure DIC, cough, dyspnea, pulmonary edema, ocular irritation, conjunctivitis, headache, substernal chest pain, fever (low grade), tachypnea, tachycardia, dermal irritation

Treatment Strategy

Decontamination:

Dermal: Remove contaminated clothing and wash area with soap and water

Inhalation: Administer 100% humidified oxygen

Ocular: Remove contact lenses and irrigate copiously with water or normal saline

Supportive therapy: Corticosteroid use is of uncertain benefit. Avoid net positive fluid balance; loop diuretics can aid in controlling fluid balance. Partial liquid ventilation may be useful. Monitor urinary fluoride in severe cases (urinary fluoride level over 3 mg/L is an indication of exposure)

Additional Information Colorless gas; ten times as toxic as phosgene gas

Lethal inhalation exposure: ~0.6 ppm

Phenibut

CAS Number 35568-37-7

Synonyms Beta-phenyl-gamma-aminobutyric Acid HCl; Fenibut; Phenylgam

Use Neuropsychotropic drug developed in Russia in the 1960s; has anxiolytic effects; usually ingested as a powder

Mechanism of Toxic Action A beta-phenyl derivative of the inhibitory central nervous sytem neuromediator GABA with GABA(A) and GABA(B) agonist effects. Also stimulates dopamine receptors.

Adverse Reactions Sedation, lethargy
Usual Dosage About 250 mg 3 times/day
Treatment Strategy
Decontamination: Oral activated charcoal may be useful within 1 hour of ingestion
Supportive care: Naloxone is not effective; flumazenil has unknown effectiveness; respiratory depression and cardiac effects are minimal
Specific References
Lapin I, "Phenibut (Beta-phenyl-GABA): A Tranquilizer and Nootropic Drug," *CNS Drug Rev*, 2001, 7(4):471-81.

Phenyldichloroarsine

CAS Number 696-28-6
UN Number 1556
Synonyms Dichlorophenylarsine; FDA; PD; Phenylarsinedichloride; TL69
Military Classification Blister Agent
Mechanism of Toxic Action An organic dithiol-arsenical that is most often in a liquid state and is odorless and colorless. In animals, it binds to glutathione and erythrocytes in red blood cells. Upon contact with water, can form hydrogen chloride. Phenylarsenious oxide is a hydrolysis product; a highly potent mucosal and dermal irritant.
Clinically Resembles Poison ivy
Signs and Symptoms of Acute Exposure Lacrimation, conjunctivitis, cough, vomiting, dermal burns, blisters, contact dermatitis
Immediate ocular effects; dermal effects can be delayed for $^1/_2$-1 hour
Treatment Strategy
Decontamination:
Dermal: Remove contaminated clothing; wash area copiously with soap and water
Inhalation: Administer 100% humidified oxygen and remove from site
Ocular: Irrigate with water or saline for at least 20 minutes. Arsenic poisoning is unlikely with eye exposure exclusively.
Ingestion: Dilute with milk or water
Supportive therapy: Obtain arsenic level; if blood arsenic level is >7 mcg/100 mL or urine is >0.7 mg/L, consider treating as per arsenic exposure. Treat dermal burns with standard therapy; debridement of devitalized skin should occur with topical antibiotics for burns. Monitor and treat for arsenic exposure. Hypotension can be treated with intravenous crystalloid solutions and placement of the patient in the Trendelenburg position. Dopamine or norepinephrine are the vasopressor agents of choice. Use aggressive pain management.
Environmental Persistency Minimal
Personal Protection Equipment Chemical protective clothing with positive-pressure self-contained breathing apparatus (SCBA)
Additional Information
Related to mustard gas, lewisite, arsenic
Colorless gas or liquid with a weak odor; stable in storage; decomposes immediately by wet clothing; insoluble in water; with water contact, PD slowly decomposes to hydrochloric acid. No food chain bioconcentration.
Action on metal: None with principle agent; corrodes metals when it reacts with water due to acid formation
Mean detectable concentration (nasal/throat irritation): 0.9 mg/min/m^3
Mean incapacitating dose (vomiting): 16 mg/min/m^3
Ocular injury: 633 mg/min/m^3
Median lethal dose: 2600 mg - min/m^3
Mean capacitating dose: As a vomiting agent - 16 mg - min/m^3; as a blistering agent - 1800 mg - min/m; eye injury - 633 mg - min/m^3

About 30% as toxic to the eyes and 90% of skin blistering action as distilled mustard (HD)
Liquid density: 1.65 at 20°C
Specific gravity: 1.6516 at 19°C
Vapor density: 7.7
LD50 (skin - guinea pig): 4 mg/kg

Phosgene

Related Information Phosgene - Patient Information Sheet
CAS Number 75-44-5
UN Number 1076
Synonyms Agent CG; Carbon Oxychloride; Carbonic Dichloride; Carbonyl Chloride; Chloro-formyl Chloride; Diphosgene; Phosgen
Military Classification Pulmonary Agent; Choking Agent
CDC Classification for Diagnosis
  Biologic: No biologic marker exists for phosgene exposure.
  Environmental: Confirmation of phosgene in environmental samples is not available.
CDC Case Classification
  Suspected: A case in which a potentially exposed person is being evaluated by healthcare workers or public health officials for poisoning by a particular chemical agent, but not specific credible threat exists.
  Probable: A clinically compatible case in which a high index of suspicion (credible threat or patient history regarding location and time) exists for phosgene exposure, or an epidemiologic link exists between this case and a laboratory-confirmed case.
  Confirmed: A clinically compatible case in which laboratory tests (not available for phosgene) have confirmed exposure. The case can be confirmed if laboratory testing was not performed because either a predominant amount of clinical and nonspecific laboratory evidence of a particular chemical was present of a 100% certainty of the etiology of the agent is known.
Use In the manufacture of isocyanates, aniline dyes, plastics, and insecticides. Initially prepared in 1812, but used extensively in World War I gas warfare (approximately 1.3 million victims and 90,000 fatalities); product of combustion of volatile chlorine compounds
Mechanism of Toxic Action Direct cytotoxicity through diamide formation that crosslink cell components. Reacts readily with hydroxyl, sulfhydryl, and ammonia groups found in albumin, amino acids, and vitamins; pulmonary irritant with low water solubility thus primarily affecting lung parenchyma on inhalation (delayed alveolar injury)
Adverse Reactions
  Cardiovascular: Hypotension, bradycardia, sinus bradycardia, angina, Raynaud's phenom-enon, leukocytosis
  Endocrine & metabolic: Hypovolemia
  Hematologic: Methemoglobinemia, leukocytosis (atypical lymphocytosis), hemolysis (>200 ppm)
  Ocular: Conjunctivitis, corneal opacification
  Respiratory: Respiratory distress, pulmonary edema ("pink foam"), acute respiratory distress syndrome, tachypnea, dyspnea
Toxicodynamics/Kinetics Metabolism: Hydrolyzes slowly to produce carbon dioxide and hydrochloric acids
Pregnancy Issues No adverse effects reported
Clinically Resembles Chlorine
Diagnostic Test/Procedure
  Electrolytes, Blood
  Pulse Oximetry

Signs and Symptoms of Acute Exposure Asthenia, chest pain, cough, dermal burns, dyspnea, headache, hemoptysis, nausea, ocular irritation, throat burning, vomiting; pulmonary edema within 1-2 hours (high dose), 4-6 hours (moderate dose), 8-24 hours (low dose)

Admission Criteria/Prognosis Patients with known or suspected phosgene exposure >25 ppm/minute, should be observed for at least 12 hours. Patients with development of bronchospasm or pulmonary edema should probably be admitted into an intensive care unit. Survival after 48 hours is usually associated with recovery.

Treatment Strategy
Decontamination:
Dermal: Remove contaminated clothing. Irrigate copiously with saline or water (especially with diphosgene exposure).
Ocular: Irrigate with water or saline
Oral: Lavage within 1 hour. Irrigate exposed areas. Steroids may be helpful. Do not use diuretics.
Respiratory: Administer 100% humidified oxygen.
Supportive therapy: Bedrest; monitor and treat for pulmonary edema. Support of pulmonary function is the mainstay of therapy with mechanical ventilation and positive end-expiratory pressure. The use of nebulized sodium bicarbonate, leukotriene-receptor antagonists, ibuprofen, methylprednisolone, cyclophosphamide, colchicine, and N-acetylcysteine (I.V. or nebulized) has met with some success in experimental models, but is of uncertain clinical benefit. Even so, if exposure to over 50 ppm of phosgene is considered, suggested regimen for pulmonary edema prophylaxis include: ibuprofen 800 mg P.O., methylprednisolone 1 g I.V. or dexamethasone phosphate 10 mg by aerosol, aminophylline 5 mg/kg I.V. or P.O. followed by 1 mg/kg every 8-12 hours to achieve serum aminophylline level of 10-20 mcg/mL, terbutaline 0.25 mg SubQ, N-acetylcysteine 10-20 mL of a 20% solution by aerosolization, oxygen as needed. Use of parenteral prednisolone (1 g I.V.), aerosolized dexamethasone, and theophylline may be useful, but has not been studied in humans. Prophylactic administration of prednisolone (250 mg I.V.), ibuprofen, or N-acetylcysteine (20 mL of a 20% solution) has also not been clinically demonstrated to prevent pulmonary edema. Inverse-ratio ventilation using lower tidal volume mechanical ventilation (6 mL/kg) and plateau pressures of 30 cm water may decrease mortality. While positive end expiratory pressures should be considered, oxygen toxicity should be monitored. Use of perfluorocarbon partial liquid ventilation and exogenous surfactant in chemical-induced ARDS is investigational.

Rescuer Contamination Low

Environmental Persistency Low; air concentrations are reduced by rain or fog
Environmental half-life of phosgene in soil and water: 3 minutes to 1 hour; however, it may persist in shelters or buildings; air persistency 30 minutes to 1 hour

Additional Information Vapor contact of >3 ppm phosgene with wet skin may lead to erythema. Phosgene levels >330 ppm/minute can be fatal. Prolonged exposure at 3 ppm for 3 hours can be fatal. Levels >4 ppm can cause ocular irritation and 5 ppm can cause cough. Liquid phosgene is a frostbite hazard.
Pulmonary edema developing within 4 hours is a bad prognosis; usually it will develop within 4-8 hours postexposure. Olfactory paralysis may occur after a short time of initial exposure. Blindness may occur. On chest x-ray, blurred enlargement of hilar area may be the earliest finding.
Colorless gas; latent period may last 3 days. Estimated worldwide production is >5 billion pounds. Odor threshold: 0.5 ppm.
Odor of newly mown hay, corn, or moldy hay. At low doses, it has a freshly mowed grass odor; higher doses give a pungent odor. Vapor density: >4. TLV-TWA: 0.1 ppm; IDLH: 2 ppm; PEL-TWA: 0.1 ppm
Soluble in water; rapidly hydrolyzed in water to hydrochloric acid and carbon dioxide. Does not bioconcentrate.

Specific References
Belson MG, Schier JG, and Patel MM, "Case Definitions for Chemical Poisoning," *MMWR Recomm Rep*, 2005, 54(RR-1):1-24.
Borak J and Diller WF, "Phosgene Exposure: Mechanisms of Injury and Treatment Strategies," *J Occup Environ Med*, 2001, 43(2):110-9.
Kales SN and Christiani DC, "Acute Chemical Emergencies," *N Engl J Med*, 2004, 350(8): 800-8.

Phosgene Oxime

Synonyms "Nettle Gas"; Agent CX; CX; Dichloroformoxine; Oximedichloroforoxime
Military Classification Vesicant
Use Chemical warfare, first synthesized in Germany in 1929; often combined with VX
Mechanism of Toxic Action A highly irritating halogenated oxime that causes dermal or mucosal corrosion, possibly through necrotizing effect of chlorine, oxime, or carbonyl group.
Toxicodynamics/Kinetics Absorption: Dermal/inhalation; not ocular
Pregnancy Issues No data
Clinically Resembles Nettle stings
Signs and Symptoms of Acute Exposure Rapid onset; causes intense, immediate eye pain, lacrimation, conjunctivitis, temporary blindness. When phosgene oxime comes into contact with skin, through liquid or vapor, there is extreme dermal pain followed by blanching with surrounding erythema within 30-60 seconds. A wheal forms in about 30 minutes and the original blanched area turns brown within 24 hours. Scabs with skin desquamation occurs within 7 days. The scab generally falls off within 21 days. Skin can take 1-3 months to heal with pruritus being prominent during this time period. Does not really produce blisters; can also cause upper airway irritation resulting in dyspnea, cough, and delayed pulmonary edema (delayed as long as 24 hours).
Treatment Strategy
Decontamination: Move patient rapidly out of area - remove clothing
Dermal: Irrigate with copious amounts of water or saline. Chloranine or phenol towelettes are **not** effective. Burns should be treated with standard therapy.
Ocular: Copious irrigation with 0.9% saline or water; monitor runoff of irrigation to a pH of 7 and continue irrigation for at least 30 minutes. Assess for solid particles. Topical mydriatic cycloplegics and antibiotics should be given (no systemic toxicity occurs following only ocular exposure). Use of 0.5% normal sodium chloride eye drops hourly or artificial tears (four times daily) with soft contact lenses may facilitate epithelization.
Personal Protection Equipment Full NBC clothing/respirator
Rescuer Contamination High possibility prior to decontamination
Environmental Persistency Not persistent; rapidly hydrolyzed in alkaline solutions
Additional Information Colorless solid or liquid with a high vapor pressure
Vapor density (as compared to air): 3.9
Odor: Disagreeable
Soluble in water
Melting point: 35°C to 40°C; concentrations <8% have little effect; can penetrate rubber
Median lethal dosage:
Inhalation: 1500-2000 mg/min/m^3
Skin: 25 mg/kg

Phosphine

Related Information Phosphine - Patient Information Sheet
CAS Number 7803-51-2
UN Number 2199

PHOSPHINE

Synonyms Delicia; Detia Gas EX-B; Hydrogen Phosphide; PH3; Phosphoreted Hydrogen; Phosphorus Trihydride

Use Silicon crystal treatment in semiconductor industry, contaminant in acetylene, grain fumigation, and rat poisons. Produced by reaction of hydrogen with metal phosphides.

Mechanism of Toxic Action Protoplasmic toxicity inhibits cytochrome C oxidase pathway primarily in myocardium

Adverse Reactions

Cardiovascular: Tachycardia, cardiovascular collapse, hypotension, flutter (atrial), angina, sinus tachycardia, arrhythmias (ventricular), chest tightness

Central nervous system: Panic attacks, coma, seizures, dizziness

Hepatic: Hepatic toxicity (late)

Neuromuscular & skeletal: Rhabdomyolysis, intention tremor

Renal: Oliguria, renal failure

Respiratory: Pulmonary edema, shortness of breath, dyspnea

Toxicodynamics/Kinetics Onset of action: Rapid, within 3 hours of exposure

Pregnancy Issues No teratogenic effects from acute exposures are known

Applies to Grain Fumigation; Rat Poison

Signs and Symptoms of Acute Exposure Abdominal pain, ataxia, cardiac arrhythmia, chest pain, cough, diplopia, dizziness, drowsiness, dyspnea, fibrillation (atrial), garlic-like breath, irritability, jaundice, seizures, tachycardia (ventricular), tremor, vomiting, diarrhea

Admission Criteria/Prognosis

Mild exposure: Asymptomatic individuals can be discharged after 6 hours of observation

Serious exposure: Patients should be observed for pulmonary edema or liver damage for 72 hours; survival after 4 days is usually associated with recovery

Treatment Strategy

Decontamination: Inhalation: Oxygenation; if ingested, lavage within 1 hour with 1:10,000 dilution of potassium permanganate to reduce availability of phosphine; activated charcoal may be useful

Supportive therapy: Calcium gluconate and magnesium sulfate may be useful for cardiac arrhythmia. Magnesium sulphate may be particularly useful at a dose of 3 g (continuous I.V. infusion) over 3 hours followed by 6 g/day for the next 3-5 days; may be unresponsive to vasopressin. Ventricular arrythmia can be treated with trimetazidine (20 mg orally, twice daily). Inverse-ratio ventilation using lower tidal volume mechanical ventilation (6 mL/kg) and plateau pressures of 30 cm water may decrease mortality. While positive end expiratory pressures should be considered, oxygen toxicity should be monitored. Use of perfluorocarbon partial liquid ventilation and exogenous surfactant in chemical-induced ARDS is investigational.

Enhancement of elimination: Hemodialysis may be useful in cases of renal damage.

Rescuer Contamination No risk of secondary contamination with phosphine gas; risk of secondary contamination with solid phosphine exposure

Personal Protection Equipment Positive pressure self-contained breathing apparatus is recommended (probable at air concentrations >10 ppm). Phosphine gas exposure does not require chemical-protective gear. Solid phosphine exposure requires rubber gloves, aprons, goggles, and footwear protection.

Reference Range Blood and urine phosphorus levels are not reliable.

Additional Information Adult deaths have been described with ingestion of 4 g of zinc phosphide or 500 mg aluminum phosphide. Colorless gas; flammable; garlic odor

Odor (garlic) threshold: 2 ppm

IDLH: 50 ppm

Gas density: 1.17

Slightly water soluble (0.3%)

May ignite spontaneously with air, oxidizer

STEL: 1 ppm

Potentially lethal air concentration: About 10 ppm

Phosphine does not appear to exhibit cumulative effects

Specific References

Lauterbach M, Solak E, Kaes J, et al, "Epidemiology of Hydrogen Phosphide Exposures in Humans Reported to the Poison Center in Mainz, Germany, 1983-2003," *Clin Toxicol*, 2005, 43:575-81.

Quinuclidinyl Benzilate

CAS Number 6581-06-2

Synonyms Agent 15; Agent 3; Agent Buzz; BZ; QNB

Military Classification Incapacitating Agent

CDC Classification for Diagnosis

Biologic: A case in which BZ is detected in urine (29), as determined by CDC.

Environmental: No method is available for detecting BZ in environmental samples.

CDC Case Classification

Suspected: A case in which a potentially-exposed person is being evaluated by healthcare workers or public health officials for poisoning by a particular chemical agent, but no specific credible threat exists.

Probable: A clinically compatible case in which a high index of suspicion (credible threat or patient history regarding location and time) exists for BZ exposure, or an epidemiologic link exists between the case and a laboratory-confirmed case.

Confirmed: A clinically compatible case in which laboratory tests on biologic samples have confirmed exposure. The case can be confirmed if laboratory testing was not performed because either a predominant amount of clinical and nonspecific laboratory evidence of a particular chemical was present or a 100% certainty of the etiology of the agent is known.

Use Incapacitating agent

Mechanism of Toxic Action An anticholinergic glycolate (3-fold and 25-fold more potent than scopolamine and atropine, respectively) which competitively inhibits acetylcholine in the muscarinic receptors in the peripheral nervous system producing parasympathetic blockade. It is believed to be 25 times more potent than atropine.

Adverse Reactions

Cardiovascular: Tachycardia, flushing

Central nervous system: Anticholinergic syndrome, confusion, delirium (may last 2-3 days), hyperthermia, seizures, dystonia, coma, psychosis, paranoia, visual hallucinations, incoordination

Gastrointestinal: Foul breath odor, xerostomia, ileus

Genitourinary: Urinary retention

Neuromuscular & skeletal: Tremors, rhabdomyolysis

Ocular: Mydriasis (may last 2-3 days), photophobia, conjunctivitis, blurred vision

Respiratory: Tachypnea

Toxicodynamics/Kinetics

Onset of effect: Within 1 hour

Peak effect: 4-10 hours

Duration of effect: 48-72 hours

Absorption: Dermal: 5% to 10% (effects may be delayed); usually found in propylene glycol vehicle

Metabolism: Two major metabolites (3-quinuclidinol and benzilic acid)

Elimination: Renal

Clinically Resembles Thyrotoxicosis, heat stroke

Signs and Symptoms of Acute Exposure At high doses, produces a toxic delirium that may render victims incapable of purposeful actions. Confused mental status may last several days.

Treatment Strategy

Decontamination:

Dermal: Remove contaminated clothes/jewelry; wash with soap and water

Inhalation: 100% humidified oxygen

Ocular: Remove contact lens; irrigate copiously with water or saline

Oral: Activated charcoal/gastric lavage within 2 hours

Supportive therapy:

Agitation/seizures: Diazepam: I.V.:

Children: 0.1 mg/kg

Adults: 2-10 mg **or**

Lorazepam: I.V.:

Children: 0.05 mg/kg

Adults: 1-2 mg. Avoid phenothiazines.

Tachyarrhythmias: Physostigmine: I.V.:

Children: 0.02 mg/kg up to 0.5 mg over several minutes; can repeat at 5-minute intervals up to 2 mg

Adults: 1-4 mg over 5 minutes; can repeat in 20 minutes

Total reversal of anticholinergic effects is uncertain. Due to its lack of central effects, pyridostigmine and neostigmine are not useful. Beta adrenergic blockers also can be used to treat supraventricular tachyarrhythmias. Monitor for hyperthermia or rhabdomyolysis. Monitor fluid status. Full recovery usually occurs in 96 hours.

Personal Protection Equipment Maximum protection with M9 mask and hood, M3 butyl rubber suit, M2A1 butyl boots, and M3 or M4 butyl gloves. Outgassing can occur.

Additional Information

Precursor chemicals may include 3-hydroxy-1-methylpiperidine (CAS: 3554-74-3), 3-Quinuclidinol (CAS: 1619-34-7), 3-Quinuclidone (CAS: 1619-34-7), and methyl benzilate (CAS: 76-89-1)

Symptomatic dose: $\sim$2.5 mcg/kg; less than 1 mcg/kg will likely not cause any effect

Incapacitating dose for 50% of adults: 6.2 mcg/kg

Lethal concentration in air needed to incapacitate 50% of unprotected exposed individuals: $\sim$200,000 mg/min/m^3

Increased lethality in hot weather

Odorless, crystalline white powder slightly soluble in water; pH: 3.64

For more information contact: U.S. Army Medical Research Institute of Chemical Defense (USAMRICD) at Aberdeen Proving Ground, Maryland (410-436-3628).

Specific References

Belson MG, Schier JG, and Patel MM, "Case Definitions for Chemical Poisoning," *MMWR Recomm Rep*, 2005, 54(RR-1):1-24.

Gundry CS, "Four Common Myths about Chemical and Biological Terrorism," *Inside Homeland Security*, 2007, 5(1):8-13.

Sarin

CAS Number 107-44-8

Synonyms GB; Isopropyl Methylphosphonofluoridate

Military Classification Chemical Warfare Weapon

Mechanism of Toxic Action Inhibits hydrolysis of acetylcholine by acetylcholinesterase binding with this enzyme. Irreversible binding (aging) takes place. This results in acetylcholine excess at the neuronal synapse. May penetrate blood brain barrier and effect GABA transmission. Similar to organophosphate agent; inhibits the enzyme acetylcholinesterase thus resulting in acetylcholine excess at the neuronal synapse; may penetrate blood brain barrier and thus affect GABA transmission.

Adverse Reactions

Cardiovascular: Sinus bradycardia, tachycardia, heart rhythm disturbances

Central nervous system: Seizures, ataxia, coma, headache, fatigue, memory loss, nightmares, excess cholinergic toxicity, muscarinic activity, nicotinic activity

Endocrine & metabolic: Hypokalemia has been reported in GB intoxication but the mechanism is unclear

Gastrointestinal: Diarrhea, nausea, salivation, vomiting, abdominal pain

Genitourinary: Involuntary urination

Ocular: Lacrimation, early mydriasis rapidly followed by persistent miosis (may take 1-2 months for pupillary response to normalize), dim or blurred vision, eye pain

Respiratory: Rhinorrhea, bronchospasm, cough, tachypnea, dyspnea, wheezing, bronchorrhea; death usually results from respiratory failure

Toxicodynamics/Kinetics

Onset of action:

Inhalation: Within 5 minutes

Dermal: Minutes to 1 hour

Half-life: Aging of sarin-acetylcholine complex ~5 hours

Pregnancy Issues Data do not suggest teratogenicity

Clinically Resembles Jatropha multifida plant poisoning, organophosphate insecticides, nicotine poisoning

Treatment Strategy

Adjunctive treatment: Diazepam for seizures, when intubation and mechanical ventilation is used and when >6 mg atropine is given (or 3 MK I kits)

Pretreatment/prophylaxis: Pyridostigmine

Decontamination:

Dermal: Remove all contaminated clothing. Wash with copious amounts of water. A 1% to 5% hypochlorite solution (household bleach diluted) with alkaline soap may be used. Immediate field/outdoor decontamination is suggested. Avoid decontamination inside the healthcare facility. Do not delay antidotal intervention in lieu of decontamination.

Inhalation: Remove from exposure site and remove contaminated clothes. Administer 100% humidified oxygen. Protect airway. If unconscious, consider intubation with mechanical ventilation.

Ocular: Copious irrigation with water or saline

Supportive therapy: Atropine is the mainstay of therapy. Depending upon severity of exposure, 10-20 mg cumulatively over the first 2-3 hours. Atropine should be titrated to bronchial secretions, **not** ocular signs. Pralidoxime should be administered 1-2 g I.V. within 6 hours of exposure over 10 minutes, with repeat dose in 1 hour if weakness persists, then every 4-12 hours afterwards as needed. Pralidoxime is most effective if given 3 hours postsarin exposure. If Mark I kits are available, acute treatment suggestions: mild exposure 1 kit, mild to moderate exposure 2 kits (2nd kit 5 minutes after the first), moderate to severe exposure 3 kits (each kit separated by 5 minutes). Long term I.V. atropine and 2-PAM may be required. The best oximes for acetylcholinesterase reactivation appear to be HI-6 and obidoxime along with oxime K027 and K033.

Special populations: Children are especially vulnerable to these compounds owing to their smaller mass, smaller airway diameter, higher respiratory rate, and minute volumes. Due to their short stature, they are exposed to chemicals that settle close to the ground.

Pediatric dose: Atropine .05 mg/kg I.V./I.M., 2-PAM 25-50 mg/kg I.V./I.M., diazepam 0.2 to 0.5 mg/kg I.V.

Environmental Persistency Sarin is the most volatile of the "G" agents, posing a significant vapor hazard. Vapors can off-gas from contaminated clothes or surfaces posing a secondary contamination risk. May persist for 1 day in soil. Evaporation time is about 2 hours. Miscible in water and solvents; vapor density: 4.86; liquid density: 1.10 g/mL

Personal Protection Equipment Protective equipment for healthcare responders is critical

Respiratory protection: Pressure-demand, self-contained breathing apparatus (SCBA) is recommended

Skin protection: Chemical protective clothing and butyl rubber (M3 and M4 Norton) with chemical goggles and face shield

Reference Range Cholinesterase activity <10% of normal is consistent with severe poisoning

Additional Information In the face of appropriate treatment, few sequellae are expected. Neuropathy and ataxia were noted in less than 10% of victims of the Japanese sarin attacks

(1994 and 1995) and resolved within 3 months of exposure. However, insomnia, PTSD, (<8% incidence after 5 years) nightmares may persist.

Specific References

American Academy of Pediatrics Committee on Environmental Health and Committee on Infectious Diseases, "Chemical-Biological Terrorism and Its Impact on Children: A Subject Review," *Pediatrics*, 2000, 105(3 Pt 1):662-70.

Arnold JF, "CBRNE - Nerve Agents, G-Series: Tabun, Sarin, Soman," http://www.emedicine.com/EMERG/topic898.htm. Last accessed October 12, 2003.

Ben Abraham R, Weinbroum AA, Rudick V, et al, "Perioperative Care of Children With Nerve Agent Intoxication," *Paediatr Anaesth*, 2001, 11(6):643-9.

Benitez FL, Velez-Daubon, and Keyes DC, "CBRNE - Nerve Agents, V-Series: Ve, Vg, Vm, Vx," http://www.emedicine.com/emerg/topic899.htm. Last accessed October 3, 2003.

Buckley NA, Roberts D, and Eddleston M, "Overcoming Apathy in Research on Organophosphate Poisoning." *BMJ*, 2004, 329(7476):1231-3 (review).

Corvino TF, Nahata MC, Angelos MG, et al, "Availability, Stability, and Sterility of Pralidoxime for Mass Casualty Use," *Ann Emerg Med*, 2006, 47(3):272-7.

Eyer P, Hagedorn I, Klimmek R, et al, "HLo 7 Dimethanesulfonate, A Potent Bispyridinium-Dioxime Against Anticholinesterases," *Arch Toxicol*, 1992, 66(9):603-21.

Giardino NJ, "Modeling Sarin Inhalation Exposure and Subsequent Dose to Civilian Population Groups," *The Forensic Examiner*, 2004, 13(1):11-3.

Gur I, Bar-Yishay E, and Ben-Abraham R, "Biphasic Extrathoracic Cuirass Ventilation for Resuscitation," *Am J Emerg Med*, 2005, 23(4):488-91.

Henretig FM, Cieslak TJ, and Eitzen EM Jr, "Biological and Chemical Terrorism," *J Pediatr*, 2002, 141(3):311-26. Erratum in: *J Pediatr*, 2002, 141(5):743-6.

Karalliedde L, Wheeler H, Maclehose R, et al, "Possible Immediate and Long-Term Health Effects Following Exposure to Chemical Warfare Agents," *Public Health*, 2000, 114(4):238-48.

Kassa J, "Comparison of the Effects of BI-6, A New Asymmetric Bipyridine Oxime, With HI-6 Oxime and Obidoxime in Combination With Atropine on Soman and Fosdrine Toxicity in Mice," *Ceska Slov Farm*, 1999, 48(1):44-7.

Kassa J, "Review of Oximes in the Antidotal Treatment of Poisoning by Organophosphorus Nerve Agents," *J Toxicol Clin Toxicol*, 2002, 40(6):803-16.

Kuca K, Bielavsky J, Cabal J, et al, "Synthesis of a New Reactivator of Tabun-Inhibited Acetylcholinesterase," *Bioorg Med Chem Lett*, 2003, 13(20):3545-7.

Kuca K and Kassa J, "*In Vitro* Reactivation of Acetylcholinesterase Using the Oxime K027," *Vet Hum Toxicol*, 2004, 46(1):15-8.

Kuca K, Jun D, "Reactivation of Sarin-Inhibited Pig Brain Acetylcholinesterase Using Oxime Antidotes," *J Med Toxicol*, 2006, 2(4):141-146.

LeJeune KE, Dravis BC, Yang F, et al, "Fighting Nerve Agent Chemical Weapons With Enzyme Technology," *Ann N Y Acad Sci*, 1998, 864:153-70.

Moreno G, Nocentini S, Guggiari M, et al, "Effects of the Lipophilic Biscation, Bis-Pyridinium Oxime BP12, on Bioenergetics and Induction of Permeability Transition in Isolated Mitochondria," *Biochem Pharmacol*, 2000, 59(3):261-6.

Noort D, Benschop HP, and Black RM, "Biomonitoring of Exposure to Chemical Warfare Agents: A Review," *Toxicol Appl Pharmacol*, 2002, 184(2):116-26 (review).

"Poison Warfare (Nerve) Gases," http://www.emergency.com/nervgas.htm. Last accessed October 10, 2003.

Rotenberg JS and Newmark J, "Nerve Agent Attacks on Children: Diagnosis and Management," *Pediatrics*, 2003, 112(3 Pt 1):648-58.

Rousseaux CG and Dua AK, "Pharmacology of HI-6, an H-series Oxime," *Can J Physiol Pharmacol*, 1989, 67(10):1183-9.

Salem H and Sidell FR, "Nerve Gases," *Encyclopedia of Toxicology*, Vol 1, Wexler P, ed: Academic Press, 2002, 380-5.

Salzman MS, Haroz R, Bartrand TA, et al, "The Tokyo Subway Sarin Attack Revisited," *Clin Toxicol (Phila)*, 2005, 43:684.

Shiloff JD and Clement JG, "Comparison of Serum Concentrations of the Acetylcholinesterase Oxime Reactivators HI-6, Obidoxime, and PAM to Efficacy Against Sarin (Isopropyl Methyl-phosphonofluoridate) Poisoning in Rats," *Toxicol Appl Pharmacol*, 1987, 89(2):278-80.

Velez-Daubon L, Benitez FL, Keyes DC, "CBRNE - Nerve Agents, Binary: GB2, VX2," *Emedicine*, 2003, http://www.emedicine.com/EMERG/topic 900.htm. Last accessed October 13, 2003.

Wei G, Chang A, and Hamilton RJ, "Nerve Agent Antidote Kits Enable Nurses to Treat More Mass-Casualty Patients Than Multidose Vials," *Acad Emerg Med*, 2006, 13(5):S177.

Yanagisawa N, Morita H, Nakajima T, "Sarin experiences in Japan: acute toxicity and long-term effects," *J Neurol Sci*, 2006, 249, 76-85.

Sodium Azide

CAS Number 12136-89-9; 26628-22-8
UN Number 1687
Synonyms Azium; Azomide
Use Shell detonators in explosive industry; found as principle agent (350-600 g) for providing nitrogen for the rapid expansion (in 0.05 seconds) of automobile air bags; preservative for laboratory reagents (concentration ~1 mg/mL); nematocide; herbicide; used in explosives industry
Mechanism of Action Mucosal irritant; may inhibit oxidative phosphorylation; can cause vasodilitation
Adverse Reactions
  Cardiovascular: Asystole, hypotension, initial bradycardia followed by tachycardia, chest pain, arrhythmias (atrial/ventricular), myocardial depression, congestive heart failure, vasodilation, cardiomyopathy
  Central nervous system: Hypothermia, hyperthermia, headache, agitation, seizures, coma
  Dermatologic: Dermal burns
  Endocrine & metabolic: Polydipsia, metabolic acidosis
  Gastrointestinal: Diarrhea, nausea, vomiting, abdominal cramps
  Hematologic: Leukocytosis
  Neuromuscular & skeletal: Weakness, hyporeflexia, paresthesia
  Ocular: Photophobia, lacrimation, keratitis, corneal burn, mydriasis
  Respiratory: Hyperventilation, tachypnea, dyspnea, pulmonary edema
  Miscellaneous: Diaphoresis
Toxicodynamics/Kinetics
  Absorption: Inhalation, dermal or ingestion
  Metabolism: Converted to nitric oxide
Pregnancy Issues No clear teratogenic effects in humans
Monitoring Parameters Arterial blood gas, ECG, creatine phosphokinase, electrolytes, chest x-ray
Clinically Resembles Cyanide, fluoroacetate
Admission Criteria/Prognosis Admit any ingestion >40 mg in adults or any symptomatic patient 2 hours postexposure; any patient with metabolic acidosis should be admitted
Treatment Strategy
  Decontamination:
    Oral: Activated charcoal
    Dermal: Flush with water
    Inhalation: Administer 100% humidified oxygen
    Ocular: Copious irrigation with saline or water
  Supportive therapy: I.V. sodium bicarbonate (1-3 mEq/kg) for acidosis; phenobarbital is probably the most effective agent to treat seizures. Hypotension can be treated with crystalloid solution (10-20 mL/kg) and placement in Trendelenburg position. Vasopressors

(dopamine or norepinephrine) can be used for resistant cases. The use of sodium nitrite or hyperbaric oxygen is of theoretical benefit, with human data lacking in efficacy.
Enhancement of elimination: Extracorporeal removal is of no benefit. Exchange transfusion does not appear to be beneficial.

Personal Protection Equipment Chemical protective clothing and positive pressure self-contained breathing apparatus should be considered

Reference Range Postmortem blood levels (following ingestion of sodium azide): 8-262 mg/L

Additional Information Fatal oral dose: 13 mg/kg

Oral dose of 0.5 mcg/kg can result in reduction of blood pressure; positive ferric chloride (10% to 20%) test of gastric aspirate can occur (red precipitate)

Rescuer can become mildly toxic (headache, nausea) from expired air or gastric aspirate of sodium azide toxic patients (due to hydrazoic acid).

Odorless, colorless, highly explosive. Specific gravity: 1.846. TLV-ceiling: 0.11 ppm

Byproducts of sodium azide detonation include sodium hydroxide and nitrogen.

Other chemical constituents in automobile air bags include 2,4-dinitrotoluene, boron, potassium nitrate, nitrocellulose and cupric oxide.

Specific References

Cooper H and Thomas T, "Ocular Injuries Related to Airbag Use," *Am J Emerg Med*, 2004, 22(2):135-7.

Duma SM, Rath AL, Jernigan MV, et al, "The Effects of Depowered Airbags on Eye Injuries in Frontal Automobile Crashes," *Am J Emerg Med*, 2005, 23(1):13-9.

Martin TG and Robertson WO, "Laboratory Workplace Coffee Tampering With Sodium Azide," *J Toxicol Clin Toxicol*, 2004, 42(5):748-9.

Nordt SP, Molloy M, Ryan J, et al, "Burns From Automobile Airbags," *J Emerg Med*, 2003, 25(2):201-2.

Sodium Monofluoroacetate

CAS Number 62-74-8

UN Number 2629

Synonyms Fluoroacetic Acid (Sodium Salt); SMFA; Sodium Fluoroacetate, 1080, Furatol, Fratol

CDC Classification for Diagnosis

Biologic: No biologic marker for sodium monofluoroacetate is available.

Environmental: Detection of sodium monofluoroacetate in environmental samples, as determined by FDA.

CDC Case Classification

Suspected: A case in which a potentially exposed person is being evaluated by healthcare workers or public health officials for poisoning by a particular chemical agent, but no specific credible threat exists.

Probable: A clinically compatible case in which a high index of suspicion (credible threat or patient history regarding location and time) exists for a sodium monofluoroacetate exposure, or an epidemiologic link exists between this case and a laboratory-confirmed case.

Confirmed: A clinically compatible case with laboratory confirmation from environmental samples. The case can be confirmed if laboratory testing was not performed because either a predominant amount of clinical and nonspecific laboratory evidence of a particular chemical was present or a 100% certainty of the etiology of the agent is known.

Use Rodenticide (banned in U.S. in 1972); used for coyote control in Mexico; possible agroterrorism agent or water-bourne terrorism agent

Mechanism of Toxic Action Metabolized to fluorocitrate, which then blocks Kreb cycle metabolism by inhibiting the mitochondrial enzyme aconitase

Adverse Reactions

Cardiovascular: Prolonged QT intervals on EKG (28%), ventricular fibrillation, atrial fibrillation (16%), hypotension, ventricular arrhythmias, increased t-wave amplitude, asystole

Central nervous system: Auditory hallucinations, seizures, tetany, coma, ataxia
Endocrine & metabolic: Hypocalcemia
Gastrointestinal: Vomiting, salivation, nausea
Hepatic: Hepatic necrosis
Neuromuscular & skeletal: Paresthesia, hypertonicity, carpopedal spasms
Ocular: Nystagmus, blurred vision, an ocular irritant
Renal: Renal failure (acute)
Respiratory: Hemorrhagic pulmonary edema (may be delayed for 24-72 hours), hyperventilation

Toxicodynamics/Kinetics
Absorption: Through gastrointestinal tract, eyes, and lungs; dermal absorption does not appear to occur if skin is intact
Elimination: Fecal/urine

Clinically Resembles Gastroenteritis, sodium azide

Applies to Compound 1080®

Signs and Symptoms of Acute Exposure Typically begin 30-90 minutes after exposure (but may be delayed for 20 hours); coma, hallucinations, hypotension, muscle spasms, mydriasis, myoglobinuria, paresthesia, seizures, sinus tachycardia, stupor. Late findings include renal and/or hepatic necrosis.

Admission Criteria/Prognosis Any symptomatic patient, any electrolyte abnormalities, acidosis, or suspected ingestions >2 mg/kg should be considered for admission into an intensive care unit; poor prognostic signs include hypotension, metabolic acidosis, increased serum creatinine and ventricular tachycardia development

Treatment Strategy
Decontamination: Lavage within 1 hour with magnesium sulfate or activated charcoal. Do not use water for lavage. Dermal: Remove contaminated clothing and jewelry and irrigate skin with soap and water. Ocular: Irrigate with water or saline
Supportive therapy: Treat seizures with phenobarbital or diazepam, intravenous calcium (10-20 mL of 10% calcium chloride in adults or 10-20 mL/kg in children); or 0.1-0.2 mL/kg of 10% calcium gluconate up to 20 mL should be given for evidence of hypocalcium (prolonged QT interval, tetany). Aggressive airway management/intubation may be necessary. Monitor fluid levels and renal function (myoglobinuria). Mephentermine may be more effective than levarterenol in treating hypotension. Digitalis can be utilized in pulmonary edema. Procainamide may be useful for ventricular arrhythmias. Glycerol monoacetate effective in animals; clinical use not as promising. Acetamide (500 mL of 10% solution in D_5W infused over 30 minutes every 4 hours) or ethanol use is also controversial (to inhibit the conversion of fluoroacetate to fluorocitrate). Acetamide or ethanol therapy has been used in experimental models, but this modality has not been substantiated in humans.
Enhancement of elimination: Multiple dose of activated charcoal may be effective

Environmental Persistency Low

Reference Range Autopsy urinary levels of 368 mg/L associated with lethality 17 hours after exposure to 465 mg of sodium fluoroacetate
Mean urinary sodium monofluoroacetate level of 368 mg/L associated with fatal poisoning

Additional Information Approximately 2-10 mg/kg is fatal.
A white, odorless water soluble powder usually associated with a blue dye; initially developed for chemical warfare
Vinegar taste only when in solution; water soluble. When heated, toxic sodium oxide and fluoride fumes are emitted.
Ethanol treatment may aggravate hypokalemia.
Fluoroacetamide is less toxic; trifluoroacetic acid used in high performance liquid chromatography is an irritant and can be dermally absorbed, but is not as toxic
IDLH: 2.5 mg/m^3
STEL: 0.15 mg/^{3}

Specific References

Belson MG, Schier JG, and Patel MM, "Case Definitions for Chemical Poisoning," *MMWR Recomm Rep*, 2005, 54(RR-1):1-24.

Höjer J, Hung HT, Du NT, et al, "An Outbreak of Severe Rodenticide Poisoning in North Vietnam Caused by Illegal Fluoroacetate," *J Toxicol Clin Toxicol*, 2003, 41(5):646.

Norris WR, Temple WA, Eason CT, et al, "Sorption of Fluoroacetate (Compound 1080) by Colestipol, Activated Charcoal and Anion-Exchange Resins *In Vitro*, and Gastrointestinal Decontamination in Rats," *Vet Hum Toxicol*, 2000, 42(5):269-75.

Robinson RF, Griffith JR, Wolowich WR, et al, "Intoxication With Sodium Monofluoroacetate (Compound 1080)," *Vet Hum Toxicol*, 2002, 44(2):93-5.

Soman

CAS Number 96-64-0

Synonyms GD; Pinacolyl Methylphosphonofluoridate

Military Classification Chemical Warfare Weapon - Nerve Agent

Mechanism of Toxic Action Inhibits hydrolysis of acetylcholine by acetylcholinesterase binding with this enzyme. Irreversible binding (aging) takes place. This results in acetylcholine excess at the neuronal synapse. May penetrate blood brain barrier and effect GABA transmission. Similar to organophosphate agent; inhibits the enzyme acetylcholinesterase thus resulting in acetylcholine excess at the neuronal synapse; may penetrate blood brain barrier and thus affect GABA transmission

Adverse Reactions - Symptoms may be delayed for up to 18 hours post exposure

Cardiovascular: Sinus bradycardia, tachycardia, heart rhythm disturbances

Central nervous system: Seizures, ataxia, coma, headache, fatigue, memory loss, nightmares, excess cholinergic toxicity, muscarinic activity, nicotinic activity

Gastrointestinal: Diarrhea, nausea, salivation, vomiting, abdominal pain

Genitourinary: Involuntary urination

Ocular: Lacrimation, early mydriasis rapidly followed by persistent miosis (may take 1-2 months for pupillary response to normalize), dim or blurred vision, eye pain

Respiratory: Rhinorrhea, bronchospasm, cough, tachypnea, dyspnea, wheezing, bronchorrhea; death usually results from respiratory failure

Toxicodynamics/Kinetics

Onset of action:

Inhalation: Within minutes

Dermal: Within minutes

Half-life: Aging of soman-acetylcholine complex ~2-5 minutes

Pregnancy Issues Data do not suggest teratogenicity

Clinically Resembles Jatropha multifida plant poisoning, organophosphate insecticides, nicotine poisoning

Treatment Strategy

Adjunctive treatment: Diazepam for seizures, when intubation and mechanical ventilation is used and when >6 mg atropine is given (or 3 MK I kits)

Pretreatment/prophylaxis: Pyridostigmine

Decontamination:

Dermal: Remove all contaminated clothing. Wash with copious amounts of water. A 1% to 5% hypochlorite solution (household bleach diluted) with alkaline soap may be used. Immediate field/outdoor decontamination is suggested. Avoid decontamination inside the healthcare facility. Do not delay antidotal intervention in lieu of decontamination.

Inhalation: Remove from exposure site and remove contaminated clothes. Administer 100% humidified oxygen. Protect airway. If unconscious, consider intubation with mechanical ventilation.

Ocular: Copious irrigation with water or saline

Supportive therapy: Atropine is the mainstay of therapy. Depending upon severity of exposure, 10-20 mg cumulatively over the first 2-3 hours. Atropine should be titrated to bronchial secretions, **not** ocular signs. Pralidoxime should be administered 1-2 g I.V. over 10 minutes, with repeat dose in 1 hour if weakness persists, then every 4-12 hours afterwards as needed. Pralidoxime is most effective if given 3 hours postsoman exposure. If Mark I kits are available, acute treatment suggestions: mild exposure 1 kit, mild to moderate exposure 2 kits (2nd kit 5 minutes after the first), moderate to severe exposure 3 kits (each kit separated by 5 minutes). Long term I.V. atropine and 2-PAM may be required.

Special populations: Children are especially vulnerable to these compounds owing to their smaller mass, smaller airway diameter, higher respiratory rate, and minute volumes. Due to their short stature, they are exposed to chemicals that settle close to the ground. Pediatric dose: Bolus: Atropine .05 mg/kg I.V./I.M., 2-PAM 25-50 mg/kg I.V./I.M., diazepam 0.2 to 0.5 mg/kg I.V.

Environmental Persistency Volatile substance with moderate persistence. Vapors can off-gas from contaminated clothes or surfaces posing a secondary contamination risk.

Personal Protection Equipment Protective equipment for healthcare responders is critical
Respiratory protection: Pressure-demand, self-contained breathing apparatus (SCBA) is recommended
Skin protection: Chemical protective clothing and butyl rubber (M3 and M4 Norton) with chemical goggles and face shield

Reference Range Cholinesterase activity <10% of normal is consistent with severe poisoning

Additional Information Soman is more difficult to treat than the other "G" agents. Pralidoxime is not as effective with soman owing to the fact it has a rapid onset of action; irreversible aging occurs within minutes. Although there are other more effective oxime antidotes, they are not widely available. Immediate decontamination and antidotal therapy is essential. Atropine and 2-PAM remain the treatment of choice. Long term atropine may be necessary. In the face of appropriate treatment, few sequelae are expected. However, insomnia, posttraumatic stress disorder, and nightmares may persist.

Specific References

American Academy of Pediatrics Committee on Environmental Health and Committee on Infectious Diseases, "Chemical-Biological Terrorism and Its Impact on Children: A Subject Review," *Pediatrics*, 2000, 105(3 Pt 1):662-70.

Arnold JF, "CBRNE - Nerve Agents, G-Series: Tabun, Sarin, Soman," http://www.emedicine.com/EMERG/topic898.htm. Last accessed October 12, 2003.

Ben Abraham R, Weinbroum AA, Rudick V, et al, "Perioperative Care of Children With Nerve Agent Intoxication," *Paediatr Anaesth*, 2001, 11(6):643-9.

Benitez FL, Velez-Daubon, and Keyes DC, "CBRNE - Nerve Agents, V-Series: Ve, Vg, Vm, Vx," http://www.emedicine.com/emerg/topic899.htm. Last accessed October 3, 2003.

Corvino TF, Nahata MC, Angelos MG, et al, "Availability, Stability, and Sterility of Pralidoxime for Mass Casualty Use," *Ann Emerg Med*, 2006, 47(3):272-7.

Giardino NJ, "Modeling Sarin Inhalation Exposure and Subsequent Dose to Civilian Population Groups," *The Forensic Examiner*, 2004, 13(1):11-3.

Gur I, Bar-Yishay E, and Ben-Abraham R, "Biphasic Extrathoracic Cuirass Ventilation for Resuscitation," *Am J Emerg Med*, 2005, 23(4):488-91.

Henretig FM, Cieslak TJ, and Eitzen EM Jr, "Biological and Chemical Terrorism," *J Pediatr*, 2002, 141(3):311-26. Erratum in: *J Pediatr*, 2002, 141(5):743-6.

Karalliedde L, Wheeler H, Maclehose R, et al, "Possible Immediate and Long-Term Health Effects Following Exposure to Chemical Warfare Agents," *Public Health*, 2000, 114(4):238-48.

Kassa J, "Comparison of the Effects of BI-6, A New Asymmetric Bipyridine Oxime, With HI-6 Oxime and Obidoxime in Combination With Atropine on Soman and Fosdrine Toxicity in Mice," *Ceska Slov Farm*, 1999, 48(1):44-7.

Kassa J, "Review of Oximes in the Antidotal Treatment of Poisoning by Organophosphorus Nerve Agents," *J Toxicol Clin Toxicol*, 2002, 40(6):803-16.

Kuca K, Bielavsky J, Cabal J, et al, "Synthesis of a New Reactivator of Tabun-Inhibited Acetylcholinesterase," *Bioorg Med Chem Lett*, 2003, 13(20):3545-7.

Lallement G, Masqueliez C, Baubichon D, et al, "Early Changes in MAP2 Protein in the Rat Hippocampus Following Soman Intoxication," *Drug Chem Toxicol*, 2003, 26(4):219-29.

LeJeune KE, Dravis BC, Yang F, et al, "Fighting Nerve Agent Chemical Weapons With Enzyme Technology," *Ann N Y Acad Sci*, 1998, 864:153-70.

Moreno G, Nocentini S, Guggiari M, et al, "Effects of the Lipophilic Biscation, Bis-Pyridinium Oxime BP12, on Bioenergetics and Induction of Permeability Transition in Isolated Mitochondria," *Biochem Pharmacol*, 2000, 59(3):261-6.

"Poison Warfare (Nerve) Gases," http://www.emergency.com/nervgas.htm. Last accessed October 10, 2003.

Rotenberg JS and Newmark J, "Nerve Agent Attacks on Children: Diagnosis and Management," *Pediatrics*, 2003, 112(3 Pt 1):648-58.

Rousseaux CG and Dua AK, "Pharmacology of HI-6, an H-series Oxime," *Can J Physiol Pharmacol*, 1989, 67(10):1183-9.

Salem H and Sidell FR, "Nerve Gases," *Encyclopedia of Toxicology*, Vol 1, Wexler P, ed: Academic Press, 2002, 380-5.

Shiloff JD and Clement JG, "Comparison of Serum Concentrations of the Acetylcholinesterase Oxime Reactivators HI-6, Obidoxime, and PAM to Efficacy Against Sarin (Isopropyl Methylphosphonofluoridate) Poisoning in Rats," *Toxicol Appl Pharmacol*, 1987, 89(2):278-80.

Velez-Daubon L, Benitez FL, Keyes DC, "CBRNE - Nerve Agents, Binary: GB2, VX2," *Emedicine*, 2003, http://www.emedicine.com/EMERG/topic 900.htm. Last accessed October 13, 2003.

Wei G, Chang A, and Hamilton RJ, "Nerve Agent Antidote Kits Enable Nurses to Treat More Mass-Casualty Patients Than Multidose Vials," *Acad Emerg Med*, 2006, 13(5):S177.

Sulfuric Acid

CAS Number 7664-93-9; 8014-95-7 (oleum)

UN Number 1830; 1831; 1832

Synonyms Battery Acid; Dihydrogen Sulfate; Oil of Vitriol; Oleum (Fuming Sulfuric Acid); Sulfur Acid; Sulphine Acid; Vitriol Brown Oil

Commonly Includes Automotive batteries; used in fur and leather industries; component in smog; major component in acid rain; formed from sulfur trioxide and water

Use In the manufacture of acetic acid, hydrochloric acid, hydrolysis of cellulose; used for metal cleaning; primarily used in phosphate fertilizer production; it is also used in the manufacture of dyes, explosives, petroleum, paper pulp, ore processing. Toilet bowl cleaners (which contain sodium bisulfite) produce sulfuric acid upon contact with water.

Mechanism of Toxic Action Acid oxidizer; corrosive to the skin, eyes, mucous membranes, gastrointestinal and respiratory tracts; chars tissue by removing water

Adverse Reactions

Cardiovascular: Shock, vascular collapse, chest pain, angina

Central nervous system: Headache

Gastrointestinal: Gastritis, throat irritation

Ocular: Lacrimation, iritis, cataracts, glaucoma

Renal: Renal failure

Respiratory: Tachypnea, bronchoconstriction, ARDS, laryngeal edema, cough, dyspnea

Toxicodynamics/Kinetics

Absorption: Can be absorbed

Metabolism: Dissociates into hydronium and sulfate

Elimination: Renal (as sulfate)

Applies to Dyes (Manufacture); Explosives (Manufacture); Fertilizer Production (Phosphate); Metal Cleaning; Ore Processing; Paper Pulp; Petroleum (Manufacture); Toilet Bowl Cleaners

Signs and Symptoms of Acute Exposure Choking, corneal burns, cough, discoloration of teeth, dyspnea, hemoptysis/hematemesis. Esophageal/gastric burns are rare, but ingestion of 15-50 mL (of a 26.4 to 35.4 normal solution) can cause severe gastroesophageal burns and gastric perforation. Dermal contact with >10% concentration is caustic. Dermal contact at >50% total body surface area can be fatal as dermal burns with coagulative necrosis develop. Corneal exposure at concentrations >1.25% can result in severe ocular damage

Admission Criteria/Prognosis Admit any symptomatic patient, or dermal contact >30% total body surface area over extremities or trunk

Treatment Strategy
   Decontamination: **Do not** induce emesis. Treat inhalation injuries with supplemental oxygen. Activated charcoal is not effective. Dilute with cold milk, cornstarch, or large amounts of cold water; endoscopy for severe mouth burns
   Ocular: Irrigate copiously with saline (preferred over water due to less heat production). Irrigation may need to last for 3 hours. Continue to irrigate until runoff exhibits a neutral pH. Topical mydriatics and antibiotics may be useful.
   Dermal: Irrigate affected area with copious amounts of water. Remove jewelry and clothing
   Enhancement of elimination: Hemodialysis if renal failure develops

Personal Protection Equipment Full chemical protective clothing (gloves, boots, rubber over clothing, and goggles) with positive pressure self-contained breathing apparatus.

Reference Range Normal blood sulfate concentration range: 0.8-1.2 mg/dL

Additional Information Lethal dose: 135 mg/kg
   Clear, colorless gas; odorless except when heating (choking odor)
   Reacts violently with water or alcohol; will corrode or dissolve metals
   Miscible in water
   pH of a 1N solution: 0.3; 0.1N: 1.2; 0.01N: 2.1
   Vapor density: 3.4
   TLV-TWA: 1 mg/m^3
   TLV-STEL: 3 mg/m^3
   IDLH: 80 mg/m^3
   PEL-TWA: 1 mg/m^3
   Background sulfate levels in North American lakes are 20-40 µeq/L, although eastern North American lakes can have concentrations as high as 100 µeq/L.

Tabun

CAS Number 77-81-6

Synonyms GA; N-Dimethylphosphoramideocyanidate; O-Ethyl N

Military Classification Chemical Warfare Weapon

Mechanism of Toxic Action Inhibits hydrolysis of acetylcholine by acetylcholinesterase binding with this enzyme. Irreversible binding (aging) takes place. This results in acetylcholine excess at the neuronal synapse. May penetrate blood brain barrier and effect GABA transmission. Similar to organophosphate agent; inhibits the enzyme acetylcholinesterase thus resulting in acetylcholine excess at the neuronal synapse; may penetrate blood brain barrier and thus affect GABA transmission.

Adverse Reactions
   Cardiovascular: Sinus bradycardia, tachycardia, heart rhythm disturbances
   Central nervous system: Seizures, ataxia, coma, headache, fatigue, memory loss, nightmares, excess cholinergic toxicity, muscarinic activity, nicotinic activity
   Gastrointestinal: Diarrhea, nausea, salivation, vomiting, abdominal pain
   Genitourinary: Involuntary urination
   Ocular: Lacrimation, early mydriasis rapidly followed by persistent miosis (may take 1-2 months for pupillary response to normalize), dim or blurred vision, eye pain

Respiratory: Rhinorrhea, bronchospasm, cough, tachypnea, dyspnea, wheezing, bronchor-rhea; death usually results from respiratory failure

Toxicodynamics/Kinetics
Onset of action:
Inhalation: Within 5 minutes
Dermal: Minutes to 1 hour
Half-life: Aging of tabun-acetylcholine complex ~14 hours

Pregnancy Issues Data do not suggest teratogenicity

Clinically Resembles Jatropha multifida plant poisoning, organophosphate insecticides, nicotine poisoning

Treatment Strategy
Adjunctive treatment: Diazepam for seizures, when intubation and mechanical ventilation is used and when >6 mg atropine is given (or 3 MK I kits)
Pretreatment/prophylaxis: Pyridostigmine
Decontamination:
Dermal: Remove all contaminated clothing. Wash with copious amounts of water. A 1% to 5% hypochlorite solution (household bleach diluted) with alkaline soap may be used. Immediate field/outdoor decontamination is suggested. Avoid decontamination inside the healthcare facility. Do not delay antidotal intervention in lieu of decontamination.
Inhalation: Remove from exposure site and remove contaminated clothes. Administer 100% humidified oxygen. Protect airway. If unconscious, consider intubation with mechanical ventilation.
Ocular: Copious irrigation with water or saline
Supportive therapy: Atropine is the mainstay of therapy. Depending upon severity of exposure, 10-20 mg cumulatively over the first 2-3 hours. Atropine should be titrated to bronchial secretions, **not** ocular signs. Pralidoxime should be administered 1-2 g I.V. over 10 minutes, with repeat dose in 1 hour if weakness persists, then every 4-12 hours afterwards as needed. Pralidoxime is most effective if given 3 hours post-tabun exposure. If Mark I kits are available, acute treatment suggestions: mild exposure 1 kit, mild to moderate exposure 2 kits (2nd kit 5 minutes after the first), moderate to severe exposure 3 kits (each kit separated by 5 minutes). Newer developed oximes (K027, K048) may prevent neurotoxicity of Tabum in rodert models. Long term I.V. atropine and 2-PAM may be required.
Special populations: Children are especially vulnerable to these compounds owing to their smaller mass, smaller airway diameter, higher respiratory rate, and minute volumes. Due to their short stature, they are exposed to chemicals that settle close to the ground.
Pediatric dose: Atropine .05 mg/kg I.V./I.M. (titrate to cessation of bronchial secretions), 2-PAM 25-50 mg/kg I.V./I.M., and diazepam 0.2 to 0.5 mg/kg I.V. (for seizure control)

Environmental Persistency Volatile substance with moderate persistence. Vapors can off-gas from contaminated clothes or surfaces posing a secondary contamination risk. May persist in soil; half-life is up to 36 hours.

Personal Protection Equipment Protective equipment for healthcare responders is critical
Respiratory protection: Pressure-demand, self-contained breathing apparatus (SCBA) is recommended
Skin protection: Chemical protective clothing and butyl rubber (M3 and M4 Norton) with chemical goggles and face shield

Reference Range Cholinesterase activity <10% of normal is consistent with severe poisoning

Additional Information In the face of appropriate treatment, few sequelae are expected. However, insomnia and nightmares may persist.
Vapor density: 5-6; liquid density: 1.08 g/mL

Specific References
American Academy of Pediatrics Committee on Environmental Health and Committee on Infectious Diseases, "Chemical-Biological Terrorism and Its Impact on Children: A Subject Review," *Pediatrics*, 2000, 105(3 Pt 1):662-70.

Arnold JF, "CBRNE - Nerve Agents, G-Series: Tabun, Sarin, Soman," http://www.emedicine.com/EMERG/topic898.htm. Last accessed October 12, 2003.

Ben Abraham R, Weinbroum AA, Rudick V, et al, "Perioperative Care of Children With Nerve Agent Intoxication," *Paediatr Anaesth*, 2001, 11(6):643-9.

Benitez FL, Velez-Daubon, and Keyes DC, "CBRNE - Nerve Agents, V-Series: Ve, Vg, Vm, Vx," http://www.emedicine.com/emerg/topic899.htm. Last accessed October 3, 2003.

Corvino TF, Nahata MC, Angelos MG, et al, "Availability, Stability, and Sterility of Pralidoxime for Mass Casualty Use," *Ann Emerg Med*, 2006, 47(3):272-7.

Eyer P, Hagedorn I, Klimmek R, et al, "HLo 7 Dimethanesulfonate, A Potent Bispyridinium-Dioxime Against Anticholinesterases," *Arch Toxicol*, 1992, 66(9):603-21.

Giardino NJ, "Modeling Sarin Inhalation Exposure and Subsequent Dose to Civilian Population Groups," *The Forensic Examiner*, 2004, 13(1):11-3.

Henretig FM, Cieslak TJ, and Eitzen EM Jr, "Biological and Chemical Terrorism," *J Pediatr*, 2002, 141(3):311-26. Erratum in: *J Pediatr*, 2002, 141(5):743-6.

Howland MA, "Pralidoxime - Antidotes in Depth," *Goldfrank's Toxicological Emergencies*, 6th ed, Norwalk, CT: Appleton and Lange, 1998, 1361-5.

Karalliedde L, Wheeler H, Maclehose R, et al, "Possible Immediate and Long-Term Health Effects Following Exposure to Chemical Warfare Agents," *Public Health*, 2000, 114(4):238-48.

Kassa J, "Comparison of the Effects of BI-6, A New Asymmetric Bipyridine Oxime, With HI-6 Oxime and Obidoxime in Combination With Atropine on Soman and Fosdrine Toxicity in Mice," *Ceska Slov Farm*, 1999, 48(1):44-7.

Kassa J, "Review of Oximes in the Antidotal Treatment of Poisoning by Organophosphorus Nerve Agents," *J Toxicol Clin Toxicol*, 2002, 40(6):803-16.

Kassa J, Kunesova G, "A Comparison of the Potency of Newly Developed Oximes (K027, K048) and Commonly Used Oximes (HI-6) to Counteract Tabun-Induced Neurotoxicity in Rats," *J Appl Toxicol*, 2006, 26:309-316.

Kuca K, Bielavsky J, Cabal J, et al, "Synthesis of a New Reactivator of Tabun-Inhibited Acetylcholinesterase," *Bioorg Med Chem Lett*, 2003, 13(20):3545-7.

LeJeune KE, Dravis BC, Yang F, et al, "Fighting Nerve Agent Chemical Weapons With Enzyme Technology," *Ann N Y Acad Sci*, 1998, 864:153-70.

Moreno G, Nocentini S, Guggiari M, et al, "Effects of the Lipophilic Biscation, Bis-Pyridinium Oxime BP12, on Bioenergetics and Induction of Permeability Transition in Isolated Mitochondria," *Biochem Pharmacol*, 2000, 59(3):261-6.

"Poison Warfare (Nerve) Gases," http://www.emergency.com/nervgas.htm. Last accessed October 10, 2003.

Rotenberg JS and Newmark J, "Nerve Agent Attacks on Children: Diagnosis and Management," *Pediatrics*, 2003, 112(3 Pt 1):648-58.

Rousseaux CG and Dua AK, "Pharmacology of HI-6, an H-series Oxime," *Can J Physiol Pharmacol*, 1989, 67(10):1183-9.

Salem H and Sidell FR, "Nerve Gases," *Encyclopedia of Toxicology*, Vol 1, Wexler P, ed: Academic Press, 2002, 380-5.

Shiloff JD and Clement JG, "Comparison of Serum Concentrations of the Acetylcholinesterase Oxime Reactivators HI-6, Obidoxime, and PAM to Efficacy Against Sarin (Isopropyl Methyl-phosphonofluoridate) Poisoning in Rats," *Toxicol Appl Pharmacol*, 1987, 89(2):278-80.

Velez-Daubon L, Benitez FL, and Keyes DC, "CBRNE - Nerve Agents, Binary: GB2, VX2," *Emedicine*, 2003, http://www.emedicine.com/EMERG/topic 900.htm. Last accessed October 13, 2003.

Wei G, Chang A, and Hamilton RJ, "Nerve Agent Antidote Kits Enable Nurses to Treat More Mass-Casualty Patients Than Multidose Vials," *Acad Emerg Med*, 2006, 13(5):S177.

Tetramethylenedisulfotetramine

CAS Number 80-12-6
Synonyms 4-2-4; Tetramine; TETS

THALLIUM SULFATE

Use Rodenticide (found in China); banned in the United States

Mechanism of Toxic Action Noncompetitive gamma-aminobutyric acid (GABA) antagonist through neuronal chloride channel blockade

Toxicodynamics/Kinetics Absorption: Oral/inhalation; not well absorbed dermally

Signs and Symptoms of Acute Exposure
Tachycardia, seizures (generalized), coma, mydriasis; onset of symptoms may be 30 minutes to 13 hours
Mild symptoms include headache, nausea, vomiting, dizziness, anorexia

Additional Information
Oral human LD_{50}: 100 mcg/kg
White powder 100 times more toxic than potassium cyanide. Fatalities usually occur within 3 hours.
Odorless/tasteless powder that easily dissolves in water

Specific References
Whitlow KS, Belson M, Barrueto F, et al, "Tetramethylenedisulfotetramine: Old Agent and New Terror," *Ann Emerg Med*, 2005, 45:609-13.

Thallium Sulfate

CAS Number 7440-28-0

CDC Classification for Diagnosis
Biologic: A case in which elevated spot urine thallium levels are detected (reference level: <0.5 mcg/L), as determined by a commercial laboratory.
Environmental: Detection of thallium in environmental samples, as determined by NIOSH or FDA.

CDC Case Classification
Suspected: A case in which a potentially exposed person is being evaluated by healthcare workers or public health officials for poisoning by a particular chemical agent, but no specific credible threat exists.
Probable: A clinically compatible case in which a high index of suspicion (credible threat or patient history regarding location and time) exists for thallium exposure, or an epidemiologic link exists between this case and a laboratory-confirmed case.
Confirmed: A clinically compatible case in which laboratory tests of biologic and environmental samples have confirmed exposure. The case can be confirmed if laboratory testing was not performed because either a predominant amount of clinical and nonspecific laboratory evidence of a particular chemical was present or a 100% certainty of the etiology of the agent is known.

Use Primarily in semiconduction industry in manufacture of switches and closures; radiopharmaceutical agent banned by the EPA as a pesticide in 1972

Mechanism of Toxic Action Thallium distributes intracellularly like potassium; it is a cellular toxin; inhibits mitochondrial oxidative phosphorylation, disrupts protein synthesis, alters heme metabolism

Adverse Reactions Neurologic, dysphagia, cardiac, hepatic, renal, dermatologic, gastrointestinal, pulmonary, salivation, chorea (extrapyramidal), xerostomia, sinus bradycardia, QRS prolongation, sinus tachycardia

Toxicodynamics/Kinetics
Absorption: Through skin and gastrointestinal tract
Distribution: V_d: 1-5 L/kg; following intoxication, highest concentration found in kidney and urine with lesser amounts in intestine, thyroid, testes, pancreas, skin, bone, spleen
Half-life: 2-4 days
Elimination: Large amount excreted in urine within first 24 hours after which fecal elimination predominates

Pregnancy Issues Crosses placenta, fetal exposure in 3rd trimester resulted in fetal alopecia and nail changes

Signs and Symptoms of Acute Exposure Alopecia, ataxia, bradycardia, cardiac arrhythmias, cerebral edema, coma, degeneration of central and peripheral nervous systems, tremor, delirium, dementia, disorientation, drowsiness, fatty infiltration, feces discoloration (black), fever (poor prognostic sign), garlic-like breath, gastroenteritis, hyperesthesia of palm and sole, lacrimation, liver necrosis, Mees' lines, metallic taste, motor/sensory neuropathy, mydriasis, nausea, nephritis, optic neuropathy, paresthesia, pulmonary edema, seizures, tachycardia, vomiting. GI symptoms occur within 1 day. Neurological symptoms occur 4-6 days postingestion.

Admission Criteria/Prognosis Any symptomatic patient or development of hypocalcemia should be admitted; suspected ingestions should also be admitted

Treatment Strategy
Decontamination:
Oral: Lavage (with 1% sodium iodine to convert thallium sulfate to thallium iodine) is preferable over ipecac. Activated charcoal or polystyrene sulfonate may be effective. Prussian blue (not commercially available for therapeutic use and not FDA approved) exchanges thallium for potassium and is fecally excreted; dosage: 250 mg/kg orally in 4 divided doses; continue therapy until thallium urinary level falls to <0.5 mcg/24 hours. Acetylcysteine treatment is investigational.
Dermal: Remove contaminated clothing; wash with soap and water
Ocular: Irrigate with water or saline
Enhancement of elimination: Multiple dosing of activated charcoal may be effective. Potassium chloride has been recommended in the past to increase elimination (when given after 48-72 hours), but may result in increased CNS toxicity. Forced saline diuresis (500 mL/hour of urine output) is recommended. Hemodialysis or hemoperfusion may theoretically be effective, but has not been demonstrated to be beneficial in humans. Furosemide can enhance urinary thallium excretion in animal models.

Reference Range Blood levels >30 mcg/L are indicative of thallium exposure; toxicity associated with urine levels >20 mcg/L. Serum thallium and cerebrospinal fluid levels of 8700 mcg/L and 1200 mcg/L, respectively, have been associated with fatality.

Additional Information Estimated lethal oral dose: Humans: 10-12 mg/kg
Odorless, tasteless, radiopaque
Chelating agents are not effective. Dithiocarb worsens symptoms by causing central redistribution of thallium. Monitor for hypocalcemia in the acute phase. Assay blood by flame atomic absorption.
The water soluble salts of thallium (sulfate, acetate, malonate, and carbonate) are more toxic than the sulfide and iodide (less water soluble).
Estimated daily thallium intake in a 70-kg adult: Drinking water: 2 mcg; Food: 5 mcg; Inhalation: 3.4 ng
Thallium is present in concentrations of 0.024 mcg/g of cigarettes and 0.06-0.17 mcg/g in cigar stubs
Thallium contaminated heroin has been documented in France.
Average ambient urban air concentration of thallium: 0.04 ng/m^3; thallium is present in earth's crust at concentrations of 0.3-0.7 ppm
PEL-TWA: 0.1 mg/m^3; TLV-TWA: 0.1 mg/m^3; IDLH: 20 mg/m^3

Specific References
Belson MG, Schier JG, and Patel MM, "Case Definitions for Chemical Poisoning," *MMWR Recomm Rep*, 2005, 54(RR-1):1-24.
Cumpston KL, Burk M, Burda A, et al, "Darkness on the Edge of Town: A Rural Family Maliciously Poisoned by Thallium," *J Toxicol Clin Toxicol*, 2003, 41(5):739.
Hentschel H, Fukula E, and Bertram G, "Unrecognized Severe Thallium Poisoning," *J Toxicol Clin Toxicol*, 2001, 39:287.

Hervé Y, Arouko H, Rams A, et al, "Acute Collective Thallotoxicosis: Three Cases," *J Toxicol Clin Toxicol*, 2000, 38(2):255-6.

Hoffman RS, "Thallium Poisoning: Past, Present, and Future," *J Toxicol Clin Toxicol*, 2001, 39:237.

Hoffman RS, "Thallium Toxicity and the Role of Prussian Blue in Therapy," *Toxicol Rev*, 2003, 22(1):29-40.

Hoffman RS, Stringer JA, Feinberg RS, et al, "Comparative Efficacy of Thallium Adsorption by Activated Charcoal, Prussian Blue, and Sodium Polystyrene Sulfonate," *J Toxicol Clin Toxicol*, 1999, 37(7):833-7.

Mercurio-Zappala M, Hardej D, Hoffman RS, et al, "Using Cell Culture to Assess Thallium Neurotoxicity: A Preliminary Study," *J Toxicol Clin Toxicol*, 2004, 42(5):826.

Mulkey JP and Oehme FW, "Are 2,3-Dimercapto-1-propanesulfonic Acid or Prussian Blue Beneficial in Acute Thallotoxicosis in Rats?" *Vet Hum Toxicol*, 2000, 42(6):325-9.

Rusyniak DE, Furbee RB, and Kirk MA, "Thallium and Arsenic Poisoning in a Small Midwestern Town," *Ann Emerg Med*, 2002, 39(3):307-11.

Rusyniak DE, Kao LW, Nanagas KA, et al, "Dimercaptosuccinic Acid and Prussian Blue in the Treatment of Acute Thallium Poisoning in Rats," *J Toxicol Clin Toxicol*, 2003, 41(2):137-42.

Senecal PE and Chalut D, "A Darker Shade of Prussian Blue: The Difficult Quest for the Thallium (Tl) Antidote," *J Toxicol Clin Toxicol*, 2000, 38(5):554.

Sharma AN, Nelson LS, and Hoffman RS, "Cerebral Spinal Fluid (CSF) Analysis in Fatal Thallium Poisoning," *J Toxicol Clin Toxicol*, 2001, 39(3):237-8.

Stringer JA, Feinberg RS, and Hoffman RS, "Comparative Efficacy of Thallium Adsorption by Activated Charcoal, Prussian Blue and Sodium Polystyrene Sulfonate," *J Toxicol Clin Toxicol*, 1998, 36(5):529.

Thermite

Military Classification Incendiary Agent

Use Explosive device

Mechanism of Toxic Action Powdered or granulated aluminum and powdered iron oxide mixture that reacts violently when heated (at $-2200°C$). Reaction produces aluminum oxide and elemental iron.

Adverse Reactions Dermal burns, ocular burns, dyspnea, respiratory distress

Clinically Resembles Burns, metal fume fever

Treatment Strategy

Decontamination: Dermal: Flush with copious amounts of water; administer 100% humidified oxygen; aggressive airway management/intubation may be necessary

Supportive therapy: Thermite itself has no inherent systemic toxicity. Remove all contaminated clothing. Evaluate for carbon monoxide exposure and treat with standard burn protocol to maintain urine output at 1-2 mL/kg/hour. Topical antibiotics such as mafenide can be used in burn wound dressing. Take precautions against infection. Monitor for associated traumatic injury.

Rescuer Contamination Take precautions against flammables that have not been completely ignited or extinguished.

Personal Protection Equipment Level A suits not protective of thermal injury

Additional Information When combined with binders, the material is called thermate

Specific References

Arnold JL, Halpern P, Tsai MC, et al, "Mass Casualty Terrorist Bombings: A Comparison of Outcomes by Bombing Type," *Ann Emerg Med*, 2004, 43(2):263-73.

Mendelson JA, "Some Principles of Protection Against Burns From Flame and Incendiary Munitions," *J Trauma*, 1971, 11(4):286-94.

Triethylamine

CAS Number 121-44-8
Synonyms TEA; Diethylaminoethane; TEN
Use A chemical intermediate for resins, rubber, and quaternary ammonium compounds. Used as a corrosion inhibitor. Cold box sand core (foundry) workers can be exposed.
Mechanism of Toxic Action A mucosal irritant which may exhibit inhibition of monoamine oxidase activity resulting in hyperadrenergic activity. Aqueous solutions have a ptt of about 10.
Pharmacodynamics/Kinetics
  Absorbtion: Orally, dermally, and by inhalation
  Metabolism: To triethylamine-n-oxide (TEAO)
  Half-life: 3-4 hours
  Excretion: Renally (93%)
Signs and Symptoms of Acute Exposure
  Headache, nausea, ocular irritation, corneal edema, visual blurriness (lasts 2-4 hours), dizziness, syncope, anxiety, dermal burns, dermatitis
Treatment Strategy
  Decontamination:
    Oral: Dilute with 4-8 oz of milk or water in an adult (up to 4 oz in children)
    Dermal: Remove contaminated clothing; wash with soap and water
    Ocular: Irrigate with saline or water for at least 20 minutes; remove contact lenses
Reference Range A triethylamine air concentration level of 4.1 mg/m^3 corresponds to a urinary concentration of 36 mmol/mol creatinine. An air level of 10 mg/m^3 corresponds to a urine and plasma triethylamine level of 65 mmol/mol creatinine and 1.9 mmol/L, respectively
Additional Information
  TLV-TWA: 1 ppm
  Colorless liquid with an ammonia-like odor
  Four hours of exposure (at air concentration levels of 40.6 mg/m^3) can result in corneal edema and blurred vision (foggy vision or "blue haze")

VR

CAS Number 159939-87-4
Synonyms Russian V Gas; Russian VX; RVX
Military Classification Nerve Agent
Clinically Resembles Jatropha multifida plant poisoning, organophosphate insecticides, nicotine poisoning
Treatment Strategy
  Adjunctive treatment: Diazepam for seizures, when intubation and mechanical ventilation is used and when >6 mg atropine is given (or 3 MK I kits)
  Pretreatment/prophylaxis: Pyridostigmine
  Decontamination:
    Dermal: VR as an oily liquid is a significant dermal hazard. Remove all contaminated clothing. Wash with copious amounts of water. Dilute ammonia, alcohol, ether or acetate may be used to wash oily liquid from the skin. A 1% to 5% hypochlorite solution (household bleach diluted) with alkaline soap may be used. Immediate field/outdoor decontamination is suggested. Avoid decontamination inside the healthcare facility. Do not delay antidotal intervention in lieu of decontamination.
    Inhalation: Remove from exposure site and remove contaminated clothes. Administer 100% humidified oxygen. Protect airway. If unconscious, consider intubation with mechanical ventilation.
    Ocular: Copious irrigation with water or saline

Supportive therapy: Atropine is the mainstay of therapy. Depending upon severity of exposure, 10-20 mg cumulatively over the first 2-3 hours. Atropine should be titrated to bronchial secretions, **not** ocular signs. Pralidoxime should be administered 1-2 g I.V. over 10 minutes, with repeat dose in 1 hour if weakness persists, then every 4-12 hours afterwards as needed. Pralidoxime is most effective if given 3 hours post-VR exposure. If Mark I kits are available, acute treatment suggestions: mild exposure 1 kit, mild to moderate exposure 2 kits (2nd kit 5 minutes after the first), moderate to severe exposure 3 kits (each kit separated by 5 minutes). Long term I.V. atropine and 2-PAM may be required.

Special populations: Children are especially vulnerable to these compounds owing to their smaller mass, smaller airway diameter, higher respiratory rate, and minute volumes. Due to their short stature, they are exposed to chemicals that settle close to the ground.

Pediatric dose: Atropine .05 mg/kg I.V./I.M. (titrate to cessation of bronchial secretions), 2-PAM 25-50 mg/kg I.V./I.M., and diazepam 0.2 to 0.5 mg/kg I.V. (for seizure control)

Additional Information A colorless liquid organophosphate with oily consistency developed in Russia (Novocheboksarsk) in the 1970s; VR is structurally very similar to its British counterpart, VX. HI-6 may be the most effective oxine and exposed individuals may require more atropine. Seizures also may be quite prevalent.

VX

CAS Number 20820-80-8

Military Classification Chemical Warfare Weapon - Nerve Agent

Mechanism of Toxic Action Inhibits hydrolysis of acetylcholine by acetylcholinesterase binding with this enzyme. Irreversible binding (aging) takes place. This results in acetylcholine excess at the neuronal synapse. May penetrate blood brain barrier and effect GABA transmission. Similar to organophosphate agent; inhibits the enzyme acetylcholinesterase thus resulting in acetylcholine excess at the neuronal synapse; may penetrate blood brain barrier and thus affect GABA transmission.

Adverse Reactions

Cardiovascular: Sinus bradycardia, tachycardia, heart rhythm disturbances

Central nervous system: Seizures, ataxia, coma, headache, fatigue, memory loss, nightmares, excess cholinergic toxicity, muscarinic activity, nicotinic activity

Gastrointestinal: Diarrhea, nausea, salivation, vomiting, abdominal pain

Genitourinary: Involuntary urination

Ocular: Lacrimation, mydriasis/miosis may present as a delayed sign, dim or blurred vision, eye pain

Respiratory: Rhinorrhea, bronchospasm, cough, tachypnea, dyspnea, wheezing, bronchorrhea; death usually results from respiratory failure

Toxicodynamics/Kinetics

Onset of action:

Inhalation: Within 5 minutes

Dermal: Minutes to 1 hour

Half-life: Aging of VX-acetylcholine complex −1-2 days

Pregnancy Issues Data do not suggest teratogenicity

Clinically Resembles Jatropha multifida plant poisoning, organophosphate insecticides, nicotine poisoning

Treatment Strategy

Adjunctive treatment: Diazepam for seizures, when intubation and mechanical ventilation is used and when >6 mg atropine is given (or 3 MK I kits)

Pretreatment/prophylaxis: Pyridostigmine

Decontamination:

Dermal: VX as an oily liquid is a significant dermal hazard. Remove all contaminated clothing. Wash with copious amounts of water. Dilute ammonia, alcohol, ether or acetate may be used to wash oily liquid from the skin. A 1% to 5% hypochlorite solution (household bleach diluted) with alkaline soap may be used. Immediate field/outdoor decontamination is suggested. Avoid decontamination inside the healthcare facility. Do not delay antidotal intervention in lieu of decontamination.

Inhalation: Remove from exposure site and remove contaminated clothes. Administer 100% humidified oxygen. Protect airway. If unconscious, consider intubation with mechanical ventilation.

Ocular: Copious irrigation with water or saline

Supportive therapy: Atropine is the mainstay of therapy. Depending upon severity of exposure, 10-20 mg cumulatively over the first 2-3 hours. Atropine should be titrated to bronchial secretions, **not** ocular signs. Pralidoxime should be administered 1-2 g I.V. over 10 minutes, with repeat dose in 1 hour if weakness persists, then every 4-12 hours afterwards as needed. Pralidoxime is most effective if given 3 hours post-VX exposure. If Mark I kits are available, acute treatment suggestions: mild exposure 1 kit, mild to moderate exposure 2 kits (2nd kit 5 minutes after the first), moderate to severe exposure 3 kits (each kit separated by 5 minutes). Long term I.V. atropine and 2-PAM may be required.

Special populations: Children are especially vulnerable to these compounds owing to their smaller mass, smaller airway diameter, higher respiratory rate, and minute volumes. Due to their short stature, they are exposed to chemicals that settle close to the ground.

Pediatric dose: Atropine .05 mg/kg I.V./I.M. (titrate to cessation of bronchial secretions), 2-PAM 25-50 mg/kg I.V./I.M., and diazepam 0.2 to 0.5 mg/kg I.V. (for seizure control)

Environmental Persistency Oily substance that is persistent for weeks or longer. Vapors can off-gas from contaminated clothes or surfaces posing a secondary contamination risk.

Personal Protection Equipment Protective equipment for healthcare responders is critical

Respiratory protection: Pressure-demand, self-contained breathing apparatus (SCBA) is recommended

Skin protection: Chemical protective clothing and butyl rubber (M3 and M4 Norton) with chemical goggles and face shield

Reference Range Cholinesterase activity <10% of normal is consistent with severe poisoning

Additional Information In the face of appropriate treatment, few sequelae are expected. However, insomnia, PTSD, nightmares may persist.

Specific References

American Academy of Pediatrics Committee on Environmental Health and Committee on Infectious Diseases, "Chemical-Biological Terrorism and Its Impact on Children: A Subject Review," *Pediatrics*, 2000, 105(3 Pt 1):662-70.

Arnold JF, "CBRNE - Nerve Agents, G-Series: Tabun, Sarin, Soman," http://www.emedicine.com/EMERG/topic898.htm. Last accessed October 12, 2003.

Ben Abraham R, Weinbroum AA, Rudick V, et al, "Perioperative Care of Children With Nerve Agent Intoxication," *Paediatr Anaesth*, 2001, 11(6):643-9.

Benitez FL, Velez-Daubon, and Keyes DC, "CBRNE - Nerve Agents, V-Series: Ve, Vg, Vm, Vx," http://www.emedicine.com/emerg/topic899.htm. Last accessed October 3, 2003.

Corvino TF, Nahata MC, Angelos MG, et al, "Availability, Stability, and Sterility of Pralidoxime for Mass Casualty Use," *Ann Emerg Med*, 2006, 47(3):272-7.

Eyer P, Hagedorn I, Klimmek R, et al, "HLo 7 Dimethanesulfonate, A Potent Bispyridinium-Dioxime Against Anticholinesterases," *Arch Toxicol*, 1992, 66(9):603-21.

Giardino NJ, "Modeling Sarin Inhalation Exposure and Subsequent Dose to Civilian Population Groups," *The Forensic Examiner*, 2004, 13(1):11-3.

Henretig FM, Cieslak TJ, and Eitzen EM Jr, "Biological and Chemical Terrorism," *J Pediatr*, 2002, 141(3):311-26. Erratum in: *J Pediatr*, 2002, 141(5):743-6.

Howland MA, "Pralidoxime - Antidotes in Depth," *Goldfrank's Toxicological Emergencies*, 6th ed, Norwalk, CT: Appleton and Lange, 1998, 1361-5.

Karalliedde L, Wheeler H, Maclehose R, et al, "Possible Immediate and Long-Term Health Effects Following Exposure to Chemical Warfare Agents," *Public Health*, 2000, 114(4): 238-48.

Kassa J, "Comparison of the Effects of BI-6, A New Asymmetric Bipyridine Oxime, With HI-6 Oxime and Obidoxime in Combination With Atropine on Soman and Fosdrine Toxicity in Mice," *Ceska Slov Farm*, 1999, 48(1):44-7.

Kassa J, "Review of Oximes in the Antidotal Treatment of Poisoning by Organophosphorus Nerve Agents," *J Toxicol Clin Toxicol*, 2002, 40(6):803-16.

Kuca K, Bielavsky J, Cabal J, et al, "Synthesis of a New Reactivator of Tabun-Inhibited Acetylcholinesterase," *Bioorg Med Chem Lett*, 2003, 13(20):3545-7.

Moreno G, Nocentini S, Guggiari M, et al, "Effects of the Lipophilic Biscation, Bis-Pyridinium Oxime BP12, on Bioenergetics and Induction of Permeability Transition in Isolated Mitochondria," *Biochem Pharmacol*, 2000, 59(3):261-6.

"Poison Warfare (Nerve) Gases," http://www.emergency.com/nervgas.htm. Last accessed October 10, 2003.

Rotenberg JS and Newmark J, "Nerve Agent Attacks on Children: Diagnosis and Management," *Pediatrics*, 2003, 112(3 Pt 1):648-58.

Rousseaux CG and Dua AK, "Pharmacology of HI-6, an H-series Oxime," *Can J Physiol Pharmacol*, 1989, 67(10):1183-9.

Salem H and Sidell FR, "Nerve Gases," *Encyclopedia of Toxicology*, Vol 1, Wexler P, ed: Academic Press, 2002, 380-5.

Shiloff JD and Clement JG, "Comparison of Serum Concentrations of the Acetylcholinesterase Oxime Reactivators HI-6, Obidoxime, and PAM to Efficacy Against Sarin (Isopropyl Methylphosphonofluoridate) Poisoning in Rats," *Toxicol Appl Pharmacol*, 1987, 89(2): 278-80.

Velez-Daubon L, Benitez FL, and Keyes DC, "CBRNE - Nerve Agents, Binary: GB2, VX2," *Emedicine*, 2003, http://www.emedicine.com/EMERG/topic 900.htm. Last accessed October 13, 2003.

Wei G, Chang A, and Hamilton RJ, "Nerve Agent Antidote Kits Enable Nurses to Treat More Mass-Casualty Patients Than Multidose Vials," *Acad Emerg Med*, 2006, 13(5):S177.

White Phosphorus

CAS Number 12185-10-3; 7723-14-0
UN Number 1381 (dry or in water); 2447 (molten)
Synonyms Phosphorus Tetramen; Red Phosphorus; Yellow Phosphorus
Military Classification Incendiary Agent
CDC Classification for Diagnosis
   Biologic: No specific test for elemental white or yellow phosphorus is available; however, an elevated serum phosphate level might indicate that an exposure has occurred. Although phosphate production is a by-product of elemental phosphorus metabolism in humans, a normal phosphate concentration does not rule out an elemental phosphorus exposure.
   Environmental: Detection of elemental phosphorus in environmental samples, as determined by NIOSH, and an elevated phosphorus level in food, as determined by FDA, might also indicate that an exposure has occurred.
   Mechanism of Action Direct tissue damage - can decrease serum calcium when systemically absorbed
CDC Case Classification
   Suspected: A case in which a potentially exposed person is being evaluated by healthcare workers or public health officials for poisoning by a particular chemical agent, but no specific credible threat exists.
   Probable: A clinically compatible case in which a high index of suspicion (credible threat or patient history regarding location and time) exists for elemental white or yellow phosphorus

exposure, or an epidemiologic link exists between this case and a laboratory-confirmed case.

Confirmed: A clinically compatible case in which laboratory tests on environmental samples are confirmatory. The case can be confirmed if laboratory testing was not performed because either a predominant amount of clinical and nonspecific laboratory evidence if a particular chemical was present or a 100% certainty of the etiology of the agent is known.

Use Fertilizers, roach poisons, rodenticides, water treatment; used in military as ammunition in motor/artillery shells, smoke screens (smoke generator)

Mechanism of Toxic Action Damages endoplasmic reticulum; inhibits fatty acid oxidation and protein synthesis; local tissue injury by oxidative/thermal damage; inhibits blood glucose regulation

Adverse Reactions

Cardiovascular: Hypotension, tachycardia, fibrillation (atrial), flutter (atrial), sinus tachycardia, cardiovascular collapse

Central nervous system: Lethargy, irritability, coma, hyperthermia

Dermatologic: Severe burns

Endocrine & metabolic: Hypoglycemia can occur on oral exposure; hypocalcemia and hyperphosphatemia associated with phosphorus burns

Gastrointestinal: GI Irritant, vomiting, abdominal cramps, nausea

Hematologic: Anemia, leukopenia, hemolysis

Hepatic: Fatty degeneration and perilobar hepatic necrosis, hepatic failure

Neuromuscular & skeletal: Degeneration and osteoporosis, fasciculations, asterixis, hemiplegia

Renal: Tubular necrosis (acute) and cortical necrosis, renal failure

Respiratory: Cough on inhalation, tachypnea, dyspnea; significant respiratory damage upon inhalation

Miscellaneous: "Phossy jaw" (degeneration and necrosis of soft tissue and teeth in oral cavity resulting in life-threatening infections); fatty deposition of muscles and liver, decreases in serum calcium, potassium, and sodium; liver/renal toxicity along with cerebral edema may occur after 5-10 days

Toxicodynamics/Kinetics

Absorption: By oral routes; yellow phosphorus well absorbed dermally and orally; phosphine gas absorbable in lungs; red phosphorus nonabsorbable

Metabolism: Oxidation and hydrolysis to hypophosphites

Elimination: Urine and feces

Pregnancy Issues Uterine hemorrhage has occurred in first trimester after a 2 mg/kg dose (an abortifacient)

Signs and Symptoms of Acute Exposure Breath odor, coma, dermal burns, flatulence, GI irritation, hypocalcemia, luminescent stool, seizures; vomiting followed by asymptomatic phase of <12 hours to 3 days with subsequent signs/symptoms of hepatic or renal failure. Chronic exposure (5+ years) results in osteoporosis and bone degeneration, commonly of jaw ("phossy jaw"), and feces discoloration (black).

Admission Criteria/Prognosis Any patient with ingestion of white or yellow phosphorus should probably be admitted

Treatment Strategy

Decontamination:

Oral: **Do not** induce emesis. Lavage within 2-3 hours with 1:5000 to 1:10,000 potassium permanganate or water. Activated charcoal may be used.

Dermal: Remove contaminated clothing; brush off phosphorus from skin and then continuously irrigate skin with water. Apply saline soaked dressings to the affected area. Avoid any lipid-based ointments. Phosphorus will fluoresce under a Wood's lamp. Remove visualized phosphorus particles with metal forceps or a WaterPik® appliance. A 1% copper sulfate solution or silver nitrate has been advocated to aid in decontamination of dermal burns, although its use is controversial. Burns may give off odor or smoke.

Inhalation: Administer 100% humidified oxygen

Ocular: Irrigate with saline for at least 15 minutes. Following saline irrigation, several drops of 3% copper sulfate solution (applied within 15 minutes) can be given to help prevent ocular burns and then remove particles mechanically.

Supportive therapy: Treat dermal burns in traditional method. Watch for secondary infections. Monitor calcium and glucose levels. Steroids are of no benefit in preventing liver injury. Morphine sulfate can be used to treat pain. Monitor fluids as third spacing and fluid loss can occur with GI or dermal injury. N-acetylcysteine may reduce hepatic toxicity when given early postexposure:

Dose regimen: 150 mg/kg in 200 mL D_5W over 15 minutes, then 50 mg/kg in 500 mL D_5W for 4 hours, then 100 mg/kg in 1000 mL D_5W for 16 hours has been utilized. Monitor for hypocalcemia.

Enhancement of elimination: Mineral oil (100 mL in adults or 1.5 mL/kg in children <12 years of age) can be used as a cathartic. Exchange transfusion may be helpful.

Personal Protection Equipment Full protective chemical clothing with positive pressure breathing apparatus; will **not** protect against thermal injury

Reference Range Normal serum phosphate concentrations: 3.0-4.5 mg/100 mL

Additional Information Lethal oral dose: 1 mg/kg

Garlic breath odor and luminescent or smoky vomitus, flatus, or stool is pathognomonic although not frequent.

Vomiting can occur after oral ingestion of 2-23 mg/kg.

Note that phosphine gas may emanate from emesis, lavage fluid, and feces from affected patients. Thus, the patient's room should be well ventilated.

Red phosphorus is not soluble and essentially not absorbed through the gastrointestinal tract and is, therefore, considered nontoxic.

Phosphine gas has a garlic odor; may be undetectable

Gas release occurs in industrial use or with moisture contamination of aluminum or zinc phosphide rodenticides.

Match phosphorus content is essentially nontoxic. Match toxicities are secondary to potassium chlorate content.

Atmospheric half-life: 5 minutes

PEL-TWA: 0.1 mg/m^3; TLV-TWA: 0.1 mg/m^3

Vapor density: 4.42

Extremely flammable - ignites spontaneously in air at 30°C

White phosphorus gives off a greenish light with white smoke

Specific References

Arnold JL, Halpern P, Tsai MC, et al, "Mass Casualty Terrorist Bombings: A Comparison of Outcomes by Bombing Type," *Ann Emerg Med*, 2004, 43(2):263-73.

Belson MG, Schier JG, and Patel MM, "Case Definitions for Chemical Poisoning," *MMWR Recomm Rep*, 2005, 54(RR-1):1-24.

SECTION IV
LABORATORY ANALYSIS

Public Health Contacts for Laboratory Testing to Confirm Exposure During a Potential or Known Chemical Terrorism Event

(Reprinted from: www.bt.cdc.gov/chemical/lab.asp
Accessed March, 2007)

Emergencies

To obtain emergency information from CDC, contact

CDC
Director's Emergency Operations Center
Atlanta, Georgia
770-488-7100
http://intra-apps.cdc.gov/od/otper/programs/deoc-main.asp

Nonemergencies

To obtain nonemergency information, contact

CDC
National Center for Environmental Health
Division of Laboratory SciencesAtlanta, Georgia
770-488-7950
http://www.cdc.gov/nceh/dls

CDC
National Center for Infectious Diseases
Bioterrorism Rapid Response and Advanced Technology Laboratory
Atlanta, Georgia
404-639-4910

CDC
National Institute of Occupational Safety and Health
Cincinnati, Ohio
800-356-4674
http://www.cdc.gov/niosh/homepage.html

Environmental Protection Agency
National Response Center
Washington, DC
800-424-8802
http://www.epa.gov/

Food and Drug Administration
Forensic Chemistry Center
Cincinnati, Ohio
513-679-2700, extension 184
http://www.fda.gov/

Laboratory Response Network
Association of Public Health Laboratories
Washington, DC
202-822-5227
http://www.bt.cdc.gov/lrn

Centers for Disease Control and Prevention
Shipping Instructions for Specimens Collected from People
Who May Have Been Exposed to Chemical-Terrorism Agents

SECTION ONE: COLLECTING AND LABELING SPECIMENS

Required Specimens
Unless otherwise directed, collect the following specimens from each person who may have been exposed:

Whole blood
- Collect blood specimens from adults only unless you receive specific instruction from CDC to collect blood from pediatric patients.
- Collect a minimum of 12 mL of blood.
- Use three 4-mL or larger vacuum-fill only (unopened), nongel, purple-top (EDTA) tubes; use four tubes if using 3-mL tubes.
- Using indelible ink, mark each purple-top tube of blood *in the order collected* (eg, # 1, # 2, # 3, # 4 [if using 3-mL tubes]).
- In addition, collect another specimen using one 3-mL or larger, vacuum-fill only (unopened), nongel, green- or gray-top tube. Allow the tube to fill to its stated capacity.

Urine
- Collect at least 25-50 mL from potentially exposed adults and children.
- Use a screw-cap plastic container; do not overfill.
- Freeze specimen as soon as possible ($-70°$ C or dry ice preferred).
- If other than "clean catch," note method of collection on the specimen cup (eg, obtained by catheterization).

Blanks
For each lot number of tubes and urine cups used for collection, provide the following to be used as blanks for measuring background contamination:
- Two (2) empty, unopened purple-top tubes.
- Two (2) empty, unopened green- or gray-top tubes.
- Two (2) empty, unopened urine cups.

Labeling Specimens

- Label specimens with labels generated by your facility and follow your facility's procedures for proper specimen labeling.
- In addition to unique patient identifiers (eg, medical records number, specimen identification number) labels should convey the collector's initials, date and time of collection so that law enforcement officials may trace the specimen to the collector should investigations lead to legal action and the collector has to testify that he or she collected the specimen.
- If you use bar-coded labels, place the labels on blood tubes and urine cups so that when these containers are upright, the bar code looks like a ladder.
- Maintain a list of names with corresponding specimen identification numbers at the collection site so that results can be reported to patients. It is recommended that you record additional data for use in the interpretation of results. Additional data may include: time of potential exposure, method of urine collection if other than "clean-catch," indication if sample was collected postmortem, and antidotes administered prior to sample collection.
- Information provided on labels and lists may prove helpful in correlating the results obtained from CDC's Rapid Toxic Screen and subsequent analysis with the people from whom the specimens were collected.

Shipping Instructions for Specimens Collected from People Who May Have Been Exposed to Chemical-Terrorism Agents

SECTION TWO: PACKAGING SPECIMENS

Packaging consists of the following components: primary receptacles (blood tubes or urine cups), secondary packaging (materials used to protect primary receptacles), and outer packaging (polystyrene foam-insulated, corrugated fiberboard shipper).

Secondary Packaging for Blood Tubes

- To facilitate processing, package all blood tubes from the same patient together.
- Place absorbent material between the blood tubes and the first layer of secondary packaging. Use enough absorbent material to absorb the entire contents of the blood tubes.
- Separate each tube of blood collected from other tubes, or wrap tubes to prevent tube-to-tube contact. Regardless of the method used, the first layer of secondary packaging must be secured with one continuous strip of evidence tape and initialed half on the tape and half on the first layer of secondary packaging by the person making the seal. Examples of some ways to do this are to
  - Pack blood tubes in a gridded box lined with absorbent material. Seal the top half of the box to the bottom half with one continuous piece of evidence tape and write your initials half on the tape and half on the box.
  - Pack a sealable polystyrene foam container or blood tube shipment sleeve and transport tube with individually wrapped tubes. Seal the polystyrene foam container or transport tube with one continuous piece of evidence tape and write your initials half on the tape and half on the container.
- Wrap and seal the first layer of secondary packaging (e.g., gridded box) with absorbent material.
- Seal one wrapped gridded box or alternative container inside a clear, leak-proof biohazard polybag equivalent to Saf-T-Pak product STP-701, STP-711, or STP-731.
- Place this bag inside a white Tyvek® outer envelope (or equivalent) and seal the opening with a continuous strip of evidence tape initialed half on the packaging and half on the evidence tape by the individual making the seal.
- According to 49 CFR 173.199(b), if specimens are to be transported by air, either the primary receptacle or the secondary packaging used must be capable of withstanding, without leaking, an internal pressure producing a pressure differential of not less than 95 kPa (0.95 bar, 14 psi). Verify in advance that the manufacturer of either the blood tube or secondary packaging used in your facility is in compliance with the pressure differential requirement.

Outer Packaging for Blood Tubes

- Use polystyrene foam-insulated, corrugated fiberboard shipper (may be available from your transfusion service or send-outs department).
- For cushioning, place additional absorbent material in the bottom of the shipper.
- Add a single layer of refrigerator packs on top of absorbent material.
- Place the packaged specimens on top of the refrigerator packs.
- Use additional cushioning material to minimize shifting while the shipper is in transit.
- Place additional refrigerator packs on top of the secondary packaging to maintain a shipping temperature of 1°C-10°C for the duration of transit.
- Place blood shipping manifest in a sealable plastic bag and put on top of packs inside the shipper.
- Keep chain-of-custody documents for your files.
- Place lid on shipper and secure with filamentous shipping tape.
- Place your return address in the upper left-hand corner of the shipper top and put CDC's receiving address in center.
- Affix labels and markings adjacent to the shipper's/consignee's address that appears on the shipper.
- Place the UN 3373 label and the words "Biological Substance, Category B" adjacent to the label on the front of the shipper.

Secondary Packaging for Urine Cups

- Separate each urine cup from other urine cups, or wrap individual urine cups to prevent contact between urine cups. Regardless of the method used, the first layer of secondary packaging must be secured with one continuous strip of evidence tape and initialed half on the tape and half on the first layer of secondary packaging by the person making the seal. Examples of some ways to do this are to—
  - Pack urine cups in a gridded box lined with absorbent material. Seal the top half of the box to the bottom half with one continuous piece of evidence tape and write your initials half on the tape and half on the box.
  - Seal individually wrapped urine cups inside a clear, leak-proof biohazard polybag equivalent to Saf-T-Pak product STP-701, STP-711, or STP-731. Secure the closure of the bag with one continuous strip of evidence tape initialed half on the tape and half on the bag by the individual making the seal.
- Place urine cups, boxed or individually wrapped and secured properly with evidence tape, in the next layer of secondary packaging. An example of acceptable material is the Saf-T-Pak Disposable 2-Part Pressure Vessel system or its equivalent.
- Secondary packaging must have its closure secured with a single strip of evidence tape initialed half on the packaging and half on the evidence tape by the person making the seal.

389

Outer Packaging for Urine Cups

- Use polystyrene foam-insulated, corrugated fiberboard shipper (may be available from your transfusion service or send-outs department).
- For cushioning, place additional absorbent material in the bottom of the shipper.
- Place a layer of dry ice on top of the absorbent material. Do not use flakes or large chunks of dry ice for shipment because large chunks have the potential for shattering urine cups during transport.
- Ensure that specimens will remain frozen or will freeze during transport.
- Place packaged urine cups in the shipper.
- Use additional absorbent or cushioning material between wrapped urine cups to minimize shifting while shipper is in transit.
- Place an additional layer of dry ice on top of samples.
- Place the urine shipping manifest in a sealable plastic bag and put on top of dry ice inside the shipper.
- Keep chain-of-custody documents for your files.
- Place lid on shipper and secure with filamentous shipping tape.
- Place your return address in the upper left-hand corner of the shipper top and put CDC's receiving address in center.
- Place the UN 3373 label and the words "Biological Substance, Category B" adjacent to the label on the front of the shipper.
- Place a Class 9/UN 1845 hazard label on the same side of the shipper as the UN 3373 marking.
- If the proper shipping name (either dry ice or carbon dioxide, solid) and Class 9/UN 1845 is not preprinted on the hazard label, add it in an area adjacent to the label.
- Note the weight of dry ice (in kg) on the preprinted area of the hazard label, or place that information adjacent to the Class 9/UN 1845 hazard label.
- Orientation arrows are not required on a shipper containing "Biological substance, category B." If you use arrows, be sure to orient the inner packaging so that closures are aligned with the arrows.
- If the shipper will be transported by a commercial air carrier, complete an airway bill. On the airway bill, note the proper shipping name and UN number for each hazardous material and identify a person responsible for the shipper per IATA packing instruction 650.

Shipping Instructions for Specimens Collected from People Who May Have Been Exposed to Chemical-Terrorism Agents

SECTION THREE: SHIPPING SPECIMENS

Follow the guidance provided in your state's chemical-terrorism comprehensive response plan. If you are directed to ship the specimens to CDC, please ship the specimens to the following address:

> **Centers for Disease Control and Prevention**
> **Attn: Charles Dodson**
> **4770 Buford Hwy.**
> **Building 110 Loading Dock**
> **Atlanta, GA 30341**
> **(770) 488-4305**

Preparing Documentation

- Since blood tubes and urine cups cannot be shipped together in the same package, prepare a separate shipping manifest for each.
- Note on shipping manifest if urine sample is collected by means other than clean catch (e.g., catheterization).
- Place each shipping manifest (with specimen identification numbers) in a plastic zippered bag on top of the specimens before closing the lid of the polystyrene foam-insulated, corrugated fiberboard shipper.
- Do not transport chain-of-custody forms with specimens. Each entity or organization handling the specimens is responsible for the specimens only during the time that it has control of the specimens.
- Each entity or organization receiving the specimens must sign-off on the chain-of-custody form of the entity or organization relinquishing the specimens to close that chain. Electronic procedures such as electronic chain-of-custody and barcode readers will expedite this process.
- When receiving specimens, each new entity or organization must begin its own chain of custody. The entity or organization relinquishing the specimens must sign its chain of custody to close the chain and indicate that they have transferred the specimens.

Note: When the person relinquishing the specimens (relinquisher) and the person receiving the specimens (receiver) are not together at the time of specimen transfer, the relinquisher must document on its chain-of-custody form that the receiver is the express courier (e.g., FedEx, Delta Dash, DHL, UPS) and must document the shipment tracking number or have the person transporting the specimens sign the chain-of-custody to indicate that he or she has taken control of the specimens. Likewise, when receivers get the specimens, they will document on their chain-of-custody form that the relinquisher is the express courier (and provide the tracking number) or have the person transporting the specimens sign the chain-of-custody form.

CDC Specimen-Collection Protocol for a Chemical-Exposure Event

For detailed instructions see CDC's Shipping Instructions for Specimens Collected for Specimens Collected from People Who May Have Been Exposed to Chemical-Terrorism Agents.

Collect blood and urine samples for each person involved in the chemical-exposure event.

Note: For children, collect only urine samples unless otherwise directed by CDC.

Blood-Sample Collection

For each person, collect blood in glass or plastic tubes in the following order: 1st: collect specimens in three (3) EDTA (purple-top) 4 mL or larger plastic or glass tubes; 2nd: collect another specimen in one (1) gray- or green-top tube. Collect the specimens by following the steps below:

1 Collect a minimum of 12 mL of blood in three (3) 4 mL or larger glass or plastic tubes. If using 3-mL tubes, use 4 tubes.

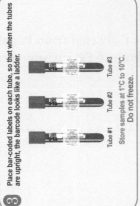

Do not use gel separators.

2 Mix contents of tubes by inverting them 5 or 6 times.

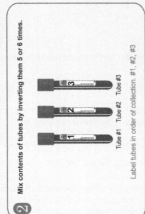

Tube #1   Tube #2   Tube #3

Label tubes in order of collection. #1, #2, #3

3 Place bar-coded labels on each tube, so that when the tubes are upright, the barcode looks like a ladder.

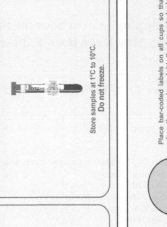

Tube #1   Tube #2   Tube #3

Store samples at 1°C to 10°C. Do not freeze.

4 After collecting samples in the purple-top tubes, collect one (1) sample in a gray- or green-top tube (gray-top tube shown). Allow the tube to fill to its stated capacity.

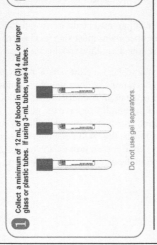

Do not use gel separators.

5 Mix contents of the tube by inverting it 5 or 6 times.

6 Place bar-coded labels on the tube, so that when the tube is upright, the barcode looks like a ladder.

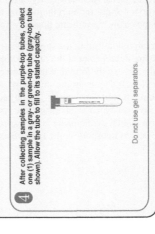

Store samples at 1°C to 10°C. Do not freeze.

Urine-Sample Collection

For each person, collect 25 mL–50 mL of urine in a screw-cap urine cup.

Label the urine cup with the appropriate bar-coded label as shown. Indicate on the cup how the sample was collected if the method was other than "clean catch" (ie, catheterization).

Freeze samples (optimally at -70°C).

Place bar-coded labels on all cups so that when the cup is upright, the barcode looks like a ladder.

Instructions for Shipping Urine Specimens to CDC after a Chemical-Exposure Event

Guidance in Accordance with Packaging Instructions International Air Transport Authority (IATA) 650 Biological Substance Category B

For detailed instructions, see CDC's *Shipping Instructions for Specimens Collected from People Who May Have Been Exposed to Chemical-Terrorism Agents.*

1. Use a gridded box or individually wrapped cups sealed with evidence tape to separate urine cups. Place absorbent material in the bottom of the box and insert the cups.

2. Use one continuous piece of evidence tape to seal the gridded box or Saf-T-Pak inner leak-proof polybag (or equivalent) containing wrapped urine cup(s). Write initials half on the evidence tape and half on the box or bag.

3. Wrap the gridded box with absorbent material and secure with tape. Seal the box inside a Saf-T-Pak inner leak-proof polybag (or equivalent).

4. Place the sealed Saf-T-Pak inner leak-proof polybag (or equivalent) inside a white Tyvek® outer envelope (or equivalent). Note: If primary receptacles do not meet the internal pressure requirement of 95 kPa, use compliant secondary packaging materials.

5. Seal the opening of this envelope with a continuous piece of evidence tape. Write initials half on the evidence tape and half on the envelope.

6. Use polystyrene foam-insulated, corrugated fiberboard shipper to ship boxes to CDC. Place absorbent pad in the bottom of the shipper.

7. Place a layer of dry ice in the bottom of the shipper on top of the absorbent material. **DO NOT** use large chunks or flakes of dry ice.

8. Place the packaged urine cups in the shipper. Use absorbent material or cushioning material to minimize shifting while box is in transit. Place additional dry ice on top of samples.

9. Place the urine shipping manifest in a sealable plastic bag and put on top of the sample boxes inside the shipper. **Keep your chain-of-custody documents for your files.** Place lid on the shipper.

10. Secure the outer container lid with filamentous shipping tape. Place your return address in the upper left-hand corner of the shipper top and put the CDC Laboratory receiving address in the center.

11. Add the UN 3373 label and the words "Biological Substance Category B" on the front of the shipper. UN 3373 is the code identifying the shipper's contents as "Biological Substance, Category B."

12. Place a Class 9/UN 1845 label on the front of the shipper. This label for dry ice MUST indicate the weight of dry ice (in kg) in the shipper and the proper name (either dry ice or carbon dioxide, solid).

13.

Send shipment via FedEx to:

Centers for Disease Control and Prevention
Attn: Charles Dodson
4770 Buford Hwy.
Building 110 Loading Dock
Atlanta, GA 30341
(770) 488-4305

Department of Health and Human Services
Centers for Disease Control and Prevention

11/2006

PAGE _____ OF_____

CENTERS FOR DISEASE CONTROL AND PREVENTION **CHEMICAL TERRORISM URINE SPECIMEN COLLECTION** **AND SHIPPING MANIFEST**	

DATE SHIPPED: _____

SHIPPED BY:_____

CONTACT TELEPHONE: _____

SIGNATURE: _____

DATE RECEIVED: _____

RECEIVED BY: _____

SIGNATURE:_____

TOTAL NUMBER OF SPECIMENS IN THIS CONTAINER:	URINE CUPS:	
TOTAL NUMBER OF BLANK URINE CUPS PROVIDED IN THIS CONTAINER:	BLANK URINE CUPS:	

COMMENTS: _____

_____

_____

_____

SHIPPING ADDRESS: CDC
ATTN: Charles Dodson
4770 Buford Hwy
Building 103 Loading Dock
Atlanta, GA 30341
(770) 488-4305

PUBLIC HEALTH CONTACTS FOR LABORATORY TESTING

PAGE _____ OF_____

PLEASE INDICATE THE AMOUNT OF URINE COLLECTED IN THE UC COLUMN UC = URINE CUP		
Patient/Victim ID Label	UC (Amount)	Comments:
		_____ _____ _____
		_____ _____ _____
		_____ _____ _____
		_____ _____ _____
		_____ _____ _____

NOTE: Please include 2 empty urine cups from each lot number collected for
background contamination measurement.

Instructions for Shipping Blood Specimens to CDC after a Chemical-Exposure Event

Guidance in Accordance with Packaging Instructions International Air Transport Authority (IATA) 650 Biological Substance Category B

For detailed instructions see CDC's *Shipping Instructions for Specimens Collected from People Who May Have Been Exposed to Chemical-Terrorism Agents.*

① Place purple- and gray- or green top tubes by patient number into gridded-type box lined with an absorbent pad. If using an alternative packaging method, pack all tubes from the same patient together while preventing tube-to-tube contact.

② Seal gridded box or alternative secondary container with one continuous piece of evidence tape. The individual making the seal must initial half on the tape and half on the packaging.

③ Wrap gridded box in absorbent pad and tape to seal. Seal gridded box or alternative container inside a Saf-T-Pak clear inner, leak-proof polybag (or equivalent).

④ Place the sealed Saf-T-Pak inner leak-proof polybag (or equivalent) inside a white Tywek® outer envelope (or equivalent). Note: If primary receptacles do not meet the internal pressure requirement of 95 kPa, use compliant secondary packaging materials.

⑤ Seal the opening of this envelope with a continuous piece of evidence tape. Write initials half on the evidence tape and half on the envelope.

⑥ Use polystyrene foam-insulated, corrugated fiberboard shipper to ship boxes to CDC. Place absorbent material in the bottom of the shipper.

⑦ Place refrigerator packs in a single layer on top of the absorbent material.

⑧ Place the packaged specimens in the shipper. Use cushioning material to minimize shifting while box is in transit. Place additional refrigerator packs on top of samples.

⑨ Place the blood shipping manifest in a sealable plastic bag and put on top of the sample boxes inside the shipper. Keep your chain-of-custody documents for your files. Place lid on the shipper.

⑩ Secure the shipper lid with filamentous shipping tape. Place your return address in the upper left-hand corner of the shipper top and put the CDC Laboratory receiving address in the center.

⑪ Add the UN 3373 label and the words "Biological Substance Category B" on the front of the shipper. UN 3373 is the code identifying the shipper's contents as "Biological Substance Category B."

UN3373

Biological Substance
Category B

Send shipment via FedEx to:

Centers for Disease Control and Prevention
Attn: Charles Dodson
4770 Buford Hwy.
Building 110 Loading Dock
Atlanta, GA 30341
(770) 488-4305

PAGE _____ OF_____

CENTERS FOR DISEASE CONTROL AND PREVENTION CHEMICAL TERRORISM BLOOD SPECIMEN COLLECTION AND SHIPPING MANIFEST		
DATE SHIPPED: _____		
SHIPPED BY:_____		
CONTACT TELEPHONE: _____		
SIGNATURE: _____		
DATE RECEIVED: _____		
RECEIVED BY: _____		
SIGNATURE: _____		
TOTAL NUMBER OF SPECIMENS IN THIS CONTAINER:	PURPLE-TOP TUBES:	
	GREEN/GRAY-TOP TUBES:	
TOTAL NUMBER OF BLANK TUBES PROVIDED IN THIS CONTAINER:	PURPLE-TOP TUBES:	
	GREEN/GRAY-TOP TUBES:	
COMMENTS:_____ _____ _____ _____		

SHIPPING ADDRESS: CDC
ATTN: Charles Dodson
4770 Buford Hwy
Building 103 Loading Dock
Atlanta, GA 30341
(770) 488-4305

397

PUBLIC HEALTH CONTACTS FOR LABORATORY TESTING

PAGE _____ OF_____

CENTERS FOR DISEASE CONTROL AND PREVENTION CHEMICAL TERRORISM BLOOD SPECIMEN COLLECTION AND SHIPPING MANIFEST					
PLACE A √ IN EACH BOX FOR SAMPLES SHIPPED – PLACE A X IN EACH BOX FOR SAMPLES NOT SHIPPED PLEASE INDICATE THE SIZE TUBE COLLECTED (5 OR 7 mL) IN THE COMMENTS PT = PURPLE-TOP GT = GREEN/GRAY-TOP					
Patient/Victim ID Labe	PT 1	PT 2	PT 3	GT	Comments:
					_____ _____ _____
					_____ _____ _____
					_____ _____ _____
					_____ _____ _____
					_____ _____ _____

NOTE: Please include 2 empty purple-top tubes and 2 empty green/gray-top tubes from each lot number collected for background contamination measurement.

Sentinel Laboratory Guidelines for Suspected Agents of Bioterrorism and Emerging Infectious Diseases

Avian Influenza A H5N1

American Society for Microbiology (Reprinted from www.asp.org/policy/index.asp
Accessed March, 2007)

Credits: Avian Influenza A H5N1

Subject Matter Expert, ASM

Michael J. Loeffelholz, Ph.D. D(ABMM)
ViroMed Laboratories/LabCorp
Minnetonka, MN

ASM Laboratory Protocol Working Group

Vickie Baselski, Ph.D.
University of Tennessee at Memphis
Memphis, TN

Roberta B. Carey, Ph.D.
Centers for Disease Control and Prevention
Atlanta, GA

Peter H. Gilligan, Ph.D.
University of North Carolina
Hospitals/Clinical Microbiology and Immunology Labs
Chapel Hill, NC

Larry Gray, Ph.D.
TriHealth Laboratories and
University of Cincinnati College of Medicine
Cincinnati, OH

Rosemary Humes, MS, MT (ASCP) SM
Association of Public Health Laboratories
Silver Spring, MD

Karen Krisher, Ph.D.
Clinical Microbiology Institute
Wilsonville, OR

Judith Lovchik, Ph.D.
Public Health Laboratories, NYCDOH
New York, NY

Chris N. Mangal, MPH
Association of Public Health Laboratories
Silver Spring, MD

Daniel S. Shapiro, M.D.
Lahey Clinic
Burlington, MA

Susan E. Sharp, Ph.D.
Kaiser Permanente
Portland, OR

Alice Weissfeld, Ph.D.
Microbiology Specialists, Inc.
Houston, TX

David Welch, Ph.D.
Medical Microbiology Consulting
Dallas, TX

Mary K. York, Ph.D.
MKY Microbiology Consultants
Walnut Creek, CA

Coordinating Editor:

James W. Snyder, Ph.D.
University of Louisville
Louisville, KY

I. General Information

A. Special Instructions

Do NOT perform culture on specimens if avian influenza A H5N1 is suspected, unless performed under enhanced Biosafety Level (BSL) 3 laboratory conditions. Very few laboratories have the capability to operate under enhanced BSL 3 conditions. These instructions for culturing influenza A H5N1 may change if the epidemiological and clinical characteristics of the disease, and pathogenesis of the virus change.

Sentinel laboratories may perform rapid influenza antigen tests and direct fluorescent antibody staining on respiratory specimens from suspected avian influenza A H5N1 cases, but only under BSL 2 conditions in a Class II biological safety cabinet. However, influenza A H5N1-specific reverse-transcriptase (RT)-PCR, available at LRN Reference and other limited laboratories, is the preferred method because of its high sensitivity.

The next influenza pandemic will likely result in a dramatic increase in testing orders and the temporary deterioration of production and distribution systems. As such, laboratorians should plan for shortages of basic laboratory supplies.

The U.S. Department of Health and Human Services has prepared an influenza pandemic plan that outlines testing and biosafety requirements (www.hhs.gov/pandemicflu/plan/). Additional guidance for laboratories is available at http://www2a.cdc.gov/han/archivesys/ ViewMsgV.asp?AlertNum=00246.

B. Description of Organism

Influenza A virus is a member of the family *Orthomyxoviridae*. Influenza viruses are enveloped, with a segmented, single-stranded RNA genome. This family also contains influenza B and C viruses. Point mutations in the envelope protein hemagglutinin (H), referred to as antigenic drift, result in the emergence of new strains of influenza A and B viruses and the resultant annual outbreaks and epidemics. Subtyping of influenza A virus is based on antigenic characteristics of two envelope proteins, H and neuraminidase (N). New influenza A virus subtypes emerge as the result of reassortment of H and N sequences from two different subtypes, referred to as antigenic shift. These new subtypes are responsible for influenza pandemics. There are currently 16 recognized H subtypes and 9 recognized N subtypes. While virtually all combinations of influenza A subtypes naturally infect waterfowl and shorebirds, certain subtypes infect poultry and mammalian species. Subtypes H1N1, H3N2, H2N2 and H1N2 have circulated, or are currently circulating widely, among humans. Subtype H5N1, causing highly pathogenic avian influenza, was identified in 1996 in southern China. Influenza A H5N1 is significant, though not unique, in its ability to cross normal species barriers and directly infect humans. Avian subtypes H9N2 and H7N7 are also known to cause infection in humans, so public health influenza surveillance programs monitor for emergence of any novel strains in humans. However, the wide geographical distribution of H5N1 in avian species, and the number and severity of human infections are unprecedented. If, or when, the virus reassorts to a strain transmitted readily among humans, and unless there is a dramatic decrease in the pathogenicity of the resulting virus, the result will likely be an influenza pandemic with mortality rates not seen since the 1918 pandemic.

C. Epidemiology

The classical epidemiologic cycle of influenza A virus includes wild waterfowl and shorebirds, which are naturally infected; domestic waterfowl and poultry, which acquire virus from wild birds; pigs, which serve as "mixing vessels" for both avian and mammalian adapted strains; and humans, who are susceptible to the reassorted viruses. Reassortment can also occur during human-to-human transmission. Influenza A virus also infects maritime mammals, including seals and whales, dogs, and horses. The H5N1 virus has bypassed this epidemiologic cycle, crossed normal species barriers, and is capable of being transmitted directly from poultry to humans. First identified as a cause of highly pathogenic avian influenza in southern China in 1996, the virus has since spread to Southeast Asia, Middle East, Eastern and Western Europe, and Africa. H5N1 has been found in domestic fowl and a variety of migratory and resident wild bird species. Avian influenza strains infect the intestinal tract and are shed at high titers in feces. Transmission rates are high among birds congregating at bodies of water. The presence of H5N1 in a number of migratory bird species has resulted in its rapid spread among continents. In addition to birds, the virus has also been found in several mammalian species. Felines have become infected as a result of consumption of infected dead birds. Human infections caused by H5N1 were first identified in 1997. Human H5N1 infections are the result of exposure to high viral titers in infected birds or feces. There is evidence of human-to-human transmission, yet secondary cases are very limited due to avian host specificity of H5N1. Hundreds of confirmed human cases of avian influenza H5N1 have been reported to the World Heath Organization (http://www.who.int/csr/disease/avian_influenza/country/cases_table_2006_10_03/en/index.html), and the mortality rate has been approximately 50%. Human cases have been reported from Azerbaijan, Cambodia, China, Djibouti, Egypt, Indonesia, Iraq, Thailand, Turkey, and Viet Nam.

Influenza pandemics result from the emergence of antigenically distinct subtypes of influenza A virus. During the twentieth century, influenza pandemics occurred in 1918, 1957 and 1968. The subtypes causing these pandemics all had avian origins, and adapted to high transmissibility among humans. Pathogenicity among the pandemic virus strains varied.

Influenza virus is considered a Category C biothreat agent. Agents in this category are emerging, readily available, and easily disseminated. In addition, they are capable of causing high morbidity and mortality rates http://www.bt.cdc.gov/agent/agentlist-category.asp.

D. Clinical Features

Influenza caused by H5N1 shares features with those caused by the Spanish influenza pandemic of 1918. Morbidity and mortality are severe in previously healthy, young and middle-aged persons. The innate immune response is in part responsible for pathogenesis, causing fluid accumulation in the lungs. While influenza caused by H5N1 is notable for its aggressive course and high mortality rates, evidence indicates that mild disease and asymptomatic infections occur. Symptomatic cases are characterized by high fever, cough, and lower respiratory tract symptoms (shortness of breath, pulmonary infiltrates) in virtually 100% of patients. Diarrhea occurs more frequently than with influenza caused by human-adapted subtypes. The frequency of pneumonia and diarrhea distinguish avian from seasonal influenza. Over 50% of reported H5N1 influenza cases were fatal. Death is primarily due to respiratory or multi-organ failure. Unlike human-adapted subtypes, H5N1 is found in relatively high titers in lower respiratory tract specimens, throat swabs and stool. Although it is not known if infections with a pandemic strain will have frequently demonstrable viremia, the current H5N1 virus has been isolated from serum. The Centers for Disease Control and Prevention has developed a case definition and risk assessment for human influenza caused by influenza A H5N1 (http://www2a.cdc.gov/han/archivesys/ViewMsgV.asp?AlertNum = 00246). When there is a suspect case, clinicians and infection control practitioners should work closely with state health department epidemiologists to perform a case risk assessment.

E. Treatment and Prevention

Recent isolates of influenza A H5N1 show varying resistance to the adamantanes (amantadine and rimantadine): clade 1 viruses are resistant, while the majority of clade 2 viruses are sensitive. The neuraminidase inhibitors, oseltamivir and zanamivir, are active against influenza A H5N1. However, the emergence of high-level resistance to oseltamivir during oseltamivir treatment has been demonstrated in some patients with influenza A H5N1 infections. These two patients had detectable virus at the end of a full course of treatment. An important component of avian influenza A H5N1 pandemic preparedness programs is the stockpiling of adequate supplies of neuraminidase inhibitors. Clinical trials and development efforts are currently ongoing for both inactivated and attenuated influenza A H5N1 vaccines.

II. Procedures

A. General

Sentinel laboratories should NOT inoculate specimens suspected of containing influenza A H5N1 virus into cell culture. Only laboratories capable of performing culture under BSL 3 conditions with enhancements should perform culture to evaluate a suspected influenza H5N1 case. If these criteria are met, and culture is performed, consultation with CDC and the state public health laboratory is recommended.

The use of rapid antigen tests for influenza is increasing in laboratories, point of care locations, and in physicians' offices. These tests are among the least reliable for diagnosis of influenza, and should not be used to rule out avian influenza in a suspect case, especially during the current pre-pandemic phase.

Although influenza A H5N1 is the agent of highest concern at this time, it is important to note that a pandemic could occur from other novel strains of influenza. For this reason, testing for H5N1 virus alone is not recommended and any unusual influenza viruses that cannot be subtyped should be referred to the CDC.

B. Precautions

Culture diagnosis of suspected avian influenza A H5N1 requires enhanced BSL 3 laboratory conditions. Enhancements include use of respirators, decontamination of all waste (solid and liquid), and showering of personnel before exiting. Molecular and rapid antigen testing can be performed on respiratory specimens under standard BSL 2 conditions in a Class II biological safety cabinet. A complete description of enhanced BSL 3 requirements is available at (www.hhs.gov/pandemicflu/plan/).

C. Specimens

Respiratory specimens Throat swabs and lower respiratory samples such as bronchoalveolar lavage and tracheal aspirates are the *preferred* specimens for detection of influenza A H5N1 virus. Nasal swabs and aspirates are acceptable, but may contain lower titers than throat swabs. This is an important distinction between H5N1 and seasonal, human adapted influenza A subtypes.

Collection of a nasopharyngeal swab and a throat swab from the same patient (can be submitted in the same tube of viral transport medium) would provide optimal specimens for both human adapted and avian influenza strains.

Rectal specimens In contrast to seasonal human influenza, diarrhea is a common symptom of H5N1 infections. Influenza A H5N1 viral RNA has been detected in rectal swabs by RT-PCR. However, current CDC testing recommendations (www.hhs.gov/pandemicflu/plan/) do not include rectal/stool specimens. Additionally, rectal specimens are inappropriate for human-adapted influenza strains.

Serum Influenza A H5N1-specific antibody can be detected in serum by the microneutraliza-tion assay. Paired specimens, the first collected during acute illness and the second collected 2–4 weeks later, are required for definitive diagnosis. Sentinel laboratories should contact their local or state public health laboratory, or the CDC for information on serologic testing for influenza A H5N1.

Specimen Collection and Handling
Detection of influenza A H5N1 is more likely from specimens collected within the first three days of illness onset. If possible, serial specimens should be collected over several days from the same patient to increase clinical sensitivity.

Dacron or rayon tipped swabs should be used for specimen collection, as other materials may inhibit RT-PCR. Rapid antigen detection kits provide or specify swab types. Swabs placed in viral transport medium are generally suitable for RT-PCR testing. Specific specimen requirements are provided by the testing laboratory.

The collection of lower respiratory specimens generates aerosols, and requires infection control precautions for influenza A H5N1, including the use of gloves, gown, eye protection, and a respirator rated at least N-95.

Specimens should be stored at refrigerated temperatures, unless specified otherwise by test procedures. For virus isolation, specimens should be stored at refrigerated temperatures no longer than 2 days, or frozen at $\leq 70°$, and shipped on dry ice. Follow current regulations for packaging and shipping hazardous materials (http://www.iata.org/ps/publications/9065.htm; ASM Sentinel Laboratory Guidelines for Suspected Agents of Bioterrorism/Packing and Shipping Infectious Substances, Diagnostic Specimens, and Biological Agents http://www.asm.org/ASM/files/LeftMarginHeaderList/DOWNLOADFILENAME/000000001202/Pack&Ship12-18-06.pdf

The Sentinel laboratory should contact the nearest designated LRN Reference laboratory prior to shipping specimens.

Rejection Criteria
Rejection criteria include lack of patient identification on the specimen, incomplete documentation on the requisition form, and improper specimen type or handling. Exceptions may be made, at the discretion of the laboratory director, out of public health or medical necessity. Due to the potential for false positive results and public health panic, specimens may also be rejected for H5N1-specific testing if the patient does not meet clinical case or risk assessment criteria.

D. Testing

Rapid Antigen Tests
Because rapid influenza antigen tests provide a result in 30 minutes or less, they significantly impact patient treatment and management. These tests are widely used for diagnosis of influenza in central, point-of-care, and physician office laboratories. Several rapid antigen tests are commercially available, some of which are able to distinguish between influenza A and B types. Some of these tests are CLIA waived. Rapid antigen tests are less sensitive than culture or RT-PCR. Manufacturer claimed sensitivity ranges from 40 to100% and specificity from 52-100% compared to viral culture. Because of the varied specificity, the positive predictive value of rapid influenza tests is often reduced when disease prevalence is low. Therefore, positive results outside of the influenza "season" should be interpreted with caution, and confirmed by additional tests.

While rapid antigen capture assays may detect avian influenza subtypes, including H5N1, currently available tests are not capable of distinguishing specific influenza A subtypes. Recent evidence indicates that currently available rapid antigen tests are extremely insensitive for H5N1, and should not be used to rule out avian influenza in a suspect case, especially during the current pre-pandemic phase. Rapid antigen testing can be performed on respiratory specimens from suspected avian influenza cases under standard BSL 2 conditions in a Class II biological safety cabinet.

Fluorescent Antibody Staining of Antigens

The staining of influenza antigens with fluorescent antibody is an additional rapid test. When performed directly on cells from respiratory specimens, this method can provide results in less than an hour. Availability of fluorescent antibody staining is restricted to laboratories with immunofluorescent microscopes and trained technologists able to accurately interpret fluorescent staining patterns. Since this is not a point-of-care test, the factors with the most significant impact on turn around time of test results are specimen transport to the testing laboratory, and batching of specimens. Fluorescently-labeled antibodies specific for influenza A and B viruses are available. Some commercially available influenza antibodies are provided in pools with antibodies to other common respiratory viruses. Fluorescent antibody staining is generally considered to be slightly more sensitive than rapid antigen tests. Specificity is high, but depends on well trained, experienced technologists.

Fluorescent antibody staining reagents specific for influenza A virus will detect avian influenza A H5N1. When this guideline was written there was no H5N1-specific fluorescent antibody reagent commercially available. Fluorescent antibody staining can be performed on respiratory specimens from suspected avian influenza cases under BSL 2 conditions in a Class II biological safety cabinet.

Nucleic Acid Amplification

Nucleic acid amplification methods such as RT-PCR and nucleic acid sequence-based amplification (NASBA) are becoming more commonly used for detection of influenza virus and other respiratory viruses. Using real-time, fluorescent detection of amplified product, laboratories are able to perform molecular tests in less than 3 hours. These are consistently the most sensitive methods for detection of influenza virus, including H5N1. High specificity requires judicious selection of primers and probes, optimization of amplification conditions, and interpretation of results. Continuous adherence to laboratory protocol is essential to avoid false positives due to carry-over contamination. Currently, the only Food and Drug Administration-cleared nucleic acid amplification test for influenza viruses, including H5N1, is the RT-PCR method used by LRN Reference laboratories. In addition to LRN Reference laboratories, some commercial and hospital laboratories offer nucleic acid amplification testing for influenza A H5N1. Unlike the LRN method, these tests are developed and validated in-house by each laboratory. As such, the performance characteristics of the tests may vary between laboratories.

RT-PCR testing only for the H5N1 subtype is not recommended. Specimens from suspect cases should be tested for both influenza A and B, and currently circulating influenza A subtypes in addition to H5N1.

Initial specimen processing, including addition of lysis buffer can be performed within a biological safety cabinet in a BSL 2 laboratory. Some specimen lysis buffers do not inactivate viruses. If lysis buffer that is known to inactivate the virus is used, further processing can be performed outside the biological safety cabinet.

Culture

Culture provides highly specific laboratory diagnosis of influenza, but requires fresh, refrigerated specimens for optimal sensitivity. Specimens in viral transport medium must be kept at 2-8°C and processed within 48 hours to avoid excessive decrease in viral titer. With proper specimen handling, culture is more sensitive than antigen detection methods. Historically, culture methods have been considered too slow to impact patient management. Incubation for at least 5 days is generally required to detect influenza virus in tubes cultures. Tubes are generally held for 14 days prior to reporting a negative result. Influenza virus is detected in tube cultures by the presence of cytopathic effect (CPE), adsorption or agglutination of red blood cells, and fluorescently-labeled antibodies specific for influenza A and B viruses. Spin-amplified shell vial cultures have reduced the time to detection to 1-3 days. However, this is considered by many to be too slow to impact patient treatment or isolation decisions. Culture is essential for detecting influenza infection missed by rapid testing, confirmation of non-culture results when disease prevalence is low, and to obtain isolates for characterization and surveillance.

Influenza A H5N1 will grow in cell lines commonly used for isolation of human-adapted influenza virus, including primary monkey kidney, Madin Darby canine kidney, A549, and others. Sentinel laboratories must be cognizant that H5N1 and other highly pathogenic novel subtypes can be cultivated unknowingly from unrecognized human cases of avian influenza. Do not perform culture on specimens if avian influenza A H5N1 is suspected, unless performed under enhanced BSL 3 laboratory conditions. Contact the LRN Reference laboratory for instructions on handling and shipping of specimens from suspected avian influenza cases.

Instructions for sentinel laboratories that set up viral cultures under BSL2 conditions on a specimen or patient later determined to be positive for influenza A H5N1:

- Consult with the public health laboratory or CDC. These laboratories may request specimens or cultures.
- Isolate specimens, cell culture vessels and supplies potentially contaminated with the influenza A H5N1 virus. Sterilize by autoclaving, and discard.
- Disinfect work area.
- Monitor potentially exposed staff for symptoms. Public health or medical experts may recommend quarantine and prophylactic treatment.
- Highly pathogenic avian influenza viruses, including H5N1 are select agents regulated by the Animal and Plant Health Inspection Service (APHIS) of the U.S. Department of Agriculture (USDA). If live H5N1 virus is isolated from a clinical specimen, CDC or APHIS must be notified, and the agent must be transferred or destroyed. Documentation of transfer or destruction must be maintained for three years.

Serology

Serologic test methods to detect influenza virus-specific antibodies are available, and generally performed in reference and public health laboratories. These methods include indirect fluorescent antibody (IFA), complement fixation (CF), hemagglutination inhibition (HI) and neutralization. HI and neutralization are often used with specific antigens to identify influenza virus type and subtype-specific antibody titers. The diagnostic utility of serology is limited by the need, generally, to collect both acute and convalescent sera to identify either seroconversion or a four-fold rise in antibody titer. As such, serologic methods that detect IgG responses have relatively little impact on patient management. IFA and other tests that detect IgM antibodies can detect acute infection, but sensitivity is reduced because serum IgM levels are often low due to repeated exposure to vaccine or circulating virus.

E. Interpretation/Reporting/Action

When a patient presents with suspected avian influenza, communication with the public health department is essential.

Specimens from suspected avian influenza cases should be referred to the local or state public health laboratory, or another LRN Reference laboratory. A list of emergency contacts for State Public Health Laboratories can be downloaded at http://www.asm.org/ASM/files/LeftMarginHeaderList/DOWNLOADFILENAME/000000000527/emcontactlist8-6.pdf. CDC has provided all state public health laboratories with RT-PCR protocols for molecular detection of influenza A and B viruses, and molecular subtyping for H1, H3 and H5. Contact the LRN Reference laboratory for instructions on handling and shipping of specimens from suspected avian influenza cases.

Thoroughly document the referral of specimens to the LRN Reference laboratory, as well the results reported by the laboratory. Follow institutional protocols for reporting positive test results obtained by an outside referral laboratory.

Positive influenza A H5N1 test results should be confirmed at the CDC.

References

1. Cimons, M. 2005. Many eyes on influenza virus in Asia, anticipating pandemic. ASM News. 71:420-424.
2. Committee of the World Health Organization (WHO) Consultation on Human Influenza A/H5. 2005. Avian influenza A (H5N1) infection in humans. N. Engl. J. Med. 353: 1374-1385.
3. Cox, N. J. and T. Ziegler. 2003. Influenza viruses. p.1360-1367. *In*, P. R. Murray, E. J. Barron, J. H. Jorgensen, M. A. Pfaller, R. H. Yolken (ed.), *Manual of Clinical Microbiology, 8th Ed.* ASM Press, Washington, D. C.
4. de Jong, M. D., V. C. Bach, T. Q. Phan, M. H. Vo, T. T. Tran, B. H. Nguyen, M. Beld, T. P. Le, H. K. Truong, V. V. Nguyen, T. H. Tran, Q. H. Do, and J. Farrar. 2005. Fatal avian influenza A (H5N1) in a child presenting with diarrhea followed by coma. N. Engl. J. Med. 352:686-691.
5. de Jong, M. D., C. P. Simmons, T. T. Thanh, V. M. Hien, G. J. Smith, T. N. Chau, D. M. Hoang, N. Van Vinh Chau, T. H. Khanh, V. C. Dong, P. T. Qui, B. Van Cam, Q. Ha do, Y. Guan, J. S. Peiris, N. T. Chinh, T. T. Hien and J. Farrar. 2006. Fatal outcome of human influenza A (H5N1) is associated with high viral load and hypercytokinemia. Nat Med. 12:1203-7.
6. de Jong, M.D., T. T. Tran, H. K. Truong, M. H. Vo, G. J. Smith, V. C. Nguyen, V. C. Bach, T. Q. Phan, Q. H. Do, Y. Guan, J. S. Peiris, T. H. Tran, and J. Farrar. 2005. Oseltamivir resistance during treatment of influenza A (H5N1) infection. N. Engl. J. Med. 353:2667-2672.
7. Fedorko, D. P., N. A. Nelson, J. M. McAuliffe and K. Subbarao. 2006. Performance of rapid tests for detection of avian influenza A virus types H5N1 and H9N2. J. Clin. Microbiol. 44:1596-1597.
8. Gray, L. D. and J. W. Snyder. 2006. Packing and shipping biological materials. p. 383-401. *In*, D. O. Fleming and D. L. Hunt (ed.), *Biological Safety, Principles and Practices, 4th Ed.* ASM Press, Washington, D. C.
9. Nyoman Kandun, I., H. Wibisono, E. R. Sedyaningsih, Yusharmen, W. Hadisoedarsuno, W. Purba, H. Santoso, C. Septiawati, E. Tresnaningsih, B. Heriyanto, D. Yuwono, S. Harun, S. Soeroso, S. Giriputra, P. J. Blair, A. Jeremijenko, H. Kosasih, S. D. Putnam, G. Samaan, M. Silitonga, K. H. Chan, L. L. M. Poon, W. Lim, A. Klimov, S. Lindstrom, Y. Guan, R. Donis, J. Katz, N. Cox, M. Peiris, and T. M. Uyeki. 2006. Three Indonesian Clusters of H5N1 Virus Infection in 2005. N. Engl. J. Med. 355:2186-2194.
10. Oner, A. F., A. Bay, S. Arslan, H. Akdeniz, H. A. Sahin, Y. Cesur, S. Epcacan, N. Yilmaz, I. Deger, B. Kizilyildiz, H. Karsen and M. Ceyhan. 2006. Avian Influenza A (H5N1) Infection in Eastern Turkey in 2006. N. Engl. J. Med. 355:2179-2185.
11. Yuen K. Y., P. K. Chan, M. Peiris, D. N. Tsang, T. L. Que, K. F. Shortridge, P. T. Cheung, W. K. To, E. T. Ho, R. Sung and A. F. Cheng. 1998. Clinical features and rapid viral diagnosis of human disease associated with avian influenza A H5N1 virus. Lancet. 351(9101):467-71.

Sentinel Laboratory Guidelines for Suspected Agents of Bioterrorism

Clinical Laboratory Bioterrorism Readiness Plan

American Society for Microbiology (Revised August 10, 2006)
Reprinted from www.asm.org/policy/index.asp
Accessed March 2007

Appendix

Appendix A: CDC Biosafety Level (BSL) Designations for Laboratories				
BSL	Agents	Practices	Safety Equipment	Facility
1	Not known to cause disease in healthy adults	Standard microbiological procedures	None required	Open bench top sink required
2	Associated with human disease. Hazard = autoinoculation, ingestion, or mucous membrane exposure	BSL-1 practice plus limited access, biohazard warning signs, Sharps precautions, and a biosafety manual defining waste decontamination or medical surveillance policies	Class I or II biosafety cabinet (BSC), splash guards and other devices to prevent splashes or aerosols. PPE = Lab coats, gloves, and face protection as needed	BSL-1 plus autoclave available
3	Indigenous or exotic agents with potential aerosol transmission. Disease may have serious or lethal consequences	BSL-2 practice plus controlled access, decontamination of all waste, decontamination of clothing before laundering, baseline serum	BSL-2 safety equipment plus respiratory protection as needed	BSL-2 plus physical separation from access corridors, self-closing double-door access, exhausted air not recirculated, negative airflow into the lab
4	Dangerous and exotic agents that pose high risk of life-threatening disease; aerosol transmitted	BSL-3 practices plus clothing change before entering, shower on exit, all materials decontaminated on exit from facility	All procedures conducted in Class III BSC or Class I or II BSC in combination with full-body, air-supplied, positive-pressure personnel suit	BSL-3 plus separate building or isolated zone, dedicated supply/exhaust, vacuum and decon system

408

Appendix B: Recommended BSL for BT Agents

Agent	BSL Specimen Handling	BSL Culture Handling	Specimen Exposure Risk	Recommended Laboratories Precautions
Alphaviruses	2	3	Blood, CSF. Tissue culture and animal inoculation studies should be performed at BSL-3 and are **NOT Sentinel (Level A) laboratory procedures.**	BSL-2: Activities involving clinical material collection and transport BSL-3: Activities with high potential for aerosol or droplet production
Bacillus anthracis	2	3	Blood, skin lesion exudates, CSF, pleural fluid, sputum, and rarely urine and feces	BSL-2: Activities involving clinical material collection and diagnostic quantities of infectious cultures BSL-3: All activities involving manipulations of cultures
Brucella spp.[a]	2	3	Blood, bone marrow, CSF, tissue, semen, and occasionally urine	BSL-2: Activities limited to collection, transport, and plating of clinical material BSL-3: All activities involving manipulations of cultures
Burkholderia pseudomallei	2	3	Blood, sputum, CSF, tissue, abscesses, and urine	BSL-2: Activities limited to collection, transport, and plating of clinical material BSL-3: All activities involving manipulations of cultures
Burkholderia mallei	2	3	Blood, sputum, CSF, tissue, abscesses, and urine	BSL-2: Activities limited to collection, transport, and plating of clinical material BSL-3: All activities involving manipulations of cultures
Coxiella burnetii[b]	2	3	Blood, tissue, body fluids, feces. Manipulation of tissues from infected animals and tissue culture should be performed at BSL-3 and are **NOT Sentinel laboratory procedures**	BSL-2: Activities limited to collection and transport of clinical material, including serological examinations
Clostridium botulinum[c]	2	3	Toxin may be present in food specimens, clinical material (serum, gastric, and feces). **TOXIN IS EXTREMELY POISONOUS!**	BSL-2: Activities with materials known to be or potentially containing toxin must be handled in a BSC (class II) with a lab coat, disposable surgical gloves, and a face shield (as needed). BLS-3: Activities with high potential for aerosol or droplet production

Table *(Continued)*

| Agent | BSL | | Specimen Exposure Risk | Recommended Laboratories Precautions |
	Specimen Handling	Culture Handling		
Francisella tularensis[d]	2	3	Skin lesion exudates, respiratory secretions, CSF, blood, urine, tissues from infected animals, and fluids from infected arthropods	BLS-2: Activities limited to collection, transport, and plating of clinical material BLS-3: All activities involving manipulations of cultures
Yersinia pestis[e]	2	3	Bubo fluid, blood, sputum, CSF, feces, and urine	BSL-2: Activities involving clinical material collection and diagnostic quantities of infectious cultures BSL-3: Activities with high potential for aerosol or droplet production
Smallpox[f]	4	4	Lesion fluid or crusts, respiratory secretions, or tissue	BSL-2: Packing and shipping. Do **NOT** put in cell culture.
Staphylococcal enterotoxin B	2	2	Toxin may be present in food specimens, clinical material (serum, gastric, urine, respiratory secretions, and feces), and isolates of *S. aureus.*	BSL-2: Activities involving clinical material collection and diagnostic quantities of infectious cultures
VHF[g]	4	4	Blood, urine, respiratory, and throat secretions, semen, and tissue	BSL-2: Packing and shipping. Do **NOT** put in cell culture.

[a] Laboratory-acquired brucellosis has occurred by "sniffing" cultures; aerosols generated by centrifugation; mouth pipetting; accidental parenteral inoculations; and sprays into eyes, nose, and mouth; by direct contact with clinical specimens; and when no breach in technique could be identified.

[b] Laboratory-acquired infections have been acquired from virulent phase I organisms due to infectious aerosols from cell culture and the use of embryonated eggs to propagate *C. burnetii.*

[c] Exposure to toxin is the primary laboratory hazard, since absorption can occur with direct contact with skin, eyes, or mucous membranes, including the respiratory tract. The toxin can be neutralized by 0.1 M sodium hydroxide. *C. botulinum* is inactivated by a 1:10 dilution of household bleach. Contact time is 20 min. If material contains both toxin and organisms, the spill must be sequentially treated with bleach and sodium hydroxide for a total contact time of 40 min.

[d] Laboratory-acquired tularemia infection has been more commonly associated with cultures than with clinical materials or animals. Direct skin/mucous membrane contact with cultures, parenteral inoculation, ingestion, and aerosol exposure have resulted in infection.

[e] Special care should be taken to avoid the generation of aerosols.

[f] Ingestion, parenteral inoculation, and droplet or aerosol exposure of mucous membranes or broken skin with infectious fluids or tissues are the primary hazards to laboratory workers.

[g] Respiratory exposure to infectious aerosols, mucous membrane exposure to infectious droplets, and accidental parenteral inoculation are the primary hazards to laboratory workers.

Appendix C: Specimen Selection: Bioterrorism Agents[a]

Disease/Agent	Specimen Selection	Transport & Storage	Specimen Plating and Processing				
			SBA	CA	MAC	Stain	Other
Anthrax (*Bacillus anthracis*)	Possible *Bacillus anthracis* exposure in an asymptomatic patient	Swab of anterior nares: Only to be collected if so advised by local public health authorities	≤24 h, RT	No	No	No	None	Follow public health instructions on anterior nares swab ONLY if advised to collect these.
		Vesicular stage: Collect fluid from intact vesicles on sterile swab(s). The organism is best demonstrated in this stage.	≤24 h, RT	X	X	X	Gram stain	
		Eschar stage: Without removing eschar, insert swab moistened in sterile saline beneath the edge of eschar, rotate, and collect lesion material.	≤24 h, RT	X	X	X	Gram stain	
	Cutaneous	Vesicular stage and eschar stage: collect 2 punch biopsies Place one biopsy in 10% formalin to be sent to CDC for histopathology, immunohistochemical staining, and PCR.	One punch biopsy in 10% formalin. Once in formalin, can be stored until transported to CDC	No	No	No	Performed at CDC	Arrange for transport to CDC.
		Submit second biopsy in an anaerobic transport vial for culture	Second punch biopsy in anaerobic transport vial ≤24 h, RT	X	X	X	Gram stain	
		Blood cultures: Collect 2 sets (1 set is 2 bottles) per institutional procedure for routine blood cultures.	Transport at RT. Incubate at 35–37°C per blood culture protocol	Blood culture bottles			Positive in some cases during late stages of disease	
		Purple-top tube (EDTA): for inpatients only, collect for direct Gram stain	≤24 h, RT	No	No	No	Gram stain	
		Red-top or blue-top tubes for serology; White Tube for PCR	≤24 h, 4°C	No	No	No	No	Arrange for transport to CDC.

411

Table (Continued)

Disease/Agent	Specimen Selection	Transport & Storage	Specimen Plating and Processing				
			SBA	CA	MAC	Stain	Other
Anthrax (*Bacillus anthracis*) (continued) — Gastrointestinal	Stool: Collect 5–10 g in a clean, sterile, leakproof container.	≤24 h, 4°C	Inoculate routine stool plating media plus CNA or PEA.				Minimal recovery
	Blood cultures: Collect 2 sets (1 set is 2 bottles) per institutional procedure for routine blood cultures.	Transport at RT. Incubate at 35–37°C per blood culture protocol.	Blood culture bottles			Positive in late stages of disease	
	Purple-top tube (EDTA): for inpatients only, collect for direct Gram stain.	≤2 h, RT	No	No	No	Gram stain	
	Red-top or blue-top tubes for serology; White Tube for PCR	≤24 h, 4°C	No	No	No	No	
Inhalation	Sputum: Collect expectorated specimen into a sterile, leakproof container.	≤24 h, 4°C	X	X	X	Gram stain	Minimal recovery
	Pleural fluid: Collect specimen into sterile, leakproof container.	≤24 h, 4°C	X	X	X	Gram stain	Save excess (if any) for PCR.
	Blood cultures: Collect 2 sets (1 set is 2 bottles) per institutional procedure for routine blood cultures.	Transport at RT. Incubate at 35–37°C per blood culture protocol.	Blood culture bottles			Positive in late stages of disease	
	Purple-top tube (EDTA): For inpatients only, collect for direct Gram stain.	≤2 h, RT	No	No	No	Gram stain	
	Red-top or blue-top tubes for serology; White Tube for PCR	≤24 h, 4°C	No	No	No	No	
Meningitis	Cerebrospinal fluid culture: Aseptically collect CSF per institutional procedure.	≤24 h, RT	X	X		Gram stain	May be seen in late stages of disease; consider adding broth medium such as brain heart infusion.
	Blood cultures: Collect 2 sets (1 set is 2 bottles) per institutional procedure for routine blood cultures.	Transport at RT. Incubate at 35–37°C per blood culture protocol.	Blood culture bottles			Positive in late stages of disease	

Table (Continued)

Disease/Agent	Specimen Selection	Transport & Storage	Specimen Plating and Processing				
			SBA	CA	MAC	Stain	Other
Brucellosis (Brucella melitensis, B. abortus, B. suis, B. canis)	Serum: Collect 10-12 cc (ml) of acute-phase specimen as soon as possible after disease onset. Follow with a convalescent-phase specimen obtained 21 days later.	Transport in ≤2 h, at RT. Store at -20°C.	Specimen should be stored and shipped frozen at -20°C to State Laboratory or other LRN Reference (Level B/C) laboratory.			Serologic diagnosis: 1. Single titer: ≥1:160 2. 4-fold rise 3. IgM NOTE: *B. canis* does not cross-react with standard serologic reagents.	
	Acute, sub-acute, or chronic	Blood: Collect 2 sets (1 set is 2 bottles) per institutional procedure for routine blood cultures.	Transport at RT. Incubate at 35-37°C	Blood culture bottles: Subculture at 5 days and hold 21 days.				Blood culture isolation rates vary from 15-70% depending on methods and length of incubation. Cultures should be manipulated in a biological safety cabinet. Personal protective equipment includes gloves, gown, mask, and protective faceshield. All cultures should be taped shut during incubation.
		Bone marrow, spleen, or liver: Collect per institution's surgical/pathology procedure.	≤24 h, RT	X	X		Gram stain	Cultures should be manipulated in a biological safety cabinet. Personal protective equipment includes gloves, gown, mask, and protective faceshield. All cultures should be taped shut during incubation.
Meningitis	Cerebrospinal fluid culture: Aseptically collect CSF per institutional procedure.	≤24 h, RT	X	X		Gram stain	Cultures should be manipulated in a biological safety cabinet. Personal protective equipment includes gloves, gown, mask, and protective faceshield. All cultures should be taped shut during incubation. Consider adding both medium such as brain heart infusion.
	Cerebrospinal fluid for anti-body testing	-20°	Specimen should be stored and shipped frozen at -20°C or lower temperature to State Laboratory or other LRN Reference laboratory.			None	

413

Table (Continued)

Specimen(s) of choice for confirming botulism:
1. Serum
2. Wound/tissue
3. Stool
4. Incriminated food

Disease/Agent	Specimen Selection					Transport & Storage	Specimen Handling
	Specimen type	Clinical syndrome					
		Foodborne	Infant	Wound	Intentional release (airborne)		
Botulism (botulinum toxin)	Enema fluid – 20 ml	X	X		X	4°C	Purge with a minimal amount of sterile nonbacteriostatic water to minimize dilution of toxin.
	Food sample – 10-50g	X	X		X	4°C	Foods that support C. botulinum growth will have a pH of 3.5-7.0; most common pH is 5.5-6.5. Submit food in original container, placing individually in leakproof sealed transport devices.
	Gastric fluid – 20 ml	X,A				4°C	Collect up to 20 ml.
	Intestinal fluid – 20 ml	A	A			4°C	Autopsy: Intestinal contents from various areas of the small and large intestines should be provided.
	Nasal swab (anaerobic swab)				X	RT	For aerosolized botulinum toxin exposure, obtain nasal cultures for C. botulinum and serum for mouse toxicity testing.
	Serum – 15-20mls	X,A		X	X	4°C	Serum should be obtained as soon as possible after the onset of symptoms and before antitoxin is given. Whole blood (30 ml [3 red-top or gold-top tubes]) is required for mouse toxicity testing. In infants, serum is generally not useful, since the toxin is quickly absorbed before serum can be obtained.
	Stool >25 g	X	X	X	X	4°C	Botulism has been confirmed in infants with only "pea-size" stools. Please note: Anticholinesterase given orally, as in patients with myasthenia gravis, has been shown to interfere with toxin testing.
	Vomitus – 20 ml	X				4°C	Collect up to 20 ml.
	Wound, tissue - anaerobic swab or transport system					Anaerobic swab or transport system Transport at RT	Exudate, tissue, or swabs must be collected and transported in an anaerobic transport system. Samples from an enema or feces should also be submitted, since the wound may not be the source of botulinum toxin.

Table (Continued)

Disease/Agent	Specimen Selection		Transport and Storage	Specimen Handling
	Specimen type	Foodborne	Airborne (intentional release)	
Staphylococcal enterotoxin B (From Staphylococcus aureus)	Serum – 10 ml	X	X	2-8°C	1. Obtain as soon as possible after the onset of symptoms to detect the toxin. 2. Also collect 7-14 days after onset of illness to compare acute and convalescent antibody titers. 3. Do not send whole blood, since hemolysis during transit will compromise the quality of the specimen.
	Nasal swab - dacron or rayon swab		X	2-8°C	Collect a nasal swab within 24 h of exposure by rubbing a dry, sterile swab (Dacron or rayon) on the mucosa of the anterior nares. Place in protective transport tube.
	Induced respiratory secretions		X	2-8°C	Collect sputum induced by instilling 10-25 ml of sterile saline into nasal passages into a sterile screw-top container.
	Urine – 20-30 ml	X	X	2-8°C	Collect into a sterile, leakproof container with screw-top lid.
	Stool or gastric aspirate—10-50 g	X	X	2-8°C	Collect into a sterile, leakproof container with screw-top lid.
	Postmortem 10 g	X	X	2-8°C	Obtain specimens of the intestinal contents from different levels of the small and large bowel. Place 10 g of specimen into a sterile, leakproof container with screw-top lid. Obtain serum as previously described.
	Culture isolate	X	X	2-8°C	Send S. aureus isolate for toxin testing on appropriate agar slant.
	Food specimen	X	X	2-8°C	Food should be left in its original container if possible or placed in sterile unbreakable containers and labeled carefully. Place containers individually in leakproof containers (i.e., sealed plastic bags) to prevent cross-contamination during shipment. Empty containers with remnants of suspected contaminated foods can be examined.

Table (*Continued*)

Disease/Agent	Specimen Selection	Transport & Storage	Specimen Plating and Processing				
			SBA	CA	MAC	Stain	Other
Plague (*Yersinia pestis*)	Possible *Y. pestis* exposure in asymptomatic patient: No cultures or serology indicated						Follow public health instructions if advised to collect specimens.
	Blood cultures: Collect 2 sets (1 set is 2 bottles) per institutional procedure for routine blood cultures.	Transport at RT. Incubate at 35-37°C per blood culture protocol.	Blood culture bottles			Gram stain of positive cultures	If suspicion of plague is high, obtain an additional set for incubation at RT (22-28°C) without shaking
	Tiger-top, red-top, or gold-top tube: For serology (acute and, if needed for diagnosis, convalescent serum in 14 days) Green-top (heparin) tube: For PCR	≤24 h, 4°C		No	No	No	Patients with negative cultures having a single titer, ≥1:10, specific to F1 antigen by agglutination would meet presumptive criteria.
	Bubonic: Lymph node (bubo) aspirate: Flushing with 1.0 ml of sterile saline may be needed to obtain material.	Transport at RT or 4°C if transport is delayed. Store at ≤24 h, 4°C.	X	X	X	Gram stain, Giemsa, Wright's stain	Contact LRN Reference lab or above laboratory to prepare smears for DFA.
	Tissue: Collect in sterile container with 1 to 2 drops of sterile, nonbacteriostatic saline.	Transport at RT or 4°C if transport is delayed. Store at ≤24 h, 4°C.	X	X	X	Gram stain, Giemsa, Wright's stain	Contact LRN Reference lab or above laboratory to prepare smears for DFA.
	Throat: Collect routine throat culture using a swab collected into a sterile, leakproof container.	≤24 h, 4°C	X	X	X	Gram stain	Contact LRN Reference lab or above laboratory to prepare smears for DFA.

Table (*Continued*)

Disease/ Agent	Specimen Selection	Transport & Storage	Specimen Plating and Processing					
			SBA	CA	MAC	Stain	Other	
Plague (*Yersinia pestis*) (continued)	Sputum/throat: Collect routine throat culture using a swab or expectorated sputum collected into a sterile, leakproof container.	≤24 h, 4°C	X	X	X	Gram stain	Contact LRN Reference lab or above laboratory to prepare smears for DFA.	
	Bronchial/tracheal wash: Collect per institution's procedure in an area dedicated to collecting respiratory specimens under isolation/containment circumstances, i.e., isolation chamber/"bubble."	≤24 h, 4°C	X	X	X	Gram stain	Contact LRN Reference lab or above laboratory to prepare smears for DFA.	
Pneumonic	Blood cultures: Collect 2 sets (1 set is 2 bottles) per institutional procedure for routine blood cultures.	Transport at RT. Incubate at 35-37°C per blood culture protocol.	Blood culture bottles			Gram stain of positive cultures	If suspicion of plague is high, obtain an additional set for incubation at RT (22-28°C) without shaking.	
	Tiger-top, red-top, or gold-top tube: For serology (acute and, if needed for diagnosis, convalescent serum in 14 days) Green-top (heparin) tube: For PCR	≤24 h, 4°C	No	No		No	Patients with negative cultures having a single titer, ≥1:10, specific to F1 antigen by agglutination would meet presumptive criteria.	

417

Table (*Continued*)

Disease/ Agent	Specimen Selection	Transport & Storage	Specimen Plating and Processing				
			SBA	CA	MAC	Stain	Other
Plague (*Yersinia pestis*) (continued)	Blood cultures: Collect 2 sets (1 set is 2 bottles) per institutional procedure for routine blood cultures.	Transport at RT. Incubate at 35-37°C per blood culture protocol.	Blood culture bottles			Gram stain of positive cultures	If suspicion of plague is high, obtain an additional set for incubation at RT (22-28°C) without shaking.
	Tiger-top, red-top, or gold-top tube: For serology (acute and, if needed for diagnosis, convalescent serum in 14 days) Green-top (heparin) tube: For PCR	≤24 h, 4°C		No		No	Patients with negative cultures having a single titer, ≥1:10, specific to F1 antigen by agglutination would meet presumptive criteria.
	Cerebrospinal fluid	Transport at RT. Store incubated at 35-37°C.	X	X		Gram stain	Can add broth culture at RT (22-28°C) without shaking.
Tularemia (*Francisella tularensis*)	Possible Francisella *tularensis* exposure in asymptomatic patient	No cultures or serology indicated					Follow public health instructions if advised to collect specimens.
	Oculoglandular	Conjunctival scraping	≤24 h, 4°C	X	X	Gram stain; prepare smears for DFA referral.	Add a BCYE plate and a plate selective for *Neisseria gonorrhoeae* such as modified Thayer-Martin. Manipulate cultures in a biological safety cabinet. Personal protective equipment includes gloves, gown, mask, and protective faceshield. All cultures should be taped shut during incubation.

418

Table (*Continued*)

Disease/Agent	Specimen Selection	Transport & Storage	Specimen Plating and Processing					
			SBA	CA	MAC	Stain	Other
Tularemia (*Francisella tularensis*) (continued)	Oculo-glandular (continued) Lymph node aspirate: Flushing with 1.0 ml of sterile saline may be needed to obtain material.	Transport at RT, 4°C if transport is delayed. Store at ≤24 h, 4°C.	X	X	X	Gram stain; prepare smears for DFA referral.	Add a BCYE plate and a plate selective for *Neisseria gonorrhoeae*, such as modified Thayer-Martin. Manipulate cultures in a biological safety cabinet. Personal protective equipment includes gloves, gown, mask, and protective faceshield. All cultures should be taped shut during incubation.
	Blood cultures: Collect 2 sets (1 set is 2 bottles) per institutional procedure for routine blood cultures. Growth is more likely from aerobic bottle.	Transport at RT. Incubate at 35-37°C per blood culture protocol.	Blood culture bottles; subculture the broth to BCYE plate and incubate aerobically.			Cultures should be manipulated in a biological safety cabinet. Personal protective equipment includes gloves, gown, mask, and protective faceshield. All cultures should be taped shut during incubation.	
	Ulcero-glandular	Blood cultures: Collect 2 sets (1 set is 2 bottles) per institutional procedure for routine blood cultures. Growth is more likely from aerobic bottle.	Transport at RT. Incubate at 35-37°C per blood culture protocol.	Blood culture bottles; subculture the broth to BCYE plate and incubate aerobically.			Cultures should be manipulated in a biological safety cabinet. Personal protective equipment includes gloves, gown, mask, and protective faceshield. All cultures should be taped shut during incubation.	
		Ulcer or tissue: Collect biopsy (best specimen), scraping, or swab.	≤24 h, 4°C	X	X	X	Gram stain	Add a BCYE plate and a plate selective for *Neisseria gonorrhoeae*

Table (*Continued*)

Disease/Agent		Specimen Selection	Transport & Storage	Specimen Plating and Processing				
				SBA	CA	MAC	Stain	Other
Tularemia (*Francisella tularensis*) (continued)	Ulcero-glandular (continued)	Lymph node aspirate: Flushing with 1.0 ml of sterile saline may be needed to obtain material.	Transport at RT; 4°C if transport is delayed. Store at ≤24 h, 4°C.	X	X	X	Gram stain; prepare smears for DFA referral.	such as modified Thayer-Martin. Prepare smears for DFA referral. Manipulate cultures in a biological safety cabinet. Personal protective equipment includes gloves, gown, mask, and protective faceshield. All cultures should be taped shut during incubation.
		Sputum/throat: Collect routine throat culture using a swab or expectorated sputum collected into a sterile, leakproof container.	≤24 h, 4°C	X	X	X	Gram stain	
		Bronchial/tracheal wash: Collect per institution's procedure in an area dedicated to collecting respiratory specimens under isolation/containment circumstances, i.e., isolation chamber/"bubble."	≤24 h, 4°C	X	X	X	Gram stain	
	Pneumonic	Blood cultures: Collect 2 sets (1 set is 2 bottles) per institutional procedure for routine blood cultures. Growth is more likely from aerobic bottle.	Transport at RT. Incubate at 35-37°C per blood culture protocol.	Blood culture bottles; Subculture the broth to BCYE plate and incubate aerobically.				Cultures should be manipulated in a biological safety cabinet. Personal protective equipment includes gloves, gown, mask, and protective faceshield. All cultures should be taped shut during incubation.
		2 Red-top or gold-top tubes: For PCR and serology (acute and, if needed for diagnosis, convalescent serum in 14 days)	≤2 h RT, ≤24 h, 4°C			No		Positive serology test would meet presumptive criteria. Confirmation requires culture identification or a 4-fold rise in titer.

Disease/Agent	Specimen Selection and Transport	Specimen Handling

Smallpox (Variola virus)	See CDC document "Specimen Collection and Transport Guidelines" for detailed instructions http://www.bt.cdc.gov/agent/smallpox/response-plan/index.asp#guidec (Click on Guide D)
	
	NOTE: Only recently, successfully *vaccinated* personnel (within 3 years) wearing appropriate barrier protection (gloves, gown, and shoe covers) should be involved in specimen collection for suspected cases of smallpox. Respiratory protection is not needed for personnel with recent, successful vaccination. Masks and eyewear or faceshields should be used if splashing is anticipated.
	
	If unvaccinated personnel must be utilized to collect specimens, only those without contraindications to vaccination should be utilized, as they would require immediate vaccination if the diagnosis of smallpox is confirmed. Fit-tested N95 masks should be worn by unvaccinated individuals caring for suspected patients.
	
	Biopsy specimens
	Rash: Scabs / Vesicular fluid
	Posterior tonsillar tissue swab: Swab
	Blood: Use plastic tubes
	Autopsy: Portions of skin containing lesions, liver, spleen, lung, lymph nodes, and/or kidney
	
	See CDC document "Specimen Collection and Transport Guidelines" for detailed instructions (Guide D).
		1. **A suspected case of smallpox should be reported immediately to the respective Local and State Health Departments for review.**
		
		2. And if, after review, smallpox is still suspected, one of the following should be contacted immediately:
		
		  A. PDC Emergency Response Hotline (24 hours): 770-488-7100
		
		  B. Poxvirus Section, Division of Viral and Rickettsial Diseases, NCID, CDC, Atlanta, Georgia 30333. Laboratory: 404-639-4931
		
		  C. Bioterrorism Preparedness and Response Program, NCID, CDC: 404-639-0385 or 404-639-2468. (8 am to 5 pm weekdays)
		
		NOTE: Approval must be obtained prior to the shipment of potential smallpox patient clinical specimens to CDC
		
		3. At this time, review the packaging/shipping requirements with CDC and request assistance in coordinating a carrier for transport/shipment.
		
		4. **Hand carry all specimens and do not send specimens via pneumatic tube system.**
		
		5. Do not attempt viral cultures: this is a Biosafety Level 4 agent, and this could result in a very unsafe situation in which there is a significant amount of infectious virus.

421

Table (*Continued*)

Disease/agent	Specimen Selection	Transport & Storage	Specimen Handling
VHF (Various viruses including Ebola, Marburg, Lassa, Machupo, Junin, Guanarito, Sabia, Crimean-Congo hemorrhagic fever, Rift Valley fever, Omsk hemorrhagic fever, Kyasanur Forest disease virus, and others)	Serum for antibody testing: Collect blood in red-top or gold-top tubes. Obtain convalescent serum at least 14 days after acute specimen is obtained. Use a Vacutainer or other sealed sterile dry tube for blood collection. Viral culture, blood: Collect serum, heparinized plasma (green-top tube), or whole blood during acute febrile illness. Throat wash specimens: Mix with equal volume of viral transport medium. Urine: Mix with equal volume of viral transport medium. CSF, tissue, other specimens	Transport within ~2 h, at RT. Store at –20°C to –70°C. Transport at RT. Store 4°C or frozen on dry ice or liquid nitrogen. Transport on wet ice. Store at –40°C or colder. Transport on wet ice. Store at –40°C or colder. As per discussion with CDC	Specific handling conditions are currently under development. Contact CDC to discuss proper collection and handling. 1. Double-bag each specimen. 2. Swab the exterior of the outside bag with disinfectant *before* removal from the patient's room. 3. Do not use glass tubes. 4. Hand carry all specimens and do not send specimens via pneumatic tube system. **NOTE:** Disposable equipment and sharps go into rigid containers containing disinfectant that are then autoclaved or incinerated. Double-bag refuse. The exterior of the outside bag is to be treated with disinfectant and then autoclaved or incinerated. **Do not attempt tissue culture isolation.** This is only to be done in a Biosafety Level 4 facility. In laboratory: 1. Strict barrier precautions are to be used. Personal protective equipment includes gloves, gown, mask, shoe covers, and protective faceshield. 2. Handle specimens in biological safety cabinet if possible. 3. Consider respiratory mask with HEPA filter. 4. Specimens should be centrifuged at low speed.
	Blood cultures: If clinical and travel history warrants, collect 2 sets (1 set is 2 bottles) of blood cultures per institutional procedure for routine blood cultures.	Transport at RT. Incubate at 35–37°C per blood culture protocol.	Bacteremia with disseminated intravascular coagulation and malaria due to *Plasmodium falciparum* are two life-threatening and treatable clinical entities that can present with prominent clinical findings of hemorrhage and fever in a patient with a travel history to areas with VHF. Handle with precautions noted above. Continue to use the same precautions as above.
	Malaria smear of peripheral blood: If clinical and travel history warrants	Lavender-top tube at RT	

Table (*Continued*)

Disease/Agent	Specimen Selection	Transport & Storage	Specimen Handling
Q. fever (*Coxiella burnetii*)	Serum: Collect 10 ml of serum (red-top, tiger-top, or gold-top tube) as soon as possible after onset of symptoms (acute) and with a follow-up specimen (convalescent) at >14 days for serological testing.	Transport within ~6 h, at 4°C. Store at −20° to −70	**Do not attempt tissue culture isolation**, as that could result in a very unsafe situation in which there is a significant amount of infectious organism.
	Blood: Collect blood in EDTA (lavender) or sodium citrate (blue) and maintain at 4°C for storage and shipping for PCR or special cultures. If possible, collect specimens prior to anti-microbial therapy.	Transport within ~6 h, at 4°C. Store at 4°C.	Sentinel laboratories should consult with State Public Health Laboratory Director (or designate) prior to or concurrent with testing if *C. burnetii* is suspected by the attending physician.
	Tissue, body fluids, others, including cell cultures and cell supernatants: Specimens can be kept at 2-8°C if transported within 24 h. Store frozen at −70°C or on dry ice.	Transport within <24 h, at 2-8°C. Store at −70°C or on dry ice.	Serology is available through commercial reference as well as public health laboratories.
Alphaviruses (Includes Eastern equine, Western equine, Venezuelan equine encephalitis viruses and others)	Serum: Collect 10 ml of serum (red-top, tiger-top, or gold-top tube) as soon as possible after onset of symptoms (acute) and with a follow-up specimen (convalescent) at ≥14 days for serological testing.	Transport within ~6 h, at 4°C. Store at −20°C to −70°C	**Do not attempt tissue culture isolation**, as that could result in a very unsafe situation in which there is a significant amount of infectious organism.
	Blood: Collect blood in EDTA (lavender) or sodium citrate (blue) and maintain at 4°C for storage and shipping for PCR or special studies.	Transport within ~6 h, at 4°C. Store at 4°C.	
	Cerebrospinal fluid: Specimens (greater than 1 ml) can be kept at 2-8°C if transported within 24 h. If frozen, store at −70°C and transport on dry ice.	Transport on wet ice. If already frozen, store at −70°C and transport on dry ice.	
	Tissue, body fluids, others, including cell cultures and cell supernatants: Specimens can be kept at 2-8°C if transported within 24 h. If frozen, store at −70°C and transport on dry ice.	Transport on wet ice. If already frozen, store at −70°C and transport on dry ice.	

Table (Continued)

Disease/Agent	Specimen Selection	Transport & Storage	Specimen Plating and Processing					
			SBA	CA	MAC	PC	Stain	Other
Melioidosis and glanders (Burkholderia pseudomallei and Burkholderia mallei)	Possible Burkholderia pseudomallei or Burkholderia mallei exposure in asymptomatic patient	No cultures or serology indicated						Follow public health instructions if advised to collect specimens.
	Clinical illness	Bone marrow		X			Gram stain	B. pseudomallei is a small gram-negative bacillus that may demonstrate bipolar morphology on stain. B. mallei is a small gram-negative coccobacillus. Incubation should be at 35 to 37°C, ambient atmosphere; CO_2 incubation is acceptable
	Blood cultures: Collect 2 sets (1 set is 2 bottles) per institutional procedure for routine blood cultures OR collect lysis-centrifugation (e.g., Isolator) blood cultures.	Transport within ≤2 h, at RT. Store ≤24 h, at 4°C	Blood culture bottles OR Collect lysis-centrifugation (e.g., Isolator) blood cultures and plate to:					Cultures should be manipulated in a biological safety cabinet. Personal protective equipment includes gloves, gown, mask, and protective face shield. All cultures should be taped shut during incubation. Incubation should be at 35 to 37°C, ambient atmosphere; CO_2 incubation is acceptable.
		Transport at RT. Incubate at 35-37°C per blood culture protocol.		X				

Table (*Continued*)

Disease/Agent	Specimen Selection	Transport & Storage	Specimen Plating and Processing					
			SBA	CA	MAC	PC	Stain	Other
Melioidosis and glanders (*Burkholderia pseudomallei* and *Burkholderia mallei*) (continued)	Respiratory specimens, abscess material, wound specimens, urine	Transport within ≤2 h, at RT. Store ≤24 h, at 4°C	X	X	X	X	Gram stain	If the laboratory has *B. cepacia* selective agar medium, it has been useful in isolation of *B. pseudomallei* for specimens in which indigineous microflora is likely to be encountered. Ashdown medium is a selective medium specifically designed for recovery of *B. pseudomallei*. This medium is not likely to be available in most Sentinel laboratories. Incubation should be at 35 to 37°C; ambient atmosphere; CO_2 incubation is acceptable.
	Clinical illness (continued) Serum: Red-top or gold-top tube for both acute and convalescent (obtained 14 days after the acute specimen)	Transport within ~6 h, at 4°C.Store at −20°C to −70°C.	Obtain if serologic diagnosis of *B. pseudomallei* infection is being considered.					

[a] Abbreviations: A, autopsy; BCYE, buffered charcoal-yeast extract agar; C, centigrade; CA, chocolate agar; CNA, colistin-nalidixic acid agar; DFA, direct fluorescent antibody; MAC, MacConkey agar; PEA, phenylethyl alcohol blood agar; RT, room temperature; VHF, viral hemorrhagic fever; PC, selective medium for *Burkholderia cepacia*.

SENTINEL LABORATORY GUIDELINES

Appendix D: Agent Characteristics Summary: Microorganisms

Characteristic	B. anthracis	Y. pestis	Burkholderia Pseudomallei and B. mallei	F. tularensis	Brucella spp.	Variola Virus (Smallpox)
Gram stain morphology	• Large gram-positive rod • Nonmotile • From blood agar: no capsule, central to subterminal spores that do not enlarge the cell • From blood: capsule, no spores	• Plump gram-negative rod • Gram stain: ± bipolar or "safety pin" appearance • Wright-Giemsa: bipolar or "safety pin" appearance	• B. pseudomallei: small gram-negative rod • B. mallei: small gram-negative coccobacillus • Gram stain: ± bipolar or "safety pin" appearance (B. pseudomallei) • Wright-Giemsa: bipolar or "safety pin" appearance (B. pseudomallei)	• Minute GNCB • Poorly staining • Smaller than Haemophilus influenzae • Pleomorphic	• Tiny GNCB • Faintly staining	
Growth	• Standard conditions • Extremely rapid	• 28°C optimal, without agitation • 35-37°C more slowly	• 35-37°C • Ambient atmosphere, though CO_2 is acceptable	• Aerobic conditions • Growth is best on media containing cysteine, such as BCYE, but will often grow initially on chocolate or BA	• Grows in blood culture media • Can require blind subculturing	• Grows in most cell lines • Unusual or unrecognizable CPE

Table *(Continued)*

Characteristic	B. anthracis	Y. pestis	Burkholderia Pseudomallei and B. mallei	F. tularensis	Brucella spp.	Variola Virus (Smallpox)
Colonial morphology (BA)	• Nonhemolytic • Ground glass • Irregular/wavy edges • Tenacious • "Beaten egg whites" when touched with loop	• Pinpoint at 24-48 h • "Fried egg" or "hammered copper" or shiny at 48-72 h • Nonhemolytic	*B. pseudomallei:* • SBA: small, smooth creamy colonies in first 1 to 2 days, gradually changing after a few days to dry, wrinkled colonies similar to *Pseudomonas stutzeri* *B. mallei:* • SBA: smooth, gray, translucent colonies without pigment	• Does not pass well on BA	• Small colonies • Punctate after 48 h • Nonhemolytic	• CPE can be passed

Tests	• Cat (+)	• Cat (+)
		• Ox (−)
		• Urease (−)
		• MAC: Lac (−)
		• Indole (−)
			• Cat (+)
			• Colistin (10 µg) and polymyxin B (300 U) (R)
			• Motility (+)
			  B. pseudomallei
			• Motility (−) *B. mallei*
			• Indole (−)
			• Oxidase (+)
			  B. pseudomallei
			• Oxidase (+/−)
			  B. mallei
			• MAC: Lac (−)
			  (*B. pseudomallei*)
			• MAC: Lac (−) or NG (*B. mallei*)
				• Cat wk (+)
				• Ox (−)
				• Urease (−)
				• β-Lac (−)
				• Satellite (−)
				• MAC: NG
					• Ox (+)
					• Urease (+), though some are negative
					• Satellite (−)
					• MAC: Poor to NG	

427

Appendix E: Bioterrorism Agent Clinical Summary

Disease	Virulence Factor(s)	Infective Dose (ID)	Incubation Period	Duration of Illness	Person-to-Person Transmission[e]	Isolation Precautions for Hospitalized[f]	Persistence of Organism
Inhalation anthrax	Exotoxin[a] capsule	Lower limit unknown, ID2 estimated at 9 spores[b]	1-6 days	3-5 days	No	Standard	>40 yr
Brucellosis	LPS;[c] PMN survival	10-100 organisms	5-60 days (usually 1-2 mo)	Weeks to months	Via breast milk[g] and sexually[h] (rare)	Standard	Water/soil, ~10 wk
Botulism	Neurotoxin	0.001 µg/kg is LD_{50} for type A	6 h to 10 days (usually 1-5 days)	Death in 24-72 h; lasts months if not lethal	No	Standard	Food/water, ~weeks
Glanders	Little studied, possible antiphagocytic capsule	Low	10-14 days via aerosol	Death in 7-10 days in septicemic form	YES (low)	Standard	Very stable
Melioidosis	Possibly LPS, exotoxin, intracellular survival, antiphagocytic capsule	Low	2 days to 26 yr	Days to months	YES (rare)[j]	Standard	Very stable in water/soil
Pneumonic plague	V and W antigens LPS (endotoxin) F1 antigen[d]	<100 organisms	2-3 days	1-6 days	YES (high)	Droplet[f]	Soil, up to 1 yr
Q fever	Intracellular survival LPS (endotoxin)	1-10 organisms	10-40 days	~2 wk (acute), months to years (chronic)	Rare[j]	Standard	Very stable
Smallpox		10-100 particles	7-17 days	~4 wk	YES (high)	Airborne[f]	Very stable
Staphylococcal enterotoxin B	Superantigen	0.0004 µg/kg incapacitation; LD_{50} is 0.02 µg/kg	3-12 h after inhalation	Hours	No	Standard	Resistant to freezing

Table (Continued)

Disease	Virulence Factor(s)	Infective Dose (ID)	Incubation Period	Duration of Illness	Person-to-Person Transmission[e]	Isolation Precautions for Hospitalized[f]	Persistence of Organism
Tularemia	Intracellular survival	10-50 organisms	2-10 days	≥2 wk	Single case report during autopsy	Standard	Moist soil, ~months
VHF	Varies with virus	1-10 particles	4-21 days	7-16 days	YES (moderate)	Airborne and contact[f]	Unstable

[a] *B. anthracis* exotoxin or exotoxins consist of three components: the **edema factor** and **lethal factor** exert their effect within cells by interacting with a common transport protein designated **"protective antigen"** (so named because, when modified, it contributes to vaccine efficacy). Expression of toxic factors is mediated by one plasmid, and that of the capsule(D-glutamic acid polypeptide) is mediated by a second plasmid.Strains repeatedly subcultured at 42°C become avirulent as a result of losing virulence-determining plasmids, which is thought to be the basis for Pasteur's attenuated anthrax vaccine used at Pouilly-le-Fort in 1881.

[b] The estimate that nine inhaled spores would infect 2% of the exposed human population is based on data from Science **266**:1202-1208, 1994. The dose needed to infect 50% of the exposed human population may be 8,000 or higher.

[c] The major virulence factor for brucellosis appears to be an endotoxic lipopolysaccharide (LPS) among smooth strains.Pathogenicity is related to an LPS containing poly *N*-formyl perosamine O chain, Cu-Zn superoxide cismutase, erythrulose phosphate dehydrogenase, intracellular survival stress-induced proteins, and adenine and guanine monophosphate inhibitors of phagocyte functions.

[d] The V and W antigens and the F1 capsular antigens are only expressed at 7°C and not at the lower temperature of the flea (20 to 25°C).

[e] Periods of communicability are as follows: for **inhalation anthrax and botulism, none;** no evidence of person-to-person transmission; **pneumonic plague,** 72 h following initiation of appropriate antimicrobial therapy or until sputum culture is negative; **smallpox, approximately 3 weeks;** usually corresponds with the initial appearance of skin lesions to their final disappearance and is most infectious during the first week of rash via inhalation of virus released from oropharyngeal lesion secretions of the index case; **VHF, varies with virus, but at minimum, all for the duration of illness,** and for Ebola/Marburg transmission through semen may occur up to 7 weeks after clinical recovery.

[f] Guidelines for isolation precautions in hospitals can be found in Infect.Control Hosp. Epidemiol. **17**:53-80, 1996, in addition to the standard precautions that apply to all patients.

[g] Published reports of possible transmission of brucellosis via human breast milk may be found in Int.J.Infect.Dis. **4**:55-56, 2000; Ann. Trop. Paediatr. **10**:305-307, 1990; J.Infect. **26**:346-348, 1993; and Trop. Geogr.Med. **40**:151-152, 1988.

[h] Published reports of possible sexual transmission of brucellosis can be found in Lancet **i**:773, 1983; Aten Primaria **8**:165-166, 1991; Lancet **337**:848-849, 1991; Lancet **347**:1763, 1996; Lancet **337**:14-15, 1991; Infection **11**:313-314, 1983; and Lancet **348**:615, 1996.

[i] See Lancet **337**:1290-1291, 1991.

[j] Published reports of possible sexual transmission of Q fever can be found in Clin.Infect.Dis. **22**:1087-1088, 1996; and Clin. Infect.Dis. **33**:399-402, 2001.

Field Identification and Decontamination of Toxins

Chad Tameling, BS

SET Environmental, Inc., Wheeling, Illinois

Contributions by Bijan Saeedi, MS

SET Environmental, Inc., Wheeling, Illinois

Hazardous materials emergency responders are faced with the challenge of identifying toxins n the field (biological, radiological, chemical) and cleaning them off of people, buildings, and the environment in as quick and efficient manner possible. Ramifications of not correctly identifying and decontaminating toxins could be life-threatening as well as monetarily damaging.

Over the years, SET has developed an operating procedure to achieve the goals of 1) quickly and accurately identifying the toxin and 2) effectively decontaminating people, property, and environment in a time-sensitive manner.

Basic theory of hazardous materials emergency response dictates that when responders arrive on the scene, prior to "going to work," they must know the identity of the toxin. Protective equipment selection, decontamination, remediation, and disposal technologies are dictated by toxin identification. Unfortunately, the luxury of knowing toxin identity is not always the case.

There are 3 categories of toxin identification methods SET implements in hazardous materials emergency response: 1) natural characteristics, 2) field characterization, and 3) mobile laboratory identification.

Natural Characteristics This is the quickest (and most crude) method of toxin identification, and is mostly applicable to chemical agents. Employing 2 of the 5 senses (sight, smell), emergency responders are sometimes able to get an idea of toxin identity. Please note, if an emergency responder is close enough to the scene to smell an unknown toxic release, he is too close! However, there are many instances when an incidental odor may be a telltale sign of toxin identity. Many chemical agents have characteristic odors (freshly mowed straw smell of phosgene, pungent choking odor of corrosives, sweet-smelling odor of aromatic compounds, garlic-like odor of arsine gas) that may lead to a tentative identification or general categorization of an unknown toxin.

Many chemicals have characteristic colors in their sedentary state or when they are reacting. When an emergency responder pulls up to the scene, he may observe brown smoke (nitrogen dioxide gas from reacting nitric acid) or see a bright yellow/orange powder (potential explosive compounds) or a green vapor cloud (chlorine gas leak) and be able to tentatively identify or categorize the unknown toxin. Although the natural characteristic method is simple and sometimes effective, it is not always conclusive. That is why the modern emergency responder employs field characterization techniques.

Field Characterization This method employs a combination of wet chemistry and real-time direct read instrumentation to identify the unknown toxin. Please note, if an emergency responder is close enough to the unknown toxin to employ field characterization, he should be donned in appropriate PPE (personal protection equipment) and have a decontamination system established (please see later sections). A well-equipped hazardous materials emergency response team should have a compliment of direct-read instrumentation including, but not limited to the following:
- Radiation survey meters (alpha, beta radiation at a minimum)
- Organic vapor analyzers (PID, FID)
- 4- or 5-gas meters (for oxygen, % flammability and toxic gas readings)

- Mercury vapor analyzer
- Inorganic gas (chlorine) detector

Direct-read instrumentation is complemented with wet chemistry. When responders are able to approach an unknown toxin, they can also use wet chemistry techniques to identify chemical families or characteristics of the material. In the past 2 years, with the recent focus on WMD and terrorism, there have been many manufacturers of "test kits" that can claim to positively identify unknown chemical agents. These test kits are still in the developmental stage, and have many false readings and/or interferences. There are other test kits on the market that test for a particular agent (VX, botulism). These kits are more accurate, however, they have a shelf-life and a responder could fill a 12-foot straight truck just with test kits if he had to get every agent-specific kit on the market. SET's approach is to employ simpler tests and use these findings, in conjunction with a solid understanding of chemistry, to characterize or identify unknown chemical toxins.

SET responds with the following wet chemistry field tests:

- pH
- Oxidizer potential
- Flammability
- Cyanide
- Sulfide
- Water solubility
- Sugar
- Chlorine
- Nitrate/nitrite/nitrogen

While these methods are sound and effective, they do not address today's threat of biological or natural occurring toxins. To meet this challenge, SET had to incorporate instrumentation for identification of unknown toxins.

Instrumentation Identification For many years prior to the WMD phenomenon, SET utilized Fourier transform infrared spectroscopy (FT-IR) and gas chromatograph/mass spectroscopy (GC/MS) as primary tools in positively identifying inorganic and organic solids, liquids, and gases. FT-IR spectra is used for identification of functional groups such as alcohols, amines, esters, and so on. The fingerprint portion of FT-IR spectrum is used to exactly identify a compound. The spectrum of an unknown compound can be searched against libraries of thousands of spectrums within seconds using a fast computer. Sometimes the size of a sample is so small that a good spectrum cannot be obtained using conventional FT-IR spectroscopy. FT-IR microscope equipped with nitrogen-cooled MCT (mercury cadmium telluride) detector is capable of giving a good spectrum of a sample as small as few microns.

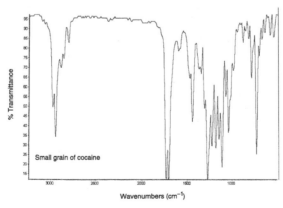

Small grain of cocaine

% Transmittance

Wavenumbers (cm⁻¹)

GC-MS is used for complex mixtures. In GC-MS, compounds are separated and identified by their mass fragments (mass spectrum). Following is an example of the mass spectrum of chloropicrin (trichloronitromethane).

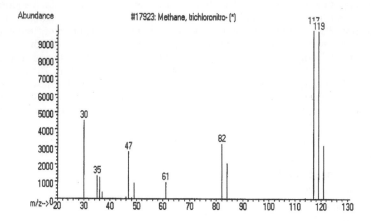

A graphic example of FT-IR spectrum obtained with microscope follows.

This instrumentation, combined with wet chemistry tests and other information obtained during the field survey, has enabled emergency responders to positively identify toxins in the field or in the laboratory setting for years.

Immediately following the World Trade Center attacks, the anthrax epidemic heightened the public awareness and concern about the threat of biological agents. The standard practice of preparing a sample, culturing, and growing it for 2-3 weeks was not acceptable. Businesses faced with the possibility of a bioagent attack in their factories or offices stood to lose tens of thousands of dollars each hour they were shut down!

SET responded by utilizing polymerase chain reaction (PCR) technology. Rapid replication of an unknown sample's DNA allowed to test against positive controls for well-known bioterrorist agents within hours rather than weeks. PCR testing in the field is relatively new, and there are a limited number of bioagent tests available, but it is the quickest, most accurate method available today. The positive result needs to be verified by conventional culture growth in the lab.

An example may help illustrate SET's methodology. The first example was recent news in the Chicago area when police arrested Dr. Chaos for storing sodium cyanide in the Chicago Metra train tunnels. The FBI had already obtained the sample. The sample originally thought to be cocaine came to SET as an unknown white powder. The following tests were administered:

- Water-soluble?
- pH of the material is around 8, not corrosive.
- FTIR spectra presented key functional group peak of cyanide at 2088 wave number (see attached spectra) identifying cyanide. In addition to cyanide, the sample contained sodium carbonate.
- Flame test of material presented bright orange flame identifying sodium salt.
- Wet chemistry test using pyridine barbituric acid confirmed the presence of cyanide.
- Unknown toxin is identified as sodium cyanide.

432

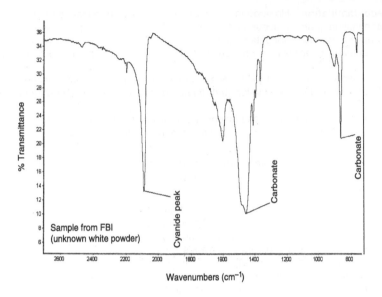

Wavenumbers (cm^{-1})

Nonintrusive Identification of Compounds

Sometimes a suspicious container is too dangerous to open for sampling; in this case, other methods may be employed.

To identify a substance through a metal shell, a method such as isotopic neutron spectroscopy (INS) is used. The metal shell is emitted by a neutron source (isotope of Californium 252) then the emitting ray from chemical contents is detected with a sensitive detector such as a nitrogen-cooled germanium detector.

For nonintrusive sampling through glass or plastic or even a powder in medical gel capsule, FT-Raman spectroscopy may be used.

FT-Raman Spectrum of Cyclohexane

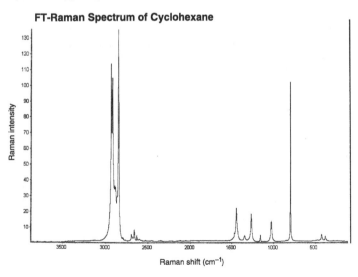

Raman shift (cm^{-1})

433

Field Decontamination   Hazardous materials emergency response operations mandate a decontamination plan in accordance with 29 CFR 1910.120(q). Emergency responders should have a standard decontamination plan that can be adjusted to site-specific situations prior to entering a hazardous environment. The basis of the plan should encompass the following:

1. What type of equipment will be needed to implement the decontamination?
2. What type of decontamination and decontamination agent will be effective against the toxin?
3. How many decontamination stations are appropriate for the situation.
4. Considerations when setting up a decontamination site.

Decontamination Equipment Selection   There are many brands and suppliers of decontamination equipment such as showers and absorbent pads. Emergency responders should have the following equipment:

- 6-mL poly sheeting for staging areas and ground cover
- Decon pools (plastic pools from K-Mart work well)
- Absorbent pads (universal and hydrocarbon)
- Low-pressure hand sprayers with wands
- Scrub brushes (soft/stiff bristle)
- Hand scrapers and putty knives
- Drums (55-gallon and 5-gallon poly and steel)
- 55-gallon capacity 2-mL poly bags (to hold contaminated personal effects)
- Duct tape (lots of it)
- Decon showers (make your own or buy online)
- Hose (garden hose, fire hose)
- Fittings for various water supplies (garden hose, Chicago, fire hydrant)
- Chemical decontamination agents (hypochlorite, citric acid, bicarbonate, surfactant)
- Collapsible bladders for holding decon rinse
- Submersible pumps for pumping decon rinse out of decon stations
- Stretchers/back boards/foldable cots for nonambulatory patients

Types of Decontamination and Decontamination Agents   There are two basic methods of decontamination: physical removal and chemical removal. There are many suggestions by OSHA, NIOSH, and EPA about gas/vapor sterilization, halogen stripping, thermal degradation, etc, but many of these methods are not practical from a time and money standpoint.

Physical decontamination (mechanically removing contaminant) is most applicable for gross decontamination of gooey tar-like material. High-volume water streams from a fog nozzle or pressure wash tip may mechanically remove material but produce overspray and lots of run off to be managed. Typical physical decontamination consists of:

- Scraping off goop with a tongue depressor or putty knife
- Wiping areas with absorbent pads
- Doffing contaminated clothing or PPE

Drums should be staged on 6-mL poly sheeting at the decontamination station. Contaminated PPE, clothes, and absorbent pads should be placed in the drums and sealed for proper disposal.

Chemical decontamination (inactivate contaminants by chemical detoxification or disinfection) is most applicable for highly soluble contaminants. The most readily available material (and probably most widely effective) is hypochlorite solution. The military uses two different concentrations of chlorine solution in decontamination procedures. A 0.5% solution is used for skin decontamination, and a 5% solution is used for equipment. Examples of other chemical decontamination agents are:

- Surfactants (ionic or nonionic)
- Detergents
- Citric acid solution
- Sodium bicarbonate solution
- Kerosene/light fuel oils
- D-limonene/alpha-pinene solution

Chemical decontamination solutions can be administered via hand sprayers or showers at appropriate stations. Showers produce much more runoff than hand sprayers and will require much more chemical since it is applied at a faster rate. Do not let victims freeze if outside temperatures are cold. Also, cold water from a fire hydrant can cause hypothermia in victims even in warm weather.

How Many Decontamination Stations Are Appropriate for the Situation? In the mid 80s and early 90s, emergency responders would try to establish decontamination as described by OHSA 29CFR 1910.120(q)(2)(vii) and spend 4-6 hours setting up decon pools, drop pads, and safety showers before any work in the hot zone occurred. In today's post 9/11 terrorism era, the 19-step decontamination stations are too cumbersome for time-sensitive situations where protecting life and property is critical.

EPA Suggested Maximum Decontamination Layout for Level A Protection

EPA has a suggested minimum decontamination layout, which is more applicable to the emergency responder.

435

Minimum Decontamination Layout for Level C Protection

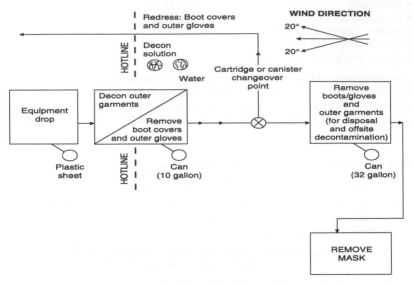

Experience and cost benefit analysis of spending time to decontaminate protective suits for reuse vs doffing protective suits and disposing of them has proven that disposable suits made much more sense on a time and money basis. For that reason, SET Environmental, Inc. has used a modified version of the EPA minimum decontamination layout. This modified decontamination layout works on the principle of "keeping it simple" and working with a 3-step process.

3-Step Process
1. Equipment drop and gross mechanical decontamination.
2. Quick physical decon by doffing suits and placing them in 55-gallon drums to be sealed and sent for offsite disposal.
3. Remove breathing apparatus (SCBA or APR) and place on 6-mL poly for staging, and finally remove inner gloves. Breathing apparatus can be decontaminated for reuse, and the inner glove drum shall be sealed and sent for offsite disposal.

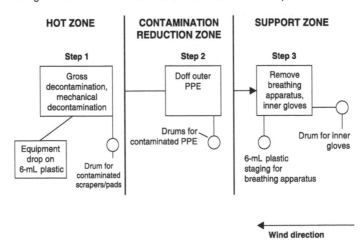

If necessary, Step 1 can be modified by adding a chemical decontamination stage. Sludge-like materials or acutely toxic materials may require dissolving/detoxification/neutralization before the responder should attempt to doff his suit. Chemical decontamination agents are applied with hand sprayers, collected, and pumped into portable tanks or drums for off-site disposal.

Emergency responders are able to progress through this decontamination scenario with little or no help from "decon techs" who traditionally had to be at each station to support decon operations. This allows more trained personnel to work the incident rather than man decon station.

The 3-step process can be adapted for emergency responders decontaminating non-ambulatory patients as well.

1. Gross decontamination/physical decontamination (remove clothes).
2. Chemical decontamination, addition of detoxification/neutralization chemicals to patients using hand sprayers.
3. Final rinse, using walk-through decontamination showers, which allow emergency responders to carry a litter/gurney/backboard with a patient on it.

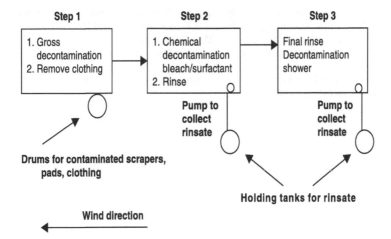

This scenario requires "decon techs" at each station, and also requires personnel to carry patients on stretchers through the system.

Considerations for Setting Up a Decontamination Site Rule number one in emergency response is to stage uphill and upwind of a hazardous materials incident.

1. The decontamination pad should be a safe distance from any potential hazardous exposures, secondary devices, or structural collapse.
2. Consider the topography of the location for natural drainage patterns or natural pools that could be used for runoff collection.
3. Are there buildings (parking garages, car washes) that could be used as shelters or triage/staging areas for incoming patients?
4. The closest utilities, such as water and electric, should be identified for potential use.
5. Establish staging area for additional resources (vehicles, ambulance) that coordinate with traffic-flow patterns to and from the site.

Verification of Decontamination In an emergency situation, people are moving quickly, working long hours, fatigued, and may not completely decontaminate absolutely everything.

This poses the threat that a residual hazard may be present. A rule of thumb is the more porous the material, the harder it is to decontaminate. Nonporous material (plastics, metal, enamel) can be readily cleaned. Materials like leather are going to readily absorb chemicals and slowly leach vapors over time. Equipment should be tested prior to being put back in service:

1. Articles should be placed in 2-mL plastic bags and sealed and allowed to sit in the sun for and heat for at least 30 minutes.
2. After 30 minutes, the bags should be slightly opened and an air monitor placed in the bags to read vapors in the bag. Wipe samples can be taken for certain contaminants and sent to a lab for analysis.
3. If residual contamination is found, the materials should be disposed of, or if the materials are critical, additional decontamination can be performed.
4. Porous materials (leather, canvas) can be additionally decontaminated by being soaked in water heated to 122°F to 131°F for 4-6 hours and then air dried without excess heat.
5. Metal or plastic can be decontaminated with a bleach slurry or hot soapy water. After 30 minutes, flush with water and aerate the item outdoors for several hours.

References

Emergency Response to Terrorism Job Aid, by Federal Emergency Management Agency, United States Fire Administration, National Fire Academy, and United States Department of Justice Office of Justice Programs, May 2000, 1st Ed.

Illinois Emergency Management Agency Chem-Bio Handbook, 1998, Jane's Information Group.

NIOSH/OSHA/USGC/EPA Occupational Safety and Health Guidance Manual for Hazardous Waste Site Activities, October 1985.

Acetylcholinesterase, Red Cell, and Serum

Synonyms Cholinesterase I; Cholinesterase; Erythrocytic; Cholinesterase; True; Erythrocyte Cholinesterase

Use Red cell cholinesterase is measured to diagnose organophosphate and carbamate toxicity and to detect atypical forms of the enzyme. Cholinesterase is irreversibly inhibited by organophosphate insecticides and reversibly inhibited by carbamate insecticides. Serum or plasma pseudocholinesterase is commonly used to measure acute toxicity, while erythrocyte levels are better for chronic exposure. (Serum level returns to normal before red cell levels.) Half-life of serum pseudocholinesterase is ~8 days, whereas that of acetylcholinesterase in red blood cells is 3 months. Persons with an atypical form of the enzyme (with low enzyme activity) exhibit prolonged apnea following the use of certain suxamethonium-type muscle relaxants in anesthesia (succinylcholine sensitivity - AA phenotype). These atypical forms may be detected by the use of fluoride or dibucaine inhibition.

The assay was used to evaluate sarin toxicity (*vide infra*).

Applies to Carbamate Toxicity; Nerve Agent Exposure; Organophosphate Toxicity; Succinyl-choline Sensitivity

Abstract The **red cell** enzyme (**true cholinesterase**) is specific for the substrate acetylcholine. The **serum** enzyme (**pseudocholinesterase**) hydrolyzes other choline esters.

Limitations Values decrease as erythrocytes age. Values are higher in younger red blood cells and reticulocytosis, and may mask the effect of acetylcholinesterase inhibition. Activity in red blood cells may not always provide a good index of intoxication with acetylcholine inhibitors.

Pseudocholinesterase in serum is the indicated test for succinylcholine sensitivity.

Specimen Red blood cells

Container Green top (heparin) tube or heparinized capillary tubes

Storage Instructions Stable at 4°C to 25°C for 1 week.

Reference Range Not well established, varies with method, age, sex, and use of oral contraceptives. Typical value: 30-50 units/g Hb.

Methodology Methods are based on determination of the rate of hydrolysis of an ester catalyzed by the enzyme acetylcholinesterase and include colorimetry, fluorometry, spectro-photometry-based systems.

Additional Information Cholinesterase activity is low at birth and higher in adult males than females. Because of the many constituent amino acids, many molecular variants are possible. The RBC level is increased in hemolytic states such as the thalassemias, spherocytosis, hemoglobin SS, and acquired hemolytic anemias. It is decreased in paroxysmal nocturnal hemoglobinuria and in relapse of megaloblastic anemia and it returns to normal with therapy. It is not widely regarded as useful as a test for paroxysmal nocturnal hemoglobinuria.

In patients poisoned by systemic insecticides (eg, organophosphates, carbamates, and nerve agents), both RBC acetylcholinesterase and plasma cholinesterase are usually inhibited. The effect on the serum enzyme is more marked, however, and serum levels are usually utilized in diagnosis and assessment of recovery. Recovery is best determined by looking for a plateau in erythrocyte cholinesterase activity. Toxic potential may vary, plasma versus red cell cholinesterase, such that in some cases erythrocyte levels may be needed for diagnosis and/or monitoring. If there is suspicion that a decrease in cholinesterase activity may not relate to the inhibitor effect of an organophosphate, then red cell level of acetylcholinesterase should be obtained. If both serum and RBC levels are significantly decreased, findings are those of exogenous toxic effect.

AchE activity use was studied following a terrorist attack in Tokyo subways in which terrorists used sarin. Systemic poisoning was apparently less likely to evolve when pupil size was normal on arrival. Miosis was perceived as a more sensitive index of exposure to sarin vapor then RBC AchE.

Anion Gap, Blood

Synonyms Electrolyte Gap; Gap; Ion Gap

Use Extensively used for quality control in the laboratory, the widest clinical application of the anion gap is in the diagnosis of types of metabolic acidosis. Unmeasured cations include calcium and magnesium. Unmeasured anions include protein, PO_4^{3-}, SO_4^{2-}, and organic acids. Organic acidosis includes lactic acidosis and ketoacidosis.

A marked elevation of anion gap, >30 mM/L, bears a strong implication of metabolic acidosis. Increased anion gaps are found in states such as renal failure and toxic exposures, ie, cyanide). A >30 mM/L gap increase commonly is secondary to lactic acidosis or ketoacidosis, but can also be caused by rhabdomyolysis or nonketotic hyperglycemic coma.

In diabetic ketoacidosis, plasma glucose is high, often much >300 mg/dL (SI: >16.7 mM/L), pH is <7.3, and ketones are found in blood and urine. Increased serum osmolality and increased calculated osmolality (osmolar gap) are found, and serum sodium is often decreased.

In alcoholic ketoacidosis glucose may be increased, normal or low, but a high alcohol level may be found, and amylase and uric acid may be increased.

Applies to Urinary Anion Gap

Test Includes A calculation from electrolytes, sodium, potassium, HCO_3^-, and chloride to ascertain quantities of unmeasured cations and anions

Abstract The anion gap is useful in evaluation of patients with acid-base abnormalities. The sum of anions and cations must be equal in the blood.

Limitations Minor differences in formula are used by different laboratories. A spurious increase may follow excessive exposure of the sample to room air, as well as, underfilling the Vacutainer® tube. Some gaps remain unexplained. Not all metabolic abnormalities are detected by abnormal gaps (eg, isopropanol ingestion is accompanied by normal gap, but ketone bodies are positive). There are a number of causes of normal anion gap acidosis associated with hyperchloremia. Anion gap is unsuitable as a quick screen for lactic acidosis. Still useful, the anion gap should not replace assay for lactate, creatinine, ketone bodies, or osmolality. In one study, only 66% of patients with an anion gap of 20-29 mM/L could be proven to have an organic acidosis.

Specimen Blood

Container Red top tube

Reference Range 6-16 mM/L (SI: 6-16 mM/L); slight differences may be established in different laboratories. Until recently electrolytes were done mostly by flame photometry. As ion-selective electrodes have come into wider use, reference ranges for anion gap will probably change.

Methodology Calculation: $(Na^+ + K^+) - (Cl^- + HCO_3^-)$ or $Na^+ - (Cl^- + HCO_3^-)$ = anion gap; actually determined by the difference between concentrations of anions and cations.

Additional Information Anion gap represents approximately the sum of the unmeasured anions, charges of which, with chloride and HCO_3^-, balance sodium. (Measured anions are chloride and bicarbonate. Measured cations are sodium.)

Anion gap high ("unmeasured anions"): With pH high: extracellular volume contraction; massive transfusion (with renal failure and/or volume contraction); carbenicillin, penicillin (large doses), salts of organic acids such as citrate. With pH low: uremia: most common cause; abnormal anion gap in uremia is usually seen only when creatinine is >4.0 mg/dL (SI: >354 μmol/L). Uremic acidosis is rare without hyperphosphatemia. Nonketotic hyperglycemic coma and rhabdomyolysis may cause high anion gap metabolic acidosis. Lactic acidosis and diabetic or alcoholic ketoacidosis characteristically fall into this group. With normal osmolal gap: salicylate and paraldehyde toxicity; with increased osmolal gap: methanol and ethylene glycol toxicity.

High anion gap metabolic acidosis without elevated lactic acid or acetone; consider: ketoacidosis with negative or slightly positive "acetone" if the patient is hypoxic and/or has alcoholic ketoacidosis, such ketoacidosis may be life-threatening; salicylate toxicity; methanol toxicity (paint thinners); ethylene glycol toxicity (antifreeze) - urinary sediment contains abundant calcium oxalate and/or hippurate crystals; paraldehyde intoxication (may have positive ketone reactions); toluene toxicity (transmission fluid, paint thinner inhalation or sniffing).

Anion gap low: Caused by retained unmeasured anions. The most common cause is hypoalbuminemia (eg, in nephrosis, cirrhosis), dilution, hypernatremia, very marked hypercalcemia, very severe hypermagnesemia, IgG myeloma and polyclonal gamma globulin increases - hyperviscosity with certain lab instruments, lithium toxicity, bromism (low anion gap may not be present). Decreased anion gap with spurious hyperchloremia and with hyponatremia is reported in hyperlipidemia. Dilution of extracellular fluid may cause a decreased gap. The finding of a low anion gap is perceived as an unreliable diagnostic parameter and may indicate potential laboratory error.

Normal anion gap may occur with **metabolic acidosis,** causes have been published. They include diarrhea, renal tubular acidosis, hyperalimentation, ureteroileostomy, ureterosigmoidostomy, external drainage of pancreaticobiliary fluids, NH_4Cl, and other drugs.

The urinary anion gap is used in the diagnosis of hyperchloremic metabolic acidosis and evaluation of renal potassium wasting.

Anthrax Detection

Use Detect the presence of *Bacillus anthracis* in clinical specimens from humans, grazing animals including cattle, goats, and sheep.

Test Includes Aerobic culture and Gram stain of specimen

Abstract Anthrax is caused by the bacteria *Bacillus anthracis*, and is one of the most serious biological agents that may be used as a weapon. The natural reservoir for this gram-positive nonmotile, nonhemolytic *Bacillus* is the soil. It exists in an infected host as a vegetative bacillus, and in the environment as a spore. Spores can remain viable for decades, and are the usual infective form. Bacterial culture for aerobic organisms will provide proper growth conditions for *Bacillus anthracis*; however, it is important to notify the laboratory if this organism is suspected.

Limitations Patients with inhalation anthrax become ill and often die rapidly. Sputum cultures are rarely positive.

Specimen Blood, sputum, cerebrospinal fluid, pleural fluid, stool, rectal swab, or cutaneous lesion fluid may be cultured; *vide infra*.

Container Sterile container or swab

Storage Instructions If transportation of swabs requires more than 1 hour, then store at 2°C to 8°C.

Causes for Rejection Nasal swabs and environmental samples (ie, powder) should not be tested by a routine clinical laboratory. Nasal swabs should **not** be used routinely to determine diagnosis or therapy.

Turnaround Time Negative results are typically reported as follows: blood and CSF: 5-7 days; sputum and wounds: 2 days. Preliminary morphologic information for positive cultures is usually available within 24 hours. Suspicious isolates should be forwarded to a state or local public health laboratory for complete identification. Analytic test time for rapid-cycle real-time PCR assays may be only 20-30 minutes.

Methodology Inoculation of sheep's blood agar, incubation of media at 35°C in ambient or CO_2-enhanced atmospheric conditions. Cultures are nonhemolytic.

Blood, impression smears from a lesion, or rarely, cerebrospinal fluid reveal short chains of large, encapsulated gram-positive bacilli.

Enzyme-linked immunosorbent assay (ELISA) and polymerase chain reaction (PCR) are available at reference laboratories. Immunohistochemical staining methods are available.

Rapid-cycle real-time PCR for anthrax has recently been described. LightCycler PCR instrumentation permits anthrax detection following autoclaving.

Additional Information Anthrax is a zoonotic infection that may present as cutaneous, pulmonary, or gastrointestinal disease. Although rare, human anthrax does occur throughout the world, including the U.S., and has usually been acquired by contact with contaminated herbivores or animal products such as hides, wool, or other fractions. Means of transmission include contact with inoculation of minor lesions, meat ingestion, handling, or inhalation.

Industrial processing of animal hair accounted for 153 (65%) of 236 anthrax cases reported tio the CDC from 1955-1999. While most of these cases were cutaneous anthrax, 10% were inhalation anthrax.

Methods to prevent inhalation anthrax from animal hide/hair exposure:

- Work with hides that are tanned or treated (air drying does not destroy B anthracis spores)
- Regular hand washing with soap and warm water
- Wearing durable protective gloves
- Working in a well-ventilated workplace
- Spores can be inactivated by heating hides to an internal temperature of 158°F (70°C) or placement in boiling water for over 30 minutes
- Remove clothing before leaving workplace and launder
- Clean workplace with high-efficiency particulate air vacuum
- Avoid vigorous shaking or beating hides

Because of its potential application as an agent of biological warfare, laboratories throughout the U.S. should have procedures in place to make a preliminary identification and to notify the proper authorities if a suspicious isolate is recovered. It is recommended that laboratories utilize biosafety level 2 facilities and practices for handling specimens and cultures for *Bacillus anthracis*.

Patients who progressed to the fulminant phase have a mortality rate of 97% (regardless of the treatment they received).

Specific References

Beatty ME, Ashford DA, Griffin PM, et al, "Gastrointestinal Anthrax: Review of the Literature," *Arch Intern Med*, 2003, 163(20):2527-31.

Ben-Noun LL, "Figs - The Earliest Known Ancient Drug for Cutaneous Anthrax," *Ann Pharmacother*, 2003, 37(2):297-300.

Blank S, Moskin LC, and Zucker JR, "An Ounce of Prevention Is a Ton of Work: Mass Antibiotic Prophylaxis for Anthrax, New York City, 2001," *Emerg Infect Dis*, 2003, 9(6):615-22.

Bottei EM, "It's Ricin and VX and Anthrax. Oops, Our Bad - It's Baking Powder," *J Toxicol Clin Toxicol*, 2004, 42(5):794.

Cantrell FL and Carlson T, "Impact of Biological/Chemical Terrorism on Poison Centers," *J Toxicol Clin Toxicol*, 2003, 41(5):681.

CDC Centers for Disease Control, "Inhalation Anthrax Associated With Dried Animal Hides - Pennsylvania and New York City," *MMWR*, 2006, 55(10):280-2.

Cetaruk EW, "Anthrax As a Biological Weapon: Recent Experiences in the United States," *J Toxicol Clin Toxicol*, 2002, 40(3):242-3.

Crupi RS, Asnis DS, Lee CC, et al, "Meeting the Challenge of Bioterrorism: Lessons Learned From West Nile Virus and Anthrax," *Am J Emerg Med*, 2003, 21(1):77-9.

American Medical Association; American Nurses Association-American Nurses Foundation; Centers for Disease Control and Prevention; Center for Food Safety and Applied Nutrition, Food and Drug Administration; Food Safety and Inspection Service, US Department of Agriculture, "Diagnosis and Management of Foodborne Illnesses: A Primer for Physicians and Other Health Care Professionals," *MMWR Recomm Rep*, 2004, 53(RR-4):1-33.

Eachempati SR, Flomenbaum N, and Barie PS, "Biological Warfare: Current Concerns for the Health Care Provider," *J Trauma*, 2002, 52(1):179-86.

Fine AM, Wong JB, Fraser HS, et al, "Is It Influenza or Anthrax? A Decision Analytic Approach to the Treatment of Patients With Influenza-Like Illnesses," *Ann Emerg Med*, 2004, 43(3):318-28.

Forrester MB and Stanley SK, "Calls About Anthrax to the Texas Poison Center Network in Relation to the Anthrax Bioterrorism Attack in 2001," *Vet Hum Toxicol*, 2003, 45(5):247-8.

Franz DR and Zajtchuk R, "Biological Terrorism: Understanding the Threat, Preparation, and Medical Response," *Dis Mon*, 2002, 48(8):493-564.

Holty JC, Bravata DM, Liu H, et al, "Systematic Review: A Century of Inhalational Anthrax Cases From 1900 to 2005," *Ann Intern Med*, 2006, 144:270-80.

"Human Anthrax Associated With an Epizootic Among Livestock - North Dakota," 2000, *MMWR Morb Mortal Wkly Rep*, 2001, 50(32):677-80.

Hupert N, Bearman GM, Mushlin AI, et al, "Accuracy of Screening for Inhalational Anthrax After a Bioterrorist Attack," *Ann Intern Med*, 2003, 139(5 Pt 1):337-45.

Inglesby TV, O'Toole T, Henderson DA, et al, "Anthrax As a Biological Weapon, 2002: Updated Recommendations for Management," *JAMA*, 2002, 287(17):2236-52.

Lanska DJ, "Anthrax Meningoencephalitis," *Neurology*, 2002, 59(3):327-34.

Leitman S, "Plasmapheresis of Anthrax Vaccinees for Production of An," *Crisp Data Base National Institutes of Health*, 2002.

Martin BL, Collins LC, Nelson MR, et al, "Immediate Hypersensitivity to Anthrax Vaccine: Three Cases," *J Allergy Clin Immunol*, 2000, 105:349.

Muniz AE, "Lymphocytic Vasculitis Associated With the Anthrax Vaccine: Case Report and Review of Anthrax Vaccination," *J Emerg Med*, 2003, 25(3):271-6.

Nicas M, Neuhaus J, and Spear RC, "Risk-Based Selection of Respirators Against Infectious Aerosols: Application to Anthrax Spores," *J Occup Environ Med*, 2000, 42(7):737-48.

Partridge R, Alexander J, Lawrence T, et al, "Medical Counterbioterrorism: The Response to Provide Anthrax Prophylaxis to New York City US Postal Service Employees," *Ann Emerg Med*, 2003, 41(4):441-6.

Roche K, McKay CA, and Bayer MJ, "Hospital Antidote Stocking Subsequent to the 2001 Terror Acts," *J Toxicol Clin Toxicol*, 2003, 41(5):719.

Rusnak J, Boudreau E, Bozue J, et al, "An Unusual Inhalational Exposure to *Bacillus anthracis* in a Research Laboratory," *J Toxicol Clin Toxicol*, 2004, 46(4):313-4.

Schultz CH, "Chinese Curses, Anthrax, and the Risk of Bioterrorism," *Ann Emerg Med*, 2004, 43(3):329-32.

Sikka R, Khalid M, Chae E, et al, "Impact of the Anthrax Bioterrorism Incidents of 2001 on Antibiotic Utilization," *Ann Emerg Med*, 2004, 44:S92.

Sox HC, "A Triage Algorithm for Inhalational Anthrax," *Ann Intern Med*, 2003, 139(5 Pt 1): 379-81.

Sternbach G, "The History of Anthrax," *J Emerg Med*, 2003, 24(4):463-7.

Subbarao IA, Johnson C, Bond WF, et al, "Creation and Pilot Testing of the Advanced Bioterrorism Triage Card," *Ann Emerg Med*, 2004, 44:S93.

Swanson-Biearman B and Krenzelok EP, "Delayed Life-Threatening Reaction to Anthrax Vaccine," *J Toxicol Clin Toxicol*, 2001, 39(1):81-4.

Swanson ER and Fosnocht DE, "Anthrax Threats: A Report of Two Incidents From Salt Lake City," *J Emerg Med*, 2000, 18(2):229-32.

"Use of Anthrax Vaccine in the United States: Recommendations of the Advisory Committee on Immunization Practices," *J Toxicol Clin Toxicol*, 2001, 39(1):85-100.

Wills BK, Leikin J, Weidner K, et al, "Analysis of Suspicious Powders in Northern Illinois Following the Post 9/11 Anthrax Scare," *Clin Toxicol (Phila)*, 2005, 43:696.

Arsenic, Blood

Synonyms As

Use Blood arsenic is for the diagnosis of acute poisoning only; use urine, hair, or nails for chronic poisoning. Acute arsenic poisoning may be signaled by the abrupt onset of vomiting and diarrhea.

Applies to Hair Analysis; Heavy Metal Screen, Arsenic

Abstract Arsenic is a toxic heavy metal. It exists in various forms. Arsine gas, As^{3+}, and As^{5+} are toxic forms; organic forms are not much less toxic. Arsenic is found in soil and rocks. Major sources of human exposure are arsenic in food resulting from broad use of arsenical insecticides and from drinking water, especially well water. Acute arsenic toxicity follows accidental ingestion, industrial accidents, suicide or homicide. For children, 2 mg/kg body weight can cause lethal arsenic poisoning. Adamsite and Lewisite, war gases of World War I, are arsenic compounds. In certain areas of the United States, fresh water supplies contain up to 1.4 mg/L arsenic, substantially in excess of the acceptable limit of 0.01 mg/L.

Limitations Short half-life in blood, 4-6 hours. Serum or plasma is the least useful specimen, except in acute poisoning.

Volume 30 mL

Minimum Volume 10 mL

Container Trace metal-free certified EDTA tubes

Causes for Rejection Containers not metal-free

Reference Range <70 mcg/L (SI: 0.93 µmol/L)

Critical Values Poisoning: ≥100 mcg/L (SI: ≥1.33 µmol/L)

Methodology Inductively-coupled plasma-mass spectrometry (ICP-MS), electrothermal atomic absorption spectrometry (AA)

Additional Information
- Half-life: 4-6 hours
- Volume of distribution: 0.2 L/kg

Arsenic can be absorbed through the gastrointestinal tract, by inhalation, and by penetration of the skin.

Arsine gas (AsH_3), combining with the globin chain in red cells, causes hemolysis with hemoglobinuria and hematuria. Acute renal failure may cause death.

Arsenic intoxication causes hypotension, tachycardia, conduction blocks, dysrhythmias, changes of mental status, rhabdomyolysis, pulmonary edema, encephalopathy, seizures, neuropathy, hepatic and renal dysfunction, hemolytic anemia, and bone marrow toxicity. Arsenic trioxide has provided remission of acute promyelocytic leukemia. Cell Therapeutics Inc. issued a "Dear Health Care Provider" letter reminding clinicians that QTc prolongation with torsade de pointes arrhythmia and sudden death have been associated with the use of arsenic trioxide.

Arsenic, Hair, Nails

Synonyms As; Quantitative; As^{3+} (As III); As^{5+} (As V)

Use Diagnose chronic arsenic exposure and intoxication. Complications of chronic exposure include carcinomas of skin; associations with carcinomas of lung and with transitional cell carcinoma of bladder and kidney.

Applies to Hair Analysis; Mee's Lines

Abstract Chronic toxicity manifests cutaneous changes, including alopecia, hyperkeratosis, hyperpigmentation, and tumors including basal and squamous cell carcinoma. See Arsenic, Serum or Plasma.

Limitations Hairs and nails do not detect recent exposure.

Specimen Clean hair or nails, ≥0.5 g

Volume 0.5 g

Container Clean envelope or heavy metal-free screw top plastic container

Collection Extreme care is necessary to avoid surface contamination. Hair from the nape of the neck indicates more recent ingestion (usually within 3 months), while axillary or pubic hair provides evidence of earlier exposure (6-12 months before). Toenails are preferable to fingernails; they are less prone to surface contamination.

Special Instructions Hair should be clean, free of oil and tonic; clip close. Nails should be thoroughly washed, dried, and clipped close to cuticle.

Reference Range Up to 1 mcg/g (SI: 0.13 nmol/g)

Critical Values Values >100 mcg/g (SI: >13.4 nmol/g) of hair are considered toxic; values >5 mcg/g of hair are consistent with exposure

Methodology Inductively-coupled plasma-mass spectrometry (ICP-MS), electrothermal atomic absorption spectrometry (AA)

Additional Information Chronic exposure to arsenic commonly involves insecticides, industrial sources, or contamination of food, water, soil, or medications. Evidence of chronic arsenic poisoning may be manifested 2-8 weeks after ingestion.

Arsenic, Urine

Synonyms As; Quantitative; Urine; As^{3+} (As III); As^{5+} (As V)

Use Evaluate recent exposure to arsenic, arsenic toxicity

Applies to Arsenate; Arsenic, Gastric Content; Arsenite; Dimethylarsine; DMA; MMA; Monomethylarsine

Test Includes Organic forms of arsenic can also be evaluated.

Abstract This toxic heavy metal appears in urine and stools. Its excretion rate in urine is used to determine toxicity. Arsenic exists in various inorganic and organic forms, of which arsine gas, As^{3+}, and As^{5+} are most toxic. As^{3+} and As^{5+} are partially detoxified to monomethyl-arsine (MMA) and dimethylarsine (DMA) and excreted in urine.

Limitations Spot levels, if normal, may not rule out arsenic poisoning. Seafood, particularly shellfish, can increase urinary As to as much as 2000 mcg/L.

Patient Preparation Patient should avoid seafood for 48 hours before collection is begun. Organic arsenic may be found especially in shellfish, cod, and haddock. Seafood may contain arsenic as high as 10 mg/lb.

Specimen 24-hour urine

Volume Entire collection

Container Acid-washed plastic container, no preservative, no metal cap or insert

Storage Instructions Refrigerate or freeze.

Reference Range Ranges for urine in organic arsenic levels can be variable among different laboratories. A general guideline is given: normal: <120 mcg/L (SI: >1.59 μmol/L); chronic exposure: 100-200 mcg/L (SI: 1.3-2.6 μmol/L).

Critical Values Toxic: >850 mcg/L (SI: >11.3 μmol/L) (inorganic)

Methodology Atomic absorption spectrometry (AA), inductively-coupled plasma-mass spectrometry (ICP-MS). Chromatography before analysis is used to distinguish between organic, nontoxic forms and inorganic, toxic forms.

Additional Information 25 mL acidified gastric washing is acceptable for arsenic analysis; gastric content normally contains no arsenic. Random urine samples are acceptable. Arsenic is radiopaque. Abdominal x-rays may prove helpful, but usually are not.

Specific References

Wang RY, Paschal DC, Osterloh J, et al, "Urinary-Speciated Arsenic Levels From Selected U.S. Regions," *J Toxicol Clin Toxicol*, 2004, 42(5):770-1.

Autopsy

Synonyms Necropsy; Postmortem Examination

Use A teaching instrument of great value, the autopsy remains a component of good medical care. A vital clinical quality control measure, it is a definitive monitor of quality of care given to the patient or by the medical system. The autopsy supports recognition of emerging medical entities, such as acquired immunodeficiency syndrome (AIDS), toxic shock syndrome, sudden infant death syndrome, hantavirus infection, Legionnaire disease, and Ebola virus infection, and provides recognition of changes and complications taking place in old ones.

Validated for over a century, the autopsy is a great deal more than an educational tool. It determines the cause and manner of death, determines severity of disease, detects unsuspected diseases, and represents an effort to preserve the quality of medical practice and to support excellence in medicine (quality assurance). It enhances medical knowledge, provides insight to understand pathogenesis and recognition of hereditary/familial diseases relevant to genetic counseling and possibly pertinent to surviving relatives, and permits diagnosis of contagious diseases and exposure to toxic entities. The autopsy provides valid comparisons between premortem and postmortem diagnoses. It produces valid statistics not otherwise available. Autopsies provide information on certain environmental or occupational exposures, provide data relevant to the public health and vital statistics, yield information on the success or failure of therapy, and provide data for evaluation of new procedures and new drugs not obtained by other means. Postmortem examination provides answers to clinically unresolved problems.

Autopsies support communication between the attending physician and the decedent's family and diminish family grief over unanswered questions.

Contraindications Improperly completed written permission

Applies to Cause of Death; Coroner's Case; Death Certificate; Disease Reporting; Medical Examiner's Case; Quality Assurance in the Practice of Medicine; Vital Statistics

Abstract The expression "autopsy" means to see for oneself, or to see with one's own eyes. The procedure provides analysis of cause of death; the nature, extent, and type of disease(s); the effects of therapy; and indications of genetic influences. It is an unparalleled support of quality assessment. It determines, for instance, the primary site of tumor, source of bleeding, presence of familial disorders, and major and minor findings that may or may not have been anticipated, may or may not have required treatment, and may or may not have contributed to the patient's demise.

Limitations Autopsies may be time-consuming, often are expensive but are not regularly reimbursed. The need for relevance has been expressed and emphasis on partial autopsies has been emphasized.

Specimen Specimens for culture, urine and blood for toxicology if indicated, solid tissues if indicated for drugs and toxins (brain, kidney, liver, striated muscle)

Container Gray top (sodium fluoride) tube for alcohol, drugs, carbon monoxide

Turnaround Time A 30-day standard presently exists.

Special Instructions Consent: A properly signed autopsy permit is usually required before a hospital necropsy can be performed. It should be informed and uncoerced. A valid permit must contain the signature of the highest ranking survivor in the next-of-kin lineage. A commonly used decreasing order of responsibility: spouse, adult children, parents, adult brothers and sisters, relatives, and then anyone who will accept responsibility for the body for purposes of burial. (State laws are not uniform.) Permits include name of decedent, date of birth, hospital number, date and time of death, authorizing signature(s) with relationship to the decedent, date and time permission given, and extent (complete autopsy or specific designated restrictions). Clinical information is always desirable, including the major clinical diagnoses, surgical procedures, recognized complications, a summary of the clinical cause and terminal episode. Compliance with appropriate state laws is, of course, necessary. Witnesses are often required to sign autopsy permits. It is usually the responsibility of the attending physician to obtain permission for the autopsy from the next-of-kin.

Moral and religious concerns have been effectively summarized.

It is desirable as well as courteous for the attending physician to discuss the clinical particulars with the pathologist before the dissection is begun. In regard to physician attendance at autopsy, some of the most important clinicopathologic correlations and in-depth investigations occur in autopsies at which clinicians attend. Topics are often discussed that are not in the medical record, and morphologic findings may provide immediate feedback to a clinician in a way unavailable to him/her from even a lengthy autopsy report. The importance of clinical history to guide postmortem examination has been recognized at least since the 18th Century. Clinician attendance at autopsy is always rewarding to everyone involved. Discussions between the pathologist and the floor nursing staff may prove invaluable before necropsies on hospital inpatients are begun. Particularly in autopsies on individuals who are not hospital inpatients, critical clinical facts are likely to be unavailable, fragmentary, or completely unknown. The presence of the attending physician at such autopsies may compensate to some degree for the lack of a well worked up chart. Except in Coroner's cases, it is usually the responsibility of the attending physician to complete the death certificate.

Methodology Gross organ examination, often with microscopy, with subsequent special procedures as indicated. Useful guides are published. Extensive methods and procedures are available. They include radiography, angiography, cytopathology, immunology, and molecular techniques.

Aerobic and anaerobic blood cultures and lung cultures are often taken. Tissue and catheter insertion sites may be cultured. Special microbiologic techniques (eg, DNA probes, PCR) are available. Collection of a tube of blood is recommended. It can be used to test for HIV, hepatitis C or B, or other agents or substances.

Toxicology and many other procedures are performed as indicated.

Blood and bile dried on a filter paper card can be used to provide postmortem metabolic screening for a subset of cases of sudden death, including instances of sudden infant death syndrome.

Additional Information

THE AUTOPSY MAY SUPPORT RISK MANAGEMENT

It may eliminate suspicion and provide reassurance to families; it may provide facts instead of conjecture, and support malpractice defense and reduce medicolegal claims as well as improve the quality of care.

QUALITY ASSURANCE AND THE AUTOPSY

Overreliance on contemporary procedures including imaging techniques occasionally leads to missed diagnoses. Of 176 autopsies in a published series, 44.9% uncovered one or more undiagnosed causes of death. The most often misdiagnosed or missed entities included infections, infarcts, and malignancies. The most frequently overlooked immediate causes of death in another study included pulmonary embolism and gastrointestinal hemorrhage. In a Belgian paper, the most frequent class I missed major diagnoses were fungal infections, cardiac tamponade, abdominal hemorrhage, and myocardial infarct.

Discrepancies between antemortem and postmortem diagnoses lead to skepticism about critical health statistics that are not based on autopsy findings. Cabot had questioned the accuracy of death certificates in 1912; such questions have persisted in the United Kingdom, in the United States, and in a number of other Western countries for good reasons. The vast inaccuracy of death certificates can lead to wrong priorities and wrong policies derived from faulty statistics.

It is not currently possible to predict which autopsy cases will have high yields. Overwhelming data document the need for the autopsy in quality assurance. Ideally cases to be autopsied should be randomly selected to exclude bias.

Guidelines for request for postmortem examination somewhat overlap forensic considerations and may include:
- death due to chemical, biological, or radiation attack
- unanticipated death
- obscure cause of death

- death occurring while the patient is being treated with a new therapy
- intraoperative or intraprocedural death
- death occurring within 48 hours after surgery or an invasive diagnostic procedure
- death related to pregnancy or within 7 days of delivery
- death during a psychiatric admission
- death in admitted infants and children

MEDICAL EXAMINER AND CORONER SYSTEMS; FORENSIC PATHOLOGY
In possible medical legal cases, the Medical Examiner or Coroner should be contacted before any suggestion regarding autopsy permission is made to the family of the deceased. Cases falling under the jurisdiction of the Medical Examiner or Coroner usually include all unnatural deaths including sudden unexpected or unexplained death, death due to accident and violence (death following injury immediately or delayed for an indefinite time), cases of suspected homicide or suicide, unusual or suspicious circumstances including poisoning, and cases falling in the public interest (ie, possible threat to public health such as meningitis or other contagious diseases). The Medical Examiner's or Coroner's Office should be notified in the event of all deaths in which a physician was not recently attending the patient (usually defined as the 48 hours prior to death) or in which the personal physician is unwilling to sign the death certificate. The Medical Examiner's or Coroner's office should be notified of unexpected deaths of infants and children and of all drug deaths, save for anesthetic deaths in most jurisdictions, or complications of diagnostic or therapeutic procedures. Deaths in custody (jail, psychiatric facilities, and other custodial situations) should elicit notification. Deaths due to unlawful termination of pregnancy, whether self-induced or otherwise, require notification. Deaths related to disease, injury, or toxins related to employment require notification. Such deaths need not be immediate. Investigation may be indicated for a body to be cremated or buried at sea, for unclaimed bodies, for deaths of operators of public conveyances while performing their duties, and for unexplained deaths of public officials. Persons who have knowledge of such deaths are usually required to notify the medical examiner or Coroner. Failure to do so may be a misdemeanor. Transfer of a body from another jurisdiction without medical death certification may be illegal, depending on local statutes and regulations.
The autopsy may resolve insurance questions, including addressing diagnoses of suicide or homicide.

HAZARDS OF THE AUTOPSY
Hazards of necropsies include tuberculosis, hepatitis B, acquired immune deficiency syndrome (AIDS), and Jakob-Creutzfeldt disease. Formaldehyde does not kill the etiologic agent of Jakob-Creutzfeldt disease and it may not kill mycobacteria promptly. So-called "universal precautions" can be observed but are time-consuming and do not provide absolute guarantees of safety. The CDC has recommended level 2 biosafety conditions, including hoods, for autopsy suites. A superb table of infectious agents of potential high risk is included in Collins and Hutchins.

MEDICAL EXAMINERS, CORONERS, AND BIOLOGIC TERRORISM

A Guidebook for Surveillance and Case Management From *MMWR*, "Recommendations and Reports," June 11, 2004/53(RR08); 1-27 [in public domain]

Biosafety Concerns

Autopsy Risks Biosafety is critical for autopsy personnel who might handle human remains contaminated with biologic terrorism agents. Tularemia, viral hemorrhagic fevers, smallpox, glanders, and Q fever have been transmitted to persons performing autopsies (i.e., prosectors); certain infections have been fatal. Infections can be transmitted at autopsies by percutaneous inoculation (i.e., injury), splashes to unprotected mucosa, and inhalation of infectious aerosols. All of the Category A pathogens are potentially transmissible to autopsy personnel, although the degree of risk varies considerably among these organisms.

Additionally, autopsies of persons who die as the result of terrorism-related infections might expose autopsy personnel to residual surface contamination with infectious material. For example, botulinum toxin has the potential to be inhaled by autopsy personnel if it is present on the body surface at the time of examination. Heavy surface contamination of the body is unlikely because of the incubation period for the majority of infectious agents and the likelihood that a victim will have bathed and changed clothes after exposure and before becoming symptomatic and dying. However, if such residual material (e.g., powder) is present, examination and specimen collection should be undertaken by using appropriate biosafety procedures to protect autopsy and analytic laboratory personnel from possible exposure to more concentrated infectious material.

Because human remains infected with unidentified biologic terrorism pathogens might arrive at autopsy without warning, basic protective measures described in this report should be maintained for all contact with potentially infectious materials. In addition to these measures, certain high-risk activities (e.g., use of oscillating saw) are known to increase the potential for worker exposure and should be performed with added safety precautions.

Autopsy Precautions Existing guidelines for biosafety and infection control for patient care are designed to prevent transmission of infections from living patients to care providers, or from laboratory specimens to laboratory technicians. Although certain biosafety and infection-control guidelines are applicable to the handling of human remains, inherent differences exist in transmission mechanisms and intensity of potential exposures during autopsies that require specific consideration.

As with any contact involving broken skin or body fluids when caring for live patients, certain precautions must be applied to all contact with human remains, regardless of known or suspected infectivity. Even if a pathogen of concern has been ruled out, other unsuspected agents might be present. Thus, all human autopsies must be performed in an appropriate autopsy room with adequate air exchange by personnel wearing appropriate personal protective equipment (PPE). All autopsy facilities should have written biosafety policies and procedures; autopsy personnel should receive training in these policies and procedures, and the annual occurrence of training should be documented.

Standard Precautions are the combination of PPE and procedures used to reduce transmission of all pathogens from moist body substances to personnel or patients. These precautions are driven by the nature of an interaction (e.g., possibility of splashing or potential of soiling garments) rather than the nature of a pathogen. In addition, transmission-based precautions are applied for known or suspected pathogens. Precautions include the following:
- airborne precautions — used for pathogens that remain suspended in the air in the form of droplet nuclei and that can transmit infection if inhaled;
- droplet precautions — used for pathogens that are transmitted by large droplets traveling 3–6 feet (e.g., from sneezes or coughs) and are no longer transmitted after they fall to the ground; and
- contact precautions — used for pathogens that might be transmitted by contamination of environmental surfaces and equipment.

All autopsies involve exposure to blood, a risk of being splashed or splattered, and a risk of percutaneous injury. The propensity of postmortem procedures to cause gross soiling of the immediate environment also requires use of effective containment strategies. All autopsies generate aerosols; furthermore, postmortem procedures that require using devices (e.g., oscillating saws) that generate fine aerosols can create airborne particles that contain infectious pathogens not normally transmitted by the airborne route.

PPE For autopsies, Standard Precautions can be summarized as using a surgical scrub suit, surgical cap, impervious gown or apron with full sleeve coverage, a form of eye protection (e.g., goggles or face shield), shoe covers, and double surgical gloves with an interposed layer of cut-proof synthetic mesh. Surgical masks protect the nose and mouth from splashes of body fluids (i.e., droplets >5 μm); they do not provide protection from airborne pathogens. Because of the fine aerosols generated at autopsy, prosectors should at a minimum wear N-95 respirators for all autopsies, regardless of suspected or known pathogens. However, because of the efficient generation of high concentration aerosols by mechanical devices in the autopsy setting, powered air-purifying respirators (PAPRs) equipped with N-95 or high-efficiency particulate air (HEPA) filters should be considered. Autopsy personnel who cannot wear N-95 respirators because of facial hair or other fit limitations should wear PAPRs.

Autopsy Procedures   Standard safety practices to prevent injury from sharp items should be followed at all times. These include never recapping, bending, or cutting needles, and ensuring that appropriate puncture-resistant sharps disposal containers are available. These containers should be placed as close as possible to where sharp items are used to minimize the distance a sharp item is carried. Filled sharps disposal containers should be discarded and replaced regularly and never overfilled.

Protective outer garments should be removed when leaving the immediate autopsy area and discarded in appropriate laundry or waste receptacles, either in an antechamber to the autopsy suite or immediately inside the entrance if an antechamber is unavailable. Handwashing is requisite upon glove removal.

Engineering Strategies and Facility Design Concerns   Air-handling systems for autopsy suites should ensure both adequate air exchanges per hour and correct directionality and exhaust of airflow. Autopsy suites should have a minimum of 12 air exchanges/hour and should be at a negative pressure relative to adjacent passageways and office spaces. Air should never be returned to the building interior, but should be exhausted outdoors, away from areas of human traffic or gathering spaces (e.g., air should be directed off the roof) and away from other air intake systems. For autopsies, local airflow control (i.e., laminar flow systems) can be used to direct aerosols away from personnel; however, this safety feature does not eliminate the need for appropriate PPE.

Clean sinks and safety equipment should be positioned so that they do not require unnecessary travel to reach during routine work and are readily available in the event of an emergency. Work surfaces should have integral waste-containment and drainage features that minimize spills of body fluids and wastewater.

Biosafety cabinets should be available for handling and examination of smaller infectious specimens; however, the majority of available cabinets are not designed to contain a whole body. Oscillating saws are available with vacuum shrouds to reduce the amount of particulate and droplet aerosols generated. These devices should be used whenever possible to decrease the risk of dispersing aerosols that might lead to occupationally acquired infection.

Vaccination and Postexposure Prophylaxis   Vaccines are available that convey protection against certain diseases considered to be potentially terrorism-associated, including anthrax, plague, and tularemia. However, these vaccines are not recommended for unexposed autopsy workers at low risk. Consistent application of standard safety practices should obviate the need for vaccination for *B. anthracis* and *Y. pestis*. In 2003, the U.S. Department of Health and Human Services (DHHS) initiated a program to administer vaccinia (smallpox) vaccine to first responders and medical personnel. In this context, persons who might be called on to assess remains or specimens from patients with smallpox should be included among this group.

The administration of prophylactic antibiotics to autopsy workers exposed to potentially lethal bacterial pathogens is sometimes appropriate. For example, autopsy personnel exposed to *Y. pestis* aerosols should consider receiving such treatment regardless of vaccination status. Similarly, because tularemia can result from infection with a limited number of organisms, an exposure to *F. tularensis* should also prompt consideration of antimicrobial prophylaxis. However, decisions to use antimicrobial postexposure prophylaxis should be made in consultation with infectious disease and occupational health specialists, with consideration made of vaccination status, nature of exposure, and safety and efficacy of prophylaxis.

Decontamination of Body-Surface Contaminants If human remains with heavy, residual surface contamination (i.e., visible) must be assessed, they should be cleansed before being brought to the autopsy facility and after appropriate samples have been collected in the field. Surface cleaning should be performed with an appropriate cleaning solution (e.g., 0.5% hypochlorite solution or phenolic disinfectant) used according to manufacturer's instructions. If the number of remains requiring autopsy is limited (i.e., one or two), cleaning of heavily contaminated remains can be undertaken in an autopsy facility that has the infrastructure, capacity, and hazardous materials (HAZMAT)-trained personnel to perform the cleaning safely. Heavily contaminated remains should not be brought to facilities where patient care is performed. Both personnel carrying contaminated remains and personnel occupying areas through which remains are being carried should wear PPE. HAZMAT personnel should perform large-scale decontamination outdoors in a controlled setting. To ensure mutual understanding of the roles and responsibilities of HAZMAT and death-investigation personnel in situations with contaminated remains, ME/Cs should develop response protocols with HAZMAT personnel before such an event occurs.

Waste Handling Liquid waste (e.g., body fluids) can be flushed or washed down ordinary sanitary drains without special procedures. Pretreatment of liquid waste is not required and might damage sewage treatment systems. If substantial volumes are expected, the local wastewater treatment personnel should be consulted in advance. Solid waste should be appropriately contained in biohazard or sharps containers and incinerated in a medical waste incinerator.

Storage and Disposition of Corpses The majority of potential biologic terrorism agents (*B. anthracis*, *Y. pestis*, or botulinum toxin) are unlikely to be transmitted to personnel engaged in the nonautopsy handling of a contaminated cadaver. However, such agents as the hemorrhagic fever viruses and smallpox virus can be transmitted in this manner. Therefore, Standard Precautions should be followed while handling all cadavers before and after autopsy.

When bodies are bagged at the scene of death, surface decontamination of the corpse-containing body bags is required before transport. Bodies can be transported and stored (refrigerated) in impermeable bags (double-bagging is preferable), after wiping visible soiling on outer bag surfaces with 0.5% hypochlorite solution. Storage areas should be negatively pressured with 9-12 air exchanges/hour.

The risks of occupational exposure to biologic terrorism agents while embalming outweigh its advantages; therefore, bodies infected with these agents should not be embalmed. Bodies infected with such agents as *Y. pestis* or *F. tularensis* can be directly buried without embalming. However, such agents as *B. anthracis* produce spores that can be long-lasting and, in such cases, cremation is the preferred disposition method. Similarly, bodies contaminated with highly infectious agents (e.g., smallpox and hemorrhagic fever viruses) should be cremated without embalming. If cremation is not an option, the body should be properly secured in a sealed container (e.g., a Zigler case or other hermetically sealed casket) to reduce the potential risk of pathogen transmission. However, sealed containers still have the potential to leak or lose integrity, especially if they are dropped or are transported to a different altitude.

ME/Cs should work with local emergency management agencies, funeral directors, and the state and local health departments to determine, in advance, the local capacity (bodies per day) of existing crematoriums, and soil and water table characteristics that might affect interment. For planning purposes, a thorough cremation produces approximately

3-6 pounds of ash and fragments. ME/Cs should also work with local emergency management agencies to identify sources and costs of special equipment (e.g., air curtain incinerators, which are capable of high-volume cremation) and the newer plasma incinerators, which are faster and more efficient than previous incineration methods. The costs of such equipment and the time required to obtain them on request should be included in state and local terrorism preparedness plans.

ME/C's Role in Biologic Terrorism Surveillance

ME/Cs should be a key component of population-based surveillance for biologic terrorism. They see fatalities among persons who have not been examined initially by other physicians, emergency departments, or hospitals. In addition, persons who have been seen first by other health-care providers might die precipitously, without a confirmed diagnosis, and therefore fall under medicolegal jurisdiction. Autopsies are a critical component of surveillance for fatal infectious diseases, because they provide organism-specific diagnoses and clarify the route of exposure. With biologic terrorism-related fatalities, organisms identified in autopsy tissues can be characterized by strain to assist in the process of criminal attribution.

Models for ME surveillance for biologic terrorism mortality include sharing of daily case dockets with public health authorities (e.g., King County, Washington, and an active symptom-driven case acquisition and pathology syndrome-based public health reporting system developed in New Mexico). Different areas of responsibility exist for ME/Cs regarding their role in effective surveillance for possible terrorism events. The following steps should be taken in local jurisdictions to enable ME/Cs to implement biologic terrorism surveillance:

• Death-investigation laws should be changed to enable ME/Cs to assume jurisdiction and investigate deaths that might constitute a public health threat, including those threats that are probably communicable.

• Any unexplained deaths possibly involving an infectious cause or biologic agent should be investigated to make etiology (organism)-specific diagnoses.

• Uniform standards for surveillance should be used. For example, the Med-X system developed in New Mexico uses a set of antemortem symptoms to determine autopsy performance. The system's syndromic approach to postmortem diagnosis allows alerting of public health authorities to specific constellations of autopsy findings that could represent infectious agents before the specific agent is identified. Diseases caused by biologic terrorism agents are rare.

• Electronic information and data systems should be designed to allow rapid recognition of excess mortality — incorporating the ability to assess possible commonalities among cases — and rapid communication/notification of such information to public health authorities who can use the information for effective response.

• Close working relationships should be developed between ME/Cs and local or state health departments to facilitate two-way communication that includes alerts to ME/Cs of possible outbreaks or clusters of nonfatal infectious diseases, which might have unrecognized fatal cases, and appropriate reporting by ME/Cs to public health authorities of notifiable disease conditions. Additionally, public health authorities should notify ME/Cs of the epidemiology of biologic terrorism-associated and other emerging infectious diseases in their community.

ME/C's Role in Data Collection, Analysis, and Dissemination

For public health surveillance, criminal justice, and administrative purposes, ME/Cs should promptly, accurately, and thoroughly collect, document, electronically store, and have available for analysis and reporting, case-specific death-investigation information. Initially, depending upon local resources and legal restrictions, all aspects of data management and use might not need to occur in-house. Recognizing that numerous entities use medicolegal death-investigation data, ME/Cs should establish collaborations with public health and law enforcement professionals to achieve the goal of complete, accurate, and timely case-specific death-investigation data. Advance planning and policy development should also clarify to whom such data may be released and under which circumstances. To facilitate this process, the following steps should be taken:

- Death-investigation information should be documented on standard forms that are consistent in content, at a minimum, with the Investigator's Death Investigation Report Form (IDIRF) and Certifier's Death Investigation Report Form (CDIRF).
- Death-investigation data should be stored in an electronic database consistent with, at a minimum, the content outlined in the Medical Examiner/Coroner Death Investigation Data Set (MCDIDS). These data elements should be updated periodically.
- Electronic death-investigation data sets should include the results of laboratory tests that are performed in the case in question.
- Entry of data into an electronic database should be prompt so that the database is current.
- Electronic databases should allow searching for and grouping of cases by disease or injury and circumstances of death.
- Electronic death-investigation data should be stored in open, nonproprietary formats so that it can be shared as needed.
- Death-investigation records should be stored in accordance with state or local regulations. Ideally, these records should be stored in perpetuity in a format that ensures future retrieval. The format or media of electronic records might require periodic updating.
- Mechanisms should be in place to ensure that electronic death investigation data can be shared with public health authorities, law enforcement agencies, and other death-investigation agencies while providing for appropriate confidentiality and control of the release of information to authorized personnel or organizations only.
- ME/Cs should have specific policies that outline the organizations and agencies that are authorized to receive death-investigation information and the conditions in which such information may be released.
- Policies and mechanisms should be in place to avoid releasing death-investigation information inappropriately and to avoid withholding information that should be available to the public.
- ME/C offices should consider establishing links with state/local public health agencies, academic institutions, or other health organizations to promote epidemiologic analysis and use of their medicolegal death-investigation data in an ongoing manner. Certain ME/C offices have determined that employing a staff epidemiologist is beneficial.

Jurisdictional, Evidentiary, and Operational Concerns

Federal Role On February 28, 2003, Homeland Security Presidential Directive 5 (HSPD-5) modified federal response policy. Under the new directive, the Secretary of Homeland Security is the principal federal official for domestic incident management. Pursuant to the Homeland Security Act of 2002 (Public Law 107-296), the Secretary of the U.S. Department of Homeland Security (DHS) is responsible for coordinating federal operations within the United States to prepare for, respond to, and recover from terrorist attacks, major disasters, and other emergencies. The Secretary will coordinate the federal government's resources used in response to or recovery from terrorist attacks, major disasters, or other emergencies if and when any one of the following four conditions applies: 1) a federal department or agency acting under its own authority has requested the assistance of the Secretary; 2) the resources of state and local authorities are overwhelmed and federal assistance has been requested by the appropriate state and local authorities; 3) more than one federal department or agency has become substantially involved in responding to the incident; or 4) the Secretary has been directed to assume responsibility for managing the domestic incident by the President.

HSPD-5 further stipulates that the U.S. Attorney General, through the FBI, has lead federal responsibility for criminal investigations of terrorist acts or terrorist threats by persons or groups inside the United States, or directed at U.S. citizens or institutions abroad, where such acts are within the federal criminal jurisdiction of the United States. The FBI, in cooperation with other federal departments and agencies engaged in activities to protect national security, will also coordinate the activities of the other members of the law enforcement community to detect, prevent, preempt, and disrupt terrorist attacks against the United States. In the event of a weapons of mass destruction (WMD) threat or incident, the local FBI field office special agent in charge (SAC) will be responsible for leading the federal criminal investigation and law enforcement actions, acting in concert with the principal federal officer (PFO) appointed by the U.S. Department of Homeland Security and state and local officials.

The FBI has a WMD coordinator in each of the agency's 56 field offices (Appendix B). These persons are responsible for pre-event planning and preparedness, as well as responding to WMD threats or incidents. ME/Cs are encouraged to contact their local FBI WMD coordinator before an incident to clarify roles and responsibilities, and ME/Cs should contact the coordinator in any case where concerns or suspicions exist of a potential WMD-related death.

The FBI assertion of jurisdiction at the scene of a terrorist event would not necessarily usurp (or relieve) ME/Cs from their statutory authority and responsibility to identify decedents and determine cause and manner of death. Such an arrangement is consistent with the performance of medicolegal death investigation where other federal crimes are involved. ME/Cs who conduct terrorism-associated death investigations should be prepared to present their medicolegal death investigation findings in federal court.

Public Health Agency Authority   State public health laws might establish the health department's specific authority to control certain aspects of operations, personnel, or corpses in a public health emergency. For example, the Center for Law and the Public's Health at Georgetown and Johns Hopkins Universities, at the request of CDC, has created a model state emergency health powers act for adoption by states. Different states have either enacted versions of this act or are in the process of introducing similar legislative bills. ME/Cs should know specifically how existing state laws might provide for the health department to take control and dictate the disposition of human remains (burial or cremation). A state's emergency health powers act might also provide for
- mandatory medical examinations for ME/C personnel;
- isolation and quarantine of the public or ME/C personnel;
- vaccination against and treatment for illnesses among ME/Cs; and
- control, use, and destruction of facilities.

ME/Cs and health departments should work together as part of the emergency planning process to determine which emergency health powers might be established by the health department and under what circumstances these might be invoked for each potential biologic terrorism agent. Determining how health departments and ME/C operations can best interact, including documenting concerns regarding the availability of death-investigation personnel and the control and disposition of human remains, should be emphasized. ME/Cs should take part in community exercises to clarify and practice their role in the emergency response process.

General Operations   In the majority of terrorism-associated scenarios, ME/Cs are responsible for identifying remains and determining the cause and manner of death. To that end, ME/Cs might need to enlist additional local, state, or federal assistance while maintaining primary responsibility for death investigation. ME/Cs should request this assistance from the local or state emergency operations center (EOC), as appropriate. The probable source of federal assistance is the Disaster Mortuary Operational Response Team (DMORT). However, DMORT has not yet developed capacity to respond to events precipitated by the release of biologic agents (further details regarding DMORT and other federal agencies are discussed in following sections).

Where possible, postmortem examinations for identifying remains and determining cause and manner of death should occur within the local or state jurisdiction where victims have died. Local resources dictate whether the statutory ME/C system can accomplish this with existing personnel and within existing facilities, or whether additional local, state, or federal assistance is necessary. Moving substantial numbers of human remains, particularly those contaminated by a biologic agent (known or unknown) to locations considerably distant from the scenes of death is neither feasible nor safe. Two potential strategies can be used to augment the biosafety capacity of local agencies having limited resources. One strategy would be to develop a mobile Biosafety Level 3 autopsy laboratory. Another strategy would be to develop regional Biosafety Level 3 autopsy centers that can handle cases from surrounding jurisdictions or states. A combination of the two approaches will probably achieve the best coverage of national needs.

Postmortem Examinations and Evidence Collection A large-scale biologic event might create more fatalities than combined local, state, and federal agencies can store and examine. Small or rural jurisdictions might be overwhelmed by a relatively limited number of fatalities, whereas larger state or city ME/C offices could conceivably process greater numbers of human remains. No formulas exist that can be used to determine in advance the autopsy rate and the extent of autopsy that might be needed. In the event of a biologic event, ME/Cs should perform complete autopsies on as many cases as feasible on the basis of case volume and biosafety risks. These autopsies should meet the standards that forensic pathologists usually meet for homicide cases. Conferring with the FBI and appropriate prosecutorial authorities early in the process will ensure that appropriate documentary and diagnostic maneuvers are employed that will support the criminal justice process. Similarly, interacting with public health authorities early in the death-investigation process should ensure that appropriate diagnostic evaluations are conducted to support the public health investigation and response.

After the etiologic agent has been determined, certain (or all) other potentially related fatalities can be selectively sampled to confirm the presence of the organism in question. ME/Cs should coordinate the decision to transition from complete autopsies to more limited examinations with law enforcement and public health professionals. Selective sampling could include skin swabs and needle aspiration of blood or other body fluids, tissues for culture, or biopsies of a particular tissue or organ for histologic diagnostic tests (e.g., immunohistochemical procedures and electron microscopy). The required specimens from a limited autopsy and the diagnostic procedures employed will be dictated by the nature of the biologic agent. Guidelines for targeted organs or tissues for culture or analysis were discussed previously. As with all homicides, chain-of-custody for specimens should be maintained at all times.

Whenever a complete autopsy is performed, the goals should be to 1) establish the disease process and the etiologic agent; 2) determine that the agent or disease is indeed the cause of death; and 3) reasonably rule out competing causes of death. When limited autopsies or external examinations are performed, ME/C personnel should

- identify the deceased;
- document the appearance of the body;
- establish that the presenting clinical symptoms and signs are consistent with the alleged etiologic agent;
- confirm the presence of the etiologic agent in the body;
- state with reasonable probability that the alleged agent was the underlying cause of death (e.g., inhalational anthrax infection); and
- state with reasonable probability the likely immediate cause of death (e.g., pneumonia, meningitis, or mediastinitis).

Forming a reasonably sound medical opinion regarding cause and manner of death can be accomplished with knowledge of the presenting syndrome and circumstantial events, external examination of the body, and testing of appropriate specimens to document the etiologic agent. For example, in a confirmed smallpox outbreak, identifying the deceased, externally examining the body and photographing the lesions, and obtaining samples from the lesions for culture or electron microscopy might be adequate.

Biologic evidence obtained at autopsy can be sent to local or state health department laboratories, and other physical evidence can be sent to the usual crime laboratory, unless otherwise instructed by the FBI. Laboratories within LRN, as described previously, are responsible for coordinating the transfer of evidence or results to the FBI, U.S. Attorney General, or local and state legal authorities, as appropriate. Consistent with routine practice, ME/Cs should document all evidence transfers adequately.

Cause and Manner of Death Statements   Death certificates are not withheld from the public record, even when the cause of death is terrorism-related. The cause of death section should be used to fully explain the sequence of the cause of death (e.g., "hemorrhagic mediastinitis due to inhalational anthrax"). If death resulted from a terrorism event, the manner of death should be classified as homicide. The "how injury occurred" section on the death certificate should be completed, and it should reflect how the infectious agent was delivered to the victim (e.g., "victim of terrorism; inhaled anthrax spores delivered in mail envelope"). The place of injury should be the statement of where (i.e., geographic location) the agent was received.

Reimbursement for Expenses and Potential Funding Sources

Additional funding for ME/Cs might be needed for either preparedness or use during an actual biologic terrorism event. ME/Cs should prepare financially for potential future terrorist events that might be similar to the anthrax attacks of October–November 2001. In crisis situations, funding is retroactive but no less a concern.

Preparedness funding can support multiple activities, including training of ME/Cs for large-scale terrorism events. Certain activities involving training of ME/Cs have occurred through DMORT, a program authorized by the DHHS Office of Emergency Preparedness to rapidly mobilize ME/Cs to respond to incidents of mass fatality. Preparedness funding can also support surveillance activities in ME/C offices. As part of the Bioterrorism Preparedness and Response cooperative agreements with state health departments, CDC has provided funding to New Mexico and other states that are pursuing ME/C surveillance systems as an enhancement to their traditional surveillance systems. The New Mexico Office of the Medical Investigator has been a recipient of this funding through the New Mexico Department of Health since the inception of the cooperative agreement program. This funding has supported development of specialized surveillance techniques for deaths caused by potential agents of biologic terrorism and recognition of ME/Cs as a key resource for all phases — early detection, case characterization, incident response and recovery — of a public health emergency response. CDC encourages pursuit of this enhanced (ME/C) surveillance capacity through cooperative agreements with states, if the state has made adequate progress with other critical capacity goals.

ME/Cs might obtain preparedness funding by integrating their response activities into the existing EOCs that have been established at selected state and county levels (integration of ME/C offices into this framework is discussed in Communications and the Incident Command System). When ME/C offices are integrated into the emergency response system, ME/Cs have an opportunity to make emergency management officials aware of ME/C emergency responsibilities and resource needs.

The sources of funding for consequence management, including medicolegal death investigation, will depend on the scope of the terrorism event. In events with a limited number of deaths, funding for activities related to the detection and diagnosis might remain at the office level. Because terrorism deaths are homicides, these deaths will contribute to an office's jurisdictional workload, and future planning for preparedness funding should be considered. Certain ME/C offices are already a part of the local or state public health department or are already affiliated with an EOC. ME/C offices, health departments, and

EOCs are strongly encouraged to forge links for effective preparedness and response and to participate in joint training exercises to maximize preparedness funding.

In events with multiple deaths, a federal emergency might be declared. As long as ME/Cs' offices are officially working through the state or local EOC, certain expenses associated with the response (e.g., cost of diagnostic testing) can be submitted to the Federal Emergency Management Agency (FEMA) for reimbursement. In the majority of localities, these requests for resources required for appropriate response during an event should be submitted through local emergency management agencies that are part of state and local EOCs. Costs will probably be covered by the agency that has jurisdiction over the disaster (e.g., FEMA). In cases where a presidential disaster declaration is made, testing costs, victim identification, mortuary services, and those services that are covered by the National Disaster Medical System (a mutual aid network that includes DHHS, the Department of Defense, and FEMA) are reimbursable under Emergency Support Function 8 (Health and Medical) of the Federal Response Plan (FRP).

Under FRP, FEMA covers 75% of reimbursement costs; the remaining 25% are covered by the state through emergency funds or in-kind reimbursement. FEMA also supports state emergency funds through the DHHS electronic payments management system. In an emergency, all requests for reimbursement flow from their point of origin, in this case from an ME/C, through the state EOC/emergency management agency, to FEMA. Before an event, ME/Cs should clarify the procedures to follow to ensure that they will be reimbursed for expenses incurred as part of their emergency response.

DMORT

DMORT is a national program that includes volunteers, divided into 10 regional teams responsible for supporting death investigation and mortuary services in federal emergency response situations involving natural disasters and mass fatalities associated with transportation accidents or terrorism. Team members are specialists from multiple forensic disciplines, funeral directors, law enforcement agents, and administrative support personnel. Each team represents a FEMA region. DMORT members are activated through DHS after mass fatalities or events involving multiple displaced human remains (e.g., a cemetery washout after a flood).

The primary functions of DMORT include the identification of human remains, evidence recovery from the bodies, recovery of human remains from the scene, and assisting with operation of a family assistance center. Whenever possible, identification of the bodies is made by using commonly accepted scientific methods (e.g., fingerprint, dental, radiograph, or DNA comparisons).

Upon activation, DMORT members are federal government employees. When DMORT is activated, representatives from DHS are also sent to manage the logistics of deployment. The FBI most commonly staffs the fingerprint section of the morgue. The Armed Forces DNA Identification Laboratory in Rockville, Maryland, has traditionally performed DNA analyses; the arrangements for this testing are negotiated separately with the local ME/C.

After a request for DMORT assistance has been made, one of two Disaster Portable Morgue Units (DPMUs) and DMORT staff are sent to the disaster site. DPMUs contain specialized equipment and supplies, prestaged for deployment within hours to a disaster site. DPMUs include all of the equipment required for a functional basic morgue with designated workstations and prepackaged equipment and supplies. DPMUs can operate at Biosafety Level 2, but do not have the ventilatory capacity necessary to protect prosectors and other nearby persons from airborne pathogens. DPMUs also contain equipment for site search and recovery, pathology, anthropology, radiology, photography, and information resources, as well as office equipment, wheeled examination tables, water heaters, plumbing equipment, electrical distribution equipment, personal protective gear, and temporary partitions and supports. DPMUs do not have the materials required to support microbiologic sampling. When a DPMU is deployed, members of the DPMU team (i.e., a subset of DMORT) are sent to the destination to unload the DPMU equipment and establish and maintain the temporary morgue. Additional equipment is required locally after DMORT activation. At a minimum, this equipment includes a facility in which to house the morgue equipment, a forklift to move the DPMU equipment into the temporary morgue facility, and refrigerated trucks to hold human remains.

ME/Cs can request DMORT response after a mass fatality or after an incident resulting in the displacement of a substantial number of human remains. ME/Cs should follow state protocols for DMORT requests. Typically, ME/Cs contact the state governor's office, which then requests DMORT from DHS. The request should include an estimate of how many deaths occurred (if known), the condition of the bodies (if known), and the location of the incident. When deployed, DMORT supports ME/Cs in the jurisdiction where the incident occurred. All medicolegal death investigation records created by DMORT are given to ME/Cs at the end of the deployment, and ME/Cs are ultimately responsible for all of the identifications made and the documents created pertaining to the incident.

DMORT-WMD Team   The DMORT-WMD team is composed of national rather than regional volunteers. The primary focus of DMORT-WMD is decontamination of bodies when death results from exposure to chemicals or radiation. DMORT-WMD is developing resources to respond to a mass disaster resulting from biologic agents. However, this team might have difficulty in responding to such an event if the deaths occur in multiple locations.

The major forensic disciplines (i.e., forensic dentistry, forensic anthropology, and forensic pathology) as well as funeral directors, law enforcement, criminalists, and administrative support persons are represented on the DMORT-WMD team. Members of DMORT-WMD undergo specialized training that focuses on chemical and radiologic decontamination of human remains. The DMORT-WMD unit has separate equipment, stored separately from the DPMU, including PPE (up to and including level A suits), decontamination tents, and equipment to gather contaminated water. DMORT-WMD teams are requested and deployed in the same manner as general DMORTs.

Communications and the Incident Command System

ME/Cs are key members of the biologic terrorism detection and management response team in any community and should be integrated into the comprehensive communication plan during any terrorism-associated event. Routine and consistent communication among ME/Cs and local and state laboratories, public health departments, EOCs, communication centers, DMORT, and other agencies, is critical to the success of efficient and effective biologic terrorism surveillance, fatality management, and public health and criminal investigations. Planning for different emergency scenarios and participation in disaster response exercises are necessary to ensure effective response to a terrorism event.

Each state and certain counties have some type of emergency operation center that has been organized to provide a coordinated response during a terrorism event. ME/Cs should verify their jurisdiction's EOC contact point and work with them periodically regarding concerns related to preparedness and response.

All EOCs follow the Incident Command System (ICS), an internationally recognized emergency management system that provides a coordinated response across organizations and jurisdictions. The ICS structure allows for individual EOC decision making and different information flow in each state. ME/Cs should determine how the EOC functions in their jurisdiction.

Each ICS is composed of a managing authority that directs the response of health department, law enforcement, and emergency management officials during a planning exercise, emergency, or major disaster. In addition to assessing the incident and serving as the interagency contact, ICS also coordinates the response to information inquiries and the safety monitoring of assigned response personnel. The ICS organizational framework, includes planning, operations, logistics, and finance/administration sections. ME/Cs are most likely to participate in the operations team, which makes tactical decisions regarding the incident response and implements those activities defined in action plans. This team might also include public health, emergency communications, fire, law enforcement, EMS, and state emergency management agency staff.

During a suspected terrorism event, ME/Cs should be responsible for the following actions to facilitate communication:

• Promptly inform laboratory, public health, and law enforcement personnel of findings of investigations of suspected biologic terrorism-related deaths as well as personnel needs and new developments. To expedite information exchange, ME/Cs should familiarize themselves with the appropriate contact persons and agencies for response in their jurisdictions.

• Answer the EOCs' requests to collect and report data in a timely manner.

• Coordinate communication of their activities with the state emergency management agency and EOCs for their jurisdiction to avoid release of confidential or speculative information directly to the public or media.

Disclaimer References to non-CDC sites on the Internet are provided as a service to *MMWR* readers and do not constitute or imply endorsement of these organizations or their programs by CDC or the U.S. Department of Health and Human Services. CDC is not responsible for the content of pages found at these sites. URL addresses listed in *MMWR* were current as of the date of publication.

Table 1. Matrix of Autopsy Pathologic Syndromes and Potential Terrorism-Related Illnesses* or Agents

Illness or Agent	Autopsy Pathologic Syndrome
Plague, tularemia, Q fever, inhaled staphylococcal enterotoxin B, ricin	Community-acquired pneumonia: diffuse alveolar damage (ARDS)
Smallpox, viral hemorrhagic fevers, T-2 mycotoxins	Diffuse rash
Plague, tularemia, anthrax, viral hemorrhagic fevers, T-2 mycotoxins	Sepsis syndromes (i.e., disseminated intra-vascular coagulopathy [DIC]
Anthrax	Hemorrhagic mediastinitis or meningitis
Brucellosis, viral hemorrhagic fevers	Hepatitis, fulminant hepatic necrosis
Venezuelan equine encephalitis and other equine encephalomyelitis agents	Encephalitis, meningitis
Viral hemorrhagic fever (Lassa)	Pharyngitis, epiglottitis and other upper airway infections
Cutaneous anthrax, bubonic plague, tularemia	Soft tissue infections — cellulitis, abscess, necrotizing fasciitis
Escherichia coli and *Shigella colitis*, gastrointestinal anthrax	Hemorrhagic colitis

* Adapted from Med-X, New Mexico Surveillance Program.

U.S. death investigation systems by state, 2001

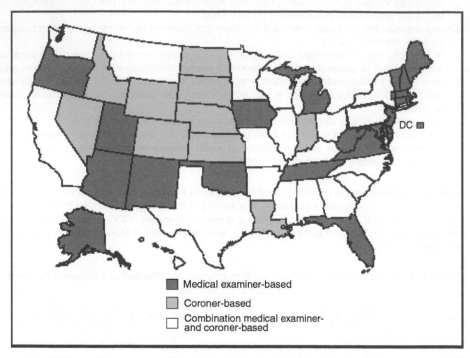

Medical Examiners, Coroners, and Biologic Terrorism: A Guidebook for Surveillance and Case Management

Classification of Biologic Terrorism Agents

Category A Agents
- Variola major (smallpox)
- *Bacillus anthracis* (anthrax)
- *Yersinia pestis* (plague)
- *Clostridium botulinum* toxin (botulism)
- *Francisella tularensis* (tularemia)
- Hemorrhagic fever viruses, including
  — Filoviruses including Ebola and Marburg hemorrhagic fever
  — Arenaviruses, including Lassa (Lassa fever) and Junin (Argentine hemorrhagic fever) and related viruses

Category B Agents
- *Coxiella burnetii* (Q fever)
- *Brucella* species (brucellosis)
- *Burkholderia mallei* (glanders)
- Alphaviruses including Venezuelan encephalomyelitis and eastern and western equine encephalomyelitis viruses
- Ricin toxin from *Ricinus communis* (castor beans)
- Epsilon toxin of *Clostridium perfringens*
- *Staphylococcus* enterotoxin B
- Food- and waterborne pathogens
  — *Salbmonella* species
  — *Shigella dysenteriae*
  — *Escherichia coli* O157:H7
  — *Vibrio cholerae*
  — *Cryptosporidium parvum*

Category C Agents
- Nipah virus
- Hantaviruses
- Tickborne hemorrhagic fever viruses
- Tickborne encephalitis viruses
- Yellow fever virus
- Multidrug-resistant *Mycobacterium tuberculosis*

Medical Examiners, Coroners, and Biologic Terrorism: A Guidebook for Surveillance and Case Management

Process for submitting specimens containing suspected Gategory A, B, or C* biologic agents (except smallpox virus) for testing within the Laboratory Response Network (LRN)

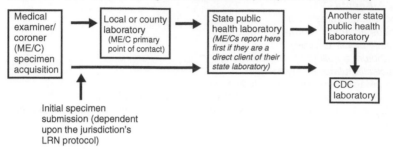

*Note: Dependent upon the LRN-designated capacity (Level A, sentinel; Level B, core; Level C, advanced), laboratory confirmation might occur on-site or require referral to the next higher-level laboratory for confirmatory testing or correct biocontainment.

Specific References

Burton EC and McPhee SJ, "Autopsy Overview," *Autopsy Performance and Reporting*, 2nd ed, Chapter 1, Collins KA and Hutchins GM, eds, Northfield, IL: College of American Pathologists, 2003, 3-12.

Collins KA and Cina SJ, "Ancillary Studies for Autopsy Pathology," *Autopsy Performance and Reporting*, 2nd ed, Chapter 24, Collins KA and Hutchins GM, eds, Northfield, IL: College of American Pathologists, 2003, 229-38.

Collins KA and Hutchins GM, *Autopsy Performance and Reporting*, 2nd ed, Northfield, IL: College of American Pathologists, 2003.

DiMaio VJ and DiMaio DJ, *Forensic Pathology*, 2nd ed, Boca Raton, FL: CRC Press, 2001.

Finkbeiner WE, Ursell PC, and Davis RL, *Autopsy Pathology: A Manual and Atlas*, New York, NY: Churchill Livingstone, 2003.

Hutchins GM, Berman JJ, Moore GW, et al, "Practice Guidelines for Autopsy Pathology," *Arch Pathol Lab Med*, 1999, 123(11):1085-92.

Lundberg GD, "Now Is the Time to Emphasize the Autopsy in Quality Assurance," *JAMA*, 1988, 260(23):3488.

Mayo Reference Services, "The Metabolic Autopsy: Postmortem Screening in Cases of Sudden, Unexpected Death," *Mayo Communiqué*, 2003, 28(9):1.

Nichols L, Dunn J, and Presnell SE, "Sampling of Microorganisms During the Autopsy," *Autopsy Performance and Reporting*, 2nd ed, Chapter 25, Collins KA and Hutchins GM, eds, Northfield, IL: College of American Pathologists, 2003, 239-44.

Nichols WS and Geller SA, "High Risk Autopsy Cases," *Autopsy Performance and Reporting*, 2nd ed, Chapter 12, Collins KA and Hutchins GM, eds, Northfield, IL: College of American Pathologists, 2003, 93-104.

Noite KB, "Medical Examiners and Bioterrorism," *Am J Forensic Med Pathol*, 2000, (4): 419-420.

Pellegrino ED, "Moral and Religious Concerns About the Autopsy," *Autopsy Performance and Reporting*, 2nd ed, Chapter 4, Collins KA and Hutchins GM, eds, Northfield, IL: College of American Pathologists, 2003, 27-36.

Randall BB, Fierro MF, Froede RC, et al, "Forensic Pathology," *Autopsy Performance and Reporting*, 2nd ed, Chapter 7, Collins KA and Hutchins GM, eds, Northfield, IL: College of American Pathologists, 2003, 55-64.

Roosen J, Frans E, Wilmer A, et al, "Comparison of Premortem Clinical Diagnoses in Critically III Patients and Subsequent Autopsy Findings," *Mayo Clin Proc*, 2000, 75(6):562-7.

Volmar KE, "History of the Autopsy Technique," *Autopsy Performance and Reporting*, 2nd ed, Chapter 3, Collins KA and Hutchins GM, eds, Northfield, IL: College of American Pathologists, 2003, 17-25.

Wagner SA, *Color Atlas of the Autopsy*, Boca Raton, FL: CRC Press, 2003.

Babesiosis Serology

Synonyms Nantucket Fever Serological Test

Use *Babesia* species serological test is used to diagnose babesiosis

Abstract *Babesia microti* and *Babesia divergens* are tick-borne intraerythrocytic protozoans, which can cause symptoms resembling those of *Plasmodium falciparum*. Like malaria, babesiosis causes hemolytic anemia. Asplenic, immunocompromised, and elderly subjects are especially at risk, but immunocompetent persons can develop the disease. These organisms also cause disease in cattle, including Texas fever. Babesiosis is enzootic in Southern New England, Southern New York, Wisconsin, and Minnesota.

Limitations False reactivity may be seen in patients with malaria. Patients exposed to *Babesia* on the West Coast of the U.S. should be tested for antibodies to the *Babesia* WA1 species rather than the *Babesia microti* antigen due to the lack of cross-reactivity between these organisms.

Specimen Serum, blood for smear

Volume 10 mL

Container Red top tube

Storage Instructions Refrigerate at 4°C.

Reference Range Negative

Critical Values Titers >1:128 are considered consistent with infection. A fourfold increase in titer establishes diagnosis.

Methodology Indirect immunofluorescent antibody (IFA), serum. Note: Intraerythrocytic ring forms and tetrads can be identified in the peripheral blood film. The ring forms resemble those of *P. falciparum* malaria, but the rare tetrad forms are diagnostic of babesiosis.

Additional Information The geographic distribution of babesiosis is worldwide. Babesiosis is transmitted in nature by hard-bodied ticks. The ticks become infected by feeding on various infected vertebrate animals (cattle, deer, moles, and mice). *Babesia* can also be transmitted to humans by blood transfusions. *Babesia divergens* is the most common species reported in Europe while *B. microti* is the agent most frequently identified in the U.S. Recently, two variants of *Babesia* species have been reported in the U.S. One variant is from the states of Washington and California (WA1 type) and the other is from Missouri (MO-1).

Babesiosis is particularly severe in patients who have undergone splenectomy (and who lack the RBC "pitting" function of the spleen). It may be potentially life-threatening in immunosuppressed persons or those of advanced age. Infections in individuals with normal spleens may often be asymptomatic. The Northern deer tick, *Ixodes scapularis* (*Ixodes dammini*), which transmits one of the *Babesia* species (*B. microti*, a rodent parasite) also transmits Lyme disease. Subjects with either disease should be considered for the other. Concurrent babesiosis and ehrlichiosis has been reported.

Pantanowitz et al have recently outlined helpful diagnostic features in the peripheral blood film.

Bacterial Culture, Aerobes

Synonyms Aerobic Bacterial Culture

Use Recover and identify obligately aerobic and facultatively anaerobic bacteria suspected of causing infections and provide a guide for therapy

Applies to *Neisseria* sp; *Pseudomonas* spp; *Staphylococcus aureus*; Bacterial Culture, Abscess; Enterobacteriaceae Culture

Test Includes Culture of aerobic or facultative anaerobes contributing to an infectious process. Antimicrobial susceptibility testing is commonly needed. Cultures for fungi and mycobacteria should be requested if indicated. Bacterial culture on tissue, abscess material, or sterile fluids should include a Gram stain as well as bacterial cultures.

Abstract Culture for aerobic bacteria utilizes methods capable of detecting obligately aerobic (those incapable of reproducing in the absence of oxygen) and facultatively anaerobic (those capable of reproducing in the presence or absence of oxygen) organisms. Obligately anaerobic organisms (those incapable of reproducing in the presence of oxygen) will not be recovered in aerobic cultures.

The overwhelming majority of bacterial pathogens are facultatively anaerobic organisms (eg, all streptococci, enterococci, staphylococci, members of the family Enterobacteriaceae, *Haemophilus influenzae*, *Pasteurella multocida*, *Vibrio* spp). There are relatively few obligately aerobic bacteria that are human pathogens; those that are commonly encountered include *Pseudomonas* spp and most *Neisseria* spp.

Most abscesses are caused by mixed bacterial growth that includes both aerobic and anaerobic bacteria. Initial antibiotic coverage should be broad-spectrum to cover these possibilities. Routine culture will not detect slow-growing microorganisms such as mycobacteria.

A useful overview of aerobic microbiology is that provided by Pezzlo.

Limitations It is often difficult to distinguish contaminating organisms from etiologic agents of infection. Organisms most likely to contaminate specimens include *Corynebacterium* species, coagulase-negative staphylococci, alpha-hemolytic streptococci, *Propionibacterium* species, and *Bacillus* species. These organisms are not invariably contaminants, however, and may be pathogenic in certain settings. Occasionally, organisms may grow slowly, be difficult to isolate in pure culture, or be difficult to identify. Fastidious bacteria may not be recovered despite significant efforts to collect and properly submit a specimen. A Gram stain should be performed on certain specimens (sputum, tissue, abscess material, etc) if sufficient material is obtained, to provide early presumptive information and to help interpret culture results.

Patient Preparation For collection of abscess material or tissue, the site is decontaminated (surgical soap and 70% isopropyl alcohol) to eliminate potentially contaminating aerobic and anaerobic bacteria that colonize many surfaces. Adjacent skin surfaces must not be touched.

Specimen Body site, tissue, or fluid associated with infection; fluid, pus, or other material properly obtained from an abscess

Container Specimens may be aspirated into a syringe and capped; all air should be expelled from the syringe prior to transport. Clinical material may be transferred from the syringe to commercially available vials. Specimens may also be transferred to sterile containers. Specimens in syringes with needles are not acceptable because of concerns about needlestick injuries. Swabs are inferior specimens.

Collection Contamination with normal flora must be avoided. Ideally, pus obtained by needle aspiration through an intact surface, which has been aseptically prepared, is put directly into a transport device or transported directly to the laboratory in the original syringe. Sampling of open lesions is enhanced by deep aspiration using a sterile needle and syringe. Curettings of the base of an open lesion may also provide a good yield. Irrigation should be done with nonbacteriostatic sterile saline. Pulmonary samples are obtained by transtracheal percutaneous needle aspiration by trained physicians or by use of a special sheathed catheter. If swabs must be used, two should be collected; one for culture and one for Gram stain. Specimens collected and transported in syringes should be transported to the laboratory within 30 minutes of collection.

The portion of the biopsy specimen submitted for culture should be separated from the portion submitted for histopathology by the surgeon or pathologist, utilizing sterile technique. Bedside inoculation of sterile body fluid into blood culture bottles improves sensitivity.

Causes for Rejection Specimens exposed to air, refrigerated specimens, or those having excessive delay in transit have a suboptimal yield; specimens in fixative

Turnaround Time Varies with specimen. Final reports of negative results are typically available as follows: sputum and wounds: 2 days; aseptically obtained body fluids: 5 days; tissue and abscess material: 2 days. Preliminary information from positive cultures is usually available within 24 hours; speciation and antimicrobial susceptibility is often available

24 hours later, but certain organisms require more time to identify. If actinomycosis is suspected clinically, the specimen may be held for 2 weeks.

Methodology Inoculation of microbiological media, incubation of media at temperatures varying from 25°C to 42°C (usually 35°C) in ambient or CO_2-enhanced atmospheric conditions

Additional Information *Staphylococcus aureus* causes community- and hospital-acquired infections that are increasing in frequency. The emergence of multidrug-resistant strains has made therapy more difficult. These gram-positive cocci cause diseases, which include toxic shock syndrome, staphylococcal scalded skin syndrome, and food poisoning. Metastatic infections include spread to the skeleton, kidneys, and lungs. *S. aureus* is among the most common pathogens causing sepsis.

S. aureus is the most frequent organism in osteomyelitis. Other causes include coagulase-negative staphylococci, *Proprionibacterium*, *Enterobacteriaceae*, *Pseudomonas aeruginosa*, streptococci, anaerobic bacteria, *Salmonella*, *Streptococcus pneumoniae*, *Bartonella henselae*, *Pasteurella multocida*, *Eikenella corrodens*, *Aspergillus*, *Mycobacterium avium* complex, *Mycobacterium tuberculosis*, *Candida albicans*, *Brucella*, *Coxiella burnetii*, and various fungi.

Pneumonia caused by gram-negative bacilli may be community- or hospital-acquired. The former patient group is composed almost entirely of individuals with chronic diseases (eg, chronic obstructive pulmonary disease, alcoholism, or malignant disease). The second (nosocomial) group is principally found in subjects whose pneumonia is secondary (eg, the postoperative state). Pneumonias caused by gram-negative bacilli are commonly related to aspiration. Blood cultures are useful in diagnosis, especially in those with community-acquired infections. The yield of positive cultures from effusion fluid is about 30%. Bronchoalveolar lavage or bronchial brush techniques are useful.

Bacterial Culture, Anaerobes

Use Recover and identify obligately anaerobic bacteria suspected of causing infections. Cultures from a variety of clinical settings should be accomplished to provide both aerobic and anaerobic culture (eg, those from biopsies and body fluids, abscesses, wounds, and bites).

Contraindications It is usually not relevant to seek anaerobes in acute cholecystitis, acute osteomyelitis, acute otitis media, acute sinusitis, appendicitis, bronchitis, cystitis, meningitis, pharyngitis, primary peritonitis, pyelonephritis, or superficial skin lesions.

Applies to Bacterial Culture, Abscess; Body Fluid Culture

Test Includes Culture for anaerobic bacteria; Gram stain is usually performed. Cultures for fungi and mycobacteremia should be requested if indicated.

Abstract Culture for anaerobic bacteria utilizes methods capable of detecting obligate anaerobes (those incapable of reproducing in the absence of oxygen). Facultatively anaerobic (those capable of reproducing in the presence of oxygen) organisms are also recovered in these cultures, but the primary purpose of anaerobic cultures is to recover obligate anaerobes. Anaerobic bacteria most likely to be identified include *Bacteroides* sp, *Prevotella* sp, *Fusobacterium* sp, *Peptostreptococcus* sp, other streptococci, *Gemella morbillorum*, *Staphylococcus saccharolyticus*, *Veillonella* sp, *Actinomyces* sp, *Eubacterium* sp, *Proprionibacterium* sp, *Bifidobacterium* sp, *Lactobacillus* sp, and *Clostridium* sp.

Limitations Anaerobic bacterial cultures are intended to recover most common obligately anaerobic bacterial pathogens. Some obligate anaerobes die after very brief exposure to oxygen and are very difficult to recover in culture. Anaerobic bacteria may contaminate clinical specimens that are not collected aseptically; it is often challenging to distinguish contaminants from etiologic agents of infection. Specimens in anaerobic transport containers are suboptimal for fungus culture.

BLOOD GASES AND pH, ARTERIAL

Specimen Abscess, blood, aseptically obtained body fluid (eg, pleural, peritoneal, synovial), wounds, etc are appropriate specimens
Volume 0.5 mL to 10 mL or small piece of tissue from biopsy site
Container Capped syringe; biopsy; anaerobic transport media, blood culture media
Collection Swabs usually cause unacceptable exposure of anaerobes to oxygen and have a propensity to dry out. Some anaerobes will be killed by contact with oxygen for only a few seconds. Swabs are considered inferior specimens; if they must be used, two are advised: one for Gram stain, the other for culture.
Causes for Rejection Specimens from sites in which anaerobic bacteria are normal flora (eg, throat, rectal swabs, urine, bronchial washes, cervico-vaginal mucosal swabs, sputums) are unacceptable for anaerobic culture. Specimens that have not been appropriately protected from atmospheric oxygen cannot yield accurate results.
Turnaround Time Negative cultures are typically reported as follows: blood cultures: 5-7 days; aseptically obtained body fluids: 5 days; swab specimen: 2-5 days.
Methodology Inoculation of microbiological media suitable for recovering anaerobes and incubation of media at 35°C in an atmosphere lacking oxygen. Biochemical, gas-liquid chromatography (GLC), DNA hybridization, and RNA homology and sequencing define anaerobic genera.
Additional Information Serious anaerobic infections are often due to mixed flora, which are pathologic synergists. Anaerobes frequently recovered from closed postoperative wound infections include *Bacteroides fragilis*, ~50%; *Prevotella melaninogenica* (previously *Bacteroides melaninogenicus*), ~25%; *Peptostreptococcus prevotii*, ~15%; and *Fusobacterium* species, ~25%. Anaerobes are seldom recovered in pure culture (10% to 15% of cultures). Aerobes and facultative bacteria, when present, are frequently found in lesser numbers than the anaerobes. Anaerobic infection is most commonly associated with operations involving opening or manipulating the bowel or a hollow viscus (eg, appendectomy, cholecystectomy, colectomy, gastrectomy, bile duct exploration, etc). The ratio of anaerobes to facultative species is normally about 10:1 in the mouth, vagina, and sebaceous glands and at least 1000:1 in the colon.
Anaerobic bacteria can be involved in all types of bacterial infections, since these bacteria consist of a major part of the indigenous flora of humans. The most common anaerobe encountered in human infections is *Bacteroides fragilis*. A number of other anaerobes also cause human disease (*Prevotella* sp, *Porphyromonas* sp, *Fusobacterium* sp, *Veillonella* sp). Tetanus and botulism are serious diseases due to toxins produced by anaerobes.

Bacterial Culture, Blood

Use Isolate, identify, and determine antimicrobial susceptibility of pathogenic organisms causing bacteremia
Contraindications Use of a 2% iodine preparation is contraindicated in patients sensitive to iodine. Green soap may be substituted for the iodine or alcohol acetone alone may be used. See Patient Preparation
Test Includes Isolation of both aerobic and anaerobic microorganisms and antimicrobial susceptibility testing on all significant isolates
Abstract A blood culture is one of the most significant procedures that a laboratory performs. Positive results should be promptly called to the ordering physician. New automated blood culture systems provide continuous monitoring, which allows detection of positive cultures 24 hours/day. A major problem of blood cultures is the possibility of contamination with normal skin flora. Other factors relevant to detection of microbial pathogens in blood include the volume of blood cultured, the number of separate cultures, the extent of dilution, the types of media, the devices selected, the presence of unusual or fastidious organisms, and the presence of antibiotics.

Limitations Transient bacteremia caused by brushing teeth, bowel movements, or scratching of the skin may cause a positive blood culture, but usually will not cause all three sets to be positive. Blood culture contamination produces false-positive results and subjects patients to the side effects of inappropriate therapy. The most common bacterial contaminants in blood cultures include coagulase-negative staphylococci, *Corynebacterium*, viridans streptococci, and *Bacillus* species.

Prior antibiotic therapy may cause negative blood cultures or delayed growth. Blood cultures from patients suspected of having *Brucella*, tularemia, or *Leptospira* must be requested as special cultures. Consultation with the laboratory for the recovery of these organisms prior to collection of the specimen is recommended. When patients with infective endocarditis have negative bacterial blood cultures, the possibility of a fungal infection should be considered. Yeast often are isolated from routine blood cultures. However, if fungi are specifically suspected, a separate fungal blood culture should be drawn along with each of the routine blood culture specimens. *Mycobacterium avium-intracellulare* (MAI) is recovered from blood of immunocompromised patients, particularly those with acquired immunodeficiency syndrome (AIDS). Special procedures are required for the recovery of these organisms (ie, lysis filtration concentration or use of a special mycobacteria blood culture medium).

Patient Preparation Blood cultures are sometimes contaminated by skin flora. Such contamination can be markedly reduced by careful attention to skin preparation and antisepsis prior to collection of the specimen.

After location of the vein by palpation, the venipuncture site should be cleansed with 70% alcohol (isopropyl or ethyl) and then swabbed in a circular motion concentrically from the center outward using tincture of iodine or a povidone iodine solution. The iodine should be allowed to dry before the venipuncture is undertaken. If palpation is required during the venipuncture, the glove covering the palpating finger tip should be disinfected. In iodine-sensitive patients, a double alcohol, green soap, or acetone alcohol preparation may be substituted.

Alcoholic chlorhexidine has also been recommended for skin preparation.

Aftercare Iodine used in the skin preparation should be carefully removed from the skin after venipuncture.

Specimen Venous blood. The yield of positives is not increased by culturing arterial blood, even in endocarditis.

Volume 10 mL; 5 mL per bottle

Minimum Volume Pediatrics, 1 mL in vented bottle. If more than 1 mL is available, divide equally between two bottles.

Container Bottles of trypticase soy broth or other standard medium

Sampling Time Ideally, two to three sets of blood cultures should be collected per febrile episode; collection of each set should be separated by at least 1 hour from the previous specimen. Such intervals provide maximum recovery of microorganisms in patients with intermittent bacteremia, and documentation of persistent bacteremia in patients with intravascular infections (eg, endocarditis, intravenous catheter site infections). If multiple sets must be collected simultaneously, draw two sets initially from separate sites, and collect a third set at least 1 hour later. Although three blood culture sets provide optimal yield, the cost-effectiveness of this approach has been challenged and some individuals propose collection of only two sets per febrile episode.

Collection Blood cultures should be drawn prior to initiation of antimicrobial therapy. If more than one culture is ordered, the specimens should be drawn from separately prepared sites. A syringe and needle, transfer set, or pre-evacuated set of tubes containing culture media may be used to collect blood. Collection tubes should be held below the level of the venipuncture to avoid reflux. A sample volume of 10-30 mL in adults or 1-5 mL in children is collected for each set. The likelihood of recovering a pathogen increases as the volume of blood sampled increases; however, drawing of more than three blood culture sets per bacteremic episode rarely increases yield. If a syringe and needle or transfer set is used, the top of the blood culture bottle should also be aseptically prepared. See table.

Blood Culture Collection		
Clinical Disease Suspected	**Culture Recommendation**	**Rationale**
Sepsis, meningitis, osteomyelitis, septic arthritis, bacterial pneumonia	Two to three sets of cultures, each 10-30 mL for adults	Assure sufficient sampling in cases of intermittent or low level bacteremia. Minimize the confusion caused by a positive culture resulting from transient bacteremia or skin contamination.
Fever of unknown origin (eg, occult abscess, empyema, typhoid fever, etc)	Two to three sets of cultures - one from each of two prepared sites, the second or third drawn after a brief time interval (30 minutes). If cultures are negative after 24-48 hours obtain two more sets, preferably prior to an anticipated temperature rise.	The yield after four sets of cultures is minimal. A maximum of three sets per patient per day for 3 consecutive days is recommended.
Endocarditis		
Acute	Obtain two to three blood culture sets within 2 hours, then begin therapy.	95% to 99% of acute endocarditis patients (untreated) will yield a positive in one of the first three cultures.
Subacute	Obtain two to three blood culture sets on day 1, repeat if negative after 24 hours. If still negative or if the patient had prior antibiotic therapy, repeat again.	Adequate sample volume despite low level bacteremia or previous therapy should result in a positive yield.
Immunocompromised host (eg, AIDS)		
Septicemia, fungemia mycobacteremia	Obtain two to three sets of cultures from each of two prepared sites; consider lysis concentration technique to enhance recovery for fungi and mycobacteria.	Low levels of fungemia and mycobacteremia frequently encountered.
Previous antimicrobial therapy		
Septicemia, bacteremia; monitor effect of antimicrobial therapy	Obtain two to three sets of cultures from each of two prepared sites; consider use of antimicrobial removal device (ARD) or increased volume >10 mL/set.	Recovery of organisms is enhanced by dilution, increased sample volume, and removal of inhibiting antimicrobials.

Interpretation of results can be enhanced by collecting blood cultures from more than one site and after a time interval (1 hour). Cultures should be taken as early as possible in the course of a febrile episode.

Storage Instructions Specimens collected in tubes with SPS® (sodium polyanetholesulfonate) should be processed without delay. The specimen should be transferred to appropriate culture media to avoid any possible decrease in yield due to storage or prolonged contact with SPS®. Culture bottles from some automated systems can sit at room temperature for several hours.

Turnaround Time Common laboratory procedure is to issue a final negative culture report after 5-7 days. A preliminary positive culture report based upon Gram stain and primary subculture is usually available at 24-72 hours.

Special Instructions The requisition should indicate current antibiotic therapy, clinical diagnosis, and relevant history.

Reference Range Negative

Critical Values Positive cultures are immediately phoned to the nursing station or physician.

Methodology Aerobic and anaerobic culture in broth media with detection of bacterial growth by a variety of methods. The antimicrobial removal device procedure (ARD) includes use of an adsorbent resin in the aerobic bottle. Resin-containing bottles are also available for several automated detection systems. In the lysis centrifugation procedure, blood is lysed and centrifuged using a Wampole Isolator™ tube or similar method. The sediment is inoculated to media appropriate for growing aerobic and anaerobic bacteria, fungi, and mycobacteria. A method of continuously monitoring media for bacterial growth is available from several commercial sources.

Additional Information Blood culture should be collected from patients with community-acquired pneumonia. The most important single test for diagnosis of infective endocarditis is the blood culture. The diagnosis of bacterial meningitis is accomplished by blood culture as well as culture and examination of cerebrospinal fluid. Most children with bacterial meningitis are initially bacteremic.

Sequential blood cultures in nonendocarditis patients using a 20-mL sample resulted in an 80% positive yield after the first set, a 93% yield after the second set, and a 98% yield after the third set. The volume of blood cultured seems to be more important than the specific culture technique being employed by the laboratory. The isolation of coagulase-negative staphylococci (CNS) poses a critical and difficult clinical dilemma. Although CNS are the most commonly isolated organisms from blood cultures, only a few (6.3%) of the isolates represent "true" clinically significant bacteremia. Conversely, CNS are well recognized as a cause of infections involving prosthetic devices, cardiac valves, CSF shunts, dialysis catheters, and indwelling vascular catheters. Ultimately, the physician is responsible for determination of whether an organism is a contaminant or a pathogen. The decision is based on both laboratory and clinical data. Patient data including patient history, physical examination, body temperatures, clinical course, and laboratory data (ie, culture results, white blood cell count, and differential) are relevant. Clinical experience and judgment may play a significant role in resolution of this clinical dilemma. Various sources of contamination include the patient's own skin flora, transient benign bacteremias, and perhaps, disinfection materials.

Recovery of mycobacteria, atypical mycobacteria, and *Legionella* may also be enhanced by lysis filtration.

The use of antimicrobial removal devices (ARD) or resin bottles to attempt to increase the yield of blood cultures drawn from patients on antimicrobial therapy is controversial. Some microorganisms are occasionally not recovered with the use of ARD blood cultures. It is, therefore, advised that at least one culture in a series of three be requested without the use of the ARD bottles. ARD blood cultures are substantially more expensive than routine blood cultures.

Blood and Fluid Precautions, Specimen Collection

Synonyms Isolation Patients; Precautions for Specimen Collection; Precautions; Specimen Collection

Abstract Exposure to blood and body fluid-borne pathogens can be significantly reduced by following universal precautions. It is also well known that universal precautions are not strictly followed among all healthcare workers.

Patient Preparation The Occupational Safety and Health Administration (OSHA) Final Rule requires that the risk to healthcare workers of accidental exposure to infection be minimized. By careful planning and thoughtful attention to detail, an appropriate and representative specimen can be safely collected.

Before entering the isolation room or drawing area:

- Check orders and assemble the equipment needed for this patient.
- Read the isolation sign on the door or patient's chart. It will explain the type of isolation and what you must wear and do. Follow these directions carefully.
- Find out if it is necessary to take a tourniquet and/or a plastic holder into the room. Many times these items will be there already.
- Take in the minimum equipment needed: tourniquet; plastic holder; evacuated tube needle; alcohol sponges; evacuated blood collection tubes or blood culture media; glass slides (if a blood smear is to be made).

In the room:

- Ask the patient to state his/her name to confirm patient identification. Check wristband.
- Put on gloves.
- Place paper towels on table and place your equipment on these towels.
- Obtain blood samples in the usual manner, avoiding any unnecessary contact with the patient and the bed.
- After obtaining blood samples, leave tourniquet and plastic holder in room and discard needle in proper container.
- Place several clean paper towels on the table, one on top of the other. If the outside of the tubes is contaminated, follow established laboratory decontamination procedures.
- If blood smears were made, place smears on two clean paper towels. When ready to leave, wrap smears and tubes in the top paper towel and discard the bottom paper towel.
- Label specimens for proper identification as directed by institutional policy. Label specimens for infectious hazards in a distinctive manner as required by institutional policy. Since the implementation of universal blood and body fluid precautions for all patients, special labeling for specific patients may be eliminated, depending upon institutional policies and local regulations. In any case, universal precautions must be observed.
- Wash hands.
- Bring specimens to the laboratory.

Collection All specimens of blood and body fluids should be put in a well-constructed container with a secure lid to prevent leaking during transport.

Special Instructions

Precautions for laboratories: Universal precautions should be followed at all times. Blood and other body fluids from all patients should be considered infective. To supplement universal blood and body fluid precautions, the following precautions are recommended for healthcare workers in clinical laboratories.

All persons collecting and processing blood and body fluid specimens should wear gloves. Masks, protective eyewear, and laboratory coats or gowns should be worn if contact with blood or body fluids is anticipated. Gloves should be changed and hands washed after completion of specimen processing.

For routine procedures, such as histologic and pathologic studies or microbiologic culturing, a biological safety cabinet is not necessary. However, biological safety cabinets (class I or II) should be used whenever procedures are conducted that have a high potential for generating droplets. These include activities such as blending, sonicating, and vigorous mixing. Mechanical pipetting devices must be used for manipulating all liquids in the laboratory. Mouth pipetting must not be done.

Use of needles and syringes should be limited to situations in which there is no alternative, and the recommendations for preventing injuries with needles outlined under universal precautions must be followed.

Laboratory work surfaces should be decontaminated with an appropriate chemical germicide after a spill of blood or other body fluids and when work activities are completed.

Contaminated materials used in laboratory tests should be decontaminated before reprocessing or be placed in bags and disposed in accordance with institutional policies for disposal of infective waste.

Scientific equipment that has been contaminated with blood or other body fluids should be decontaminated and cleaned before being repaired in the laboratory or transported to the manufacturer.

All persons must wash their hands after completing laboratory activities and should remove personal protective equipment before leaving the laboratory.

Implementation of universal blood and body fluid precautions for all patients eliminates the need for warning labels on specimens, since blood and other body fluids from all patients should be considered infective. OSHA rules require "Biohazard" labeling or color coding of containers of regulated waste, refrigerators and freezers containing blood or other potentially infectious material, and containers used to store, transport, or ship such materials.

Additional Information

Universal Precautions: Since medical history and examination cannot reliably identify all patients infected with HIV or other blood-borne pathogens, blood and body fluid precautions should be consistently used for all patients. This approach, recommended by CDC and referred to as "universal blood and body fluid precautions" or "universal precautions," must be used in the care of all patients as a result of OSHA's Final Rule.

All healthcare workers must take precautions to prevent injuries caused by needles, scalpels, and other sharp instruments or devices during procedures; when cleaning used instruments; during disposal of used needles; and when handling sharp instruments after procedures. To prevent needlestick injuries, needles must not be recapped, purposely bent or broken by hand, removed from disposable syringes, or otherwise manipulated by hand. After they are used, disposable syringes and needles, scalpel blades, and other sharp items must be placed in puncture-resistant containers for disposal; the puncture-resistant containers should be located as close as possible to the area of use. Large-bore reusable needles should be placed in a puncture-resistant container for transport to the reprocessing area. Recapping of syringes, if absolutely necessary, may be done by a one-handed method that employs the use of a recapping block.

Although saliva has not been implicated in HIV transmission, to minimize the need for emergency mouth-to-mouth resuscitation, mouthpieces, resuscitation bags, or other ventilation devices should be available for use in areas in which the need for resuscitation is predictable.

Healthcare workers who have exudative lesions or weeping dermatitis should refrain from all direct patient care and from handling patient care equipment until the condition resolves.

Pregnant healthcare workers are not known to be at greater risk of contracting HIV infection than healthcare workers who are not pregnant; however, if a healthcare worker develops HIV infection during pregnancy, the infant is at risk of infection resulting from perinatal transmission. Because of this risk, pregnant healthcare workers should be especially familiar with and strictly adhere to precautions to minimize the risk of HIV transmission.

Sterilization and Disinfection: Standard sterilization and disinfection procedures for patient care equipment currently recommended for use in a variety of healthcare settings, including hospitals, medical and dental clinics and offices, hemodialysis centers, emergency care facilities, and long-term nursing care facilities, are adequate to sterilize or disinfect instruments, devices, or other items contaminated with blood or other body fluids from persons infected with blood-borne pathogens including HIV.

Cleaning and Decontaminating Spills of Blood or Other Body Fluids: Chemical germicides that are approved for use as "hospital disinfectants" and are tuberculocidal when used at recommended dilutions can be used to decontaminate spills of blood and other body fluids. Strategies for decontaminating spills of blood and other body fluids in a patient care setting are different than for spills of cultures or other materials in clinical, public health, or research laboratories. In patient care areas, visible material should first be removed and then the area should be decontaminated. With large spills of cultured or concentrated infectious agents in the laboratory, the contaminated area should be flooded with a liquid germicide before cleaning, then decontaminated with fresh germicidal agent. In both settings, gloves should be worn during the cleaning and decontaminating procedures.

HIV is inactivated rapidly after being exposed to commonly used chemical germicides at concentrations that are much lower than used in practice. Embalming fluids (formalin preparations) are similar to the types of chemical germicides that have been tested and found to completely inactivate HIV. Formalin may not rapidly inactivate hepatitis B virus nor quickly kill bacteria. It is a slow-acting antiseptic agent requiring 18 hours or more to kill microorganisms. In addition to commercially available chemical germicides, a solution of sodium hypochlorite (household bleach) prepared daily is an inexpensive and effective germicide. Concentrations ranging from -500 ppm (1:100 dilution of household bleach) sodium hypochlorite to 5000 ppm (1:10 dilution of household bleach) are effective, depending on the amount of organic material (eg, blood, mucus) present on the surface to be cleaned and disinfected. Disinfecting surfaces in cases of known Jakob-Creutzfeld agent may require full strength bleach. Commercially available chemical germicides may be more compatible with certain medical devices that might be corroded by repeated exposure to sodium hypochlorite.

Housekeeping: Environmental surfaces such as walls and floors are not associated with transmission of infections to patients or healthcare workers. Therefore, extraordinary attempts to disinfect or sterilize such environmental surfaces are unnecessary. Cleaning and removal of soil should be done routinely.

Infective Waste: There is no epidemiologic evidence to suggest that most hospital waste is any more infective than residential waste. Moreover, there is no epidemiologic evidence that hospital waste has caused disease in the community as a result of improper disposal. Therefore, identifying wastes for which special precautions are indicated is largely a matter of judgment about relative risk of disease transmission. The most practical approach to the management of infective waste is to identify those wastes with potential for causing infection during handling and disposal and for which some special precautions appear prudent. Hospital wastes for which special precautions are required include microbiology laboratory waste, pathology waste, blood specimens or blood products, and other potentially infectious material. Any item that has had contact with blood, exudates, or secretions may be potentially infective. Infective waste, in general, should either be incinerated or should be autoclaved before disposal in a sanitary landfill. Bulk blood, suctioned fluids, excretions, and secretions may be carefully poured down a drain connected to a sanitary sewer. Sanitary sewers may also be used to dispose of other infectious wastes capable of being ground and flushed into the sewer.

Survival of HIV in the Environment: The most extensive study on the survival of HIV after drying involved greatly concentrated HIV samples (ie, 10 million tissue culture infectious doses/mL). This concentration is at least 100,000 times greater than that typically found in the blood or serum of patients with HIV infection. HIV was detectable by tissue culture techniques 1-3 days after drying, but the rate of inactivation was rapid. Studies performed at CDC have also shown that drying HIV causes a rapid (within several hours) 1-2 log (90% to 99%) reduction in HIV concentration. In tissue culture fluid, cell-free HIV could be detected up to 15 days at room temperature, up to 11 days at 37°C (98.6°F), and up to 1 day if the HIV was cell-associated. HIV can be isolated from peripheral blood mononuclear cells and plasma for up to 48 hours after sample collection.

Risk to Healthcare Workers of Acquiring Hepatitis or HIV in Healthcare Settings: Comparative risks of needlestick transmission are estimated by the rule of threes: Hepatitis B is transmitted in 30% of exposures, hepatitis C in 3%, and HIV-1 in 0.3%. An estimated 2.7 million people in the U.S. have active hepatitis C infection. The risks associated with occupational mucous membrane and cutaneous exposures are likely to be substantially smaller.

Blood Collection Tube Information

Synonyms Blood Container Description; Tubes for Blood Collection; Vacutainer® Tube Description

Additional Information The table describes most commonly used color codes, optimum and minimum volumes required, and additives contained in common vacuum draw tubes. In the last 5-10 years, due to the advent of different anticoagulant combinations and other additives, tube color codes have changed significantly and are becoming increasingly confusing. International Standard Organization (ISO), along with blood collection tube manufacturers is making efforts to standardize color codes. It is important to be certain that a tube is filled with the prescribed minimum volume in order to avoid spurious results due to an inappropriate anticoagulant to specimen ratio.

Tube Codes		
Color	**Optimum Volume/Minimum Volume**	**Additive**
Blue	4.5 mL/4.5 mL	Sodium citrate
Blue/navy	7 mL/3 mL	No additive (for trace metals)
		Heparin (for trace metals)
Culture (yellow)	8.3 mL/8.3 mL	SPS
FSP (blue)	2 mL/2 mL	Thrombin, trypsin inhibitor
Gray	5 mL/5 mL	Potassium oxalate, sodium fluoride
	7 mL/7 mL	
Green	10 mL/3.5 mL	Heparin
Lavender	7 mL/2 mL	EDTA
Orange	10 mL/NA	Thrombin
Red	10 mL/NA	None
Red/gray (gel)	10 mL/NA	Inert barrier material; clot activator
Yellow	5 mL/NA	ACD
Yellow/black	7 mL	Thrombin
Pediatric Tubes		
Blue	2.7 mL/2.7 mL	Sodium citrate
Culture (yellow)	3.3 mL/3.3 mL	SPS
Green	2 mL/2 mL	Heparin
Lavender	2 mL/0.6 mL	EDTA
	3 mL/0.9 mL	
	4 mL/1 mL	
Red	2 mL/NA	None
	3 mL/NA	
	4 mL/NA	

Blood Gases and pH, Arterial

Synonyms ABGs; Arterial Blood Gases; Gases; Arterial

Use Blood gas and pH testing are done to evaluate oxygen and carbon dioxide gas exchange, respiratory function including hypoxia and acid-base status. They are clinically indicated in a wide variety of medical and surgical situations involving cardiorespiratory, metabolic, and central nervous system disturbances.

The context in which the specimen is drawn is pivotal relevant to its significance (eg, pH 7.10 drawn in the immediate postictal state may be of little consequence, but is ominous in methanol intoxication).

Applies to Allen Test; Base Excess; FiO_2; HCO_3^-; Oxygen Saturation; PaO_2; pCO_2; pH; pO_2; TCO_2

Test Includes Measured results include pH, pCO_2 ($PaCO_2$), and pO_2 (PaO_2). Calculated values include, among others, total carbon dioxide (TCO_2), bicarbonate (HCO_3^-), oxygen saturation, and base excess.

Abstract The measurement in arterial blood of pH, pCO_2, pO_2 and the calculation of HCO_3^-, TCO_2, and O_2 saturation, are used to evaluate oxygen and carbon dioxide exchange, respiratory function, and acid-base balance. Arterial blood is preferred for these determinations due to its superior uniformity throughout the body, but venous pH is extremely similar in most situations and is more easily obtained.

Limitations Arterial puncture is a hazardous procedure and may be extremely difficult in some individuals. Complications of arterial puncture potentially include hematoma, bleeding, arterial occlusion and insufficiency, infection, and, very rarely, gangrene.

Markedly elevated leukocytes and/or platelets in a blood gas specimen will significantly alter blood gas and pH results regardless of specimen transport/storage methods. Point-of-care or *in vivo* methods of analysis are helpful in these situations.

By some instruments, O_2 saturation is calculated from oxyhemoglobin and total hemoglobin. The reported value may be misleading when nonfunctional hemoglobins (COHb, MetHb, or sulfhemoglobin) are present or other hemoglobins with different dissociation curves are present. Calculations commonly assume body temperature of 37°C.

Variability of results occurs; changes in pO_2 in isolated reports must be interpreted cautiously and in light of data trends, oxygen delivery, and the patient's clinical appearance. Such variation occurs without change in FiO_2 or the patient's clinical status.

Arterial gases and pH are of little value in treatment decisions for carbon monoxide poisoning.

Patient Preparation Patient should be supine, relaxed. The patient's temperature, breathing pattern, and concentration of inspired air (FiO_2) should be recorded.

Aftercare Observe for bleeding. Pressure must be applied to puncture site for at least 10-15 minutes; longer times are required for anticoagulated patients.

Specimen Whole blood (arterial)

Container Heparinized blood gas syringe (plastic or glass) or via an indwelling arterial line. Gases are capable of dissolving in plastic, which may alter results in some situations. Mahoney et al reported clinically significant increases in pO_2 levels in whole blood stored in iced plastic syringes for 30 minutes. Specimens stored in iced glass syringes did not change significantly.

Collection Very small diameter needles are used. Specimen is drawn into air-free heparinized syringe, then stoppered. The radial artery is frequently used after the Allen test, which assesses the presence of normal collateral circulation. The brachial artery is the second choice. The specimen should be transported to the laboratory immediately and analysis should be prompt. If testing will be delayed by more than 10-15 minutes, the specimen should be placed in a slurry of ice chips and water. Mode of oxygen delivery (quantity of therapeutic oxygen or room air) and patient's temperature must be indicated. Rapid changes may occur if collected immediately after exercise. Avoid excessive heparin. Strict anaerobiosis must be maintained.

Storage Instructions Testing should occur immediately; therefore, specimens should not require storage. However, if testing must be delayed more than 10-15 minutes, the specimen should be cooled to about 0°C by placing the syringe into a slurry of ice chips and water. Delay of analysis should not exceed 1 hour. Specimens for critical alveolar-arterial oxygen tension or shunt fraction samples must be analyzed immediately (within 10 minutes) to minimize changes in gas tensions. The following *in vitro* changes occur in blood gas parameters: pH decreases by <0.01 pH units/hour at 4°C, pCO_2 increases by about 0.5 mm Hg/hour at 4°C, and pO_2 decreases negligibly (<3 mm Hg/hour) if collected in a glass syringe and stored in an ice water slurry.

Causes for Rejection Specimen not received correctly iced, air bubbles or clots in syringe, unsealed/open syringe

Special Instructions Sample obtained just after a change in inspired oxygen concentration (FiO_2) (eg, room air or quantity of therapeutic oxygen delivered) is likely to generate confusing results. Normally, arterial blood gases and pH will achieve steady-state levels within a minute or two after a change in FiO_2 or alveolar ventilation, but in certain disease states (eg, lung disease) the time necessary to achieve equilibrium can be as long as 20-30 minutes.

Reference Range
- Arterial pH: 7.35-7.45
- TCO_2: 23-29 mmol/L

- pCO_2: 35-45 mm Hg
- pO_2: newborns: 60-70 mm Hg, adults: 80-95 mm Hg
- O_2 saturation: 95% to 99%

Such intervals must be interpreted in light of the FiO_2 and other variables.

Possible Panic Range pH: <7.2, >7.55; pCO_2: <20 mm Hg, >60 mm Hg; pO_2: <40 mm Hg

Methodology Selective electrodes measuring pH, pCO_2, and pO_2

Additional Information A pH value <7.25, without elevation of pCO_2, may indicate need for a lactate determination.

Potassium leaves the intracellular fluids in acidemia, leading to hyperkalemia.

The acute respiratory distress syndrome (ARDS) is usually initially characterized by respiratory alkalosis and hypoxemia.

Assessment of acid-base status of tissues in patients in circulatory failure may be misleading if only arterial blood gas data is available. Adrogué presents data supporting the need for information on mixed venous as well as arterial gases in care of critically ill patients.

Botulism, Diagnostic Procedure

Synonyms *Clostridium botulinum* Toxin Identification Procedure; Infant Botulism; Toxin Identification

Use Diagnose infant botulism, classic botulism in adults

Contraindications Due to the difficulty in performance of the diagnostic test and because of the extensive epidemiological studies initiated upon receipt of the specimen, State Department of Health Laboratories require specific clinical symptomatology for infant botulism and therefore should be consulted early to optimize handling of the suspect case.

Abstract A neurotoxin, botulin, may be produced by *C. botulinum* in foods which have been improperly preserved. Characteristics of this type of food poisoning include vomiting and abdominal pain, disturbances of vision, motor function and secretion, mydriasis, ptosis, dry mouth, and cough.

Limitations The toxin from *C. botulinum* binds almost irreversibly to individual nerve terminals; thus, serum and cerebrospinal fluid specimens may yield false-negative results. Can detect as little as 0.03 ng of botulinum toxicity.

Specimen Vomitus, serum, stool, gastric aspirates, cerebrospinal fluid or autopsy tissue; food samples

Volume 50 g vomitus or feces, gastric washings, cerebrospinal fluid or autopsy tissue, 15-20 mL serum; infant botulism: 25 g feces

Container Sterile wide-mouth, leakproof, screw-cap jar; red top tube

Storage Instructions Keep refrigerated at 4°C except for unopened food samples.

Special Instructions The laboratory must be notified prior to obtaining specimen in order to prepare for transport of the specimen to the State Health Laboratory or Center for Disease Control.

Reference Range No toxin identified, no *Clostridium botulinum* isolated

Methodology Toxin neutralization test in mice

Specific References

Armada M, Love S, Barrett E, et al, "Foodborne Botulism in a Six-Month-Old Infant Caused by Home-Canned Baby Food," *Ann Emerg Med*, 2003, 42(2):226-9.

Fox CK, Keet CA, and Strober JB, "Recent Advances in Infant Botulism," *Pediatr Neurol*, 2005, 32(3):149-54.

Richardson WH, Frei SS, and Williams SR, "A Case of Type F Botulism in Southern California," *J Toxicol Clin Toxicol*, 2003, 41(5):653.

Thompson JA, Filloux FM, Van Orman CB, et al, "Infant Botulism in the Age of Botulism Immune Globulin," *Neurology*, 2005, 64(12):2029-32.

Carboxyhemoglobin, Blood

Synonyms Carbon Monoxide; CO; COHb

Use Determine the extent of carbon monoxide poisoning

Applies to Methylene Chloride; Oximeters, Pulse

Test Includes COHb is sometimes included in Blood Gases, but may be ordered as a separate test.

Abstract A byproduct of incomplete combustion of hydrocarbons, carbon monoxide (CO) is a colorless, tasteless, and odorless gas. It binds tightly to hemoglobin (Hb) to form COHb, reducing oxygen-carrying capacity of blood. The affinity of Hb for CO is 200-250 times that of oxygen. It also binds to myoglobin and cytochrome oxidase. Carbon monoxide poisoning is seen from smoke inhalation, suicide attempt and accidental exposure. It is the most common cause of poisoning death in the U.S., accounting for thousands of emergency department visits and some 800 deaths annually. This test measures hemoglobin-bound carbon monoxide.

Limitations CO binds to cytochrome oxidase, interfering with cellular respiration. Thus, although COHb assays provide information on exposure to CO, they do not always consistently correlate with symptoms or prognosis.

Patient Preparation In suspected carbon monoxide poisoning, the specimen should be collected immediately.

Specimen Whole blood, venous or arterial

Minimum Volume 1 mL whole blood

Container Green top (heparin) tube or lavender top (EDTA) tube, depending upon laboratory methods

Collection Keep tube capped

Storage Instructions Refrigerate immediately after collection. Do not remove cap. Carboxyhemoglobin is stable 4 months in a filled, well-capped tube.

Reference Range
- **Nonsmoker:** <3%
- **Smoker:** 1-2 packs/day: 4% to 5%, >2 packs/day: 8% to 10%

Carboxyhemoglobin in the newborn may run to 10% to 12%. Carbon monoxide is a metabolic product of hemoglobin catabolism. The increased turnover of hemoglobin in the newborn together with decreased efficiency of the infant's respiratory system may and does lead to higher levels of carboxyhemoglobin.

Critical Values Exposure to CO concentrations 80-140 ppm for 1-2 hours can lead to COHb results of 3% to 6%; in some patients, even these levels can precipitate angina and cardiac arrhythmias. Healthy individuals can tolerate levels of 10% without any symptoms. Toxic concentration is 20%; lethal is >50%.

Possible Panic Range Disturbance of judgment, headache, and dizziness occur at 10% to 30%; coma at 50% to 60%; fatality occurs at 30% to 60% or more, and rapid death at level of 80%.

Methodology Spectrophotometry; gas chromatography is used to measure CO.

Additional Information A danger of missed diagnosis of CO intoxication is continued exposure of the patient and others to a toxic environment. The cherry red color of CO poisoning is not consistently seen. CO intoxication may contribute to the risk of myocardial infarction. The half-life of carboxyhemoglobin at room air is −6 hours. The half-life with 100% O_2 administration, at atmospheric pressure, is 80 minutes. With O_2 at three atmospheres, the half-life is 24 minutes. Use of hyperbaric oxygen in the treatment of carbon monoxide poisoning is more effective than normobaric oxygen therapy (Weaver et al 2002).

Specific References

Centers for Disease Control and Prevention (CDC), "Carbon Monoxide Poisonings Resulting From Open Air Exposures to Operating Motorboats - Lake Havasu City, Arizona, 2003," *MMWR Morb Mortal Wkly Rep*, 2004, 53(15):314-18.

Chee KJ, Suner S, Partridge RA, et al, "Noninvasive Carboxyhemoglobin Monitoring: Screening Emergency Department Patients for Carbon Monoxide Exposure," *Acad Emerg Med*, 2006, 13(5):S177.

Chain-of-Custody Protocol

Synonyms Chain-of-Evidence Form; Specimen Chain-of-Custody Protocol

Use Maintain sample integrity in the collection, handling, and storage of urine or other samples

Applies to Bullets; Medical Legal Specimens

Abstract A procedure to ensure sample integrity from collection through transport, receipt, sampling, and analysis. It is associated with a Chain-of-Custody form. Similar forms are used (chain-of-evidence) for other forensic materials such as guns, bullets, chemicals, etc.

Specimen Usually urine for drugs-of-abuse-related monitoring; blood for alcohol testing

Container Plastic urine cup with locking lid covered by seal, which is signed or initialed (if for drugs of abuse)

Causes for Rejection Sample container not sealed or labeled

Special Instructions Form requires signature of sample donor as well as that of the person receiving the sample at the collection site.

Reference Range Normal: all seals intact and Chain-of-Custody form completed.

Additional Information A written record of specimen transfer from patient, to analyst, to storage and disposal is maintained on all specimens covered by chain-of-custody. All drug screens, blood alcohols, most bullets, or any other tests or objects that have medicolegal significance should be accompanied by Chain-of-Custody and a written release form.

Chemical Terrorism Event

Source: Centers for Disease Control and Prevention (CDC), Morbidity and Mortality Weekly Report (MMWR), 2005 - http://www.cdc.gov

Specimen Collection

For each patient, collect the following blood tubes: 3 EDTA (purple top) and 1 gray or green top tube in that order.
   3, 5, or 7 mL purple top tube #1 - Metals
   3, 5, or 7 mL purple top tube #2 - Cyanide
   3, 5, or 7 mL purple top tube #3 - Chemical adducts w/Hgb & albumin

For pediatric patients, collect urine only unless otherwise directed by the CDC.
   Purple top tube #1 - Place a "1" on tube #1 with indelible ink
If collecting in 3 mL purple top tubes, collect a 4th purple top tube.
Mix the EDTA tubes by inverting 5-6 times.

Label each EDTA tube with the appropriate label/Place a single strip of evidence tape over the top of the tube, making sure it is secured on both sides of the tube. Initial tube so that your initials are half on the tube and half on the tape.
   "-PT1" on purple top tube #1 - Metals
   "-PT2" on purple top tube #2 - Cyanide
   "-PT3" on purple top tube #3 - Chemical adducts w/Hgb & albumin

*Make sure that tube #1 is the first tube drawn, tube #2 is the second tube drawn, etc, and that the correct label is put on each tube!

Store blood samples at 4°C.

*Make sure the gray or green top tube is collected after all the purple top tubes.
   3, 5, or 7 mL gray or green top tube - Volatile organic compounds

Mix the tube by inverting 5-6 times.

Label either the gray or green top tube with the appropriate label. Place a single strip of evidence tape over the top of the tube as shown, making sure it is secured on both sides of the tube. Initial tube so that your initials are half on the tube and half on the tape.
   "-GRT" on gray or green top tube - Volatile organic compounds

Store blood samples at 4°C.

In addition to the blood tubes, collect *at least* 25 mL or urine in a screw cap urine cap for each patient.
   Military nerve agents
   Organophosphate pesticides
   Incapacitating agents and drugs of abuse
   Ricin and saxitoxin
   Sulfur and nitrogen mustard
   Lewisite
   Heavy metals (Hg, As, Sb, Ba, Be, Cd, Ca, Co, Pb, Mo, Pt, Tl, W, U, Se)
   Creatinine correction

For pediatric patients, collect urine only unless otherwise directed by the CDC.

Label the urine cup with the appropriate label. Place a single strip of evidence tape across the top of the urine cup, making sure it is secured on both sides of the cup. Initial the cup and tape so that your initials are half on the cup and half on the tape.
"-UC"

Please note the placement of the labels. Labels should be placed on all tubes/cups so that when the tube/cup is standing upright, the bar code looks like a ladder. Also note the placement of your initials.

Freeze urine specimens at $\leq 70^{\circ}$C.

www.bt.cdc.gov/labissues/pdf/chemspecimencollection.pdf

Shipping Instructions - Blood

Separate tubes by type (purple or green/gray).

Pack tubes separated by type in a gridded box. Prevent glass-to-glass contact.

Wrap each box with absorbent material (ie, Chux pad) and seal with waterproof, tamper-evident forensic evidence tape.

Write initials half on the evidence tape and half on the absorbent material.

Place one wrapped box in the clear Saf-T-Pak® waterproof bag and seal. Note: If primary receptacles have not been shown to meet the 95kPa, use compliant secondary packaging materials.

Place this bag inside the white Tyvek Saf-T-Pak® bag and seal.

Seal the opening of this bag with waterproof, tamper-evident forensic evidence tape. Write initials half on the evidence tape and half on the bag.

After all bagged boxes have been packaged, they need to be put in a shipper to be sent to the CDC or other LRN-C Laboratory. Use Styrofoam-insulated, corrugated fiberboard shippers.

Place absorbent material such as a Chux pad in the bottom of the shipper.

Place cold packs in a single layer on top of the absorbent material.

Place the packaged boxes in the shipper, using plenty of cushioning material to minimize shifting while in transit.

Place additional cold packs on top of samples.

Place the Blood Shipping Manifest in a sealable plastic bag and put on top of the sample boxes inside the shipper. Place the styrofoam lid on the cooler.

Secure the outer container lid with filamentous shipping tape. Place your shipping address in the upper left hand corner of the box top and the CDC or other LRN-C Laboratory receiving address in the center.

Add the UN3373 label. If the label bears the words "Diagnostic Specimens," there is no need to rewrite this statement on the container. If "Diagnostic Specimens" is not on the UN3373 label, stamp or write "Diagnostic Specimens" on the container's lid.

www.bt.cdc.gov/labissues/pdf/chemspecimenshipping-blood.pdf

Shipping Instructions - Urine

Wrap each urine cup separately in absorbent material and seal with waterproof, tamper-evident forensic evidence tape. Write initials half on the evidence tape and half on the absorbent material.

OR

Use a gridded box to separate urine cups. Place sealed urine cups into the box. Urine cups do not need to be wrapped individually with absorbent material.

Separated urine cups should be wrapped with absorbent material. Evidence tape is used to secure the seal. Write initials half on the evidence tape and half on the absorbent material.

Place wrapped cups in the waterproof inner Saf-T-Pak® bag and seal. Note: If primary receptacles have not been shown to meet 95 kPa, use compliant secondary packaging materials.

Place this bag inside the white Tyvek Saf-T-Pak® bag and seal.

Seal the opening of this bag with waterproof, tamper-evident forensic evidence tape.

Write initials half on evidence tape and half on bag.

After all urine cups have been bagged, they need to be put in a shipper to be send to the CDC or other LRN-C Laboratory.

Use styrofoam-insulated shippers.

Place absorbent material such as a Chux pad in the bottom of the shipper.

Place a layer of dry ice in the bottom of the cooler. DO NOT use large chunks.

Place the packaged urine cups in the shipper using plenty of absorbent material or cushioning material to minimize shifting while in transit.

Place additional dry ice on top of samples.

Place the Urine Shipping Manifest in a sealable plastic bag and put on top of the samples inside the shipper.

Place the styrofoam lid on the cooler.

Secure the outer container lid with filamentous shipping tape. Place your address in the upper left-hand corner of the box top and the CDC or other LRN-C Laboratory receiving address in the center.

Add the UN3373 label. If the label bears the words "Diagnostic Specimens," there is no need to rewrite this statement on the container. If "Diagnostic Specimens" is not on the UN3373 label, stamp or write "Diagnostic Specimens" on the container's lid.

Place a Class 9 label on the container. This label MUST indicate the amount of dry ice in the container, UN1845 and the proper name (either "Carbon Dioxide, Solid" or D"Dry Ice").

www.bt.cdc.gov/labissues/pdf/chemspecimenshipping-urine.pdf

Public Health Contacts for Laboratory Testing to Confirm Exposure During a Potential or Known Chemical Terrorism Event

Source: Centers for Disease Control (CDC), "Recommendations and Reports,"
MMWR, 2005, 54(RR01):25.

Emergencies

To obtain emergency information from CDC, contact:
CDC
Director's Emergency Operations Center
Atlanta, GA
770-488-7100
http://intra-apps.cdc.gov/od/otper/programs/deoc-main.asp

Nonemergencies

To obtain nonemergency information, contact:
CDC
National Center for Environmental Health
Division of Laboratory Sciences
Atlanta, GA
770-488-7950
http://www.cdc.gov/nceh/dls

CDC
National Institute of Occupational Safety and Health
Cincinnati, OH
800-356-4674
http://www.cdc.gov/nciosh/homepage.html

CDC
National Center for Infectious Diseases
Bioterrorism Rapid Response and Advanced Technology Laboratory
Atlanta, GA
404-639-4910

Environmental Protection Agency
National Response Center
Washington, DC
800-424-8802
http://www.epa.gov

Food and Drug Administration
Forensic Chemistry Center
Cincinnati, OH
513-679-2700, ext 184
http://www.fda.gov

Laboratory Response Network
Association of Public Health Laboratories
Washington, DC
202-822-5227
http://www.bt.cdc/lrn

Complete Blood Count

Synonyms CBC

Use Evaluate anemia, leukemia, reaction to inflammation and infections, peripheral blood cellular characteristics, state of hydration and dehydration, polycythemia, hemolytic disease of the newborn; manage chemotherapy decisions

Test Includes The components of a complete blood count (CBC) vary in different laboratories. Tests commonly included are: WBC count, differential count, Hct, Hb, RBC count, WBC and RBC morphology, RBC indices, platelet estimate, platelet count, RDW, and histograms. Variations on the theme include automated 5-part WBC differentials: granulocytes, monocytes, lymphocytes, eosinophils, basophils, and additional RBC and platelet indices. In addition, current analyzers have reticulocyte capability including determination of a set of reticulocyte indices.

Abstract The CBC is a profile of tests rather than a single test. It is the standard, broadly inclusive, usually automated test for evaluation of RBC, WBC, and platelets.

Limitations Hemoglobin (and thus the derived MCH and MCHC) may be falsely high if the plasma is lipemic or if the white count is >50,000 cells/mm^3. "Spun" (manual centrifuged) microhematocrits are ~3% higher (due to plasma trapping) compared to automated hematocrit levels. The increase is especially pronounced in cases of polycythemia (increased Hct levels) and when the cells are hypochromic and microcytic. The spun Hct level (as compared to automated instruments' calculated level) may be 12% higher at Hct levels of 70% and MCV of 48 fL with decrease in change to 3% higher at Hct levels of 70% with MCV of 100 fL. Cold agglutinins (high titer) may cause spurious macrocytosis and low RBC count. This results when RBC couplets are "seen" and processed as single cells by the detection circuitry. Keeping the blood warm and warming the diluent prior to and during counting can correct this problem. Cryoproteinemia (cryoglobulinemia) may cause pseudoleukocytosis or pseudothrombocytosis. Malaria may be a cause of pseudoreticulocytosis.

Specimen Whole blood

Volume 5 mL

Minimum Volume 1 mL

Container Lavender top (EDTA) tube. International Council for Standardization in Hematology recommendation is for use of K$_2$-EDTA, 1.5-2.2 mg/mL of blood as anticoagulant for blood cell counting and sizing.

Collection Mix specimen 10 times by gentle inversion. If specimen is not brought to the laboratory immediately refrigeration is required. If the anticipated delay in arrival is more than 4 hours, two blood smears should be prepared immediately after the venipuncture and submitted with the blood specimen.

Storage Instructions EDTA-anticoagulated sample should be analyzed within 6 hours at room temperature and within 24 hours when stored at 4°C. Blood cell parameters are stable for up to 24-48 hours (WBC differential is stable for 24 hours) at 4°C.

Causes for Rejection Improper tube, clotted specimen, hemolyzed specimen, dilution of blood with I.V. fluid

Turnaround Time If the analyzer is operational, a stat result may be available within 5-10 minutes.

Special Instructions Blood specimen and diluent may require prewarming to obtain meaningful results if cold agglutinins are present.

Reference Range See table.

Mean ($\pm$1 SD) Reference Intervals for Hematologic Values

Age	Hb (g/dL)	Hct (%)	RBC ($\times 10^6/mm^3$) ($\times 10^{12}/L$)	MCV (fL)	MCH (pg)	MCHC (g/dL)	WBC ($\times 10^3/mm^3$) ($\times 10^9/L$)
Birth (cord blood)	17.1$\pm$1.8	52$\pm$5	4.64$\pm$0.5	113$\pm$6	37$\pm$2	33$\pm$1	
1 d	19.4$\pm$2.1	58$\pm$7	5.30$\pm$0.5	110$\pm$6	37$\pm$2	33$\pm$1	
2-6 d	19.8$\pm$2.4	66$\pm$8	5.40$\pm$0.7	122$\pm$14	37$\pm$4	30$\pm$3	
14-23 d	15.7$\pm$1.5	52$\pm$5	4.92$\pm$0.6	106$\pm$11	32$\pm$3	30$\pm$2	
24-37 d	14.1$\pm$1.9	45$\pm$7	4.35$\pm$0.6	104$\pm$11	32$\pm$3	31$\pm$3	
40-50 d	12.8$\pm$1.9	42$\pm$6	4.10$\pm$0.5	103$\pm$11	31$\pm$3	30$\pm$2	
2-3.5 mo	11.3$\pm$1.0	37$\pm$4	3.81$\pm$0.5	98$\pm$9	30$\pm$3	30$\pm$2	
5-10 mo	11.6$\pm$0.7	38$\pm$3	4.28$\pm$0.5	91$\pm$8	27$\pm$3	30$\pm$2	
1-3 y	11.9$\pm$0.6	39$\pm$2	4.45$\pm$0.4	87$\pm$7	27$\pm$2	30$\pm$2	
3-5 y	12.3$\pm$0.8	36$\pm$3	4.4$\pm$0.3	81$\pm$5	28$\pm$2	34$\pm$1	7.7$\pm$2.2
6-8 y	12.7$\pm$0.9	37$\pm$2	4.5$\pm$0.3	83$\pm$5	28$\pm$2	34$\pm$1	7.5$\pm$2.0
9-11 y	13.1$\pm$0.9	38$\pm$2	4.6$\pm$0.4	83$\pm$5	28$\pm$2	34$\pm$1	7.0$\pm$1.8
12-14 y							
Male	13.8$\pm$1.0	40$\pm$3	4.8$\pm$0.4	84$\pm$5	29$\pm$2	34$\pm$1	7.0$\pm$1.8
Female	13.2$\pm$1.0	39$\pm$2	4.5$\pm$0.3	86$\pm$5	29$\pm$2	34$\pm$1	7.1$\pm$2.1
15-17 y							
Male	14.7$\pm$1.0	43$\pm$3	5.0$\pm$0.4	87$\pm$5	30$\pm$2	34$\pm$1	7.2$\pm$2.0
Female	13.3$\pm$1.0	39$\pm$3	4.5$\pm$0.3	88$\pm$5	30$\pm$2	34$\pm$1	7.7$\pm$2.1
18-64 y							
Male	15.2$\pm$1.1	44$\pm$3	5.0$\pm$0.4	89$\pm$5	31$\pm$2	34$\pm$1	7.4$\pm$2.1
Female	13.5$\pm$1.1	40.5$\pm$3	4.4$\pm$0.4	90$\pm$6	30$\pm$2	34$\pm$1	7.2$\pm$2.1
65-74 y							
Male	14.8$\pm$1.4	44$\pm$4	4.8$\pm$0.5	91$\pm$6	31$\pm$2	34$\pm$1	7.1$\pm$2.0
Female	13.7$\pm$1.2	40.5$\pm$3	4.5$\pm$0.4	90$\pm$6	31$\pm$2	34$\pm$1	6.8$\pm$2.4

Adapted from Johnson TR, "How Growing Up Can Alter Lab Values in Pediatric Laboratory Medicine," *Diag Med* (special issue), 1982, 5:13-8; and Second National Health and Nutrition Examination Survey (NHANES), "Hematological and National Biochemistry Reference Data for Persons 6 Months - 74 Years of Age," *DHHS Publication No (PHS) 83-1682*, Hyattsville, MD: Public Health Service, Dec 1982.

Above values are reference intervals for the population defined in NHANES II (1976-1980, see above). These intervals are not necessarily identical to "normal intervals" for local populations, but are sufficiently broad such that mean$\pm$2 SD should include most (at least 95%) of normal subjects. NHANES III data (from a 6-year study, 1988-1994) is available, in part, as it pertains to iron deficiency. See Looker AC, Dallman PR, Carroll MD, et al, "Prevalence of Iron Deficiency in the United States," *JAMA*, 1997, 277(12):973-6. NHANES III Hb cutoff values were calculated as the mean Hb of the reference group minus 1.645 SD (corresponds to the 5th percentile value for a variable such as Hb that has a gaussian distribution). Thus, adult males (ages 20-49) had Hb level of 15.30$\pm$0.97 with a cutoff value <13.7 g/100 mL. Adult females (20-49 years) had Hb level of 13.48$\pm$0.91 with a cutoff value <12.0 g/100 mL.

Critical Values Hematocrit: <18% or >54%; hemoglobin: <6.0 g/dL or >18.0 g/dL; WBC on admission: <2500/mm^3 or >30,000/mm^3; platelets: <20,000/mm^3 or >1,000,000/mm^3

Methodology Varies considerably between institutions. Most laboratories have high capacity multichannel instruments in place (available from multiple commercial sources). The majority measure RBC and WBC variables on the basis of changes in electrical impedance as cells and platelets are pulled through a tiny aperture with subsequent computer processing of electrical signals. Accuracy (with proper standardization) and precision (usually in the 0.5% to 2% range) is significantly improved over older manual and semiautomated methods. Some instruments count impulses as cells flow across a laser beam.

Additional Information Presence of one or more of the following may be indications for further investigation: hemoglobin < 10 g/dL, hemoglobin > 18 g/dL, MCV > 100 fL, MCV < 80 fL, MCHC > 37%, WBC > 20,000/mm³, WBC < 2000/mm³, presence of sickle cells, significant spherocytosis, basophilic stippling, stomatocytes, significant schistocytosis, oval macrocytes, tear drop red blood cells, eosinophilia (>10%) monocytosis (>15%), nucleated red blood cells in other than the newborn, malarial organisms or the possibility of malarial organisms, hypersegmented (five or more nuclear segments) PMNs, agranular PMNs, Pelger-Huët anomaly, Auer rods, Döhle bodies, marked toxic granulation, mononuclears in which apparent nucleoli are prominent (blast type cells), presence of metamyelocytes, myelocytes, promyelocytes, neutropenia, presence of plasma cells, peculiar atypical lymphocytes, significant increase or decrease in platelets. Some quantitative elements of the CBC are related to each other, normally, such that examination of the results of any individual analysis allow for the application of a simple but effective case individualized quality control maneuver. The RBC count, hemoglobin, and hematocrit may be interpreted by applying a "rule of three." If red cells are normochromic/normocytic, the RBC count times three should approximately equal the hemoglobin and the hemoglobin multiplied by three should approximate the hematocrit. If there is significant deviation from these relationships, check for supporting abnormalities in RBC indices and in the peripheral blood smear; in addition, patient identification may be a problem. If patient transfusion can be excluded, then RBC indices should vary little consecutively from day to day.

Anemias have been classified on the basis of their MCV and RDW (RBC heterogeneity). This classification has been especially helpful in the separation of iron deficiency from thalassemia. Heterozygous thalassemia (thalassemia minor) when associated with normal hemoglobin has a normal RDW (13.4±1.2%) while RDW is high with iron deficiency (16.3±1.8%). RDW will be increased slightly in cases of thalassemia with slight anemia. RDW may not always separate cases of iron deficiency from thalassemia minor unless a higher cutoff value (RDW of 17.0%) is used. A recently proposed algorithm uses ethnic background and MCV to provide a "high index of suspicion" for detection of thalassemia trait when dealing with multicultural populations. RDW is not clinically useful in distinguishing the anemia of chronic disease from iron deficiency. In some 30% to 50% of patients with anemia of chronic disease, red cells are hypochromic and microcytic, often with decreased serum iron, iron binding capacity and transferrin saturation even with demonstrably adequate iron stores.

Histogram of MCHC is of value in the diagnosis of hereditary spherocytosis and the differentiation of α- from β-thalassemic red cells. RDW is an insensitive parameter for the diagnosis of vitamin B_{12} deficiency, as well as for the diagnosis of folate deficiency, and the RDW has no value in separating alcohol-related macrocytosis from B_{12}/folate deficiency. In a hospitalized urban patient population, zidovudine treatment of AIDS may be the most common cause of macrocytosis.

The RDW in healthy pregnant women may rise during the last 4-6 weeks before onset of labor, possibly due to increased bone marrow activity. This change could serve to indicate impending parturition.

Most recent generation automated hematology analyzers include leukocyte differential determination and a system of "flags" to indicate presence of abnormal, atypical, and possible immature granulocytes and/or blasts. Numerous studies evaluating performance of flagging systems have been published. Evaluation of this WBC discriminate function is important in detection and characterization of leukemia.

Recently available analyzers incorporate the ability to perform reticulocyte counts, nucleated red blood cell determinations, CD4:CD8, CD64, immature granulocyte, and variant lymphocyte counts. A flow cytometric analysis of platelets, "ImmunoPlt" assay is based in part on CD61 monoclonal antibody labeling (implemented on the Cell-Dyn 4000). It is especially suited for analysis of thrombocytopenic specimens (interference by nonplatelet particles is decreased).

Detailed descriptions of current analyzers have been published and include instruments offered by the following manufacturers:

- Abbott Diagnostics (Cell-Dyn 3200, 3700, and 4000)
- ABX Diagnostics Inc (Pentra 60^{c+}, 120 Retic Hematology Analyzer)
- Bayer® Diagnostics (Advia 120 Hematology System)
- Beckman Coulter, Inc (Coulter GEN-S, HmX, STK-S with Reticulocytes, MAXM with Reticulocytes)
- Roche Diagnostics Corp (Sysmex SF-3000/SF-Alpha, 9500/SE-Alpha II, SE-9500R/SE-Alpha IIR/HST, XE 2100/XE Alpha II/HST)

Specific References

Rodriguez RM, Abdullah R, Miller R, et al, "A Pilot Study of Cytokine Levels and White Blood Cell Counts in the Diagnosis of Necrotizing Fasciitis," *Am J Emerg Med*, 2006, 24:58-61.

Cyanide, Blood

Synonyms CN$^-$; Hydrocyanic Acid; Potassium or Sodium Cyanide

Use Establish the diagnosis of cyanide poisoning

Abstract This highly toxic substance is one of the oldest poisons known. It binds to cytochrome oxidase and prevents cellular respiration, causing tissue hypoxia, severe lactic acidosis, high anion gap metabolic acidosis, and death. Pharmacokinetic estimates vary widely, probably depending on the circumstances of poisoning.

Specimen Whole blood, since cyanide is concentrated in erythrocytes. Venous blood may appear bright red. Serum may be used in some settings.

Volume 10 mL

Minimum Volume 7 mL plasma or 5 mL serum

Container Lavender top (EDTA) tube preferred; red top (for serum)

Sampling Time Stat

Storage Instructions Fill tube to capacity and keep tightly closed; analyze as soon as possible.

Reference Range

Whole blood cyanide:
- Nonsmoker: 0.016 mg/L (SI: 0.61 µmol/L)
- Smoker: 0.041 mg/L (SI: 1.57 µmol/L)

Serum cyanide:
- Nonsmoker: 0.004 mg/L (SI: 0.15 µmol/L)
- Smoker: 0.006 mg/L (SI: 0.23 µmol/L)
- Toxic: >0.1 mg/L (SI: >3.84 µmol/L)

Methodology Photometric, ion-specific potentiometry

Additional Information Cyanide is found in insecticides, rodenticides, vermicides, metal polishes, and electroplating baths. Other sources include ore refining, laetrile, synthetic rubber manufacturing, and the seeds of cherries, plums, peaches, apricots, pears, apples, crab apples, chokeberries, and lima beans. Some cyanide poisoning occurs among victims of fires, since plastic construction materials produce cyanide from combustion. In such situations, carbon monoxide poisoning may coexist with cyanide poisoning. Fires that involve urea foam insulation may produce hydrocyanic acid, which may be inhaled. Symptoms of toxicity include headache, agitation, vomiting, and confusion. A scent of bitter almonds is suggestive, but not all individuals can detect it.

Disseminated Intravascular Coagulation Screen

Synonyms Consumptive Coagulopathy Screen; DIC Screen; Screen for Disseminated Intravascular Coagulation

Use Diagnose DIC in patients with an underlying disorder known to cause DIC and/or with clinical suspicion of DIC

Applies to Fibrinolysis; Schistocytes

Abstract The most useful panel of tests to screen for disseminated intravascular coagulation (DIC) includes D-dimer or fibrin degradation products (FDP), prothrombin time (PT), activated partial thromboplastin time (PTT), platelet count, examination of peripheral blood smear, and fibrinogen. These tests are not, however, specific for DIC.

Limitations D-dimer and FDP are positive with physiologic clot formation and lysis, and they may be positive in liver disease because they are normally cleared by the liver. Therefore, DIC can be difficult to diagnose in the presence of liver disease in some cases. Results should be reviewed in relation to the clinical situation.

Specimen Plasma (and whole blood for platelet count and peripheral blood smear)

Volume Approximately 15 mL blood, multiple specimens

Container Blue top (sodium citrate) tube; lavender top (EDTA) tube for platelet count and peripheral blood smear

Collection Routine venipuncture. Draw blue top before lavender top (EDTA) tube. If a red top is drawn, draw red top tube before blue top tube. Immediately invert tube gently at least 4 times to mix. Blue top tubes must be appropriately filled. Deliver tubes immediately to the laboratory.

Storage Instructions Blue top tube: separate plasma from cells as soon as possible; plasma may be stored on ice for up to 4 hours; otherwise store frozen

Causes for Rejection Blue top tube received more than 4 hours after collection, blue top tube not filled, clotted specimens

Turnaround Time 1-2 hours (often less than 1 hour if requested stat)

Reference Range See individual tests.

Methodology See individual tests.

Additional Information DIC is a common acquired coagulation disorder resulting from excessive activation of the coagulation system, usually due to massive tissue injury, sepsis, or certain pregnancy complications. The normal anticoagulant and fibrinolytic systems are overwhelmed and cannot contain the coagulation activation, which becomes systemic, resulting in disseminated microvascular thrombi. Thrombosis consumes platelets, coagulation factors, and the natural anticoagulants, which consequently become depleted. The decrease in coagulation factors causes PT and PTT prolongations, and may lead to bleeding. Depletion of platelets also contributes to the bleeding risk. The fibrinolytic system is activated to dissolve the fibrin thrombi, resulting in consumption of plasminogen as it is converted into plasmin, and the formation of fibrin degradation products (FDP) including D-dimers as plasmin degrades fibrin clots. FDP can contribute to bleeding, because they impair fibrin clot formation and interfere with platelet function. In acute DIC, the most obvious clinical symptom is bleeding, although the insidious underlying disseminated microvascular thrombosis may lead to tissue ischemia and consequently multiorgan failure. The key laboratory findings are elevated D-dimers or FDP, prolonged PT and/or PTT, and decreased or decreasing platelets and fibrinogen. Repeat testing may be needed to show that fibrinogen and/or platelets are decreasing over time. Fibrinogen is decreased in $\sim$50% of acute DIC cases; the PT is prolonged in $\sim$70%; and the PTT is prolonged in $\sim$50% of acute DIC cases. Thus, it is important to note that these tests can be normal in a substantial percentage of DIC cases. D-dimer or FDP should be positive in DIC.

Chronic DIC may develop when the activation of the coagulation system is low-grade and prolonged, as occurs with malignancy, retained dead fetus, aneurysm, or hemangioma.

486

The clinical features and laboratory findings in chronic DIC can be much more subtle than with acute DIC. Fibrinogen and platelet levels are commonly elevated, because they can increase during acute phase reactions in response to illness (including malignancy), injury, or other conditions. The PT and PTT may actually be short, possibly due to increased circulating activated coagulation factors. Large-vessel thrombosis can occur in chronic DIC of malignancy. The main laboratory abnormality for acute or chronic DIC is positive D-dimers or FDP, neither of which are specific for DIC.

Schistocytes are present on the peripheral blood smear in 50% or more of acute DIC cases. Schistocytes are generated by microangiopathic hemolysis of red blood cells severed by flowing through fibrin strands. A large number of other coagulation tests may be abnormal in acute or chronic DIC, but their clinical utility for DIC diagnosis remains uncertain. These include decreases in the natural anticoagulant proteins antithrombin, protein C, and protein S; prolonged thrombin time; elevated markers of coagulation activation (eg, prothrombin fragment 1.2, fibrinopeptide A, fibrinopeptide B, fibrin monomers, thrombin-antithrombin complexes), and the appearance of markers of fibrinolysis (plasmin-antiplasmin complexes, decreased plasminogen, and antiplasmin). The thrombin time is often prolonged because of decreased fibrinogen and/or elevated FDP. Elevated FDP interfere with fibrin polymerization, prolonging the thrombin time. Plasma markers of platelet activation, such as platelet factor 4 and beta-thromboglobulin, may also be detected. None of these markers are specific for DIC.

Treatment of the underlying condition is the primary treatment of DIC, along with supportive care including transfusions if needed for bleeding. Heparin use is controversial. New strategies, such as antithrombin concentrates or recombinant activated protein C, are under investigation.

Electrolytes, Blood

Synonyms Plasma Electrolytes; Serum Electrolytes

Use Monitor electrolyte status; screen water balance; diagnose respiratory and metabolic acid-base balance; evaluate hydrational status, diarrhea, dehydration, ketoacidosis in diabetes mellitus and other disorders; evaluate alcoholism and other toxicity states

Test Includes Sodium, potassium, chloride; often total CO_2, but in some laboratories pH and pCO_2 are measured and bicarbonate (HCO_3) and CO_2T are calculated. Anion gap is reported with electrolytes by some laboratories.

Limitations Hemolysis and prolonged contact of serum with cells produces elevation of potassium. The usual order for "electrolytes" does not include magnesium, osmolality, phosphorus, or lactic acid.

Specimen Blood

Volume 10 mL

Minimum Volume 3 mL serum or plasma

Container Red top tube or green top (heparin) tube

Collection Best to collect without tourniquet if possible. Do not allow patient to clench-unclench his/her hand.

Storage Instructions Do not freeze.

Methodology Ion-selective electrodes (ISE) or flame photometry are used for sodium and potassium. Sodium and potassium may also be measured by inductively-coupled plasma emission spectrometry.

Additional Information May be performed on heparinized plasma but not on EDTA plasma. Knowledge of pertinent clinical criteria allows for more cost effective ordering of blood electrolytes in patients in whom the information will be clinically significant. The anion gap, calculated $Na^+ - (Cl^- + HCO_3^-)$, provides useful information for interpreting acid-base disorders and may be useful for establishing a differential diagnosis in some conditions.

Specific References

Desai B and Seaberg DC, "The Utility of Routine Electrolytes and Blood Cell Counts in Patients With Chest Pain," *Am J Emerg Med*, 2001, 19(3):196-8.

Elkharrat D, Chevret S, LeCorre A, et al, "Appropriateness of Blood Count and Serum Electrolytes Prescribed in an Emergency Department," *Ann Emerg Med*, 2000, 35(5):S16.

Walsh CR, Larson MG, Leip EP, et al, "Serum Potassium and Risk of Cardiovascular Disease: The Framingham Heart Study," *Arch Intern Med*, 2002, 162(9):1007-12.

Electromyography

Synonyms Electrodiagnostic Study; EMG

Use In the neurologic literature, EMG has been performed in a wide variety of clinical situations, many of which are experimental or highly specialized in nature. Common indications for EMG in general practice include the following.

Evaluation of the patient with clinical features of primary muscle disease (symmetric and proximal weakness, muscle atrophy, intact sensory system, etc). Examples include:

- muscular dystrophy
- glycogen storage disease
- myotonia
- inflammatory myopathies (systemic lupus, sarcoidosis, infectious myopathies)
- polydermatomyositis
- alcoholic myopathy
- endocrine myopathies, and others

Evaluation of the patient with lower motor neuron disease, including:

- suspected peripheral nerve lesions, such as diffuse peripheral neuropathies, spinal root lesions, and trauma
- suspected disease of the anterior horn cells (characterized by asymmetric weakness, muscle atrophy, fasciculations), as in amyotrophic lateral sclerosis or poliomyelitis

Evaluation of botulism

Assessment of the patient with suspected upper motor neuron disease, when prior imaging studies are inconclusive. This includes occult lesions of the corticospinal tract (syringomyelia, tumor) and, less commonly, lesions of the cerebral tract (tumor, CVA). Evaluation of the patient with suspected neuromuscular junction disease (NMJ). This includes myasthenia gravis and the paraneoplastic Eaton-Lambert syndrome. Conventional EMG, as described here, is not the diagnostic test of choice for myasthenia gravis. However, other forms of EMG such as single fiber EMG and repetitive stimulation tests are highly specific for NMJ disease. Assessment of the patient with severe and persistent muscle cramps. Serial documentation of response to therapy for known cases of myopathy or neuropathy. Identification of significantly diseased muscle groups to help guide muscle biopsy (if clinical examination is not inconclusive). EMG is less useful in:

- the restless legs syndrome
- transient, self-resolving muscle cramps
- uncomplicated cases of polymyalgia rheumatica unless the diagnosis is in doubt or underlying myositis is suspected
- routine cases of fibrositis/fibromyalgia (EMG abnormalities have recently been documented in the medical literature but needle examination is not routinely indicated)

Adverse Reactions Local discomfort at the site of needle insertion is common. This is often mild in severity and has no significant sequelae. Pneumothorax has been rarely documented in the literature. This was associated with needle examination of paraspinal muscles.

Transient bacteremia has been reported, but routine antibiotic prophylaxis for patients with high-risk cardiac lesions is not generally recommended.

Contraindications The following situations represent relative contraindications:
- severe coagulopathy, including hemophilia and marked thrombocytopenia. It should be noted that EMG has been performed safely with platelet counts as low as 20,000/mm^3, but this is not recommended.
- systemic anticoagulation (eg, intravenous heparin, oral Coumadin®)
- patients with an unusual susceptibility to systemic infections (EMG has been known to cause transient bacteremia)
- patients undergoing a muscle biopsy after EMG require special consideration. It is well known that needle insertion and manipulation during EMG may cause local microscopic tissue damage on a traumatic basis. Histologically, this damage may be confused with a focal myopathy. Thus, some experts avoid detailed needle examinations of the specific muscle group which will be biopsied (although EMG testing of surrounding muscle groups is frequently performed).

Test Includes Insertion of a needle electrode into skeletal muscle to measure electrical activity and assess physiologic function. In this procedure, percutaneous, extracellular needle electrodes are placed into a selected muscle group. Muscle action potentials (AP) are detected by these electrodes, amplified, and displayed on a cathode ray oscilloscope. In addition, fluctuations in voltage are heard as "crackles" over a loudspeaker, permitting both auditory and visual analysis of the muscle APs. Testing is performed with the muscle at rest, with a mild voluntary contraction, and with maximal muscle contraction (where recruitment pattern and interference are noted). Unlike nerve conduction studies, EMG does not involve external electrical stimulation. Muscle APs (normal or abnormal) are physiologically generated. In various diseases of the motor system, typical electrical abnormalities may be present: increased insertional activity, abnormal motor unit potentials, fibrillations, fasciculations, positive sharp waves, decreased recruitment pattern, and others. EMG assesses the integrity of upper motor neurons, lower motor neurons, the neuromuscular junction, and the muscle itself. However, EMG is seldom diagnostic of a particular disease entity. Its major use lies in differentiating between the following disease classes: primary myopathy, peripheral motor neuron disease, and disease of the neuromuscular junction. As with nerve conduction velocity studies (with which EMG is usually paired), EMG should be considered an extension of the history and physical examination.

Patient Preparation Details of procedure are reviewed with the patient. Considerable patient apprehension often accompanies "needle tests" and calm reassurance from the medical team will go far in allaying such anxieties. Aspirin products should be discontinued 5-7 days beforehand. Nonsteroidal agents should also be stopped several days in advance. Routine medications may be taken on the morning of the examination. If coagulopathy is suspected, appropriate hematologic tests should be ordered (PT/PTT, CBC, bleeding time, etc). If a primary muscle disease is suspected, creatine phosphokinase (CPK) level should be drawn prior to needle examination. Routine EMG testing may cause minor elevations in CPK up to one and one-half times baseline. However, striking elevations in CPK, as seen with polymyositis or muscular dystrophy, are not associated with EMG testing.

Aftercare No specific activity restrictions are necessary. Patient may resume previous activity level.

Special Instructions Requisition from ordering physician should include brief clinical history, tentative neurologic diagnosis, and the specific limb(s) or muscle group(s) in question. Physician should also state whether nerve conduction studies are desired, although in some centers these may be added at the neurologist's discretion.

Critical Values Abnormalities in one or more of the previously stated parameters may be seen.

Insertional activity: Increased in both neurogenic disorders (eg, lower motor nerve disease) and myogenic disorders (eg, polymyositis), and thus is considered nonspecific. Decreased insertional activity is less common, but may be associated with far advanced denervation or myopathy, especially when muscle is replaced by fat or collagen. A distinctive insertional pattern is seen with myotonia, an unusual neurologic disorder, and is termed "myotonic discharge."

Abnormal activity at rest: Instead of the electrical silence that characterizes the muscle at rest, spontaneous action potentials in single muscle fibers ("fibrillation potentials") may be observed in several disease states.

Fibrillations are seen 1-3 weeks after destruction of a lower motor neuron. Denervated muscle fibers develop heightened chemosensitivity and individual muscle fibers contract spontaneously. The phenomenon of "positive sharp waves" may also be seen. Fibrillations may also occur with severe polymyositis when extensive areas of necrosis interrupt nerve innervation. The naked eye is unable to perceive fibrillations.

Fasciculations represent random contractions of a full motor unit, often visible through the skin. (A motor unit is comprised of an anterior horn cell, axon, neuromuscular junction, and the numerous muscle fibers supplied by the axon.) Fasciculations may be benign, with no other EMG abnormalities observed. They may also be associated with amyotrophic lateral sclerosis, other anterior horn cell diseases, nerve root compression, herniated nucleus pulposus syndrome, acute polyneuropathy, and others.

Abnormalities in the motor unit potential (MUP): Individual motor unit potentials are distinguishable during a submaximal muscle contraction. Abnormalities in amplitude, shape (number of phases, serrations, configuration), and duration are possible. Increased MUP amplitude is seen in lower motor neuron disease but is normal in upper motor neuron disease. Decreased MUP amplitude is characteristic of polymyositis and other myopathies and duration of the MUP is also decreased.

Characteristic EMG findings of botulism include normal nerve conduction velocity, normal sensory nerve function, a pattern of brief, small amplitude motor potentials, and an incremental response or facilitation to repetitive stimulation (often only seen at 50 Hz)

Abnormalities in interference pattern: The normal "full recruitment" pattern seen during maximal muscle contraction is often compromised in disease states. In myogenic lesions, such as polymyositis and myotonia, the amplitude of the MUPs is significantly decreased but the recruitment pattern is normal (ie, the number of activated motor units is normal but the number of muscle fibers per motor unit is diminished). In LMN lesions, the number of motor units recruited is decreased. In severe cases of neuropathy, the maximum interference pattern resembles that of a single MUP, with individual potentials visible. Amplitude of MUPs may be normal. These findings are summarized in the previous figure.

Equipment EMG is performed in a specially equipped procedure room, usually reserved for electrodiagnostic studies. Basic instrumentation includes:

• needle electrodes; these may be monopolar (sharpened, coated steel wires), coaxial, or bipolar (two wires within a needle). These needles record electrical activity from muscle fibers directly contacting the tip, as well as fibers within a several millimeter radius.

• amplifier with filters

• cathode ray oscilloscope with the vertical axis measuring voltage, the horizontal axis measuring time. This device usually has an audio amplifier and loudspeaker, which converts APs to sound energy.

• data storage apparatus (eg, magnetic tape recorder)

Methodology A brief neurologic examination is performed prior to the start of the procedure. For EMG of the extremities, patient lies recumbent on the examination table. When paraspinal muscles are tested, the patient adopts a prone position. No intravenous sedatives or pain medications are used. Local anesthesia is also not required despite the generous number of needle insertions. The skin is cleansed thoroughly with alcohol pads, as necessary. Patient is instructed to relax as much as possible. The following steps are carried out.

• Recording needle electrode is inserted percutaneously into the muscle under considera-tion. The initial electrical activity of the muscle, as seen on the oscilloscope screen and heard over the loudspeaker, is termed the insertional activity.

• Next, the needle is held stationary and the muscle action potentials during voluntary relaxation are recorded.

• Patient performs a mild, submaximal contraction of the test muscle. The summed muscle action potentials - the motor unit potential - are observed on the oscilloscope.

• Finally, a maximal muscle contraction is carried out. The compound action potentials generated during this maneuver are studied for interference and recruitment pattern.

Needle examination is, by nature, a slow and labor-intensive process. Numerous muscles must be tested individually including both symptomatic and clinically asymptomatic muscles. Within a specific muscle, several independent sites may need to be examined, particularly when the muscle has a large surface area. A number of myopathic processes are focal and sampling errors even within an individual muscle are possible (ie, disease process may effect proximal portion of a muscle, sparing distal fibers).

Results are interpreted by neurologist or physiatrist with preliminary impression written in chart immediately. A formal, typed report is completed several days later. The fundamental principles underlying test interpretation are as follows.

• Insertional activity: Immediately upon needle insertion, there is a brief burst of electrical activity lasting <300 msec. This "insertional activity" is heard over the loudspeaker and may be increased or decreased in various disease states.

• Electrical activity at rest: Muscle tissue is normally silent at rest. No action potentials are seen on the oscilloscope.

• Minimal muscle contraction: When a minimal contraction is performed, several motor unit potentials (MUPs) are activated. Several individual APs are normally visible on the oscilloscope at a rate of 4-5/second. The idealized configuration of a single MUP is depicted in the following figure (see normal column).

• Full voluntary contraction: As the strength of the muscle contraction increases, further muscle units are "recruited". On the oscilloscope the APs appear more disorganized and individual APs can no longer be recognized. At the peak of a contraction the "complete recruitment pattern" is seen, which represents a compilation of motor unit potentials firing asynchronously. The normal interference pattern is considered "full," that is, the amplitude of APs is high (≤5 mV) and firing rate is fast (40/second).

EMG FINDINGS

LESION \ EMG Steps	NORMAL	NEUTOGENIC LESION		MYOGENIC LESION		
		Lower Motor	Upper Motor	Myopathy	Myotonia	Polymyositis
1 Insertional Activity	Normal	Increased	Normal	Normal	Myotonic Discharge	Increased
2 Spontaneous Activity	—	Fibrillation Positive Wave	—	—	—	Fibrillation Positive Wave
3 Motor Unit Potential	0.5-1.0 mV 5-10 ms	Large Unit Limited Recruitment	Normal	Small Unit Early Recruitment	Myotonic Discharge	Small Unit Early Recruitment
4 Interference Pattern	Full	Reduced Fast Firing Rate	Reduced Slow Firing Rate	Full Low Amplitude	Full Low Amplitude	Full Low Amplitude

Idealized EMG findings, normal, neurogenic lesions, and myogenic lesions. Reproduced with permission from Kimura J, Chapter 13, "Types of Abnormality," *Electrodiagnosis in Diseases of Nerve and Muscle: Principles and Practice*, 2nd ed, Philadelphia, PA: FA Davis, 1989, 263.

Specific References

Adams RD and Victor M, *Principles of Neurology: Companion Handbook*, 6th ed, New York, NY: McGraw-Hill Book Co, 1997.

Aminoff MJ, *Electromyography in Clinical Practice*, 4th ed, New York, NY: Churchill-Livingstone, 1999.

Martin JB and Hauser SL, "Approach to the Patient With Neuromuscular Disease," *Harrison's Principles of Internal Medicine*, 14th ed, Fauci AS, Braunwald E, Isselbacher KJ, et al, eds, New York, NY: McGraw-Hill Book Co, 1998, 2277-93.

Fibrinogen

Synonyms Factor I

Use One of several tests performed in a DIC panel, a prolonged PT or PTT evaluation, and an evaluation of a patient with an unexplained bleeding history

Applies to Acute Phase Reactant; Afibrinogenemia; Dysfibrinogenemia; Plasmin; Sedimentation Rate; Thrombin

Abstract Fibrinogen is converted into fibrin clot by thrombin. Fibrinogen levels < 100 mg/dL can be associated with bleeding. Acquired decreases in fibrinogen (eg, with liver dysfunction or DIC) are much more common than hereditary deficiencies.

Limitations Heparin concentrations > 0.6 units/mL can falsely decrease the result with the Clauss method (described below). Usual therapeutic doses of heparin do not significantly affect PT-based methods. The Ellis method is more sensitive to heparin than the Clauss method. Some reagents contain hexadimethrine bromide (Polybrene) to neutralize heparin, allowing fibrinogen to be measured in specimens containing heparin. Fibrin degradation products (FDP) >30-100 mcg/mL may decrease fibrinogen values with the Clauss method. Hirudin or argatroban anticoagulation may falsely decrease fibrinogen levels levels with the Clauss and Ellis method, and possibly the PT-based method.

Specimen Plasma

Minimum Volume 4.5 mL

Container Blue top (sodium citrate) tube

Collection Routine venipuncture. If multiple tests are being drawn, draw blue top tubes after any red top tubes but before any lavender top (EDTA), green top (heparin), or gray top (oxalate/fluoride) tubes. Immediately invert tube gently at least 4 times to mix. Tubes must be appropriately filled. Deliver tubes immediately to the laboratory.

Storage Instructions Separate plasma from cells as soon as possible. Store plasma at room temperature for up to 2 hours, at 2°C to 8°C for up to 4 hours, or store frozen.

Causes for Rejection Specimen received more than 4 hours after collection, tubes not filled, clotted specimen

Turnaround Time Less than 1 day

Reference Range Approximately 150-400 mg/dL

Methodology

Functional (activity) assays: The majority of clinical laboratories use the Clauss method, which is essentially a dilute thrombin time. A high concentration of thrombin is added to dilute patient plasma, which converts fibrinogen into fibrin clot. The clotting time is inversely proportional to the amount of fibrinogen in the sample. In the Ellismethod, a lower amount of thrombin is added to undiluted patient plasma and change in turbidity is measured in a spectrophotometer. In the PT-based method, thromboplastin (tissue factor with phospholipid) is added to undiluted patient plasma to generate endogenous thrombin, and light scatter or turbidity is measured. The measured optical change (before and after fibrin clot formation) is proportional to the amount of fibrinogen in the sample.

Antigen assays (immunoassays) for fibrinogen measure the quantity of fibrinogen without assessing fibrinogen function. This method is not routinely indicated and is usually a send-out test (see Additional Information for its use in dysfibrinogenemia evaluations).

Additional Information Fibrinogen decreases with liver disease, due to decreased hepatic synthesis. However, fibrinogen may be normal or even elevated until late stages of hepatic disease. Fibrinogen decreases in DIC due to excessive thrombin generation, which converts fibrinogen into fibrin. The plasma fibrinogen level, however, is not a sensitive marker for DIC. Presence of a high fibrinogen level in patients with DIC is associated with a poor prognosis. Fibrinogen also decreases with thrombolytic therapy and fibrinolysis because plasmin breaks down fibrinogen in addition to fibrin.

Fibrinogen becomes elevated during acute phase reactions and during pregnancy, smoking, and physical inactivity. As with certain other acute phase reactants (eg, C-reactive protein), elevated fibrinogen has been associated with an increased risk of myocardial infarction. There is evident complexity involving gene to gene and gene to environment interactions in the genesis and modulation of cardiovascular risk. Thrombogenicity may relate to characteristics of the fibrin network (thin fibers and small pores with restricted entry of fibrinolytic enzymes) also affected by fibrinogen and factor XIII polymorphisms.

Hereditary deficiencies of fibrinogen are rare. The PT and PTT may be prolonged. Bleeding symptoms may include bruising, epistaxis, menorrhagia, bleeding with surgery, trauma, dental extractions, and postpartum, and bleeding in the gastrointestinal or genitourinary tract. Miscarriage and poor wound healing are also complications of fibrinogen deficiency. Umbilical stump bleeding and bleeding with circumcision may be noted in newborns with afibrinogenemia. Intracranial hemorrhage has been reported with afibrinogenemia. In general, deficiencies of fibrinogen tend to be milder than factor VIII or IX deficiencies (hemophilia).

There are three major types of fibrinogen deficiency. The homozygous quantitative form, called afibrinogenemia, results in a severe quantitative deficiency of fibrinogen and an increased risk for bleeding. The heterozygous form of this deficiency is hypofibrinogenemia, with less severe reductions in the fibrinogen level and little or no bleeding. Fibrinogen consists of two copies of each of three polypeptide chains called α, β, and γ. Among the afibrinogenemia mutations that have been characterized thus far, most have been found in the α-fibrinogen chain gene.

Dysfibrinogenemia is a qualitative fibrinogen deficiency, characterized by the production of dysfunctional fibrinogen. Many different mutations are known to cause hereditary dysfibrinogenemia. Most patients with hereditary dysfibrinogenemia are heterozygous. Rare homozygous cases have been reported. Dysfibrinogenemia patients are usually asymptomatic or have mild bleeding, but severe bleeding has been reported. Interestingly, some dysfibrinogenemia cases are associated with thrombosis, with or without bleeding. Dysfibrinogenemia has an estimated prevalence of 0.8% in patients with venous thrombosis. Arterial thrombosis is less frequent than venous thrombosis in these patients. Acquired forms of dysfibrinogenemia, of uncertain clinical significance, can be seen with liver disease or acute phase reactions with generation of high levels of fibrinogen (Galanakis D, personal communication 1999). The thrombin time and Reptilase® time, which measure the clotting time during the conversion of fibrinogen into fibrin, are often prolonged in dysfibrinogenemia. The PT and PTT may also be prolonged. In dysfibrinogenemia, assays that measure fibrinogen function show lower levels than assays that measure fibrinogen quantity (immunological or "antigen" assays), because fibrinogen function is impaired but fibrinogen quantity is not. This potentially diagnostic disparity between functional and antigen levels may be less pronounced with PT-based functional fibrinogen assays than with Clauss-based functional assays.

Hantavirus Serology

Synonyms Muerto Canyon Strain Virus

Use Confirm the diagnosis of hantavirus pulmonary syndrome

Applies to Hantavirus Pulmonary Syndrome

Test Includes Detection of IgM and IgG antibody specific for the Muerto Canyon strain of hantavirus (nucleocapsid proteins)

Abstract An outbreak of severe respiratory illness associated with respiratory failure, shock, and high mortality was recognized in May, 1993 in the southwestern part of the U.S. The cause of the illness was identified as a unique hantavirus now known as the Muerto Canyon strain, and the disease is now called hantavirus pulmonary syndrome (HPS). Since the recognition of this disease, other cases have been recognized in 17 states, with most of the cases occurring west of the Mississippi. HPS begins with nonspecific symptoms such as fever and myalgia, which is followed in 3-6 days by progressive cough and shortness of breath. Common findings during this later stage include tachypnea, tachycardia, fever, and hypotension. Bilateral abnormalities on the chest radiograph are detected, and pleural effusions are common. Hemoconcentration, thrombocytopenia, prolonged activated partial thromboplastin time, an increased proportion of immature granulocytes on the peripheral blood smear, leukocytosis, and elevated levels of serum lactate dehydrogenase and aspartate aminotransferase are found. Serum antibodies are detectable at the time of clinical presentation.

Limitations Assays for the detection of antibody to hantavirus are experimental and none have been approved by the Food and Drug Administration for use in the United States. All requests for testing must be sent to the CDC.

Specimen Serum from acute phase of illness

Container Red top tube

Storage Instructions Serum can be stored at 4°C up to 1 week; serum should be stored at −70°C after 1 week and during shipping

Special Instructions Specimens should be sent to the CDC through state health departments.

Reference Range No detectable hantavirus IgM or less than a fourfold increase in IgG specific for the N and G1 proteins of the Muerto Canyon virus.

Methodology Western blot; enzyme-linked immunosorbent assay (ELISA)

Additional Information Hantaviruses are single-stranded RNA viruses of the family Bunyaviridae. The Muerto Canyon strain of hantavirus has been found in a proportion of the deer mouse (*Peromyscus maniculatus*) population, which is prevalent in the western U.S. Thus, this rodent species is thought to be the reservoir for the etiologic agent of HPS. Recommendations for prevention include avoidance of contact with the deer mouse and excreta from deer mice. Currently, no evidence exists for person-to-person transmission of HPS.

HPS can also be diagnosed by detection of hantavirus antigen in tissue by immunohistochemistry with a monoclonal antibody reactive with conserved hantaviral nucleoproteins. In addition, hantaviral nucleotide sequences can be detected in tissue using a reverse transcriptase polymerase chain reaction.

Heavy Metal Screen, Blood

Synonyms Metals, Blood; Poisonous Metals, Blood; Toxic Metals, Blood

Use Screen for heavy metal poisoning

Test Includes Antimony, arsenic, bismuth, boron, cadmium, cobalt, copper, lead, mercury, thallium

Abstract Used principally to detect arsenic, cadmium, mercury, and lead poisoning.

Specimen Whole blood (EDTA) plus serum

Volume 20 mL

Container Special metal-free tube and red top tube

Storage Instructions Refrigerate: do not spin down.

Special Instructions Check with laboratory performing the assay to determine what elements will be detected and for special instructions.

Methodology Atomic absorption spectrometry (AA), inductively-coupled plasma (ICP)

Heavy Metal Screen, Urine

Synonyms Metal Screen; Metals, Toxic; Poisonous Metals, Urine; Toxic Metals, Urine

Use Screen for heavy metal poisoning and toxic exposure

Test Includes Arsenic, mercury, lead (could also include nickel and cadmium)

Specimen 24-hour urine

Volume Entire collection

Minimum Volume 150 mL aliquot

Container Plastic, acid-washed urine container (preferably polyethylene), no preservative, 20-25 mL 6 N HCl (low metal content)

Storage Instructions Refrigerate

Reference Range Arsenic: <50 mcg/L; lead: <80 mcg/L; mercury: <20 mcg/L; nickel: <25 mcg/L; cadmium: <10 mcg/L

Methodology Atomic absorption spectrometry (AA), inductively-coupled plasma (ICP)

Additional Information See Heavy Metal Screen, Blood for further information.

Identification DNA Testing

Synonyms DNA Analysis for Parentage Evaluation; DNA Fingerprinting; DNA Testing; Genetic Identification by DNA Fingerprinting; Paternity Testing by DNA Testing; RFLP Analysis for Parentage Evaluation

Use The analysis of highly polymorphic regions of human DNA can clarify the relationships between individuals and verify the identify of unknown individuals (such as suspects in criminal investigations or unidentified victims of murder).

Test Includes Identification of individuals by using DNA polymorphic regions

Abstract Progress in the field of DNA technology and Human Genome Project has resulted in tremendous knowledge about the genetic material that makes each individual unique. The DNA from both maternal and paternal sources may be normal, but will have slight variations in character. These variations can be detected and used to map heredity much like the variations in blood group antigens and the human leukocyte antigen (HLA) system. By using between 20-30 different polymorphic sites on different chromosomes, identity or parentage can be established with up to 99.99% exclusion probability. Other DNA identification applications include identification of suspects in forensic cases, and origin and migration history of modern humans.

Patient Preparation Patient should receive no transfusions 90 days prior to testing.

Specimen Peripheral whole blood, tissue, amniotic fluid, semen, or cultured cells; dried blood, hair and skin scrapings are frequently used in forensic cases.

Volume Two tubes of peripheral blood in ACD tubes or lavender top tubes (EDTA); 0.1 g to 1 g of tissue; 2 to 5 mL of semen; two T25 flasks of cells

Minimum Volume 0.1 g tissue

Collection Blood should be collected in a yellow top (ACD) tube or lavender top (EDTA) tube; tissue should be frozen at $-70°C$; amniotic cells, fibroblasts, or lymphocytes should be grown in appropriate media in T25 tissue culture flasks. Collection in heparin tubes should be avoided as heparin interferes in polymerase chain reaction. Dried blood, hair and skin scrapings are collected in a plastic sealable bag.

Storage Instructions Store tissue at $-70°C$ or on dry ice. Peripheral blood should be stored and shipped at 4°C. Do **not** freeze blood. Dried blood, hair and skin scraping can be stored at room temperature.

Causes for Rejection If the tissue specimen thaws during transport to the laboratory or before shipping, DNA may not be obtained from the specimen. Blood samples that have been frozen and thawed will yield low quality DNA. Specimens inadequately identified will be rejected.

Turnaround Time 2-4 weeks. Samples of DNA can be stored for an unlimited amount of time.

Reference Range The test provides a 99.99% exclusion probability.

Methodology DNA is released and isolated from the white blood cells, tissue, or cultured cells by lysing. The DNA is then digested with various restriction enzymes and electrophoresed through an agarose gel. DNA is then transferred to a solid support such as a nylon membrane and hybridized with a radioactive or fluorescent DNA probe. After washing the unhybridized DNA probe off the membrane, the target DNA is exposed to X-ray or fluorescence sensitive film to detect the polymorphic regions of DNA. Certain regions of the human genome show a high degree of polymorphism in that >85% of the population show heterogeneity. These regions are highly informative in determining DNA identification. Upon digestion with different restriction enzymes, the size and pattern of the DNA fragments vary with each individual, an inherited trait. If the appropriate family members are tested, the inheritance pattern can be established, and applied to determination of the paternity of a child or the zygosity of twins. There is application to the identification of an unknown criminal or victim.

Sufficient material to provide a source of DNA may not be available at the scene of the crime. When the quantity of DNA (eg, from dried blood, hair and skin scraping) is very small, the polymerase chain reaction is used to amplify DNA before identification by digestion with restriction enzymes.

Additional Information The genetic material of humans is highly polymorphic, and an individual's genotype represents a unique pattern that determines that person's identity and heredity. The only exception to this rule is identical twins, since they are derived from a single fertilized egg and hence have the same DNA profile.

Healthcare professionals should exercise care in the collection and storage of specimens for DNA testing to prevent contamination and to preserve evidence that may have crucial legal significance.

Lactic Acid, Whole Blood, or Plasma

Synonyms Blood Lactate; Lactate, Blood

Use The differential diagnosis of type A lactic acidosis includes hypoxemia (eg, carbon monoxide, anemia, methemoglobinemia, respiratory failure), hypotension, shock, decreased perfusion, and strenuous exercise. Type B may be caused by ethanol, methanol, ethylene glycol, phenformin, cyanide, nitroprusside, salicylate, nalidixic acid, streptozocin, diabetes, liver failure, renal failure, infection, systemic malignancy, and inborn errors of metabolism.

Lactate determination is generally indicated if anion gap is >20 mmol/L and if pH is <7.25 and the pCO_2 is not elevated.

When lactate is <45 mg/dL (SI: <5.0 mmol/L), suspect carbohydrate infusions, exercise, diabetic ketoacidosis, or ethanol. When lactate is >45 mg/dL, suspect shock, severe anemia, severe congestive failure, or systemic malignancy.

Contraindications Absence of acidosis is **not** a contraindication for this test.

Applies to Biotin; D-Lactate; Metformin; Oxygen Transport; Phenformin

Abstract Lactate (L-lactic acid) is formed from pyruvate in glycolysis. Strenuous exercise can produce a 10- or 15-fold increase in venous plasma lactate within several seconds. Blood lactate is lowest during fasting and reaches the upper end of the reference interval in the postprandial state. The D-isomer of lactic acid does not occur in human metabolism, but may be present in rare instances (see Additional Information). Lactic acidosis, with elevated blood lactate, occurs in two clinical contexts. Type A lactic acidosis is due to hypoxia and is the more common form; type B lactic acidosis may be due to drugs, inborn errors of metabolism, severe liver disease, or a metabolic myopathy.

Limitations Assays for L-lactate provide no information about the isomer, D-lactate (see Additional Information).

Patient Preparation Mannitol interference is described for a whole-blood analysis method.

Specimen Whole blood, arterial or venous, or plasma, *vide infra*. Arterial blood is preferred, since contraction of muscles can cause increase in lactate in venous blood.

Container Gray top (sodium fluoride) tube; heparinized syringe, heparin-containing tube, anaerobic draw, depending upon available instrumentation

Collection Avoid hand-clenching, and if possible use of a tourniquet. A tourniquet or a patient clenching and unclenching his/her hand will lead to build-up of potassium and lactate from the hand muscles.

Lactate is commonly needed with or as stat follow-up to venous or arterial pH. Serial determinations are often valuable. Send specimen on wet ice.

Storage Instructions Centrifuge immediately and take off plasma (unless laboratory uses a whole blood method). Keep plasma on ice or at 2°C to 8°C, analyze promptly.

Causes for Rejection Specimen not received on ice

Turnaround Time Use of whole blood improves turnaround time.

Special Instructions Keep tube on ice until delivered. Tube must be processed within 15 minutes of being drawn.

Reference Range See table.

Whole Blood Lactate		
	mmol/L*	mg/dL*
Venous		
At rest	0.5-1.3	5-11
In hospital	0.9-1.7	8-15
Arterial		
At rest	0.36-0.75	3-7
In hospital	0.36-1.25	3-11
Plasma Lactate		
	mmol/L†	mg/dL†
Venous	0.5-2.2	4.5-19.8
Arterial	0.5-1.6	4.5-14.4

*Sacks DB, "Carbohydrates," *Tietz Textbook of Clinical Chemistry*, 3rd ed, Burtis CA and Ashwood ER, eds, Philadelphia, PA: WB Saunders Co, 1999, 789.
†Painter PC, Cope JY, and Smith JL, "Reference Information for the Clinical Laboratory," *Tietz Textbook of Clinical Chemistry*, 3rd ed, Burtis CA and Ashwood ER, eds, Philadelphia, PA: WB Saunders Co, 1999, 1822.

Critical Values There is an inverse relationship between hyperlactatemia and survival. Lactate >36 mg/dL (4 mmol/L) is a strong predictor of need for hospital admission from Emergency Department (ED) as well as predictor of mortality.

Possible Panic Range ≥45.0 mg/dL

Methodology Enzymatic; other methods include gas chromatography (GC), amperometric, enzymatic, substrate-specific electrode. Whole blood analysis by trilayer-biosensors is available.

Additional Information Most assays for lactic acid measure only the L-isomer of lactic acid, and all the discussion above relates to the L-isomer. The D-isomer of lactic acid, while not produced in human metabolism, can, on rare occasions, be absorbed from gastrointestinal microorganisms. Serious disease from D-lactic acidosis has been reported after intestinal bypass operations. D-lactic acidosis is rare, and difficult to diagnose. D-lactic acid can be measured by gas-liquid chromatography and by microbiologic assay.

Normal L-lactate occurs with high D-lactate in D-lactic acidosis. Metabolic acidosis following jejunoileal bypass for obesity, related to altered gastrointestinal flora, may develop in subjects who develop dysarthria, cerebellar ataxia, and confusion as well, in whom D-lactate is the causative anion.

Liver Disease: Laboratory Assessment, Overview

Synonyms Liver Profile

Use Evaluate hepatobiliary disease, hepatoma, autoimmune hepatitis and cirrhosis, including biliary cirrhosis; investigate otherwise unexplained increases in such tests as AST, ALT, alkaline phosphatase, or prolongation of prothrombin time; work up possible alcoholism.

A number of useful tests include those to work up pancreatitis, such as serum and urine amylase and serum lipase.

The two most common causes of persistent abnormalities of liver function in Western countries are hepatitis C and heavy alcohol consumption. Use of blood alcohol determinations for investigation of liver disease has been advocated. Other tests useful in alcoholism include carbohydrate deficient transferrin, MCV, albumin, and folate levels, triglycerides, GGT, and bilirubin. AST increment is greater than that of ALT in alcoholic hepatitis; the AST:ALT ratio is >2.0 and AST is not increased more than 250 units/L. Prothrombin time is often helpful. Alcoholism causes malnutrition, including deficiencies of vitamins, including thiamine and vitamin A as well as folate. It leads to hyperlacticacidemia, hyperuricemia, ketosis, and hyperlipidemia. Alcoholics are vulnerable to a wide variety of substances, solvents, and medications including acetaminophen. Alcoholism may lead to a variety of complications including cirrhosis, gastritis, malnutrition, pancreatitis, and cardiomyopathy.

Cholestasis and biliary tract: alkaline phosphatase, GGT, total bilirubin, conjugated bilirubin, eosinophil count, urine bile (as part of urinalysis) are used. In extrahepatic biliary obstruction the serum alkaline phosphatase is increased two to three times or more while AST remains <300 units/L. Very high alkaline phosphatase levels may be found with intrahepatic cholestasis, such that alkaline phosphatase, which is high out of proportion to the severity of jaundice, may indicate an intrahepatic disease. Viral, alcoholic, or drug-related cholestatic hepatitis may give rise to chemistry tests indistinguishable from those of extrahepatic obstruction. Primary biliary cirrhosis, primary sclerosing cholangitis, overlap syndrome, and autoimmune cholangiopathy sometimes represent difficult differential diagnosis.

Liver excretory function: urine urobilinogen, total bilirubin, conjugated bilirubin.

In **viral hepatitis**, ALT is greater than AST, but in chronic liver diseases the AST:ALT ratio is less useful. In cirrhosis caused by hepatitis B and other agents, the AST:ALT ratio may be >1.0. In acute viral hepatitis, AST is usually three to five times or more higher (as multiples of the upper limit of normal) than LD; in cases that clinically resemble hepatitis, but in which LD equals or exceeds AST, LD isoenzymes may be useful. LD_4 and LD_5 are the hepatic fractions. An isomorphic pattern, if detected, may suggest infectious mononucleosis, CMV infection, neoplasm, or cirrhosis/alcoholism, depending on clinical setting. Transaminases increase in viral and drug-induced hepatitis and peak within 7-14 days, usually in the low thousands range, returning to normal in about 6 weeks in uncomplicated viral hepatitis. At onset they are generally more than 10 times the upper limit of the reference interval. Alkaline phosphatase is only moderately elevated. Acute viral hepatitis includes at least five separate disease entities.

Appropriate positive serological tests support a diagnosis of viral hepatitis, while negative ones provide support for other disorders including drug-induced hepatitis. Resolution of liver disease with removal of the offending agent enhances the latter diagnosis. Ecstasy, a synthetic amphetamine, may cause hepatic damage resembling acute viral hepatitis.

Chronic hepatitis includes chronic hepatitis B, C, and D and autoimmune hepatitis. It clinically may resemble alcoholic and nonalcoholic steatohepatitis. A review of its classification, criteria, and grading is available. Nonalcoholic steatohepatitis (NASH), a common cause of liver disease in Western countries, is associated with type 2 diabetes, hyperlipidemia, jejunoileal bypass, and drugs including amiodarone and perhexilene. The only clinical manifestation of hepatic steatosis and NASH may be moderate increases of aminotransferase concentrations. The ratio of ALT to AST (ALT:AST ratio or De Ritis ratio) is >1, but <1.0 in alcoholic liver disease (it is normally <1). Liver biopsy is recommended.

Immunologic stimulation: Protein electrophoresis: features suggestive of cirrhosis but not always present in that disease include low albumin, low $alpha_2$, polyclonal or oligoclonal gammopathy, and beta/gamma bridging. Oligoclonal gammopathy is found in <50% of cases of autoimmune hepatitis. HLA phenotypes may be relevant.

Autoimmune hepatitis, found predominantly in young to middle-aged women, is recognized by the presence of increased immunoglobulins and the presence of autoantibodies. More than 80% have hypergammaglobulinemia. Two other major groups of autoimmune liver diseases are recognized, primary biliary cirrhosis and primary sclerosing cholangitis. The distinction between autoimmune hepatitis and chronic viral hepatitis with hyperglobulinemia and/or autoantibodies is sometimes blurred. Wilson disease also enters this differential diagnosis. In autoimmune hepatitis, classical laboratory features include elevation of serum aminotransferases, bilirubin and alkaline phosphatase, and increased gamma globulins, especially IgG with autoantibodies (ANA and antismooth muscle antibodies in autoimmune hepatitis type 1). In type 2 autoimmune hepatitis, anti-liver-kidney-microsomal 1 antibodies are detected. Soluble liver antigen and smooth muscle antibody are anticipated in type 3. Liver biopsy is recommended for disease confirmation.

In ischemic hepatitis, as develops in instances of cardiac failure and hypotension, the transaminases may be >10,000 IU/L, and vast increases of LDH may be found. Such increased transaminase concentrations may be seen as well with acetaminophen overdose and with herpes simplex hepatitis.

In the presence of liver disease with hemolysis, Wilson disease must be considered. With hypoceruloplasminemia, hypocupremia, and hypercupruria, a diagnosis of Wilson disease is expedited, but liver biopsy for microscopy including rhodamine and orcein stains and tissue copper analysis are worthwhile.

Hemochromatosis: Mild abnormalities in liver profile tests (AST, ALT, ALP) may occur in hemochromatosis. Iron, IBC, transferrin, ferritin, and molecular testing are available. Liver biopsy is considered essential for confirmation of the diagnoses of Wilson disease and of many cases of hemochromatosis, save that biopsy may be unnecessary in patients younger than age 40 with normal results.

Liver disease related to alpha$_1$-antitrypsin deficiency is associated with periodic acid-Schiff positive diastase-resistant globules and an abnormal α_1-antitrypsin phenotype, PiZZ homozygotes, and several other alleles. (The MM phenotype (PiMM) is normal.)

Hepatic functional reserve: Both albumin and prothrombin time are useful in evaluation of the liver, but they are nonspecific. Albumin reflects hepatic synthesis and nutritional status but is lost in a variety of gastrointestinal and renal diseases.

Liver metabolic function: Serum ammonia may increase in liver necrosis and cirrhosis as well as in Reye syndrome.

Hepatoma (carcinoma, liver) and other tumors: Useful assays include alkaline phosphatase, GGT, total LD, CEA, and HB$_s$Ag. Alpha$_1$-fetoprotein may increase moderately in nonmalignant liver diseases; rising or high levels may indicate hepatoma. Such clinical laboratory tests do not prove tumor without imaging and biopsy. Other primary tumors of liver, benign and malignant, are found.

Metastatic tumors are the most common malignant neoplasms of the liver. The most frequent primaries include carcinoma of the stomach, colon, pancreas, esophagus, lung, and breast. The liver is a site of spread of malignant lymphomas.

Applies to Acetaminophen; ALT:AST Ratio; AST:ALT Ratio; De Ritis Ratio; Ecstasy; LFTs; Liver Function Tests, So-Called

Test Includes The differential diagnosis of liver disease often requires albumin; bilirubin, total; bilirubin, direct; alkaline phosphatase; aspartate aminotransferase (AST/SGOT); alanine aminotransferase (ALT/SGPT); protein, total; and often GGT (GGTP). It may also include serum protein electrophoresis, prothrombin time, and hepatitis serological tests when indicated. Alpha$_1$-antitrypsin phenotype and quantitation on occasion explains cases otherwise difficult to classify but is rarely, if ever, included in liver profiles. Other tests that may be helpful in evaluation of liver disease include immunoglobulins IgG, IgA, IgM; ANA; antimitochondrial antibody; smooth muscle antibody; liver/kidney microsomal type 1 antibodies and alpha-fetoprotein. Ammonia is useful in selected cases (eg, Reye syndrome, urea cycle disorders, and certain organic acidurias). Lactate dehydrogenase and LD isoenzymes may be useful in the differential diagnosis of icterus.

Abstract Characterization of liver disease requires correlation of the medical history, physical examination, laboratory test results, and, when indicated, the liver biopsy. Classification of liver disease includes hepatocellular, cholestatic (intra- or extrahepatic) and infiltrative states. Clinical assessment includes patient history and physical examination. Family history is especially relevant in cases of hemolytic anemia, Gilbert syndrome, Dubin-Johnson syndrome, Wilson disease, hemochromatosis, and α_1-antitrypsin deficiency.

Limitations Some types of hepatic injury may not be accompanied by very much increase in enzymes (eg, injury from ethionine or phosphorus).

The enzymes that are sometimes called "LFTs" ("liver function tests") often lack specificity for the liver when used alone, and reflect injury or disease of other organs as well. Use of LDH isoenzymes supports specificity: LDH_5 reflects disease of liver or of striated muscle, and CK arises from injury to the latter.

More specialized tests are often necessary to establish an etiologic diagnosis. Liver biopsy is often needed to provide precise diagnosis in subjects with subacute to chronic abnormalities. See table.

Groups of Serum Enzymes According to Their Sensitivities				
Enzyme	Cholestasis*	Hepatocellular Necrosis	Chronic Injury	Injury of Other Organs[†]
Group I Cholestasis > Hepatic Injury				
ALP, GGT	↑↑↑	↑	↑	±
Group II Hepatic Injury > Cholestasis				
A - Extrahepatic and Hepatic Disease				
AST, LDH	↑	↑↑↑	↑	↑
B - More Specificity for the Liver				
ALT	↑	↑↑↑	↑	↑
C - Still Greater Specificity for Hepatic Injury/Disease				
LDH_5	↑	↑↑↑	↑	
Group III Insensitivity to Liver Injury				
CK	Normal	Normal	Normal	↑

5'N = 5'-nucleotidase; ALP = alkaline phosphatase; ALT = alanine aminotransferase; AST = aspartate aminotransferase; CK = creatine kinase; LAP = leucine aminopeptidase; LDH_5 = least anodic isoenzyme of lactic dehydrogenase; LDH = lactase dehydrogenase; ↑ = increased; ↑↑↑ = markedly increased; ± = little change.
*Obstructive jaundice or intrahepatic cholestasis.
[†]Cardiac or skeletal muscle, brain, or kidney.
Modified from Zimmerman HJ, *Hepatotoxicity: The Adverse Effects of Drugs and Other Chemicals on the Liver*, 2nd ed, Philadelphia, PA: Lippincott Williams and Wilkins, 1999.

Storage Instructions Protect specimens for bilirubin from light.

Causes for Rejection Hemolysis interferes with certain tests.

Special Instructions The specimens should be handled with extra precaution, especially if there is a greater than usual possibility of viral hepatitis.

Additional Information Relevant topics in the Disease Index include Acetaminophen Toxicity, Carcinoma (Liver), Cirrhosis, Cirrhosis (Primary Biliary), Hemochromatosis, Hepatic Necrosis/Failure, Hepatitis, Hepatitis (Autoimmune), Jaundice, and Wilson Disease.

Mercury, Blood

Synonyms Cinnabar; Hg, Blood; Organic Mercury; Quicksilver

Use Evaluate for mercury toxicity

Applies to Methylmercury

Abstract Elemental mercury is toxic if inhaled or injected, but may safely pass through the digestive tract if small amounts are swallowed. Blood analysis for mercury is used principally for evaluation of toxicity from organic mercury, found in wood preservatives, paints, fungicides, cosmetics, foods, seeds, and in contaminated fish. The mercury catastrophe in Minamata Bay, Japan, involved ingestion of methylmercury-contaminated fish.

Limitations Once exposure ceases, blood levels may not provide a good indicator of remaining body burden. There are marked variations by different investigators in mercury levels considered toxic.

Specimen Whole blood

Volume 7 mL

Container Special metal-free EDTA tube

Causes for Rejection Failure to collect blood in a special metal-free container or exposure to metal containing dusts

Special Instructions Whole blood is analyzed.

Reference Range Whole blood: 0.6-59 mcg/mL (SI: 3-294 nmol/L). Normally whole blood mercury is <10 mcg/L, but this number varies widely in a large population of healthy unexposed individuals.

Critical Values >50 mcg/L (SI: 250 nmol/L) if exposure is to alkyl mercury compounds and >200 mcg/L (SI: 1000 nmol/L) if exposure is to Hg^{+2} compounds.

Methodology Electrothermal atomic absorption (AA)

Additional Information Organic methyl mercury is an important environmental mercurial contaminant. It was discovered that elemental mercury could be oxidized into inorganic mercury (Hg^{+2}) and then into organic mercury (methylmercury) from industrial wastes. Organic mercury then accumulates in large amounts in predator fish, and thus, into the human food chain. Ingestion of mercury-laden fish in Minimata Bay led to severe neurologic deficits.

The major physical forms of mercury to which humans are exposed are (elemental) mercury vapor and methylmercury compounds. Severe acute poisoning leads to pulmonary distress with acute pneumonitis, hemoptysis, and cyanosis. Mechanical ventilation is often required. Permanent sequelae may include irreversible lung impairment and pulmonary fibrosis.

Half-life of inorganic mercury is 24 days and of methylmercury (organic mercury) is 54 days. Hair analysis may be used for poisoning or chronic exposure.

Specific References

Centers for Disease Control and Prevention (CDC), "Measuring Exposure to an Elemental Mercury Spill - Dakota County, Minnesota, 2004," *MMWR Morb Mortal Wkly Rep*, 2005, 18;54(6):146-9.

Gray T, Baker B, Lintner C, et al, "Chronic Consumption of Fish Associated With Mercury Toxicity," *J Toxicol Clin Toxicol*, 2004, 42(5):807-8.

Mercury, Urine

Synonyms Hg, Urine

Use Urine mercury is best used for evaluation of inorganic mercury exposure.

Abstract See Mercury, Blood.

Limitations Organic mercury is found mostly in red cells and is lipid soluble. Urine mercury may not be useful for evaluating pure organic mercury poisoning. Once exposure ceases, urine levels may not represent a good indicator of remaining body burden.

Specimen 24-hour urine

Volume Entire collection

Minimum Volume 50 mL aliquot

Container Plastic (preferably polyethylene) acid-washed container, no preservative

Storage Instructions Store in special metal-free container.

Causes for Rejection Failure to collect sample in a special metal-free container or exposure to metal-containing dusts

Reference Range 24-hour urine: <20 mcg/L (SI: <100 nmol/L)

Critical Values Significant exposure is indicated when daily urine mercury reaches 50 mcg/day.

Possible Panic Range >150 mcg/L (SI: >748 nmol/L); symptoms are found with levels >600 mcg/L (>3000 nmol/L); lethal urine levels: >800 mcg/L (SI: >3992 nmol/L).

Methodology Electrothermal atomic absorption (AA)

Additional Information See Mercury, Blood

Specific References

Azizz-Baumgartner E, Luber G, Jones R, et al, "Mercury Exposure Assessment in a Nevada Middle School - 2004," *J Toxicol Clin Toxicol*, 2004, 42(5):808-9.

Centers for Disease Control and Prevention (CDC), "Measuring Exposure to an Elemental Mercury Spill - Dakota County, Minnesota, 2004," *MMWR Morb Mortal Wkly Rep*, 2005, 18;54(6):146-9.

Nehls-Lowe H, Stanton N, and Stremski E, "Air Sampling and Urine Analysis Following Decontamination of an Elemental Mercury (Hg) Spill in a High School Chemistry Lab," *J Toxicol Clin Toxicol*, 2004, 42(5):825.

Methemoglobin, Blood

Synonyms MetHb; NADH-MetHb Reductase

Use Evaluate cyanosis, especially in the presence of normal arterial gases; evaluate polycythemia and hemoglobinopathies; work up dyspnea and headache; work up "poppers" and "sniffers"; evaluate drug or chemical toxicity, since most instances of methemoglobinemia are so acquired; monitor patients on high dose nitrate therapy; measurement in CSF may detect small cerebral and subdural hematomas.

Applies to Cytochrome b_5 Reductase

Abstract This pigment is hemoglobin in which the iron is in the trivalent state. It cannot act as an oxygen carrier.

Limitations Sulfhemoglobin, methylene blue, and Evans blue dye may interfere. Methemoglobin exhibits pH sensitivity.

Specimen Blood

Volume 5 mL

Minimum Volume 1 mL whole blood

Container Green top (heparin) tube

Storage Instructions Keep tube on ice. pH dependent. Should be run within 8 hours, or false-negatives may occur. Run as promptly as possible after draw. Studies have shown up to 10% drop in 4 hours, up to 16% drop in 8 hours, in samples kept on ice. Such studies have not been extensive. May be drawn into sodium fluoride-containing tubes and immediately frozen at 0°C to −4°C prior to analysis.

Reference Range Up to 1.5% of total hemoglobin. Smokers have a slightly higher percent methemoglobin than do nonsmokers.

Possible Panic Range Headache and other symptoms occur at levels >30%. Methemoglobinemia can be fatal, particularly >70% saturation levels.

Methodology Spectrophotometry; Hb M variants are best detected by electrophoresis because spectrophotometry is unreliable due to their abnormal ferrihemoglobin spectra.

Additional Information Methemoglobin is an inactive, oxidized form of hemoglobin resulting in decreased oxygen-carrying capacity of blood. Concentrations of methemoglobin of >10% to 15% of hemoglobin will cause cyanosis. Sulfhemoglobin will interfere with methemoglobin determined by the above method. Methemoglobinemia may be hereditary or acquired. Polycythemia is occasionally present as a compensatory mechanism. Elevations of methemoglobin lead to dyspnea and headache, and can be lethal. Most instances of methemoglobinemia are acquired, from drugs and chemicals. Nitro and amino groups are especially involved, eg, aniline and derivatives, nitrites, nitroglycerin, nitrate salts in burn patients, dapsone (perhaps the most common cause of drug-induced methemoglobinemia), acetophenetidin, phenacetin and some sulfonamides, chlorates, quinones, large doses of ferrous sulfate, and many other drugs and some intestinal bacteria. Well water containing nitrate is the most common cause of methemoglobinemia in the newborn. Methemoglobinemia has been reported after exposure to automobile exhaust fumes.

Hereditary methemoglobinemia is uncommon. It may be due to a deficiency of red cell NADH-methemoglobin reductase (diaphorase, also termed cytochrome b_5 reductase), which has an autosomal recessive mode of inheritance. It may also be the result of presence of certain hemoglobinopathies, members of the Hb M family including Hb M Saskatoon, Boston, Iwate, Hyde Park, and Milwaukee. These have autosomal dominant mode of inheritance and may be associated with clinical cyanosis. Hb Seattle and other hemoglobinopathies also show increase in the *in vitro* rate of methemoglobin formation. A recently identified new hemoglobin variant, Hb Warsaw, is also characterized by elevated blood levels of methemoglobin.

A study of postmortem methemoglobin levels showed a range of 0.8% to 57% in individuals who, clinically, should have had normal antemortem concentrations. There was no correlation with antemortem circumstances, autopsy findings, or interval of time from death to autopsy. Chocolate brown color to arterial blood indicates methemoglobin concentration is >10%. Brown skin color, tachycardia, and tachypnea occur with levels >30%.

Lipemia will give falsely elevated methemoglobin levels.

Specific References

Babbitt CJ and Garrett JS, "Diarrhea and Methemoglobinemia in an Infant," *Pediatr Emerg Care*, 2000, 16(6):416-7.

Cruz-Landeira A, Bal MJ, Quintela O, et al, "Determination of Methemoglobin and Total Hemoglobin in Toxicological Studies by Derivative Spectrophotometry," *J Anal Toxicol*, 2002, 26(1):67-72.

Lee D, Shih RD, Axtell S, et al, "Methemoglobinemia in Smoke Inhalation Victims," *Acad Emerg Med*, 2000, 7(5):497.

Matteucci MJ, Reed WJ, and Tanen DA, "Failure of Sodium Thiosulfate to Reduce Methemoglobin Produced *In Vitro*," *Acad Emerg Med*, 2001, 8(5):443.

Wallace KL and Curry SC, "Postcollection Rise in Methemoglobin Level in Frozen Blood Specimens," *J Toxicol Clin Toxicol*, 2002, 40(1):91-4.

Wallace KL, Curry SC, and Brooks D, "Postcollection Rise in Methemoglobin Level in Frozen Blood," *J Toxicol Clin Toxicol*, 2000, 38(5):540-1.

Zilker T, "Methemoglobin and Intoxications," *J Toxicol Clin Toxicol*, 2001, 39(3):249-50.

Methemoglobin, Whole Blood

Synonyms MetHb

Use Evaluate causes of cyanosis. Examination for metHb is used to monitor patients on high dose nitrate therapy. Its measurement in CSF may detect small cerebral and subdural hematomas.

Applies to Carboxyhemoglobin; Reduced Hemoglobin; Sulfhemoglobin

Abstract Methemoglobin (metHb) is a form of hemoglobin in which the iron has been oxidized from the normal ferrous (Fe^{++}) to the ferric (Fe^{+++}) state. MetHb cannot bind oxygen to act as an oxygen carrier.

Specimen Whole blood

Volume 5 mL

Minimum Volume 1 mL heparinized whole blood

Container Green top (heparin) tube

Storage Instructions Keep tube on ice. Run as promptly as possible after draw because there is significant decrease with time. Studies have shown up to 10% drop in 4 hours, up to 16% drop in 8 hours, in samples kept on ice.

Reference Range Up to 1% of total hemoglobin. Smokers have slightly higher values than nonsmokers.

Possible Panic Range Headache and other symptoms occur at levels >30%. Methemoglobinemia levels >70% may be lethal.

Methodology Co-oximetry, spectrophotometry. Hb M variants are best detected by electrophoresis.

Additional Information MetHb is an inactive, oxidized form of hemoglobin (Hb), which does not contribute to the oxygen-carrying capacity of blood. Therefore, arterial %O_2 saturation will be inappropriately low for a given inhaled air oxygen concentration and p_aO_2, if the calculation for %O_2 saturation is based on total Hb. Concentrations of metHb of over 10% to 25% of Hb cause cyanosis. The most common cause of cyanosis is the presence of excessive reduced Hb, which becomes clinically apparent (as a bluish discoloration of skin and mucous membranes) when the capillary level is >5 g/dL. Cyanosis appears with metHb concentrations of 10% to 25%, but symptoms are minimal until metHb rises to 35% to 40%, at which patients may experience fatigue, dizziness, dyspnea, headache, and tachycardia. At the 60% level, lethargy and stupor may occur; levels >70% in adults may be fatal.

Methemoglobinemia may be hereditary or acquired. Polycythemia is occasionally present as a compensatory mechanism. Most instances of methemoglobinemia are acquired from drugs and chemicals. Nitro and amino groups are especially involved, eg, aniline and derivatives, nitrites, nitroglycerin, nitrate salts in burn patients, flutamide, metoclopramide, phenazopyridine, dapsone (perhaps the most common cause of drug-induced methemoglobinemia), phenacetin, acetophenetidin, prilocaine, some sulfonamides, sulfones, chlorates, primaquine, quinones, large doses of ferrous sulfate, and many other drugs and some intestinal bacteria.

Hereditary methemoglobinemia is uncommon. The most common phenotype is due to a deficiency of red cell NADH-methemoglobin reductase (diaphorase, also termed cytochrome b_5 reductase), which is inherited as an autosomal recessive trait. Homozygotes have metHb levels of 15% to 20%. Heterozygotes are apt to develop toxic methemoglobinemia when exposed to substances that can oxidize hemoglobin iron. Certain hemoglobin variants (particularly M variant) with abnormal tendency to stabilize Fe^{+++} cause methemoglobinemia.

Methylene Blue Stain, Stool

Synonyms Fecal Leukocyte Stain

Use Examine fecal specimens for the presence of leukocytes as an indicator of inflammatory diarrhea (eg, invasive enteric infection). Diarrhea for more than 4 weeks is an indication for evaluation.

Applies to Leukotest® (Fecal Lactoferrin)

Test Includes Methylene blue, Gram, or Wright stain of stool smear

Abstract Evaluation for fecal leukocytes is a part of the initial evaluation of chronic diarrhea. In general, the presence of fecal leukocytes provides indication for stool culture. Absence of fecal leukocytes, however, does not exclude bacterial diarrhea or the need for stool culture.

Limitations Ten percent to 15% of stools that yield an invasive bacterial pathogen on culture have an absence of fecal leukocytes. Fecal leukocytes are present in idiopathic inflammatory bowel disease.

Patient Preparation Collect specimen prior to barium procedures.

Specimen Fresh random stool, rectal swab

Volume 1 rectal swab, 1 mL stool

Container Cup specimen is more sensitive than swab specimen in detection of fecal leukocytes.

Storage Instructions Refrigerate

Reference Range No predominance of yeast, cocci in clusters, or leukocytes

Methodology Smear of stool (preferably mucus) with one drop methylene blue, coverslip, and observe for the presence of leukocytes.

Additional Information Conditions associated with marked fecal leukocytes, blood, and mucus include predominantly bacterial infections including invasive *E. coli*, shigellosis, salmonellosis, *Helicobacter*, *Yersinia* infection, ulcerative colitis, and cases of antibiotic-associated colitis and pseudomembranous colitis. *Salmonella typhi* may evoke a monocyte response. Conditions associated with modest numbers of fecal leukocytes include early shigellosis involving small bowel, and cases of antibiotic-associated colitis. In amebiasis, stool leukocytes are variable. Diarrhea can be watery or bloody. Conditions associated with an absence of fecal leukocytes include toxigenic bacterial infection including *Vibrio cholerae*, giardiasis, and viral infections. The methylene blue stain for polymorphonuclear leukocytes has a high sensitivity (85%) and specificity (88%) for bacterial diarrhea (*Shigella*, *Salmonella*, *Helicobacter*). Positive predictive value is 59%. Negative predictive value is 97%. Combined with a history of abrupt onset, more than four stools per day, and no vomiting before the onset of diarrhea, the stool methylene blue stain for fecal polymorphonuclear leukocytes is a very effective presumptive diagnostic test for bacterial diarrhea. A positive occult blood test may also be suggestive of acute bacterial diarrhea being more sensitive (79% vs 42%) than the fecal leukocyte test (in detection of invasive bacteria in pediatric patients). The occult blood test in this setting, however, lacks specificity and has been found to have a positive predictive value of only 24%.

Neither absence of fecal occult blood and/or leukocytes should pre-empt the use of culture. When both tests were positive, there was a sensitivity of 81% and specificity of 74% for bacterial diarrhea.

Slightly over 50% of patients with collagenous colitis have demonstrable fecal leukocytes.

A commercially available latex agglutination screening test (Leukotest®) for the leukocyte marker lactoferrin has several advantages over the microscopic-based detection of fecal leukocytes. Both tests, however, are insufficiently sensitive to be used as screening tests for such pathogens as *Campylobacter*, *Salmonella*, or *Shigella* spp.

Nicotine, Serum, or Plasma

Use Work-up of acute poisoning in children ingesting cigarettes or cigarette butts

Abstract Nicotine, one of the most toxic of all poisons, is a neural stimulant found in most tobacco products including transdermal patches and Nicorette® gum.

Specimen Serum or plasma

Container Red top tube

Storage Instructions Store at −20°C for 1 week.

Reference Range Serum concentrations >50 ng/mL may be associated with toxicity. Mean plasma concentration after smoking one cigarette: 5-30 ng/mL.

Possible Panic Range Plasma concentrations >13,000 ng/mL have been associated with **fatality**.

Methodology Gas chromatography (GC), high performance liquid chromatography (HPLC), immunoassay

Additional Information
Half-life: 24-84 minutes
Volume of distribution: 1.0 L/kg
Acute toxicity may include, but is not limited to: nausea, cyanosis, insomnia, hyponatremia, blurred vision, hyperventilation, nystagmus, dementia, abdominal pain, apnea, and respiratory depression. Chronic exposure to nicotine in the form of cigarette smoke also has other deleterious effects including exacerbation of atherogenesis.

Organophosphate Pesticides, Urine, Blood, or Serum

Use Determine occupational, accidental, and intentional poisoning. Insecticides are among the most toxic pesticides and cause most pesticide intoxication.

Applies to Carbamate Toxicity; Insecticides; Pesticides; Pralidoxime

Test Includes Azinphos-methyl, carbophenthion, chlorpyrifos, coumaphos, diazinon, dichlorvos, dimethoate, ethion, fenchlorphos, fenthion, fonofos, malathion, metasystox, methyl parathion, mevinphos, p-nitrophenol, paraoxon, parathion, phorate, terbufos

Abstract The organophosphates and the carbamates are the insecticide groups. Organophosphorus insecticide poisoning is a major global health problem with ~3 million poisonings or more and 220,000 deaths annually. These agents cause poisoning by irreversibly inhibiting acetylcholinesterase. Organophosphate poisoning causes CNS intoxication and polyneuropathy.

There are two cholinesterase enzymes in blood: acetylcholinesterase (also called true cholinesterase) in red cells; and "pseudocholinesterase" (acylcholine acylhydrolase) in serum. Organophosphates inhibit both enzymes, but the toxic effect on the serum enzyme, pseudocholinesterase, is more rapid and intense, so that it is somewhat more useful in the initial diagnosis of organophosphate toxicity. Red cell acetylcholinesterase is often preferred for evaluating chronic organophosphate exposure; red cell acetylcholinesterase levels normalize more slowly than do serum pseudocholinesterase values.

Limitations Correlation between severity of toxicity and the amount of pesticide metabolite in urine is poor.

Specimen Urine and blood or serum

Container Lavender top (EDTA) tube or red top tube

Storage Instructions Freeze sample if analysis cannot be performed immediately.

Methodology High performance thin layer chromatography (HPTLC). Urinary metabolites, the di-alkyl-phosphates, can be measured in urine by gas chromatography.

Additional Information Treatment of organophosphate poisoning includes atropine treatment and suctioning of oral secretions as required until atropinization is achieved. (Atropine is a physiologic antidote, used to treat muscarinic effects.) Pralidoxime (Protopam®, 2-PAM), a specific antidote, should be administered to seriously ill organophosphate-poisoned patients. If induction of paralysis with muscle-relaxing agents is required for intubation, succinylcholine should be avoided because of potential prolonged duration of paralysis secondary to pseudocholinesterase inhibition by the organophosphate.

The chemical terrorist agents Sarin, Soman, Tabun, and VX are organophosphate nerve gases.

Osmolality, Calculated, Serum, or Plasma

Use Hyponatremia is a laboratory finding that raises the possibility of an osmolal disturbance and prompts additional laboratory studies. Since some, but by no means all, patients with hyponatremia are at risk for acute cerebral edema, prompt diagnosis is essential. The following table details typical findings in various clinical conditions having hyponatremia as a common denominator.

Typical Serum Sodium, Osmolality, and Effective Osmolality (Tonicity) in Different Clinical States*

Condition	Serum Sodium mmol/L	Blood Glucose mmol/L (mg/dL)	Serum Urea Nitrogen mmol/L (mg/dL)	Mannitol or Ethanol mmol/L	Osm_c mmol/kg H_2O	Osm_m mmol/kg H_2O	Osmolal Gap mmol/kg H_2O	Effective Osmolality mmol/kg H_2O[†]	Risk of Cerebral Edema[‡]
Normal	140	5 (90)	5 (14)	0	290	290	0	285 (normal)	None
Hyponatremia (without abnormal amounts of other solutes)	120	5 (90)	5 (14)	0	250	250	0	245 (low)	Increased
Pseudohyponatremia (eg, from extreme hypertriglyceridemia)	120	5 (90)	5 (14)	0	250	290	40	285 (normal)	Unchanged
Hyponatremia caused by severe hyperglycemia	120	75 (1350)	5 (14)	0	320	320	0	315 (high)	Variable[§]
Hyponatremia caused by retention of mannitol	120	5 (90)	5 (14)	75[¶]	250	325	75	320 (high)	Decreased
Hyponatremia together with high serum urea nitrogen	120	5 (90)	45 (126)	0	290	290	0	245 (low)	Increased
Hyponatremia together with high blood ethanol level	120	5 (90)	5 (14)	40[#]	250	290	40	245 (low)	Increased
Hypernatremia	160	5 (90)	5 (14)	0	330	330	0	325 (high)	Decreased

*Osm_c indicates calculated osmolality (calculated as 2[Na] + SUN/2.8 + GLU/18, where SUN indicates serum urea nitrogen, and GLU glucose); Osm_m indicates measured osmolality (assume that normal osmol gap is zero). Osmolal gap is $Osm_m - Osm_c$.

[†]Effective osmolality (tonicity) is that portion of osmolality inducing transmembrane water movement; the cited values for effective osmolality were calculated by subtracting the contributions of urea and ethanol (if present) from the measured osmolality.

[‡]Immediate risk of the osmotic type of cerebral edema before treatment (the more acute the hyponatremia, the greater the risk, since osmotic adaptation is less advanced).

[§]Effect on intracellular fluid volume depends on clinical circumstances.

[¶]Neglecting the correction factor for serum water content. 75 mmol of mannitol per liter corresponds to approximately 1365 mg/dL. Substances (eg, urea and ethanol) that easily cross cell membranes contribute to measured osmolality, but not to tonicity.

[#]Neglecting the correction factor for serum water content. 40 mmol of ethanol per liter corresponds to approximately 184 mg/dL.

Adapted from Oster JR and Singer I, "Hyponatremia, Hypo-osmolality, and Hypotonicity," *Arch Intern Med*, 1999, 159(4):333-6.

Causes of Increased Osmolal Gap, Overview
Exogenous Causes
Acetone
Methanol
Ethanol
Isopropanol
Ethylene glycol
Propylene glycol
Ethyl ether
Dimethyl sulfoxide*
Glycine*
Mannitol*
Osmotic contrast dyes
Diseases
Chronic renal failure
Lactate acidosis?
Alcoholic ketoacidosis
Diabetic ketoacidosis
Hyperlipidemia,[†] hyperglobulinemia[†]
Artifactual
Lavender top (EDTA[‡]) tube 15 mOsm/L[§]
Gray top (sodium fluoride - potassium oxalate) tube 150 mOsm/L[§]
Blue top (citrate) tube 10 mOsm/L[§]
Green top (lithium heparin) tube 6 mOsm/L[§]

*These agents are given by intravenous infusion.
[†]Sodium was measured by flame emission spectrophotometry.
[‡]EDTA indicates ethylenediaminetetraacetic acid.
[§]Values are reported as calculated contributions.
Adapted from Osterloh JD, Kelly TJ, Khayam-Bashi H, et al, "Discrepancies in Osmolal Gaps and Calculated Alcohol Concentrations," *Arch Pathol Lab Med*, 1996, 120(7):637-41.

Applies to Isopropyl Alcohol Intoxication; Methanol; Osmolal Gap; Pseudohyponatremia

Test Includes Sodium, urea nitrogen (BUN), glucose

Abstract Serum osmolality may be directly measured, or predicted, using one of several equations based on the concentrations of serum sodium, glucose, and urea. The purpose of each equation is the necessity, now unique to the U.S., of converting the conventionally reported glucose and urea results into molar (or SI) units. A formula widely used to calculate osmolality in mOsm/kg H_2O, is:

Calculated osmolality = 2[Na] + BUN/2.8 + GLU/18

in which BUN is the serum urea nitrogen in mg/dL; and GLU is the plasma or serum glucose in mg/dL. Dividing the BUN by 2.8 converts the measurement from mg/dL to mmol/L. The same is true for dividing the glucose by 18. Multiplying the sodium, already in mmol/L, by 2 is done to account for the osmotically active anions (mostly chloride and bicarbonate) associated with the sodium. The osmolal gap is the arithmetic difference between the measured osmolality and the calculated osmolality.

The sodium value used in the equation should be one measured without dilution by a sodium-selective ion electrode. This avoids using the erroneous value (euphemistically called pseudohyponatremia), which results from a sodium measurement made after dilution by flame photometry on a specimen with an increased concentration of lipid or protein.

Limitations Considerable variability occurs in the normal osmolal gap.

Patient Preparation Patient ideally should be fasting for 8 hours, a setting not usually possible when the need for investigation of osmolality arises.

Specimen Serum or plasma

Container Red top tube, green top (heparin) tube

Collection Keep one green top tube on ice, should a pH be needed.

Causes for Rejection Gross hemolysis

Reference Range
Calculated osmolality: 290 mOsm/kg H_2O
Measured osmolality: up to age 60 years: 275-295 mOsm/kg H_2O, older than 60 years: 280-301 mOsm/kg H_2O
Osmolal gap: 9.0 ($\pm$6.4) mOsm/kg H_2O

Possible Panic Range A gap >20 mOsm/kg H_2O (SI: >20 mmol/kg H_2O)

Osmolality, Serum

Synonyms Serum Osmolality

Use Serum osmolality is used to evaluate electrolyte and water balance, hyperosmolar status, hydration/dehydration status, acid-base balance, seizures, antidiuretic hormone function, liver disease, and hyperosmolar coma. Osmolality is proportional to the concentration of particles in solution. Freezing point depression serum osmolality with calculated osmolal gap is useful in screening for and approximating the serum concentrations of certain low molecular weight toxins, such as ethanol, ethylene glycol, isopropanol, and methanol, especially as a rapid approximation for emergent situations. See Limitations.

High serum osmolality may result from hypernatremia, dehydration, hypovolemia, hyperglycemia, mannitol therapy, azotemia, and ingestion of ethanol, methanol, or ethylene glycol. Thus, osmolality has a role in toxicology and in coma evaluation. Very low birth weight infants may have elevated serum osmolality for the first week of life.

Low serum osmolality may be secondary to overhydration, hyponatremia, and the syndrome of inappropriate antidiuretic hormone secretion (SIADH).

Serum osmolality measurements do not measure the fraction of serum that is water. Osmolality measurement by freezing point depression is also indifferent to permeability of solutes to cell membranes.

Applies to Osmolal Gap; Osmolality:Serum Ratio; Sodium, Serum:Osmolality, Serum Ratio; Urine:Serum Osmolality Ratio

Replaces Osmolarity

Abstract The osmolality of a solution is the number of particles in a liter of solution. Osmolality is independent of particle size or charge.
Ethanol ingestion is a common cause of increased osmolality and is responsible for most osmolal gap testing.

Limitations When vapor pressure osmometry is used, volatile solutes (eg, alcohols and glycols) may remain in the vapor phase and not be detected.

Specimen Serum

Container Red top tube

Collection Pediatrics: Blood drawn from heelstick

Storage Instructions Refrigerate or freeze serum if not run within 4 hours.

Reference Range 275-295 mOsm/kg H_2O (SI: 275-295 mmol/kg H_2O)
Urine:serum osmolality ratio: 1.0-3.0 with fluid restriction: 3.0-4.7
Sodium, serum:osmolality:serum ratio: 0.43-0.50

Possible Panic Range <265 mOsm/kg H_2O (SI: <265 mmol/kg H_2O), >320 mOsm/kg H_2O (SI: >320 mmol/kg H_2O). Result of 385 mOsm/kg H_2O (SI: 385 mmol/kg H_2O) may reflect stupor in hyperglycemia. Values 400-420 mOsm/kg H_2O (SI: 400-420 mmol/kg H_2O) may reflect grand mal seizures. Values >420 mOsm/kg H_2O (SI: >420 mmol/kg H_2O) may be lethal.

Methodology Freezing point depression (more often used) or vapor pressure elevation

Additional Information Measured osmolality usually exceeds the calculated osmolality. If measured osmolality is >15 mOsm/kg H$_2$O (SI: >15 mmol/kg H$_2$O) greater than calculated, the differential diagnosis includes: methanol, ethylene glycol, ethanol, or other toxicity; shock and trauma. Elevated serum osmolality with normal sodium suggests hyperglycemia, uremia, or alcoholism. Both serum and urine values and calculated osmolality (see Osmolality, Calculated, Serum, or Plasma) are sometimes required for accurate diagnosis.

After overnight dehydration, urine:serum osmolality ratio is usually ≥3. Even with fluid restriction, the ratio is 0.2-0.7 in diabetes insipidus. It is usually normal in neurogenic polyuria without fluid restriction and is increased with fluid restriction.

The expression "urinary:plasma ratio" is also used. Laboratory criteria for hypovolemia include elevation of urea nitrogen:creatinine ratio >25 and/or increased serum osmolality and sodium concentration. Serum sodium and osmolality increase in dehydration; the serum sodium:serum osmolality ratio remains within normal limits.

Q Fever Serology

Use Support the diagnosis of Q fever; the most common symptoms include fever, chills, and headache.

Test Includes Detection of antibody specific for *Coxiella burnetii*, the organism responsible for Q fever

Abstract *Coxiella burnetii*, originally called *Rickettsia burneti*, is a pleomorphic coccobacillus with a gram-negative cell wall, that does not stain with the Gram stain. It is an obligate intracellular pathogen. The primary reservoirs for Q fever are cattle, sheep, and goats. The organism withstands heat and drying, survives on inanimate surfaces, and can persist in the environment long after an infected animal has vacated an area. It is highly infectious and its distribution is worldwide.

A broad clinical spectrum exists, from subclinical to fatal. Transmission is by inhalation of the organism. Person to person transmission has been documented but it is rare.

This is a potential agent of bioterrorism.

Limitations Reagents prepared from fresh isolates (phase I organisms) react differently from those from multiply-passaged organism, a laboratory artifact (phase II). Enzyme-linked immunosorbent assay (ELISA) is more sensitive than indirect immunofluorescent antibody (IFA) or complement fixation (CF) assay. Serology tests for the diagnosis of Q fever have not been standardized. Cross reactions with *Legionella* have been described.

Specimen Serum

Volume 10 mL

Minimum Volume 2 mL serum

Container Red top tube

Sampling Time Acute and convalescent samples are recommended, the latter 2-4 weeks from onset.

Reference Range Titer: <1:10; comparison of acute and convalescent titers is of greatest diagnostic value

Critical Values Titer: ≥1:10; fourfold or greater increase in titer provides evidence of recent infection

Methodology Complement fixation (CF), indirect immunofluorescent antibody (IFA), enzyme-linked immunosorbent assays (ELISA)

Additional Information Q fever should be considered in the presence of fever with negative blood cultures. Serology is preferable to culture for detection of infection with *C. burnetii*. Nucleic acid amplification has also been used to detect *C. burnetii* in blood specimens of infected patients. Sera from patients with Q fever do not react in the Weil-Felix test with *Proteus* antigen. Convalescent sera react best with phase II organism (see Limitations), but sera from chronic persistent infection react best with phase I organisms. Chronic Q fever is uncommon and tends to affect patients with valvular heart disease.

Shiga Toxin Test, Direct

Use Detect toxin-producing strains after they have been isolated, directly from patient specimens such as stool, directly from food (especially meat), or from fecal specimens of food animals such as cattle.

Applies to *E. coli* O157:H7; Verocytotoxins

Test Includes Detection of *E. coli* O157:H7 by assay for the Shiga-like toxin directly in stool or after isolation of a bacterial isolate

Abstract In the late 1970s, the *E. coli* serotype O157:H7 was recognized as a human pathogen and was associated with severe bloody diarrhea and the hemolytic-uremic syndrome. This serotype of *E. coli* bacteria produces a Shiga-like toxin that is similar to the Shiga toxin expressed by *Shigella dysenteriae*. The production of the toxin, or the genes encoding the toxin, can be detected by a variety of methods, and some are commercially available.

Limitations Results from such assays should be confirmed by the isolation of a Shiga-toxin producing *E. coli*. Some of the commercial assays may have false-positive results with *Pseudomonas aeruginosa*.

Specimen Stool, food, or bacterial isolate

Container Sterile container

Reference Range No toxin detected

Critical Values Positive detection of Shiga-like toxin

Methodology Enzyme immunoassays (EIA), latex agglutination, genetic detection methods

Additional Information Certain strains of *E. coli* have the ability to produce toxins (verotoxins or Shiga-like toxins) that closely resemble the Shiga toxin of *Shigella dysenteriae*. Since *E. coli* isolates that produce the Shiga-like toxin are associated with severe hemorrhagic colitis and/or the life-threatening hemolytic uremic syndrome (HUS), any laboratory detection of this toxin should be considered clinically important and should be immediately reported to the attending physician and the state health department. The detection of this toxin or the genes required for toxin production can be used to predict the presence of this bacteria.

Smallpox Diagnostic Procedures

Synonyms Variola Major/Variola Minor Diagnostic Procedures

Use Diagnose papulovesicular eruptions in a setting of possible bioterrorism

Abstract The last natural case of smallpox occurred in Somalia in 1977. Two known stocks of variola virus are known to be retained, one by the Centers for Disease Control and Prevention (CDC) in Atlanta and one by the Vector Institute in Novosibirsk, Russia. Routine smallpox vaccination was stopped in the U.S. in 1972, and elsewhere since about 1982. Thus, a large pool of susceptible individuals now exists.

Oral and airway lesions occur first. Macules become papules, then vesicles, then pustules. The disease appears first on the face and extremities, and lesions follow on the trunk and extremities.

Since smallpox bears a potential for human-to-human transmission, it has a terrifying potential as a bioterrorist weapon.

Specimen Skin lesion scrapings, vesicular or pustular fluid, crusts, scabs, blood, tonsillar swabbings, or saliva sent to CDC. The CDC should be contacted anytime, day or night, at 770-488-7100. Public health officials should be contacted. Agents such as variola can be handled only in biosafety level 4 laboratories.

Container Vacutainer® tube

Collection Specimen collection should be done by a previously vaccinated individual, or one who is vaccinated that day. Mask and gloves should be worn.

Paired sera samples are needed for serologic testing to recognize recent infection.

Storage Instructions Storage of biopsies, scabs, vesicular fluid for virology: $-20°C$ to $-70°C$. Transport conditions: $4°C$ in 6 hours or less.

Special Instructions Guidelines for the collection and shipping of specimens are available at http://www.cdc.giv/smallpox.

Methodology Immunohistochemistry for viral antigen; polymerase chain reaction (PCR) for Orthopoxvirus genetic material; live-cell cultures with nucleic acid identification; growth on chorioallantoic membranes; serologic testing detecting IgM responses. Complement fixation (CF), immunofluorescence, Ouchterlony techniques, and hemadsorption with chicken erythrocytes can be used.

Electron microscopy for recognition of virions can be used.

The inclusion bodies of smallpox, designated Guarnieri bodies, are found in intraepidermal lesions characterized by multilocular vesicles with ballooning and reticular degeneration. Guarnieri bodies or the virions on electron microscopy do not discriminate between Orthopoxvirus species.

Additional Information The differential diagnosis of papulovesicular and maculopapular eruptions has recently been published. The differential diagnosis of smallpox includes chickenpox, erythema multiforme, and herpes virus infections. Lesions of chickenpox begin on the trunk.

Variola virus lacks animal reservoirs.

The possibility of bioterrorism has generated need for education (eg, the April 25, 2002 and January 30, 2003 issues of the *N Engl J Med*).

Specific References

Miller M, Kosmala-Runkle D, Coon T, et al, "ECG Findings in Patients With Vaccinia-Related Myopericarditis," *J Toxicol Clin Toxicol*, 2004, 42(5):803.

Thallium, Urine, or Blood

Use Diagnose thallium exposure and toxicity, including alopecia with neuropathy, which may resemble that of Guillain-Barré syndrome, and abdominal colic.

Applies to Mee's Lines

Abstract A by-product of lead smelting, thallium salts are used in photomultiplier tubes, infrared detectors and transmitters, lens and glass making, and in rockets and flares. It has been, and in some areas remains, a component of insecticides, pesticides, and rodenticides. Thallium has been excluded as a rodenticide in the U.S. since 1972. It may be found in grain.

Limitations There are marked variations in heavy metal levels considered toxic by different investigators.

Patient Preparation The patient should be instructed to use a plastic bedpan or urinal if necessary, not metal.

Specimen 24-hour urine, serum

Volume Entire collection

Container Special metal-free EDTA tube or metal-free plastic urine container.

Causes for Rejection Specimen allowed to contact metal or dusts with metal, use of nonmetal-free containers

Reference Range Urine: <2 mcg/L (SI: <9.8 nmol/L); serum: <0.5 mcg/L (SI: <24.5 nmol/L)

Critical Values Serum value in most normal individuals is <10 mcg/L or 10 ng/mL (SI: 49 nmol/L); daily urine excretion in most normal individuals is <10 mcg/day (SI: 49 nmol/day). Spot urinary thallium concentration in normal unexposed individuals is <1.5 mcg/L.

Possible Panic Range Blood levels >100 mcg/L (SI: 490 nmol/L) or urine values >200 mcg/L (SI: 978 nmol/L)

Methodology Graphite furnace atomic absorption spectrometry (GFAAS), inductively coupled plasma-mass spectrometry (ICP-MS)

Additional Information Thallium is almost 100% absorbed from the GI tract and, like potassium, distributes throughout the body. Symptoms of thallium poisoning initially begin with generalized nausea, vomiting, abdominal pain, diarrhea, and gastrointestinal bleeding. Other nonspecific clinical findings include polyneuritis, encephalitis, delirium, ophthalmologic symptoms, convulsions, shock, and coma. Alopecia and painful ascending peripheral neuropathy are the most characteristic components of a thallium "toxidrome." Because of the delayed development of alopecia (several weeks after poisoning), the diagnosis of thallotoxicosis is often initially overlooked until alopecia appears. Mee's lines (transverse white lines on the nails) appear on the hands and feet 1 month after exposure, but are not specific for thallium poisoning. Other dermatological findings include crusted eczematous lesions, hypohydrosis, anhydrosis, palmar erythema, painful glossitis, stomatitis, and hair discoloration.

Thiocyanate, Serum, Plasma, or Urine

Synonyms Ethyl and Methyl Thiocyanate (Thanite® and Lethane®); Potassium Thiocyanate (KSCN)

Use Follow exposure to cyanide; evaluation of clearance of thiocyanate; monitor nitroprusside toxicity and smoking or nonsmoking compliance

Applies to Cyanide; Nipride®; Nitroprusside

Abstract Thiocyanate is a relatively inert metabolite of the antihypertensive drug, nitroprusside. It is also a product of cyanide metabolism, a by-product of cigarette smoking, and is found in the serum and urine of individuals consuming cassava beans.

Limitations Because of rapid metabolism of the drug, results are usually meaningless in the clinical setting of acute CN^- exposure by the time they are reported.

Specimen Serum or plasma, urine

Volume 7 mL

Container Red top tube, lavender top (EDTA) tube, plastic urine container

Reference Range Serum, nonsmoker: 1-4 mcg/mL (SI: 0.02-0.07 mmol/L), smoker: up to 10 mcg/mL (SI: up to 0.17 mmol/L); urine: nonsmoker: 1-4 mg/24 hours (SI: 0.02-0.07 mmol/day), smoker: 7-17 mg/24 hours (SI: 0.12-0.30 mmol/day)

Possible Panic Range Serum: >35 mcg/mL (SI: >0.60 mmol/L); 200 mcg/mL (SI: 3.44 mmol/L) is lethal

Methodology Photometry, chromatography, ion chromatography with fluorescence and ultraviolet detection

Additional Information Thiocyanate toxicity may occur with long-term nitroprusside use (7-10 days with normal renal function and 3-6 days with renal impairment). When thiosulfate is given to treat cyanide toxicity, thiocyanate toxicity may occur. Toxic manifestations may include psychotic behavior, agitation, and convulsions.

Home fires in which urea foam insulation burns produce hydrocyanic acid and formaldehyde.

Tularemia Diagnostic Procedures

Synonyms *Francisella tularensis* Diagnostic Procedures; Rabbit Fever Diagnostic Procedures

Use Investigation of illness characterized by an ulcerative lesion at a site of inoculation, with regional lymphadenopathy, fever, and pneumonia. Primary pneumonic tularemia follows inhalation. Ulceroglandular, glandular, oculoglandular, oropharyngeal, intestinal, pneumonic, and typhoidal disease types are recognized. Serologic testing is useful for retrospective diagnosis.

Abstract *Francisella tularensis* is a zoonotic bacteria and is found in a number of small mammals, including rabbits. It is transmitted to man by contact with such animals, by inhalation, or by insect bites. The diagnosis of tularemia is often made in the absence of a positive culture for *F. tularensis*. Detection of increased titers of specific antibodies and culture remains the most definitive procedure.

If *F. tularensis* is used as a biological weapon, it will cause febrile illness 3-5 days later. Disease presents as pneumonia, followed by respiratory failure and shock. The abrupt onset of illness among large numbers of individuals with rapid disease progression would lead to consideration of possible bioterrorism.

Limitations There is serologic cross reactivity with *Brucella* species, *Proteus* OX-19, and *Yersinia* species. IgM and IgG titers may remain elevated (1:20-1:80) for over a decade after infection, limiting the value of unpaired specimens; single titers may be misleading.

Specimen Serum for serology; culture and Gram stain from blood, sputum, pleural fluid, lymph node or lymph node aspiration material, pharyngeal washings, skin or mucosal lesions.

Volume 7 mL

Minimum Volume 1 mL serum

Container Red top tube for serology

Sampling Time Paired sera collected 2-3 weeks apart are recommended for serology.

Storage Instructions Refrigeration is recommended if transportation will require longer than 1 hour.

Special Instructions *F. tularensis* is a risk to laboratory personnel and extra precautions are indicated.

Reference Range Agglutination titer: <1:40; ELISA: <1:500; less than a fourfold increase in paired titer

Critical Values A single serologic result with a titer ≥1:160 in a patient having clinical tularemia or a fourfold rise in titer is diagnostic.

Methodology Agglutination, hemagglutination, enzyme-linked immunosorbent assay (ELISA), antigen detection assays, immunoblotting, pulsed-field gel electrophoresis, direct fluorescent antibody, immunohistochemistry, PCR, culture, Gram stain. It is an intracellular, nonmotile, aerobic, faintly staining, tiny gram-negative coccobacillus, or pleomorphic rod. The fastidious organism requires cysteine for growth.

Additional Information In the United States, there are approximately 100-200 tularemia cases reported annually with 1-4 deaths. Antibodies to *F. tularensis* develop 2-3 weeks after infection and peak in 4-5 weeks. Rising titers over a 2-week interval are the best indicator of recent infection.

Although laboratory diagnosis of tularemia is often established by serologic methods, *F. tularensis* may be recovered in culture from a variety of clinical specimens. Isolation is hazardous and sometimes difficult. Blood cultures are often negative. If tularemia is clinically suspected, contact the laboratory so that appropriate precautions can be taken.

Uncrossmatched Blood, Emergency

Synonyms Emergency Blood; Emergency Transfusion; Universal Donor Blood; Urgent Transfusion

Use Blood replacement in exsanguinating emergency, massive acute blood loss.

Contraindications Do not use group O whole blood even in emergencies, for patients of other types, because the anti-A and anti-B can cause hemolysis of the recipient's RBCs. In life-threatening trauma or bleeding when the patient's type is unknown, group O RBCs should be used. Whenever possible, Rh-negative RBCs should be used in females of childbearing age.

Applies to Emergency Issue of Uncrossmatched Blood; Exsanguinating Emergency; Massive Acute Blood Loss

Test Includes No testing is needed in the Transfusion Service if group O, Rh negative blood is issued. ABO group and Rh type can be completed in 5-10 minutes for issue of group-specific blood. As soon as possible, complete antibody screen and crossmatch and other serologic tests (eg, antibody identification) if indicated.

Abstract The administration of a blood component, usually red cells, before the completion of routine pretransfusion testing, in an emergency situation when a delay in transfusion would harm the patient. Risks include those a physician must evaluate (eg, degree of athero-sclerosis, nature of the disease, level of oxygenation, heart rate, blood pressure, and control or lack of control of bleeding).

Limitations All parties involved need to understand that blood issued in life-threatening emergencies is clearly more dangerous than that in controlled circumstances.

Patient Preparation Proper patient identification is necessary. Emergency Department (ER) may use special or temporary identification. Care must be taken to follow transfusion protocols, especially when multiple trauma victims are being treated simultaneously, as errors in sample/patient identification may lead to fatal ABO hemolytic transfusion reactions.

Specimen Venous or arterial blood

Container One red top tube or one lavender top (EDTA) tube

Collection (Of sample from intended recipient): Identify patient by wristband(s) or other system specially set up for identification of unconscious or noncommunicating patients in emergencies. Label tube specimen with the same information, including identification number from the wristband; label requisition form with identification number. Requisition should be signed by collector, indicating that patient's identity has been verified. Positive identification of patient sample is important, even in an emergency. If it is impossible to get a blood sample, record this fact.

Turnaround Time Although uncrossmatched O Rh-negative red blood cells can be issued immediately if available, ABO and Rh type can be done in only 5-10 minutes. Antibody screen and crossmatch require as much as 1 hour, longer if antibodies are detected. The process is much quicker if patient has already had a type and screen.

Special Instructions An emergency request for uncrossmatched blood should include name of physician requesting blood and signature of person authorized, name and location of patient, and nature of emergency. There should also be a statement that the situation was sufficiently urgent to require release of blood before completion of testing.

Critical Values 30% to 40% loss of blood volume is associated with increased signs of shock, and >40% relates to severe shock.

Methodology When issuing uncrossmatched blood, apply a label indicating uncrossmatched status.

EMERGENCY RELEASE COMPATIBILITY TESTING INCOMPLETE

Additional Information There is no such thing as a "universal donor." Group O blood lacks A and B but has antigens of other blood group systems, any of which may be a problem for a given patient. Group O RBCs can be transfused when the blood type is unknown. A blood sample can be ABO and Rh typed in 5-10 minutes. Thus, the patient may then receive type-specific RBCs when their blood type is determined. If the patient's indirect antiglobulin test (screen for unexpected antibodies) is negative, transfusion of uncrossmatched but type-specific/compatible blood carries a very low risk of being incompatible. Although volume can be made up temporarily with plasma expanders, possible adverse effects of albumin solutions have been noted in critically ill patients.

Virus Detection by DFA

Synonyms Direct Detection of Virus; Direct Fluorescent Antibody Test for Virus; Viral Direct Detection by Fluorescent Antibody; Virus Fluorescent Antibody Test

Use Rapid diagnosis of HSV, VZV, RSV, parainfluenza, influenza, and rabies infections

Applies to Herpes Simplex Virus, Direct Detection; Influenza Virus, Direct Detection; Measles Virus, Direct Detection; Mumps Virus, Direct Detection; Parainfluenza Virus, Direct Detection;

Rabies Virus, Direct Detection; Respiratory Syncytial Virus, Direct Detection; Varicella-Zoster Virus, Direct Detection

Test Includes Direct (nonculture) detection of virus-infected cells

Limitations Physicians must specify the particular viruses they suspect or desire to be tested. Only a few viruses (HSV, VZV, RSV, influenza, parainfluenza, measles, mumps, and rabies virus) can be detected in this manner. It is possible for the test to be negative in the presence of viral infection. Expertly trained and experienced personnel, excellent quality reagents and adequate numbers of cells are required. Contact laboratory prior to requesting test to determine if laboratory offers this/these tests.

Specimen Impression smears of tissues, lesion scrapings and swabs, frozen sections, cell suspensions, upper respiratory tract swabs

Causes for Rejection Insufficient material, slides broken or badly scratched, fixative used on slide preparation (generally, the laboratory is responsible for fixing specimens)

Turnaround Time Less than 1 day

Special Instructions Make at least four impression smears or place four frozen sections on four separate slides. Cell suspensions should be centrifuged, resuspended to slight turbidity, and applied to prewelled slides.

Reference Range
No virus detected

Methodology Monoclonal antibody reagents and immunofluorescence microscopy are used to detect viruses/viral antigens in specimen cells.

Additional Information Generally, this test is not as sensitive as cell culture.

Specific References
Drew WL, "Controversies in Viral Diagnosis," *Rev Infect Dis*, 1986, 8(5):814-24.
Drew WL, "Diagnostic Virology," *Clin Lab Med*, 1987, 7(4):721-40.
Smith TF, "Rapid Methods for the Diagnosis of Viral Infections," *Laboratory Medicine*, 1987, 18:16-20.

Volatile Screen, Blood, or Urine

Synonyms Toxicology, Volatiles

Use Evaluate methanol and isopropanol toxicity, and alcohol drug abuse

Applies to Acetone; Ethanol; Isopropanol; Methanol

Test Includes Determination of volatiles by gas chromatography including acetone, ethanol, isopropanol, and methanol.

Abstract This screening profile measures ethanol and other possible volatiles.

Limitations Urine levels do not correlate well with blood levels.

Specimen Serum or plasma, urine, gastric fluid

Volume 7 mL blood, 25 mL urine, 25 mL gastric washing

Minimum Volume 3 mL serum

Container Red top tube, gray top (sodium fluoride) tube; tightly stoppered container for urine and gastric fluid

Collection All containers should be tightly stoppered. The gray (oxalate/fluoride) tube top is recommended for medicolegal collections and if storage is prolonged. Sodium fluoride (50 mg) can be added as a preservative to urine and gastric samples. Other anticoagulants (eg, heparin EDTA) are acceptable.

Reference Range None detected

Possible Panic Range Blood: acetone, methanol, isopropanol >50 mg/dL (SI: acetone: >8.6 mmol/L, methanol: >15.6 mmol/L, isopropanol: >8.32 mmol/L); ethanol: >200 mg/dL (SI: >43.4 mmol/L); urine: acetone, methanol, isopropanol >50 mg/dL (SI: acetone: >8.6 mmol/L, methanol: >15.6 mmol/L, isopropanol: 8.32 mmol/L); ethanol: >160 mg/dL (SI: >34.7 mmol/L)

Methodology Gas chromatography (GC)

Additional Information Both methanol and isopropanol are more intoxicating than ethanol. Like ethanol, methanol exhibits zero order elimination kinetics. Methanol is converted to formaldehyde and formic acid, which causes retinal damage leading to metabolic acidosis and blindness. Treatment may include infusion of ethanol to inhibit methanol metabolism or treatment with alcohol dehydrogenase inhibitors.

In most laboratories, ethylene glycol is not part of volatile screen. This fact should be made clear to hospital staff and ordering physicians.

Warming, Blood

Use For very rapid, massive transfusion (>50 mL/kg/hour in adults or 15 mL/kg/hour in children), patients with severe cold agglutinin disease and infants undergoing exchange transfusion.

Contraindications When moderate volumes of blood are given at ordinary rates, warming is unnecessary. It is probably unnecessary also in most patients with cold agglutinin disease or paroxysmal cold hemoglobinuria who are not seriously ill.

Abstract Warming should take place, when necessary, using an FDA-approved device during passage through the transfusion set. A visible thermometer and a warning system are required.

Limitations Uncontrolled warming of donor blood can severely damage RBCs. Although red cells must be heated to 44°C or higher to be damaged, it is probably best not to allow warming above 40°C. Warming must be done so as not to cause hemolysis.

Additional Information Relatively large volumes of blood at refrigerator temperature, infused rapidly, can cause hypothermia and cardiac arrest. Blood subjected to excessive heat (ie, above 44°C) may be lethal. A quality assurance protocol is essential for all blood warmers and is required by accrediting agencies.

West Nile Virus Diagnostic Procedures

Use Evidence of encephalitis, myelitis, meningoencephalitis, fever, muscle weakness and pain, fever, and acute often asymmetrical flaccid paralysis may indicate WNV infection. Usually a self-limited dengue-like disease, other clinical observations include roseolar or maculopapular rash, nausea and vomiting, lymphadenopathy, and polyarthropathy.

Abstract Although severe neurologic disease develops in 1 of 150 infections, most West Nile virus (WNV) infections are subclinical. Advanced age is a risk factor. Immunocompromised individuals may be at high risk for death. Encephalitis is more common than meningitis/ meningoencephalitis. A poliomyelitis-like syndrome occurs as well. The agent is an arbovirus principally spread by *Culex* mosquito vectors and by birds, especially crows and jays.

WNV IgM and IgG antibody concentrations on enzyme-linked immunoassay can be confirmed by plaque-reduction neutralization assay. IgM testing is the method of choice for diagnostic disease testing. PCR is the test of choice for donor screening.

Limitations Cross-reactions may occur in subjects who have recently been vaccinated against related flaviviruses (eg, yellow fever, Japanese encephalitis) or recently infected with one of these agents (eg, dengue fever); but cross-reactivity was not reported to represent a problem in acute-phase samples tested for IgM antibody. Antibody tests fail to identify recently infected individuals who are yet to become seropositive, but who are potential blood donors.

Specimen Serum, plasma, cerebrospinal fluid (CSF), brain tissue (humans), mosquitoes, avian samples. CSF IgM or the plaque reduction neutralization test provides definitive diagnosis.

Sampling Time Specimens drawn before 8 days after onset of symptoms may be serologically negative. When WNV-specific IgM is detected in serum but not cerebrospinal fluid, serum from acute and convalescent stages is needed. In that setting, increase in titer by a factor ≥4 is required for confirmation of diagnosis.

Turnaround Time The plaque reduction neutralization test usually requires up to 8 days to test for both WNV and St Louis encephalitis virus.

Special Instructions WNV is a biosafety level 3 pathogen.

The laboratory should know the date of onset and dates of sample collection to assess whether to anticipate IgM or IgG reactions. Travel history is relevant, and vaccination history is important.

Methodology IgM capture and IgG enzyme-linked immunosorbent assays (ELISA), microsphere immunoassay. IgM antibody may be found without IgG virus-specific IgG in the first week of illness. Switch to IgG antibody is reported following 4-5 days of illness and is found earlier in CSF. A positive serum result can be followed by WNV IgM in CSF by ELISA, by WNV RNA in CSF, by fourfold increase in IgG, or by virus isolation. ELISA serum titers can be confirmed with plaque-reduction neutralization.

The plaque reduction neutralization test is considered a gold standard confirmation method. It is sensitive but slow and potentially hazardous to laboratory workers.

The presence of WNV RNA is confirmatory by reverse transcriptase-PCR testing, but its sensitivity in CSF is only 57%, in serum only 14%. The sensitivity of PCR testing is less than that of IgM capture ELISA methodology. Screening of donors for blood donations with nucleic acid-based assays is advocated.

WNV may be identified by immunocytochemistry and immunofluorescent assays.

Additional Information Monitoring for West Nile virus activity includes mosquitoes, chickens, wild birds, and susceptible mammals, including horses as well as humans. The virus can survive in donated blood and can be transmitted by red cell, platelet and fresh frozen plasma transfusions. Transmission has been conclusively linked to organ transplantation. Other means of transmission include breast-feeding, transplacental transmission, and laboratory acquisition.

The antibody response to WNV infection by IgM antibody capture enzyme-linked immunosorbent assay is positive in most subjects 7-8 days after onset. Such IgM response in serum or cerebrospinal fluid is detectable within 4 days of onset in about 75%. Such IgM antibody persists for a year or more. Detection of WNV-specific IgM in cerebrospinal fluid is confirmatory of current infection, while detection in serum is thought to provide only a diagnosis of probable infection.

Yersinia pestis Diagnostic Procedures

Synonyms Plague Diagnostic Procedures

Use Although serological studies can confirm the diagnosis of plague, bacteriologic methods, ELISA, and the new rapid diagnostic test are preferable; *vide infra*.

Applies to *Yersinia pestis* Culture

Abstract *Yersinia pestis* (*Pasteurella pestis* until 1970) is the etiologic agent of plague. Epidemic bubonic plague has been described historically. It was responsible for the deaths of 25% or more of Europe's population in the Middle Ages, the Black Death of the 14th century and subsequent epidemics. Currently, epidemics occur throughout the world with at least 2000 cases reported annually. The cycle can be stable (enzootic) or epidemic (epizootic) in rodents, squirrels, and prairie dogs. Fleas become infected and carry the organisms. In the U.S., plague occurs west of the 100th meridian (North Dakota to Texas).

The major forms of plague include lymphadenitis (bubonic plague), and secondary and primary pneumonic disease. A present concern is the possibility that *Y. pestis* may be used as a biological weapon. Used by bioterrorists, *Y. pestis* would probably be aerosolized, causing pneumonic plague. A terrorist attack might be recognized by a sudden outbreak of severe pneumonia, possibly with sepsis.

Limitations F1 antigen may diffuse into the Cary-Blair transport medium, leading to false negatives. Isolation and ELISA are specific but their sensitivity is impaired by deterioration and/or contamination during transportation and antibiotic therapy prior to sampling.

Specimen Serum for serology; sputum, blood, or aspirated material from lymph nodes for culture and immunofluorescence; sputum, serum, urine, and bubo aspirates for rapid diagnostic test

Volume 10 mL

Minimum Volume 2 mL

Container Red top tube for serology; specimens for microbiology can be transported in Cary-Blair agar.

Storage Instructions Acidified serum for serology may be stored in a refrigerator.

Turnaround Time A rapid diagnostic test has recently been described by Chanteau et al. It is a hand-held immunochromatographic dipstick assay. Cultures may grow slowly, and should be held for 7 days before discard. Hemagglutinating antibodies generally appear toward the end of the first week of disease.

Special Instructions In cases of suspected plague, the Centers for Disease Control and Prevention, Vector-borne Infectious Diseases, Fort Collins, CO (970) 221-6400 should be contacted at once. Sera must be inactivated and absorbed with sheep erythrocytes prior to testing. Sputum, aspirates, and tissue are hazardous.

Reference Range Serology: titer ≤1:10

Critical Values Serology: A fourfold or greater increase in titer or a single titer >1:16 indicates exposure to *Y. pestis* and should be immediately reported to the physician and the CDC.

Methodology F1 antigen is found in blood, sputum, and bubo specimens from infected subjects. It is specific to *Y. pestis*. Methods include immunocapture ELISA for F1 antigen, passive hemagglutination on acute and convalescent serum for fraction 1 (F1) antigen; direct antigen staining in urine, testing for IgM and IgG antibodies by PCR, direct immunofluorescence, and tissue immunostaining. Culture: *vide infra*. The new rapid diagnostic test uses monoclonal antibodies to the F1 antigen. Gram, Wright, Giemsa, or Wayson stains, directly or after mouse inoculation.

Additional Information A hemagglutination titer ≥1:16 is presumptive evidence of an immunologic response to *Yersinia pestis*. Seeing the stained organism in clinical material (aspiration of a bubo, sputum) can also make diagnosis, using conventional stains or fluorescent antibody. Appearing as gram-negative bacilli or as coccobacilli, Wright, Giemsa, or Wayson preparations may demonstrate bipolar staining. Blood, bubo (lymph node) aspiration, sputum, bronchial washings, and other materials can be stained and cultured. The organism grows at 25°C to 28°C on blood or MacConkey agar. It grows in brain-heart broth at 37°C. Identification of the F1 capsular antigen from organisms grown at 37°C is diagnostic.

SECTION V
RADIATION

The Reality of Radiation Risks: A Recent Case of Polonium 210 as a Weapon of Assassination and Subsequent Public Health Threat

R.B. McFee, DO, MPH, FACPM

While radiation terrorism seems a remote threat, the fact is radioactive materials suitable for malevolent usage are ubiquitous and readily available. The case that occurred recently in the United Kingdom, and the chapter that follows, underscore the need to be well trained in response to and vigilant about the threats of radiation/nuclear agents.

"The chilling reality is that nuclear materials and technologies are more accessible now than at any other time in history." Former Director – Central Intelligence, John Deutch.[1]

This reality became all too evident in November 2006 when former KGB agent Alexander Litvinenko was murdered by poisoning with radioactive polonium – 210 in the United Kingdom.[2-5]

He was not the only person exposed in this saga of espionage, murder and intrigue. At least 12 people tested positive for contamination as of January 2007.[3] Moreover, fears of polonium-210 contamination have led literally thousands of people to contact the National Health Services (NHS) Direct helpline, which was established in the aftermath of the assassination. The British initiated an investigation, which uncovered traces of [210]Po at a restaurant and bar, both in Mayfair – a posh section of London. Some traces had been found on British Airways (BA) aircraft; fortunately none of the 1700 passengers on BA flights or 250 patrons of the restaurants and bar have become sick or were contaminated. A former KGB agent and former Russian army officer both tested positive for [210]Po in Moscow; both have received medical treatment.[2,3]

Of all the Weapons of Mass Destruction (WMD) agent categories – chemical, biological, radiologic, nuclear and explosive that exist as terrorist threats – radiological and nuclear agents represent the ultimate lethal mass casualty as well as psychological threat, yet remain the most underemphasized and least prepared for among these weapons.[1,6-9] Almost half of hospitals lack a plan for nuclear terrorism.[10] Even when they exist, few have been practiced. This is not just a domestic problem. According to the British Medical Journal, while the United Kingdom has one of the longest established systems of public protection after radiation incidents, such plans are not well known by health professionals, making them thus less than ideally placed to advise or protect their patients.[11] A Medline® literature review of WMD related articles revealed most publications and preparedness efforts emphasize biological and chemical weapons with a glaring lack of guidance on radiation by comparison. Responses to recent incidents involving radiation indicate that most general practitioners are uncertain about the health consequences of exposure to ionizing radiation and the medical management of exposed patients.[12-16]

This lack of emphasis on radiation preparedness became evident when Mr. Litvinenko presented to a British hospital in early November. Clinicians initially believed his symptoms were due to thallium exposure but eventually determined the toxicant was [210]Po.[2-5] By then the likelihood of treatment success decreased significantly; he died three weeks later.

There are several treatment options available to treat the radiation poisoned patient, from colony stimulating factors to increase blood cell counts and decorporative antidotes (chelators), these must be initiated early. Currently available chelators – domestic and foreign include Dimercaptosuccinic acid (DMSA), mono-N—(1-butyl)-meso-2,3-dimercapto succinamide (Mi-BDMA), N,N'-di (2-hydroxyethyl_ethylenediamine-N,N'-biscarbodithioate (HOEtTTC), N-(2,3-dimercaptopropyl) phthalamidic acid (DMPA) as well as British Anti-Lewisite (BAL) - oxytiol or oxathiol (Russian); each demonstrating varying degrees of capabilities in reducing body burden and toxicity.[17,18] Under laboratory conditions HOEtTTC appears to be superior to other options.[17,19] A Russian radiation expert asserted that oxytiol and other countermeasure based upon electrochemical characteristics of cyclic acetals and other similarly acting thiols, dithiols and azathiols could have conferred a survival benefit to Mr. Litvinenko.[5,20,21] Not all chelators are available globally; guidance from on site radiation expertise, and REAC-TS are essential given rapid treatment is a critical element to survival. A cornerstone to accurate diagnosis is obtaining a thorough history and integrating such information with an index of suspicion. The delay in correctly diagnosing Mr. Litvinenko underscores the need to consider radiation and exotic toxicants, *especially* in the context of a victim who has been involved in dangerous occupations that might predispose the patient to increased risk from unusual exposures – information that should have been revealed as a result of a thorough, timely history.

Polonium was discovered and chemically separated by Marie and Pierre Curie; Madame Curie named it after her native Poland. There are over 20 isotopes of Po; ^{210}Po is the most stable form.[4] ^{210}Po was an important component during the early development of nuclear weapons; it has a 138 day half life (t ½) and is an alpha emitter. While alpha particles cannot travel great distances in air, and can personal protective equipment, intact skin and clothes are effective barriers; unlike other radioactive elements, ^{210}Po is relatively safe to transport.[4,7,8] However, if such high energy particles are inhaled, ingested or inserted into abraded skins or wounds, they can damage tissue.[7,8,21] Within minutes the cells lining the gastrointestinal tract of Mr. Litvinenko would begin to die and slough off, which would cause nausea, pain and severe internal bleeding.[7-10,21] ^{210}Po would effect other systems and, unless early decorporative (chelation) treatment is initiated to lower the body burden and separate the poison from the patient, significant morbidity and ultimately death can be anticipated.

Po can be used to produce radioisotope thermoelectron generators (RTGs), which are used to produce electricity to operate satellites and unmanned space vehicles. However owing to the short t ½ of Po, it is being replaced by plutonium, as a longer lived alpha emitter. In recent years obtaining quantities of polonium requires neutron bombardment of bismuth. As a poison, it is highly toxic; ^{210}Po is several orders of magnitude more toxic, on a milligram per milligram basis than hydrogen cyanide. It is estimated that one gram of ^{210}Po could kill 50 million and sicken another 50 million.

To decrease our vulnerability, the medical community should have a better understanding of radiation hazards, the newer treatments available such as Prussian blue, Zn-DTPA and other emerging countermeasures, in addition to knowing how to rapidly access the expertise of the Radiation Emergency Assistance Center – Training Site (REAC-TS) – a 24-hour emergency service that can be reached at 865.576.3131 and 865.576.1005.[7,8,22]

Whether thought of as first responders or last preemptors, emergency medical professionals need to be able to identify rapidly a possible radiologic event, control potential public anxiety or outright panic and protect themselves with appropriate personal protective equipment.[1] Responding to a radiation incident requires advanced planning, training, heightened awareness and information sharing across performance cultures tasked with preparedness and public protection.[22]

References

1. McFee RB, Leikin JB, Kiernan K. Preparing for an era of weapons of mass destruction (WMD) – Are we there yet? Why we should all be concerned, Part II. Vet Human Tox 2004; 46 (6): 347-351.
2. Day M. Former spy's death causes public health alert. BMJ 2006; 333:1117
3. Dyer O. More cases of polonium-210 contamination are uncovered in London. BMJ 2007; 334:65.
4. Kaplan K, Maugh TH. Polonium-210's Quiet Trail of Death. Jan. 1, 2007 http://ww.mjwcorp.com/rad_dose_assessments_poloniumarticle.php. Last accessed 1/27/07

5. Komarov S. Litvinenko Could Have Been Saved. Moscow News http://english.mn.ru/english/issue.php?2006-48-11. Last accessed 1/27/07.
6. McFee RB. Preparing for an era of weapons of mass destruction (WMD) – Are we there yet? Why we should all be concerned, Part I Vet Human Tox 2002; 44:193-199.
7. Leikin JB, McFee RB, Walter F, et al. A primer for nuclear terrorism. Dis Mon 2003; 49: 485-516.
8. Leikin JB, McFee RB, Walter F. Chemical and nuclear agents of terrorism. In Hessl S ed. Clinics in Occupational and Environmental Medicine Law Enforcement 2003; 3: 477-505.
9. Timins JK, Lipoti JA. Radiological terrorism. NJ Med 2003; 100: 14-21.
10. Dirty bomb threat puts spotlight on unprepared Eds: do you have a plan? ED Manag 2002; 14:97-100.
11. Turai I, Veress K, Gunalp B, Souchkevitch G. Medical response to radiation incidents and radionuclear threats. BMJ 2004; 328:568-572.
12. Meineke V, van Beuningen D. Sohns TG, Fleidner TM. Medical management principles for radiation accidents Mil Med 2003; 168: 219-222.
13. International Atomic Energy Agency and World Health Organization – Follow up of delayed health consequences of acute accidental radiation exposure; lessons to be learned from their medical management. Viena IAEA 2002;129(IAEA Tec Doc 1300).
14. Turai I, Crick M, Ortiz-Lopez P, et al. Response to radiologicasl accidents: the rolr of the International Atomic Energy Agency. Eadiotprot 2001;36:459-75.
15. Turai I, Veress K. Radiation accidents:occurrence types consequences, medical manage Thomas ment, and the lessons to be learned. Centr Europ J Occup Enviro Med 2001; 7: 3-14.
16. Gower- Thomas K, Lewis M, Shiralkar S, etal. Doctors knowledge of radiation exposure is deficient. BMJ 2002;324:919.
17. Rencova J, Volf V, Jones MM, Singh PK. Mobilization and detoxification of polonium – 210 in rats by 2,3 dimercaptosuccinic acid and its derivatives. Int J Radiat Biol 2000 Oct; 76 (10): 1409-15.
18. Bogdan GM, Aposhian HV. N-(2,3-dimercaptopropyl)phthalamidic acid (DMPA) increases polonium – 210 excretion. Biol Met 1990; 3 (3-4): 232-6.
19. Rencova J, Volf V, Jones MM, et al. Bis-dithiocarbamates: effective chelating agents for mobilization of polonium-210 from rat. Int J Radiat Biol 1995 Feb; 67(2): 229-34.
20. Roy RK, Bagaria P, Naik S, Kavala V, Patel BK. Chemoselectiveities in acetalization, thioacetalization, oxathioacetalization and azathioacetalization. J Phys Chem A Mol Spectrosc Kinet Environ Gen Theory 2006 Feb 16; 110 (6): 2181 – 7.
21. Skomorkhova TN, Borisov VP. Eliminatory effect of oxathiol upon exposure to fumes of metallic mercury 203 Hg. Gig Tr Prof Zabol 1982 Mar; (3): 46-47.
22. McFee RB, Leikin JB. Radiation Terrorism: The unthinkable possibility, the ignored reality. JEMS 2005 April: 78-92.

A Primer for Nuclear Terrorism Preparedness

Jerrold B. Leikin, MD

Robin B. McFee DO, MPH

Frank G. Walter, MD. FACEP, FACMT, FAACT

Richard G. Thomas, Pharm D. ABAT

Keith Edsall MBBS, MRCS, LRCP, MSc

Adapted from Elsevier with permission.

Table I. Classification of Radiological & Nuclear Terrorism	
Name	Example
Simple radiological device	Placing high gamma source (Ir, Cs, Co) in busy public place
Radiological dispersal device	Spreading radioactive material with conventional explosion ("dirty bomb")
Nuclear reactor sabotage	Disabling nuclear reactor's cooling system
Improvised nuclear device	Detonating homemade nuclear device in a city
Nuclear weapon	Detonating one kiloton, suitcase-sized, nuclear weapon pirated from one of nuclear powers of world

Preparedness efforts for mass destruction (WMD) and terrorism of a population have emphasized biological and chemical agents.[1-4]

Medical training concerning radiation poisoning remains underemphasized in spite of the potential risk radiation terrorism poses when one considers radiation sources are ubiquitous, and available from medical, industrial, and military sources.[4-6] (Table I) Such widespread proliferation of nuclear materials increases the probability that some form of radiation weapon will be used. The potential for a radiation event can result from terrorism, as occurred in the Chechen threat to use a radiation dispersal device (RDD) using Cesium – 137 in Moscow, or from the mishandled disposal of a medical radiotherapy device in Brazil, which was responsible for the worst radiation accident in the Western Hemisphere.[4] Chernobyl and scattered other high profile events although relatively rare, demonstrate that when they do occur, the potential for multiple casualties, and long-term illness is great. An RDD is any device that causes the purposeful dissemination of radioactive material across an area without a nuclear detonation.[4,9] (Table II) This type of device can be produced by blowing up a radioactive source with conventional explosives. An RDD would cause conventional blast related casualties that have become contaminated with radionuclides.[4] An alternate device – an simple dispersal device (SDD) is a source of radioactive material that does not rely on an explosion to cause harm, instead relying on the intrinsic radioactivity of the source material to affect individuals. In this era of terrorism, clinicians may be the first to identify sentinel cases, and thus need to think in terms of complex presentations from multiple and simultaneous injury sources, and communicate suspicions, concerns and observations to appropriate members of response teams. Although most major metropolitan areas have designated nuclear response hospitals, the location of an RDD may occur closer to a community hospital than an tertiary care facility. The medical management of patients who can present with suspected exposure, no exposure but significant psychological trauma, or a major radiological emergency, with or without contamination, depends upon trained personnel with appropriate equipment. Our concerns as health care professionals must go beyond Cold War thinking in terms of a nuclear holocaust, and expand to include

Table II. Classification: Simple Radiological Device & "Dirty Bomb" Sources														
	Major Radionuclides												
Sources	C 14	Co 60	Cs 137	Ga 67	H 3	I 125	I 131	Ir 192	P 32	Tc 99	Tl 201	U 238	Xe 133
Medical Diagnostic & Therapeutic Facilities		X		X			X		X	X	X		X
Research Labs	X	X			X	X						X	
Industrial Imaging & Sterilizing Facilities		X	X					X					

527

preparedness against "dirty bombs," such as RDD, as well as increase our knowledge in managing industrial incidents.

A) Introduction to Radiological Casualties

Managing a patient exposed to or contaminated by radioactive material is directed toward rapid recognition, prevention of secondary contamination, patient decontamination, and appropriate intervention, referral, and follow-up.[7,8,9]

When dealing with radiation problems the type of radiation source is less important than the dose it is producing. If the dose produced is high enough, regardless of the source, biological damage will result. Unfortunately, the body cannot see, smell, hear, taste or feel radiation. A victim of a radiological or nuclear event can receive a lethal dose of radiation without any warning signs or symptoms until the early signs of radiation sickness develop.[8]

Managing a victim is guided in part by whether the patient was contaminated or irradiated (or both). A contaminated patient is one that has radioactive material on the body and/or clothes, and thus poses the risk of spreading this contaminant to others and the environment. Irradiated patients pose no threat to health care providers, and can be managed without special protective equipment.

Patients can be affected by four radiation exposure scenarios:

1. **Irradiation**
   Irradiation injury is radiation exposure without contamination. There is no need to delay treatment – no decontamination is necessary. For example, patients who receive diagnostic X-rays are irradiated, but not radioactive. We do not decontaminate nor isolate patients after they receive diagnostic X-rays.

2. **External contamination**
   Radionuclides are present on external body surfaces, and may be transferable. Precautions must be taken to avoid contaminating health care workers and facilities.

3. **Internal contamination**
   Radionuclides can be deposited internally by inhalation, ingestion, injection, or wounds. These particles can then become incorporated into the body.

4. **Combined injury**
   RDD or nuclear explosions can create blast or mechanical injury, burns and radiation injury. Treatment of life-threatening, nonradiation related injuries should not be delayed in favor of treating radiation related illness.

It is important to mention, although beyond the scope here, the psychological impact associated with any radiation event will include profound acute and chronic stress disorders, possible somatization responses, and large numbers of walking, worried but physically well patients.[10]

B) Nuclear Radiation

Ionizing Radiation

Ionizing radiation consists of electromagnetic energy or energetic particles emitted from a radioactive nuclide source. Ionizing radiation is able to strip electrons from atoms causing chemical change resulting in biological damage. Ionizing radiation can occur naturally, can be generated or produced from radioactive atoms.

The basic building block of all matter is the atom. The atom consists of a central nucleus and around this nucleus are defined shells of electrons orbiting around.

The main part of the atom of concern is the nucleus. The nucleus has two basic building blocks, firstly the neutron as its name implies is neutral, secondly the proton which is positively charged. Positively charged particles will tend to repel each other therefore, to hold this nucleus together there is a special nuclear glue which has mass, (this will become apparent later called "binding energy"). This binding energy has three important characteristics. It is very strong, acts over a very short range and is independent of charge. Each element is defined by the number of protons it has in its nucleus, known as the Atomic Number.

For a stable element the ratio of protons to neutrons is one proton to 1.2 neutrons. A radioactive nucleus has an imbalance of this ratio and it is usually due to an excess of neutrons.

The unstable nucleus can become stable by altering its components in the following ways. A neutron can change itself into a proton by the emission of a negative electron this negative electron is called a Beta particle.

> Neutron = Proton + -ve Electron

A proton can change itself into a neutron with the emission of a positive electron called a positron. Note that a positron cannot survive on its own and is annihilated. This occurs when a positron meets a negative electron, both particles are changed into energy producing two gammas of 0.5 mev (mega electron volts).

> Proton = Neutron + +ve electron.

So as one can see the unstable nucleus can interchange its component parts reaching the ratio of one proton to 1.2 neutrons. Note that if the proton number changes then a completely different element is formed.

One now asks the question is the new nucleus that is formed having the correct proton to neutron ratio stable? The answer to this question is no. Why? To answer this question one must consider the nuclear glue (binding energy). As a result of the particles that have been ejected from the nucleus there is an excess of nuclear glue which has mass, this excess mass is changed into electromagnetic energy of very short wavelength and this is where gamma irradiation is derived. Thus the nucleus finally becomes stable.

Unstable heavy elements have the capability of ejecting larger particles consisting of two protons and two neutrons. (Table III) These particles are known as Alpha particles.

Therefore the types of radiation that are of concern are:

Alpha particles
Beta particles
Neutrons
Gamma Rays
X-Rays

Table III. Atomic Structure		
Selected Particles	**Atomic Mass (amu = daltons)**	**Charge**
Proton	1	+1
Neutron	1	0
Alpha (α) Particle	4	+2
Beta (β) particles		
Electron	1/1,823	−1
Positron	1/1,823	+1

The following discusses properties of the various types of ionizing radiation:

Alpha Particles

Alpha particles consist of two protons that are positively charged, and two neutrons. Having two protons, which are positively charged, they are very reactive. Alpha particles can travel 2-3 cms in air and only microns into tissue. They can be stopped completely by a piece of paper therefore will be stopped by clothing. If an alpha particle lands on the skin this is not a problem. The reason for this is the outer layers of skin are dead and are several microns thick. The alpha particle generally will not penetrate through the dead layers of skin to get to the live layers to cause damage. Therefore an alpha particle is not an external hazard to the skin. On the other hand, if an Alpha particle gets inside the body by, ingestion, inhalation, or via a wound. In this situation the alpha particle is adjacent to live tissue and could cause serious damage to the immediate tissue within a few microns in depth. Therefore Alpha particles are an internal hazard only.

Beta Particles

Beta particles are high-energy negative electrons that are emitted from a decaying nucleus. They can travel a meter or so through air and millimeters into skin. They have a spectrum of energies depending on the radioactive isotope concerned. These energies are measured in mega electron volts (Mev). The higher the electron volts the deeper the penetration e.g. Beta particle of energy of 0.1 Mev will penetrate 0.15 cms into tissue. Beta particle with energy of 5 Mev will penetrate 5cms into tissue. If these beta particles are allowed to remain on the skin for long periods of time they could cause severe burns of the skin and also the anterior compartment of the eye. They are also a hazard if they are internally deposited into the body.

Gamma Rays

Gamma rays are very high-energy rays of short wavelength that are very penetrating. hey can travel many meters in air and many centimeters into tissue. Protective clothing will not protect you against Gamma rays, but your clothing will prevent contamination of your skin from gamma emitting radioactive isotopes. Gamma rays are not contaminating.

Neutrons

Neutrons have no electrical charge. They are very penetrating and cause damage via two main pathways. The first pathway is by collision therefore imparting energy to other particles (i.e. the "billiard ball" effect). The second pathway is by neutron capture. Certain elements have the affinity to capture neutrons. One element of concern is sodium. When the body is exposed to a flux of neutrons our body Sodium 23 has the affinity to capture a neutron to become Sodium 24, which is radioactive and is one of the occasions where an individual becomes radioactive.

X-rays

X-rays are similar to gamma rays, but they have lower energy. X-rays are produced as the result of high energy electrons passing a positive nucleus. As the electron slows down, it gives off X-rays which is also called Bremsstrahlung radiation (braking radiation). X-rays, like gamma rays, are very penetrating; X-rays are not contaminating.

It should be noted that in addition to the type, source and threat of radioactivity that some elements present, these same substances may pose an additional toxic threat depending upon their state of matter, external temperature, and circumstances of exposure, by being corrosive, or explosive. Cesium for example is an alkali metal, and can explode if exposed to water. Cesium hydroxide is a strong base and attacks glass. In fact it is the most electropositive and alkaline element and one of the few metals that is liquid at or around room temperature. Cesium reacts with ice at temperatures above $-116°C$. Therefore the responder should attempt to identify the radionuclide source, the element or chemical involved, and associated critical information such as flash point, type of radiation emission (beta, gamma, alpha) and other characteristics that may pose a risk to patient and health care worker. Cesium is found in nature as a single stable isotope – Cesium 133. However there are 30 other radioactive isotopes ranging from Cs 114 to Cs 145, with half lives ranging from .57 seconds (Cs 114) to 3×10^6 years (Cs 135). Cesium 137 is commonly found in medical and industrial applications, has a relatively long half – life and thus could be utilized with TNT or other explosive chemicals to create a "dirty" RDD device, or left unshielded as an SDD. Cs 137 emits both gamma and beta radiation. It is completely absorbed by the lungs and GI tract, as well as internalized via wounds. It is soluble in most forms and is metabolically handled like a potassium analog, with urinary excretion. Deaths have occurred due to Cesium Acute Radiation Syndrome (ARS).

Therefore, in this era of terrorism, explosives not only create blast and thermal damage, they now can be mixed with toxic chemicals or radionuclides. The clinician must be aware of the spectrum of risk that these chemicals and events can present, and the requisite assets necessary to manage such exigencies. Although the ensuing radiation risk could be minimal, the terrorism value of spreading even a low yield, low threat source of radiation would be quite effective in terms of public concern and possible panic.

The clinician must be aware of the spectrum of risk that these chemicals and events can present, and the requisite assets necessary to manage such exigencies. Although the ensuing radiation risk could be minimal, the terrorism value of spreading even a low yield, low threat source of radiation would be quite effective in terms of public concern and possible panic.

Radiation hazard is related to the strength of the source (radioactivity) and the total radioactive energy absorbed per unit volume of tissue (dose). Activity is the number of radioactive disintegrations per second, which is referred to as the Curie or Becquerel (Bq). 1 curie $= 3.7 \times 10^{10}$ Bq. Biologic injury from a given type of ionizing radiation depends upon the amount of radioactive energy deposited into a physiologically critical volume of tissue. This is quantified by units of absorbed dose.

Time

Exposure and absorption are dose and time dependent. By limiting the time of exposure you limit the exposure. Therefore time with potentially contaminated patients should be kept to a minimum until they are decontaminated.

Distance

Keeping the maximum distance between oneself and the source will limit the exposure. The concept of the inverse square law applies i.e. if you double the distance between yourself and the source the exposure will be reduced to a quarter of its original value. The reduction of the exposure is exponential.

For example, if an individual is standing two feet away from a source and is receiving a dose of 160 Rads, by doubling the distance to four feet the dose will be reduced to 40 Rads i.e. one quarter. If the distance is then doubled again i.e. at eight feet the dose will be reduced to 10 Rads. Therefore by increasing the distance from two feet to eight feet the dose has reduced from 160 Rads to 10 Rads. (Table IV). Also, if possible, removing the patient from the irradiated contaminated site will reduce the dose to the patient.

Table IV. Radiological Protection: Increased Distance Example	
Distance From Source (Feet)	Dose (Rad)
2	160
4	40
8	10

Shielding (TableV)

Alpha

A single piece of paper will stop alpha completely. Hospital protective clothing will completely protect you against alpha particles.

Beta

To shield against Beta particles you require a light material such as aluminum or thick plastic. Materials of low atomic number are chosen to shield against Beta particles to minimize the production of Bremsstrahlung radiation (X-rays). Your hospital protective clothing will only partially protect you from Beta irradiation.

Gamma

Gamma irradiation is electro magnetic radiation, unlike particulate radiation, it cannot be completely stopped only attenuated. However, with shielding, it is markedly attenuated and the remaining radiation would be at a very level.

Shielding is provided by lead or concrete. However, in an emergency these materials may be difficult to get at short notice, therefore the setting up of barriers or limiting the time may be your best option.

Neutrons

Generally materials of high hydrogen content and low atomic number are used to shield against fast neutrons such as paraffin, wax, water and for thermal neutrons Boron or Cadmium. Neutron radiation is the result of either nuclear fission (e.g. a nuclear bomb or reactor core), or nuclear fusion (e.g. a neutron bomb). In either case, patients are not going to be contaminated with neutron radiation nor will health care workers be exposed to it. Thus shield is usually unnecessary.

Table V. Radiological Protection: Radiation Shielding						
Name	Type	Mass (amu)	Charge	Penetration	Adequate Shielding	Main Problems
Alpha	Particle	4	+2	cm thru air Stratum corenum (dead outer skin layer) only (20 μm)	A single piece of paper Universal precautions PPE Regular clothing	Contamination
Beta	Particle	1/1,823	−1 or +1	m thru air mm thru tissue	Aluminum foil Thick plastic	Contamination
Neutron	Particle	1	0	m thru air cm in or thru body	Hydrogen content: Water Polystyrene Parrafin wax	Irradiation or neutron capture: $^{23}Na + n = ^{24}Na$
X-ray	Ray	0	0	m thru air Thru body	Lead Concrete	Irradiation
Gamma	Ray	0	0	100m thru air Thru body	Thick lead Thick concrete Depleted Uranium (U238)	Irradiation

Quantity

When dealing with contaminated clothing, do not store but discard it in the emergency department. This would result in producing a radioactive source in the department, thereby increasing radiation exposure of staff. Therefore, it is important that the clothing of contaminated patients is removed from the emergency department to a designated area outside of the hospital.

Radiation Units (Table VI)

The basic unit for measuring dose is the RAD (Radiation Absorbed Dose) This is defined as the deposition of 0.01 Joules of energy deposited in one kilogram of tissue. To quantify the amount of damage that is suspected from a radiation exposure RADS are converted into REM (Roentgen Equivalent Man). The REM is adjusted to reflect the type of radiation absorbed and the likely damage produced. For most purposes, consider the REM & RAD to be equivalent.

The United States is the only country in the world that uses the REM & RAD, the rest of the world uses the international units named the Gray and the Sievert. The reason this is mentioned is that all publications and new text books on radiation medicine are published in the new international units. The conversion of the units are: 100 RAD = 1 Gray and 100 Rem = 1 Sievert.

Background Radiation

Radioactivity has existed in the earths crust for millions of years. We live in a radiation environment being exposed from the earths crust, cosmic radiation and man made radiation. The highest contribution of man made radiation coming from hospital X-ray and nuclear departments. We are all radioactive; our bodies contain the radioactive isotopes Carbon 14 and Potassium 40.

The average radiation background, both natural and man made for the United States is 360 mRem/year; 80% natural and 20% manmade.

It is important to realize that background radiation will change from area to area and day to day. Comparing granite mountainous regions to the flat plains the background radiation is much higher in the mountainous regions. Therefore it is important to know what the background radiation level is in your area at the time of an incident because you are looking for anything above background.

Putting radiation doses into prospective, some examples are shown on Table VII.

Table VI. Radiation Dose					
Unit	Meaning	Symbol	Value	System	Usual Setting
RAD	Radiation Absorbed Dose	rad	0.01 J/kg	USA	"Accident" Incident
REM	Radiation Equivalent Man	rem	Biological damage from 0.01 J/kg of X-rays	USA	Occupational
Gray	Eponym	Gy	1 J/kg = 100 rad	SI	"Accident" Incident
Sievert	Eponym	Sv	Biological damage from 1 Gy of X-rays = 100 rem	SI	Occupational

Table VII. Radiation Doses	
Natural and man-made radiation	360 mRem/yr.
Chest X-ray	10 mRem
Flight LA to Paris	4.8 mRem
Smoking 1.5 ppd	16 Rem/yr
Mild acute radiation sickness	200 Rem

C) Exposure, Contamination & Incorporation.

With respect to a radiation incident regardless of the source or the type of incident three types of radiation injury could occur. External radiation, contamination with radioactive material and incorporation of the radioactive material into body tissues, to this could be added the complication of trauma. (Table VIII & Table IX) It is important to realize that any combination of these could happen at the same instant.

External Exposure

External radiation occurs when the whole of the body or part of the body is exposed to penetrating radiation from an external source (namely gamma and X-rays from an external source). It is important to realize in this situation the patient could be seriously injured from radiation but the patient is no danger to the medical staff. Following this type of exposure the patient is not radioactive and can be treated accordingly.

To emphasize this concept, this is similar to when you go to an X-ray department for a chest X-ray. When the X-ray has been taken you do not become radioactive, the same applies to someone who has been exposed to penetrating radiation.

Table VIII. Summary: Exposure vs. Contamination					
Incident	Radiation Location	Source Type	Physical State	Patient Decon Necessary	Secondarily Contaminating
Exposure	External	Rays (energy)	None	No	No
External Contamination	External (skin surface)	Particles (matter)	Solid Liquid	Yes Yes	Yes Yes
Internal Contamination	Internal	Particles (matter)	Solid Liquid Gas	Yes Yes No	Yes Yes No

Table IX. Chemical vs. Radiological Terminology		
Chemical	Vs.	Radiological
Absorption	vs.	Internal contamination
Distribution	vs.	Incorporation
Metabolism (catabolism)	vs.	Incorporation
Elimination	vs.	Decorporation

Contamination

Contamination can either be external or internal. External contamination is as a result of radioactive materials in the form of a solid liquid or a gas released into the atmosphere precipitating on the individuals. The exposed surfaces of the skin can become contaminated namely the hair, neck, face and hands. Obviously the clothes will be contaminated as well, but by taking the clothes off, with this simple procedure you can remove 90% of the contamination. Internal contamination is achieved by three main routes, inhalation, ingestion and via wounds resulting in internal radiation of the organs of entry.

Incorporation

Incorporation refers to uptake of radioactive materials into organs tissue and cells. In general materials that are distributed around the body depends on their physical, chemical forms and solubility. Incorporation cannot occur unless internal contamination has occurred. Also if you know what element or elements you are dealing with you have a reasonable idea what the target organ will be, however a word of caution in the accident situation you will know what radioactive isotope or isotopes you are dealing with as all radioactive sources are registered.

BUT IN THE TERRORIST SITUATION INITALLY YOU WILL NOT KNOW WHAT RADIO-ACTIVE SOUCE OR SOURCES YOU ARE DEALING WITH.

These four types of events can occur individually, or as a result of the combination of any of the three. Also to this you could add the complications of trauma and illness.

DEFINITION OF NUCLEAR TERRORISM: The use of radioactive material in various forms to cause the maximum amount of disruption, panic, injury and fear into the general population.

There are five groups of terrorist threat that can be considered:

- Simple radiological device
- Radiological dispersal device
- Reactor
- Improvised nuclear device
- Nuclear weapon

1. It is emphasized that the most plausible threat would be the first two; the reason being is that there are hundreds of thousands of radioactive sources used in industry and medicine around the world.

   Much information can be gained from previous worldwide radiation accidents in preparing and planning for nuclear terrorism.

Simple Radiological Device (SDD)
This is the deliberate act of placing a high energy source or spreading radioactive material in a highly populated areas e.g. airports, train stations, ports, sports venue etc. exposing individuals to various levels of radiation.

In 1987 in Goiania, Brazil, two thieves stole a Cesium 137 hospital therapy source in its shielding and sold it for scrap metal. The source consisted of 1375 curies of Cs 137 unfortunately the source was broken up and shared amongst various individuals. At the time the individuals concerned were not aware it was a radioactive source. The incident was not detected for fifteen days. The medical response and clean up phase took several months to complete. In this situation there was both exposure and contamination. The overall findings were as follows:

- 112,800 surveyed for contamination.
- 249 Contaminated.

- 120 Externally contaminated on clothes and shoes.
- 129 Externally and internally contaminated.
- 20 Required specific hospital treatment.
- 14 Developed bone marrow depression.
- Treated with GM-CSF.
- 4 Died due to hemorrhage and infection.

There have been other cases around the world of stolen radioactive sources resulting in serious radiation injuries (e.g. Juarez in Mexico1983 involving a Cobalt 60 source (450 Ci), and recently in Thailand, again involving Cobalt 60 sources.)

In the last couple of years there have been two incidences of stolen sources. First 16 Brachiotherapy sources of Cesium 137 were stolen from a hospital in North Carolina. Second, in Florida, where an industrial radiography source of Iridium 192 was stolen. These sources have not been found up until the present day.

Cesium 137 is ubiquitous as a medical and industrial radiation source. A common industrial usage is in the construction of highways; it is used in devices that measure the density of asphalt. During the last 18 months, several such devices in the Southeast United States have been stolen or become missing – whereabouts remain unknown. In addition to the radiation risk that Cesium 137 poses, this substance is highly toxic – explosive when exposed to air and water, must be stored in special containment, and can cause significant dermal injury upon direct contact. Hospitals, industry and scientific research facilities are especially vulnerable to the theft of such materials compared to military instillations. As such, occupations utilizing such materials should implement safeguards and surveillance systems to ensure that these radiation sources remain fully accounted for.

Radiological Dispersal Device (Table II)

This is where explosives are attached to a radioactive isotope and detonated. This could result in large areas of local contamination and also contamination of the people in the area and of responding emergency services. It could also lead to death of several people as a result of the explosion and significant exposure to radiation of individuals close to the scene of the incident. In this type of event a nuclear reaction does not take place.

Cesium 137 has been identified as a potential radiation source for such a "dirty" device. In this era of terrorism, responders must be aware of the secondary risk that a potential radiation source may pose an additional risk to responding to a potential explosive threat. The risk of a secondary device as well as a 'dirty' primary device must always be considered.

Reactor (Table X)

In the western world the probability of a terrorist event to a reactor is low, but recent events highlight the possibility that this could happen. This low probability is due to the high security surrounding a reactor and the safety systems incorporated into a reactor. There is extensive shielding around a reactor therefore very significant amounts of explosive would be required to breach the reactor core.

Table X. Classification: Nuclear Sabotage & Nuclear Weapons Fallout												
	Major Radionuclides											
Sources	Am	Cf	Ce 144	Co 60	Cs 137	Cm	H 3	I 131	Pu 239	P o	Sr 90	U
Nuclear Reactor Sabotage Fallout			X	X	X		X	X	X		X	
Improvised Nuclear Device & Nuclear Weapon Fallout	X	X	X		X	X	X	X	X	X	X	X

Terrorist events could compose of use of very large amounts of explosives but it would be very difficult for terrorist to breach the security cordon, there is the possibility of a jumbo jet crashing into a reactor or a nuclear pond of used reactor cores but again security is very high for this type of scenario to happen. Recently published results from computer and engineering scenarios studies suggest the construction of most reactors would sustain a direct hit from a commercial aircraft flying below 300 MPH into a reactor. However, there have been some scientists that hold these results into question.

Most people are aware of the reactor accidents at Three Mile Island and Chernobyl. The amount of radiation released from Three Mile Island was very small indeed and no radiation injuries developed.

The Chernobyl reactor incident was caused as a result of a technical design fault. This accident resulted in two explosions with fires and melt down of the core leading to serious wide spread contamination of the environment.

Considering the early phase of the accident this lead to the following:

- 237 were hospitalized.
- 134 developed the Acute Radiation Syndrome.
- 28 died within the first three months.
- The initial explosion killed 2.
- 1 died from heart failure.
- 250,000 were permanently evacuated from the area.

The main radioactive isotopes causing health problem were Cs 137 and I 131.

Considering the grave seriousness of the accident it is commendable that the response teams actions lead to a low overall mortality.

Improvised Nuclear Device (Table XI)

This is a device that would produce a nuclear yield. This type of device, if successful, would produce a nuclear yield similar to that of Hiroshima & Nagasaki with the release of radiation, blast and thermal pulses together with significant radioactive fallout. Construction of such a device would be very difficult because of sophisticated engineering and expertise required. Realistically a terrorist organization might be able to produce a partial yield producing reduced effects. A realistic situation is that the conventional explosive would detonate and blow the device apart, therefore resulting in environmental contamination with weapon material such as Plutonium or Uranium.

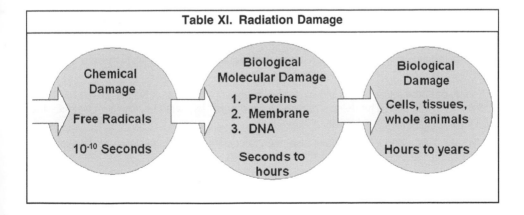

Table XI. Radiation Damage

Chemical Damage

Free Radicals

10^{-10} Seconds

Biological Molecular Damage

1. Proteins
2. Membrane
3. DNA

Seconds to hours

Biological Damage

Cells, tissues, whole animals

Hours to years

Nuclear Weapon

The probability of stealing a nuclear weapon in the Western World is very remote because of the high security. However a Russian general has stated publicly that between 50 to100 one kiloton nuclear weapons are unaccounted for in the former Soviet Union. These weapons are potential "suitcase bombs" and represent an increasing threat.

The consequences of the detonation of a one-kiloton nuclear weapon within the first minute would result in the following:

- Blast range would reach a distance of 400 yards.
- Thermal radiation would reach the distance as the blast.
- Nuclear radiation would reach half a mile i.e. gamma & neutrons.

The radioactive fallout could produce very high exposure rates up to half a mile. The rule of sevens applies. After 7 hours the dose is reduced by a tenth, 7×7 i.e. 49 hours the dose is reduced by another tenth, $7 \times 7 \times 7$ (i.e. 343hrs) the dose is reduced by another tenth. Hence the concept of sheltering for a period of two weeks.

The electromagnetic pulse, which really applies to a high aerial burst, would result in damage to solid state electronic equipment including hospital solid state defibrillators, electrocardiograms ventilators etc.

E) Radiation Cellular Damage (Table XI)

The basic building block of all tissue is the cell. Damage to the main components of the cell namely, the cell membrane, complex proteins and especially the DNA can lead to serious malfunction of the cell or even death.

The chemical damage caused by free radicals is immediate, it happens in 10^{-10} seconds.

The mechanism of damage is via two pathways, the direct and the indirect pathway. The direct pathway accounts for 20% of the damage and the indirect pathways accounts for 80% of the damage. The main target within the cell is the chromosomal DNA. The direct pathway causes single strand or double strand breaks in the chromosomes DNA which in some circumstances can be lethal to the cell. The indirect pathway is by ionizing body water into the hydrogen ion and the hydroxyl ion, it is the hydroxyl ion that causes the damage to the DNA. This damage can be repaired, in some cases modified or in some instances be beyond repair.

The biological expression of this damage at the cellular level takes seconds to hours. The clinical expression of this damage can take hours to years. High acute doses can express the Acute Radiation Syndrome within hours. Lower doses that are not capable of producing the acute radiation syndrome and even with very low doses, have the probability of producing cancer 20 to 30 years later. Other problems that can be produced weeks to months after exposure include cataract formation requiring an acute dose of 200 rem to the lens of the eye, neutrons are more potent in producing this effect.

Effects on some fetuses were shown in the survivors of Hiroshima & Nagasaki who received a dose over 50 Rads. The defects at birth showed small for dates i.e. low birth weight, small head circumference and mental retardation.

Hypothyroidism and infertility have also been associated with acute doses of radiation.

F) Acute Radiation Syndrome (A.R.S.)

To understand the Acute Radiation Syndrome two concepts need to be discussed. Firstly the LD 50/60 and secondly cell death.

Our body senses cannot detect radiation, after the dropping of the bombs in Hiroshima & Nagasaki which resulted in deaths and horrific injuries, experiments were carried out on animals to find out what was the required lethal dose to kill 50% of the animals in a set period of time. This was defined as the LD 50/30 i.e. the lethal radiation dose required to kill 50% of the animal population within a 30 day period. The killing effect depended on health, age, environment and state of health of the various animal species. The LD50 for animals varied from species to species. With humans the LD50 takes longer to express itself, therefore for humans we use the LD50/60 that is the lethal dose required to kill 50% of the population within a 60 day period. Unfortunately with humans we can only use the data from individuals who have died as a result of radiation accidents and the total number that have been killed over the last 50 years is approximately 130. Therefore the LD50/60 for humans is a range, not an individual number. The LD 50/60 for man is between 250-450 Rads.

Radiation cell death is also an important consideration. The definition of cell death is defined as, the dose required to stop cell division and not the killing of the cell outright. It is found that a much lower dose is required to stop cell division compared to killing the cell outright. Why is this important? If for example we consider the hematopoietic system and consider a red blood stem cell, this one stem cell has the capability of producing over a million red blood cells. By stopping this one stem cell dividing in the long term there is a loss of over a million red blood cells. The importance of this is that a sub lethal dose produces this effect. Therefore when considering the acute effects of radiation we are concerned with rapidly dividing cell populations. Two important organ systems which have rapidly dividing cell lines, are the heamatopoietic system and the gastrointestinal syndrome.

The Acute Radiation Syndrome follows a predictable course over a few hours to several weeks after exposure to ionizing radiation. The onset of clinical effects is dependant on dose, the higher the dose, the shorter the time interval with respect to the presentation of clinical signs and symptoms. The most important factors to produce the A.R.S. are:

- High Dose
- High dose rate
- Penetrating radiation
- Whole body exposure

Other factors that should be considered are sex, age, existing medical conditions and genetic considerations. It is important to realize the source term does not matter i.e. radioactive isotope industrial or medical, weapon, reactor etc. If the dose is high enough it will produce the same effects.

A.R.S. Signs and Symptoms

The signs and symptom of the A.R.S. occur in four distinct phases:

- Prodromal Phase.
- Latent Phase.
- Illness Phase.
- Recovery or Death phase.

Prodromal Phase (Table XII)
Onset is related to total dose of radiation received can be minutes to hours post exposure. Bouts of nausea, vomiting and anorexia lasting for 2-4 days at doses below approximately 500 Rads.Above this level of dose the symptoms may be persistent. After very high doses of radiation, additional symptoms such as prostration, fever, respiratory symptoms, conjunctivitis, erythema and increased excitability may develop. Development of gastro intestinal symptoms within 2 hours of exposure portend a serious and usually fatal outcome. (Table XIII)

Table XII. Prodromal Phase: Approximate ARS Degree & Dose				
Signs & Symptoms	Mild (1-2 Gy)	Moderate (2-4 Gy)	Severe (4-6 Gy)	Lethal (>8 Gy)
Vomiting Onset Incidence (%)	≥2 h 10-50	1-2 h 70-90	<1 h 100	<10 min 100
Diarrhea Onset Incidence (%)	None — —	None — —	Mild 3-8 h <10	Heavy <1 h ~100
Temperature Onset Incidence (%)	Normal — —	<38.5°C 1-3 h 10-80	Fever ≥38.5°C 1-2 h 80-100	High Fever <1 h 100
Headache Onset Incidence (%)	Slight — —	Mild — —	Moderate 4-24 h 50	Severe 1-2 h 80-90
Level of Consciousness Onset Incidence (%)	Normal — —	Normal — —	Normal — —	Unconscious for sec-min Sec-min 100% ≥50 Gy

Latent Phase

At doses of 200-300 Rads or so, after two to four days the signs and symptoms regress and a latent period develops lasting 2-3 weeks. As the dose increases the latent phase becomes shorter. (Table XIV)

During this time critical cell populations lymphocytes, leucocytes and platelets are decreasing as a result of bone marrow insult. Higher doses will affect the gastro intestinal cells leading to the gastrointestinal syndrome. At very high doses there is no latent phase at all.

Illness Phase

This is the period where overt illness develops, which may be characterized by nausea, vomiting, bleeding, diarrhea and other symptoms. (Table XV)

Recovery or Death Phase

This follows the illness phase in which either recovery can take weeks or months or where the dose of radiation has been so high that death ensues.

Therefore whole body exposure of doses between 100-800 Rads will produce the hematopoietic Syndrome. Doses between 800-3000 Rads will produce the gastro intestinal

Table XIII. Prodromal Phase Predicts Survival Prognosis of ARS		
Prognosis	Whole-Body Radiation Dose	Prodromal Phase Key Manifestations
Survival likely (probable)	<2 Gy	No signs or symptoms to mild nausea & vomiting (N/V) for a few hours
Survival possible	2-8 Gy	Early, persistent N/V for 24-48 h
Survival unlikely (improbable)	>8 Gy	Early, severe, persistent N/V, diarrhea, & shock. Loss of consciousness signals poor prognosis. Coma heralds death. Death often ≤ 72 h.

Table XIV. Latent Phase: Approximate ARS Degree & Dose

Signs, Symptoms & Lab	Mild (1-2 Gy)	Moderate (2-4 Gy)	Severe (4-6 Gy)	Lethal (>8 Gy)
Epilation (hair loss) Onset	None —	Moderate ≥15 days	Moderate 11-21 days	Complete <10 days
Lymphocyte Count (10⁹cells/L) on Day 6	0.8-1.5	0.5-0.8	0.3-0.5	0.0-0.1
Latency Phase Duration (days)	21-35	18-28	8-18	0

Table XV. Illness (Critical) Phase: Approximate ARS Degree & Dose

Signs, Symptoms, & Lab	Mild (1-2 Gy)	Moderate (2-4 Gy)	Severe (4-6 Gy)	Lethal (>8 Gy)
Onset (days)	21-35	18-28	8-18	0
Fatigue	Yes	Yes	Yes	Yes
Infections	No	Yes	Yes	Yes
Bleeding	No	Yes	Yes	Yes
Platelet Count (10⁹/L)	60-100	30-60	25-35	<20
Shock	No	No	No	Yes
Coma	No	No	No	Yes
Lethality (%) Onset (weeks)	0 N/A	0-50 6-8	20-70 4-8	100 by 1-2 weeks 1st day

syndrome and doses over 3000 Rads will produce the cardiovascular/central nervous (CV/CNS) syndrome.

Hematopoietic Syndrome (Table XVI) This system shows the earliest indication of the severity of the radiation exposure, through the rapid decline of the lymphocyte cell line, also decline of granulocytes, platelets, thrombocytes and recticulocytes. The systemic effects leads to immuno dysfunction, increased infectious complications, anemia, hemorrhage and impaired wound healing.

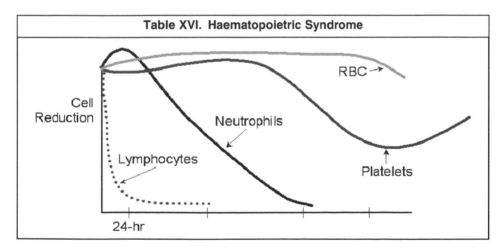

Table XVI. Haematopoietric Syndrome

Initially the absolute lymphocyte count is crucial. Lymphocyte count should be measured every 4-6 hours for the first 48 hours and an Andrews curve can be plotted. The average lymphocyte count is approximately 2500 cells per ml.If the total lymphocyte count is above1200 cells per ml, it is unlikely the patient has received a lethal dose of radiation. If the cell count is between 300-1200 cells per ml a significant exposure has occurred. The patient should be hospitalized isolated with barrier protection. Lymphocyte counts below 300 cells per ml are classified as critical, this warrants consideration for the use of Colony Stimulating Factors (CSF) and/ or stem cell transfusions, these important decisions usually being made by a hematologist. Although the lymphocytes rapidly drop in the first 48 hours note that the neutrophils initially rise this is the steroid trauma response. If surgery cannot be performed in the first 48 hours, it should be delayed for 2 to 3 months or until the bone marrow has recovered. It is important to realize that infection remains a leading cause of death in radiation exposed individuals due to the profound injury to hematopoietic colonies and the subsequent immune compromise. Strict attention to infection control and promotion of health defenses are critical during definitive care measures.

Gastro Intestinal System   The classical symptoms of the Gastro Intestinal Syndrome (G.I.T), are nausea vomiting and diarrhoea these are seen with doses over 800 rads. Damage is to the specialized epithelial cell lining of the small intestine. The time of onset of the vomiting and diarrhea is important. Patient who vomit in the first 2-6 hours is suggestive of a very high dose. This can lead to loss of fluids and opportunistic infection leading to septicaemia as a result of loosing the integrity of the villi cells lining the small intestine which act as a barrier. Persistent high fever and bloody diarrhea are serious signs. Death can occur within one to two weeks.

A word of caution, be aware of psychological nausea and vomiting. This has been recorded at accidents with people who have thought they have been irradiated.

Cerebrovascular/Central Nervous System   Unfortunately with this syndrome there is no recovery. Doses of 3000 Rads and above produce this syndrome this is most probably due to micro vascular leaks within the brain. Early vomiting and diarrhea occur within minutes. Patients tend to develop confusion, disorientation, convulsions, cerebral edema and coma. Also they develop hyperpyrexia and usually die within a couple of days.

G) Local Radiation to the Skin

Most irradiated casualties do not receive whole body exposure but only partial body exposure. This is due to part of their bodies being shielded. The skin is a good indicator of dose in A.R.S. by the types of lesions that are produced. The lesions that are produced are. (Table XVII)

- Epilation (Loss of body hair)
- Erythema
- Dry Desquamation
- Moist Desquamation
- Necrosis

It is important to realize that the patient may not present to you on the first day. In two serious world radiation accidents from contamination and exposure, the time intervals of realizing the accidents were 15 and 17 days respectively.

Epilation
This is the loss of body hair. (e.g. arms, legs, chest etc.) To produce this effect you require a dose of 300 RADS (3 Gys). Epilation normally takes 17-21 days to express itself.

Erythema
Late erythema develops 2-3 weeks after exposure. The dose to produce this is 300 to 600 RADS (3 Gys to 6 Gys).

Table XVII. Local Radiation Injury	
5 Skin Lesions	**Dose (Gy)**
• **Epilation** (hair loss) • 17-21 days	3

• **Erythema** (redness)
  • Early (Initial)
    • Within hours, not minutes
      • Within minutes implies chemical burn
    • ≤24-48 h
  • Examine hourly for 24 h
    • Take photos & note time
  • Latent LRI
    • No erythema
    • Between 2 days & 2-3 weeks
  • Later (second or main)
    • Onset: ≥2-3 weeks
	6

• **Dry desquamation** (peeling skin)
  • No blisters
  • Onset: Weeks 2-4
	10

• **Moist desquamation** (blistered, peeling skin)
  • Blisters
    • Small blisters
    • Large blisters
  • Onset: Weeks 2-8
	>15
	>25

• **Necrosis** (death)
  • Dead tissue
  • Gangrene
  • Autoamputation
  • Onset: Days-months
	>50

Early erythema which may appear in a few hours indicates a dose in the region of 600 RADS (6 Gys). Note sometimes this erythema disappears only to reoccur later. Therefore take photographs to record the erythema. A word of caution erythema occurring within a few minutes is most probably due to the chemical form of the contaminant and not due to irradiation.

Dry Desquamation

This is dry peeling of the skin. The dose to produce this effect is in the region of 1000 RADS (10 Gy). This expression normally takes 2-4 weeks.

West Desquamation

This results in the formation of small blisters or large bullae. The dose for small blister formation is in the region of 1500 to 2500 RADS (15 to 25 Gys). For large blister formation the dose would be greater than 2500 Rads (>25 Gys). The expression of blister formation can take from 2-8 weeks depending on dose.

Necrosis

To produce this effect the dose is greater than 5000 RADS (>50 Gys) Necrosis can take from a few days to months to express itself.

An accident involved a 28 Ci Iridium-192 source which a patient put it in his back pocket. This resulted in several blisters within 5 hours, one large bullae within 18 hours. Within one week the patient developed a large necrotic ulcer on the right buttock. The dose at 1 cm depth was calculated to be over 50,000 RADS (500 Gys).

H) Radiation Triage

It is important to mention at this stage that an irradiated contaminated casualty is not a medical emergency. The triage of casualties from a radiological or nuclear incident should focus first on trauma and other physical injuries. If you did nothing with a patient that received a very high dose of radiation, the individual would take days to die. Therefore your FIRST PRIORITY IS TO STABILIZE THE PATIENT AND ONLY WHEN THIS IS DONE, DOES ONE CONSIDER THE CONTAMINATION IRRADIATION PROBLEM. A.B.Cs. (airway, breathing, circulation) always take priority.

Since the degree of radiation injury will not be initially apparent, triage criteria will need to be based on associated injuries and complaints.

Patients who have received large doses of radiation will present with signs and symptoms already describe under the acute radiation syndromes.

Patients who have suffered trauma as well irradiation have an increased chance of dying compared to a single insult. Patients with burns, wounds and fractures who has received an acute dose of irradiation over 200 Rads (2Gy). Effort must be made to close wounds, stabilize fractures, cover burns and perform surgical stabilizing treatment, within the first 48 hrs after injury. After 48 hours, surgery should be delayed for 2-3 months.

Once radiological survey and decontamination is complete, patients can be divided into three main groups.

Survival probable less than 100 RADS (1 Gy)
Survival possible 200-800 RADS (2-8 Gy)
Survival improbable greater than 800 RADS (>8Gy)

Survival Probable

These patients will have very few signs and symptoms, there may be nausea and vomiting which general subsides after a few hours. Once the initial full blood differential has been taken and you are satisfied that the lymphocyte and granulocyte counts are acceptable, these patients can be sent home and treated on an outpatient basis. Ideally blood should be taken for chromosome analysis and send it to a cytogenetics laboratory.

Survival Possible

These patients present with nausea and vomiting lasting about 24–48 hrs which recedes and the patient goes into the latent phase. If the lymphocyte count drops below 50% of normal these patients should be reversed barriered and managed by a Hematologist. These patients will develop bone marrow suppression, which could lead to aplasia and pancytopenia that could be fatal. Blood replacement products, antibiotics, antiviral and antifungal agent maybe necessary. Colony Stimulating Factors (CSF) or stem cell transplants from bone marrow, peripheral blood, or fetal cord blood maybe also be required. Urgent stem cell transplantation may need to be considered for exposures exceeding 4 Gys.

Survival Improbable

These patients have received an acute whole body exposure of over 800 RADS (8 Gy). The prognosis for these patients is poor. Recent accidents have recorded success in the 800-1200 Rads (8-12 Gys) range, as far as the heamopoietic syndrome is concerned but unfortunately, these patients developed severe lung damage a few months later and died. In the mass casualty situation these patient are made as comfortable as possible, survival is improbable.

I) Preparing the Hospital & Staff to Receive Irradiated Casualties/Contaminated Casualties

In the emergency department you are accustomed to dealing with chemical casualties on a weekly basis. Only slight modifications are required in your emergency plan. A key person you must add to your team is a Nuclear Health Physicist. This person will be invaluable to you with respect to radiation problems. Early notification by EMS will facilitate preparation for receiving contaminated patients. Train regularly as a team. Initially the Health Physicist might not be available, hence the need for training.

Ideally you should divide your emergency department into a dirty and clean area. Remove all unnecessary equipment from the dirty area, if there is not time, cover equipment with drapes. If X-rays are required use a portable X-ray machine.

Contaminated clothing of the patients should be doubled bagged, using polythene bags, labeled and moved out of the emergency department to a place designated by the health physicist.

Protective clothing required by hospital staff in the emergency department is listed in (Table XVIII).

Dosimeters should also be provided for each member of staff.[11] These include a quartz fiber dosimeter (QFD) worn at the neck of the surgical gown for easy access and a thermoluminescent dosimeter (TLD) which is worn under the surgical gown (Table XIX). Your Health Physicist will assist you with these. The reason for wearing dosimeters is because our body senses cannot detect radiation. The quartz fiber dosimeter can be read immediately therefore enabling control of individual doses.

With respect to staff, only commit staff when needed.[12] Rotate staff as much as possible to share the committed dose. Golden Rule is no eating, drinking or smoking in a contaminated environment. Another very important point which is often missed make sure you use the rest room before entering the dirty radiation environment because you might be there for a long period of time.

Instrumentation is very important you should be able to use an instrument in an emergency and differentiate between alpha, beta and gamma radiations (Table XX).

Note that Geiger counter does not count alpha particles. Common errors with instruments are: Flat batteries therefore check instruments on a regular basis (spare batteries should be obtainable).

- Forgetting to measure background on the day.
- Instrument on the wrong switching position.
- Reading the wrong scale.
- Monitoring too fast.
- Alpha monitoring holding the probe too far away

Table XVIII. Protective Clothing
• Surgical greens & gown
• Surgical cap, mask and overshoes
• Surgical face shield
• Wear two pairs of surgical gloves, tape inner set of surgical gloves to surgical gown.
• Tape over the opening on the back of the surgical gown
• Tape over shoes to surgical trousers.
• Wear a plastic apron.

Table XIX. Labeled Dosimeters
• QFD and Charger
• TLD
• Finger TLD
• Electronic Readout Dosimeter
• Personal Air Sampler

Table XX. Suggested Equipment for Radiation Exposures
• Alpha detector
• Decontamination stretcher with drainage collection tank
• Direct reading dosimeters plus charges
• Dosimeters/film badges (see Table XIX)
• Lead storage containers
• Level A biohazard suits
• Level B biohazard suits
• Level C biohazard suits
• N 95 masks
• Neoprene industrial gloves
• Plastic (Polyethylene) floor/wall covering
• Rolls of absorbent paper with plastic backing
• Self contained Breathing Apparatus (SCBA)
• Tyvex Suits
• Ventilation filter paper

- Putting a surgical glove over alpha instrument probe, if alpha will not penetrate a piece of paper it certainly will not penetrate a surgical glove. However you can put a surgical glove over a beta gamma probe to protect it from contamination.
- Switch off the audible alarm, always use the instrument headphones provided, this will prevent the patient panicking.

These detectors should be checked regularly with appropriate staff training.

J. Dealing with an Irradiated Contaminated Casualty[13-18]

You most probably will not get any warning of these patients arriving. In a terrorist event you will not know what the radioactive source term is, you might be dealing with large or small amounts of contamination, generally large amounts of contamination are at the incident site. The patient might have received significant dose of radiation as well but this depends on the source. It is important to re-emphasize that infection remains a leading cause of death in radiation exposed individuals due to the profound injury to hematopoietic colonies and the subsequent immune compromise. Strict attention to infection control and the promotion of host defenses is critical.

Depending on the type of radiological or nuclear incident, victims may be contaminated and irridiated. Once the patient has been medically stabilized, attention should be given to the presence and level of contamination. Patients can be internally contaminated by ingesting, inhaling or having a wound created by or exposed radioactive materials, or an externally contaminated victim who did not have respiratory protection.[4] However, the majority of radioactive material from an event will remain in the environment of the incident. By using simple survey techniques, victim contamination can be detected and quantified immediately. If contamination is detected on the clothing, carefully have the victim remove their outer clothing inside out or cut off the clothing. Double bag the clothes in polythene bags and label with the patient's name. Place bags in large plastic trash barrels and remove them to a designated area out of the emergency department. After removing the clothes, re-survey the patient's body completely, paying special attention to the exposed parts of the body, namely, head, neck, hands, and legs. Be careful to avoid contaminating the survey instrument by covering it in a small disposable piece of plastic. For any victims with positive survey results, obtain a history of exposure to a radionuclide from a medical diagnostic procedure during the previous 72 hours. Antidotes or chelating agents can be useful in altering the absorption of radioactive materials in the GIT depending upon the toxicokinetics, and solubility of the substance. Cesium is rapidly absorbed in the GIT whereas plutonium is not. Ion exchange resins such as Prussian blue, under IND (investigational new drug) can decrease GI uptake of certain radionuclides, such as Cesium.[4,6] Diethylenetriaminepentaacetic acid (DTPA-also an IND) can chelate heavy metal radionuclides, including radionuclides that are heavier than uranium with the exception of neptunium. Ca DTPA should be used in the initial 24 hours post exposure, followed by Zn DTPA. DTPA is available through REAC/TS at Methodist Medical Center 1.865.481.1000. The DTPA from nuclear medicine departments should not be used interchangeably with the above.[6] Dimercaprol can chelate arsenic, and other elements, and can be considered for treating internal exposures from radioisotopes of those same elements. Consultation with REAC/TS or staff health physicist is recommended depending upon the exposure.

General Medical Considerations (4,6)

Radiation events can be managed with modern techniques and technology.

1. It is important to have on staff or readily available a health physicist and to develop protocols to respond to a radiation victim. Skills based training are essential to familiarize and enhance your facility response. The Radiation Emergency Assistance Center and Training Site (REAC/TS) is operated by the Department of Energy and is available on a 24 hour basis. 1. 615. 576.3131 or 1.615. 481.1000.
2. Treat the patient, before the poison! Treatment of serious medical problems takes precedence over radiation related issues (Table XXI). When the threat of contamination is unknown, responders should wear protective clothing and respirators.
3. Follow the ABC's (Airway, Breathing, Circulation)
4. Realize the two most radiosensitive systems are the hematopoietic and gastrointestinal systems.
   a. Gastroenteritis can result in significant fluid losses
   b. Leukopenia can result in significant infections
5. Enhance elimination of the toxin
6. Utilize appropriate
   a. Infection control – antimicrobials, colony stimulating factors (Table XXII).
   b. Antidotes and pharmacotherapy – such as chelating agents or blocking drugs (Table XXIII).
7. System and dose specific intervention
   a. Slit lamp examination to document cataracts
   b. Liver function tests and coagulation studies
   c. Chromosomal changes in lymphocytes are considered highly sensitive indicators or radiation exposures, including as little as 10 rad.
   d. Use of the Andrews Lymphocyte Nomogram or Modified USSR Classification for prognosis
8. Decontamination

As stated before the patient must be stabilized first, then and only then do you consider the contamination problem. When taking a past medical history an important question to ask is, Has the patient had a recent nuclear medicine procedure? There have been the odd occasions where this type of patient has presented. Also, it is important to ascertain if there is any history of renal disease, since most drugs that are used to excrete the incorporated

Table XXI. Medical Management Based on Phase & Degree of ARS					
ARS Phase	ARS Degree & Dose				
	Mild (1-2 Gy)	Moderate (2-4 Gy)	Severe (4-6 Gy)	Very Severe (6-8 Gy)	Lethal (>8 Gy)
Prodromal	Decon if contaminated with particles (matter)				
	Outpatient labs & observation	Inpatient general hospital labs & observation. Transfer to specialty hospital, if necessary.	Inpatient treatment in specialty hospital.	Inpatient treatment in specialty hospital	Palliative symptom treatment

radioactive isotopes are excreted via the kidneys. Once you are satisfied that the patient is stable, then start to monitor the patient. If you find contamination on the clothing, stop monitoring and carefully cut the clothes off. Placing the clothes carefully into a double polythene bag which must be labeled with the patients name and remove these clothes out of the emergency department to the designated area. Then re-monitor the patients body completely, paying special attention to the exposed parts of the body namely, head, neck, and hands and also exposed legs with respect to females.

Decontamination of intact skin generally is achieved with soap and water, work from the periphery to the center it might require a few washings to remove the contamination. It is important to drape the area of concern with waterproof surgical sheets to prevent further spread of contamination. Save all washings from the patient for later analysis. If there is difficulty in removing the contamination from the intact skin you can try a light abrasive such as a puff pad which ladies use to remove their make up or a mixture of ground corn and washing powder worked into a paste. If the skin becomes red and angry stop the abrasive technique cover area with a dry dressing and seek expert advice.

With respect to wounds mask around the wound with waterproof surgical drapes. Use sterile normal saline to irrigate the wound with the aid of a 50 ml syringe. The wound might require irrigation several times. Save all washings from the wound including any debris that has been removed. If the wound is still contaminated then refer to a surgeon. The objective is to decontaminate down to background levels.

Table XXII. Infection Prevention (4,6)

I. Reduce pathogen infections
    a. Low microbial content food
    b. Low microbial content water
    c. Reverse isolation
    d. Precise isolation precautions – frequent hand washing
 	
II. Treat neutropenic phase of illness
    a. Selective gut decontamination that preserves commensels but suppresses aerobes
    b. Sucralfate or misoprostel may prevent gastric hemorrhage
    c. Early oral feeding
 	
III. Hematopoietic growth factors
    a. Filgrastim – granulocyte colony-stimulating factor
    b. Sargramostim – granulocyte macrophage colony stimulating factor
    c. Selected use of fresh, irradiated platelets and blood products
    d. Match antimicrobials to susceptibility patterns at the health care facility
    e. Beware of life-threatening gram negative bacterial infections
    f. Aggressive empiric antimicrobial use
    g. Early identification of infection and associated nidus

Table XXIII. Antidotes/Common Radionuclide (4-6)	
Cesium 137	Prussian blue (ferric hexacyanoferrate) which adsorbs cesium in the GI tract. It may enhance elimination
Iodine 131	Potassium iodide (KI) blocks thyroid uptake
Plutonium 239	Diethylenetriamine pentaacetic acid (DTPA) – can be used as a chelator as well as for wounds.
	EDTA is also a suitable chelator
	Aluminum hydroxide antacids may bind Plutonium in the GIT
Radium 226	Immediate Lavage with 10% magnesium sulfate followed by saline and magnesium purgatives
	Ammonium chloride may increase fecal elimination
Strontium 90	Aluminum hydroxide antacids may bind Strontium in the GIT
	Aluminum phosphate can decrease absorption by 85%.
	Ammonium chloride can acidify the urine and enhance excretion
	Barium sulfate may reduce Sr absorption
	Stable strontium can competitively inhibit the metabolism and increase excretion of radiostrontium
Tritium	Oral fluids will reduce the biological half time from 12 to approximately 6 days. Caution – do not overhydrate patients
Uranium	Sodium bicarbonate renders the uranyl ion less nephrotoxic
	Diuretics

Decontamination of hair, this can be done in two ways if the patient is restricted to a gurney, the patients hair can be washed in the barber position over a sink. If the patient is ambulant then the hair can be washed over a sink or the patient could bend over in a shower shampoo the hair once or twice then shower the patient. The reason for shampooing the hair first is that this is the most likely site for contamination, this prevents the contamination trickling down the body reducing the risk of contamination getting trapped in body hair or body crevasses. Ideally one should save this contaminated water but it depends on the number of patients you are required to shower.

K) Biological Sampling

Nasals swabs must be taken off every patient one from each nostril. Note the time that the swabs were taken post incident. Take a throat swab, save all sputum samples that are produced. Save all urine samples and all fecal samples the latter is the most difficult one to get patients to provide. It must be stressed to the patient how important this sample is with respect to measuring insoluble radioactive isotopes that have been ingested or inhaled. Initial wound swab must be taken, save all irrigation fluid and swabs used and any debris. Save all decontaminated fluid used to decontaminate skin.

Blood

Do a full complete count and white cell differential looking at especially the lymphocyte and neutrophil count every 4 hours. If you suspect the patient has received a significant dose of irradiation, send blood off for chromosome dicentric count to a laboratory that has this capability. For neutrons, send blood to a nuclear lab that is able to identify Sodium 24 (Na 24).

Table XXIV. Care of the Dead

• Put in a body bag
• Do not use your own morgue
• Ideally use a refrigerator truck, not a freezing truck
   • Preserve forensic data
• Do not cremate contaminated bodies
   • Radioactive matter is not destroyed by fire
• Try to respect religious beliefs

In a terrorist event you will not know initially what isotope or isotopes you are dealing with. In this situation it is advisable to send the clothes of the first couple of individuals and if they have wounds, their initial wound dressings to a nuclear laboratory to identify what isotope or isotopes you may be dealing with. You should also inform the hospital pharmacy of the isotopes you are dealing with, so that they can acquire the relevant drug that is required. Remember that some of the drugs used in radiation emergencies are not easily available.

L) Consideration of the Dead (Table XXIV)

It is important to have a plan to facilitate the dead. Bodies that are contaminated should not be placed in the hospital morgue, because you do not want to loose you pathology facilities due to radioactive contamination. The requirement is a temporary morgue. The ideal solution to this, is the use of a mobile chilling unit, as used in the food industry, note that the requirement is a chilling unit, not a freezer unit, other wise some forensic evidence could be lost, remember we are dealing with a crime. This unit could be strategically placed in the hospital grounds.

If these bodies cannot be completely decontaminated then they must be buried. They cannot be cremated because you cannot destroy nuclear material by fire. This could present with religious problems within certain religion denominations e.g. some religions insist that the body must be cremated within the first 24 hour of death. Therefore letters of dispensation should be in place from the relevant religious heads for dispensation in this type of situation.

M) Treatment of Radiation Contaminated Injuries

Hospital staff should be prepared for any type of nuclear accident, e.g. terrorist event, reactor accident or an industrial source to name a few, remember, regardless of the source term, if the dose is high enough the same biological effects will be produced.

Patients who have radioactive isotopes incorporated in their wounds RADIATION CONTAMINATED INJURIES require more specialized treatment. Consultation with specialist in hematology, oncology, radiation, and infectious disease should be obtained. Access to nuclear departments with whole body monitors and other nuclear facilities will also be required.

Treatment for internally contaminated patient is determined by the patient's medical condition, past medical history, biological sample results and whole body monitoring results.

Medications that are used depends on the isotope that one is dealing with, an excellent reference is the NCRP Report No 65 and the Safety Series 47, IAEA (International Atomic Energy Authority).

550

Conclusion

With casualties from a radiological or nuclear event, the full spectrum of clinical personnel will be in involved in the treatment of these patients in one way or another. It is important to remember that the nuclear health physicist is a key member of your emergency team. With respect to deep radiation specialists advice, contact the Radiation Emergency Assistance Center and Training Site (REAC/TS) operated by the department of energy, at Oak Ridge Tennessee which is the World Health Organization (WHO) & International Atomic Energy Authority (IAEA) collaboration center. This center has 24 hour emergency cover throughout the year and can be contacted by telephone at 865-576-3131 weekdays and 865-576-1005 weekends and after hours. REAC/TS provide week long courses for doctors, nurses, paramedics, and emergency planners emphasizing the practical side of decontamination.

Addendum The preceding was adapted from the Advanced Hazmat Life Support manual.[19] Advanced Hazmat Life Support (AHLS), a 2-day continuing education program, held its first course in October 1999, and has continued to grow into an international program. AHLS is based in the United States at the University of Arizona Emergency Medicine Research Center and is co-sponsored by the American Academy of Clinical Toxicology. AHLS trains medical personnel including paramedics, physicians, nurses, pharmacists and toxicologists in the medical management of people exposed to hazardous materials, including nuclear, biological and chemical terrorism.

Courses are offered throughout the U.S., Canada and in Australia. Check our website for more information and a listing of all courses at www.ahls.org or contact the International Headquarters at 520-626-2305, ahlsinfo@aemrc.arizona.edu for more information.

References
1. Waeckerle JF, Seamans S, Whiteside M, et al: Executive Summary: Developing Objectives, Content, and Competencies for the Training of Emergency Medical Technicians, Emergency Nurses to Care for Casualties Resulting From Nuclear, Biological, or Chemical (NBC).
2. Nussbaum RH, Kohnlein W: Ionizing Radiation. In Greenberg MI (ed): Occupational, Industrial, and Environmental Toxicology. St. Louis, Missouri, Mosby Publishing, 1997, 517-530.
3. Harley NH: Effects of Radiation and Radioactive Materials. In Klaassen CD (ed): Casarett and Doull's Toxicology. The Basic Science of Poisons. New York, New York, McGraw Hill, 6th Edition, 2001, pp 914-917.
4. Medical Management of Radiological Casualties Handbook. 2nd Edition. Col. D. G. Jarrett Editor. Armed Forces Radiobiology Research Institute. April 2003. Bethesda, Maryland.
5. Radiation. Fun Fong, Dirk Schrader, Show Fong. In Fords Clinical Toxicology, 1st ed., Copyright © 2001 W. B. Saunders Company. Pages 782-788.
6. Radiation (Ionizing) E. T Wythe In Poisoning and Drug Overdose 3rd Edition. Kent R Olson, Editor. 1999 Appleton and Lange, Stamford, Connecticut. Pages280-282.
7. White SR; Hospital and Emergency Department Preparedness for Biologic, Chemical and Nuclear Terrorism. Clin Occup Environ Med 2003: 2(2) 405-426.
8. Mettler F., Volez, G.; Major Radiation Exposure-What to Expect and How to Respond. New Engl J. Med 436.L 1554-61, 2002.
9. Sullivan, Jr., Stewart C.: Weapons of Mass Destruction Nuclear Agents. Clin Occup Environ Med 2(2): 327-337, 2003.
10. Lees-Haley, PR: Psychologic Responses to Bioterrorist Attacks. Clin Occup Environ Med. 2(2): 361-377,2003.
11. Greenburg MR, Farrell TP, Hendrickson RG: Routine Screening for EnvironmentalRadiation by first Responders Explosions and Fires. Ann Emerg Med 41 (3); 426-427,2003.

12. Thorne, C.D., Curbow B., Oliver, M. et al: Terrorism Preparedness Training for Nonclinical Workers Empowering Them to Take Action. J Occup EnViron Med; 45: 333-337,2003.
13. Weisdorf D, Chao N, Waselenko JK et al, "Acute Radiation Injury: Contingency Planning for Triage, Supportive Care and Transplantation," *Biol Blood Marrow Transplant* 2006: June 12(6): 672-682.
14. Weisdorf D, Apperley J, Courmelon P et al, "Radiation and Transplantation," *Biol Blood Marrow Transplant* 2007, Jan 13 Suppl 1: 103-106.
15. Dainiak N, Waselenko JK, Armitage JO et al, "The Hematologist and Radiation Casualties," *Hematology Am Soc Educ Program* 2003: 473-496.
16. Waselenko JK, MacVittie TJ, Blakely WF et al, "Medical Management of the Acute Radiation Syndrome: Recommendations of the Strategic National Stockpile Radiation Working Group," *Ann Intern Med* 2004 140(12):1037-1051.
17. Flynn DF, Goans RE, "Nuclear Terrorism: Triage and Medical Management for Radiation and Combined-Injury Casualties," *Surg Clin North Am* 2006 86(3):601-636.
18. McFee R, Leikin JB, "Radiation Terrorism," *JEMS* April 2005: 78-92.
19. Walter, F.G., Klein R., Thomas, R.G.: Toxic Terrorism. In Walter F.G, Meislin HW (eds): Advanced Hazmat Life Support Provider Manual. Tuscon, Arizona; Arizona Board of Regents, 2nd Edotopm. 2000, pp 279-304.

Americium

CAS Number 7440-35-9

Synonyms Am

Use Product of nuclear weapon detonation

Mechanism of Toxic Action Alpha radiation; significant respiratory absorption with minimal gastrointestinal absorption. Primary toxicity involves skeletal and liver deposition with bone marrow suppression. Also emits gamma rays (weakly)

Pregnancy Issues No apparent adverse effect

Treatment Strategy Supportive therapy: DTPA: 1 gram intravenously in 250 mL of isotonic saline or D_5W. If given one hour after exposure, 50% of body burden can be decreased.

Additional Information Isotope is AM-241 discovered by Dr. Glenn Seaborg in 1944. A man-made metal produced when plutonium atoms absorb neutrons in nuclear reactors and in nuclear weapons detonations. A silver white crystal solid with a half-life of 432.7 years

Specific References

Jarrett DG, *Medical Management of Radiological Casualties Handbook*, 1st ed, Military Medical Operations Office, Armed Forces Radiobiology Research Institute, Bethesda, MD, 1999.

Pizzarello DJ, "Radiation," *Goldfrank's Toxicologic Emergencies*, Goldfrank LR, ed, Stamford, CT: Appleton & Lange, 1968.

"Radiation Disasters and Children: American Academy of Pediatrics Policy Statement - Committee on Environmental Health," *Pediatrics*, 2003, 111(6):1455-66.

"Radiation Threats," http://www.imc-la.com/cbr/L1d-m.html. Last accessed 10/10/03.

Waselenko JK, MacVittie TJ, Blakely WF, et al, "Medical Management of the Acute Radiation Syndrome: Recommendations of the Strategic National Stockpile Radiation Working Group," *Ann Intern Med*, 2004, 140(12):1037-51.

Yow RB, "Radiation Poisoning," *Clinical Management of Poisoning and Drug Overdose*, 2nd ed, Haddad LM and Winchester JF, Philadelphia, PA: WB Saunders Co, 2001, 624-35.

Cesium

CAS Number 7440-46-2

Synonyms Ce137; Ce137m; Ce133; Caesium

Use Medical therapy; moisture-density gauges in construction industry; leveling gauges and thickness gauges in drilling industry; use in medical and research facilities

Mechanism of Toxic Action Beta and gamma emission - decays to barium-137m

Toxicity: Concentrated in muscle and soft tissues; secreted into the gut and reabsorbed. Beta particles can cause severe burns. Cesium is metabolically handled like a potassium analog, with urinary excretion. Cesium can cause acute radiation syndrome.

Toxicodynamics/Kinetics Half-life: 30 years

Overdosage Treatment

Decontum: Ocular: Irrigate with water or saline

Oral: Activated Charcoal is not useful

Prussian blue: Oral or nasogastric tube: 150-250 mg/kg daily (or 1 gram in 100 to 200 ml of water orally) in 2-4 divided doses. It is often dissolved in 50 mL of 15% mannitol to prevent constipation. Treatment for cesium should be for a minimum of 30 days.

If given within ten minutes, Prussian blue can reduce gut absorption of cesium by 40%, and reduce half-life by 43% to 69%.

Signs and Symptoms of Acute Exposure Depending upon exposure and dose of radiation, acute radiation syndrome (ARS) can occur

Additional Information There are 11 major radioactive isotopes of cesium. Only three have half-lives long enough to warrant concern: cesium-134, cesium-135 and cesium-137. Each of

these decays by emitting a beta particle, and their half-lives range from about 2 to 2 million years. Of these three, the isotope of most concern is cesium-137, which has a half-life of 30 years. Its decay product, barium-137m (the "m" means metastable) stabilizes itself by emitting an energetic gamma ray with a half-life of about 2.6 minutes. It is this decay product that makes cesium an external hazard; a hazard without being taken into the body.
Liquid at room temperature but bonds easily with chlorides to form a crystalline powder

Decay Properties of the Radioactive Isotopes of Cesium				
Isotope	Half-life (years)	Decay Mode	Intensity Percent	Beta Particle Energy (MeV)
^{134}Cs	2.062	$\beta_1{}^-$	27	0.02309
		$\beta_2{}^-$	2.5	0.1234
		$\beta_3{}^-$	70	0.2101
^{137}Cs	30	$\beta_1{}^-$	94.6	0.1734
		$\beta_2{}^-$	5.4	0.4246

Specific References

"Cesium," *Human Health Fact Sheet, Environmental Assessment Division, Argonne National Laboratory*, http://www.stoller-eser.com/Quarterlies/2001Q1/cesium.htm. Last accessed 10/12/03.

Dainiak N, Waselenko JK, Armitage JO, et al, "The Hematologist and Radiation Casualties," *Hematology (Am Soc Hematol Educ Program)*, 2003, 473-96 (review).

Jarrett DG, *Medical Management of Radiological Casualties Handbook*, 1st ed, Military Medical Operations Office, Armed Forces Radiobiology Research Institute, Bethesda, MD, 1999.

Lydon TJ, DeRoos FJ, and Perrone J, "Pulseless Ventricular Tachycardia Secondary to Alternative Cancer Therapy With Cesium Chloride," *J Toxicol Clin Toxicol*, 2004, 42(5):731.

"Radiation Disasters and Children: American Academy of Pediatrics Policy Statement - Committee on Environmental Health," *Pediatrics*, 2003, 111(6):1455-66.

"Radiation Threats," http://www.imc-la.com/cbr/L1d-m.html. Last accessed 10/10/03.

Ring JP, "Radiation Risks and Dirty Bombs," *Health Phys*, 2004, 86(2 Suppl):S42-7.

Robison WL, Conrado CL, Bogen KT, et al, "The Effective and Environmental Half-Life of 137Cs at Coral Islands at the Former US Nuclear Test Site," *J Environ Radioact*, 2003, 69(3):207-23.

Waselenko JK, MacVittie TJ, Blakely WF, et al, "Medical Management of the Acute Radiation Syndrome: Recommendations of the Strategic National Stockpile Radiation Working Group," *Ann Intern Med*, 2004, 140(12):1037-51.

Yow RB, "Radiation Poisoning," *Clinical Management of Poisoning and Drug Overdose*, 2nd ed, Haddad LM and Winchester JF, Philadelphia, PA: WB Saunders Co, 2001, 624-35.

Cobalt

CAS Number 10198-40-0

Synonyms Co60

Use Radiotherapy in hospitals (sealed gamma sources); industrial radiography; component - improvised nuclear device to enhance fallout danger. Also found in food irradiation facilities as well as medical and research facilities.

Mechanism of Toxic Action High-energy gamma rays and 0.31-MeV beta rays.
Toxicity:
Depending upon exposure and dose of radiation, acute radiation syndrome (ARS)
Beta and gamma radiation - whole body radiation risk
People may ingest cobalt-60 with food and water that has been contaminated, or may inhale it in contaminated dust, where it is rapidly absorbed. Once in the body, some cobalt-60 is

quickly eliminated in the feces. The rest is absorbed into the blood and tissues, mainly the liver, kidney, and bones. Absorbed cobalt leaves the body slowly, mainly in the urine.

All ionizing radiation, including that of cobalt-60, is known to cause cancer. Therefore, exposures to gamma radiation from cobalt-60 result in an increased risk of cancer. The magnitude of the health risk depends on the quantity of cobalt-60 involved and on exposure conditions.

Data are lacking about how Co60 is absorbed from wounds. Wound contamination is generally not a serious clinical problem from the viewpoint of hindering the healing of the wound (requires 50 to 100 rads/day) or from causing acute necrosis (requires 200 or more rads/day). It is of concern because of the psychological effect on the individual and possible long-term effects from internal contamination

Adverse Reactions

Cardiovascular: Cardiomyopathy

Dermatologic: Contact or allergic dermatitis, nonimmunologic contact urticaria, eczema

Endocrine & metabolic: Metabolic acidosis, hypothyroidism, goiter, hyperlipidemia

Gastrointestinal: Nausea, vomiting, diarrhea

Hematologic: Anemia or polycythemia (not noted with inhalation exposure), methemoglobinemia, neutropenia

Ocular: Optic atrophy, conjunctivitis

Respiratory: Wheezing, interstitial fibrosis, giant cell pneumonitis, pulmonary fibrosis, bronchospasm

Renal: Goodpasture's syndrome

Toxicodynamics/Kinetics

Absorption: Variable GI absorption, absorbed through skin and inhalation

Half-life: 5-10 years

Elimination: Lungs and kidneys

Signs and Symptoms of Acute Exposure Hypothyroid goiter, hypertension, cardiomegaly, cyanosis, polycythemia, pruritus

Treatment Strategy

Decontamination: External contamination requires removal from/of contaminant, irrigation with saline or water. Internal contamination may be treated with chelation by penicillamine. Lavage within 1 hour/activated charcoal in oral ingestion. Iron can decrease oral absorption.

Supportive therapy: Sodium bicarbonate for metabolic acidosis

Enhanced elimination: Calcium disodium EDTA (50 mg/kg/day for 5 days). BAL, N-acetylcysteine or 10 mg/kg/dose every 8 hours for 5 days followed by 10 mg/kg/dose every 12 hours for 14 days. 2,3 dimercaptosuccinic acid (DMSA, succimer) while maintaining adequate hydration, have also been shown to increase fecal excretion. Hemodialysis may also be useful.

Additional Information

A hard, brittle gray metal with a bluish tint; can be magnetized similar to iron

Clothing can be decontaminated with ultrasonic cleaners, vacuuming, or dilute solutions of nitric, hydrochloric, or sulfuric acid.

The most common radioactive isotope of cobalt is cobalt-60. Cobalt-60 undergoes radioactive decay with the emission of beta particles and strong gamma radiation. It ultimately decays to nonradioactive nickel. The half-life of cobalt-60 is 5.27 years.

A technique called "whole-body counting" can detect gamma radiation emitted by cobalt-60 in the body. A variety of portable instruments can directly measure cobalt-60 on the skin or hair. Other techniques include measuring the level of cobalt-60 in soft tissues (such as organs) and in blood, bones, milk, or feces. Normal levels in urine ≤ 2.2 mcg/L; plasma ≤ 1.2 mcg/L

Specific References

Jarrett DG, *Medical Management of Radiological Casualties Handbook*, 1st ed, Military Medical Operations Office, Armed Forces Radiobiology Research Institute, Bethesda, MD, 1999.

Pizzarello DJ, "Radiation," *Goldfrank's Toxicologic Emergencies*, Goldfrank LR, ed, Stamford, CT: Appleton & Lange, 1968.

"Radiation Disasters and Children: American Academy of Pediatrics Policy Statement - Committee on Environmental Health," *Pediatrics*, 2003, 111(6):1455-66.

"Radiation Threats," http://www.imc-la.com/cbr/L1d-m.html. Last accessed 10/10/03.

Waselenko JK, MacVittie TJ, Blakely WF, et al, "Medical Management of the Acute Radiation Syndrome: Recommendations of the Strategic National Stockpile Radiation Working Group," *Ann Intern Med*, 2004, 140(12):1037-51.

Yow RB, "Radiation Poisoning," *Clinical Management of Poisoning and Drug Overdose*, 2nd ed, Haddad LM and Winchester JF, Philadelphia, PA: WB Saunders Co, 2001, 624-35.

Depleted Uranium

Synonyms DU

Use By-product of uranium used in tank armor and munitions; as counterweights on wing parts in helicopters and airplanes

Mechanism of Toxic Action DU has both chemical and mild radioactive effects. DU emits alpha, beta, and gamma radiation. Inhaled uranium compounds may be metabolized and excreted in the urine, which may result in kidney injury or failure. DU fragments in wounds can become encapsulated and subsequently metabolized, resulting in whole body distribution - especially bone and kidney.

Toxicodynamics/Kinetics
Half-life:
  U-234: 246,000 years
  U-235: 700 million years
  U-238: 4.47 billion years
Elimination: Fecal (99%)

Pregnancy Issues Crosses the placenta according to lab tests

Treatment Strategy I.V. infusion of isotonic 1.4% sodium bicarbonate enhances renal elimination. Tubular diuretics may be useful. Wound fragments should be removed when the benefit exceeds the risk. Dialysis may be necessary depending upon clinical.
  Laboratory tests include: 24-hour urine uranium bioassay, renal function testing, beta-2-microglobulin, creatinine clearance, liver-function studies

Environmental Persistency DU munitions have been abandoned in multiple battlefields. Long-term exposure risk is related to the dose or radiation. DU is of most risk to persons in the military and in the vicinity of detonated DU munitions. Short-term exposure without wounding is unlikely to cause acute short-term effects.

Additional Information Silvery white metal; detectable with an end window Geiger-Mueller counter; does not absorb through skin; uranium limit in drinking water is 30 mcg/L

Specific References

Jarrett DG, *Medical Management of Radiological Casualties Handbook*, 1st ed, Military Medical Operations Office, Armed Forces Radiobiology Research Institute, Bethesda, MD, 1999.

Pizzarello DJ, "Radiation," *Goldfrank's Toxicologic Emergencies*, Goldfrank, LR, ed, Stamford, CT: Appleton & Lange, 1968.

"Radiation Disasters and Children: American Academy of Pediatrics Policy Statement - Committee on Environmental Health," *Pediatrics*, 2003, 111(6):1455-66.

"Radiation Threats," http://www.imc-la.com/cbr/L1d-m.html. Last accessed 10/10/03.

Waselenko JK, MacVittie TJ, Blakely WF, et al, "Medical Management of the Acute Radiation Syndrome: Recommendations of the Strategic National Stockpile Radiation Working Group," *Ann Intern Med*, 2004, 140(12):1037-51.

Yow RB, "Radiation Poisoning," *Clinical Management of Poisoning and Drug Overdose*, 2nd ed, Haddad LM and Winchester JF, Philadelphia, PA: WB Saunders Co, 2001, 624-35.

Iodine

CAS Number 10043660
Synonyms Iodine 129; Iodine 131; Iodine 132; Iodine 134; Iodine 135
U.S. Brand Names Iodex [OTC]; Iodoflex™; Iodosorb®
Use Product of nuclear weapon (plutonium or uranium) detonation. Normal fission product in uranium fuel rods; found following destruction of nuclear reactor
Mechanism of Toxic Action Most of the radiation is in the form of beta rays with some gamma radiation. High respiratory and gastrointestinal absorption. Primary toxicity is to the thyroid gland. Thyroid uptake concentrates radioactive iodine, which locally irradiates the tissue.
Toxicodynamics/Kinetics
Half-life:
I^{131}: 8 days
I^{129}: 15.7 million years
Elimination: Renal
Signs and Symptoms of Acute Exposure Conjunctival irritation, nasal/throat irritation, thyroid nodules/thyroid cancer
Treatment Strategy
Prophylaxis: Sodium iodide (NaI) or potassium iodide (KI) 130 mg will prevent uptake if taken prior to the event. KI effectiveness is greatest when administered immediately before the exposure to prevent radioiodine from reaching the thyroid, which is near 100% but diminishes to less than 10% if started 24 hours postexposure.
Supportive therapy: Oral:
Children 3-18 years <150 lbs: Administer 65 mg/day while remaining in a radioactive iodine-contaminated environment. Children >150 lbs may receive an adult dose.
Adults:
Postexposure: 300 mg NaI or KI daily for 7-10 days will prevent thyroid uptake. Propylthiouracil 100 mg every 8 hours for 8 days or methimazole 10 mg every 8 hours for 2 days; then 5 mg every 8 hours for 6 days may also be used.
Environmental Persistency Postreactor accident winds will determine fallout pattern.
Specific References

Jarrett DG, *Medical Management of Radiological Casualties Handbook*, 1st ed, Military Medical Operations Office, Armed Forces Radiobiology Research Institute, Bethesda, MD, 1999.
Pizzarello DJ, "Radiation," *Goldfrank's Toxicologic Emergencies*, Goldfrank LR, ed, Stamford, CT: Appleton & Lange, 1968.
"Radiation Disasters and Children: American Academy of Pediatrics Policy Statement - Committee on Environmental Health," *Pediatrics*, 2003, 111(6):1455-66.
"Radiation Threats," http://www.imc-la.com/cbr/L1d-m.html. Last accessed 10/10/03.
Waselenko JK, MacVittie TJ, Blakely WF, et al, "Medical Management of the Acute Radiation Syndrome: Recommendations of the Strategic National Stockpile Radiation Working Group," *Ann Intern Med*, 2004, 140(12):1037-51.
Yow RB, "Radiation Poisoning," *Clinical Management of Poisoning and Drug Overdose*, 2nd ed, Haddad LM and Winchester JF, Philadelphia, PA: WB Saunders Co, 2001, 624-35.

Phosphorus

CAS Number 14596373
Synonyms P32; P33
Use Medical and research facilities
Toxicokinetics
Half-life: P32 - 14 days
P33 - 25 days
Elimination - renal

Mechanism of Toxic Action Beta radiation; high respiratory and gastrointestinal absorption; completely absorbed from all sites. Primary toxicity for rapidly dividing cells. Deposited in the bone marrow. Local irradiation causes cell damage. A Beta emitter oral ingestions with water.

Treatment Strategy Decontamination: Lavage

Additional Information Can be detected with the beta shield open on a beta-gamma detector - shielding should be with Thick greater than $^3/_8$ inch plexiglass, acrylic, lucite, plastic or wood.

Specific References

Jarrett DG, *Medical Management of Radiological Casualties Handbook*, 1st ed, Military Medical Operations Office, Armed Forces Radiobiology Research Institute, Bethesda, MD, 1999.

"Radiation Disasters and Children: American Academy of Pediatrics Policy Statement - Committee on Environmental Health," *Pediatrics*, 2003, 111(6):1455-66.

"Radiation Threats," http://www.imc-la.com/cbr/L1d-m.html. Last accessed 10/10/03.

Waselenko JK, MacVittie TJ, Blakely WF, et al, "Medical Management of the Acute Radiation Syndrome: Recommendations of the Strategic National Stockpile Radiation Working Group," *Ann Intern Med*, 2004, 140(12):1037-51.

Plutonium

CAS Number 13981163; 7440-07-5

UN Number 2918

Synonyms Plutonium Metal; Plutonium-236 Isotope; Plutonium-237 Isotope; Plutonium-238 Isotope; Plutonium-239 Isotope; Plutonium-240 Isotope; Plutonium-241 Isotope; Plutonium-242 Isotope; Plutonium-243 Isotope; PU

Use Primary fissionable material in nuclear weapons; primary radioactive contaminant in nuclear weapons accidents (from uranium)

Mechanism of Toxic Action

Alpha and gamma radiation. The primary radiation is from inhalation in the form of alpha particles; plutonium is not an external irradiation hazard.

Dermal: Can cause dermal burns, also tissue necrosis on contact. Plutonium can enter wounds, otherwise limited absorption.

GI absorption: Depends upon chemical state of plutonium. Generally not significant. The metal is not absorbed.

Pulmonary: Primary toxicity is from inhalation. Particles ≤ 5 microns remain in the lung and are metabolized based upon salt solubility. These particles cause local irradiation damage.

Toxicodynamics/Kinetics

Absorption: No GI absorption

Half-life: 180-200 years; PU-239: 24,400 years

Elimination: Feces and urine

Treatment Strategy

Decontamination: Remove contaminated clothes - store in specialized containment. Isolate all contaminants including irrigation run-off. Wound absorption is variable but washing intact skin is effective.

Dermal: Plutonium may be washed from intact skin. Irrigate skin with water; scrub gently with soft brush or sponge being careful to avoid abrading skin. 5% hypochlorite (dilute household bleach) can be used.

Ocular: Copious irrigation with water or saline

Medication: Nebulized or I.V.: calcium diethylenetriamine pentacetic acid (CaDTPA) administered using 1 g diluted in 250 mL of 5% dextrose in water infused over a 1-1$^1/_2$-hour period every 5 days. Especially useful when administered within 24 hours postexposure followed by ZnDTPA 1 g/daily (IND). Monitor urine levels. EDTA (calcium Versenate®) can also be used, although DTPA is more effective. DTPA is also useful for open wounds, in addition to inhalation or ingestion. CaDTPA is available through Radiation Emergency Assistance Center at 615-576-1004.

Additional Information
Silver gray metal, which becomes yellow when exposed to air
Patients should be monitored by use of a Geiger-Mueller counter. Of note, not all hospital monitors detect alpha particles. Personnel should be familiar with detection devices. Universal precautions and barrier protection is sufficient personal protective equipment for alpha articles.

Specific References
Dainiak N, Waselenko JK, Armitage JO, et al, "The Hematologist and Radiation Casualties," *Hematology (Am Soc Hematol Educ Program)*, 2003, 473-96 (review).

Jarrett DG, *Medical Management of Radiological Casualties Handbook*, 1st ed, Military Medical Operations Office, Armed Forces Radiobiology Research Institute, Bethesda, MD, 1999.

Khokhryakov VF, Suslova KG, Kudryavtseva TI, et al, "Relative Role of Plutonium Excretion With Urine and Feces From Human Body," *Health Phys*, 2004, 86(5):523-7.

"Radiation Disasters and Children: American Academy of Pediatrics Policy Statement - Committee on Environmental Health," *Pediatrics*, 2003, 111(6):1455-66.

"Radiation Threats," http://www.imc-la.com/cbr/L1d-m.html. Last accessed 10/10/03.

Scott BR and Peterson VL, "Risk Estimates for Deterministic Health Effects of Inhaled Weapons Grade Plutonium," *Health Phys*, 2003, 85(3):280-93.

Waselenko JK, MacVittie TJ, Blakely WF, et al, "Medical Management of the Acute Radiation Syndrome: Recommendations of the Strategic National Stockpile Radiation Working Group," *Ann Intern Med*, 2004, 140(12):1037-51.

Yow RB, "Radiation Poisoning," *Clinical Management of Poisoning and Drug Overdose*, 2nd ed, Haddad LM and Winchester JF, Philadelphia, PA: WB Saunders Co, 2001, 624-35.

Polonium

CAS Number 13981-52-7 (polonium 210)
Synonyms Radium F
Use Polonium 210 is a radon daughter found in uranium miners; an alpha particle emitter
Adverse Reactions Radiation sickness
Ocular: Cataracts
Respiratory: Lung cancer
Pharmacodynamics/Kinetics
Half-life: 138 days
Pregnancy Issues
Fetus may be exposed to about 8% of maternal dose
Additional Information
Low melting point
Overdose Treatment
Overdosage Treatment
Decontamination:
Oral: Gastric Lavage with saline; Activated Charcoal is *Not* useful
Supportive: Internal contamination may respond to succimer or BAL. (BAL Dose: 3 to 5 mg per kg in every four hours for two days followed by 2.5 to 3 mg/kg in every six hours for 2 days then 2.5 to 3 mg/kg in every 12 hours for one week)
Specific References
Rencova J, Volf V, Jones MM, Singh PK, "Mobilization and Detoxification of Polonium-210 in Rats by 2,3-Dimercaptosuccinic Acid and Its Derivatives," *Int J Radiat Biol*, 2000, 76(10), 1409-15.

Radium

CAS Number 7440-14-4 (Radium 226 and radium 228)
Use Medical and research facilities
Mechanism of Toxic Action Alpha and beta radiation
Adverse Reactions
  Hematologic: Leukopenia, anemia, aplastic anemia
  Hepatic: Cirrhosis
  Ocular: Cataracts
Toxicodynamics/Kinetics
  Half-life:
    Radium 223: 11 days in soil by alpha decay
    Radium 224: 4 days by alpha decay
    Radium 226: 1620 years by alpha decay
    Radium 228: 6 years by beta decay
Treatment Strategy
  Decontamination:
    Dermal: Wash with soap and water
    Ocular: Irrigate with water or saline
    Symptomatic and supportive care
Environmental Persistency Residence atmosphere time 1-10 days
Additional Information Bone sarcomas, breast cancer, and liver cancer have been associated with radium exposure. Chronic myeloid leukemia can occur.
Specific References

De Lorenzo RA, "When It's Hot, It's Hot," *Ann Emerg Med*, 2005, 45(6):653-4.

Jarrett DG, *Medical Management of Radiological Casualties Handbook*, 1st ed, Military Medical Operations Office, Armed Forces Radiobiology Research Institute, Bethesda, MD, 1999.

Koenig KL, Goans RE, Hatchett RJ, et al, "Medical Treatment of Radiological Casualties: Current Concepts," *Ann Emerg Med*, 2005, 45(6):643-52.

Lloyd RD, Taylor GN, and Miller SC, "Does Low Dose Internal Radiation Increase Lifespan?" *Health Phys*, 2004, 86(6):629-32.

"Radiation Disasters and Children: American Academy of Pediatrics Policy Statement - Committee on Environmental Health," *Pediatrics*, 2003, 111(6):1455-66.

"Radiation Threats," http://www.imc-la.com/cbr/L1d-m.html. Last accessed 10/10/03.

Waselenko JK, MacVittie TJ, Blakely WF, et al, "Medical Management of the Acute Radiation Syndrome: Recommendations of the Strategic National Stockpile Radiation Working Group," *Ann Intern Med*, 2004, 140(12):1037-51.

Yow RB, "Radiation Poisoning," *Clinical Management of Poisoning and Drug Overdose*, 2nd ed, Haddad LM and Winchester JF, Philadelphia, PA: WB Saunders Co, 2001, 624-35.

Strontium

CAS Number 7440-24-6; 7789-06-2 (Strontium chromate); 10042-76-9 (Strontium nitrate)
UN Number 1507 (Strontium nitrate)
Synonyms Sr
Use Product of nuclear weapon detonation; direct fission product (uranium daughter); power source for space vehicles, weather stations, and navigational beacons
Mechanism of Toxic Action Depending upon exposure and dose of radiation, acute radiation syndrome (ARS); beta and gamma radiation; limited respiratory absorption and moderate gastrointestinal absorption. Primary toxicity is to the bone; up to 50% of a dose will

be deposited in bone. Strontium behaves like calcium and concentrates in bone and teeth. Main isotope is Sr-90, which emits beta particles and decays to yttrium-90.

Toxicodynamics/Kinetics Half-life: 29.1 years; yttrium-90: 64 hours

Signs and Symptoms of Acute Exposure Linked to bone cancer and leukemia, ocular/dermal burns (yttrium-90), Flushing

Treatment Strategy Symptomatic and supportive care

Additional Information
Silver gray metal, which changes to yellow when exposed to air
EPA drinking water limit of 90 Sr is 8 pCi/L

Specific References

Jarrett DG, *Medical Management of Radiological Casualties Handbook*, 1st ed, Military Medical Operations Office, Armed Forces Radiobiology Research Institute, Bethesda, MD, 1999.

Kirrane BM, Hoffman RS, and Nelson LS, "Massive Strontium Ferrate Ingestion Does Not Produce Acute Toxicity," *Clin Toxicol (Phila)*, 2005, 43:749.

"Radiation Disasters and Children: American Academy of Pediatrics Policy Statement - Committee on Environmental Health," *Pediatrics*, 2003, 111(6):1455-66.

"Radiation Threats," http://www.imc-la.com/cbr/L1d-m.html. Last accessed 10/10/03.

Waselenko JK, MacVittie TJ, Blakely WF, et al, "Medical Management of the Acute Radiation Syndrome: Recommendations of the Strategic National Stockpile Radiation Working Group," *Ann Intern Med*, 2004, 140(12):1037-51.

Yow RB, "Radiation Poisoning," *Clinical Management of Poisoning and Drug Overdose*, 2nd ed, Haddad LM and Winchester JF, Philadelphia, PA: WB Saunders Co, 2001, 624-35.

Tritium

CAS Number 10028-17-8

Synonyms H₃, Thymidine

Use Component of nuclear weapons; luminescent for gun sights and muzzle velocity detectors

Mechanism of Toxic Action Tritium is a beta particle emitter that poses the greatest risk in closed spaces. Tritium is absorbed through the skin, ingested, or inhaled. Distribution in the body equals that of water. It is excreted in urine. Urine samples will be positive within an hour of a significant exposure.

Toxicodynamics/Kinetics Half-life: 10-12 days

Treatment Strategy Supportive therapy: Increasing oral fluids will reduce the half-life by 50%

Specific References

"Radiation Disasters and Children: American Academy of Pediatrics Policy Statement - Committee on Environmental Health," *Pediatrics*, 2003, 111(6):1455-66.

"Radiation Threats," http://www.imc-la.com/cbr/L1d-m.html. Last accessed 10/10/03.

Waselenko JK, MacVittie TJ, Blakely WF, et al, "Medical Management of the Acute Radiation Syndrome: Recommendations of the Strategic National Stockpile Radiation Working Group," *Ann Intern Med*, 2004, 140(12):1037-51.

Yow RB, "Radiation Poisoning," *Clinical Management of Poisoning and Drug Overdose*, 2nd ed, Haddad LM and Winchester JF, Philadelphia, PA: WB Saunders Co, 2001, 624-35.

Uranium

CAS Number 7440-61-1

UN Number 2979

Use Weapons grade material; component in munitions

Mechanism of Toxic Action Alpha, beta and gamma radiation; depending upon exposure and dose of radiation, acute radiation syndrome (ARS). Inhaled uranium may be metabolized and excreted in the urine. Urinary levels of 100 mcg/dL may cause renal failure. Soluble salts of uranium are readily absorbed while the metal form is not.

Toxicodynamics/Kinetics Half-life: Billions of years

Pregnancy Issues Other than effects of radiation, rodent studies demonstrate increased risk of cleft palate and skeletal variations at doses >1 mg/kg/day

Signs and Symptoms of Acute Exposure Dermal/ocular burns, acute radiation sickness, lacrimation, conjunctivitis, shortness of breath, cough, nausea, vomiting, lymphatic cancer

Treatment Strategy
Decontamination:
Dermal: Promptly remove clothing; wash skin with soap and water
Inhalation: Give 100% humidified oxygen
Ocular: Irrigate with copious amounts of water or saline

Additional Information
Laboratory evaluation: Urinalysis, 24-hour urine for uranium bioassay, renal and hepatic function test, β-2-microglobulin and creatinine clearance
Typical uranium concentrations:
Native soil: 0.9-9 ppm
Ground water: 0.1-40 ppm
IDLH: 10 mg/m^3
Essentially insoluble in water
See table.

Concentrations of Uranium in Some Foods	
Food	Uranium Concentration (ng/g raw weight)
Whole grain products	1.45*
Potatoes	2.66-2.92*; 15-18[†]
Carrots	7.7[†]
Root vegetables	0.94-1.20*
Cabbage	4.7[†]
Meat	0.58-1.32*; 20[†]
Poultry	0.14-0.42*
Beef	14[†]
Beef liver	26[†]
Beef kidney	70[†]
Eggs	0.23*; 9.6[†]
Dairy products	0.08-0.31*
Cow milk	4[†]
Milk	1-2[†]
Fresh fish	0.43-0.85*; 11[†]
Shellfish	9.5-31.0*
Welsh onion	69[†]
Flour	0.25-0.68*
Wheat bread	19[†]
Baked products	1.32-1.5*; 12[†]
Polished rice	1.43-6.0*; 15[†]
Macaroni	0.4-0.63*
Tea	5[†]
Coffee	6[†]
Parsley	60[†]
Red pepper	5[†]
Mustard	0.2[†]

Table *(Continued)*	
Concentrations of Uranium in Some Foods	
Food	Uranium Concentration (ng/g raw weight)
Table salt	40[†]
Canned vegetables	0.09-0.18*
Fruit juices	0.04-0.12*
Canned fruits	0.18-0.29*
Fresh fruits	0.71-1.29*
Dried beans	1.5-3.67*
Fresh vegetables	0.52-0.92*

From: U.S. Department of Health and Human Services, "Toxicological Profile for Uranium," Agency for Toxic Substances and Disease Registry, September 1999.
*Reference: National Council on Radiation Protection and Measurements
[†]Reference: Environmental Protection Agency

Average Uranium Concentrations in Drinking Water for States Where Concentration Exceeds 1 pCi/L

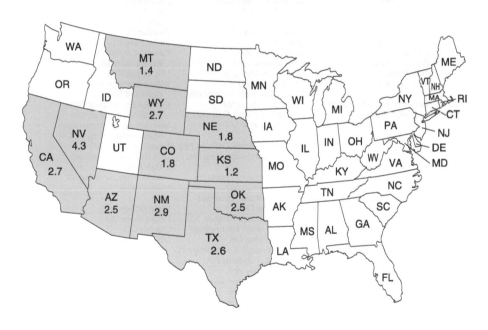

Concentration >1 pCi/L

Concentration ≤1 pCi/L

Source: NCRP 1984b

Department of Energy
Major Offices, Facilities, and Laboratories

Operations Offices (o)
Oakland CA
Idaho Falls ID
Chicago IL
Las Vegas NV
Albuquerque NM
Savannah SC
Oak Ridge TN
Richland WA

Laboratories / National Laboratories (♦)
Lawrence Berkeley Natl Lab, Berkeley CA
Lawrence Livermore Natl Lab, Berkeley CA
Stanford Linear Accelerator, Menlo Park CA
Natl Renewable Energy Lab, Golden CO
Idaho National Energy and Env Lab, ID
Fermi Natl Accelerator Lab, Batavia IL
Argonne Natl Lab, Argonne IL
Ames Lab, Ames IA
MIT Bates Lab, Cambridge MA
Princeton Plasma Physics Lab, NJ
Inhalation Toxicology Res Inst, Albuquerque NM

Los Alamos Natl Lab, Los Alamos NM
Sandia Natl Labs, Albuquerque NM
Livermore CA
Brookhaven Natl Lab, Brookhaven NY
Oak Ridge Natl Lab, Oak Ridge TN
T Jefferson Natl Accelerator, Newport News VA
Pacific Northwest Natl Lab, Richland WA

Special Purpose Offices (✛)
Energy Technology Eng Ctr, LA, CA
Naval Petroleum Reserves 1&2, Kern Co, CA
Naval Oil Shale Reserves 1&3, Rifle CO
Yucca Mountain Project, NV
Natl Petroleum Technology Office, Tulsa OK
Federal Energy Technology Centers, Pittsburg PA, Morgantown WV
Oak Ridge Inst for Science & Education, TN
Naval Oil Shale Reserve 2, Vernal UT
Naval Petroleum Reserve 3, Casper WY

Facilities (+)
Grand Junction CO
Rocky Flats CO
Pinellas FL
Idaho Falls ID
Kansas City Plant, Kansas City MO
Las Vegas NV
Waste Isolation Pilot Plant, Carlsbad NM
Fernald Env Management Project, Cincinnati OH
Miamisburg Environmental Management Project, OH
Savannah River SC
Oak Ridge Reservation, Oak Ridge TN
Pantex Plant, Amarillo TX
Hanford WA

Field Offices (#)
Rocky Flats CO
Golden CO
Ohio (4 sites in OH, 1 site in New York)

Specific References

Archer VE, Coons T, Saccomanno G, et al, "Latency and the Lung Cancer Epidemic Among United States Uranium Miners," *Health Phys*, 2004, 87(5):480-9.

Dainiak N, Waselenko JK, Armitage JO, et al, "The Hematologist and Radiation Casualties," *Hematology (Am Soc Hematol Educ Program)*, 2003, 473-96 (review).

DeVol TA and Woodruff RL Jr, "Uranium in Hot Water Tanks: A Source of TENORM," *Health Phys*, 2004, 87(6):659-63.

Iyengar GV, Kawamura H, Dang HS, et al, "Contents of Cesium, Iodine, Strontium, Thorium, and Uranium in Selected Human Organs of Adult Asian Population," *Health Phys*, 2004, 87(2):151-9.

Jarrett DG, *Medical Management of Radiological Casualties Handbook*, 1st ed, Military Medical Operations Office, Armed Forces Radiobiology Research Institute, Bethesda, MD, 1999.

McDiarmid MA, Squibb K, and Engelhardt SM, "Biologic Monitoring for Urinary Uranium in Gulf War I Veterans," *Health Phys*, 2004, 87(1):51-6.

"Radiation Disasters and Children: American Academy of Pediatrics Policy Statement - Committee on Environmental Health," *Pediatrics*, 2003, 111(6):1455-66.

"Radiation Threats," http://www.imc-la.com/cbr/L1d-m.html. Last accessed 10/10/03.

Waselenko JK, MacVittie TJ, Blakely WF, et al, "Medical Management of the Acute Radiation Syndrome: Recommendations of the Strategic National Stockpile Radiation Working Group," *Ann Intern Med*, 2004, 140(12):1037-51.

Yow RB, "Radiation Poisoning," *Clinical Management of Poisoning and Drug Overdose*, 2nd ed, Haddad LM and Winchester JF, Philadelphia, PA: WB Saunders Co, 2001, 624-35.

APPENDIX

Algorithmic Approach to Case Classification

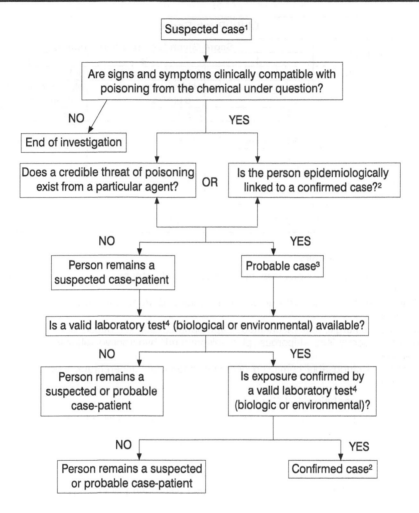

[1]Suspected case: A case in which a protentially exposed person is being evaluated by healthcare workers or public health officials for poisoning by a particular chemical agent, but no specific credible threat exists.

[2]Confirmed case: A clinically compatible case with laboratory confirmation by using either biologic or environmental samples. The case can be confirmed if laboratory testing was not performed because either a predominant amount of clinical and nonspecific laboratory evidence of a particular chemical was present or a 100% certainty of the etiology of the agent is known.

[3]Probable case: A clinically compatible case in which a high index of suspicion (ie, a credible threat) exists for exposure to a particular agent, or a case with an epidemiologic link to a laboratory-confirmed case.

[4]Valid laboratory test: A biologic and environmental laboratory test that has been analytically validated.

Assessment of Liver Function

Child-Pugh Score			
Component	Score Given for Observed Findings		
	1	2	3
Encephalopathy grade[1]	None	1-2	3-4
Ascites	None	Mild or controlled by diuretics	Moderate or refractory despite diuretics
Albumin (g/dL)	>3.5	2.8-3.5	<2.8
Total bilirubin (mg/dL)	<2 (<34 micromoles/L)	2-3 (34-50 micromoles/L)	>3 (>50 micromoles/L)
or			
Modified total bilirubin[2]	<4	4-7	>7
Prothrombin time (seconds prolonged)	<4	4-6	>6
or			
INR	<1.7	1.7-2.3	>2.3

[1]Encephalopathy Grades

Grade 0: Normal consciousness, personality, neurological examination, electroencephalogram

Grade 1: Restless, sleep disturbed, irritable/agitated, tremor, impaired handwriting, 5 cps waves

Grade 2: Lethargic, time-disoriented, inappropriate, asterixis, ataxia, slow triphasic waves

Grade 3: Somnolent, stuporous, place-disoriented, hyperactive reflexes, rigidity, slower waves

Grade 4: Unarousable coma, no personality/behavior, decerebrate, slow 2-3 cps delta activity

Alternative Encephalopathy Grades

Grade 1: Mild confusion, anxiety, restlessness, fine tremor, slowed coordination

Grade 2: Drowsiness, disorientation, asterixis

Grade 3: Somnolent but rousable, marked confusion, incomprehensible speech, incontinent, hyperventilation

Grade 4: Coma, decerebrate posturing, flaccidity

[2]Modified total bilirubin used to score patients who have Gilbert's syndrome or who are taking indinavir.

Child-Pugh Classification

Class A (mild hepatic impairment): Score 5-6
Class B (moderate hepatic impairment): Score 7-9
Class C (severe hepatic impairment): Score 10-15

References

Centers for Disease Control and Prevention, "Report of the NIH Panel to Define Principles of Therapy of HIV Infection and Guidelines for the Use of Antiretroviral Agents in HIV-Infected Adults and Adolescents," March 23, 2004, located at (URL) http://www.aidsinfo.nih.gov.

U.S. Department of Health and Human Services Food and Drug Administration, "Guidance for Industry, Pharmacokinetics in Patients With Impaired Hepatic Function: Study Design, Data Analysis, and Impact on Dosing and Labeling," May 2003, located at (URL) http://www.fda.gov/cder/guidance/3625fnl.pdf.

Clinical Syndromes Associated with Food-borne Diseases

Clinical Syndromes	Incubation Period (h)	Causes	Commonly Associated Vehicles
Nausea and vomiting	<1-6	*Staphylococcus aureus* (pre-formed toxins, A, B, C, D, E)	Ham, poultry, cream-filled pastries, potato and egg salad, mushrooms
		Bacillus cereus (emetic toxin)	Fried rice, pork
		Heavy metals (copper, tin, cadmium, zinc)	Acidic beverages
Histamine response and gastrointestinal (GI) tract	<1	Histamine (scombroid)	Fish (bluefish, bonito, mackerel, mahi-mahi, tuna)
Neurologic, including paresthesia and GI tract	0-6	Tetrodotoxin, ciguatera	Puffer fish
			Fish (amberjack, barracuda, grouper, snapper)
		Paralytic compounds	Shellfish (clams, mussels, oysters, scallops, other mollusks)
		Neurotoxic compounds	Shellfish
		Domoic acid	Mussels
		Monosodium glutamate	Chinese food
Neurologic and GI tract manifestations	0-2	Mushroom toxins (early onset)	Mushrooms
Moderate-to-severe abdominal cramps and watery diarrhea	8-16	*B. cereus* enterotoxin	Beef, pork, chicken, vanilla sauce
		Clostridium perfringens enterotoxin	Beef, poultry, gravy
	16-48	Caliciviruses	Shellfish, salads, ice
		Enterotoxigenic *Escherichia coli*	Fruits, vegetables
		Vibrio cholerae 01 and 0139	Shellfish
		V. cholerae non-01	Shellfish
Diarrhea, fever, abdominal cramps, blood and mucus in stools	16-72	*Salmonella*	Poultry, pork, eggs, dairy products, including ice cream, vegetables, fruit
		Shigella	Egg salad, vegetables
		Campylobacter jejuni	Poultry, raw milk
		Invasive *E. coli*	
		Yersinia enterocolitica	Pork chitterlings, tofu, raw milk
		Vibrio parahaemolyticus	Fish, shellfish
Bloody diarrhea, abdominal cramps	72-120	Enterohemorrhagic *E. coli*	Beef (hamburger), raw milk, roast beef, salami, salad dressings
Methemoglobin poisoning	6-12	Mushrooms (late onset)	Mushrooms
Hepatorenal failure	6-24	Mushrooms (late onset)	Mushrooms

Table   (*Continued*)

Clinical Syndromes	Incubation Period (h)	Causes	Commonly Associated Vehicles
Gastrointestinal then blurred vision, dry mouth, dysarthria, diplopia, descending paralysis	18-36	*Clostridium botulinum*	Canned vegetables, fruits and fish, salted fish, bottled garlic
Extraintestinal manifestations	Varied	Brainerd disease	Unpasteurized milk
		Brucella	Cheese, raw milk
		Group A *Streptococcus*	Egg and potato salad
		Listeria monocytogenes	Cheese, raw milk, hot dogs, cole slaw, cold cuts
		Trichinella spiralis	Pork
		Vibrio vulnificus	Shellfish

Adapted from "Report of the Committee on Infectious Diseases," *1997 Red Book*®, 24th ed.

Creatinine Clearance Estimating Methods in Patients with Stable Renal Function

These formulas provide an acceptable estimate of the patient's creatinine clearance **except** in the following instances.
- Patient's serum creatinine is changing rapidly (either up or down).
- Patients are markedly emaciated.

In above situations, certain assumptions have to be made.
- In patients with rapidly rising serum creatinine (ie, >0.5-0.7 mg/dL/day), it is best to assume that the patient's creatinine clearance is probably <10 mL/minute.
- In emaciated patients, although their actual creatinine clearance is less than their calculated creatinine clearance (because of decreased creatinine production), it is not possible to easily predict how much less.

Infants   Estimation of creatinine clearance using serum creatinine and body length (to be used when an adequate timed specimen cannot be obtained). **Note:** This formula may not provide an accurate estimation of creatinine clearance for infants younger than 6 months of age and for patients with severe starvation or muscle wasting.

$$Cl_{cr} = K \times L/S_{cr}$$

where:
Cl_{cr} = creatinine clearance in mL/minute/1.73 m^2
K = constant of proportionality that is age-specific

Age	K
Low birth weight ≤1 y	0.33
Full-term ≤1 y	0.45
2-12 y	0.55
13-21 y female	0.55
13-21 y male	0.70

L = length in cm
S_{cr} = serum creatinine concentration in mg/dL

Reference
Schwartz GJ, Brion LP, and Spitzer A, "The Use of Plasma Creatinine Concentration for Estimating Glomerular Filtration Rate in Infants, Children and Adolescents," *Pediatr Clin North Am*, 1987, 34(3):571-90.

Children (1-18 years)
Method 1: (Traub SL and Johnson CE, *Am J Hosp Pharm*, 1980, 37(2):195-201)

$$Cl_{cr} = \frac{0.48 \times (height)}{S_{cr}}$$

where:
Cl_{cr} = creatinine clearance in mL/min/1.73 m^2
S_{cr} = serum creatinine in mg/dL
Height = height in cm

Method 2: Nomogram (Traub SL and Johnson CE, *Am J Hosp Pharm*, 1980, 37(2):195-201)

Children 1-18 Years

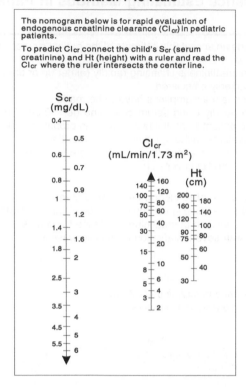

The nomogram below is for rapid evaluation of endogenous creatinine clearance (Cl_{cr}) in pediatric patients.

To predict Cl_{cr} connect the child's S_{cr} (serum creatinine) and Ht (height) with a ruler and read the Cl_{cr} where the ruler intersects the center line.

S_{cr}
(mg/dL)

Cl_{cr}
(mL/min/1.73 m²)

Ht
(cm)

Adults (18 years and older)

Method 1: (Cockroft DW and Gault MH, *Nephron*, 1976, 16:31-41)

Estimated creatinine clearance (Cl_{cr}) (mL/min):

$$Male = \frac{(140 - age) \times BW (kg)}{72 \times S_{cr}}$$
$$Female = male \times 0.85$$

Note: Use of actual body weight (BW) in obese patients (and possibly patients with ascites) may significantly overestimate creatinine clearance. Some clinicians prefer to use an adjusted ideal body weight (IBW) in such cases [eg, IBW + 0.4(ABW − IBW)], especially when calculating dosages for aminoglycoside antibiotics.

Method 2: (Jelliffe RW, *Ann Intern Med*, 1973, 79:604)

Estimated creatinine clearance (Cl_{cr}) (mL/min/1.73 m²):

$$Male = \frac{98 - 0.8 (age - 20)}{S_{cr}}$$
$$Female = male \times 0.90$$

Fluid and Electrolyte Requirements in Children

Maintenance Fluids (Two methods):

Surface area method (most commonly used in children >10 kg): 1500-2000 mL/m^2/day

Body weight method

<10 kg	100 mL/kg/day
11-20 kg	1000 mL + 50 mL/kg (for each kg >10)
>20 kg	1500 mL + 20 mL/kg (for each kg >20)

Maintenance Electrolytes (See specific electrolyte in Alphabetical Listing of Drugs for more detailed information)

Sodium: 3-4 mEq/kg/day **or** 30-50 mEq/m^2/day

Potassium: 2-3 mEq/kg/day **or** 20-40 mEq/m^2/day

Dehydration Fluid Therapy

Goals of therapy:
- Restore circulatory volume to prevent shock (10% to 15% dehydration)
- Restore combined intracellular and extracellular deficits of water and electrolytes within 24 hours
- Maintain adequate water and electrolytes
- Resolve homeostatic distortions (eg, acidosis)
- Replace ongoing losses

Analysis of the Severity of Dehydration by Physical Signs			
Clinical Sign	**Mild***	**Moderate***	**Severe***
Pre-illness body weight	5% loss	10% loss	15% loss
Skin turgor	↓	Tenting	Tenting
Mucous membranes	Dry	Very dry	Parched
Skin color	Pale	Gray	Mottled
Urine output	↓	↓↓	Azotemic
Blood pressure	Normal	Normal, ↓	↓↓
Heart rate	Normal, ↑	↑	↑↑
Fontanelle (<7 mo)	Flat	Soft	Sunken
CNS	Consolable	Irritable	Lethargic/coma
*Postpubertal children and adults experience the same symptoms with mild, moderate, and severe dehydration associated with 3%, 6%, and 9% losses in body weight, respectively.			

Restoration of Circulatory Volume (10% to 15% dehydration estimate)

Fluid boluses of 20 mL/kg using crystalloid (eg, normal saline) or 10 mL/kg colloid (eg, 5% albumin) administered as rapidly as possible; repeat until improved circulation (eg, warm skin, decreased heart rate (towards normal), improved capillary refill time, urine output restored).

Classification of Dehydration (based upon the serum sodium concentration)

Isotonic	130-150 mEq/L
Hypotonic	<130 mEq/L
Hypertonic	>150 mEq/L

Estimated Water & Electrolyte Deficits in Dehydration (moderate to severe)				
Type of Dehydration	Water (mL/kg)	Na+ (mEq/kg)	K+ (mEq/kg)	Cl^- and HCO_3^- (mEq/kg)
Isotonic	100-150	8-10	8-10	16-20
Hypotonic	50-100	10-14	10-14	20-28
Hypertonic	120-180	2-5	2-5	4-10

Adapted from *Current Pediatric Diagnosis & Treatment*, 10th ed, 1991.

Water deficit may also be calculated (in isotonic dehydration):

Water deficit (mL) = $\dfrac{\%\ \text{dehydration} \times \text{wt (kg)} \times 1000\ \text{g/kg}}{100}$

Assessment of Water Loss in Relation to Serum Na^+ Concentrations Degree of Dehydration as % Body Weight			
Serum Na^+	Mild	Moderate	Severe
Isotonic	5%	10%	15%
Hypotonic ($Na^+ < 130$)	4%	6%	8%
Hypertonic ($Na^+ > 150$)	7%	12%	17%

Example of Fluid Replacement (assume 10-kg infant with 10% isotonic dehydration)			
	Water	Sodium (mEq)	Potassium (mEq)
Maintenance	1000 mL	40	20
Deficit*	1000 mL	80	80
Total	2000 mL	120	100

*Reduce this total by any fluid boluses given initially.

First 8 hours:	Replace $^1/_3$ maintenance water	=	330 mL
	Replace $^1/_2$ deficit water	=	500 mL
	Total		830 mL/8 h = 103 mL/h
	Replace $^1/_2$ of Na^+ & K^+ = 60 mEq sodium/803 mL; 50 mEq potassium/803 mL (It is suggested that the maximum potassium initially used is 40 mEq/L and is **not** started until urine output has been established.)		
	The actual order would appear as: $D_5\,^1/_2$NS at 103 mL/hour for 8 hours; add 40 mEq/L KCl after patient voids.		
Second 16 hours:	Replace $^2/_3$ maintenance water	=	670 mL
	Replace $^1/_2$ deficit water	=	500 mL
	Total		1260 mL/16 h = 79 mL/h
	Replace remainder of sodium and potassium.		
	The actual order would appear as: $D_5\,^1/_3$NS with KCl 40 mEq/L at 79 mL/hour for 16 hours. (Use of $^1/_4$NS may be more desirable for convenience.)		

Analysis of Ongoing Losses

Electrolyte Composition of Biological Fluids (mEq/L)				
Fluid Type	Sodium	Potassium	Chloride	Total HCO$_3^-$
Stomach	20-120	5-25	90-160	0-5
Duodenal drainage	20-140	3-30	30-120	10-50
Biliary tract	120-160	3-12	70-130	30-50
Small intestine Initial drainage	100-140	4-40	60-100	30-100
Small intestine Established drainage	4-20	4-10	10-100	40-120
Pancreatic	110-160	4-15	30-80	70-130
Diarrheal stool	10-25	10-30	30-120	10-50

Adapted from *Current Pediatric Diagnosis & Treatment*, 9th ed, Appleton & Lange, 1987.

Because of the wide range of normal values, specific analyses are suggested in individual cases.

Alterations of Maintenance Fluid Requirements	
Fever	Increase maintenance fluids by 5 mL/kg/day for each degree of temperature above 38°C
Hyperventilation	Increase maintenance fluids by 10-60 mL/100 kcal BEE (basal energy expenditure)
Sweating	Increase maintenance fluids by 10-25 mL/100 kcal BEE (basal energy expenditure)
Hyperthyroidism	Variable increase in maintenance fluids: 25%-50%
Renal disease	Monitor and analyze output; adjust therapy accordingly
Renal failure	Maintenance fluids are equal to insensible losses (300 mL/m^3) + urine replacement (mL for mL)
Diarrhea	Increase maintenance fluids on a mL/mL loss basis

Oral Rehydration

Due to the high worldwide incidence of dehydration from infantile diarrhea, effective, inexpensive oral rehydration solutions have been developed. In the U.S., a typical effective solution for rehydration contains 50-60 mEq/L sodium, 20-30 mEq/L potassium, 30 mEq/L bicarbonate or its equivalent, and sufficient chloride to provide electroneutrality. Two percent to 3% glucose facilitates electrolyte absorption and short-term calories. The following table describes the electrolyte/sugar content of commonly used oral rehydration solutions.

Composition of Frequently Used Oral Electrolyte Replacement Solutions

Solutions	% CHO	Na⁺ (mEq/L)	K⁺ (mEq/L)	Cl⁻ (mEq/L)	HCO⁻ (mEq/L)
Normal saline		154		154	
Ringer's lactate		130	4	109	28
Dextrose 5% in 0.25% NaCl	5% glucose	38		38	
WHO solution	2% glucose or 4% sucrose	90	20	80	30 citrate
WHO solution, modified	2% glucose	55	25	30	50
Rehydralyte®	2.5% glucose	75	20	65	30 citrate
Ricelyte®	3% carbohydrate	50	25	45	34 citrate
Resol®	2% glucose	50	20	50	34 citrate
Rice water	2.5% carbohydrate	90	20	30	80
Pedialyte® (Ross)	2.5% glucose	45	20	35	30 citrate
Pedialyte® Freezer Pops	2.5% glucose	45	20	35	30 citrate
Enfamil Enfalyte®	3% glucose	50	25	45	34 citrate
Kao Lectrolyte	2% glucose	50	20		30 citrate
Naturalyte	2.5% glucose	45	20	35	30 citrate
Oralyte	2.5% glucose	45	20	35	48 citrate
Gatorade®	2.6% glucose, 2% fructose	23.5	<1	17	
Apple juice	3.2% glucose, 1.3% sucrose, 7.5% fructose	<1	25		
Orange juice 1:3 (dilution with water)	1% glucose, 1.2% fructose	<1	50		50 citrate
Grape juice	1.6% glucose, 2.1% fructose	0.2-0.7	8-11		8
One package cherry gelatin dissolved in 4 cups water		24	Needs added K⁺		
Coca-Cola®		1.6	<1		13.4 citrate
Pepsi-Cola®		6.5	0.8		
Beef broth		120	10		
Chicken broth		250	8		

Adapted from Aranda-Michel, J and Giannella RA, "Acute Diarrhea" A Practical Review," *Am J Med,* 1999, 106:670-6.

Initial Isolation and Protective Action Distances Table

From *2000 Emergency Response Guidebook*, Chicago, IL: LabelMaster, 2000, 311-370

Introduction

The Table of Initial Isolation and Protective Action Distances suggests distances useful to protect people from vapors resulting from spills involving dangerous goods, which are considered toxic by inhalation (TIH), including certain chemical warfare agents, or which produce toxic gases upon contact with water. The Table provides first responders with initial guidance until technically qualified emergency response personnel are available. Distances show areas likely to be affected during the first 30 minutes after materials are spilled and could increase with time.

The **Initial Isolation Zone** defines an area SURROUNDING the incident in which persons may be exposed to dangerous (upwind) and life threatening (downwind) concentrations of material. The **Protective Action Zone** defines an area DOWNWIND from the incident in which persons may become incapacitated and unable to take protective action and/or incur serious or irreversible health effects. The Table provides specific guidance for small and large spills occurring day or night.

Adjusting distances for a specific incident involves many interdependent variables and should be made only by personnel technically qualified to make such adjustments. For this reason, no precise guidance can be provided in this document to aid in adjusting the table distances; however, general guidance follows.

Factors That May Change the Protective Action Distances

The guide for a material clearly indicates the evacuation distance required to protect against fragmentation hazard. If the material becomes involved in a FIRE, the toxic hazard may become less important than the fire or explosion hazard.

If more than one tank car, cargo tank, portable tank, or large cylinder involved in the incident is leaking, LARGE SPILL distances may need to be increased.

For material with a protective action distance of 11.0+ km (7.0+ miles), the actual distance can be larger in certain atmospheric conditions. If the dangerous goods vapor plume is channeled in a valley or between many tall buildings, distances may be larger than shown in the Table due to less mixing of the plume with the atmosphere. Daytime spills in regions with known strong inversions or snow cover, or occurring near sunset, accompanied by a steady wind, may require an increase in protective action distance. When these conditions are present, airborne contaminants mix and disperse more slowly and may travel much farther downwind. In addition, protective action distances may be larger for liquid spills when either the material or outdoor temperature exceeds 30°C (86°F).

Materials that react with water to produce significant toxic gases are included in the Table of Initial Isolation and Protective Action Distances. Note that some materials that are TIH (eg, bromine trifluoride, thionyl chloride, etc) produce additional TIH materials when spilled in water. For these materials, 2 entries are provided in the Table of Initial Isolation and Protective Action Distances. If it is not clear whether the spill is on land or in water, or in cases where the spill occurs both on land and in water, choose the larger Protective Action Distance. Following the Table of Initial Isolation and Protective Action Distances is a table that lists the materials, which, when spilled in water, produce toxic gases and the toxic gases that these water reactive materials produce.

When a water-reactive TIH-producing material is spilled into a river or stream, the source of the toxic gas may move with the current or stretch from the spill point downstream for a substantial distance.

Certain chemical warfare agents have been added to the Table of Initial Isolation and Protective Action Distances. The distances shown were calculated using worst-case scenarios for these agents when used as a weapon.

Protective Action Decision Factors to Consider

The choice of protective options for a given situation depends on a number of factors. For some cases, evacuation may be the best option; in others, sheltering in-place may be the best course. Sometimes, these two actions may be used in combination. In any emergency, officials need to quickly give the public instructions. The public will need continuing information and instructions while being evacuated or sheltered in-place.

Proper evaluation of the factors listed below will determine the effectiveness of evacuation or in-place protection. The importance of these factors can vary with emergency conditions. In specific emergencies, other factors may need to be identified and considered as well. This list indicates what kind of information may be needed to make the initial decision.

The Dangerous Goods

- Degree of health hazard
- Amount involved
- Containment/control of release
- Rate of vapor movement

The Population Threatened

- Location
- Number of people
- Time available to evacuate or shelter in-place
- Ability to control evacuation or shelter in-place
- Building types and availability
- Special institutions or populations (eg, nursing homes, hospitals, prisons)

Weather Conditions

- Effect on vapor and cloud movement
- Potential for change
- Effect on evacuation or protection in-place

Protective Actions

Protective Actions are those steps taken to preserve the health and safety of emergency responders and the public during an incident involving releases of dangerous goods. The Table of Initial Isolation and Protective Action Distances predicts the size of downwind areas that could be affected by a cloud of toxic gas. People in this area should be evacuated and/or sheltered in-place inside buildings.

Isolate Hazard Area and Deny Entry means keep everybody away from the area if they are not directly involved in emergency response operations. Unprotected emergency responders should not be allowed to enter the isolation zone. This "isolation" task is done first to establish control over the area of operations. This is the first step for any protective actions that may follow. See the Table of Isolation and Protective Action Distances for more detailed information on specific materials.

Evacuate means move all people from a threatened area to a safer place. To perform an evacuation, there must be enough time for people to be warned, to get ready, and to leave an area. If there is enough time, evacuation is the best protective action. Begin evacuating people nearby and those outdoors in direct view of the scene. When additional help arrives, expand the area to be evacuated downwind and crosswind to at least the extent recommended in this chapter. Even after people move to the distances recommended, they may not be completely safe from harm. They should not be permitted to congregate at such distances. Send evacuees to a definite place, by a specific route, far enough away so they will not have to be moved again if the wind shifts.

Shelter In-Place means people should seek shelter inside a building and remain inside until the danger passes. Sheltering in-place is used when evacuating the public would cause greater risk than staying where they are, or when an evacuation cannot be performed. Direct the people inside to close all doors and windows and to shut off all ventilating, heating, and cooling systems. In-place protection may not be the best option if (a) the vapors are flammable; (b) if it will take a long time for the gas to clear the area; or (c) if buildings cannot be closed tightly. Vehicles can offer some protection for a short period if the windows are closed and the ventilating systems are shut off. Vehicles are not as effective as buildings for in-place protection.

It is vital to maintain communications with competent persons inside the building so that they are advised about changing conditions. Persons protected-in-place should be warned to stay far from windows because of the danger from glass and projected metal fragments in a fire and/or explosion.

Every dangerous goods incident is different. Each will have special problems and concerns. Action to protect the public must be selected carefully. These pages can help with initial decisions on how to protect the public. Officials must continue to gather information and monitor the situation until the threat is removed.

Background on the Initial Isolation and Protective Action Distance Table

Initial Isolation and Protective Action Distances were determined for small and large spills occurring day or night. The overall analysis was statistical in nature and utilized state-of-the-art emission rate and dispersion models; statistical release data from the U.S. DOT HMIS (Hazardous Materials Incident Reporting System) database; 5 years of meteorological observations from over 120 locations in U.S., Canada, and Mexico; and the most current toxicological exposure guidelines.

For each chemical, thousands of hypothetical releases were modeled to account for the statistical variation in both release amount and atmospheric conditions. Based on this statisticsl sample, the 90% percentile Protective Action Distance for each chemical and category was selected to appear in the table. A brief description of the analysis is provided below. A detailed report outlining the methodology and data used in the generation of the Initial Isolation and Protective Action Distances may be obtained from the U.S. Department of Transportation, Research and Special Programs Administration.

Release amounts and emission rates into the atmosphere were statistically modeled based on (1) data from U.S. DOT HMIRS database, (2) container types and sizes authorized for transport as specified in 49 CFR§172.101 and Part 173, (3) physical properties of the materials involved, and (4) atmospheric data from a historical database. The emission model calculated the release of vapor due to evaporation of pools on the ground, direct release of vapors from the container, or a combination of both, as would occur for liquefied gases, which can flash to form both a vapor/aerosol mixture and an evaporating pool. In addition, the emission model also calculated the emission of toxic vapor by-products generated from spilling water-reactive chemicals in water. Spills that involve releases of approximately 200 liters or less are considered small spills, while spills that involve quantities >200 liters are considered large spills.

Downwind dispersion of the vapor was estimated for each case modeled. Atmospheric parameters affecting the dispersion, and the emission rate, were selected in a statistical fashion from a database containing hourly meteorological data from 120 cities in U.S., Canada, and Mexico. The dispersion calculation accounted for the time-dependent emission rate from the source as well as the density of the vapor plume (ie, heavy gas effects). Since atmospheric mixing is less effective at dispersing vapor plumes during nighttime, day and night were separated in the analysis. In the table, "Day" refers to time periods after sunrise and before sunset, while "Night" includes all hours between sunset and sunrise.

Toxicological short-term exposure guidelines for the chemicals were applied to determine the downwind distance to which persons may become incapacitated and unable to take protective action or may incur serious health effects. Toxicological exposure guidelines were chosen from (1) emergency response guidelines, (2) occupational health guidelines, or (3) lethal concentrations determined from animal studies, as recommended by an independent panel of toxicological experts from industry and academia.

How to Use the Table of Initial Isolation and Protective Action Distances

1. The responder should already have:
   - identified the material by its ID number and name; (if an ID number cannot be found, use the name of material index in the blue-bordered pages [of the 2000 Emergency Response Guidebook] to locate that number).
   - found the 3-digit guide for that material in order to consult the emergency actions recommended jointly with this table
   - noted the wind direction
2. Look in this table for the ID number and name of the material involved in the incident. Some ID numbers have more than one shipping name listed - look for the specific name of the material. (If the shipping name is not known and the table lists more than one name for the same ID number, use the entry with the largest protective action distances.)
3. Determine if the incident involves a SMALL or LARGE spill and if DAY or NIGHT. Generally, a small spill is one that involves a single, small package (eg, a drum containing up to approximately 200 liters), a small cylinder, or a small leak from a large package. A large spill is one that involves a spill from a large package, or multiple spills from many small packages. Day is any time after sunrise and before sunset. Night is any time between sunset and sunrise.
4. Look up the initial ISOLATION distance. Direct all persons to move, in a crosswind direction, away from the spill to the specified distance - in meters or feet.

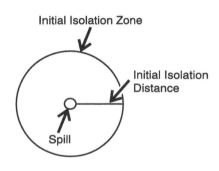

Initial Isolation Zone

Initial Isolation Distance

Spill

5. Look up the initial PROTECTIVE ACTION DISTANCE shown in the Table. For a given dangerous goods, spill size, and whether day or night, the Table gives the downwind distance - in kilometers and miles - for which protective actions should be considered. For practical purposes, the Protective Action Zone (ie, the area in which people are at risk of harmful exposure) is a square, whose length and width are the same as the downwind distance shown in the Table.

6. Initiate protective actions to the extent possible, beginning with those closest to the spill site and working away from the site in the downwind direction. When a water-reactive TIH-producing material is spilled into a river or stream, the source of the toxic gas may move with the current or stretch from the spill point downstream for a substantial distance.

The shape of the area in which protective actions should be taken (the Protective Action Zone) is shown in this figure. The spill is located at the center of the small circle. The larger circle represents the INITIAL ISOLATION Zone around the spill.

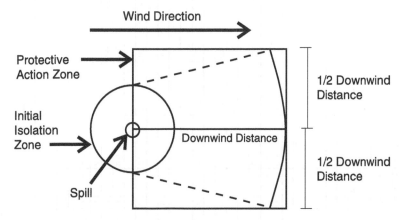

Note: See Introduction to the Table of Initial Isolation and Protective Action Distances for factors that may increase or decrease Protective Action Distances.

Call the emergency response telephone number listed on the shipping paper, or the appropriate response agency as soon as possible for additional information on the material, safety precautions, and mitigation procedures.

Table of Initial Isolation and Protective Action Distances

ID #	Name of Material	SMALL SPILLS (From a small package or small leak from a large package)						LARGE SPILLS (From a large package or from many small packages)					
		First ISOLATE in all directions		Then, PROTECT persons downwind during -				First ISOLATE in all directions		Then, PROTECT persons downwind during -			
				DAY		NIGHT				DAY		NIGHT	
		m	ft	km	mi	km	mi	m	ft	km	mi	km	mi
1005	Ammonia, anhydrous	30	100	0.2	0.1	0.2	0.1	60	200	0.5	0.3	1.1	0.7
	Ammonia, anhydrous, liquefied												
	Ammonia, solution, with >50% ammonia												
	Anhydrous ammonia												
	Anhydrous ammonia, liquefied												
1008	Boron trifluoride	30	100	0.2	0.1	0.6	0.4	215	700	1.6	1.0	5.1	3.2
	Boron trifluoride, compressed												
1016	Carbon monoxide	30	100	0.2	0.1	0.2	0.1	125	400	0.6	0.4	1.8	1.1
	Carbon monoxide, compressed												
1017	Chlorine	30	100	0.3	0.2	1.1	0.7	275	900	2.7	1.7	6.8	4.2
1023	Coal gas	30	100	0.2	0.1	0.2	0.1	60	200	0.3	0.2	0.5	0.3
	Coal gas, compressed												
1026	Cyanogen	30	100	0.3	0.2	1.1	0.7	305	1000	3.1	1.9	7.7	4.8
	Cyanogen, liquefied												
	Cyanogen gas												
1040	Ethylene oxide	30	100	0.2	0.1	0.2	0.1	60	200	0.5	0.3	1.8	1.1
	Ethylene oxide with nitrogen												
1045	Fluorine	30	100	0.2	0.1	0.5	0.3	185	600	1.4	0.9	4.0	2.5
	Fluorine, compressed												
1048	Hydrogen bromide, anhydrous	30	100	0.2	0.1	0.5	0.3	125	400	1.1	0.7	3.4	2.1
1050	Hydrogen chloride, anhydrous	30	100	0.2	0.1	0.6	0.4	185	600	1.6	1.0	4.3	2.7
1051	AC (when used as a weapon)	60	200	0.2	0.1	0.5	0.3	460	1500	1.6	1.0	3.9	2.4
1051	Hydrocyanic acid, aqueous solutions, with >20% hydrogen cyanide	60	200	0.2	0.1	0.5	0.3	400	1300	1.3	0.8	3.4	2.1
	Hydrocyanic acid, liquefied												
	Hydrogen cyanide, anhydrous, stabilized												
	Hydrogen cyanide, stabilized												
1052	Hydrogen fluoride, anhydrous	30	100	0.2	0.1	0.6	0.4	125	400	1.1	0.7	2.9	1.8
1053	Hydrogen sulfide	30	100	0.2	0.1	0.3	0.2	215	700	1.4	0.9	4.3	2.7
	Hydrogen sulfide, liquefied												
	Hydrogen sulphide												
	Hydrogen sulphide, liquefied												

Table (Continued)

ID #	Name of Material	SMALL SPILLS (From a small package or small leak from a large package)						LARGE SPILLS (From a large package or from many small packages)					
		First ISOLATE in all directions		Then, PROTECT persons downwind during -				First ISOLATE in all directions		Then, PROTECT persons downwind during -			
				DAY		NIGHT				DAY		NIGHT	
		m	ft	km	mi	km	mi	m	ft	km	mi	km	mi
1062	Methyl bromide	30	100	0.2	0.1	0.3	0.2	95	300	0.5	0.3	1.4	0.9
1064	Methyl mercaptan	30	100	0.2	0.1	0.3	0.2	95	300	0.8	0.5	2.7	1.7
1067	Dinitrogen tetronide	30	100	0.2	0.1	0.5	0.3	305	1000	1.3	0.8	3.9	2.4
	Dinitrogen tetroxide, liquefied												
	Nitrogen dioxide												
	Nitrogen dioxide, liquefied												
	Nitrogen peroxide, liquid												
	Nitrogen tetroxide, liquid												
1069	Nitrosyl chloride	30	100	0.3	0.2	1.4	0.9	365	1200	3.5	2.2	9.8	6.1
1071	Oil gas	30	100	0.2	0.1	0.2	0.1	30	100	0.3	0.2	0.5	0.3
	Oil gas, compressed												
1076	CG (when used as a weapon)	155	500	1.3	0.8	3.2	2.0	765	2500	7.2	4.5	11.0+	7.0+
1076	Diphosgene	60	200	0.2	0.1	0.5	0.3	95	300	1.0	0.6	1.9	1.2
1076	DP (when used as a weapon)	60	200	0.3	0.2	1.0	0.6	185	600	1.6	1.0	4.5	2.8
1076	Phosgene	95	300	0.8	0.5	2.7	1.7	765	2500	6.6	4.1	11.0	6.9
1079	Sulfur dioxide	30	100	0.3	0.2	1.1	0.7	185	600	3.1	1.9	7.2	4.5
	Sulfur dioxide, liquefied												
	Sulphur dioxide												
	Sulphur dioxide, liquefied												
1082	Trifluorochloroethylene	30	100	0.2	0.1	0.2	0.1	30	100	0.3	0.2	0.8	0.5
	Trifluorochloroethylene, inhibited												
1092	Acrolein, inhibited	60	200	0.5	0.3	1.6	1.0	400	1300	3.9	2.4	7.9	4.9
1098	Allyl alcohol	30	100	0.2	0.1	0.2	0.1	30	100	0.3	0.2	0.6	0.4
1135	Ethylene chlorohydrin	30	100	0.2	0.1	0.3	0.2	60	200	0.6	0.4	1.3	0.8
1143	Crotonaldehyde, inhibited	30	100	0.2	0.1	0.2	0.1	30	100	0.3	0.2	0.8	0.5
	Crotonaldehyde, stabilized												
1162	Dimethyldichlorosilane (when spilled in water)	30	100	0.2	0.1	0.3	0.2	125	400	1.1	0.7	2.9	1.8
1163	1,1-Dimethylhydrazine	30	100	0.2	0.1	0.2	0.1	60	200	0.5	0.3	1.1	0.7
	Dimethylhydrazine, unsymmetrical												
1182	Ethyl chloroformate	30	100	0.2	0.1	0.3	0.2	60	200	0.6	0.4	1.4	0.9
1185	Ethyleneimine, inhibited	30	100	0.3	0.2	0.8	0.5	155	500	1.4	0.9	3.5	2.2
1238	Methyl chloroformate	30	100	0.3	0.2	1.1	0.7	155	500	1.6	1.0	3.4	2.1
1239	Methyl chloromethyl ether	30	100	0.2	0.1	0.6	0.4	125	400	1.1	0.7	2.7	1.7

Table (Continued)

ID #	Name of Material	SMALL SPILLS (From a small package or small leak from a large package)						LARGE SPILLS (From a large package or from many small packages)						
		First ISOLATE in all directions		Then, PROTECT persons downwind during -				First ISOLATE in all directions		Then, PROTECT persons downwind during -			
				DAY		NIGHT				DAY		NIGHT	
		m	ft	km	mi	km	mi	m	ft	km	mi	km	mi
1242	Methyldichlorosilane (when spilled in water)	30	100	0.2	0.1	0.2	0.1	60	200	0.5	0.3	1.6	1.0
1244	Methylhydrazine	30	100	0.3	0.2	0.8	0.5	125	400	1.1	0.7	2.7	1.7
1250	Methyltrichlorosilane (when spilled in water)	30	100	0.2	0.1	0.3	0.2	125	400	1.1	0.7	2.9	1.8
1251	Methyl vinyl ketone	155	500	1.3	0.8	3.4	2.1	915	3000	8.7	5.4	11.0+	7.0+
	Methyl vinyl ketone, stabilized												
1259	Nickel carbonyl	60	200	0.6	0.4	2.1	1.3	215	700	2.1	1.3	4.3	2.7
1295	Trichlorosilane (when spilled in water)	30	100	0.2	0.1	0.3	0.2	125	400	1.3	0.8	3.2	2.0
1298	Trimethylchlorosilane (when spilled in water)	30	100	0.2	0.1	0.2	0.1	95	300	0.8	0.5	2.3	1.4
1340	Phosphorus pentasulfide, free from yellow or white phosphorus (when spilled in water)	30	100	0.2	0.1	0.5	0.3	155	500	1.3	0.8	3.2	2.0
	Phosphorus pentasulphide, free from yellow or white phosphorus (when spilled in water)												
1360	Calcium phosphide (when spilled in water)	30	100	0.2	0.1	0.8	0.5	215	700	2.1	1.3	5.3	3.3
1380	Pentaborane	155	500	1.3	0.8	3.7	2.3	765	2500	6.6	4.1	10.6	6.6
1384	Sodium dithionite (when spilled in water)	30	100	0.2	0.1	0.2	0.1	30	100	0.3	0.2	1.1	0.7
	Sodium hydrosulfite (when spilled in water)												
	Sodium hydrosulphite (when spilled in water)												
1397	Aluminum phosphide (when spilled in water)	30	100	0.2	0.1	0.8	0.5	245	800	2.4	1.5	6.4	4.0
1412	Lithium amide (when spilled in water)	30	100	0.2	0.1	0.2	0.1	95	300	0.8	0.5	1.9	1.2
1419	Magnesium aluminum phosphide (when spilled in water)	30	100	0.2	0.1	0.8	0.5	215	700	2.1	1.3	5.5	3.4
1432	Sodium phosphide (when spilled in water)	30	100	0.2	0.1	0.5	0.3	155	500	1.4	0.9	4.0	2.5

Table (Continued)

		SMALL SPILLS (From a small package or small leak from a large package)						LARGE SPILLS (From a large package or from many small packages)					
		First ISOLATE in all directions		Then, PROTECT persons downwind during -				First ISOLATE in all directions		Then, PROTECT persons downwind during -			
				DAY		NIGHT				DAY		NIGHT	
ID #	Name of Material	m	ft	km	mi	km	mi	m	ft	km	mi	km	mi
1433	Stannic phosphides (when spilled in water)	30	100	0.2	0.1	0.8	0.5	185	600	1.6	1.0	4.7	2.9
1510	Tetranitromethane	30	100	0.3	0.2	0.5	0.3	60	200	0.6	0.4	1.3	0.8
1541	Acetone cyanohydrin, stabilized (when spilled in water)	30	100	0.2	0.1	0.2	0.1	95	300	0.8	0.5	2.1	1.3
1556	MD (when used as a weapon)	30	100	0.3	0.2	0.8	0.5	125	400	1.3	0.8	3.5	2.2
1556	Methyldichloroarsine	30	100	0.2	0.1	0.3	0.2	60	200	0.5	0.3	1.0	0.6
1556	PD (when used as a weapon)	30	100	0.2	0.1	0.2	0.1	30	100	0.2	0.1	0.3	0.2
1560	Arsenic chloride / Arsenic trichloride	30	100	0.2	0.1	0.3	0.2	60	200	0.6	0.4	1.4	0.9
1569	Bromoacetone	30	100	0.2	0.1	0.3	0.2	95	300	0.8	0.5	1.9	1.2
1580	Chloropicrin	60	200	0.5	0.3	1.3	0.8	185	600	1.8	1.1	4.0	2.5
1581	Chloropicrin and methyl bromide mixture / Methyl bromide and chloropicrin mixtures	30	100	0.2	0.1	0.5	0.3	125	400	1.3	0.8	3.1	1.9
1581	Methyl bromide and >2% chloropicrin mixture, liquid	30	100	0.3	0.2	1.1	0.7	215	700	2.1	1.3	5.6	3.5
1582	Chloropicrin and methyl chloride mixture / Methyl chloride and chloropicrin mixtures	30	100	0.2	0.1	0.8	0.5	95	300	1.0	0.6	3.2	2.0
1583	Chloropicrin, absorbed	60	200	0.5	0.3	1.3	0.8	185	600	1.8	1.1	4.0	2.5
1583	Chloropicrin mixture, n.o.s.	30	100	0.3	0.2	1.1	0.7	215	700	2.1	1.3	5.6	3.5
1589	CK (when used as a weapon)	60	200	0.6	0.4	2.4	1.5	400	1300	4.0	2.5	8.0	5.0
1589	Cyanogen chloride, inhibited	60	200	0.5	0.3	1.8	1.1	275	900	2.7	1.7	6.8	4.2
1595	Dimethyl sulfate / Dimethyl sulphate	30	100	0.2	0.1	0.2	0.1	30	100	0.3	0.2	0.6	0.4
1605	Ethylene dibromide	30	100	0.2	0.1	0.2	0.1	30	100	0.3	0.2	0.5	0.3
1612	Hexaethyl tetraphosphate and compressed gas mixture	30	100	0.2	0.1	0.2	0.1	30	100	0.3	0.2	1.4	0.9
1613	Hydrocyanic acid, aqueous solution, with not more than 20% hydrogen cyanide (when "Inhalation Hazard" is on a package or shipping paper)	30	100	0.2	0.1	0.2	0.1	125	400	0.5	0.3	1.3	0.8

INITIAL ISOLATION AND PROTECTIVE ACTION DISTANCES TABLE

Table (Continued)

ID #	Name of Material	SMALL SPILLS (From a small package or small leak from a large package)						LARGE SPILLS (From a large package or from many small packages)					
		First ISOLATE in all directions		Then, PROTECT persons downwind during -				First ISOLATE in all directions		Then, PROTECT persons downwind during -			
				DAY		NIGHT				DAY		NIGHT	
		m	ft	km	mi	km	mi	m	ft	km	mi	km	mi
1614	Hydrogen cyanide, aqueous solution, with not more than 20% hydrogen cyanide (when "Inhalation Hazard" is on a package or shipping paper)												
1614	Hydrogen cyanide, anhydrous, stabilized (absorbed)	60	200	0.2	0.1	0.5	0.3	400	1300	1.3	0.8	3.4	2.1
	Hydrogen cyanide, stabilized (absorbed)												
1647	Ethylene dibromide and methyl bromide mixture, liquid	30	100	0.2	0.1	0.2	0.1	30	100	0.3	0.2	0.5	0.3
	Methyl bromide and ethylene dibromide mixture, liquid												
1660	Nitric oxide	30	100	0.3	0.2	1.3	0.8	155	500	1.3	0.8	3.5	2.2
	Nitric oxide, compressed												
1670	Perchloromethyl mercaptan	30	100	0.2	0.1	0.3	0.2	60	200	0.5	0.3	1.1	0.7
1680	Potassium cyanide (when spilled in water)	30	100	0.2	0.1	0.3	0.2	95	300	0.8	0.5	2.6	1.6
1689	Sodium cyanide (when spilled in water)	30	100	0.2	0.1	0.3	0.2	95	300	1.0	0.6	2.6	1.6
1694	CA (when used as a weapon)	30	100	0.2	0.1	0.5	0.3	155	500	1.6	1.0	4.2	2.6
1695	Chloroacetone, stabilized	30	100	0.2	0.1	0.3	0.2	60	200	0.6	0.4	1.3	0.8
1697	CN (when used as a weapon)	30	100	0.2	0.1	0.5	0.3	125	400	1.1	0.7	3.2	2.0
1698	Adamsite (when used as a weapon)	60	200	0.3	0.2	1.1	0.7	185	600	2.3	1.4	5.1	3.2
	DM (when used as a weapon)												
1699	DA (when used as a weapon)	60	200	0.3	0.2	1.1	0.7	185	600	2.3	1.4	5.1	3.2
1703	Tetraethyl dithiopyrophosphate and gases, in solution	30	100	0.3	0.2	1.1	0.7	365	1200	3.7	2.3	6.9	4.3
	Tetraethyl dithiopyrophosphate and gases, mixtures												
1703	Tetraethyl dithiopyrophosphate and gases, mixtures, or in solution (LC50 more than 200 ppm but not more than 5000 ppm)	30	100	0.2	0.1	0.5	0.3	125	400	0.8	0.5	2.9	1.8

INITIAL ISOLATION AND PROTECTIVE ACTION DISTANCES TABLE

Table (Continued)

ID #	Name of Material	SMALL SPILLS (From a small package or small leak from a large package)						LARGE SPILLS (From a large package or from many small packages)						
		First ISOLATE in all directions		Then, PROTECT persons downwind during -				First ISOLATE in all directions		Then, PROTECT persons downwind during -			
				DAY		NIGHT				DAY		NIGHT	
		m	ft	km	mi	km	mi	m	ft	km	mi	km	mi
1703	Tetraethyl dithiopyrophosphate and gases, mixtures, or in solution (LC50 not more than 200 ppm)	30	100	0.3	0.2	1.1	0.7	365	1200	3.7	2.3	6.9	4.3
1705	Tetraethyl pyrophosphate and compressed gas mixtures	30	100	0.3	0.2	1.3	0.8	400	1300	4.0	2.5	7.2	4.5
1705	Tetraethyl pyrophosphate and compressed gas mixtures (LC50 more than 200 ppm but not more than 5000 ppm)	30	100	0.2	0.1	0.5	0.3	125	400	0.8	0.5	2.9	1.8
1705	Tetraethyl pyrophosphate and compressed gas mixtures (LC50 not more than 200 ppm)	30	100	0.3	0.2	1.3	0.8	400	1300	4.0	2.5	7.2	4.5
1714	Zinc phosphide (when spilled in water)	30	100	0.2	0.1	0.8	0.5	185	600	1.8	1.1	5.1	3.2
1716	Acetyl bromide (when spilled in water)	30	100	0.2	0.1	0.3	0.2	95	300	0.8	0.5	2.3	1.4
1717	Acetyl chloride (when spilled in water)	30	100	0.2	0.1	0.3	0.2	95	300	1.0	0.6	2.7	1.7
1722	Allyl chlorocarbonate / Allyl chloroformate	155	500	1.3	0.8	2.7	1.7	610	2000	6.1	3.8	10.8	6.7
1724	Allyltrichlorosilane, stabilized (when spilled in water)	30	100	0.2	0.1	0.3	0.2	125	400	1.0	0.6	2.9	1.8
1725	Aluminum bromide, anhydrous (when spilled in water)	30	100	0.2	0.1	0.3	0.2	95	300	1.0	0.6	2.7	1.7
1726	Aluminum chloride, anhydrous (when spilled in water)	30	100	0.2	0.1	0.2	0.1	60	200	0.5	0.3	1.6	1.0
1728	Amyltrichlorosilane (when spilled in water)	30	100	0.2	0.1	0.2	0.1	60	200	0.5	0.3	1.6	1.0
1732	Antimony pentafluoride (when spilled in water)	30	100	0.2	0.1	0.6	0.4	155	500	1.6	1.0	3.7	2.3
1736	Benzoyl chloride (when spilled in water)	30	100	0.2	0.1	0.2	0.1	30	100	0.3	0.2	1.1	0.7
1741	Boron trichloride	30	100	0.2	0.1	0.3	0.2	60	200	0.6	0.4	1.6	1.0
1744	Bromine / Bromine, solution	60	200	0.3	0.2	1.1	0.7	185	600	1.6	1.0	4.0	2.5

INITIAL ISOLATION AND PROTECTIVE ACTION DISTANCES TABLE

Table (Continued)

ID #	Name of Material	SMALL SPILLS (From a small package or small leak from a large package)						LARGE SPILLS (From a large package or from many small packages)					
		First ISOLATE in all directions		Then, PROTECT persons downwind during -				First ISOLATE in all directions		Then, PROTECT persons downwind during -			
				DAY		NIGHT				DAY		NIGHT	
		m	ft	km	mi	km	mi	m	ft	km	mi	km	mi
1745	Bromine pentafluoride (when spilled on land)	60	200	0.5	0.3	1.3	0.8	245	800	2.3	1.4	5.0	3.1
1745	Boron pentafluoride (when spilled in water)	30	100	0.2	0.1	0.8	0.5	215	700	1.9	1.2	4.2	2.6
1746	Bromine trifluoride (when spilled on land)	30	100	0.2	0.1	0.3	0.2	60	200	0.3	0.2	0.8	0.5
1746	Bromine trifluoride (when spilled in water)	30	100	0.2	0.1	0.6	0.4	185	600	2.1	1.3	5.5	3.4
1747	Butyltrichlorosilane (when spilled in water)	30	100	0.2	0.1	0.2	0.1	60	200	0.5	0.3	1.8	1.1
1749	Chlorine trifluoride	60	200	0.5	0.3	1.6	1.0	335	1100	3.4	2.1	7.7	4.8
1752	Chloroacetyl chloride (when spilled on land)	30	100	0.2	0.1	0.5	0.3	95	300	0.8	0.5	1.6	1.0
1752	Chloroacetyl chloride (when spilled in water)	30	100	0.2	0.1	0.2	0.1	60	200	0.3	0.2	1.3	0.8
1754	Chlorosulfonic acid (when spilled on land)	30	100	0.2	0.1	0.2	0.1	30	100	0.2	0.1	0.5	0.3
1754	Chlorosulfonic acid (when spilled in water)	30	100	0.2	0.1	0.2	0.1	60	200	0.5	0.3	1.4	0.9
1754	Chlorosulfonic acid and sulfur trioxide mixture (when spilled on land)	60	200	0.3	0.2	1.1	0.7	305	1000	2.1	1.3	5.6	3.5
	Chlorosulfonic acid and sulfur trioxide mixture (when spilled in water)												
1754	Chlorosulphonic acid (when spilled on land)	30	100	0.2	0.1	0.2	0.1	30	100	0.2	0.1	0.5	0.3
1754	Chlorosulphonic acid (when spilled in water)	30	100	0.2	0.1	0.2	0.1	60	200	0.5	0.3	1.4	0.9
1754	Chlorosulphonic acid and sulphur trioxide mixture (when spilled on land)	60	200	0.3	0.2	1.1	0.7	305	1000	2.1	1.3	5.6	3.5
	Chlorosulphonic acid and sulphur trioxide mixture (when spilled in water)												

Table   (Continued)

ID #	Name of Material	SMALL SPILLS (From a small package or small leak from a large package)						LARGE SPILLS (From a large package or from many small packages)					
		First ISOLATE in all directions		Then, PROTECT persons downwind during -				First ISOLATE in all directions		Then, PROTECT persons downwind during -			
				DAY		NIGHT				DAY		NIGHT	
		m	ft	km	mi	km	mi	m	ft	km	mi	km	mi
	Sulphur trioxide and chlorosulfonic acid mixture (when spilled on land)												
	Sulphur trioxide and chlorosulfonic acid mixture (when spilled in water)												
	Sulphur trioxide and chlorosulphonic acid mixture (when spilled on land)												
	Sulphur trioxide and chlorosulphonic acid mixture (when spilled in water)												
1758	Chromium oxychloride	30	100	0.2	0.1	0.2	0.1	60	200	0.3	0.2	1.3	0.8
1777	Fluorosulfonic acid (when spilled in water)	30	100	0.2	0.1	0.2	0.1	60	200	0.5	0.3	1.4	0.9
	Fluorosulphonic acid (when spilled in water)												
1801	Octyltrichlorosilane (when spilled in water)	30	100	0.2	0.1	0.3	0.2	95	300	0.8	0.5	2.4	1.5
1806	Phosphorus pentachloride (when spilled in water)	30	100	0.2	0.1	0.3	0.2	125	400	1.0	0.6	2.9	1.8
1809	Phosphorus trichloride (when spilled on land)	30	100	0.2	0.1	0.6	0.4	125	400	1.1	0.7	2.7	1.7
1809	Phosphorus trichloride (when spilled in water)	30	100	0.2	0.1	0.3	0.2	125	400	1.1	0.7	2.6	1.6
1810	Phosphorus oxychloride (when spilled on land)	30	100	0.2	0.1	0.5	0.3	95	300	0.8	0.5	1.8	1.1
1810	Phosphorus oxychloride (when spilled in water)	30	100	0.2	0.1	0.3	0.2	95	300	1.0	0.6	2.6	1.6
1818	Silicon tetrachloride (when spilled in water)	30	100	0.2	0.1	0.3	0.2	125	400	1.3	0.8	3.4	2.1
1828	Sulfur chlorides (when spilled on land)	30	100	0.2	0.1	0.3	0.2	60	200	0.5	0.3	1.0	0.6
1828	Sulphur chlorides (when spilled in water)	30	100	0.2	0.1	0.2	0.1	60	200	0.6	0.4	2.3	1.4

591

Table (Continued)

ID #	Name of Material	SMALL SPILLS (From a small package or small leak from a large package)						LARGE SPILLS (From a large package or from many small packages)						
		First ISOLATE in all directions		Then, PROTECT persons downwind during -				First ISOLATE in all directions		Then, PROTECT persons downwind during -			
				DAY		NIGHT				DAY		NIGHT	
		m	ft	km	mi	km	mi	m	ft	km	mi	km	mi
1828	Sulphur chlorides (when spilled on land)	30	100	0.2	0.1	0.3	0.2	60	200	0.5	0.3	1.0	0.6
1828	Sulphur chlorides (when spilled in water)	30	100	0.2	0.1	0.2	0.1	60	200	0.6	0.4	2.3	1.4
1829	Sulfur trioxide	60	200	0.3	0.2	1.1	0.7	305	1000	2.1	1.3	5.6	3.5
	Sulfur trioxide, inhibited												
	Sulfur trioxide, stabilized												
	Sulfur trioxide, uninhibited												
	Sulphur trioxide												
	Sulphur trioxide, inhibited												
	Sulphur trioxide, stabilized												
	Sulphur trioxide, uninhibited												
1831	Oleum	60	200	0.3	0.2	1.1	0.7	305	1000	2.1	1.3	5.6	3.5
	Oleum, with not less than 30% free sulfur trioxide												
	Oleum, with not less than 30% free sulphur trioxide												
	Sulfuric acid, fuming												
	Sulfuric acid, fuming, with not less than 30% free sulfur trioxide												
	Sulphuric acid, fuming												
	Sulphuric acid, fuming, with not less than 30% free sulphur trioxide												
1834	Sulfuryl chloride (when spilled on land)	30	100	0.2	0.1	0.2	0.1	30	100	0.3	0.2	0.6	0.4
1834	Sulfuryl chloride (when spilled in water)	30	100	0.2	0.1	0.2	0.1	125	400	1.1	0.7	2.4	1.5
1834	Sulphuryl chloride (when spilled on land)	30	100	0.2	0.1	0.2	0.1	30	100	0.3	0.2	0.6	0.4
1834	Sulphuryl chloride (when spilled in water)	30	100	0.2	0.1	0.2	0.1	125	400	1.1	0.7	2.4	1.5
1836	Thionyl chloride (when spilled on land)	30	100	0.2	0.1	0.5	0.3	60	200	0.5	0.3	1.1	0.7
1836	Thionyl chloride (when spilled in water)	30	100	0.2	0.1	1.0	0.6	335	1100	3.2	2.0	7.1	4.4

Table (Continued)

ID #	Name of Material	SMALL SPILLS (From a small package or small leak from a large package)						LARGE SPILLS (From a large package or from many small packages)						
		First ISOLATE in all directions		Then, PROTECT persons downwind during -				First ISOLATE in all directions		Then, PROTECT persons downwind during -			
				DAY		NIGHT				DAY		NIGHT	
		m	ft	km	mi	km	mi	m	ft	km	mi	km	mi
1838	Titanium tetrachloride (when spilled on land)	30	100	0.2	0.1	0.2	0.1	30	100	0.3	0.2	0.8	0.5
1838	Titanium tetrachloride (when spilled in water)	30	100	0.2	0.1	0.3	0.2	125	400	1.1	0.7	2.9	1.8
1859	Silicon tetrafluoride	30	100	0.2	0.1	0.5	0.3	60	200	0.5	0.3	1.6	1.0
	Silicon tetrafluoride, compressed												
1892	ED (when used as a weapon)	30	100	0.3	0.2	0.8	0.5	125	400	1.3	0.8	2.6	1.6
1892	Ethyldichloroarsine	30	100	0.2	0.1	0.3	0.2	60	200	0.5	0.3	1.0	0.6
1898	Acetyl iodide (when spilled in water)	30	100	0.2	0.1	0.2	0.1	60	200	0.6	0.4	1.6	1.0
1911	Diborane	30	100	0.2	0.1	0.3	0.2	95	300	1.0	0.6	2.7	1.7
	Diborane, compressed												
1923	Calcium dithionite (when spilled in water)	30	100	0.2	0.1	0.2	0.1	30	100	0.3	0.2	1.1	0.7
	Calcium hydrosulfite (when spilled in water)												
	Calcium hydrosulphite (when spilled in water)												
1939	Phosphorus oxybromide (when spilled in water)	30	100	0.2	0.1	0.3	0.2	95	300	0.6	0.4	1.9	1.2
	Phosphorus oxybromide, solid (when spilled in water)												
1953	Compressed gas, flammable, poisonous, n.o.s. (Inhalation Hazard Zone A)	185	600	1.8	1.1	5.6	3.5	915	3000	10.8	6.7	11.0+	7.0+
1953	Compressed gas, flammable, poisonous, n.o.s. (Inhalation Hazard Zone B)	30	100	0.3	0.2	1.1	0.7	305	1000	3.1	1.9	7.7	4.8+
1953	Compressed gas, flammable, poisonous, n.o.s. (Inhalation Hazard Zone C)	30	100	0.2	0.1	1.0	0.6	215	700	2.1	1.3	5.6	3.5
1953	Compressed gas, flammable, poisonous, n.o.s. (Inhalation Hazard Zone D)	30	100	0.2	0.1	0.6	0.4	185	600	1.6	1.0	4.3	2.7

INITIAL ISOLATION AND PROTECTIVE ACTION DISTANCES TABLE

Table (Continued)

ID #	Name of Material	SMALL SPILLS (From a small package or small leak from a large package)						LARGE SPILLS (From a large package or from many small packages)					
		First ISOLATE in all directions		Then, PROTECT persons downwind during -				First ISOLATE in all directions		Then, PROTECT persons downwind during -			
				DAY		NIGHT				DAY		NIGHT	
		m	ft	km	mi	km	mi	m	ft	km	mi	km	mi
1953	Compressed gas, flammable, toxic, n.o.s. (Inhalation Hazard Zone A)	185	600	1.8	1.1	5.6	3.5	915	3000	10.8	6.7	11.0+	7.0+
1953	Compressed gas, flammable, toxic, n.o.s. (Inhalation Hazard Zone B)	30	100	0.3	0.2	1.1	0.7	305	1000	3.1	1.9	7.7	4.8
1953	Compressed gas, flammable, toxic, n.o.s. (Inhalation Hazard Zone C)	30	100	0.2	0.1	1.0	0.6	215	700	2.1	1.3	5.6	3.5
1953	Compressed gas, flammable, toxic, n.o.s. (Inhalation Hazard Zone D)	30	100	0.2	0.1	0.6	0.4	185	600	1.6	1.0	4.3	2.7
1953	Compressed gas, poisonous, flammable, n.o.s.	185	600	1.8	1.1	5.6	3.5	915	3000	10.8	6.7	11.0+	7.0+
1953	Compressed gas, poisonous, flammable, n.o.s. (Inhalation Hazard Zone A)												
1953	Compressed gas, poisonous, flammable, n.o.s. (Inhalation Hazard Zone B)	30	100	0.3	0.2	1.1	0.7	305	1000	3.1	1.9	7.7	4.8
1953	Compressed gas, poisonous, flammable, n.o.s. (Inhalaton Hazard Zone C)	30	100	0.2	0.1	1.0	0.6	215	700	2.1	1.3	5.6	3.5
1953	Compressed gas, poisonous, flammable, n.o.s. (Inhalation Hazard Zone D)	30	100	0.2	0.1	0.6	0.4	185	600	1.6	1.0	4.3	2.7
1953	Compressed gas, toxic, flammable, n.o.s.	185	600	1.8	1.1	5.6	3.5	915	3000	10.8	6.7	11.0+	7.0+
1953	Compressed gas, toxic, flammable, n.o.s. (Inhalation Hazard Zone A)												
1953	Compressed gas, toxic, flammable, n.o.s. (Inhalation Hazard Zone B)	30	100	0.3	0.2	1.1	0.7	305	1000	3.1	1.9	7.7	4.8
1953	Compressed gas, toxic, flammable, n.o.s. (Inhalation Hazard Zone C)	30	100	0.2	0.1	1.0	0.6	215	700	2.1	1.3	5.6	3.5
1953	Compressed gas, toxic, flammable, n.o.s. (Inhalation Hazard Zone D)	30	100	0.2	0.1	0.6	0.4	185	600	1.6	1.0	4.3	2.7
1953	Liquefied gas, flammable, poisonous, n.o.s.	185	600	1.8	1.1	5.6	3.5	915	3000	10.8	6.7	11.0+	7.0+

Table (Continued)

ID #	Name of Material	SMALL SPILLS (From a small package or small leak from a large package) — First ISOLATE in all directions — m	ft	Then, PROTECT persons downwind during - DAY km	DAY mi	NIGHT km	NIGHT mi	LARGE SPILLS (From a large package or from many small packages) — First ISOLATE in all directions — m	ft	Then, PROTECT persons downwind during - DAY km	DAY mi	NIGHT km	NIGHT mi
1953	Liquefied gas, flammable, poisonous, n.o.s. (Inhalation Hazard Zone A)	30	100	0.3	0.2	1.1	0.7	305	1000	3.1	1.9	7.7	4.8
1953	Liquefied gas, flammable, poisonous, n.o.s. (Inhalation Hazard Zone B)	30	100	0.2	0.1	1.0	0.6	215	700	2.1	1.3	5.6	3.5
1953	Liquefied gas, flammable, poisonous, n.o.s. (Inhalation Hazard Zone C)	30	100	0.2	0.1	0.6	0.4	185	600	1.6	1.0	4.3	2.7
1953	Liquefied gas, flammable, poisonous, n.o.s. (Inhalation Hazard Zone D)	185	600	1.8	1.1	5.6	3.5	915	3000	10.8	6.7	11.0+	7.0+
1953	Liquefied gas, flammable, toxic, n.o.s. (Inhalation Hazard Zone A)	30	100	0.3	0.2	1.1	0.7	305	1000	3.1	1.9	7.7	4.8
1953	Liquefied gas, flammable, toxic, n.o.s. (Inhalation Hazard Zone B)	30	100	0.2	0.1	1.0	0.6	215	700	2.1	1.3	5.6	3.5
1953	Liquefied gas, flammable, toxic, n.o.s. (Inhalation Hazard Zone C)	30	100	0.2	0.1	0.6	0.4	185	600	1.6	1.0	4.3	2.7
1953	Liquefied gas, flammable, toxic, n.o.s. (Inhalation Hazard Zone D)	185	600	1.8	1.1	5.6	3.5	915	3000	10.8	6.7	11.0+	7.0+
1953	Poisonous gas, flammable, n.o.s.	185	600	1.8	1.1	5.6	3.5	915	3000	10.8	6.7	11.0+	7.0+
1953	Poisonous liquid, flammable, n.o.s.	155	500	1.3	0.8	3.4	2.1	915	3000	8.7	5.4	11.0+	7.0+
1955	Compressed gas, poisonous, n.o.s.	430	1400	4.2	2.6	8.4	5.2	915	3000	11.0+	7.0+	11.0+	7.0+
1955	Compressed gas, poisonous, n.o.s. (Inhalation Hazard Zone A)	60	200	0.5	0.3	1.6	1.0	430	1400	4.0	2.5	9.8	6.1
1955	Compressed gas, poisonous, n.o.s. (Inhalation Hazard Zone B)	30	100	0.3	0.2	1.3	0.8	215	700	3.1	1.9	7.2	4.5
1955	Compressed gas, poisonous, n.o.s. (Inhalation Hazard Zone C)	30	100	0.2	0.1	0.6	0.4	185	600	1.6	1.0	4.3	2.7
1955	Compressed gas, poisonous, n.o.s. (Inhalation Hazard Zone D)	430	1400	4.2	2.6	8.4	5.2	915	3000	11.0+	7.0+	11.0+	7.0+
1955	Compressed gas, toxic, n.o.s.	430	1400	4.2	2.6	8.4	5.2	915	3000	11.0+	7.0+	11.0+	7.0+
1955	Compressed gas, toxic, n.o.s. (Inhalation Hazard Zone A)	60	200	0.5	0.3	1.6	1.0	430	1400	4.0	2.5	9.8	6.1
1955	Compressed gas, toxic, n.o.s. (Inhalation Hazard Zone B)	30	100	0.3	0.2	1.3	0.8	215	700	3.1	1.9	7.2	4.5
1955	Compressed gas, toxic, n.o.s. (Inhalation Hazard Zone C)	30	100	0.2	0.1	0.6	0.4	185	600	1.6	1.0	4.3	2.7

Table (Continued)

ID #	Name of Material	SMALL SPILLS (From a small package or small leak from a large package)						LARGE SPILLS (From a large package or from many small packages)					
		First ISOLATE in all directions		Then, PROTECT persons downwind during -				First ISOLATE in all directions		Then, PROTECT persons downwind during -			
				DAY		NIGHT				DAY		NIGHT	
		m	ft	km	mi	km	mi	m	ft	km	mi	km	mi
1955	Compressed gas, toxic, n.o.s. (Inhalation Hazard Zone D)	30	100	0.2	0.1	0.6	0.4	185	600	1.6	1.0	4.3	2.7
1955	Liquefied gas, poisonous, n.o.s.	430	1400	4.2	2.6	8.4	5.2	915	3000	11.0+	7.0+	11.0+	7.0+
1955	Liquefied gas, poisonous, n.o.s. (Inhalation Hazard Zone A)												
1955	Liquefied gas, poisonous, n.o.s. (Inhalation Hazard Zone B)	60	200	0.5	0.3	1.6	1.0	430	1400	4.0	2.5	9.8	6.1
1955	Liquefied gas, poisonous, n.o.s. (Inhalation Hazard Zone C)	30	100	0.3	0.2	1.3	0.8	215	700	3.1	1.9	7.2	4.5
1955	Liquefied gas, poisonous, n.o.s. (Inhalation Hazard Zone D)	30	100	0.2	0.1	0.6	0.4	185	600	1.6	1.0	4.3	2.7
1955	Liquefied gas, toxic, n.o.s.	430	1400	4.2	2.6	8.4	5.2	915	3000	11.0+	7.0+	11.0+	7.0+
1955	Liquefied gas, toxic, n.o.s. (Inhalation Hazard Zone A)												
1955	Liquefied gas, toxic, n.o.s. (Inhalation Hazard Zone B)	60	200	0.5	0.3	1.6	1.0	430	1400	4.0	2.5	9.8	6.1
1955	Liquefied gas, toxic, n.o.s. (Inhalation Hazard Zone C)	30	100	0.3	0.2	1.3	0.8	215	700	3.1	1.9	7.2	4.5
1955	Liquefied gas, toxic, n.o.s. (Inhalation Hazard Zone D)	30	100	0.2	0.1	0.6	0.4	185	600	1.6	1.0	4.3	2.7
1955	Methyl bromide and nonflammable, nonliquefied compressed gas mixture	30	100	0.2	0.1	0.3	0.2	95	300	0.5	0.3	1.4	0.9
1955	Organic phosphate compound mixed with compressed gas	30	100	0.3	0.2	1.3	0.8	400	1300	4.0	2.5	7.2	4.5
	Organic phosphate mixed with compressed gas												
	Organic phosphorus compound mixed with compressed gas												
1967	Insecticide gas, poisonous, n.o.s.	30	100	0.3	0.2	1.3	0.8	400	1300	4.0	2.5	7.2	4.5
	Insecticide gas, toxic, n.o.s.												
1967	Parathion and compressed gas mixture	30	100	0.2	0.1	0.3	0.2	95	300	1.0	0.6	3.2	2.0
1975	Dinitrogen tetroxide and nitric oxide mixture	30	100	0.3	0.2	1.3	0.8	155	500	1.3	0.8	3.5	2.2

Table (Continued)

ID #	Name of Material	SMALL SPILLS (From a small package or small leak from a large package)						LARGE SPILLS (From a large package or from many small packages)					
		First ISOLATE in all directions		Then, PROTECT persons downwind during -				First ISOLATE in all directions		Then, PROTECT persons downwind during -			
				DAY		NIGHT				DAY		NIGHT	
		m	ft	km	mi	km	mi	m	ft	km	mi	km	mi
	Nitric oxide and dinitrogen tetroxide mixture												
	Nitric oxide and nitrogen dioxide mixture												
	Nitric oxide and nitrogen tetroxide mixture												
	Nitrogen dioxide and nitric oxide mixture												
	Nitrogen tetroxide and nitric oxide mixture												
1994	Iron pentacarbonyl	30	100	0.3	0.2	0.6	0.4	125	400	1.1	0.7	2.4	1.5
2004	Magnesium diamide (when spilled in water)	30	100	0.2	0.1	0.2	0.1	60	200	0.5	0.3	1.3	0.8
2011	Magnesium phosphide (when spilled in water)	30	100	0.2	0.1	0.8	0.5	245	800	2.3	1.4	6.0	3.7
2012	Potassium phosphide (when spilled in water)	30	100	0.2	0.1	0.5	0.3	155	500	1.3	0.8	4.0	2.5
2013	Strontium phosphide (when spilled in water)	30	100	0.2	0.1	0.5	0.3	155	500	1.3	0.8	3.7	2.3
2032	Nitric acid, fuming	95	300	0.3	0.2	0.5	0.3	400	1300	1.3	0.8	3.5	2.2
	Nitric acid, red fuming												
2186	Hydrogen chloride, refrigerated liquid	30	100	0.2	0.1	0.6	0.4	185	600	1.6	1.0	4.3	2.7
2188	Arsine	60	200	0.5	0.3	2.1	1.3	335	1100	3.2	2.0	6.6	4.1
2188	SA (when used as a weapon)	60	200	0.8	0.5	2.4	1.5	400	1300	4.0	2.5	8.0	5.0
2189	Dichlorosilane	30	100	0.3	0.2	1.0	0.6	245	800	2.4	1.5	6.3	3.9
2190	Oxygen difluoride	430	1400	4.2	2.6	8.4	5.2	915	3000	11.0+	7.0+	11.0+	7.0+
	Oxygen difluoride, compressed												
2191	Sulfuryl fluoride	30	100	0.2	0.1	0.3	0.2	95	300	0.8	0.5	2.3	1.4
	Sulphuryl fluoride												
2192	Germane	30	100	0.2	0.1	0.8	0.5	275	900	2.7	1.7	6.6	4.1
2194	Selenium hexafluoride	30	100	0.3	0.2	1.3	0.8	245	800	2.3	1.4	6.0	3.7
2195	Tellurium hexafluoride	60	200	0.6	0.4	2.3	1.4	365	1200	3.5	2.2	7.6	4.7
2196	Tungsten hexafluoride	30	100	0.3	0.2	1.3	0.8	155	500	1.3	0.8	3.7	2.3
2197	Hydrogen iodide, anhydrous	30	100	0.2	0.1	0.5	0.3	95	300	0.8	0.5	2.6	1.6
2198	Phosphorus pentafluoride	30	100	0.3	0.2	1.1	0.7	125	400	1.1	0.7	3.5	2.2

Table (Continued)

ID #	Name of Material	SMALL SPILLS (From a small package or small leak from a large package)						LARGE SPILLS (From a large package or from many small packages)					
		First ISOLATE in all directions		Then, PROTECT persons downwind during -				First ISOLATE in all directions		Then, PROTECT persons downwind during -			
				DAY		NIGHT				DAY		NIGHT	
		m	ft	km	mi	km	mi	m	ft	km	mi	km	mi
	Phosphorus pentafluoride, compressed												
2199	Phosphine	95	300	0.3	0.2	1.3	0.8	490	1600	1.8	1.1	5.5	3.4
2202	Hydrogen selenide, anhydrous	185	600	1.8	1.1	5.6	3.5	915	3000	10.8	6.7	11.0+	7.0+
2204	Carbonyl sulfide	30	100	0.2	0.1	0.6	0.4	215	700	1.9	1.2	5.6	3.5
	Carbonyl sulphide												
2232	Chloroacetaldehyde	30	100	0.2	0.1	0.5	0.3	60	200	0.6	0.4	1.6	1.0
	2-Chloroethanol												
2334	Allylamine	30	100	0.2	0.1	0.5	0.3	95	300	1.0	0.8	2.4	1.5
2337	Phenyl mercaptan	30	100	0.2	0.1	0.2	0.1	30	100	0.3	0.2	0.6	0.4
2382	1,2-Dimethylhydrazine	30	100	0.2	0.1	0.3	0.2	60	200	0.5	0.3	1.1	0.7
	Dimethylhydrazine, symmetrical												
2407	Isopropyl chloroformate	30	100	0.2	0.1	0.3	0.2	95	300	0.8	0.5	1.9	1.2
2417	Carbonyl fluoride	30	100	0.2	0.1	1.1	0.7	125	400	1.0	0.6	3.1	1.9
	Carbonyl fluoride, compressed												
2418	Sulfur tetrafluoride	60	200	0.5	0.3	1.9	1.2	305	1000	2.9	1.8	6.9	4.3
	Sulphur tetrafluoride												
2420	Hexafluoroacetone	30	100	0.3	0.2	1.4	0.9	365	1200	3.7	2.3	8.5	5.3
2421	Nitrogen trioxide	30	100	0.2	0.1	0.2	0.1	155	500	0.6	0.4	2.1	1.3
2438	Trimethylacetyl chloride	30	100	0.2	0.1	0.2	0.1	30	100	0.3	0.2	0.8	0.5
2442	Trichloroacetyl chloride (when spilled on land)	30	100	0.2	0.1	0.3	0.2	60	200	0.6	0.4	1.4	0.9
2442	Trichloroacetyl chloride (when spilled in water)	30	100	0.2	0.1	0.2	0.1	30	100	0.3	0.2	1.3	0.8
2474	Thiophosgene	60	200	0.6	0.4	1.8	1.1	275	900	2.6	1.6	5.0	3.1
2477	Methyl isothiocyanate	30	100	0.2	0.1	0.3	0.2	60	200	0.5	0.3	1.1	0.7
2480	Methyl isocyanate	95	300	0.8	0.5	2.7	1.7	490	1600	4.8	3.0	9.8	6.1
2481	Ethyl isocyanate	215	700	1.9	1.2	4.3	2.7	915	3000	11.0+	7.0+	11.0+	7.0+
2482	n-Propyl isocyanate	125	400	1.1	0.7	2.4	1.5	765	2500	6.3	3.9	10.6	6.6
2483	Isopropyl isocyanate	185	600	1.8	1.1	3.9	2.4	430	1400	4.2	2.6	7.4	4.6
2484	tert-Butyl isocyanate	125	400	1.0	0.6	2.4	1.5	550	1800	5.3	3.3	10.3	6.4
2485	n-Butyl isocyanate	95	300	0.8	0.5	1.6	1.0	335	1100	3.1	1.9	6.3	3.9
2486	Isobutyl isocyanate	60	200	0.6	0.4	1.4	0.9	155	500	1.6	1.0	3.2	2.0
2487	Phenyl isocyanate	30	100	0.3	0.2	0.8	0.5	155	500	1.3	0.8	2.6	1.6
2488	Cyclohexyl isocyanate	30	100	0.2	0.1	0.3	0.2	95	300	0.8	0.5	1.4	0.9

Table (Continued)

ID #	Name of Material	SMALL SPILLS (From a small package or small leak from a large package)							LARGE SPILLS (From a large package or from many small packages)							
		First ISOLATE in all directions		Then, PROTECT persons downwind during -					First ISOLATE in all directions		Then, PROTECT persons downwind during -					
				DAY		NIGHT					DAY		NIGHT			
		m	ft	km	mi	km	mi		m	ft	km	mi	km	mi		
2495	Iodine pentafluoride (when spilled in water)	30	100	0.2	0.1	0.5	0.3		125	400	1.1	0.7	3.1	1.9		
2521	Diketene, inhibited	30	100	0.2	0.1	0.2	0.1		30	100	0.3	0.2	0.5	0.3		
2534	Methylchlorosilane	30	100	0.2	0.1	1.0	0.6		215	700	2.1	1.3	5.6	3.5		
2548	Chlorine pentafluoride	30	100	0.3	0.2	1.0	0.6		365	1200	3.7	2.3	8.7	5.4		
2576	Phosphorus oxybromide, molten (when spilled in water)	30	100	0.2	0.1	0.3	0.2		95	300	0.6	0.4	1.9	1.2		
2600	Carbon monoxide and hydrogen mixture	30	100	0.2	0.1	0.2	0.1		125	400	0.6	0.4	1.8	1.1		
	Carbon monoxide and hydrogen mixture, compressed															
	Hydrogen and carbon monoxide mixture															
	Hydrogen and carbon monoxide mixture, compressed															
2605	Methoxymethyl isocyanate	60	200	0.3	0.2	0.8	0.5		125	400	1.3	0.8	2.6	1.6		
2606	Methyl orthosilicate	30	100	0.2	0.1	0.2	0.1		30	100	0.3	0.2	0.6	0.4		
2644	Methyl iodide	30	100	0.2	0.1	0.3	0.2		60	200	0.3	0.2	1.0	0.6		
2646	Hexachlorocyclopentadiene	30	100	0.2	0.1	0.2	0.1		30	100	0.2	0.1	0.3	0.2		
2668	Chloroacetonitrile	30	100	0.3	0.2	0.2	0.1		30	100	0.3	0.2	0.5	0.3		
2676	Stibine	30	100	0.3	0.2	1.6	1.0		245	800	2.3	1.4	6.0	3.7		
2691	Phosphorus pentabromide (when spilled in water)	30	100	0.2	0.1	0.3	0.2		95	300	0.8	0.5	2.4	1.5		
2692	Boron tribromide (when spilled on land)	30	100	0.2	0.1	0.3	0.2		60	200	0.6	0.4	1.4	0.9		
2692	Boron tribromide (when spilled in water)	30	100	0.2	0.1	0.2	0.1		60	200	0.5	0.3	1.6	1.0		
2740	n-Propyl chloroformate	30	100	0.2	0.1	0.3	0.2		60	200	0.5	0.3	1.4	0.9		
2742	sec-Butyl chloroformate	30	100	0.2	0.1	0.2	0.1		30	100	0.3	0.2	0.6	0.4		
2742	Isobutyl chloroformate	30	100	0.2	0.1	0.2	0.1		60	200	0.3	0.2	0.8	0.5		
2743	n-Butyl chloroformate	30	100	0.2	0.1	0.2	0.1		30	100	0.3	0.2	0.5	0.3		
2806	Lithium nitride (when spilled in water)	30	100	0.2	0.1	0.2	0.1		95	300	0.8	0.5	2.1	1.3		
2810	Bis-(2-chloroethyl)ethylamine	30	100	0.2	0.1	0.2	0.1		30	100	0.2	0.1	0.3	0.2		
	Bis-(2-chloroethyl)methylamine															

Table (Continued)

ID #	Name of Material	SMALL SPILLS (From a small package or small leak from a large package)						LARGE SPILLS (From a large package or from many small packages)					
		First ISOLATE in all directions		Then, PROTECT persons downwind during -				First ISOLATE in all directions		Then, PROTECT persons downwind during -			
				DAY		NIGHT				DAY		NIGHT	
		m	ft	km	mi	km	mi	m	ft	km	mi	km	mi
	Bis-(2-chloroethyl)sulfide												
	Bis-(2-chloroethyl)sulphide												
2810	Buzz (when used as a weapon)	30	100	0.2	0.1	0.5	0.3	60	200	0.5	0.3	1.9	1.2
2810	BZ (when used as a weapon)												
2810	CS (when used as a weapon)	60	200	0.3	0.2	1.1	0.7	245	800	2.6	1.6	5.6	3.5
2810	DC (when used as a weapon)	30	100	0.2	0.1	0.8	0.5	245	800	2.3	1.4	5.3	3.3
2810	O-Ethyl S-(2-diisopropylaminoethyl) methylphosphonothiolate	30	100	0.2	0.1	0.2	0.1	30	100	0.2	0.1	0.2	0.1
2810	Ethyl N,N-dimethylphosphoramido-cyanidate	30	100	0.2	0.1	0.2	0.1	60	200	0.5	0.3	1.0	0.6
2810	GA (when used as a weapon)	30	100	0.3	0.2	0.6	0.4	155	500	1.6	1.0	3.1	1.9
2810	GB (when used as a weapon)	155	500	1.6	1.0	3.4	2.1	915	3000	11.0+	7.0+	11.0+	7.0+
2810	GD (when used as a weapon)	95	300	0.8	0.5	1.8	1.1	765	2500	6.8	4.2	10.5	6.5
2810	GF (when used as a weapon)	30	100	0.3	0.2	0.6	0.4	245	800	2.3	1.4	5.1	3.2
2810	H (when used as a weapon)	30	100	0.2	0.1	0.2	0.1	60	200	0.6	0.4	1.1	0.7
2810	HD (when used as a weapon)												
2810	HL (when used as a weapon)	30	100	0.2	0.1	0.3	0.2	95	300	1.0	0.6	1.8	1.1
2810	HN-1 (when used as a weapon)	30	100	0.2	0.1	0.2	0.1	60	200	0.6	0.4	1.3	0.8
2810	HN-2 (when used as a weapon)	30	100	0.2	0.1	0.2	0.1	60	200	0.5	0.3	1.1	0.7
2810	HN-3 (when used as a weapon)	30	100	0.2	0.1	0.2	0.1	30	100	0.2	0.1	0.3	0.2
2810	Isopropyl methylphosphonofluoridate	125	400	1.3	0.8	2.3	1.4	550	1800	5.3	3.3	8.7	5.4
2810	L (Lewisite) (when used as a weapon)	30	100	0.2	0.1	0.3	0.2	95	300	1.0	0.6	1.8	1.1
	Lewisite (when used as a weapon)												
2810	Mustard (when used as a weapon)	30	100	0.2	0.1	0.2	0.1	30	100	0.2	0.1	0.3	0.2
2810	Mustard Lewisite (when used as a weapon)	30	100	0.2	0.1	0.3	0.2	95	300	1.0	0.6	1.8	1.1
2810	Pinacolyl methylphosphonofluoridate	60	200	0.5	0.3	0.8	0.5	215	700	2.1	1.3	3.1	1.9
2810	Poisonous liquid, n.o.s. (when "Inhalation Hazard" is on a package or shipping paper)	215	700	1.9	1.2	4.3	2.7	915	3000	11.0+	7.0+	11.0+	7.0+
	Poisonous liquid, n.o.s. (Inhalation Hazard Zone A)												
2810	Poisonous liquid, n.o.s. (Inhalation Hazard Zone B)	60	200	0.5	0.3	1.3	0.8	245	800	2.3	1.4	5.0	3.1

Table (Continued)

ID #	Name of Material	SMALL SPILLS (From a small package or small leak from a large package)							LARGE SPILLS (From a large package or from many small packages)							
		First ISOLATE in all directions		Then, PROTECT persons downwind during -						First ISOLATE in all directions		Then, PROTECT persons downwind during -				
				DAY		NIGHT						DAY		NIGHT		
		m	ft	km	mi	km	mi			m	ft	km	mi	km	mi
2810	Poisonous liquid, organic, n.o.s. (when "Inhalation Hazard" is on a package or shipping paper)	215	700	1.9	1.2	4.3	2.7			915	3000	11.0+	7.0+	11.0+	7.0+
2810	Poisonous liquid, organic, n.o.s. (Inhalation Hazard Zone A)														
2810	Poisonous liquid, organic, n.o.s. (Inhalation Hazard Zone B)	60	200	0.3	0.2	1.1	0.7			185	600	1.6	1.0	4.0	2.5
2810	Sarin (when used as a weapon)	155	500	1.6	1.0	3.4	2.1			915	3000	11.0+	7.0+	11.0+	7.0+
2810	Soman (when used as a weapon)	95	300	0.8	0.5	1.8	1.1			765	2500	6.8	4.2	10.5	6.5
2810	Tabun (when used as a weapon)	30	100	0.3	0.2	0.6	0.4			155	500	1.6	1.0	3.1	1.9
2810	Tickened GD (when used as a weapon)	95	300	0.8	0.5	1.8	1.1			765	2500	6.8	4.2	10.5	6.5
2810	Toxic liquid, n.o.s. (when "Inhalation Hazard" is on a package or shipping paper)	215	700	1.9	1.2	4.3	2.7			915	3000	11.0+	7.0+	11.0+	7.0+
2810	Toxic liquid, n.o.s. (Inhalation Hazard Zone A)														
2810	Toxic liquid, n.o.s. (Inhalation Hazard Zone B)	60	200	0.5	0.3	1.3	0.8			245	800	2.3	1.4	5.0	3.1
2810	Toxic liquid, organic, n.o.s. (when "Inhalation Hazard" is on a package or shipping paper)	215	700	1.9	1.2	4.3	2.7			915	3000	11.0+	7.0+	11.0+	7.0+
2810	Toxic liquid, organic, n.o.s. (Inhalation Hazard Zone A)														
2810	Toxic liquid, organic, n.o.s. (Inhalation Hazard Zone B)	60	200	0.3	0.2	1.1	0.7			185	600	1.6	1.0	4.0	2.5
2810	Tris-(2-chloroethyl)amine	30	100	0.2	0.1	0.2	0.1			30	100	0.2	0.1	0.2	0.1
2810	VX (when used as a weapon)	30	100	0.2	0.1	0.2	0.1			60	200	0.6	0.4	1.0	0.6
2811	CX (when used as a weapon)	30	100	0.2	0.1	0.5	0.3			95	300	1.0	0.6	3.1	1.9
2826	Ethyl chlorothioformate	30	100	0.2	0.1	0.2	0.1			60	200	0.5	0.3	0.8	0.5
2845	Ethyl phosphonous dichloride, anhydrous	60	200	0.5	0.3	1.3	0.8			155	500	1.6	1.0	3.4	2.1
2845	Methyl phosphonous dichloride	60	200	0.5	0.3	1.3	0.8			245	800	2.3	1.4	5.0	3.1
2901	Bromine chloride	30	100	0.3	0.2	1.0	0.6			155	500	1.6	1.0	4.0	2.5

Table (Continued)

ID #	Name of Material	SMALL SPILLS (From a small package or small leak from a large package)						LARGE SPILLS (From a large package or from many small packages)					
		First ISOLATE in all directions		Then, PROTECT persons downwind during -				First ISOLATE in all directions		Then, PROTECT persons downwind during -			
				DAY		NIGHT				DAY		NIGHT	
		m	ft	km	mi	km	mi	m	ft	km	mi	km	mi
2927	Ethyl phosphonothioic dichloride, anhydrous	30	100	0.2	0.1	0.2	0.1	30	100	0.2	0.1	0.2	0.1
2927	Ethyl phosphorodichloridate	30	100	0.2	0.1	0.2	0.1	30	100	0.2	0.1	0.3	0.2
2927	Poisonous liquid, corrosive, n.o.s. (when "Inhalation Hazard" is on a package or shipping paper)	215	700	1.9	1.2	4.3	2.7	915	3000	11.0+	7.0+	11.0+	7.0+
2927	Poisonous liquid, corrosive, n.o.s. (Inhalation Hazard Zone A)												
2927	Poisonous liquid, corrosive, n.o.s. (Inhalation Hazard Zone B)	60	200	0.3	0.2	1.1	0.7	245	800	1.6	1.0	5.0	2.5
2927	Toxic liquid, corrosive, organic, n.o.s. (when "Inhalation Hazard" is on a package or shipping paper)	215	700	1.9	1.2	4.3	2.7	915	3000	11.0+	7.0+	11.0+	7.0+
2927	Toxic liquid, corrosive, organic, n.o.s. (Inhalation Hazard Zone A)												
2927	Toxic liquid, corrosive, organic, n.o.s. (Inhalation Hazard Zone B)	60	200	0.3	0.2	1.1	0.7	245	800	1.6	1.0	5.0	2.5
2929	Poisonous liquid, flammable, n.o.s. (when "Inhalation Hazard" is on a package or shipping paper)	155	500	1.3	0.8	3.4	2.1	915	3000	8.7	5.4	11.0+	7.0+
2929	Poisonous liquid, flammable, n.o.s. (Inhalation Hazard Zone A)												
2929	Poisonous liquid, flammable, n.o.s. (Inhalation Hazard Zone B)	30	100	0.2	0.1	0.6	0.4	125	400	1.1	0.7	2.7	1.7
2929	Poisonous liquid, flammable, organic, n.o.s. (when "Inhalation Hazard" is on a package or shipping paper)	155	500	1.3	0.8	3.4	2.1	915	3000	8.7	5.4	11.0+	7.0+
2929	Poisonous liquid, flammable, organic, n.o.s. (Inhalation Hazard Zone A)												
2929	Poisonous liquid, flammable, organic, n.o.s. (Inhalation Hazard Zone B)	30	100	0.2	0.1	0.6	0.4	125	400	1.1	0.7	2.7	1.7
2929	Toxic liquid, flammable, n.o.s. (when "Inhalation Hazard" is on a package or shipping paper)	155	500	1.3	0.8	3.4	2.1	915	3000	8.7	5.4	11.0+	7.0+

Table (Continued)

ID #	Name of Material	SMALL SPILLS (From a small package or small leak from a large package)						LARGE SPILLS (From a large package or from many small packages)					
		First ISOLATE in all directions		Then, PROTECT persons downwind during -				First ISOLATE in all directions		Then, PROTECT persons downwind during -			
				DAY		NIGHT				DAY		NIGHT	
		m	ft	km	mi	km	mi	m	ft	km	mi	km	mi
	Toxic liquid, flammable, n.o.s. (Inhalation Hazard Zone A)												
2929	Toxic liquid, flammable, n.o.s. (Inhalation Hazard Zone B)	30	100	0.2	0.1	0.6	0.4	125	400	1.1	0.7	2.7	1.7
2929	Toxic liquid, flammable, organic, n.o.s. (when "Inhalation Hazard" is on a package or shipping paper)	155	500	1.3	0.8	3.4	2.1	915	3000	8.7	5.4	11.0+	7.0+
	Toxic liquid, flammable, organic, n.o.s. (Inhalation Hazard Zone A)												
2929	Toxic liquid, flammable, organic, n.o.s. (Inhalation Hazard Zone B)	30	100	0.2	0.1	0.6	0.4	125	400	1.1	0.7	2.7	1.7
2977	Radioactive material, uranium hexafluoride, fissile (when spilled in water)	30	100	0.2	0.1	0.5	0.3	95	300	1.0	0.6	3.1	1.9
	Uranium hexafluoride, fissile containing more than 1% uranium-235) (when spilled in water)												
2978	Radioactive material, uranium hexafluoride, nonfissile or fissile excepted) (when spilled in water)	30	100	0.2	0.1	0.5	0.3	95	300	1.0	0.6	3.1	1.9
	Uranium hexafluoride, fissile excepted (when spilled in water)												
	Uranium hexafluoride, low specific activity (when spilled in water)												
	Uranium hexafluoride, nonfissile (when spilled in water)												
2985	Chlorosilanes, flammable, corrosive, n.o.s. (when spilled in water)	30	100	0.2	0.1	0.3	0.2	125	400	1.1	0.7	2.9	1.8
	Chlorosilanes, n.o.s. (when spilled in water)												
2986	Chlorosilanes, corrosive, flammable, n.o.s. (when spilled in water)	30	100	0.2	0.1	0.3	0.2	125	400	1.1	0.7	2.9	1.8
	Chlorosilanes, n.o.s. (when spilled in water)												

Table *(Continued)*

ID #	Name of Material	SMALL SPILLS (From a small package or small leak from a large package)						LARGE SPILLS (From a large package or from many small packages)					
		First ISOLATE in all directions		Then, PROTECT persons downwind during -				First ISOLATE in all directions		Then, PROTECT persons downwind during -			
				DAY		NIGHT				DAY		NIGHT	
		m	ft	km	mi	km	mi	m	ft	km	mi	km	mi
2987	Chlorosilanes, corrosive, n.o.s. (when spilled in water)	30	100	0.2	0.1	0.3	0.2	125	400	1.1	0.7	2.9	1.8
	Chlorosilanes, n.o.s. (when spilled in water)												
2988	Chlorosilanes, n.o.s. (when spilled in water)	30	100	0.2	0.1	0.3	0.2	125	400	1.1	0.7	2.9	1.8
	Chlorosilanes, water-reactive, flammable, corrosive, n.o.s. (when spilled in water)												
3023	2-Methyl-2-heptanethiol tert-Octyl mercaptan	30	100	0.2	0.1	0.2	0.1	60	200	0.5	0.3	1.1	0.7
3048	Aluminum phosphide pesticide (when spilled in water)	30	100	0.2	0.1	0.8	0.5	215	700	1.9	1.2	5.3	3.3
3049	Metal alkyl halides, n.o.s. (when spilled in water)	30	100	0.2	0.1	0.2	0.1	30	100	0.3	0.2	1.3	0.8
	Metal alkyl halides, water-reactive, n.o.s. (when spilled in water)												
	Metal aryl halides, n.o.s. (when spilled in water)												
	Metal aryl halides, water-reactive, n.o.s. (when spilled in water)												
3052	Aluminum alkyl halides (when spilled in water)	30	100	0.2	0.1	0.2	0.1	30	100	0.3	0.2	1.3	0.8
3057	Trifluoroacetyl chloride	30	100	0.3	0.2	1.4	0.9	430	1400	4.0	2.5	8.5	5.3
3079	Methacrylonitrile, inhibited	30	100	0.2	0.1	0.5	0.3	60	200	0.6	0.4	1.6	1.0
3083	Perchloryl fluoride	30	100	0.2	0.1	1.0	0.6	215	700	2.3	1.4	5.6	3.5
3122	Poisonous liquid, oxidizing, n.o.s. (when "Inhalation Hazard" is on a package or shipping paper)	155	500	1.3	0.8	3.4	2.1	915	3000	8.7	5.4	11.0+	7.0+
	Poisonous liquid, oxidizing, n.o.s. (Inhalation Hazard Zone A)												
3122	Poisonous liquid, oxidizing, n.o.s. (Inhalation Hazard Zone B)	30	100	0.2	0.1	0.6	0.4	125	400	1.1	0.7	2.7	1.7

Table (Continued)

ID #	Name of Material	SMALL SPILLS (From a small package or small leak from a large package)						LARGE SPILLS (From a large package or from many small packages)						
		First ISOLATE in all directions		Then, PROTECT persons downwind during -				First ISOLATE in all directions		Then, PROTECT persons downwind during -			
				DAY		NIGHT				DAY		NIGHT		
		m	ft	km	mi	km	mi	m	ft	km	mi	km	mi
3122	Toxic liquid, oxidizing, n.o.s. (when "Inhalation Hazard" is on a package or shipping paper)	155	500	1.3	0.8	3.4	2.1	915	3000	8.7	5.4	11.0+	7.0+
3122	Toxic liquid, oxidizing, n.o.s. (Inhalation Hazard Zone A)												
3122	Toxic liquid, oxidizing, n.o.s. (Inhalation Hazard Zone B)	30	100	0.2	0.1	0.6	0.4	125	400	1.1	0.7	2.7	1.7
3123	Poisonous liquid, water-reactive, n.o.s. (when "Inhalation Hazard" is on a package or shipping paper)	215	700	1.9	1.2	4.3	2.7	915	3000	11.0+	7.0+	11.0+	7.0+
3123	Poisonous liquid, water-reactive, n.o.s. (Inhalation Hazard Zone A)												
3123	Poisonous liquid, water-reactive, n.o.s. (Inhalation Hazard Zone B)	60	200	0.5	0.3	1.3	0.8	245	800	2.3	1.4	5.0	3.1
3123	Poisonous liquid, which in contact with water emits flammable gases, n.o.s. (when "Inhalation Hazard" is on a package or shipping paper)	215	700	1.9	1.2	4.3	2.7	915	3000	11.0+	7.0+	11.0+	7.0+
3123	Poisonous liquid, which in contact with water emits flammable gases, n.o.s. (Inhalation Hazard Zone A)												
3123	Poisonous liquid, which in contact with water emits flammable gases, n.o.s. (Inhalation Hazard Zone B)	60	200	0.5	0.3	1.3	0.8	245	800	2.3	1.4	5.0	3.1
3123	Toxic liquid, water-reactive, n.o.s. (when "Inhalation Hazard" is on a package or shipping paper)	215	700	1.9	1.2	4.3	2.7	915	3000	11.0+	7.0+	11.0+	7.0+
3123	Toxic liquid, water-reactive, n.o.s. (Inhalation Hazard Zone A)												
3123	Toxic liquid, water-reactive, n.o.s. (Inhalation Hazard Zone B)	60	200	0.5	0.3	1.3	0.8	245	800	2.3	1.4	5.0	3.1
3123	Toxic liquid, which in contact with water emits flammable gases, n.o.s. (when "Inhalation Hazard" is on a package or shipping paper)	215	700	1.9	1.2	4.3	2.7	915	3000	11.0+	7.0+	11.0+	7.0+

605

Table (Continued)

ID #	Name of Material	SMALL SPILLS (From a small package or small leak from a large package) First ISOLATE in all directions — m	ft	Then, PROTECT persons downwind during — DAY km	DAY mi	NIGHT km	NIGHT mi	LARGE SPILLS (From a large package or from many small packages) First ISOLATE in all directions — m	ft	Then, PROTECT persons downwind during — DAY km	DAY mi	NIGHT km	NIGHT mi
3123	Toxic liquid, which in contact with water emits flammable gases, n.o.s. (Inhalation Hazard Zone A)	60	200	0.5	0.3	1.3	0.8	245	800	2.3	1.4	5.0	3.1
3123	Toxic liquid, which in contact with water emits flammable gases, n.o.s. (Inhalation Hazard Zone B)	185	600	1.8	1.1	5.6	3.5	915	3000	10.8	6.7	11.0+	7.0+
3160	Liquefied gas, poisonous, flammable, n.o.s.												
3160	Liquefied gas, poisonous, flammable, n.o.s. (Inhalation Hazard Zone A)	30	100	0.3	0.2	1.1	0.7	305	1000	3.1	1.9	7.7	4.8
3160	Liquefied gas, poisonous, flammable, n.o.s. (Inhalation Hazard Zone B)	30	100	0.2	0.1	1.0	0.6	215	700	2.1	1.3	5.6	3.5
3160	Liquefied gas, poisonous, flammable, n.o.s. (Inhalation Hazard Zone C)	30	100	0.2	0.1	0.6	0.4	185	600	1.6	1.0	4.3	2.7
3160	Liquefied gas, poisonous, flammable, n.o.s. (Inhalation Hazard Zone D)	185	600	1.8	1.1	5.6	3.5	915	3000	10.8	6.7	11.0+	7.0+
3160	Liquefied gas, toxic, flammable, n.o.s.												
3160	Liquefied gas, toxic, flammable, n.o.s. (Inhalation Hazard Zone A)	30	100	0.3	0.2	1.1	0.7	305	1000	3.1	1.9	7.7	4.8
3160	Liquefied gas, toxic, flammable, n.o.s. (Inhalation Hazard Zone B)	30	100	0.2	0.1	1.0	0.6	215	700	2.1	1.3	5.6	3.5
3160	Liquefied gas, toxic, flammable, n.o.s. (Inhalation Hazard Zone C)	30	100	0.2	0.1	0.6	0.4	185	600	1.6	1.0	4.3	2.7
3160	Liquefied gas, toxic, flammable, n.o.s. (Inhalation Hazard Zone D)	430	1400	4.2	2.6	8.4	5.2	915	3000	11.0+	7.0+	11.0+	7.0+
3162	Liquefied gas, poisonous, n.o.s.												
3162	Liquefied gas, poisonous, n.o.s. (Inhalation Hazard Zone A)	60	200	0.5	0.3	1.6	1.0	430	1400	4.0	2.5	9.8	6.1
3162	Liquefied gas, poisonous, n.o.s. (Inhalation Hazard Zone B)	30	100	0.3	0.2	1.3	0.8	215	700	3.1	1.9	7.2	4.5
3162	Liquefied gas, poisonous, n.o.s. (Inhalation Hazard Zone C)	30	100	0.2	0.1	0.6	0.4	185	600	1.6	1.0	4.3	2.7
3162	Liquefied gas, poisonous, n.o.s. (Inhalation Hazard Zone D)	430	1400	4.2	2.6	8.4	5.2	915	3000	11.0+	7.0+	11.0+	7.0+
3162	Liquefied gas, toxic, n.o.s.												

Table (Continued)

ID #	Name of Material	SMALL SPILLS (From a small package or small leak from a large package)						LARGE SPILLS (From a large package or from many small packages)					
		First ISOLATE in all directions		Then, PROTECT persons downwind during -				First ISOLATE in all directions		Then, PROTECT persons downwind during -			
				DAY		NIGHT				DAY		NIGHT	
		m	ft	km	mi	km	mi	m	ft	km	mi	km	mi
	Liquefied gas, toxic, n.o.s. (Inhalation Hazard Zone A)												
3162	Liquefied gas, toxic, n.o.s. (Inhalation Hazard Zone B)	60	200	0.5	0.3	1.6	1.0	430	1400	4.0	2.5	9.8	6.1
3162	Liquefied gas, toxic, n.o.s. (Inhalation Hazard Zone C)	30	100	0.3	0.2	1.3	0.8	215	700	3.1	1.9	7.2	4.5
3162	Liquefied gas, toxic, n.o.s. (Inhalation Hazard Zone D)	30	100	0.2	0.1	0.6	0.4	185	600	1.6	1.0	4.3	2.7
3246	Methanesulfonyl chloride	95	300	0.6	0.4	2.4	1.5	245	800	2.3	1.4	5.1	3.2
	Methanesulphonyl chloride												
3275	Nitriles, poisonous, flammable, n.o.s. (when "Inhalation Hazard" is on a package or shipping paper)	30	100	0.2	0.1	0.5	0.3	60	200	0.6	0.4	1.6	1.0
	Nitriles, toxic, flammable, n.o.s. (when "Inhalation Hazard" is on a package or shipping paper)												
3276	Nitriles, poisonous, n.o.s.	30	100	0.2	0.1	0.5	0.3	60	200	0.6	0.4	1.6	1.0
	Nitriles, toxic, n.o.s.												
3278	Organophosphorus compound, poisonous, n.o.s. (when "Inhalation Hazard" is on a package or shipping paper)	60	200	0.5	0.3	1.3	0.8	245	800	2.3	1.4	5.0	3.1
	Organophosphorus compound, toxic, n.o.s. (when "Inhalation Hazard" is on a package or shipping paper)												
3279	Organophosphorus compound, poisonous, flammable, n.o.s. (when "Inhalation Hazard" is on a package or shipping paper)	60	200	0.5	0.3	1.3	0.8	245	800	2.3	1.4	5.0	3.1
	Organophosphorus compound, toxic, flammable, n.o.s. (when "Inhalation Hazard" is on a package or shipping paper)												

Table (Continued)

ID #	Name of Material	SMALL SPILLS (From a small package or small leak from a large package)						LARGE SPILLS (From a large package or from many small packages)						
		First ISOLATE in all directions		Then, PROTECT persons downwind during -				First ISOLATE in all directions		Then, PROTECT persons downwind during -			
				DAY		NIGHT				DAY		NIGHT	
		m	ft	km	mi	km	mi	m	ft	km	mi	km	mi
3280	Organoarsenic compound, n.o.s. (when "Inhalation Hazard" is on a package or shipping paper)	30	100	0.2	0.1	0.8	0.5	185	600	1.8	1.1	4.3	2.7
3281	Metal carbonyls, n.o.s.	60	200	0.6	0.4	2.1	1.3	215	700	2.1	1.3	4.3	2.7
3287	Poisonous liquid, inorganic, n.o.s. (when "Inhalation Hazard" is on a package or shipping paper)	155	500	1.3	0.8	3.7	2.3	765	2500	6.6	4.1	10.6	6.6
	Poisonous liquid, inorganic, flammable, n.o.s. (Inhalation Hazard Zone A)												
3287	Poisonous liquid, inorganic, n.o.s. (Inhalation Hazard Zone B)	60	200	0.5	0.3	1.3	0.8	245	800	2.3	1.4	5.0	3.1
3287	Toxic liquid, inorganic, n.o.s. (when "Inhalation Hazard" is on a package or shipping paper)	155	500	1.3	0.8	3.7	2.3	765	2500	6.6	4.1	10.6	6.6
3287	Toxic liquid, inorganic, n.o.s. (Inhalation Hazard Zone A)												
3287	Toxic liquid, inorganic, n.o.s. (Inhalation Hazard Zone B)	60	200	0.5	0.3	1.3	0.8	245	800	2.3	1.4	5.0	3.1
3289	Poisonous liquid, corrosive, inorganic, n.o.s. (when "Inhalation Hazard" is on a package or shipping paper)	95	300	0.6	0.4	1.8	1.1	400	1300	2.6	1.6	5.0	3.1
	Poisonous liquid, corrosive, inorganic, n.o.s. (Inhalation Hazard Zone A)												
3289	Poisonous liquid, corrosive, inorganic, n.o.s. (Inhalation Hazard Zone B)	60	200	0.3	0.2	1.1	0.7	185	600	1.6	1.0	4.9	2.5
3289	Toxic liquid, corrosive, inorganic, n.o.s. (when "Inhalation Hazard" is on a package or shipping paper)	95	300	0.6	0.4	1.8	1.1	400	1300	2.6	1.6	5.0	3.1
	Toxic liquid, corrosive, inorganic, n.o.s. (Inhalation Hazard Zone A)												
3289	Toxic liquid, corrosive, inorganic, n.o.s. (Inhalation Hazard Zone B)	60	200	0.3	0.2	1.1	0.7	185	600	1.6	1.0	4.0	2.5

Table (Continued)

ID #	Name of Material	SMALL SPILLS (From a small package or small leak from a large package)						LARGE SPILLS (From a large package or from many small packages)					
		First ISOLATE in all directions		Then, PROTECT persons downwind during -				First ISOLATE in all directions		Then, PROTECT persons downwind during -			
				DAY		NIGHT				DAY		NIGHT	
		m	ft	km	mi	km	mi	m	ft	km	mi	km	mi
3294	Hydrogen cyanide, solution in alcohol with not more than 45% hydrogen cyanide (when "Inhalation Hazard" is on a package or shipping paper)	30	100	0.2	0.1	0.3	0.2	215	700	0.6	0.4	1.9	1.2
3300	Carbon dioxide and ethylene oxide mixture, with more than 87% ethylene oxide	30	100	0.2	0.1	0.2	0.1	60	200	0.5	0.3	1.8	1.1
	Ethylene oxide mixture and carbon dioxide mixture, with more than 87% ethylene oxide												
3303	Compressed gas, poisonous, oxidizing, n.o.s.	430	1400	4.2	2.6	8.4	5.2	915	3000	11.0+	7.0+	11.0+	7.0+
	Compressed gas, poisonous, oxidizing, n.o.s. (Inhalation Hazard Zone A)												
3303	Compressed gas, poisonous, oxidizing, n.o.s. (Inhalation Hazard Zone B)	60	200	0.5	0.3	1.6	1.0	335	1100	3.4	2.1	7.7	4.8
3303	Compressed gas, poisonous, oxidizing, n.o.s. (Inhalation Hazard Zone C)	30	100	0.3	0.2	1.3	0.8	215	700	3.1	1.9	7.2	4.5
3303	Compressed gas, poisonous, oxidizing, n.o.s. (Inhalation Hazard Zone D)	30	100	0.2	0.1	0.6	0.4	185	600	1.6	1.0	4.3	2.7
3303	Compressed gas, toxic, oxidizing, n.o.s.	430	1400	4.2	2.6	8.4	5.2	915	3000	11.0+	7.0+	11.0+	7.0+
	Compressed gas, toxic, oxidizing, n.o.s. (Inhalation Hazard Zone A)												
3303	Compressed gas, toxic, oxidizing, n.o.s. (Inhalation Hazard Zone B)	60	200	0.5	0.3	1.6	1.0	335	1100	3.4	2.1	7.7	4.8
3303	Compressed gas, toxic, oxidizing, n.o.s. (Inhalation Hazard Zone C)	30	100	0.3	0.2	1.3	0.8	215	700	3.1	1.9	7.2	4.5
3303	Compressed gas, toxic, oxidizing, n.o.s. (Inhalation Hazard Zone D)	30	100	0.2	0.1	0.6	0.4	185	600	1.6	1.0	4.3	2.7
3304	Compressed gas, poisonous, corrosive, n.o.s.	430	1400	4.2	2.6	8.4	5.2	915	3000	11.0+	7.0+	11.0+	7.0+
	Compressed gas, poisonous, corrosive, n.o.s. (Inhalation Hazard Zone A)												

Table *(Continued)*

ID #	Name of Material	SMALL SPILLS (From a small package or small leak from a large package)							LARGE SPILLS (From a large package or from many small packages)							
		First ISOLATE in all directions		Then, PROTECT persons downwind during -						First ISOLATE in all directions		Then, PROTECT persons downwind during -				
				DAY		NIGHT						DAY		NIGHT		
		m	ft	km	mi	km	mi		m	ft	km	mi	km	mi
3304	Compressed gas, poisonous, corrosive, n.o.s. (Inhalation Hazard Zone B)	60	200	0.5	0.3	1.6	1.0	430	1400	4.0	2.5	9.8	6.1
3304	Compressed gas, poisonous, corrosive, n.o.s. (Inhalation Hazard Zone C)	30	100	0.3	0.2	1.3	0.8	185	600	3.1	1.9	7.2	4.5
3304	Compressed gas, poisonous, corrosive, n.o.s. (Inhalation Hazard Zone D)	30	100	0.2	0.1	0.6	0.4	185	600	1.6	1.0	4.3	2.7
3304	Compressed gas, toxic, corrosive, n.o.s.	430	1400	4.2	2.6	8.4	5.2	915	3000	11.0+	7.0+	11.0+	7.0+
3304	Compressed gas, toxic, corrosive, n.o.s. (Inhalation Hazard Zone A)												
3304	Compressed gas, toxic, corrosive, n.o.s. (Inhalation Hazard Zone B)	60	200	0.5	0.3	1.6	1.0	430	1400	4.0	2.5	9.8	6.1
3304	Compressed gas, toxic, corrosive, n.o.s. (Inhalation Hazard Zone C)	30	100	0.3	0.2	1.3	0.8	185	600	3.1	1.9	7.2	4.5
3304	Compressed gas, toxic, corrosive, n.o.s. (Inhalation Hazard Zone D)	30	100	0.2	0.1	0.6	0.4	185	600	1.6	1.0	4.3	2.7
3305	Compressed gas, poisonous, flammable, corrosive, n.o.s.	430	1400	4.2	2.8	8.4	5.2	915	3000	11.0+	7.0+	11.0+	7.0+
3305	Compressed gas, poisonous, flammable, corrosive, n.o.s. (Inhalation Hazard Zone A)												
3305	Compressed gas, poisonous, flammable, corrosive, n.o.s. (Inhalation Hazard Zone B)	60	200	0.5	0.3	1.6	1.0	430	1400	4.0	2.5	9.8	6.1
3305	Compressed gas, poisonous, flammable, corrosive, n.o.s. (Inhalation Hazard Zone C)	30	100	0.3	0.2	1.3	0.8	185	600	3.1	1.9	7.2	4.5
3305	Compressed gas, poisonous, flammable, corrosive, n.o.s. (Inhalation Hazard Zone D)	30	100	0.2	0.1	0.6	0.4	185	600	1.6	1.0	4.3	2.7
3305	Compressed gas, toxic, flammable, corrosive, n.o.s.	430	1400	4.2	2.6	8.4	5.2	915	3000	11.0+	7.0+	11.0+	7.0+

Table (Continued)

ID #	Name of Material	SMALL SPILLS (From a small package or small leak from a large package)						LARGE SPILLS (From a large package or from many small packages)					
		First ISOLATE in all directions		Then, PROTECT persons downwind during -				First ISOLATE in all directions		Then, PROTECT persons downwind during -			
				DAY		NIGHT				DAY		NIGHT	
		m	ft	km	mi	km	mi	m	ft	km	mi	km	mi
	Compressed gas, toxic, flammable, corrosive, n.o.s. (Inhalation Hazard Zone A)												
3305	Compressed gas, toxic, flammable, corrosive, n.o.s. (Inhalation Hazard Zone B)	60	200	0.5	0.3	1.6	1.0	430	1400	4.0	2.5	9.8	6.1
3305	Compressed gas, toxic, flammable, corrosive, n.o.s. (Inhalation Hazard Zone C)	30	100	0.3	0.2	1.3	0.8	185	600	3.1	1.9	7.2	4.5
3305	Compressed gas, toxic, flammable, corrosive, n.o.s. (Inhalation Hazard Zone D)	30	100	0.2	0.1	0.6	0.4	185	600	1.6	1.0	4.3	2.7
3306	Compressed gas, poisonous, oxidizing, corrosive, n.o.s. (Inhalation Hazard Zone A)	430	1400	4.2	2.6	8.4	5.2	915	3000	11.0+	7.0+	11.0+	7.0+
	Compressed gas, poisonous, oxidizing, corrosive, n.o.s. (Inhalation Hazard Zone A)												
3306	Compressed gas, poisonous, oxidizing, corrosive, n.o.s. (Inhalation Hazard Zone B)	60	200	0.5	0.3	1.6	1.0	335	1100	3.4	2.1	7.7	4.8
3306	Compressed gas, poisonous, oxidizing, corrosive, n.o.s. (Inhalation Hazard Zone C)	30	100	0.3	0.2	1.3	0.8	185	600	3.1	1.9	7.2	4.5
3306	Compressed gas, poisonous, oxidizing, corrosive, n.o.s. (Inhalation Hazard Zone D)	30	100	0.2	0.1	0.6	0.4	185	600	1.6	1.0	4.3	2.7
3306	Compressed gas, toxic, oxidizing, corrosive, n.o.s. (Inhalation Hazard Zone A)	430	1400	4.2	2.6	8.4	5.2	915	3000	11.0+	7.0+	11.0+	7.0+
	Compressed gas, toxic, oxidizing, cor-rosive, n.o.s. (Inhalation Hazard Zone A)												
3306	Compressed gas, toxic, oxidizing, corrosive, n.o.s. (Inhalation Hazard Zone B)	60	200	0.5	0.3	1.6	1.0	335	1100	3.4	2.1	7.7	4.8

INITIAL ISOLATION AND PROTECTIVE ACTION DISTANCES TABLE

Table (Continued)

ID #	Name of Material	SMALL SPILLS (From a small package or small leak from a large package)						LARGE SPILLS (From a large package or from many small packages)						
		First ISOLATE in all directions		Then, PROTECT persons downwind during -				First ISOLATE in all directions		Then, PROTECT persons downwind during -			
				DAY		NIGHT				DAY		NIGHT	
		m	ft	km	mi	km	mi	m	ft	km	mi	km	mi
3306	Compressed gas, toxic, oxidizing, corrosive, n.o.s. (Inhalation Hazard Zone C)	30	100	0.3	0.2	1.3	0.8	185	600	3.1	1.9	7.2	4.5
3306	Compressed gas, toxic, oxidizing, corrosive, n.o.s. (Inhalation Hazard Zone D)	30	100	0.2	0.1	0.6	0.4	185	600	1.6	1.0	4.3	2.7
3307	Liquefied gas, poisonous, oxidizing, n.o.s.	430	1400	4.2	2.6	8.4	5.2	915	3000	11.0+	7.0+	11.0+	7.0+
	Liquefied gas, poisonous, oxidizing, n.o.s. (Inhalation Hazard Zone A)												
3307	Liquefied gas, poisonous, oxidizing, n.o.s. (Inhalation Hazard Zone B)	60	200	0.5	0.3	1.6	1.0	335	1100	3.4	2.1	7.7	4.8
3307	Liquefied gas, poisonous, oxidizing, n.o.s. (Inhalation Hazard Zone C)	30	100	0.3	0.2	1.3	0.8	215	700	3.1	1.9	7.2	4.5
3307	Liquefied gas, poisonous, oxidizing, n.o.s. (Inhalation Hazard Zone D)	30	100	0.2	0.1	0.6	0.4	185	600	1.6	1.0	4.3	2.7
3307	Liquefied gas, toxic, oxidizing, n.o.s.	430	1400	4.2	2.6	8.4	5.2	915	3000	11.0+	7.0+	11.0+	7.0+
	Liquefied gas, toxic, oxidizing, n.o.s. (Inhalation Hazard Zone A)												
3307	Liquefied gas, toxic, oxidizing, n.o.s. (Inhalation Hazard Zone B)	60	200	0.5	0.3	1.6	1.0	335	1100	3.4	2.1	7.7	4.8
3307	Liquefied gas, toxic, oxidizing, n.o.s. (Inhalation Hazard Zone C)	30	100	0.3	0.2	1.3	0.8	215	700	3.1	1.9	7.2	4.5
3307	Liquefied gas, toxic, oxidizing, n.o.s. (Inhalation Hazard Zone D)	30	100	0.2	0.1	0.6	0.4	185	600	1.6	1.0	4.3	2.7
3308	Liquefied gas, poisonous, corrosive, n.o.s.	430	1400	4.2	2.6	8.4	5.2	915	3000	11.0+	7.0+	11.0+	7.0+
	Liquefied gas, poisonous, corrosive, n.o.s. (Inhalation Hazard Zone A)												
3308	Liquefied gas, poisonous, corrosive, n.o.s. (Inhalation Hazard Zone B)	60	200	0.5	0.3	1.6	1.0	430	1400	4.0	2.5	9.8	6.1
3308	Liquefied gas, poisonous, corrosive, n.o.s. (Inhalation Hazard Zone C)	30	100	0.3	0.2	1.3	0.8	185	600	3.1	1.9	7.2	4.5
3308	Liquefied gas, poisonous, corrosive, n.o.s. (Inhalation Hazard Zone D)	30	100	0.2	0.1	0.6	0.4	185	600	1.6	1.0	4.3	2.7

Table (Continued)

ID #	Name of Material	SMALL SPILLS (From a small package or small leak from a large package)						LARGE SPILLS (From a large package or from many small packages)					
		First ISOLATE in all directions		Then, PROTECT persons downwind during -				First ISOLATE in all directions		Then, PROTECT persons downwind during -			
				DAY		NIGHT				DAY		NIGHT	
		m	ft	km	mi	km	mi	m	ft	km	mi	km	mi
3308	Liquefied gas, toxic, corrosive, n.o.s. (Inhalation Hazard Zone A)	430	1400	4.2	2.6	8.4	5.2	915	3000	11.0+	7.0+	11.0+	7.0+
3308	Liquefied gas, toxic, corrosive, n.o.s. (Inhalation Hazard Zone B)	60	200	0.5	0.3	1.6	1.0	430	1400	4.0	2.5	9.8	6.1
3308	Liquefied gas, toxic, corrosive, n.o.s. (Inhalation Hazard Zone C)	30	100	0.3	0.2	1.3	0.8	185	600	3.1	1.9	7.2	4.5
3308	Liquefied gas, toxic, corrosive, n.o.s. (Inhalation Hazard Zone D)	30	100	0.2	0.1	0.6	0.4	185	600	1.6	1.0	4.3	2.7
3309	Liquefied gas, poisonous, flammable, corrosive, n.o.s. (Inhalation Hazard Zone A)	430	1400	4.2	2.6	8.4	5.2	915	3000	11.0+	7.0+	11.0+	7.0+
3309	Liquefied gas, poisonous, flammable, corrosive, n.o.s. (Inhalation Hazard Zone B)	60	200	0.5	0.3	1.6	1.0	430	1400	4.0	2.5	9.8	6.1
3309	Liquefied gas, poisonous, flammable, corrosive, n.o.s. (Inhalation Hazard Zone C)	30	100	0.3	0.2	1.3	0.8	185	600	3.1	1.9	7.2	4.5
3309	Liquefied gas, poisonous, flammable, corrosive, n.o.s. (Inhalation Hazard Zone D)	30	100	0.2	0.1	0.6	0.4	185	600	1.6	1.0	4.3	2.7
3309	Liquefied gas, toxic, flammable, corrosive, n.o.s. (Inhalation Hazard Zone A)	430	1400	4.2	2.6	8.4	5.2	915	3000	11.0+	7.0+	11.0+	7.0+
3309	Liquefied gas, toxic, flammable, corrosive, n.o.s. (Inhalation Hazard Zone B)	60	200	0.5	0.3	1.6	1.0	430	1400	4.0	2.5	9.8	6.1
3309	Liquefied gas, toxic, flammable, corrosive, n.o.s. (Inhalation Hazard Zone C)	30	100	0.3	0.2	1.3	0.8	185	600	3.1	1.9	7.2	4.5

INITIAL ISOLATION AND PROTECTIVE ACTION DISTANCES TABLE

ID #	Name of Material	SMALL SPILLS (From a small package or small leak from a large package)						LARGE SPILLS (From a large package or from many small packages)						
		First ISOLATE in all directions		Then, PROTECT persons downwind during -				First ISOLATE in all directions		Then, PROTECT persons downwind during -			
				DAY		NIGHT				DAY		NIGHT	
		m	ft	km	mi	km	mi	m	ft	km	mi	km	mi
3309	Liquefied gas, toxic, flammable, corrosive, n.o.s. (Inhalation Hazard Zone D)	30	100	0.2	0.1	0.6	0.4	185	600	1.6	1.0	4.3	2.7
3310	Liquefied gas, poisonous, oxidizing, n.o.s.	430	1400	4.2	2.6	8.4	5.2	915	3000	11.0+	7.0+	11.0+	7.0+
	Liquefied gas, poisonous, oxidizing, corrosive, n.o.s. (Inhalation Hazard Zone A)												
3310	Liquefied gas, poisonous, oxidizing, corrosive, n.o.s. (Inhalation Hazard Zone B)	60	200	0.5	0.3	1.6	1.0	335	1100	3.4	2.1	7.7	4.8
3310	Liquefied gas, poisonous, oxidizing, corrosive, n.o.s. (Inhalation Hazard Zone C)	30	100	0.3	0.2	1.3	0.8	185	600	3.1	1.9	7.2	4.5
3310	Liquefied gas, poisonous, oxidizing, corrosive, n.o.s. (Inhalation Hazard Zone D)	30	100	0.2	0.1	0.6	0.4	185	600	1.6	1.0	4.3	2.7
3310	Liquefied gas, toxic, oxidizing, corrosive, n.o.s.	430	1400	4.2	2.6	8.4	5.2	915	3000	11.0+	7.0+	11.0+	7.0+
	Liquefied gas, toxic, oxidizing, corrosive, n.o.s. (Inhalation Hazard Zone A)												
3310	Liquefied gas, toxic, oxidizing, corrosive, n.o.s. (Inhalation Hazard Zone B)	60	200	0.5	0.3	1.6	1.0	335	1100	3.4	2.1	7.7	4.8
3310	Liquefied gas, toxic, oxidizing, corrosive, n.o.s. (Inhalation Hazard Zone C)	30	100	0.3	0.2	1.3	0.8	185	600	3.1	1.9	7.2	4.5
3310	Liquefied gas, toxic, oxidizing, corrosive, n.o.s. (Inhalation Hazard Zone D)	30	100	0.2	0.1	0.6	0.4	185	600	1.6	1.0	4.3	2.7
3318	Ammonia solution, with more than 50% ammonia	30	100	0.2	0.1	0.2	0.1	60	200	0.5	0.3	1.1	0.7
3355	Insecticide gas, poisonous, flammable, n.o.s.	430	1400	4.2	2.6	8.4	5.2	915	3000	11.0+	7.0+	11.0+	7.0+

Table   (Continued)

ID #	Name of Material	SMALL SPILLS (From a small package or small leak from a large package)						LARGE SPILLS (From a large package or from many small packages)					
		First ISOLATE in all directions		Then, PROTECT persons downwind during -				First ISOLATE in all directions		Then, PROTECT persons downwind during -			
				DAY		NIGHT				DAY		NIGHT	
		m	ft	km	mi	km	mi	m	ft	km	mi	km	mi
3355	Insecticide gas, poisonous, flammable, n.o.s. (Inhalation Hazard Zone A)	60	200	0.5	0.3	1.6	1.0	430	1400	4.0	2.5	9.8	6.1
3355	Insecticide gas, poisonous, flammable, n.o.s. (Inhalation Hazard Zone B)	30	100	0.3	0.2	1.3	0.8	215	700	3.1	1.9	7.2	4.5
3355	Insecticide gas, poisonous, flammable, n.o.s. (Inhalation Hazard Zone C)	30	100	0.2	0.1	0.6	0.4	185	600	1.6	1.0	4.3	2.7
3355	Insecticide gas, poisonous, flammable, n.o.s. (Inhalation Hazard Zone D)	430	1400	4.2	2.6	8.4	5.2	915	3000	11.0+	7.0+	11.0+	7.0+
3355	Insecticide gas, toxic, flammable, n.o.s. (Inhalation Hazard Zone A)	60	200	0.5	0.3	1.6	1.0	430	1400	4.0	2.5	9.8	6.1
3355	Insecticide gas, toxic, flammable, n.o.s. (Inhalation Hazard Zone B)	30	100	0.2	0.1	0.2	0.1	60	200	0.5	0.3	1.1	0.7
3315	Ammonia solution, with more than 50% ammonia	30	100	0.3	0.2	1.3	0.8	215	700	3.1	1.9	7.2	4.5
3355	Insecticide gas, toxic, flammable, n.o.s. (Inhalation Hazard Zone C)	30	100	0.2	0.1	0.6	0.4	185	600	1.6	1.0	4.3	2.7
3355	Insecticide gas, toxic, flammable, n.o.s. (Inhalation Hazard Zone D)	30	100	0.2	0.1	0.2	0.1	30	100	0.2	0.1	0.6	0.4
9191	Chlorine dioxide, hydrate, frozen (when spilled in water)	30	100	0.2	0.1	0.5	0.3	185	600	1.4	0.9	4.0	2.5
9192	Fluorine, refrigerated liquid (cryogenic liquid)	30	100	0.2	0.1	0.2	0.1	125	400	0.6	0.4	1.8	1.1
9202	Carbon monoxide, refrigerated liquid (cryogenic liquid)	30	100	0.2	0.1	0.2	0.1	30	100	0.2	0.1	0.3	0.2
9206	Methyl phosphonic dichloride	30	100	0.2	0.1	0.2	0.1	30	100	0.3	0.2	0.5	0.3
9263	Chloropivaloyl chloride	30	100	0.2	0.1	0.2	0.1	30	100	0.3	0.2	0.5	0.3
9264	3,5-Dichloro-2,4,6-trifluoropyridine	30	100	0.2	0.1	0.2	0.1	30	100	0.3	0.2	0.5	0.3
9269	Trimethoxysilane	30	100	0.3	0.2	1.0	0.6	215	700	2.1	1.3	4.2	2.6

"+" means distance can be larger in certain atmospheric conditions.
See next page for Table of Water-Reactive Materials That Produce Toxic Gases.

Materials That Produce Large Amounts of Toxic-by-Inhalation (TIH) Gas(es) *When Spilled in Water*

Table of Water-Reactive Materials That Produce Toxic Gases

ID #	Guide #	Name of Material	TIH Gas(es) Produced
1162	151	Dimethyldichlorosilane	HCl
1242	139	Methyldichlorosilane	HCl
1250	155	Methyltrichlorosilane	HCl
1295	139	Trichlorosilane	HCl
1298	155	Trimethylchlorosilane	HCl
1340	139	Phosphorus pentasulfide, free from yellow and white phosphorus	H_2S
1340	139	Phosphorus pentasulphide, free from yellow and white phosphorus	H_2S
1360	139	Calcium phosphide	PH_3
1384	135	Sodium dithionite	H_2S, SO_2
1384	135	Sodium hydrosulfite	H_2S, SO_2
1384	135	Sodium hydrosulphite	H_2S, SO_2
1397	139	Aluminum phosphide	PH_3
1412	139	Lithium amide	NH_3
1419	139	Magnesium aluminum phosphide	PH_3
1432	139	Sodium phosphide	PH_3
1433	139	Stannic phosphides	PH_3
1541	155	Acetone cyanohydrin, stabilized	HCN
1680	157	Potassium cyanide	HCN
1689	157	Sodium cyanide	HCN
1714	139	Zinc phosphide	PH_3
1716	156	Acetyl bromide	HBr
1717	132	Acetyl chloride	HCl
1724	155	Allyl trichlorosilane, stabilized	HCl
1725	137	Aluminum bromide, anhydrous	HBr
1726	137	Aluminum chloride, anhydrous	HCl
1728	155	Amyltrichlorosilane	HCl
1732	157	Antimony pentafluoride	HF
1736	137	Benzoyl chloride	HCl
1745	144	Bromine pentafluoride	HF, HBr, Br_2
1746	144	Bromine trifluoride	HF, HBr, Br_2
1747	155	Butyltrichlorosilane	HCl
1752	156	Chloroacetyl chloride	HCl
1754	137	Chlorosulfonic acid	HCl
1754	137	Chlorosulfonic acid and sulfur trioxide mixture	HCl
1754	137	Chlorosulphonic acid	HCl
1754	137	Chlorosulphonic acid and sulphur trioxide mixture	HCl
1754	137	Sulfur trioxide and chlorosulfonic acid	HCl
1754	137	Sulphur trioxide and chlorosulphonic acid	HCl
1758	137	Chromium oxychloride	HCl
1777	137	Fluorosulfonic acid	HF
1777	137	Fluorosulphonic acid	HF
1801	156	Octyltrichlorosilane	HCl
1806	137	Phosphorus pentachloride	HCl
1809	137	Phosphorus trichloride	HCl
1810	137	Phosphorus oxychloride	HCl
1818	157	Silicon tetrachloride	HCl
1828	137	Sulfur chlorides	HCl, SO_2, H_2S

Table *(Continued)*

ID #	Guide #	Name of Material	TIH Gas(es) Produced
1828	137	Sulphur chlorides	HCl, SO_2, H_2S
1834	137	Sulfuryl chloride	HCl, SO_3
1834	137	Sulphuryl chloride	HCl, SO_3
1836	137	Thionyl chloride	HCl, SO_2
1838	137	Titanium tetrachloride	HCl
1898	156	Acetyl iodide	HI
1923	135	Calcium dithionite	H_2S, SO_2
1923	135	Calcium hydrosulfite	H_2S, SO_2
1923	135	Calcium hydrosulphite	H_2S, SO_2
1939	137	Phosphorus oxybromide	HBr
1939	137	Phosphorus oxybromide, solid	HBr
2004	135	Magnesium diamide	NH_3
2011	139	Magnesium phosphide	PH_3
2012	139	Potassium phosphide	PH_3
2013	139	Strontium phosphide	PH_3
2442	156	Trichloroacetyl chloride	HCl
2495	144	Iodine pentafluoride	HF
2576	137	Phosphorus oxybromide, molten	HBr
2691	137	Phosphorus pentabromide	HBr
2692	157	Boron tribromide	HBr
2806	138	Lithium nitride	NH_3
2977	166	Radioactive material, uranium hexafluoride, fissile	HF
2977	166	Uranium hexafluoride, fissile containing more than 1% uranium-235	HF
2978	166	Radioactive material, uranium hexafluoride, nonfissile or fissile excepted	HF
2978	166	Uranium hexafluoride, fissile excepted	HF
2978	166	Uranium hexafluoride, low specific activity	HF
2978	166	Uranium hexafluoride, nonfissile	HF
2985	155	Chlorosilanes, flammable, corrosive, n.o.s.	HCl
2985	155	Chlorosilanes, n.o.s.	HCl
2986	155	Chlorosilanes, corrosive, flammable, n.o.s.	HCl
2986	155	Chlorosilanes, n.o.s.	HCl
2987	156	Chlorosilanes, corrosive, n.o.s.	HCl
2987	156	Chlorosilanes, n.o.s.	HCl
2988	139	Chlorosilanes, n.o.s.	HCl
2988	139	Chlorosilanes, water-reactive, flammable, corrosive, n.o.s.	HCl
3048	157	Aluminum phosphide pesticide	PH_3
3049	138	Metal alkyl halides, n.o.s.	HCl
3049	138	Metal alkyl halides, water-reactive, n.o.s.	HCl
3049	138	Metal aryl halides, n.o.s.	HCl
3049	138	Metal aryl halides, water-reactive, n.o.s.	HCl
3052	135	Aluminum alkyl halides	HCl
9191	143	Chlorine dioxide, hydrate, frozen	Cl_2

Chemical Symbols for TIH Gases:					
Br$_2$	Bromine	HF	Hydrogen fluoride	PH$_3$	Phosphine
Cl$_2$	Chlorine	HI	Hydrogen iodide	SO$_2$	Sulfur dioxide
HBr	Hydrogen bromide	H$_2$S	Hydrogen sulfide	SO$_2$	Sulphur dioxide
HCl	Hydrogen chloride	H$_2$S	Hydrogen sulphide	SO$_3$	Sulfur trioxide
HCN	Hydrogen cyanide	NH$_3$	Ammonia	SO$_3$	Sulphur trioxide

PROTECTIVE CLOTHING

Street Clothing and Work Uniforms: These garments, such as uniforms worn by police and emergency medical services personnel, provide almost no protection from the harmful effects of dangerous goods.

Structural Fire Fighters' Protective Clothing (SFPC): This category of clothing, often called turnout or bunker gear, means the protective clothing normally worn by fire fighters during structural fire-fighting operations. It includes a helmet, coat, pants, boots, gloves, and a hood to cover parts of the head not protected by the helmet and facepiece. This clothing must be used with full-facepiece positive pressure self-contained breathing apparatus (SCBA). This protective clothing should at a minimum, meet the OSHA Fire Brigades Standard (29 CFR 1910.156). Structural fire fighters' protective clothing provides limited protection from heat and cold, but may not provide adequate protection from the harmful vapors or liquids that are encountered during dangerous goods incidents. Each guide includes a statement about the use of SFPC in incidents involving those materials referenced by that guide. Some guides state that SFPC provides limited protection. In those cases, the responder wearing SFPC and SCBA may be able to perform an expedient, that is quick "in-and-out," operation. However, this type of operation can place the responder at risk of exposure, injury, or death. The incident commander makes the decision to perform this operation only if an overriding benefit can be gained (ie, perform an immediate rescue, turn off a valve to control a leak, etc). The coverall-type protective clothing customarily worn to fight fires in forests or wildlands is not SFPC and is not recommended nor referred to elsewhere in this guide.

Positive Pressure Self-Contained Breathing Apparatus (SCBA): This apparatus provides a constant, positive pressure flow of air within the facepiece, even if one inhales deeply while doing heavy work. Use apparatus certified by NIOSH and the Department of Labor/Mine Safety and Health Administration in accordance with 42 CFR Part 84. Use it in accordance with the requirements for respiratory protection specified in OSHA 29 CFR 1910.134 (Respiratory Protection) and/or 29 CFR 1910.156) (f) (Fire Brigade Standard). Chemical-cartridge respirators or other filtering masks are not acceptable substitutes for positive pressure self-contained breathing apparatus. Demand-type SCBA does not meet the OSHA 29 CFR 1910.156 (f)(1)(i) Fire Brigade Standard.

Chemical Protective Clothing and Equipment: Safe use of this type of protective clothing and equipment requires specific skills developed through training and experience. It is generally not available to, or used by, first responders. This type of special clothing may protect against one chemical, yet be readily permeated by chemicals for which it was not designed. Therefore, protective clothing should not be used unless it is compatible with the released material. This type of special clothing offers little or no protection against heat and/or cold. Examples of this type of equipment have been described as (1) Vapor Protective Suits (NFPA 1991), also known as Totally-Encapsulating Chemical Protective (TECP) Suits or Level A* protection (OSHA 29 CFR 1910.120, Appendix A and B), and (2) Liquid-Splash Protective Suits (NFPA 1991 and 1993), also known as Level B* or C* protection (OSHA 29 CFR 1910.120, Appendix A and B). No single protective clothing material will protect you from all dangerous goods. Do not assume any protective clothing is resistant to cold and/or heat or flame exposure unless it is so certified by the manufacturer (NFPA 1991 5-3 Flammability Resistance Test and 5-6 Cold Temperature Performance Test).

*Consult glossary (of the 2000 Emergency Response Guidebook) for additional protection levels under the heading "Protective Clothing."

FIRE AND SPILL CONTROL

Fire Control

Water is the most common and generally most available fire extinguishing agent. Exercise caution in selecting a fire extinguishing method as there are many factors to be considered in an incident. Water may be ineffective in fighting fires involving some materials; its effectiveness depends greatly on the method of application.

Spill fires involving flammable liquids are generally controlled by applying a fire-fighting foam to the surface of the burning material. Fighting flammable liquid fires requires foam concentrate that is chemically compatible with the burning material, correct mixing of the foam concentrate with water and air, and careful application and maintenance of the foam blanket. There are 2 general types of fire fighting foam: regular and alcohol-resistant. Examples of regular foam are protein-base, fluoroprotein, and aqueous film-forming foam (AFFF). Some flammable liquids, including many petroleum products, can be controlled by applying regular foam. Other flammable liquids, including polar solvents (flammable liquids that are water soluble) such as alcohols and ketones, have different chemical properties. A fire involving these materials cannot be easily controlled with regular foam and requires application of alcohol-resistant foam. Polar-solvent fires may be difficult to control and require a higher foam-application rate than other flammable liquid fires (see NFPA/ANSI Standards 11 and 11A for further information). Refer to the appropriate guide to determine which type of foam is recommended. Although it is impossible to make specific recommendations for flammable liquids that have subsidiary corrosive or toxic hazards, alcohol-resistant foam may be effective for many of these materials. The emergency response telephone number on the shipping document, or the appropriate emergency response agency, should be contacted as soon as possible for guidance on the proper fire extinguishing agent to use. The final selection of the agent and method depends on many factors such as incident location, exposure hazards, size of the fire, environmental concerns, as well as the availability of extinguishing agents and equipment at the scene.

Water Reactive Materials

Water is sometimes used to flush spills and to reduce or direct vapors in spill situations. Some of the materials covered by this guide can react violently or even explosively with water. In these cases, consider letting the fire burn or leaving the spill alone (except to prevent its spreading by diking) until additional technical advice can be obtained. The applicable guides clearly warn you of these potentially dangerous reactions. These materials require technical advice since

1. water getting inside a ruptured or leaking container may cause an explosion
2. water may be needed to cool adjoining containers to prevent their rupturing (exploding) or further spread of the fires
3. water may be effective in mitigating an incident involving a water-reactive material only if it can be applied at a sufficient flooding rate for an extended period; and
4. the products from the reaction with water may be more toxic, corrosive, or otherwise more undesirable than the product of the fire without water applied

When responding to an incident involving water-reactive chemicals, take into account the existing conditions such as wind, precipitation, location, and accessibility to the incident, as well as the availability of the agents to control the fire or spill. Because there are variables to consider, the decision to use water on fires or spills involving water-reactive materials should be based on information from an authoritative source; for example, a producer of the material, who can be contacted through the emergency response telephone number or the appropriate emergency response agency.

Vapor Control

Limiting the amount of vapor released from a pool of flammable or corrosive liquids is an operational concern. It requires the use of proper protective clothing, specialized equipment, appropriate chemical agents, and skilled personnel. Before engaging in vapor control, get advice from an authoritative source as to the proper tactics.

There are several ways to minimize the amount of vapors escaping from pools of spilled liquids, such as special foams, adsorbing agents, absorbing agents, and neutralizing agents. To be effective, these vapor-control methods must be selected for the specific material involved and performed in a manner that will mitigate, not worsen, the incident.

Where specific materials are known, such as at manufacturing or storage facilities, it is desirable for the dangerous-goods response team to prearrange with the facility operators to select and stockpile these control agents in advance of a spill. In the field, first responders may not have the most effective vapor control agent for the material available. They are likely to have only water and only one type of fire-fighting foam on their vehicles. If the available foam is inappropriate for use, they are likely to use water spray. Because the water is being used to form a vapor seal, care must be taken not to churn or further spread the spill during application. Vapors that do not react with water may be directed away from the site using the air currents surrounding the water spray. Before using water spray or other methods to safely control vapor emission or to suppress ignition, obtain technical advice, based on specific chemical name identification.

CRIMINAL/TERRORIST USE OF CHEMICAL/BIOLOGICAL AGENTS

The following is intended to supply information to first responders for use in making a preliminary assessment of a situation that they suspect involves criminal/terrorist use of chemical and/or biological (CB) agents. To aid in the assessment, a list of observable indicators of the use and/or presence of a CB agent is provided in the following paragraphs.

Differences Between a Chemical and a Biological Agent

Chemical and biological agents can be dispersed in the air we breathe, the water we drink, or on surfaces we physically contact. Dispersion methods may be as simple as opening a container, using conventional (garden) spray devices, or as elaborate as detonating an improvised explosive device.

Chemical incidents are characterized by the rapid onset of medical symptoms (minutes to hours) and easily observed signatures (colored residue, dead foliage, pungent odor, dead insects and animals).

Biological incidents are characterized by the onset of symptoms in hours to days. Typically, there will be no characteristic signatures because biological agents are usually odorless and colorless. Because of the delayed onset of symptoms in a biological incident, the area affected may be greater due to the movement of infected individuals.

Indicators of a Possible Chemical Incident

Dead animals/birds/fish	Not just an occasional road kill, but numerous animals (wild and domestic, small and large), birds, and fish in the same area.
Lack of insect life	If normal insect activity (ground, air, and/or water) is missing, check the ground/water surface/shore line for dead insects. If near water, check for dead fish/aquatic birds.
Unexplained odors	Smells may range from fruity to flowery to sharp/pungent to garlic/horseradish-like to bitter almonds/peach kernels to new mown hay. It is important to note that the particular odor is completely out of character with its surroundings.
Unusual numbers of dying or sick people (mass casualties)	Health problems including nausea, disorientation, difficulty in breathing, convulsions, localized sweating, conjunctivitis (reddening of eyes/nerve agent symptoms), erythema (reddening of skin/vesicant symptoms), and death.
Pattern of casualties	Casualties will likely be distributed downwind, or if indoors, by the air ventilation system.
Blisters/rashes	Numerous individuals experiencing unexplained water-like blisters, weals (like bee stings), and/or rashes.
Illness in confined area	Different casualty rates for people working indoors vs outdoors dependent on where the agent was released.
Unusual liquid droplets	Numerous surfaces exhibit oily droplets/film; numerous water surfaces have an oily film. (No recent rain.)
Different-looking areas	Not just a patch of dead weeds, but trees, shrubs, bushes, food crops, and/or lawns that are dead, discolored, or withered. (No current drought.)
Low-lying clouds	Low-lying cloud/fog-like condition that is not consistent with its surroundings.
Unusual metal debris	Unexplained bomb/munitions-like material, especially if it contains a liquid.

Indicators of a Possible Biological Incident

Unusual numbers of sick or dying people or animals	Any number of symptoms may occur. Casualties may occur hours to days after an incident has occurred. The time required before symptoms are observed is dependent on the agent used.
Unscheduled and unusual spray being disseminated	Especially if outdoors during periods of darkness.
Abandoned spray devices	Devices may not have distinct odors.

Personal Safety Considerations

When approaching a scene that may involve CB agents, the most critical consideration is the safety of oneself and other responders. Protective clothing and respiratory protection of appropriate level of safety must be used. Be aware that the presence and identification of CB agents may not be verifiable, especially in the case of biological agents. The following actions/ measures to be considered are applicable to either a chemical or biological incident. The guidance is general in nature, not all encompassing, and its applicability should be evaluated on a case-by-case basis.

Approach and Response Strategies Protect yourself and use a safe approach (minimize any exposure time, maximize the distance between you and the item that is likely to harm you, use cover as protection, and wear appropriate personal protective equipment and respiratory protection). Identify and estimate the hazard by using indicators as provided above. Isolate the area and secure the scene; potentially contaminated people should be isolated and decontaminated as soon as possible. In the event of a chemical incident, the fading of chemical odors is not necessarily an indication of reduced vapor concentrations. Some chemicals deaden the senses giving the false perception that the chemical is no longer present.

Decontamination Measures Emergency responders should follow standard decontamination procedures (flush-strip-flush). Mass casualty decontamination should begin as soon as possible by stripping (all clothing) and flushing (soap and water). If biological agents are involved or suspected, careful washing and use of a brush are more effective. If chemical agents are suspected, the most important and effective decontamination will be that done within the first 1 or 2 minutes. If possible, further decontamination should be performed using a 0.5% hypochlorite solution (1 part household bleach mixed with 9 parts water). If biological agents are suspected, a contact time of 10 to 15 minutes should be allowed before rinsing. The solution can be used on soft tissue wounds, but must not be used in eyes or open wounds of the abdomen, chest, brain, or spine. For further information contact the agencies listed in the (2000 Emergency Response Guidebook) guidebook.

Note: The above information was developed by the Department of National Defense (Canada) and the U.S. Department of the Army, Edgewood Arsenal.

Median Heights and Weights and Recommended Energy Intake*

Age (y) or Condition	Weight		Height		REE[†]	Multiples	Average Energy Allowance (kcal)[‡]	
	(kg)	(lb)	(cm)	(in)	(kcal/d)	of REE	/kg	/d[§]
Infants								
0-0.5	6	13	60	24	320		108	650
0.5-1	9	20	71	28	500		98	850
Children								
1-3	13	29	90	35	740		102	1300
4-6	20	44	112	44	950		90	1800
7-10	28	62	132	52	1130		70	2000
Male								
11-14	45	99	157	62	1440	1.70	55	2500
15-18	66	145	176	69	1760	1.67	45	3000
19-24	72	160	177	70	1780	1.67	40	2900
25-50	79	174	176	70	1800	1.60	37	2900
51+	77	170	173	68	1530	1.50	30	2300
Female								
11-14	46	101	157	62	1310	1.67	47	2200
15-18	55	120	163	64	1370	1.60	40	2200
19-24	58	128	164	65	1350	1.60	38	2200
25-50	63	138	163	64	1380	1.55	36	2200
51+	65	143	160	63	1280	1.50	30	1900
Pregnant								+0
1st trimester								+300
2nd trimester								+300
3rd trimester								+300
Lactating								
1st 6 months								+500
2nd 6 months								+500

* From *Recommended Dietary Allowances*, 10th ed, Washington, DC: National Academy Press, 1989.
[†] Calculation based on FAO equations, then rounded.
[‡] In the range of light to moderate activity, the coefficient of variation is $\pm20\%$.
[§] Figure is rounded.

Pounds to Kilograms Conversion

1 pound = 0.45359 kilograms
1 kilogram = 2.2 pounds

lb	=	kg	lb	=	kg	lb	=	kg
1		0.45	70		31.75	140		63.50
5		2.27	75		34.02	145		65.77
10		4.54	80		36.29	150		68.04
15		6.80	85		38.56	155		70.31
20		9.07	90		40.82	160		72.58
25		11.34	95		43.09	165		74.84
30		13.61	100		45.36	170		77.11
35		15.88	105		47.63	175		79.38
40		18.14	110		49.90	180		81.65
45		20.41	115		52.16	185		83.92
50		22.68	120		54.43	190		86.18
55		24.95	125		56.70	195		88.45
60		27.22	130		58.91	200		90.72
65		29.48	135		61.24			

Reference Values for Adults

Automated Chemistry (CHEMISTRY A)		
Test	Values	Remarks
Serum/Plasma		
Acetone	Negative	
Albumin	3.2-5 g/dL	
Alcohol, ethyl	Negative	
Aldolase	1.2-7.6 IU/L	
Ammonia	20-70 mcg/dL	Specimen to be placed on ice as soon as collected
Amylase	30-110 units/L	
Bilirubin, direct	0-0.3 mg/dL	
Bilirubin, total	0.1-1.2 mg/dL	
Calcium	8.6-10.3 mg/dL	
Calcium, ionized	2.24-2.46 mEq/L	
Chloride	95-108 mEq/L	
Cholesterol, total	≤200 mg/dL	Fasted blood required - normal value affected by dietary habits. This reference range is for a general adult population.
HDL cholesterol	40-60 mg/dL	Fasted blood required - normal value affected by dietary habits.
LDL cholesterol	<160 mg/dL	If triglyceride is >400 mg/dL, LDL cannot be calculated accurately (Friedewald equation). Target LDL-C depends on patient's risk factors.
CO_2	23-30 mEq/L	
Creatine kinase (CK) isoenzymes		
CK-BB	0%	
CK-MB (cardiac)	0%-3.9%	
CK-MM (muscle)	96%-100%	
CK-MB levels must be both ≥4% and 10 IU/L to meet diagnostic criteria for CK-MB positive result consistent with myocardial injury.		
Creatine phosphokinase (CPK)	8-150 IU/L	
Creatinine	0.5-1.4 mg/dL	
Ferritin	13-300 ng/mL	
Folate	3.6-20 ng/dL	
GGT (gamma-glutamyltranspeptidase)		
Male	11-63 IU/L	
Female	8-35 IU/L	
GLDH	To be determined	
Glucose (preprandial)	<115 mg/dL	Goals different for diabetics
Glucose, fasting	60-110 mg/dL	Goals different for diabetics
Glucose, nonfasting (2-h postprandial)	<120 mg/dL	Goals different for diabetics
Hemoglobin A_{1c}	<8	
Hemoglobin, plasma free	<2.5 mg/100 mL	

Table   (*Continued*)

Test	Values	Remarks
Hemoglobin, total glycosolated (Hb A₁)	4%-8%	
Iron	65-150 mcg/dL	
Iron binding capacity, total (TIBC)	250-420 mcg/dL	
Lactic acid	0.7-2.1 mEq/L	Specimen to be kept on ice and sent to lab as soon as possible
Lactate dehydrogenase (LDH)	56-194 IU/L	
Lactate dehydrogenase (LDH) iso-enzymes		
LD₁	20%-34%	
LD₂	29%-41%	
LD₃	15%-25%	
LD₄	1%-12%	
LD₅	1%-15%	
Flipped LD₁/LD₂ ratios (>1 may be consistent with myocardial injury) particularly when considered in combination with a recent CK-MB positive result.		
Lipase	23-208 units/L	
Magnesium	1.6-2.5 mg/dL	Increased by slight hemolysis
Osmolality	289-308 mOsm/kg	
Phosphatase, alkaline		
Adults 25-60 y	33-131 IU/L	
Adults 61 y or older	51-153 IU/L	
Infancy-adolescence	Values range up to 3-5 times higher than adults	
Phosphate, inorganic	2.8-4.2 mg/dL	
Potassium	3.5-5.2 mEq/L	Increased by slight hemolysis
Prealbumin	>15 mg/dL	
Protein, total	6.5-7.9 g/dL	
SGOT (AST)	<35 IU/L (20-48)	
SGPT (ALT) (10-35)	<35 IU/L	
Sodium	134-149 mEq/L	
Transferrin	>200 mg/dL	
Triglycerides	45-155 mg/dL	Fasted blood required
Troponin I	<1.5 ng/mL	
Urea nitrogen (BUN)	7-20 mg/dL	
Uric acid		
Male	2-8 mg/dL	
Female	2-7.5 mg/dL	
Cerebrospinal Fluid		
Glucose	50-70 mg/dL	
Protein		
Adults and children	15-45 mg/dL	CSF obtained by lumbar puncture
Newborn infants	60-90 mg/dL	
On CSF obtained by cisternal puncture: About 25 mg/dL		

Table   (*Continued*)

Test	Values	Remarks
On CSF obtained by ventricular puncture: About 10 mg/dL		
Note: Bloody specimen gives erroneously high value due to contamination with blood proteins		
Urine (24-hour specimen is required for all these tests unless specified)		
Amylase	32-641 units/L	The value is in units/L and **not** calculated for total volume.
Amylase, fluid (random samples)		Interpretation of value left for physician, depends on the nature of fluid
Calcium	Depends upon dietary intake	
Creatine		
Male	150 mg/24 h	Higher value on children and during pregnancy
Female	250 mg/24 h	
Creatinine	1000-2000 mg/24 h	
Creatinine clearance (endogenous)		
Male	85-125 mL/min	A blood sample must accompany urine specimen.
Female	75-115 mL/min	
Glucose	1 g/24 h	
5-hydroxyindoleacetic acid	2-8 mg/24 h	
Iron	0.15 mg/24 h	Acid-washed container required
Magnesium	146-209 mg/24 h	
Osmolality	500-800 mOsm/kg	With normal fluid intake
Oxalate	10-40 mg/24 h	
Phosphate	400-1300 mg/24 h	
Potassium	25-120 mEq/24 h	Varies with diet; the interpretation of urine electrolytes and osmolality should be left for the physician.
Sodium	40-220 mEq/24 h	
Porphobilinogen, qualitative	Negative	
Porphyrins, qualitative	Negative	
Proteins	0.05-0.1 g/24 h	
Salicylate	Negative	
Urea clearance	60-95 mL/min	A blood sample must accompany specimen.
Urea N	10-40 g/24 h	Dependent on protein intake
Uric acid	250-750 mg/24 h	Dependent on diet and therapy
Urobilinogen	0.5-3.5 mg/24 h	For qualitative determination on random urine, send sample to urinalysis section in Hematology Lab.
Xylose absorption test		
Children	16%-33% of ingested xylose	
Feces		
Fat, 3-day collection	<5 g/d	Value depends on fat intake of 100 g/d for 3 days preceding and during collection.
Gastric Acidity		
Acidity, total, 12 h	10-60 mEq/L	Titrated at pH 7

Blood Gases

	Arterial	Capillary	Venous
pH	7.35-7.45	7.35-7.45	7.32-7.42
pCO_2 (mm Hg)	35-45	35-45	38-52
pO_2 (mm Hg)	70-100	60-80	24-48
HCO_3 (mEq/L)	19-25	19-25	19-25
TCO_2 (mEq/L)	19-29	19-29	23-33
O_2 saturation (%)	90-95	90-95	40-70
Base excess (mEq/L)	−5 to +5	−5 to +5	−5 to +5

HEMATOLOGY Complete Blood Count

Age	Hgb (g/dL)	Hct (%)	RBC (mill/mm^3)	RDW
0-3 d	15.0-20.0	45-61	4.0-5.9	<18
1-2 wk	12.5-18.5	39-57	3.6-5.5	<17
1-6 mo	10.0-13.0	29-42	3.1-4.3	<16.5
7 mo to 2 y	10.5-13.0	33-38	3.7-4.9	<16
2-5 y	11.5-13.0	34-39	3.9-5.0	<15
5-8 y	11.5-14.5	35-42	4.0-4.9	<15
13-18 y	12.0-15.2	36-47	4.5-5.1	<14.5
Adult male	13.5-16.5	41-50	4.5-5.5	<14.5
Adult female	12.0-15.0	36-44	4.0-4.9	<14.5

Age	MCV (fL)	MCH (pg)	MCHC (%)	Plts ($\times 10^3$/mm^3)
0-3 d	95-115	31-37	29-37	250-450
1-2 wk	86-110	28-36	28-38	250-450
1-6 mo	74-96	25-35	30-36	300-700
7 mo to 2 y	70-84	23-30	31-37	250-600
2-5 y	75-87	24-30	31-37	250-550
5-8 y	77-95	25-33	31-37	250-550
13-18 y	78-96	25-35	31-37	150-450
Adult male	80-100	26-34	31-37	150-450
Adult female	80-100	26-34	31-37	150-450

WBC and Differential

Age	WBC ($\times 10^3$/mm^3)	Segs	Bands	Lymphs	Monos
0-3 d	9.0-35.0	32-62	10-18	19-29	5-7
1-2 wk	5.0-20.0	14-34	6-14	36-45	6-10
1-6 mo	6.0-17.5	13-33	4-12	41-71	4-7
7 mo to 2 y	6.0-17.0	15-35	5-11	45-76	3-6
2-5 y	5.5-15.5	23-45	5-11	35-65	3-6

Table   (*Continued*)

Age	WBC ($\times 10^3/mm^3$)	Segs	Bands	Lymphs	Monos
5-8 y	5.0-14.5	32-54	5-11	28-48	3-6
13-18 y	4.5-13.0	34-64	5-11	25-45	3-6
Adults	4.5-11.0	35-66	5-11	24-44	3-6

Age	Eosinophils	Basophils	Atypical Lymphs	No. of NRBCs
0-3 d	0-2	0-1	0-8	0-2
1-2 wk	0-2	0-1	0-8	0
1-6 mo	0-3	0-1	0-8	0
7 mo to 2 y	0-3	0-1	0-8	0
2-5 y	0-3	0-1	0-8	0
5-8 y	0-3	0-1	0-8	0
13-18 y	0-3	0-1	0-8	0
Adults	0-3	0-1	0-8	0

Segs = segmented neutrophils.
Bands = band neutrophils.
Lymphs = lymphocytes.
Monos = monocytes.

Erythrocyte Sedimentation Rates and Reticulocyte Counts		
Sedimentation rate, Westergren	Children	0-20 mm/hour
	Adult male	0-15 mm/hour
	Adult female	0-20 mm/hour
Sedimentation rate, Wintrobe	Children	0-13 mm/hour
	Adult male	0-10 mm/hour
	Adult female	0-15 mm/hour
Reticulocyte count	Newborns	2%-6%
	1-6 mo	0%-2.8%
	Adults	0.5%-1.5%

Reference Values for Children

		Normal Values
Chemistry		
Albumin	0-1 y	2-4 g/dL
	1 y to adult	3.5-5.5 g/dL
Ammonia	Newborns	90-150 mcg/dL
	Children	40-120 mcg/dL
	Adults	18-54 mcg/dL
Amylase	Newborns	0-60 units/L
	Adults	30-110 units/L
Bilirubin, conjugated, direct	Newborns	<1.5 mg/dL
	1 mo to adult	0-0.5 mg/dL
Bilirubin, total	0-3 d	2-10 mg/dL
	1 mo to adult	0-1.5 mg/dL
Bilirubin, unconjugated, indirect		0.6-10.5 mg/dL
Calcium	Newborns	7-12 mg/dL
	0-2 y	8.8-11.2 mg/dL
	2 y to adult	9-11 mg/dL
Calcium, ionized, whole blood		4.4-5.4 mg/dL
Carbon dioxide, total		23-33 mEq/L
Chloride		95-105 mEq/L
Cholesterol	Newborns	45-170 mg/dL
	0-1 y	65-175 mg/dL
	1-20 y	120-230 mg/dL
Creatinine	0-1 y	≤0.6 mg/dL
	1 y to adult	0.5-1.5 mg/dL
Glucose	Newborns	30-90 mg/dL
	0-2 y	60-105 mg/dL
	Children to adults	70-110 mg/dL
Iron		
	Newborns	110-270 mcg/dL
	Infants	30-70 mcg/dL
	Children	55-120 mcg/dL
	Adults	70-180 mcg/dL
Iron binding	Newborns	59-175 mcg/dL
	Infants	100-400 mcg/dL
	Adults	250-400 mcg/dL
Lactic acid, lactate		2-20 mg/dL
Lead, whole blood		<10 mcg/dL
Lipase		
	Children	20-140 units/L
	Adults	0-190 units/L
Magnesium		1.5-2.5 mEq/L
Osmolality, serum		275-296 mOsm/kg
Osmolality, urine		50-1400 mOsm/kg

Table   (*Continued*)

	Normal Values	
Phosphorus	Newborns	4.2-9 mg/dL
	6 wk to 19 mo	3.8-6.7 mg/dL
	19 mo to 3 y	2.9-5.9 mg/dL
	3-15 y	3.6-5.6 mg/dL
	>15 y	2.5-5 mg/dL
Potassium, plasma	Newborns	4.5-7.2 mEq/L
	2 d to 3 mo	4-6.2 mEq/L
	3 mo to 1 y	3.7-5.6 mEq/L
	1-16 y	3.5-5 mEq/L
Protein, total	0-2 y	4.2-7.4 g/dL
	>2 y	6-8 g/dL
Sodium		136-145 mEq/L
Triglycerides	Infants	0-171 mg/dL
	Children	20-130 mg/dL
	Adults	30-200 mg/dL
Urea nitrogen, blood	0-2 y	4-15 mg/dL
	2 y to adult	5-20 mg/dL
Uric acid	Male	3-7 mg/dL
	Female	2-6 mg/dL
Enzymes		
Alanine aminotransferase (ALT) (SGPT)	0-2 mo	8-78 units/L
	>2 mo	8-36 units/L
Alkaline phosphatase (ALKP)	Newborns	60-130 units/L
	0-16 y	85-400 units/L
	>16 y	30-115 units/L
Aspartate aminotransferase (AST)	Infants	18-74 units/L
(SGOT)	Children	15-46 units/L
	Adults	5-35 units/L
Creatine kinase (CK)	Infants	20-200 units/L
	Children	10-90 units/L
	Adult male	0-206 units/L
	Adult female	0-175 units/L
Lactate dehydrogenase (LDH)	Newborns	290-501 units/L
	1 mo to 2 y	110-144 units/L
	>16 y	60-170 units/L

Blood Gases			
	Arterial	**Capillary**	**Venous**
pH	7.35-7.45	7.35-7.45	7.32-7.42
pCO_2 (mm Hg)	35-45	35-45	38-52
pO_2 (mm Hg)	70-100	60-80	24-48
HCO_3 (mEq/L)	19-25	19-25	19-25
TCO_2 (mEq/L)	19-29	19-29	23-33
O_2 saturation (%)	90-95	90-95	40-70
Base excess (mEq/L)	−5 to +5	−5 to +5	−5 to +5

Thyroid Function Tests		
T_4 (thyroxine)	1-7 d	10.1-20.9 mcg/dL
	8-14 d	9.8-16.6 mcg/dL
	1 mo to 1 y	5.5-16 mcg/dL
	>1 y	4-12 mcg/dL
FTI	1-3 d	9.3-26.6
	1-4 wk	7.6-20.8
	1-4 mo	7.4-17.9
	4-12 mo	5.1-14.5
	1-6 y	5.7-13.3
	>6 y	4.8-14
T_3 by RIA	Newborns	100-470 ng/dL
	1-5 y	100-260 ng/dL
	5-10 y	90-240 ng/dL
	10 y to adult	70-210 ng/dL
T_3 uptake		35%-45%
TSH	Cord	3-22 μIU/mL
	1-3 d	<40 μIU/mL
	3-7 d	<25 μIU/mL
	>7 d	0-10 μIU/mL

Renal Function Tests

Endogenous Creatinine Clearance vs Age (timed collection)

Creatinine clearance (mL/min/1.73 m^2) = (Cr$_u$V/Cr$_s$T) (1.73/A)
where:

Cr$_u$	=	urine creatinine concentration (mg/dL)
V	=	total urine collected during sampling period (mL)
Cr$_s$	=	serum creatinine concentration (mg/dL)
T	=	duration of sampling period (min) (24 h = 1440 min)
A	=	body surface area (m^2)

Age-specific normal values

5-7 d	50.6 ± 5.8 mL/min/1.73 m^2
1-2 mo	64.6 ± 5.8 mL/min/1.73 m^2
5-8 mo	87.7 ± 11.9 mL/min/1.73 m^2
9-12 mo	86.9 ± 8.4 mL/min/1.73 m^2
≥18 mo	
Male	124 ± 26 mL/min/1.73 m^2
Female	109 ± 13.5 mL/min/1.73 m^2
Adults	
Male	105 ± 14 mL/min/1.73 m^2
Female	95 ± 18 mL/min/1.73 m^2

Note: In patients with renal failure (creatinine clearance <25 mL/min), creatinine clearance may be elevated over GFR because of tubular secretion of creatinine.

Calculation of Creatinine Clearance From a 24-Hour Urine Collection
Equation 1:

$$Cl_{cr} = \frac{U \times V}{P}$$

where:
Cl_{cr} = creatinine clearance
U = urine concentration of creatinine
V = total urine volume in the collection
P = plasma creatinine concentration

Equation 2:

$$Cl_{cr} = \frac{(\text{total urine volume [mL]}) \times (\text{urine Cr concentration [mg/dL]})}{(\text{serum creatinine [mg/dL]}) \times (\text{time of urine collection [minutes]})}$$

Occasionally, a patient will have a 12- or 24-hour urine collection done for direct calculation of creatinine clearance. Although a urine collection for 24 hours is best, it is difficult to do since many urine collections occur for a much shorter period. A 24-hour urine collection is the desired duration of urine collection because the urine excretion of creatinine is diurnal and thus the measured creatinine clearance will vary throughout the day as the creatinine in the urine varies. When the urine collection is less than 24 hours, the total excreted creatinine will be affected by the time of the day during which the collection is performed. A 24-hour urine collection is sufficient to be able to accurately average the diurnal creatinine excretion variations. If a patient has 24 hours of urine collected for creatinine clearance, equation 1 can be used for calculating the creatinine clearance. To use equation 1 to calculate the creatinine clearance, it will be necessary to know the duration of urine collection, the urine collection volume, the urine creatinine concentration, and the serum creatinine value that reflects the urine collection period. In most cases, a serum creatinine concentration is drawn anytime during the day, but it is best to have the value drawn halfway through the collection period.

Amylase: Creatinine Clearance Ratio

$$\frac{\text{Amylase}_u \times \text{creatinine}_p}{\text{Amylase}_p \times \text{creatinine}_u} \times 100$$

u = urine; p = plasma

Serum BUN: Serum Creatinine Ratio

Serum BUN (mg/dL: serum creatinine (mg/dL))
Normal BUN: creatinine ratio is 10-15
BUN: creatinine ratio >20 suggests prerenal azotemia (also seen with high urea-generation states such as GI bleeding)
BUN: creatinine ratio <5 may be seen with disorders affecting urea biosynthesis such as urea cycle enzyme deficiencies and with hepatitis.

Fractional Sodium Excretion

Fractional sodium secretion (FENa) = $\text{Na}_u\text{Cr}_s/\text{Na}_s\text{Cr}_u \times 100\%$

where:
Na_u = urine sodium (mEq/L)
Na_s = serum sodium (mEq/L)
Cr_u = urine creatinine (mg/dL)
Cr_s = serum creatinine (mg/dL)

FENa <1% suggests prerenal failure
FENa >2% suggest intrinsic renal failure (for newborns, normal FENa is approximately 2.5%)
Note: Disease states associated with a falsely elevated FENa include severe volume depletion (>10%), early acute tubular necrosis, and volume depletion in chronic renal disease. Disorders associated with a lowered FENa include acute glomerulonephritis, hemoglobinuric or myoglobinuric renal failure, nonoliguric acute tubular necrosis, and acute urinary tract obstruction. In addition, FENa may be <1% in patients with acute renal failure and a second condition predisposing to sodium retention (eg, burns, congestive heart failure, nephrotic syndrome).

Urine Calcium: Urine Creatinine Ratio (spot sample)

Urine calcium (mg/dL): urine creatinine (mg/dL)
Normal values <0.21 (mean values 0.08 males, 0.06 females)
Premature infants show wide variability of calcium:creatinine ratio, and tend to have lower thresholds for calcium loss than older children. Prematures without nephrolithiasis had mean Ca:Cr ratio of 0.75 ± 0.76. Infants with nephrolithiasis had mean Ca:Cr ratio of 1.32 ± 1.03 (Jacinto JS, Modanlou HD, Crade M, et al, "Renal Calcification Incidence in Very Low Birth Weight Infants," *Pediatrics*, 1988, 81:31.)

Urine Protein:Urine Creatinine Ratio (spot sample)

P_u/Cr_u	Total Protein Excretion $(mg/m^2/d)$
0.1	80
1	800
10	8000

where:

P_u = urine protein concentration (mg/dL)

Cr_u = urine creatinine concentration (mg/dL)

Stress Replacement of Glucocorticoids

Recommendations for stress replacement of glucocorticoids vary. Because of the low risk involved with supplementation, some advocate administration of glucocorticoids for any patient who has received steroids, including topical steroids, within a year. Others reserve glucocorticoid administration for patients who have received more than a 14-day treatment of supraphysiologic steroid therapy within the past year. Yet others consider supplementation in any patient who has received corticosteroid therapy for at least 1 month in the past 6-12 months.

Steroid Status	Prednisone Dose*	Severity of Surgery	Steroid Regimen
Taking steroids	<10 mg/d	Any surgery	Additional steroid coverage not required; assume normal HPA response
	>10 mg/d	Minor surgery	25 mg hydrocortisone at induction
		Moderate surgery	Usual preoperative steroids plus 25 mg hydrocortisone at induction plus 100 mg hydrocortisone per day for 24 hours
		Major surgery	Usual preoperative steroids plus 25 mg hydrocortisone at induction plus 100 mg hydrocortisone per day for 48-72 hours
	High-dose immuno-suppression	Any surgery	Give usual immunosuppressive doses during perioperative period

*If patient receiving a different corticosteroid, please use the table below to convert to an equivalent prednisone dose.

Steroid Status	Time Off Steroid	Comments
Not currently taking steroids	<3 months	Treat as if on steroids
	>3 months	No perioperative steroids necessary

Please refer to the chart below for potency comparisons and equivalent dosing:

Glucocorticoid	Relative Potency		Equivalent Dose (mg)
	Anti-inflammatory	Mineralocorticoid	
Short Acting			
Cortisone	0.8	2	25
Hydrocortisone	1	2	20
Intermediate Acting			
Prednisone	4	1	5
Prednisolone	4	1	5
Triamcinolone	5	0	4
Methylprednisolone	5	0	4
Long Acting			
Dexamethasone	25-30	0	0.75
Betamethasone	25	0	0.6-0.75

References

Henriques HF III and Lebovic D, "Defining and Focusing Perioperative Steroid Supplementation," *Am Surg*, 1995, 61(9):809-13.

Nicholson G, Burrin JM, and Hall GM, "Peri-Operative Steroid Supplementation," *Anaesthesia*, 1998, 53(11):1091-104.

Salem M, Tainsh RE Jr, Bromberg J, et al, "Perioperative Glucocorticoid Coverage. A Reassessment 42 Years After Emergence of a Problem," *Ann Surg*, 1994, 219(4):416-25.

Weapons of Mass Destruction Preparedness Resources

Websites

American Academy of Clinical Toxicology
http://www.clintox.org

American Academy of Family Physicians
http://www.btresponse.org

American Association of Health-System Pharmacists, Emergency Preparedness-Counter-terrorism Resource Center
http://www.ashp.org/emergency

American College of Physicians - American Society of Internal Medicine, bioterrorism resources
http://www.acponline.org/bioterro/index.html

American Hospital Association, Chemical and Bioterrorism Preparedness Checklist

American Medical Association, Disaster Preparedness and Medical Response
http://www.ama-assn.org/ama/pub/category/6206.html

American Pharmaceutical Association Pharmacist Response Center
http://www.aphanet.org/pharmcare/ResponseCenter.htm

Annual Report to the President and Congress 2001 Chapter 7 - "Managing the Consequences of Domestic Weapons of Mass Destruction Incidents"
http://www.defenselink.mil/execsec/adr2001/index.html

Anthrax Vaccine Immunization Program
http://www.anthrax.osd.mil

Armed Forces Institute of Pathology (AFIP)
http://www.afip.org

Armed Forces Radiobiology Research Institute
http://www.afrri.usuhs.mil

Association for Professionals in Infection Control and Epidemiology, Inc, bioterrorism resources
http://www.apic.org/bioterror

The Beacon, National Disaster Preparedness Office, FBI

Canadian Center for Emergency Preparedness
http://www.ccep.ca

Center for Disease Control (CDC) and Prevention, Bioterrorism Preparedness and Response
http://www.bt.cdc.gov

Center for Disease Control (CDC) and Prevention, Bioterrorism Preparedness and Response Learning Resources
http://www.bt.cdc.gov/training/index.asp

Center for Disease Control (CDC) and Prevention, Bioterrorism Readiness Plan: A Template for Healthcare Facilities
http://www.cdc.gov/ncidod/hip/Bio/bio.htm

Center for Disease Control (CDC) and Prevention, Distance Learning Web Sites
http://www.cdc.gov/ncidod/hip/Bio/bio.htm

Center for Disease Control (CDC) and Prevention, National Center for Infectious Diseases
http://www.cdc.gov/ncidod/diseases/index.htm

Center for Disease Control (CDC) and Prevention, National Center for Infectious Diseases, Emerging Infectious Diseases
http://www.cdc.gov/ncidod/eid/index.htm

Center for Disease Control (CDC) and Prevention, National Laboratory Training Network
http://www.phppo.cdc.gov/nltn/default.asp

Center for Disease Control (CDC) and Prevention, Public Health Preparedness and Emergency Response
http://www.bt.cdc.gov/Agent/Agentlist.asp

Center for Disease Control (CDC) and Prevention, Public Health Training Network
http://www.phppo.cdc.gov/phtn/default.asp

Center for Disease Control (CDC) and Prevention, Recommended Protocol for Bioterrorism (BT) Event Notification
http://www.bt.cdc.gov/EmContact/Protocols.asp

Center for Disease Control (CDC) and Prevention, State Health Department Web Sites
http://www.phppo.cdc.gov/phtn/sites.asp

The Chemical and Biological Arms Control Institute
http://www.cbaci.org

The Chemical and Biological Arms Control Institute, the Dispatch (bimonthly report)
http://www.cbaci.org/dispatch.htm

Chemical and Biological Defense Information Analysis Center (CBIAC)
http://www.cbiac.apgea.army.mil

Chemical and Biological Defense Information Analysis Center (CBIAC), newsletter
http://www.cbiac.apgea.army.mil/awareness/newsletter/intro.html

DefenseLINK
http://www.defenselink.mil

Defense Threat Reduction Agency - Chem-Bio Defense
http://www.dtra.mil/cb/cb_index.html

Department of Defense (DOD), Global Emerging Infections Surveillance and Response System
http://www.geis.ha.osd.mil

Department of Defense (DOD), Nuclear, Biological, Chemical Medical Reference Site
http://www.nbc-med.org/others

Department of Defense (DOD), Office of Counterproliferation and Chemical/Biological Defense
http://www.acq.osd.mil/cp

Department of Justice (DOJ), National Center for Biomedical Research and Training (NCBRT) at Louisiana State University
http://www.doce.lsu.edu/ace/PagesNew/index.html

Department of Justice (DOJ), Office of Domestic Preparedness Support
http://www.ojp.usdoj.gov/odp

Department of Justice (DOJ), Office of Domestic Preparedness Support, Technical Support Help
http://www.ojp.usdoj.gov/odp/ta/ta.htm

Department of Justice (DOJ), Office of Domestic Preparedness Support, training courses
http://www.ojp.usdoj.gov/odp/ta/training.htm

Department of Transportation (DOT), Office of Hazardous Materials Safety
http://hazmat.dot.gov

Disaster Mortuary Operational Response Teams (DMORT)
http://www.dmort.org

Disaster Resource Guide
http://disaster-resource.com/

Domestic Preparedness (Commercial site - includes jobs, equipment, news)
http://www.domesticpreparedness.com

Domestic Preparedness - Conference and Event Listings (commercial site)
http://www.domesticpreparedness.com/members/calendar.html

The Emergency Information Infrastructure Partnership
http://www.emforum.org

The Emergency Information Infrastructure Partnership, newsletter
http://www.emforum.org/eiip/news.htm

Emergency Medical Services Magazine
http://www.emsmagazine.com/home.html

Emergency Preparedness News
http://www.bpinews.com/hs/pages/epn.cfm

Emergency Response and Research Institute
http://www.emergency.com

Emergency Response and Research Institute, Local Terrorism Planning Model
http://www.emergency.com/terrplan.htm

Energetic Materials Research and Testing Center (EMRTC), The National Domestic Preparedness Consortium, training programs
http://www.emrtc.nmt.edu/

Environmental Protection Agency (EPA), Chemical Emergency Preparedness and Prevention Office
http://www.epa.gov/swercepp, http://www.epa.gov/ceppo

Environmental Protection Agency (EPA), HAZMAT Conference Listings
http://yosemite.epa.gov/oswer/ceppoweb.nsf/content/events.htm

Environmental Protection Agency (EPA), National Response System
http://www.epa.gov/superfund/programs/er/nrs/index.htm

Federal Bureau of Investigation (FBI), The National Domestic Preparedness Office
http://www.fbi.gov/programs/ndpo/default.htm

Federal Emergency Management Agency (FEMA)
http://www.fema.gov

Federal Emergency Management Agency (FEMA), Federal Response Plan
http://www.fema.gov/rrr/frp

Federal Emergency Management Agency (FEMA), IMPACT newsletter (Region V)
http://www.fema.gov/regions/v/newsletter.shtm

Federal Emergency Management Agency (FEMA), National Fire Academy, compendium of WMD courses
http://www.usfa.fema.gov/fire-service/nfa.cfm

Federal Emergency Management Agency (FEMA), partners list (state EMAs, local agencies, national agencies)
http://www.fema.gov/about/partners.htm

Federal Emergency Management Agency (FEMA), terrorism consequence courses
http://training.fema.gov/EMIWeb/terrorismInfor/ctrt.asp

Federal Emergency Management Agency (FEMA), training programs
http://training.fema.gov/EMIWeb/EMICourses/index.asp

Health Sciences & Human Services Library, Terrorism resources for the healthcare community
http://www.hshsl.umaryland.edu/resources/terrorism.html

Henry L. Stimson Center, Chemical and Biological Weapons Nonproliferation Project
http://www.stimson.org/index.html

Hospital Emergency Incident Command System (HEICS) - San Mateo County
http://www.emsa.ca.gov/dms2/download.htm

Illinois Department of Public Health
http://www.idph.state.il.us

Illinois Emergency Management Agency
http://www.state.il.us/iema

Illinois Mobile Emergency Response Team (IMERT)
http://www.imert.org

Institute of Medicine
http://www.iom.edu

International Association of Emergency Managers
http://www.iaem.com

Johns Hopkins University, Center for Civilian Biodefense Studies
http://www.hopkins-biodefense.org

Johns Hopkins University, Center for Civilian Biodefense Studies, events List
http://www.hopkins-biodefense.org/pages/events/calevents.html

Johns Hopkins University, Center for Civilian Biodefense Studies, Biodefense Quarterly
http://www.hopkins-biodefense.org/pages/news/quarter.html

Joint Commission on Accreditation of Healthcare Organizations
http://www.jcaho.org

Joint Program Office-Biodefense (DOD office for biodefense equipment development)
http://www.jpobd.net

Joint Service Chemical Biological Information System, DOD (tracking system for equipment development)
http://206.37.238.107/jscbis/jscbis.cfm

The Journal of the American Medical Association (JAMA), Weapons of Mass Destruction Events With Contaminated Casualties: Effective Planning for Health Care Facilities (article)
http://jama.ama-assn.org/cgi/content/full/283/2/242

Journal of Terrorism and Political Violence
http://www.frankcass.com/jnls/tpv.htm

Mayo Clinic, Biological, Chemical weapons: Arm yourself with information
http://www.mayoclinic.com/invoke.cfm?id=MH00027

McGraw-Hill: Access Medicine, bioterrorism watch
http://www.accessmedicine.com/amed/public/amed_news/news_article/281.html

Medical NBC Online Information Server
http://www.nbc-med.org/ie40/Default.html

Medicine and Global Survival Magazine (MGS)
http://www.healthnet.org/index.php

Monterey Institute for International Studies, Center for Nonproliferation Studies
http://www.cns.miis.edu

Monterey Institute for International Studies, Center for Nonproliferation Studies, Federal Agency WMD Response Organizational Chart
http://www.cns.miis.edu/research/cbw.domest_p.htm

Monterey Institute for International Studies, chemical and biological weapons and resource page
http://www.cns.miis.edu/research/cbw/cbterror.htm

Monterey Institute of International Studies, The Nonproliferation Review, Center for Nonproliferation Studies
http://cns.miis.edu/pubs/npr/index.htm

The National Academies, responding first to bioterrorism
http://bob.nap.edu/shelves/first

National Academies of Science, Civilian Emergency Response to Chemical or Biological Weapons Incidents Project
http://www.nap.edu/html/terrorism

National Academies of Science, Institute of Medicine
http://www.iom.edu

National Association of County and City Health Officials, bioterrorism and emergency response program
http://www.naccho.org/project63.cfm

National Center for Complementary and Alternative Medicine, bioterrorism and CAM: What the public needs to know
http://www.nccam.nih.gov/health/alerts/bioterrorism

National Distaster Medical System, Team MA-1, Boston
http://www.ma1boston.com

The National Emergency Management Association
http://www.nemaweb.org/index.cfm

National Fire and Rescue Magazine
http://www.nfrmag.com

National Institutes of Health

http://www.nih.gov

National Library of Medicine-Medline Plus, disaster and emergency preparedness
http://www.nlm.nih.gov/medlineplus/disastersandemergencypreparedness.html

The National Response Team (HAZMAT & Chemical Spills)
http://www.nrt.org

National Strategy Review
http://www.nationalstrategy.com/nsr/current%20issue.htm

The New England Journal of Medicine, smallpox and smallpox vaccination
http://content.nejm.org/cgi/content/short/347/9/691

Occupational Safety & Health Administration (OSHA), Job Safety and Health Quarterly Magazine
http://www.osha-slc.gov/html/jshq-index.html

Overseas Security Advisory Council (OSAC)
http://www.ds-osac.org

San Francisco Department of Public Health
http://www.healthysf.org/pharmacy/education

The Terrorism Research Center (mostly international in nature)
http://www.terrorism.com

Texas A&M University, The National Emergency Rescue and Response Training Center
http://www.teexweb.tamu.edu/nerrtc

U.S. Agency for International Development (USAID) The Office of U.S. Foreign Disaster Assistance (OFDA)
http://www.usaid.gov/hum_response/ofda

U.S. Air Force Counterproliferation Center
http://www.au.af.mil/au/awc/awcgate/awc-cps.htm

U.S. Army Chemical School
http://www.wood.army.mil/usacmls

U.S. Army Medical Command
http://www.armymedicine.army.mil

U.S. Army Medical Department, links to training sites/conferences
http://www.nbc-med.org/others

U.S. Army Medical Research Institute of Chemical Defense
http://chemdef.apgea.army.mil

U.S. Army Medical Research Institute of Infectious Disease (USAMRIID)
http://www.usamriid.army.mil

U.S. Army National Guard Bureau Information, on military support, WMD teams, background studies
http://www.ngb.dtic.mil

U.S. Army Soldier and Biological Chemical Command (SBCCOM)
http://www.sbccom.apgea.army.mil

U.S. Army Soldier and Biological Chemical Command (SBCCOM), CB Quarterly
http://www.sbccom.apgea.army.mil/RDA/ecbc/quarterly/index.htm

U.S. Army Soldier and Biological Chemical Command (SBCCOM), Compendium of Weapons of Mass Distruction (WMD) courses
http://dp.sbccom.army.mil/fr/compendium/index.html

U.S. Army Soldier and Biological Chemical Command (SBCCOM), Edgewood Chemical Biological Center (ECBC)
http://www.sbccom.apgea.army.mil/RDA/ecbc

U.S. Army Soldier and Biological Chemical Command (SBCCOM), hotline
http://www.dp.sbccom.army.mil/fs/dp_hotline.html

U.S. Army Soldier and Biological Chemical Command (SBCCOM), Techical Escort Unit
http://teu.sbccom.army.mil

U.S. Coast Guard, Incident Command System, forms and guides
http://www.uscg.mil/bacarea/pm/icsforms/ics.htm

U.S. Coast Guard, The National Response Center
http://www.nrc.uscg.mil

U.S. Department of Health and Human Services (DHHS), Agency for Toxic Substances and Disease Registry (ATSDR), Public Health Service
http://www.atsdr.cdc.gov

U.S. Department of Health and Human Services (DHHS), Office of Emergency Preparedness (OEP) (manages the National Disaster Medical System)
http://www.ndms.dhhs.gov

U.S. Department of Health and Human Services (DHHS), Office of Emergency Preparedness consolidated training list
http://ndms.dhhs.gov/CT_Program/Training/training.html

U.S. Department of Health and Human Services (DHHS), Office of Emergency Preparedness planning documents/guides
http://ndms.dhhs.gov/CT_Program/Response_Planning/response_planning.html

U.S. Food and Drug Administration, bioterrorism
http://www.fda.gov/oc/opacom/hottopics/bioterrorism.html

U.S. Food and Drug Administration, Center for Food Safety and Applied Nutrition
http://vm.cfsan.fda.gov/list.html

U.S. Marine Corps, Chemical Biological Incident Response Force
http://www.mcwl.quantico.usmc.mil

U.S. Navy Chem/Bio Program
http://www.chembiodef.navy.mil/o_nav.htm

University of Pittsburgh Medical Center (UPMC) Health System, A Nation Prepared
http://prepared.upmc.com

World Health Organization (WHO)
http://www.who.int/en

World Health Organization (WHO), communicable disease surveillance and response
http://www.who.int/csr/en

Selected Contacts

Agency for Toxic Substances and Disease Registry (ATSDR)
(404) 639-6000

Center for Disease Control (CDC), Bioterrorism Emergency Hotline
(770) 488-7100

CHEMTREC
(800) 535-8200

Domestic Preparedness Help Line - email: cbhelp@sbccom.apgea.army.mil
(800) 535-0202

Emergency Nurses Association Weapons of Mass Distruction (WMD) Work Group
(800) 900-5969

Hospital Emergency Incident Command System
(916) 322-4336

National Domestic Preparedness Office (NDPO) - email: majortomleonard@aol.com
(202) 324-0275

National Pharmaceutical Stockpile Program (NPSP) Emergency Hotline
(770) 488-7516

National Response Center
(800) 424-8802

U.S. Army Medical Research Institute for Chemical Defense (USAMRICD), 3100 Ricketts Point Rd, Aberdeen Proving Ground, MD 21010
(410) 436-3628

U.S. Army Medical Research Institute for Infectious Disease (USARMIID)
(301) 619-4679

U.S. Department of Justice, Center for Domestic Preparedness
(256) 847-2134

U.S. Public Health Service
(800) 872-6367

VA Emergency Management Strategic Healthcare Group (EMSHG) EMSHG (104), Route 9, VAMC Bldg 203B Martinsburg, WV 25401
http://www.va.gov/emshg/(304) 264-4385

PATIENT INFORMATION SHEETS

Ammonia - Patient Information Sheet

This handout provides information and follow-up instructions for persons who have been exposed to ammonia gas or ammonium hydroxide solution (obtained from the CDC - Center for Disease Control - www.atsdr.cdc.gov/MHMII/mmg126-handout.pdf)

What is ammonia?
Ammonia is a colorless, highly irritating gas with a sharp, suffocating odor. It easily dissolves in water to form a caustic solution called ammonium hydroxide. It is not highly flammable, but containers of ammonia may explode when exposed to high heat. About 80% of the ammonia produced is used in fertilizers. It is also used as a refrigerant and in the manufacture of plastics, explosives, pesticides, and other chemicals. It is found in many household and industrial-strength cleaning solutions.

What immediate health effects can result from ammonia exposure?
Most people are exposed to ammonia from breathing the gas. They will notice the pungent odor and experience burning of the eyes, nose, and throat after breathing even small amounts. With higher doses, coughing or choking may occur. Exposure to high levels of ammonia can cause death from a swollen throat or from chemical burns to the lungs. Skin contact with ammonia-containing liquids may cause burns. Eye exposure to concentrated gas or liquid can cause serious corneal burns or blindness. Drinking a concentrated ammonia solution can cause burns to the mouth, throat, and stomach. Generally, the severity of symptoms depends on the degree of exposure.

Can ammonia poisoning be treated?
There is no antidote for ammonia poisoning, but ammonia's effects can be treated, and most people recover. Persons who have experienced serious signs and symptoms (such as severe or persistent coughing or burns in the throat) may need to be hospitalized.

Are any future health effects likely to occur?
A single small exposure from which a person recovers quickly is not likely to cause delayed or long-term effects. After a severe exposure, injury to the eyes, lungs, skin, or digestive system may continue to develop for 18 to 24 hours, and serious delayed effects, such as gastric perforation, chronic pulmonary obstructive disease, or glaucoma, are possible.

What tests can be done if a person has been exposed to ammonia?
Specific tests for the presence of ammonia in blood or urine generally are not useful to the doctor. If a severe exposure has occurred, blood and urine analyses, chest x-rays, and other tests may show whether the lungs have been injured. Testing is not needed in every case. If ammonia contacts the eyes, the doctor may put a special dye in the eyes and examine them with a magnifying lamp.

Where can more information about ammonia be found?
More information about ammonia can be obtained from your regional poison control center; your state, county, or local health department; the Agency for Toxic Substances and Disease Registry (ATSDR); your doctor; or a clinic in your area that specializes in occupational or environmental health. If the exposure happened at work, you may wish to discuss it with your employer, the Occupational Safety and Health Administration (OSHA), or the National Institute for Occupational Safety and Health (NIOSH). Ask the person who gave you this form for help in locating these telephone numbers.

* This section is reprinted from the CDC website, www.cdc.gov

FOLLOW-UP INSTRUCTIONS

Keep this page and take it with you to your next appointment. Follow only the instructions checked below.

[ ]	Call your doctor or the Emergency Department if you develop any unusual signs or symptoms within the next 24 hours, especially:
•	coughing
•	difficulty breathing or shortness of breath
•	wheezing or high-pitched voice
•	chest pain or tightness
•	increased pain or a discharge from exposed eyes
•	increased redness or pain or a pus-like discharge in the area of a skin burn
•	stomach pain or vomiting
[ ]	No follow-up appointment is necessary unless you develop any of the symptoms listed above.
[ ]	Call for an appointment with Dr _____ in the practice of _____ _____. When you call for your appointment, please say that you were treated in the Emergency Department at _____ Hospital by _____ and were advised to be seen again in _____ days.
[ ]	Return to the Emergency Department/ _____ Clinic on (date) _____ at _____ ____AM/PM for a follow-up examination.
[ ]	Do not perform vigorous physical activities for 1-2 days.
[ ]	You may resume everyday activities including driving and operating machinery.
[ ]	Do not return to work for _____ days.
[ ]	You may return to work on a limited basis. See instructions below.
[ ]	Avoid exposure to cigarette smoke for 72 hours; smoke may worsen the condition of your lungs.
[ ]	Avoid drinking alcoholic beverages for at least 24 hours; alcohol may worsen injury to your stomach or have other effects.
[ ]	Avoid taking the following medications:
[ ]	You may continue taking the following medication(s) that your doctor(s) prescribed for you:
[ ]	Other instructions:
•	Provide the Emergency Department with the name and the number of your primary care physician so that the ED can send him or her a record of your emergency department visit.
•	You or your physician can get more information on the chemical by contacting: _____ _____ or _____, or by checking out the following internet web sites: _____.

Signature of patient _____ Date _____

Signature of physician _____ Date _____

Arsine - Patient Information Sheet

This handout provides information and follow-up instructions for persons who have been exposed to arsine (obtained from the CDC - Center for Disease Control - www.atsdr.cdc.gov/MHMII/mmg169-handout.pdf)

What is arsine?

Arsine is a colorless, flammable gas that does not burn the eyes, nose, or throat. At high concentrations it has a garlic-like or fishy smell, but a person can be exposed to a hazardous concentration of arsine and may not be able to smell it. Arsine is widely used in the manufacturing of fiberoptic equipment and computer microchips. It is sometimes used in galvanizing, soldering, etching, and lead plating. Certain ores or metals may contain traces of arsenic. If water or acid contacts these ores or metals, they may release arsine gas at hazardous levels.

What immediate health effects can result from arsine exposure?

Breathing in arsine gas can be very harmful, even in small quantities. The main effect of arsine poisoning is to destroy red blood cells, causing anemia (lack of red blood cells) and kidney damage (from circulating red-blood-cell debris). Initially, exposed individuals may feel relatively well. Within hours after a serious exposure, the victim may develop headache, weakness, shortness of breath, and back or stomach pain with nausea and vomiting; the urine may turn a dark red, brown or greenish color. The skin may become yellow or bronze in color, the eyes red or green. Generally, the more serious the exposure, the worse the symptoms. Although arsine is related to arsenic, it does not produce the usual signs and symptoms of arsenic poisoning.

Can arsine poisoning be treated?

There is no antidote for arsine, but its effects can be treated. A doctor may give the exposed patient fluids through a vein to protect the kidneys from damage. For severe poisoning, blood transfusions and cleansing of the blood (hemodialysis) may be needed to prevent worsening kidney damage.

Are any future health effects likely to occur?

After a serious exposure, symptoms usually begin within 2-24 hours (see the Follow-up Instructions). Most people do not develop long-term effects from a single, small exposure to arsine. In rare cases, permanent kidney damage or nerve damage has developed after a severe exposure. Repeated exposures to arsine over a long period of time might cause skin or lung cancer, but this has not been studied.

What tests can be done if a person has been exposed to arsine?

Specific tests can show the amount of arsenic in urine, but this information may or may not be helpful to the doctor. Standard tests of blood, urine, and other measures of health may show whether exposure has caused serious injury to the lungs, blood cells, kidneys, or nerves. Since toxic effects of arsine poisoning may be delayed, testing should be done in all cases of suspected exposure to arsine.

Where can more information about arsine be found?

More information about arsine can be obtained from your regional poison control center; your state, county, or local health department; the Agency for Toxic Substances and Disease Registry (ATSDR); your doctor; or a clinic in your area that specializes in occupational or environmental health. If the exposure happened at work, you may wish to discuss it with your employer, the Occupational Safety and Health Administration (OSHA), or the National Institute for Occupational Safety and Health (NIOSH). Ask the person who gave you this form for help in locating these telephone numbers.

651

FOLLOW-UP INSTRUCTIONS

Keep this page and take it with you to your next appointment. Follow only the instructions checked below.

[ ]	Call your doctor or the Emergency Department if you develop any unusual signs or symptoms within the next 24-72 hours, especially:
•	unusual fatigue or weakness
•	shortness of breath
•	abnormal urine color (red or brown)
•	stomach pain or tenderness
•	unusual skin color (yellow or bronze)
[ ]	No follow-up appointment is necessary unless you develop any of the symptoms listed above.
[ ]	Call for an appointment with Dr _____ in the practice of _____. When you call for your appointment, please say that you were treated in the Emergency Department at _____ Hospital by _____ and were advised to be seen again in _____ days.
[ ]	Return to the Emergency Department/ _____ Clinic on (date) _____ at _____ AM/PM for a follow-up examination.
[ ]	Do not perform vigorous physical activities for 1-2 days.
[ ]	You may resume everyday activities including driving and operating machinery.
[ ]	Do not return to work for _____ days.
[ ]	You may return to work on a limited basis. See instructions below.
[ ]	Avoid exposure to cigarette smoke for 72 hours; smoke may worsen the condition of your lungs.
[ ]	Avoid drinking alcoholic beverages for at least 24 hours; alcohol may worsen injury to your stomach or have other effects.
[ ]	Avoid taking the following medications:
[ ]	You may continue taking the following medication(s) that your doctor(s) prescribed for you:
[ ]	Other instructions:
•	Provide the Emergency Department with the name and the number of your primary care physician so that the ED can send him or her a record of your emergency department visit.
•	You or your physician can get more information on the chemical by contacting: _____ _____ or _____, or by checking out the following internet web sites: _____

Signature of patient _____ Date _____

Signature of physician _____ Date _____

Benzene - Patient Information Sheet

This handout provides information and follow-up instructions for persons who have been exposed to benzene (obtained from the CDC - Center for Disease Control - www.atsdr.cdc.gov/MHMII/mmg3-handout.pdf)

What is benzene?
Benzene is a clear, colorless liquid with a sweet odor when in pure form. It burns readily. Benzene is obtained from crude petroleum. Small amounts may be found in products such as cigarette smoke, paints, glues, pesticides, and gasoline.

What immediate health effects can result from exposure to benzene?
Breathing benzene vapor in small amounts can cause headache, dizziness, drowsiness, or nausea. With more serious exposure, benzene may cause sleepiness, stumbling, irregular heartbeats, fainting, or even death. Benzene vapors are mildly irritating to the skin, eyes, and lungs. If liquid benzene contacts the skin or eyes, it may cause burning pain. Liquid benzene splashed in the eyes can damage the eyes. The degree of symptoms depends on the amount of exposure. Special consideration regarding the exposure of pregnant women is warranted since benzene has been shown to have a small negative effect on genes and crosses the placenta; thus, medical counseling is recommended for the acutely exposed pregnant woman.

Can benzene poisoning be treated?
There is no specific antidote for benzene, but its effects can be treated, and most exposed persons recover fully. Persons who have experienced serious symptoms may need to be hospitalized.

Are any future health effects likely to occur?
A single small exposure from which a person recovers quickly is not likely to cause delayed or long-term effects. After a severe exposure, some symptoms may take a few days to develop. Repeated exposure to benzene may cause a blood disorder (ie, aplastic anemia and pancytopenia) and cancer of blood-forming cells (ie, leukemia). Aplastic anemia and leukemia have been reported in some workers exposed repeatedly to benzene over long periods of time.

What tests can be done if a person has been exposed to benzene?
Specific tests for the presence of benzene in blood generally are not useful to the doctor. Phenol, muconic acid, or S-phenyl-N-acetyl cysteine (PhAC), breakdown products of benzene, can be measured in urine to prove benzene exposure. Other tests may show whether injury has occurred in the heart, kidneys, blood, or nervous system. Testing is not needed in every case.

Where can more information about benzene be found?
More information about benzene can be obtained from your regional poison control center; the state, county, or local health department; the Agency for Toxic Substances and Disease Registry (ATSDR); your doctor; or a clinic in your area that specializes in occupational and environmental health. If the exposure happened at work, you may wish to discuss it with your employer, the Occupational Safety and Health Administration (OSHA), or the National Institute for Occupational Safety and Health (NIOSH). Ask the person who gave you this form for help in locating these telephone numbers.

653

FOLLOW-UP INSTRUCTIONS

Keep this page and take it with you to your next appointment. Follow only the instructions checked below.

[ ]	Call your doctor or the Emergency Department if you develop any unusual signs or symptoms within the next 24 hours, especially:
•	eye and skin irritation
•	bronchial irritation, cough, hoarseness, tightness in chest
•	drowsiness, dizziness, headache, convulsions
•	irregular heart beats
[ ]	No follow-up appointment is necessary unless you develop any of the symptoms listed above.
[ ]	Call for an appointment with Dr _____ in the practice of _____ _____. When you call for your appointment, please say that you were treated in the Emergency Department at _____ Hospital by _____ and were advised to be seen again in _____ days.
[ ]	Return to the Emergency Department/ _____ Clinic on (date) _____ at _____AM/PM for a follow-up examination.
[ ]	Do not perform vigorous physical activities for 1-2 days.
[ ]	You may resume everyday activities including driving and operating machinery.
[ ]	Do not return to work for _____ days.
[ ]	You may return to work on a limited basis. See instructions below.
[ ]	Avoid exposure to cigarette smoke for 72 hours; smoke may worsen the condition of your lungs.
[ ]	Avoid drinking alcoholic beverages for at least 24 hours; alcohol may worsen injury to your stomach or have other effects.
[ ]	Avoid taking the following medications:
[ ]	You may continue taking the following medication(s) that your doctor(s) prescribed for you:
[ ]	Other instructions:
•	Provide the Emergency Department with the name and the number of your primary care physician so that the ED can send him or her a record of your emergency department visit.
•	You or your physician can get more information on the chemical by contacting: _____ _____ or _____, or by checking out the following internet web sites: _____ .

Signature of patient _____ Date _____

Signature of physician _____ Date _____

Chlorine - Patient Information Sheet

This handout provides information and follow-up instructions for persons who have been exposed to chlorine (obtained from the CDC - Center for Disease Control - www.atsdr.cdc. gov/MHMII/mmg172-handout.pdf)

What is chlorine?
Chlorine is a yellowish-green gas with a sharp, burning odor. It is used widely in chemical manufacturing, bleaching, drinking water and swimming pool disinfecting, and in cleaning agents. Household chlorine bleach contains only a small amount of chlorine but it can release chlorine gas if mixed with other cleaning agents.

What immediate health effects can be caused by exposure to chlorine?
Even small exposures to the gas may cause immediate burning of the eyes, nose, and throat, and shortness of breath, as well as coughing, wheezing, shortness of breath, and tearing of the eyes. However, once exposure is stopped, symptoms usually clear up quickly. Breathing large amounts of chlorine may cause the lining of the throat and lungs to swell, making breathing difficult. Generally, the more serious the exposure, the more severe the symptoms.

Can chlorine poisoning be treated?
There is no antidote for chlorine, but its effects can be treated and most exposed persons get well. Persons who have experienced serious symptoms may need to be hospitalized.

Are any future health effects likely to occur?
A single small exposure from which a person recovers quickly is not likely to cause delayed or long-term effects. After a serious exposure, symptoms may worsen for several hours.

What tests can be done if a person has been exposed to chlorine?
Specific tests for the presence of chlorine in blood or urine generally are not useful to the doctor. If a severe exposure has occurred, blood and urine analyses and other tests may show whether the lungs, heart, or brain has been injured. Testing is not needed in every case.

Where can more information about chlorine be found?
More information about chlorine can be obtained from your regional poison control center, your state, county, or local health department; the Agency for Toxic Substances and Disease Registry (ATSDR); your doctor; or a clinic in your area that specializes in occupational and environmental health. If the exposure happened at work, you may wish to discuss it with your employer, the Occupational Safety and Health Administration (OSHA), or the National Institute for Occupational Safety and Health (NIOSH). Ask the person who gave you this form for help in locating these telephone numbers.

FOLLOW-UP INSTRUCTIONS

Keep this page and take it with you to your next appointment. Follow only the instructions checked below.

[ ]	Call your doctor or the Emergency Department if you develop any unusual signs or symptoms within the next 24 hours, especially:
•	coughing, or wheezing
•	difficulty breathing, shortness of breath, or chest pain
•	increased pain or a discharge from injured eyes
•	increased redness or pain or a pus-like discharge in the area of a skin burn
[ ]	No follow-up appointment is necessary unless you develop any of the symptoms listed above.
[ ]	Call for an appointment with Dr _____ in the practice of _____ _____. When you call for your appointment, please say that you were treated in the Emergency Department at _____ Hospital by _____ and were advised to be seen again in _____ days.
[ ]	Return to the Emergency Department/ _____ Clinic on (date) _____ at _____AM/PM for a follow-up examination.
[ ]	Do not perform vigorous physical activities for 1-2 days.
[ ]	You may resume everyday activities including driving and operating machinery.
[ ]	Do not return to work for _____ days.
[ ]	You may return to work on a limited basis. See instructions below.
[ ]	Avoid exposure to cigarette smoke for 72 hours; smoke may worsen the condition of your lungs.
[ ]	Avoid drinking alcoholic beverages for at least 24 hours; alcohol may worsen injury to your stomach or have other effects.
[ ]	Avoid taking the following medications:
[ ]	You may continue taking the following medication(s) that your doctor(s) prescribed for you:
[ ]	Other instructions: _____
•	Provide the Emergency Department with the name and the number of your primary care physician so that the ED can send him or her a record of your emergency department visit.
•	You or your physician can get more information on the chemical by contacting: _____ _____ or _____, or by checking out the following internet web sites: _____

Signature of patient _____ Date _____

Signature of physician _____ Date _____

Hydrogen Cyanide - Patient Information Sheet

This handout provides information and follow-up instructions for persons who have been exposed to hydrogen cyanide (obtained from the CDC - Center for Disease Control - www.atsdr.cdc.gov/MHMII/mmg8-handout.pdf)

What is hydrogen cyanide?
At room temperature, hydrogen cyanide is a volatile, colorless-to-blue liquid (also called hydrocyanic acid). It rapidly becomes a gas that can produce death in minutes if breathed. Hydrogen cyanide is used in making fibers, plastics, dyes, pesticides, and other chemicals, and as a fumigant to kill rats. It is also used in electroplating metals and in developing photographic film.

What immediate health effects can be caused by exposure to hydrogen cyanide?
Breathing small amounts of hydrogen cyanide may cause headache, dizziness, weakness, nausea, and vomiting. Larger amounts may cause gasping, irregular heartbeats, seizures, fainting, and even rapid death. Generally, the more serious the exposure, the more severe the symptoms. Similar symptoms may be produced when solutions of hydrogen cyanide are ingested or come in contact with the skin.

Can hydrogen cyanide poisoning be treated?
The treatment for cyanide poisoning includes breathing pure oxygen, and in the case of serious symptoms, treatment with specific cyanide antidotes. Persons with serious symptoms will need to be hospitalized.

Are any future health effects likely to occur?
A single small exposure from which a person recovers quickly is not likely to cause delayed or long-term effects. After a serious exposure, a patient may have brain or heart damage.

What tests can be done if a person has been exposed to hydrogen cyanide?
Specific tests for the presence of cyanide in blood and urine generally are not useful to the doctor. If a severe exposure has occurred, blood and urine analyses and other tests may show whether the brain or heart has been injured. Testing is not needed in every case.

Where can more information about hydrogen cyanide be found?
More information about hydrogen cyanide can be obtained from your regional poison control center; your state, county, or local health department; the Agency for Toxic Substances and Disease Registry (ATSDR); your doctor; or a clinic in your area that specializes in occupational and environmental health. If the exposure happened at work, you may wish to discuss it with your employer, the Occupational Safety and Health Administration (OSHA), or the National Institute for Occupational Safety and Health (NIOSH). Ask the person who gave you this form for help in locating these telephone numbers.

FOLLOW-UP INSTRUCTIONS

Keep this page and take it with you to your next appointment. Follow only the instructions checked below.

[ ]	Call your doctor or the Emergency Department if you develop any unusual signs or symptoms within the next 24 hours, especially:
•	difficulty breathing, shortness of breath, or chest pain
•	confusion or fainting
•	increased pain or a discharge from your eyes
•	increased redness, pain, or a pus-like discharge in the area of a skin burn
[ ]	No follow-up appointment is necessary unless you develop any of the symptoms listed above.
[ ]	Call for an appointment with Dr _____ in the practice of _____ _____. When you call for your appointment, please say that you were treated in the Emergency Department at _____ _____ Hospital by _____ and were advised to be seen again in _____ days.
[ ]	Return to the Emergency Department/ _____ Clinic on (date) _____ at _____AM/PM for a follow-up examination.
[ ]	Do not perform vigorous physical activities for 1-2 days.
[ ]	You may resume everyday activities including driving and operating machinery.
[ ]	Do not return to work for _____ days.
[ ]	You may return to work on a limited basis. See instructions below.
[ ]	Avoid exposure to cigarette smoke for 72 hours; smoke may worsen the condition of your lungs.
[ ]	Avoid drinking alcoholic beverages for at least 24 hours; alcohol may worsen injury to your stomach or have other effects.
[ ]	Avoid taking the following medications: _____ _____ _____
[ ]	You may continue taking the following medication(s) that your doctor(s) prescribed for you: _____ _____ _____ _____
[ ]	Other instructions: _____ _____ _____ _____
•	Provide the Emergency Department with the name and the number of your primary care physician so that the ED can send him or her a record of your emergency department visit.
•	You or your physician can get more information on the chemical by contacting: _____ _____ or _____, or by checking out the following internet web sites: _____ _____.

Signature of patient _____ Date _____

Signature of physician _____ Date _____

Lewisite and Mustard-Lewisite Mixture - Patient Information Sheet

This handout provides information and follow-up instructions for persons who have been exposed to lewisite or mustard-lewisite mixture (obtained from the CDC - Center for Disease Control - www.atsdr.cdc.gov/MHMII/mmgd1-handout.pdf)

What are lewisite and mustard-lewisite mixture?
Lewisite is a chemical warfare agent that was first produced in 1918. It has not been used in warfare, although it may be stockpiled by some countries. Mustard-Lewisite Mixture is a mixture of Lewisite and Mustard. It was developed to achieve a lower freezing point for ground dispersal and aerial spraying.

What immediate health effects can be caused by exposure to lewisite and mustard-lewisite mixture?
Lewisite and Mustard-Lewisite Mixture produce pain and skin irritation immediately after exposure. Both compounds cause skin blisters and damage to the airways and eyes. They are also extremely irritating to the eyes, skin, nose, and throat. Exposure to very high levels may result in kidney and liver damage. Mustard-Lewisite Mixture can also damage the immune system.

Can lewisite and mustard-lewisite mixture poisoning be treated?
Immediate decontamination reduces symptoms. Intramuscular injection of British Anti-Lewisite (BAL) may be used to treat severe conditions but will not prevent lesions on the skin, eye, or airways. Persons who have been exposed to large amounts of Lewisite and Mustard-Lewisite Mixture will need to be hospitalized.

Are any future health effects likely to occur?
Adverse health effects, such as chronic respiratory diseases, may occur from exposure to high levels of these agents. Severe damage to the eye may be present for a long time after the exposure.

What tests can be done if a person has been exposed to lewisite or mustard-lewisite mixture?
There is no specific test to confirm exposure to Lewisite or Mustard-Lewisite Mixture; however, measurement of arsenic in the urine may help to identify exposure.

Where can more information about lewisite or mustard-lewisite mixture be found?
More information about Lewisite and Mustard-Lewisite Mixture can be obtained from your regional poison control center; the Agency for Toxic Substances and Disease Registry (ATSDR); your doctor; or a clinic in your area that specializes in toxicology or occupational and environmental health. Ask the person who gave you this form for help locating these telephone numbers.

FOLLOW-UP INSTRUCTIONS

Keep this page and take it with you to your next appointment. Follow only the instructions checked below.

[ ]	Call your doctor or the Emergency Department if you develop any unusual signs or symptoms within the next 24 hours, especially:
•	coughing, wheezing, shortness of breath, or discolored sputum
•	increased pain or discharge from injured eyes
•	increased redness, pain, or a pus-like discharge from injured skin; fever; or chills
[ ]	No follow-up appointment is necessary unless you develop any of the symptoms listed above.
[ ]	Call for an appointment with Dr _____ in the practice of _____ _____. When you call for your appointment, please say that you were treated in the Emergency Department at _____ Hospital by _____ and were advised to be seen again in _____ days.
[ ]	Return to the Emergency Department/ _____ Clinic on (date) _____ at _____AM/PM for a follow-up examination.
[ ]	Do not perform vigorous physical activities for 1-2 days.
[ ]	You may resume everyday activities including driving and operating machinery.
[ ]	Do not return to work for _____ days.
[ ]	You may return to work on a limited basis. See instructions below.
[ ]	Avoid exposure to cigarette smoke for 72 hours; smoke may worsen the condition of your lungs.
[ ]	Avoid drinking alcoholic beverages for at least 24 hours; alcohol may worsen injury to your stomach or have other effects.
[ ]	Avoid taking the following medications: _____ _____ _____
[ ]	You may continue taking the following medication(s) that your doctor(s) prescribed for you: _____ _____ _____ _____
[ ]	Other instructions: _____ _____ _____ _____
•	Provide the Emergency Department with the name and the number of your primary care physician so that the ED can send him or her a record of your emergency department visit.
•	You or your physician can get more information on the chemical by contacting: _____ _____ or _____, or by checking out the following internet web sites: _____ _____.

Signature of patient _____ Date _____

Signature of physician _____ Date _____

Nerve Agents - Patient Information Sheet

This handout provides information and follow-up instructions for persons who have been exposed to nerve agents (obtained from the CDC - Center for Disease Control - www.atsdr.cdc.gov/MHMII/mmg166-handout.pdf)

What are nerve agents?
Nerve agents are chemical warfare agents, similar to but much more potent than organophosphate insecticides. They are colorless to amber-colored, tasteless liquids that may evaporate to create a gas. GB and VX are odorless, while GA has a slight fruity odor, and GD has a slight camphor odor.

What immediate health effects can result from exposure to nerve agents?
Nerve agents are extremely toxic chemicals that attack the nervous system. As little as one drop to a few milliliters of nerve agent contacting the skin can cause death within 15 minutes. Nerve agent exposure can cause runny nose, sweating, blurred vision, headache, difficulty breathing, drooling, nausea, vomiting, muscle cramps and twitching, confusion, convulsions, paralysis, and coma. Symptoms occur immediately if you inhale nerve agent vapor but may be delayed for several hours if you get nerve agent liquid on your skin.

Can nerve agent poisoning be treated?
There are antidotes for nerve agent poisoning but they must be administered quickly after exposure. Immediate decontamination is critical and hospitalization may be needed.

Are any future health effects likely to occur?
Complete recovery may take several months. After a severe exposure with prolonged seizures, permanent damage to the central nervous system is possible.

What tests can be done if a person has been exposed to nerve agents?
Activity of a blood enzyme called acetylcholinesterase can be measured to assess exposure and recovery.

Where can more information about nerve agents be found?
More information about nerve agents can be obtained from your regional poison control center; the Agency for Toxic Substances and Disease Registry (ATSDR); your doctor; or a clinic in your area that specializes in toxicology or occupational and environmental health. Ask the person who gave you this form for help locating these telephone numbers.

661

FOLLOW-UP INSTRUCTIONS

Keep this page and take it with you to your next appointment. Follow only the instructions checked below.

[ ]	Call your doctor or the Emergency Department if you develop any unusual signs or symptoms within the next 24 hours, especially:
•	dizziness, loss of coordination, loss of memory
•	coughing, wheezing, or shortness of breath
•	nausea, vomiting, cramps, or diarrhea
•	muscle weakness or twitching
•	blurred vision
[ ]	No follow-up appointment is necessary unless you develop any of the symptoms listed above.
[ ]	Call for an appointment with Dr _____ in the practice of _____ _____. When you call for your appointment, please say that you were treated in the Emergency Department at _____ Hospital by _____ and were advised to be seen again in _____ days.
[ ]	Return to the Emergency Department/ _____ Clinic on (date) _____ at _____AM/PM for a follow-up examination.
[ ]	Do not perform vigorous physical activities for 1-2 days.
[ ]	You may resume everyday activities including driving and operating machinery.
[ ]	Do not return to work for _____ days.
[ ]	You may return to work on a limited basis. See instructions below.
[ ]	Avoid exposure to cigarette smoke for 72 hours; smoke may worsen the condition of your lungs.
[ ]	Avoid drinking alcoholic beverages for at least 24 hours; alcohol may worsen injury to your stomach or have other effects.
[ ]	Avoid taking the following medications: _____ _____ _____
[ ]	You may continue taking the following medication(s) that your doctor(s) prescribed for you: _____ _____ _____ _____
[ ]	Other instructions: _____ _____ _____ _____
•	Provide the Emergency Department with the name and the number of your primary care physician so that the ED can send him or her a record of your emergency department visit.
•	You or your physician can get more information on the chemical by contacting: _____ _____ or _____, or by checking out the following internet web sites: _____ _____.

Signature of patient _____ Date _____

Signature of physician _____ Date _____

Nitrogen Mustard (HN-1, HN-2, and HN-3) - Patient Information Sheet

This handout provides information and follow-up instructions for persons who have been exposed to nitrogen mustards (obtained from the CDC - Center for Disease Control - www.atsdr.cdc.gov/MHMII/mmg164-handout.pdf)

What are nitrogen mustards?
Nitrogen mustards are compounds that were initially developed as chemical warfare agents or pharmaceuticals. They have never been used on the battlefield. HN-2 has been used in chemotherapy.

What immediate health effects can be caused by exposure to nitrogen mustards?
Nitrogen mustards cause injury to the skin, eyes, nose and throat. Eye damage may occur within minutes of exposure. Nausea and vomiting also may occur shortly after exposure. Skin rashes, blisters, and lung damage may develop within a few hours of exposure but may take 6 hours or more. Nitrogen mustards can also suppress the immune system.

Can nitrogen mustard poisoning be treated?
There is no antidote for nitrogen mustard, but its effects can be treated and most exposed people recover. Immediate decontamination reduces symptoms. People who have been exposed to large amounts of nitrogen mustard will need to be treated in a hospital.

Are any future health effects likely to occur?
Adverse health effects, such as chronic respiratory diseases, may occur from exposure to high levels of these agents. Severe damage to the eye may be present for a long time following the exposure.

What tests can be done if a person has been exposed to nitrogen mustard?
There are no routine tests to confirm exposure.

Where can more information about nitrogen mustard be found?
More information about nitrogen mustards can be obtained from your regional poison control center; the Agency for Toxic Substances and Disease Registry (ATSDR); your doctor; or a clinic in your area that specializes in toxicology or occupational and environmental health. Ask the person who gave you this form for help locating these telephone numbers.

FOLLOW-UP INSTRUCTIONS

Keep this page and take it with you to your next appointment. Follow only the instructions checked below.

[ ]	Call your doctor or the Emergency Department if you develop any unusual signs or symptoms within the next 24 hours, especially:
•	coughing, wheezing, shortness of breath, or discolored sputum
•	increased redness, pain, or a pus-like discharge from injured skin
•	fever or chills
[ ]	No follow-up appointment is necessary unless you develop any of the symptoms listed above.
[ ]	Call for an appointment with Dr _____ in the practice of _____ _____. When you call for your appointment, please say that you were treated in the Emergency Department at _____ Hospital by _____ and were advised to be seen again in _____ days.
[ ]	Return to the Emergency Department/ _____ Clinic on (date) _____ at _____AM/PM for a follow-up examination.
[ ]	Do not perform vigorous physical activities for 1-2 days.
[ ]	You may resume everyday activities including driving and operating machinery.
[ ]	Do not return to work for _____ days.
[ ]	You may return to work on a limited basis. See instructions below.
[ ]	Avoid exposure to cigarette smoke for 72 hours; smoke may worsen the condition of your lungs.
[ ]	Avoid drinking alcoholic beverages for at least 24 hours; alcohol may worsen injury to your stomach or have other effects.
[ ]	Avoid taking the following medications: _____ _____
[ ]	You may continue taking the following medication(s) that your doctor(s) prescribed for you: _____ _____ _____ _____
[ ]	Other instructions: _____ _____ _____ _____
•	Provide the Emergency Department with the name and the number of your primary care physician so that the ED can send him or her a record of your emergency department visit.
•	You or your physician can get more information on the chemical by contacting: _____ _____ or _____, or by checking out the following internet web sites: _____ _____.

Signature of patient _____ Date _____

Signature of physician _____ Date _____

Phosgene - Patient Information Sheet

This handout provides information and follow-up instructions for persons who have been exposed to phosgene (obtained from the CDC - Center for Disease Control - www.atsdr.cdc. gov/MHMII/mmg176-handout.pdf)

What is phosgene?
At room temperature, phosgene is a colorless gas. At high concentrations, it has a suffocating odor; at low concentrations, it smells like green corn or new mown hay. It is not flammable. Phosgene is used in the manufacture of many chemicals. It is also produced when chlorine-containing chemicals burn or break down.

What immediate health effects can be caused by exposure to phosgene?
Most exposures to phosgene occur from breathing the gas. Exposure to small amounts usually causes eye, nose, and throat irritation. However, the irritating effects can be so mild at first that the person does not leave the area of exposure. Generally, the higher the exposure, the more severe the symptoms. Extended exposure can cause severe breathing difficulty, which may lead to chemical pneumonia and death. Severe breathing problems may not develop for as long as 48 hours after exposure.

Can phosgene poisoning be treated?
There is no antidote for phosgene, but its effects can be treated, and most exposed persons get well. Persons who have experienced serious symptoms may need to be hospitalized.

Are any future health effects likely to occur?
A single small exposure from which a person recovers quickly is not likely to cause delayed or long-term effects. After a serious exposure, some symptoms may take a few days to develop. Some persons who have had serious exposures have developed permanent breathing difficulty and tend to develop lung infections easily.

What tests can be done if a person has been exposed to phosgene?
Specific tests for the presence of phosgene in blood or urine generally are not useful to the doctor. If a severe exposure has occurred, chest x-rays, blood and urine analyses, and other tests may show whether the lungs or other organs have been injured. Because effects may take several days to develop, immediate and follow-up testing of lung function should be done in all cases of suspected exposure to phosgene.

Where can more information about phosgene be found?
More information about phosgene can be obtained from your regional poison control center; your state, county, or local health department; the Agency for Toxic Substances and Disease Registry (ATSDR); your doctor; or a clinic in your area that specializes in occupational and environmental health. If the exposure happened at work, you may wish to discuss it with your employer, the Occupational Safety and Health Administration (OSHA), or the National Institute for Occupational Safety and Health (NIOSH). Ask the person who gave you this form for help in locating these telephone numbers.

FOLLOW-UP INSTRUCTIONS

Keep this page and take it with you to your next appointment. Follow only the instructions checked below.

[ ]	Call your doctor or the Emergency Department if you develop any unusual signs or symptoms within the next 24 hours, especially:
•	coughing or wheezing
•	difficulty breathing or shortness of breath
•	increased pain or discharge from exposed skin or eyes
•	chest pain or tightness
[ ]	No follow-up appointment is necessary unless you develop any of the symptoms listed above.
[ ]	Call for an appointment with Dr _____ in the practice of _____ _____. When you call for your appointment, please say that you were treated in the Emergency Department at _____ Hospital by _____ and were advised to be seen again in _____ days.
[ ]	Return to the Emergency Department/ _____ Clinic on (date) _____ at _____AM/PM for a follow-up examination.
[ ]	Do not perform vigorous physical activities for 1-2 days.
[ ]	You may resume everyday activities including driving and operating machinery.
[ ]	Do not return to work for _____ days.
[ ]	You may return to work on a limited basis. See instructions below.
[ ]	Avoid exposure to cigarette smoke for 72 hours; smoke may worsen the condition of your lungs.
[ ]	Avoid drinking alcoholic beverages for at least 24 hours; alcohol may worsen injury to your stomach or have other effects.
[ ]	Avoid taking the following medications: _____
[ ]	You may continue taking the following medication(s) that your doctor(s) prescribed for you:
[ ]	Other instructions: _____
•	Provide the Emergency Department with the name and the number of your primary care physician so that the ED can send him or her a record of your emergency department visit.
•	You or your physician can get more information on the chemical by contacting: _____ _____ or _____, or by checking out the following internet web sites: _____.

Signature of patient _____ Date _____

Signature of physician _____ Date _____

Phosphine - Patient Information Sheet

This handout provides information and follow-up instructions for persons who have been exposed to phosphine or phosphides (obtained from the CDC - Center for Disease Control - www.atsdr.cdc.gov/MHMII/mmg177-handout.pdf)

What is phosphine? How are phosphides related?
Phosphine is a toxic gas that has no color and smells like garlic or fish. A serious exposure to phosphine could occur, however, even if a person does not smell it. Phosphine is used widely in the semiconductor industry. Phosphine may be encountered in grain storage silos where it has been used as a fumigant, or zinc phosphide has been put down as a rat poison. Certain pesticides containing zinc phosphide or aluminum phosphide can release phosphine when they come in contact with water or acid. The phosphine formed in the stomach when these solid phosphides are swallowed can result in phosphine poisoning.

What immediate health effects can be caused by exposure to phosphine?
Exposure to even small amounts of phosphine can cause headache, dizziness, nausea, vomiting, diarrhea, drowsiness, cough, and chest tightness. More serious exposure can cause shock, convulsions, coma, irregular heartbeat, and liver and kidney damage. Generally, the more serious the exposure, the more severe the symptoms.

Can phosphine poisoning be treated?
There is no antidote for phosphine, but its effects can be treated, and most exposed persons get well. Persons who have experienced serious symptoms may need to be hospitalized.

Are any future health effects likely to occur?
A single small exposure from which a person recovers quickly is not likely to cause delayed or long-term effects. After a severe exposure, symptoms usually begin immediately but might not appear for 72 hours or more. Some severely exposed persons have experienced long-term brain, heart, lung, and liver injury.

What tests can be done if a person has been exposed to phosphine?
There are no specific blood or urine tests for phosphine itself. Breakdown products of phosphine can be measured in urine, but the result of this test is generally not useful to the doctor. If a severe exposure has occurred, blood and urine analyses and other tests may also show whether the brain, lungs, heart, liver, or kidneys have been damaged. Testing is not needed in every case.

Where can more information about phosphine be found?
More information about phosphine and phosphides can be obtained from your regional poison control center; your state, county, or local health department; the Agency for Toxic Substances and Disease Registry (ATSDR); your doctor; or a clinic in your area that specializes in occupational and environmental health. If the exposure happened at work, you may wish to discuss it with your employer, the Occupational Safety and Health Administration (OSHA), or the National Institute for Occupational Safety and Health (NIOSH). Ask the person who gave you this form for help in locating these telephone numbers.

FOLLOW-UP INSTRUCTIONS

Keep this page and take it with you to your next appointment. Follow only the instructions checked below.

[ ]	Call your doctor or the Emergency Department if you develop any unusual signs or symptoms within the next 24 hours, especially:
•	coughing or wheezing
•	difficulty breathing or shortness of breath
•	chest pain or tightness
•	headache, dizziness, tremor, or double vision
•	difficulty walking
•	nausea, vomiting, diarrhea, or stomach pain
[ ]	No follow-up appointment is necessary unless you develop any of the symptoms listed above.
[ ]	Call for an appointment with Dr _____ in the practice of _____ _____. When you call for your appointment, please say that you were treated in the Emergency Department at _____ Hospital by _____ and were advised to be seen again in _____ days.
[ ]	Return to the Emergency Department/ _____ Clinic on (date) _____ at _____AM/PM for a follow-up examination.
[ ]	Do not perform vigorous physical activities for 1-2 days.
[ ]	You may resume everyday activities including driving and operating machinery.
[ ]	Do not return to work for _____ days.
[ ]	You may return to work on a limited basis. See instructions below.
[ ]	Avoid exposure to cigarette smoke for 72 hours; smoke may worsen the condition of your lungs.
[ ]	Avoid drinking alcoholic beverages for at least 24 hours; alcohol may worsen injury to your stomach or have other effects.
[ ]	Avoid taking the following medications: _____ _____ _____
[ ]	You may continue taking the following medication(s) that your doctor(s) prescribed for you: _____ _____ _____ _____
[ ]	Other instructions: _____ _____ _____ _____
•	Provide the Emergency Department with the name and the number of your primary care physician so that the ED can send him or her a record of your emergency department visit.
•	You or your physician can get more information on the chemical by contacting: _____ _____ or _____, or by checking out the following internet web sites: _____ _____.

Signature of patient _____ Date _____

Signature of physician _____ Date _____

Sulfur Mustard - Patient Information Sheet

This handout provides information and follow-up instructions for persons who have been exposed to sulfur mustard (obtained from the CDC - Center for Disease Control - www.atsdr.cdc.gov/MHMII/mmgd3-handout.pdf)

What are sulfur mustards?
Sulfur mustards are yellowish to brown liquids that have been used as chemical warfare agents since 1917.

What immediate health effects can be caused by exposure to sulfur mustards?
Sulfur mustards produce blistering and cell damage, but symptoms are delayed for hours. They cause damage to the skin, eyes, and respiratory tract. The eyes are the most sensitive. Nausea and vomiting may occur within the first few hours after exposure. Skin rashes, blisters, and lung damage may develop within a few hours of exposure but may take 12 to 24 hours to develop. Sulfur mustard can also suppress the immune system.

Can sulfur mustard poisoning be treated?
There is no antidote for sulfur mustard, but its effects can be treated and most exposed people recover. Immediate decontamination reduces symptoms. People who have been exposed to large amounts of sulfur mustard will need to be treated in a hospital.

Are any future health effects likely to occur?
Adverse health effects, such as chronic respiratory diseases, may occur from exposure to high levels of these agents. Severe damage to the eyes and skin may be present for a long time following the exposure.

What tests can be done if a person has been exposed to sulfur mustards?
There are no routine tests to determine if someone has been exposed to sulfur mustard. Thiodiglycol (a break-down product of mustard) may be detected in the urine up to 2 weeks following exposure; however, this test is available only in several specialized laboratories.

Where can more information about sulfur mustards be found?
More information about sulfur mustard can be obtained from your regional poison control center; the Agency for Toxic Substances and Disease Registry (ATSDR); your doctor; or a clinic in your area that specializes in toxicology or occupational and environmental health. Ask the person who gave you this form for help locating these telephone numbers.

FOLLOW-UP INSTRUCTIONS

Keep this page and take it with you to your next appointment. Follow only the instructions checked below.

[ ]	Call your doctor or the Emergency Department if you develop any unusual signs or symptoms within the next 24 hours, especially:
•	coughing, wheezing, shortness of breath, or discolored sputum
•	increased pain or discharge from injured eyes
•	Increased redness, pain, or a pus-like discharge from injured skin
•	fever or chills
[ ]	No follow-up appointment is necessary unless you develop any of the symptoms listed above.
[ ]	Call for an appointment with Dr _____ in the practice of _____ _____. When you call for your appointment, please say that you were treated in the Emergency Department at _____ Hospital by _____ and were advised to be seen again in _____ days.
[ ]	Return to the Emergency Department/ _____ Clinic on (date) _____ at _____AM/PM for a follow-up examination.
[ ]	Do not perform vigorous physical activities for 1-2 days.
[ ]	You may resume everyday activities including driving and operating machinery.
[ ]	Do not return to work for _____ days.
[ ]	You may return to work on a limited basis. See instructions below.
[ ]	Avoid exposure to cigarette smoke for 72 hours; smoke may worsen the condition of your lungs.
[ ]	Avoid drinking alcoholic beverages for at least 24 hours; alcohol may worsen injury to your stomach or have other effects.
[ ]	Avoid taking the following medications: _____
[ ]	You may continue taking the following medication(s) that your doctor(s) prescribed for you:
[ ]	Other instructions: _____
•	Provide the Emergency Department with the name and the number of your primary care physician so that the ED can send him or her a record of your emergency department visit.
•	You or your physician can get more information on the chemical by contacting: _____ _____ or _____, or by checking out the following internet web sites: _____.

Signature of patient _____ Date _____

Signature of physician _____ Date _____

INDEX

701

Kratom, 329
  mechanism of toxic action, 329
  pregnancy issues, 329
  scientific name, 329
  signs and symptoms of acute exposure, 329
  treatment strategy, 329
  use, 329

L

Laboratory analysis, 383–520
  acetylcholinesterase, red cell, and serum, 439
    container, 439
    limitations, 439
    methodology, 439
    reference range, 439
    specimen, 439
    storage instructions, 439
    synonyms, 439
    use, 439
  anion gap, blood, 440–441
    container, 440
    limitations, 440
    methodology, 440–441
    reference range, 440
    specimen, 440
    synonyms, 440
    test, 440
    use, 440
  anthrax detection, 441–443
    causes for rejection, 441
    container, 441
    limitations, 441
    methodology, 441–442
    specimen, 441
    storage instructions, 441
    test, 441
    turnaround time, 441
    use, 441
  arsenic, blood, 443–444
    causes for rejection, 444
    container, 444
    critical values, 444
    limitations, 444
    methodology, 444
    reference range, 444
    synonym, 443
    use, 443
  arsenic, hair, nails, 444–445
    collection, 444
    container, 444
    critical values, 445
    limitations, 444

    methodology, 445
    reference range, 445
    special instructions, 445
    specimen, 444
    synonyms, 444
    use, 444
  arsenic, urine, 445
    container, 445
    critical values, 445
    limitations, 445
    methodology, 445
    patient preparation, 445
    reference range, 445
    specimen, 445
    storage instructions, 445
    synonyms, 445
    test, 445
    use, 445
  autopsy, 445–462
    biologic terrorism surveillance, 452
    biosafety concerns, 448–452
    communications, 458–459
    container, 446
    contraindications, 446
    data collection, analysis, and dissemination, 452–453
    DMORT, 457–458
    forensic pathology, 448
    hazards, 448
    jurisdictional, evidentiary, and operational concerns, 453–456
    limitations, 446
    methodology, 447
    quality assurance, 447–448
    reimbursement for expenses, 456–457
    special instructions consent, 446–447
    specimen, 446
    synonyms, 445
    turnaround time, 446
    use, 445–446
  babesiosis serology, 463
    container, 463
    critical values, 463
    limitations, 463
    methodology, 463
    reference range, 463
    specimen, 463
    storage instructions, 463
    synonym, 463
    use, 463
  bacterial culture, aerobes, 463–465
    causes for rejection, 464
    collection, 464

713